Electron Density

Electron Density

Concepts, Computation and DFT Applications

Edited by

Pratim Kumar Chattaraj
Birla Institute of Technology
Mesra, India

Debdutta Chakraborty
Birla Institute of Technology
Mesra, India

WILEY

This edition first published 2024

Registered Offices
John Wiley & Sons, Inc., 111 River Street, Hoboken, NJ 07030, USA
John Wiley & Sons Ltd, The Atrium, Southern Gate, Chichester, West Sussex, PO19 8SQ, UK

For details of our global editorial offices, customer services, and more information about Wiley products visit us at www.wiley.com.

Wiley also publishes its books in a variety of electronic formats and by print-on-demand. Some content that appears in standard print versions of this book may not be available in other formats.

Library of Congress Cataloging-in-Publication Data

Names: Chattaraj, Pratim Kumar, editor. | Chakraborty, Debdutta, editor. | John Wiley & Sons, publisher.
Title: Electron density : concepts, computation and DFT applications / Pratim Kumar Chattaraj, Debdutta Chakraborty.
Description: Hoboken, NJ : Wiley, 2024. | Includes index.
Identifiers: LCCN 2024009752 (print) | LCCN 2024009753 (ebook) | ISBN 9781394217625 (hardback) | ISBN 9781394217632 (adobe pdf) | ISBN 9781394217649 (epub)
Subjects: LCSH: Electron distribution. | Density functionals.
Classification: LCC QC793.5.E626 E44 2024 (print) | LCC QC793.5.E626 (ebook) | DDC 539.7/2112–dc23/eng/20240328
LC record available at https://lccn.loc.gov/2024009752
LC ebook record available at https://lccn.loc.gov/2024009753

Cover Design: Wiley
Cover Image: © agsandrew/Shutterstock

Set in 9.5/12.5pt STIXTwoText by Straive, Chennai, India
Printed and bound by CPI Group (UK) Ltd, Croydon, CR0 4YY

C9781394217625_010724

Contents

List of Contributors

Rudra Agarwal
Department of Chemistry
School of Natural Sciences
Shiv Nadar University
Gautam Budhha Nagar, Uttar Pradesh
India

Priyanka Aggarwal
Department of Theoretical Physics
Asia Pacific Center for Theoretical Physics
POSTECH
Pohang
Korea

Leandro Ayarde-Henríquez
School of Physics
Trinity College Dublin
Dublin 2
Ireland

Paul W. Ayers
Department of Chemistry and Chemical Biology
McMaster University
Hamilton
Canada

Kanagasabai Balamurugan
Centre for High Computing
CSIR-Central Leather Research Institute (CSIR-CLRI)
Chennai
India

and

Inorganic and Physical Chemistry Laboratory
CSIR-Central Leather Research Institute (CSIR-CLRI)
Chennai
India

and

Academy of Scientific and Innovative Research (AcSIR)
Ghaziabad
India

Dali Barman
Department of Chemistry
Indian Institute of Technology Patna
Patna
India

Debojit Bhattacharya
Department of Chemistry
University of North Bengal
Siliguri, West Bengal
India

Santu Biswas
Department of Chemistry
Visva-Bharati University
Santiniketan
India

Ramon Carbó-Dorca
Institute of Computational Chemistry and Catalysis
University of Girona
Spain

and

Ronin Institute
Montclair
USA

Debdutta Chakraborty
Department of Chemistry
Birla Institute of Technology, Mesra
Ranchi, Jharkhand
India

Tanmoy Chakraborty
Department of Chemistry and Biochemistry
Sharda School of Basic Sciences and Research
Sharda University
Greater Noida
India

and

SVKM'S NMIMS (Deemed to be University)
Mumbai
India

Somesh Chamoli
Department of Chemistry
Indian Institute of Technology Bombay
Powai
Mumbai
India

Eduardo Chamorro
Departmento de Química
Centro de Química Teórica & Computational (CQT&C) and Department of Chemistry
Facultad de Ciencias Exactas
Universidad Andrés Bello
Avenida República 275
Santiago
Chile

Shamik Chanda
Department of Chemical Sciences
Indian Institute of Science Education and Research Kolkata
Mohanpur
Nadia, West Bengal
India

Pratim Kumar Chattaraj
Department of Chemistry
Birla Institute of Technology, Mesra
Ranchi, Jharkhand
India

Zhong-hua Cui
Institute of Atomic and Molecular Physics
Jilin University
Changchun
China

Snehasis Daschakraborty
Department of Chemistry
Indian Institute of Technology Patna
Patna
India

Ayan Datta
School of Chemical Sciences
Indian Association for the Cultivation of Science
Kolkata
India

Mario Duque-Noreña
Departmento de Química
Centro de Química Teórica & Computational (CQT&C) and Department of Chemistry
Facultad de Ciencias Exactas
Universidad Andrés Bello
Avenida República 275
Santiago
Chile

Achintya Kumar Dutta
Department of Chemistry
Indian Institute of Technology Bombay
Powai
Mumbai
India

Marco Franco-Pérez
Facultad de Química
Universidad Nacional Autónoma de México,
Cd. Universitaria
Ciudad de Mexico
Mexico

Shivangi Garg
Computational and Theoretical Chemistry
Group
Department of Chemistry & Centre for
Advanced Studies in Chemistry
Panjab University
Chandigarh
India

José L. Gázquez
Departamento de Química
Universidad Autónoma
Metropolitana-Iztapalapa
Ciudad de Mexico
Mexico

Debashree Ghosh
School of Chemical Sciences
Indian Association for the Cultivation of
Science
Kolkata
India

Swapan K. Ghosh
UM-DAE-Centre for Excellence in Basic
Sciences
University of Mumbai
Mumbai
India

Sourav Ghoshal
Department of Chemistry
Visva-Bharati University
Santiniketan
India

Santanab Giri
School of Applied Science and Humanities
Haldia Institute of Technology
Haldia, West Bengal
India

Neetu Goel
Computational and Theoretical Chemistry
Group
Department of Chemistry & Centre for
Advanced Studies in Chemistry
Panjab University
Chandigarh
India

Cristian Guerra
Departmento de Química
Centro de Química Teórica & Computational
(CQT&C) and Department of Chemistry
Facultad de Ciencias Exactas
Universidad Andrés Bello
Avenida República 275
Santiago
Chile

Kanika Guleria
Department of Chemistry
Indian Institute of Technology Patna
Patna
India

Manoj K. Harbola
Department of Physics
Indian Institute of Technology Kanpur
Kanpur
India

Ram Kuntal Hazra
Department of Chemistry
Physical Division
University of Delhi
Delhi
India

Xin He
Qingdao Institute for Theoretical and Computational Sciences
Institute of Frontier and Interdisciplinary Science
Shandong University
Qingdao, Shandong
P.R. China

Ruchi Jha
Advanced Technology Development Centre
Indian Institute of Technology Kharagpur
Kharagpur
India

Bharti Kapil
Department of Chemistry
Physical Division
University of Delhi
Delhi
India

Soumyajit Karmakar
Department of Chemistry
Indian Institute of Technology Kharagpur
Kharagpur
India

Masiyappan Karuppusamy
Centre for High Computing
CSIR-Central Leather Research Institute (CSIR-CLRI)
Chennai
India

and

Inorganic and Physical Chemistry Laboratory
CSIR-Central Leather Research Institute (CSIR-CLRI)
Chennai
India

and

Academy of Scientific and Innovative Research (AcSIR)
Ghaziabad
India

Harsimran Kaur
Department of Chemistry
Indian Institute of Technology Bombay
Powai
Mumbai
Maharashtra
India

Mayank Khera
Computational and Theoretical Chemistry Group
Department of Chemistry & Centre for Advanced Studies in Chemistry
Panjab University
Chandigarh
India

and

Department of Applied Science
Chandigarh College of Engineering
CGC-Jhanjheri, Mohali
India

Abhishek Kumar
Department of Physics
University of Lucknow
Lucknow, Uttar Pradesh
India

and

Department of Geology
Babasaheb Bhimrao Ambedkar University (A Central University)
Lucknow, Uttar Pradesh
India

Ashish Kumar
Department of Physics
Indian Institute of Technology Kanpur
Kanpur
India

Ratnesh Kumar
Department of Physics
University of Lucknow
Lucknow, Uttar Pradesh
India

Ravi Kumar
Computational and Theoretical Chemistry Group
Department of Chemistry & Centre for Advanced Studies in Chemistry
Panjab University
Chandigarh
India

Vijay Kumar
Center for Informatics
School of Natural Sciences
Shiv Nadar University
Gautam Budhha Nagar
Uttar Pradesh
India

and

Dr. Vijay Kumar Foundation
Gurgaon, Haryana
India

Meng Li
Key Laboratory of Chemical Biology and Traditional Chinese Medicine Research (Ministry of Education of China)
Hunan Normal University
Changsha, Hunan
P.R. China

Shubin Liu
Research Computing Center
University of North Carolina
Chapel Hill, NC
USA

and

Department of Chemistry
University of North Carolina
Chapel Hill, NC
USA

Nilangshu Mandal
School of Chemical Sciences
Indian Association for the Cultivation of Science
Kolkata
India

Sabyashachi Mishra
Department of Chemistry
Indian Institute of Technology Kharagpur
Kharagpur
India

Anirban Misra
Department of Chemistry
University of North Bengal
Siliguri, West Bengal
India

Neeraj Misra
Department of Physics
University of Lucknow
Lucknow, Uttar Pradesh
India

Sathish K. Mudedla
Inorganic and Physical Chemistry Laboratory
CSIR-Central Leather Research Institute (CSIR-CLRI)
Chennai
India

Sunita Muduli
Department of Chemistry
Indian Institute of Technology Kharagpur
Kharagpur
India

Roman F. Nalewajski
Department of Theoretical Chemistry
Jagiellonian University
Cracow
Poland

Preeti Nanda
Department of Electronics and Communication Engineering
IIIT Bhagalpur
Bhagalpur
India

Malaya K. Nayak
Theoretical Chemistry Section
Bhabha Atomic Research Centre Trombay
Powai
Mumbai
India

and

Homi Bhabha National Institute
BARC Training School Complex
Anushakti Nagar
Mumbai
India

Minh Tho Nguyen
Laboratory for Chemical Computation and Modeling
Institute for Computational Science and Artificial Intelligence
Van Lang University
Ho Chi Minh City
Vietnam

and

Faculty of Applied Technology
School of Technology
Van Lang University
Ho Chi Minh City
Vietnam

Sudip Pan
Institute of Atomic and Molecular Physics
Jilin University
Changchun
China

Shyam V.K. Panneer
Centre for High Computing
CSIR-Central Leather Research Institute (CSIR-CLRI)
Chennai
India

and

Inorganic and Physical Chemistry Laboratory
CSIR-Central Leather Research Institute (CSIR-CLRI)
Chennai
India

Maurizio A. Pantoja-Hernández
Departamento de Química
Universidad Autónoma Metropolitana-Iztapalapa
Ciudad de Mexico
Mexico

Shanti Gopal Patra
Department of Chemistry
Indian Institute of Technology Kharagpur
Kharagpur
India

and

Department of Chemistry
National Institute of Technology Silchar
Silchar
India

Atanuka Paul
Department of Chemistry
Indian Institute of Technology Kharagpur
Kharagpur
India

Patricia Pérez
Departmento de Química
Centro de Química Teórica & Computational (CQT&C) and Department of Chemistry
Facultad de Ciencias Exactas
Universidad Andrés Bello
Avenida República 275
Santiago
Chile

Anup Pramanik
Department of Chemistry
Sidho-Kanho-Birsha University
Purulia
India

Amol B. Rahane
Department of Physics
K. R. T. Arts, B. H. Commerce & A. M. Science (KTHM) College [Affiliated to Savitribai Phule Pune University (SPPU), Pune, India]
Nashik, Maharashtra
India

Biplab Rajbanshi
Department of Chemistry
Visva-Bharati University
Santiniketan
India

Prabhat Ranjan
Department of Mechatronics Engineering
Manipal University Jaipur
Jaipur
India

Mahesh K. Ravva
Department of Chemistry
SRM University – AP
Amaravati
India

Elizabeth Rincón
Facultad de Ciencias, Instituto de Ciencias Químicas
Universidad Austral de Chile
Valdivia
Chile

Chunying Rong
Key Laboratory of Chemical Biology and Traditional Chinese Medicine Research (Ministry of Education of China)
Hunan Normal University
Changsha, Hunan
P.R. China

Amlan K. Roy
Department of Chemical Sciences
Indian Institute of Science Education and Research Kolkata
Maharashtra
Nadia, West Bengal
India

Pinaki Saha
Biocomputation Group, School of Physics Engineering and Computer Science
University of Hertfordshire
UK

and

Department of Chemistry
School of Natural Sciences
Shiv Nadar University
Gautam Budhha Nagar, Uttar Pradesh
India

Supriyo Santra
School of Chemical Sciences
Indian Association for the Cultivation of Science
Kolkata
India

Pranab Sarkar
Department of Chemistry
Visva-Bharati University
Santiniketan
India

Shivalika Sharma
Department of Theoretical Physics
Asia Pacific Center for Theoretical Physics
POSTECH
Pohang
Korea

Suranjan Shil
Manipal Centre for Natural Sciences
Manipal Academy of Higher Education
Manipal, Karnataka
India

Sunny Singh
Department of Chemistry
Physical Division
University of Delhi
Delhi
India

Swapan Sinha
School of Applied Science and Humanities
Haldia Institute of Technology
Haldia, West Bengal
India

and

Department of Applied Chemistry
Maulana Abul Kalam Azad University of Technology
Haringhata, West Bengal
India

Ambrish K. Srivastava
Department of Physics
Deen Dayal Upadhyaya Gorakhpur University
Gorakhpur, Uttar Pradesh
India

Ranga Subramanian
Department of Chemistry
Indian Institute of Technology Patna
Patna
India

Venkatesan Subramanian
Centre for High Computing
CSIR-Central Leather Research Institute (CSIR-CLRI)
Chennai
India

and

Department of Chemistry
Indian Institute of Technology–Madras
Chennai
India

Nagamani Sukumar
Amrita School of AI
Amrita Vishwa Vidyapeetham
Coimbatore
641105
Tamil Nadu
India

Paweł Szarek
Navi-Chem®
Warsaw
Poland

Suresh Tiwari
Department of Chemistry
Indian Institute of Technology Patna
Patna
India

Jean-Louis Vigneresse
GéoRessources
UMR CNRS 7539
Vandœuvre-lès-Nancy
54706
France

Xinjie Wan
Key Laboratory of Chemical Biology and Traditional Chinese Medicine Research (Ministry of Education of China)
Hunan Normal University
Changsha, Hunan
P.R. China

Paulino Zerón
Departamento de Química
Universidad Autónoma Metropolitana-Iztapalapa
Ciudad de Mexico
Mexico

Dongbo Zhao
Institute of Biomedical Research
Yunnan University
Kunming, Yunnan
P.R. China

Preface

Electron density or the single particle density is a 3D function even for a many-electron system. Electron density contains all information regarding the ground state and also about some excited states of an atom or a molecule. All the properties can be written as functionals of electron density, and the energy attains its minimum value for the true density. Moreover, it is an experimental observable. It brings out a drastic reduction in computational labor due to its reduced dimensionality. Electron density helps build up various models that in turn make the understanding of complicated chemical, physical, or biological problems simpler. As we live and perceive in a 3D classical world, visualization of different systems and processes becomes much easier as opposed to that using the many-body wavefunction. Accordingly, it helps develop a classical interpretation of quantum mechanics.

This book aims to outline fundamental as well as applied aspects of state-of the-art developments in the domain of density functional theory. It is well known in the scientific literature that density functional theory constitutes the work horse for modern computational chemistry and physics. Density functional theory has been very successful in analyzing and rationalizing numerous physicochemical processes. However, there are also challenging situations where density functional theory fails to predict and/or analyze several experimentally observed phenomena. Moreover, the application of density functional theory is difficult in case of very large systems such as biological systems due to the limitation in the available computational resources. To this end, various methodological and conceptual aspects related to density functional theory have been discussed in this book. In particular, conceptual density functional theory-based reactivity descriptors are discussed in detail. Areas in which density functional theory has been successfully applied have also been discussed to give a broad overview of the field to the reader.

On the methodological and fundamental aspects, this book aims to discuss the Levy–Perdew–Sahni equation and the Kohn–Sham inversion problem within the premise of density functional theory. Several key concepts related to conceptual density functional theory-based local and global reactivity descriptors have been discussed. The relativistic treatment of many-electron systems within the premise of density functional theory has been explained in detail. In addition, the properties and applications of relativistic reduced density matrices have been discussed herein. Utilizing the Coulomb Green's function, recipes for performing many-body multiconfigurational calculations have been explored. Furthermore, application of catastrophe theory in understanding the Diels–Alder cycloaddition reaction pathways have been discussed. On the other hand, muti-scale simulation techniques have been discussed in the context of the modeling of excited-state electronic structures of biomolecules.

On the applied aspects of density functional theory, this book comprises several interesting applications. The designing principles for ultrashort H···H nonbonded contacts and ultralong

C–C bonds have been explored. Application of density functional theory in the context of analyzing the properties of materials has been discussed. The emerging topic of mechanochemistry has been explored by taking help from density-based quantum chemical formalism. Electrostatic potential and their applications in the context of analyzing several physicochemical processes are discussed. The role of electron spin density in dictating molecular magnetism has been explored via considering several diradical systems. Interesting and novel bonding scenarios have been explored and discussed via several density functional theory formalisms. Applications of density functional theory in predicting radiative efficiencies and global warming potential of some selected organic molecules are also discussed. Finally, the properties of several natural products are explored via the application of density functional theory.

Thus, this book intends to cover both fundamental and applied aspects on the state-of the-art status of density functional theory. The book would help practicing and aspiring theoretical and computational chemists/physicists to obtain a clear-cut overview on the status of density functional theory.

April 30, 2024

Pratim Kumar Chattaraj and Debdutta Chakraborty
Birla Institute of Technology Mesra, India

1

Levy–Perdew–Sahni Equation and the Kohn–Sham Inversion Problem

Ashish Kumar and Manoj K. Harbola

Department of Physics, Indian Institute of Technology Kanpur, Kanpur, India

1.1 Introduction

Density functional theory (DFT) [1–9] is a rigorous formulation for studying the ground-state of a many-electron system in terms of the corresponding density. This theory is complementary to other formulations of many-electron problems, such as those based on field-theoretic techniques [10, 11] or the model Hamiltonian approaches [12–15]. However, because of the ease of its implementation, DFT has become the de facto electronic structure theory with the ongoing development of highly accurate energy functionals. Starting with the earliest local density approximation (LDA) [2] based on the energy of homogeneous electron gas, these functionals have been developed, [16–27] guided by the various exact properties [28] they should satisfy. Obtaining these exact properties and their further investigation for different systems is thus of fundamental importance for understanding DFT well and making it applicable with high accuracy for various electronic systems. The question is this: How does one go about performing such studies? We address this in the following paragraph.

Let us start by recalling that DFT is a reformulation of quantum-mechanical many-electron problem in terms of the ground-state density. Therefore in this theory, any quantity is expressed as a functional of density. This is done by transforming the expectation value of the operator corresponding to a quantity into an expression in terms of the density. This transformation can sometimes be made exactly; however for a majority of quantities, it is not possible to do so. As a result, performing density-based calculations necessitates developing approximate expressions for such quantities in terms of the density. The best-known example of such an approximation is the exchange–correlation energy functional and the corresponding exchange–correlation potential, which is the mainstay of Kohn–Sham (KS) DFT calculations. Beginning with the LDA, a lot of effort has been put into developing approximations for better and better exchange–correlation (XC) energy functionals since the inception of DFT. This research activity is evidently vital for practical applications of theory, and continuous development of new XC energy functional has made DFT the most widely used electronic structure theory. On the other hand, to gain fundamental insights into DFT one needs to explore the connection between a density functional and its wavefunctional counterpart without necessarily expressing relevant quantities in terms of the density [29]. In such cases, one can employ the formal expression of a density functional in terms of the exact many-body wavefunction. For example, to study energy density functional for fractional numbers of electrons, a statistical mixture of pure states is used [30] to obtain the corresponding energy.

From among various exact results of DFT, our focus in this chapter is on obtaining the exact Kohn–Sham potential for a given ground-state density. This falls under the category of the inverse

Electron Density: Concepts, Computation and DFT Applications, First Edition.
Edited by Pratim Kumar Chattaraj and Debdutta Chakraborty.

problems [31–33] with the direct problem being the calculation of density from a given potential. Over the years, various methods have been proposed and employed [34–66] for doing such a calculation. We will show in this chapter that all the density-based inversion methods can be obtained using the Euler equation [67–69]

$$\left[-\frac{1}{2}\nabla^2 + v_{\text{ext}}(\mathbf{r}) + v_{\text{eff}}(\mathbf{r})\right]\sqrt{\rho} = \mu\sqrt{\rho} \tag{1.1}$$

for the density. We will begin by using this equation for the inversion and then go on to generalize it. This generalization shows the universal nature of these methods and opens the door to devising a method of one's choice suitable for a given density. In Eq. (1.1) $v_{\text{ext}}(\mathbf{r})$ is the potential in which electrons are moving; $v_{\text{eff}}(\mathbf{r})$ comprises the Pauli potential, the Hartree potential, and the exchange–correlation potential with μ being the chemical potential of the system. The expression for $v_{\text{eff}}(\mathbf{r})$ in terms of the exact wavefunction of a system has been derived by Levy–Perdew–Sahni (LPS) [67]. In the rest of this chapter, we will therefore refer to Eq. (1.1) as the LPS equation.

After demonstrating the universal nature of all density-based inversion methods, we next discuss how this understanding also leads to a general penalty method to obtain the Kohn–Sham potential for a ground-state density. Finally, using the expression of $v_{\text{eff}}(\mathbf{r})$ in terms of the wavefunction, we throw light on why methods that employ approximate wavefunctions rather than the density obtained from the same wavefunction give Kohn–Sham potential free of pathological features that are seen when these densities are employed in density-to-potential inversion. In the following, we first present the universal nature of seemingly different density-based Kohn–Sham inversion methods using Eq. (1.1) and show how these can all be generated from a single algorithm based on the LPS equation. An exciting application based on understanding universal nature is using random numbers to carry out the density-to-potential inversion. We will discuss that as well. Next, we present the general penalty method to obtain the Kohn–Sham potential from a ground-state density. Finally, we discuss how expression for $v_{\text{eff}}(\mathbf{r})$ in terms of the wavefunction can be used to find the Kohn–Sham potential of a many-electron system from an approximate wavefunction. All the methods are demonstrated by applying them to spherical systems.

1.2 One Equation $\Longrightarrow$ Several Methods; Universal Nature of Different Density-Based Kohn–Sham Inversion Algorithms [70]

1.2.1 Generating Functional $S[\rho]$ of Density-Based Kohn–Sham Inversion

In this section, we show how Eq. (1.1) can be used to obtain several different algorithms to get the Kohn–Sham potential for a given density using a generating functional $S[\rho]$ of density $\rho(\mathbf{r})$. Since the Hartree potential is well known in terms of density, obtaining the exact Kohn–Sham potential boils down to getting the exchange–correlation potential for a given density. This is done using Eq. (1.1). For this, let us first see how this equation is obtained. We begin by writing the energy functional as

$$E[\rho] = \frac{1}{8}\int \frac{|\nabla\rho|^2}{\rho}d\mathbf{r} + T_{\text{P}}[\rho] + \int v_{\text{ext}}(\mathbf{r})\rho(\mathbf{r})d\mathbf{r} + E_{\text{H}}[\rho] + E_{\text{xc}}[\rho] \tag{1.2}$$

where

$$E_{\text{H}}[\rho] = \frac{1}{2}\iint \frac{\rho(\mathbf{r})\rho(\mathbf{r}')}{|\mathbf{r}-\mathbf{r}'|}d\mathbf{r}d\mathbf{r}' \tag{1.3}$$

is the Hartree energy, $T_{\text{P}}[\rho]$ is the Pauli kinetic energy, and $E_{\text{xc}}[\rho]$ is the exchange–correlation energy. Note that $(1/8)\int |\nabla\rho(\mathbf{r})|^2/\rho(\mathbf{r})d\mathbf{r} = (1/2)\int |\nabla\sqrt{\rho(\mathbf{r})}|^2 d\mathbf{r}$ is the ground-state kinetic energy for Bosons of density $\rho(\mathbf{r})$. Thus $T_P[\rho]$ is the additional kinetic energy arising due to Fermionic nature of electrons. Now minimizing the energy functional $E[\rho]$ with respect to the density amplitude $\sqrt{\rho(\mathbf{r})}$ with the constraint that $\int \rho(\mathbf{r})d\mathbf{r} - N = 0$, one is led to Eq. (1.1) with μ as the Lagrange multiplier. Through this derivation, it is clear that the effective potential

$$v_{\text{eff}}(\mathbf{r}) = v_P(\mathbf{r}) + v_{\text{ext}}(\mathbf{r}) + v_H(\mathbf{r}) + v_{xc}(\mathbf{r}) \tag{1.4}$$

where the Pauli potential [71]

$$v_P(\mathbf{r}) = \frac{\delta T_P[\rho]}{\delta\rho(\mathbf{r})} \tag{1.5}$$

the Hartree potential

$$v_H(\mathbf{r}) = \frac{\delta E_H[\rho]}{\delta\rho(\mathbf{r})} = \int \frac{\rho(\mathbf{r}')}{|\mathbf{r}-\mathbf{r}'|}d\mathbf{r}' \tag{1.6}$$

and the exchange–correlation potential

$$v_{xc}(\mathbf{r}) = \frac{\delta E_{xc}[\rho]}{\delta\rho(\mathbf{r})} \tag{1.7}$$

For a given exact ground-state density $\rho_0(\mathbf{r})$, we can write, using Eq. (1.1), the exact exchange–correlation potential as

$$v_{xc}^0(\mathbf{r}) = \mu_0 + \frac{1}{2}\frac{\nabla^2\sqrt{\rho_0}}{\sqrt{\rho_0}} - v_{\text{ext}}(\mathbf{r}) - v_H[\rho_0] - v_P^0(\mathbf{r}) \tag{1.8}$$

Now we wish to generate $v_{xc}^0(\mathbf{r})$ iteratively starting from some approximate initial potential. Suppose in the ith iteration of solving the KS equation all the quantities are denoted by superscript i, then

$$v_{xc}^{(i)}(\mathbf{r}) = \mu^{(i)} + \frac{1}{2}\frac{\nabla^2\sqrt{\rho^{(i)}}}{\sqrt{\rho^{(i)}}} - v_{\text{ext}}(\mathbf{r}) - v_H^{(i)}(\mathbf{r}) - v_P^{(i)}(\mathbf{r}) \tag{1.9}$$

To improve the exchange–correlation potential for the next iteration, we use exact ground-state density $\rho_0(\mathbf{r})$ and write

$$v_{xc}^{(i+1)}(\mathbf{r}) = \mu^{(i)} + \frac{1}{2}\frac{\nabla^2\sqrt{\rho_0}}{\sqrt{\rho_0}} - v_{\text{ext}}(\mathbf{r}) - v_H^{(i)}(\mathbf{r}) - v_P^{(i)}(\mathbf{r}) \tag{1.10}$$

Subtracting Eq. (1.9) from Eq. (1.10), we get

$$v_{xc}^{(i+1)}(\mathbf{r}) = v_{xc}^{(i)}(\mathbf{r}) + \frac{1}{2}\frac{\nabla^2\sqrt{\rho_0}}{\sqrt{\rho_0}} - \frac{1}{2}\frac{\nabla^2\sqrt{\rho^{(i)}}}{\sqrt{\rho^{(i)}}} \tag{1.11}$$

This equation can be used to obtain the exact $v_{xc}^0(\mathbf{r})$ iteratively. To do this for a given system where $v_{ex}(\mathbf{r})$ and $\rho_0(\mathbf{r})$ are given, start with an approximate exchange–correlation potential and solve the KS equation while keeping the Hartree potential fixed to its exact value calculated from $\rho_0(\mathbf{r})$. Use the approximate density so obtained in Eq. (1.11) to get the exchange–correlation potential for the next cycle of iteration. Continue doing so until the desired convergence criteria is met.

The method presented has also been proposed independently by Finzel et al. [55]. It has been applied to atoms and other spherical systems to obtain their exact exchange–correlation potential. However, more significant aspect of derivation given is that it can be generalized so that any other function of density can be used in place of $(1/2)\nabla^2\sqrt{\rho(\mathbf{r})}/\sqrt{\rho(\mathbf{r})}$. How this is done is described next.

To get the general method, let us write Eq. (1.2) as

$$E[\rho] = S[\rho] + \tilde{T}_P[\rho] + \int v_{ext}(\mathbf{r})\rho(\mathbf{r})d\mathbf{r} + E_H[\rho] + E_{xc}[\rho] \tag{1.12}$$

where $S[\rho]$ is a functional of our choice with units of energy and $\tilde{T}_P[\rho]$ is a general "Pauli" kinetic energy functional; evidently in Eq. (1.2), $S[\rho] = (1/8)\int |\nabla\rho|^2/\rho d\mathbf{r} = (1/2)\int |\nabla\sqrt{\rho}|^2 d\mathbf{r}$. Now the exact exchange–correlation potential is written as ρ

$$v_{xc}^0(\mathbf{r}) = \mu_0 - \left.\frac{\delta S[\rho]}{\delta\rho(\mathbf{r})}\right|_{\rho_0} - v_{ext}(\mathbf{r}) - v_H[\rho_0] - \tilde{v}_P^0(\mathbf{r}) \tag{1.13}$$

Here $\left.\frac{\delta S[\rho]}{\delta\rho(\mathbf{r})}\right|_{\rho_0}$ indicates that the functional derivative has been evaluated for density ρ_0 and $\tilde{v}_P^0(\mathbf{r}) = \left.\frac{\delta \tilde{T}_P[\rho]}{\delta\rho(\mathbf{r})}\right|_{\rho_0}$.

Following the same steps as from Eq. (1.9) to (1.11), we write

$$v_{xc}^{(i)}(\mathbf{r}) = \mu^{(i)} - \left.\frac{\delta S[\rho]}{\delta\rho(\mathbf{r})}\right|_{\rho^{(i)}} - v_{ext}(\mathbf{r}) - v_H^{(i)}(\mathbf{r}) - \tilde{v}_P^{(i)}(\mathbf{r}) \tag{1.14}$$

$$v_{xc}^{(i+1)}(\mathbf{r}) = \mu^{(i)} - \left.\frac{\delta S[\rho]}{\delta\rho(\mathbf{r})}\right|_{\rho_0} - v_{ext}(\mathbf{r}) - v_H^{(i)}(\mathbf{r}) - \tilde{v}_P^{(i)}(\mathbf{r}) \tag{1.15}$$

and subtract Eq. (1.14) from Eq. (1.15) to get

$$v_{xc}^{(i+1)}(\mathbf{r}) = v_{xc}^{(i)}(\mathbf{r}) + \left.\frac{\delta S[\rho]}{\delta\rho(\mathbf{r})}\right|_{\rho^{(i)}} - \left.\frac{\delta S[\rho]}{\delta\rho(\mathbf{r})}\right|_{\rho_0} \tag{1.16}$$

Equation (1.16) provides the general algorithm to update the potential and thereby obtain the exact exchange–correlation potential for the density $\rho_0(\mathbf{r})$ using an iterative process. The steps to be followed are the same as those described for Eq. (1.11). However, the significance of Eq. (1.16) is that a functional $S[\rho]$ of our choice, which satisfies the condition discussed below, can be used in it. This opens the doors to using various functionals for updation of the potential, thereby making available various methods of doing a density-to-potential inversion, all stemming from a single equation.

Having presented the general equation for obtaining the exchange–correlation potential, we next derive the condition that the functional $S[\rho]$ should satisfy in order to be used in Eq. (1.16).

1.2.2 Condition on Generating Functional $S[\rho]$

To obtain the condition that the generating functional $S[\rho]$ satisfies, we use Lieb's [72] prescription for obtaining the universal functional $F[\rho]$ of DFT. Accordingly, for a given ground-state density ρ_0

$$F[\rho_0] = \max_{v}\left\{E[v] - \int v(\mathbf{r})\rho_0(\mathbf{r})d\mathbf{r}\right\} \tag{1.17}$$

where $E[v]$ is the exact energy for electrons moving in the external potential $v(\mathbf{r})$. Thus correct $v(\mathbf{r})$ corresponding to ground-state density $\rho_0(\mathbf{r})$ will maximize the right-hand side of Eq. (1.17) with respect to $v(\mathbf{r})$, exactly like for a given $v(\mathbf{r})$ the correct wavefunction minimizes the functional $\langle\Psi|T + v_{ee}|\Psi\rangle + \int v(\mathbf{r})\rho(\mathbf{r})d\mathbf{r}$ with minimum value being the ground-state energy for the corresponding Hamiltonian. Equation (1.17) implies that as $v(\mathbf{r})$ changes in Eq. (1.16) and gets closer to the exact potential, the quantity $E[v] - \int v(\mathbf{r})\rho_0(\mathbf{r})d\mathbf{r}$ should become larger and larger so that when the correct $v(\mathbf{r})$ reached, it attains its maximum value that is equal to the universal functional $F[\rho_0]$. For the Kohn–Sham system $v(\mathbf{r}) = v_{ext}(\mathbf{r}) + v_H(\mathbf{r}) + v_{xc}(\mathbf{r})$. In our inversion method, $v_{ext}(\mathbf{r})$ and $v_H(\mathbf{r})$ are kept fixed to their exact values, respectively, so as we go from ith to the next iteration we should have

$$E[v_{\text{ext}}(\mathbf{r}) + v_{\text{H}}(\mathbf{r}) + v_{\text{xc}}^{(i+1)}(\mathbf{r})] - \int \rho_0(\mathbf{r})(v_{\text{ext}}(\mathbf{r}) + v_{\text{H}}(\mathbf{r}) + v_{\text{xc}}^{(i+1)}(\mathbf{r}))d\mathbf{r}$$
$$-E[v_{\text{ext}}(\mathbf{r}) + v_{\text{H}}(\mathbf{r}) + v_{\text{xc}}^{(i)}(\mathbf{r})] + \int \rho_0(\mathbf{r})(v_{\text{ext}}(\mathbf{r}) + v_{\text{H}}(\mathbf{r}) + v_{\text{xc}}^{(i)}(\mathbf{r}))d\mathbf{r} \geq 0 \tag{1.18}$$

Using first-order perturbation theory, Eq. (1.18) implies

$$\int \left(v_{\text{xc}}^{(i+1)}(\mathbf{r}) - v_{\text{xc}}^{(i)}(\mathbf{r})\right) \Delta\rho(\mathbf{r})d\mathbf{r} = \int \left(\left.\frac{\delta S[\rho]}{\delta\rho(\mathbf{r})}\right|_{\rho^{(i)}} - \left.\frac{\delta S[\rho]}{\delta\rho(\mathbf{r})}\right|_{\rho_0} \right) \Delta\rho(\mathbf{r})d\mathbf{r} \geq 0 \tag{1.19}$$

or the strong condition

$$\left(\left.\frac{\delta S[\rho]}{\delta\rho(\mathbf{r})}\right|_{\rho^{(i)}} - \left.\frac{\delta S[\rho]}{\delta\rho(\mathbf{r})}\right|_{\rho_0} \right) \Delta\rho(\mathbf{r}) \geq 0 \tag{1.20}$$

Here $\Delta\rho(\mathbf{r}) = \rho^{(i)} - \rho_0(\mathbf{r})$. As long as any one of the conditions are satisfied, the functional $E[v] - \int v(\mathbf{r})\rho_0(\mathbf{r})d\mathbf{r}$ will keep becoming larger with each iteration, eventually giving the exact exchange–correlation potential for the given density $\rho_0(\mathbf{r})$. The corresponding value of $F[\rho_0]$ will be equal to the noninteracting kinetic energy for the given density $\rho_0(\mathbf{r})$. Equation (1.20) also has a simple physical interpretation: if the density in the ith iteration is larger (smaller) than the target density at a given spatial point, then in next iteration because of Eq. (1.20) the potential at that point becomes larger (smaller), thereby driving the density toward the $\rho_0(\mathbf{r})$.

1.2.3 Examples of Different Generating Functionals

We now show that all density-based methods in literature are generated from corresponding generating functionals that satisfy condition of Eq. (1.19) or of Eq. (1.20). Table 1.1 lists a few of the frequently used density-based inversion methods proposed from time to time, the corresponding functionals $S[\rho]$, and the updation of potential $v_{\text{xc}}^{(i)}$ for each iteration. It is easy to see that methods of Werden and Davidson [73], Görling [36], Gaudoin and Burke [74, 75], and Hollins et al. [54] satisfy Eq. (1.19), and for other methods convergence conditions of both Eqs. (1.19) and (1.20) are satisfied.

We also note that in applying the vLB method, one should be careful in its numerical implementation as pointed out in Ref. [44]. For example, instead of using the $v_{\text{Hxc}}(\mathbf{r})$, one may think of employing $v_{\text{xc}}(\mathbf{r})$ for potential updation and improve the exchange–correlation potential using the equation

$$v_{\text{xc}}^{(i+1)}(\mathbf{r}) = v_{\text{xc}}^{(i)}(\mathbf{r}) + \frac{\left[\rho^{(i)}(\mathbf{r}) - \rho_0(\mathbf{r})\right]}{\rho_0(\mathbf{r})} v_{\text{xc}}^{(i)}(\mathbf{r}) \tag{1.21}$$

However, Eq. (1.21) will have numerical instability. To understand this, consider the condition of convergence of updation of potential given in Eq. (1.21) using Eq. (1.19)

$$\int \frac{v_{\text{xc}}^{i}(\mathbf{r})\left[\rho^{(i)}(\mathbf{r}) - \rho_0(\mathbf{r})\right]^2}{\rho_0(\mathbf{r})} d\mathbf{r} \geq 0 \tag{1.22}$$

or Eq. (1.20)

$$\frac{v_{\text{xc}}^{(i)}(\mathbf{r})\left[\rho^{(i)}(\mathbf{r}) - \rho_0(\mathbf{r})\right]^2}{\rho_0(\mathbf{r})} \geq 0 \tag{1.23}$$

In the convergence conditions, the quantity

$$\frac{\left[\rho^{(i)}(\mathbf{r}) - \rho_0(\mathbf{r})\right]^2}{\rho_0(\mathbf{r})}$$

Table 1.1 Shown in the table are various methods of obtaining the Kohn–Sham system for a given density along with their associated functional $S[\rho]$.

Method	$S[\rho]$	Potential improvement scheme
1. Werden and Davidson [73], Görling [36], and Gaudoin and Burke [74, 75]	$-\frac{1}{2}\iint \chi_i^{-1}(\mathbf{r},\mathbf{r}')\rho(\mathbf{r})\rho(\mathbf{r}')d\mathbf{r}'d\mathbf{r}$	$v_{\rm xc}^{(i+1)}(\mathbf{r}) = v_{\rm xc}^{(i)}(\mathbf{r}) - \epsilon\int \chi_i^{-1}(\mathbf{r},\mathbf{r}')\Delta\rho(\mathbf{r}')d\mathbf{r}'$
2. van Leeuwen and Baerends [39]	$\frac{1}{2}\int \frac{v_{\rm Hxc}^{(i)}(\mathbf{r})}{\rho_0(\mathbf{r})}\rho^2(\mathbf{r})d\mathbf{r}$	$v_{\rm xc}^{(i+1)}(\mathbf{r}) = v_{\rm xc}^{(i)}(\mathbf{r}) + \epsilon\frac{\Delta\rho(\mathbf{r})}{\rho_0(\mathbf{r})}v_{\rm Hxc}^{(i)}(\mathbf{r})$
3. Wang–Parr method [38]	$\frac{1}{2}\int \frac{\rho^2(\mathbf{r})}{\sum_j \frac{1}{-\epsilon_j^{(i)}}\lvert\phi_j^{(i)}(\mathbf{r})\rvert^2}d\mathbf{r}$	$v_{\rm xc}^{(i+1)}(\mathbf{r}) = v_{\rm xc}^{(i)}(\mathbf{r}) + \frac{\Delta\rho(\mathbf{r})}{\sum_j \frac{1}{-\epsilon_j^{(i)}}\lvert\phi_j^{(i)}(\mathbf{r})\rvert^2}$
4. Kadantsev–Stott method [45]	$\frac{\alpha}{2}\int \rho^2(\mathbf{r})d\mathbf{r}$	$v_{\rm xc}^{(i+1)}(\mathbf{r}) = v_{\rm xc}^{(i)}(\mathbf{r}) + \alpha\Delta\rho(\mathbf{r})$
5. Jensen and Wasserman [56]	$\frac{\gamma}{2}\int \frac{\rho^2(\mathbf{r})}{\rho_0(\mathbf{r})}d\mathbf{r}$	$v_{\rm xc}^{(i+1)}(\mathbf{r}) = v_{\rm xc}^{(i)}(\mathbf{r}) + \gamma\frac{\Delta\rho(\mathbf{r})}{\rho_0(\mathbf{r})}$
6. Peirs, et al. [44]	$\int r^\beta\rho^n(\mathbf{r})d\mathbf{r} \quad (n=2, \beta>0)$	$v_{\rm xc}^{(i+1)}(\mathbf{r}) = v_{\rm xc}^{(i)}(\mathbf{r}) + \epsilon r^\beta\Delta\rho(\mathbf{r})$
7. Kadantsev–Stott [45]	$\int \rho^n(\mathbf{r})d\mathbf{r} \quad (n=2)$	$v_{\rm xc}^{(i+1)}(\mathbf{r}) = v_{\rm xc}^{(i)}(\mathbf{r}) + \epsilon\Delta\rho(\mathbf{r})$
8. Hollins et al. [54]	$\frac{1}{2}\iint \frac{\rho(\mathbf{r})\rho(\mathbf{r}')}{\lvert\mathbf{r}-\mathbf{r}'\rvert}d\mathbf{r}'d\mathbf{r}$	$v_{\rm xc}^{(i+1)}(\mathbf{r}) = v_{\rm xc}^{(i)}(\mathbf{r}) + \epsilon\int \frac{\Delta\rho(\mathbf{r}')}{\lvert\mathbf{r}-\mathbf{r}'\rvert}d\mathbf{r}'$

In all the cases, it is easy to they satisfy the convergence condition given by Eq. (1.19) or (1.20). Here $\Delta\rho(\mathbf{r}) = \rho^{(i)}(\mathbf{r}) - \rho_0(\mathbf{r})$ and ϵ $(0 < \epsilon < 1)$ is the mixing parameter.

is always positive. Since the exchange–correlation potential $v_{\rm xc}^{(i)}(\mathbf{r}) < 0$ for a large spatial region, the convergence conditions may not be satisfied. To avoid such type of instability, one possible way to update potential using $v_{\rm xc}(\mathbf{r})$ would be

$$v_{\rm xc}^{(i+1)}(\mathbf{r}) = v_{\rm xc}^{(i)}(\mathbf{r}) + {\rm sgn}[\rho^{(i)}, \rho_0](\mathbf{r})\left|\frac{(\rho^{(i)}(\mathbf{r}) - \rho_0(\mathbf{r}))}{\rho_0(\mathbf{r})}v_{\rm xc}^{(i)}(\mathbf{r})\right| \tag{1.24}$$

where

$${\rm sgn}[\rho^{(i)}, \rho_0](\mathbf{r}) = \begin{cases} +1, & \text{if } \rho^{(i)}(\mathbf{r}) > \rho_0(\mathbf{r}) \\ -1, & \text{if } \rho^{(i)}(\mathbf{r}) < \rho_0(\mathbf{r}) \end{cases} \tag{1.25}$$

${\rm sgn}[\rho^{(i)}, \rho_0](\mathbf{r})$ ensures the correct updation in the potential.

An alternative to the vLB method for $v_{\rm ext}(\mathbf{r}) < 0$ is obtained with

$$S[\rho] = -\frac{1}{2}\int \frac{v_{\rm ext}(\mathbf{r})}{\rho_0(\mathbf{r})}\rho^2(\mathbf{r})d\mathbf{r}$$

so that

$$v_{\rm xc}^{(i+1)}(\mathbf{r}) = v_{\rm xc}^{(i)}(\mathbf{r}) - v_{\rm ext}(\mathbf{r})\frac{[\rho^{(i)}(\mathbf{r}) - \rho_0(\mathbf{r})]}{\rho_0(\mathbf{r})} \tag{1.26}$$

The negative sign here is to ensure that for $v_{\rm ext}(\mathbf{r}) < 0$, the convergence condition of Eq. (1.20) is satisfied. For $v_{\rm ext}(\mathbf{r}) > 0$, the sign in Eq. (1.26) will be positive. We also note that for the Wang–Parr

Table 1.2 Shown in the table are various new methods of obtaining the Kohn–Sham system for a given density obtained by employing different new functional $S_1[\rho] - S_4[\rho]$.

$S[\rho]$ for different new methods	Potential improvement scheme
$S_1[\rho] = -\frac{1}{2}\int \rho^{1/2}(\mathbf{r})\nabla^2\rho^{1/2}(\mathbf{r})d\mathbf{r}$ [55, 76]	$v_{xc}^{(i+1)}(\mathbf{r}) = v_{xc}^{(i)}(\mathbf{r}) + \frac{\epsilon}{2}\left(\frac{\nabla^2\rho_0^{1/2}(\mathbf{r})}{\rho_0^{1/2}(\mathbf{r})} - \frac{\nabla^2\rho_i^{1/2}(\mathbf{r})}{\rho_i^{1/2}(\mathbf{r})}\right)$
$S_2[\rho] = \int \rho^{n+1}(\mathbf{r})d\mathbf{r} \quad (0 < n \leq 1)$	$v_{xc}^{(i+1)}(\mathbf{r}) = v_{xc}^{(i)}(\mathbf{r}) + \epsilon({\rho^{(i)}}^n(\mathbf{r}) - \rho_0^n(\mathbf{r}))$
$S_3[\rho] = \int \rho(\mathbf{r})\log\rho(\mathbf{r})d\mathbf{r}$	$v_{xc}^{(i+1)}(\mathbf{r}) = v_{xc}^{(i)}(\mathbf{r}) + \epsilon\log\frac{\rho^{(i)}(\mathbf{r})}{\rho_0(\mathbf{r})}$
$S_4[\rho] =$ $\int[\mathrm{erf}(\alpha r)\rho^{(n_1+1)}(\mathbf{r}) + (1-\mathrm{erf}(\alpha r))\rho^{(n_2+1)}(\mathbf{r})]d\mathbf{r}(\alpha > 0)$	$v_{xc}^{(i+1)}(\mathbf{r}) = v_{xc}^{(i)}(\mathbf{r}) + \epsilon\mathrm{erf}(\alpha r)({\rho^{(i)}}^{n_1}(\mathbf{r}) - {\rho_0}^{n_1}(\mathbf{r}))$ $+ \epsilon(1-\mathrm{erf}(\alpha r))({\rho^{(i)}}^{n_2}(\mathbf{r}) - {\rho_0}^{n_2}(\mathbf{r}))$

The Kohn–Sham inversion methods corresponding to functionals $S_k[\rho]$ for $k = 2-4$ satisfy the strong form of convergence condition of Eq. (1.20). The method obtained by $S_1[\rho]$ satisfies the weak form of convergence condition given by Eq. (1.19). Here, $\Delta\rho(\mathbf{r})$ and ϵ are the same as those in Table 1.1.

method, the convergence condition $\frac{[\rho^{(i)}(\mathbf{r})-\rho_0(\mathbf{r})]^2}{\sum_j \frac{1}{-\epsilon_j^{(i)}}|\phi_j^{(i)}(\mathbf{r})|^2} \geq 0$ holds only if the bound states have negative eigenvalues.

In Table 1.2, we give four new density-to-potential methods by employing different generating functionals $S[\rho]$. The potential updation scheme for $v_{xc}^{(i)}$ corresponding to functional $S_1[\rho]$ satisfies the convergence condition of Eq. (1.19), whereas others satisfy Eq. (1.20). A new feature based on our study is that different inversion schemes can be mixed to obtain a hybrid scheme. Such a hybrid scheme becomes useful if a different update criterion for modification potential is required in different regions of a given system. An example of a hybrid method is given by the generating functional $S_4[\rho]$, which is obtained by mixing the functional $S_2[\rho]$ with different values of n.

1.2.4 Application to Spherical Systems

We have applied different new inversion methods given in Table 1.2 for calculating the exchange–correlation potential of spherical systems. Systems we have considered are atoms, Hookium [77, 78], and jellium spheres [79], which have different external potentials. For atoms and Hookium, the external potential varies as $-1/r$ and r^2, respectively. On the other hand, for jellium spheres, the external potential has r^2 dependence inside the sphere and goes as $-1/r$ outside the sphere. Here r is the distance from the center of the system. For the atomic systems, we have employed the Hartree–Fock (HF) reference density of Be, Ne, Na, and Ar atoms calculated by Bunge et al. [80] using Slater basis set. For jellium spheres, we have used the electronic density of clusters of N numbers of Na atoms. For a given cluster of Na atoms, each atom contributes exactly one electron to the jellium sphere for a given cluster of atoms. We have taken the density of the jellium sphere corresponding to $N = 18, 34, 58$ atoms. These densities are obtained by solving the Harbola–Sahni (HS) formulation of DFT [5, 81–83]. For the functional $S_2[\rho]$ we have taken values of n to be 0.05, 0.45 and 0.85. To solve the Kohn–Sham equation, we have used the Herman–Skillman program [84] with suitable modifications. For every method, we take the Fermi–Amaldi self-interaction potential $-v_H^0(\mathbf{r})/N$ as the initial guess $v_{xc}^{(i=1)}(\mathbf{r})$ for the exchange–correlation potential. Here N is the number of electrons and $v_H^0(\mathbf{r})$ is the Hartree potential of given ground-state density $\rho_0(\mathbf{r})$. In asymptotic regions where density is small, we have fixed the exchange–correlation potential to its exact value $-1/r$ [85]. Throughout this chapter, we have considered the density to be small if $\rho(\mathbf{r}) \leq 10^{-6}$ for atoms and the Hookium, and $\rho(\mathbf{r}) \leq 10^{-15}$ for jellium spheres. Furthermore, to stabilize the iterative method we have employed linear mixing of potential and used

$$v_{xc}^{(i+1)}(\mathbf{r}) = (1-\varepsilon)v_{xc}^{(i)}(\mathbf{r}) + \epsilon v_{xc}^{(i+1)}(\mathbf{r}) \tag{1.27}$$

in the Kohn–Sham equation, which is equivalent to

$$v_{xc}^{(i+1)}(\mathbf{r}) = v_{xc}^{(i)}(\mathbf{r}) + \epsilon \left[\left.\frac{\delta S[\rho]}{\delta\rho(\mathbf{r})}\right|_{\rho^{(i)}} - \left.\frac{\delta S[\rho]}{\delta\rho(\mathbf{r})}\right|_{\rho_0} \right] \tag{1.28}$$

Here ϵ $(0 < \epsilon < 1)$ is known as the mixing parameter. It is taken to be 0.05 for atoms and Hookium and 0.005 for jellium spheres. To check the convergence of methods, we have calculated the quantity $\Delta_\rho = \int |\rho^{(i)}(\mathbf{r}) - \rho_0(\mathbf{r})| d\mathbf{r}$ and iterated the algorithm until $\Delta_\rho \leq 10^{-4}$.

In Figure 1.1, we have displayed the exchange potential calculated from the HF densities of atoms Be, Ne, Na, and Ar obtained by employing the functionals $S_1[\rho]$, $S_2[\rho]$, and $S_3[\rho]$. We have compared our results with the optimized effective potential (OEP) [86–88]. It is evident from the figure that apart from a little deviation near the nucleus, the calculated potential matches the corresponding exact result. Next, in Figure 1.2, we show the jellium spheres' exchange and Hookium correlation potential. The potentials obtained from functional $S_1[\rho]$ and $S_3[\rho]$ match with both systems' corresponding exact results. Since we used the density of the jellium sphere obtained from the HS potential, we took the HS potential as the exact potential for jellium spheres.

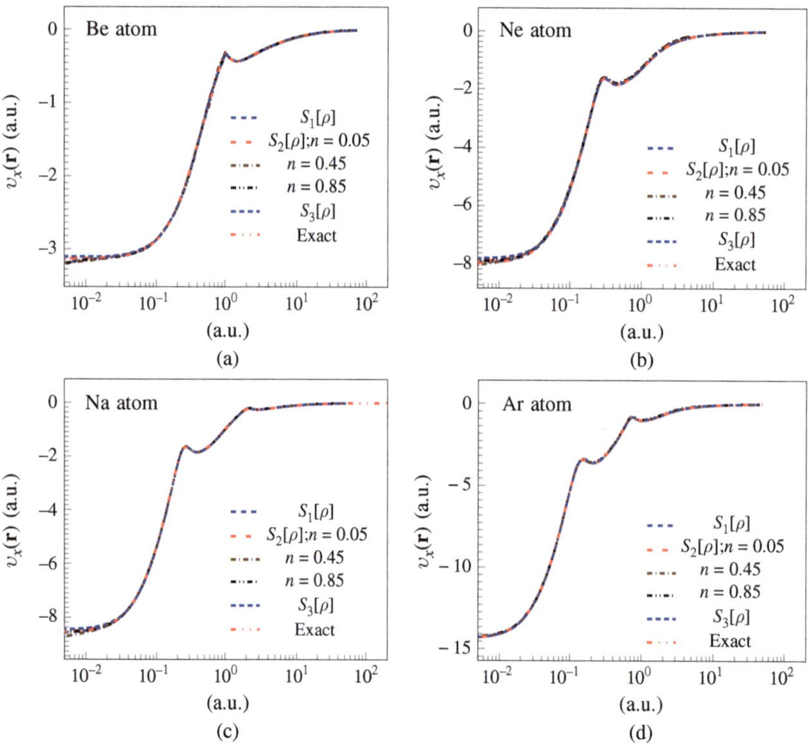

Figure 1.1 Exchange potential of Be (a), Ne (b), Na (c), and Ar (d) atoms corresponding to functionals $S_1[\rho] - S_3[\rho]$.

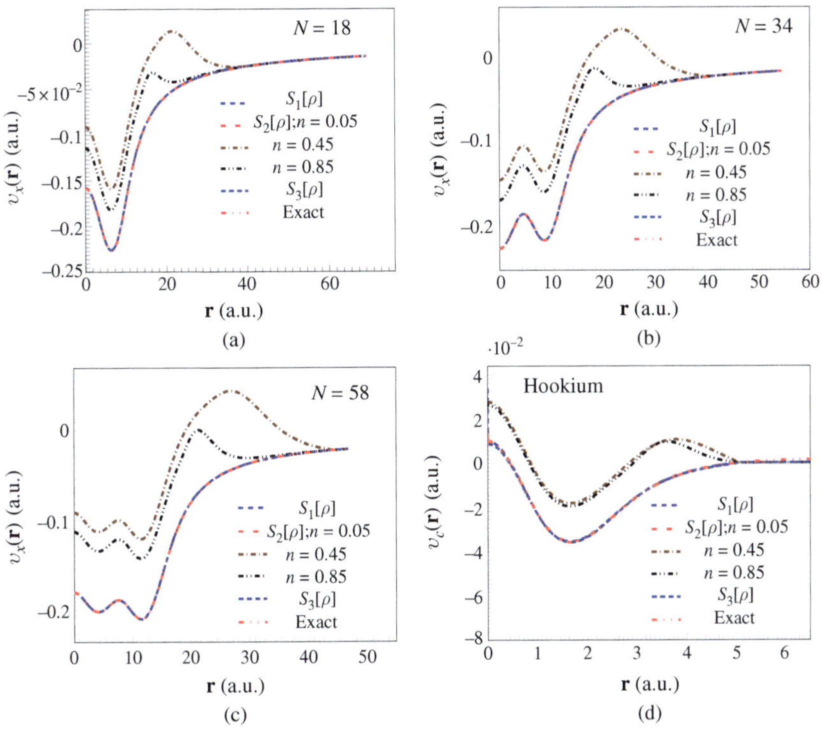

Figure 1.2 (a)–(c) Exchange potential for jellium spheres corresponding to $N = 18, 34, 58$ sodium atoms and (d) the correlation potential of Hookium atom calculated using functionals $S_1[\rho] - S_3[\rho]$.

We now discuss the results for the functional $S_2[\rho]$ for different values of n. It is evident from the figure that for large values of n (0.85 and 0.45), the calculated potential deviates from the exact result in the internal region for both systems. Deviation in the calculated potential occurs because for a large value of n, ρ^n is very small for low density in outer region. Thus, during the iterative process, it does not give sufficient correction to the potential so that it remains unaffected. However, as the value of n is made smaller and smaller, the correction in potential becomes significant. As can be seen from the figure, this is the case for $n = 0.05$, for which the calculated potential becomes close to the corresponding exact result. Using this observation, we design hybrid type functional that uses different correction functions in the different regions of the system. This then leads to proper update in the potential in each region. We demonstrate this idea through the functional $S_4[\rho]$, where we have hybridized the functional $S_2[\rho]$ for two different values $n = n_1$ and $n = n_2$ through the error function erf(αr). We have applied $S_4[\rho]$ for calculation of exchange–correlation potential taking n_1 to be small (0.05) and $n_2 = 0.45$ and 0.85. As discussed earlier, for a small value of n_1 the term ρ^{n_1} provides sufficient correction in the asymptotic region. In contrast, a large value of n_2 provides significant correction in the high-density region near the nucleus. The value of α in $S_4[\rho]$ is 0.1 for jellium spheres and 1.0 for the Hookium. In Figure 1.3, we have shown the exchange potential of $N = 34$ jellium sphere and correlation potential of Hookium obtained from hybrid functional $S_4[\rho]$. It is evident from the figure that the calculated potentials match the corresponding exact results.

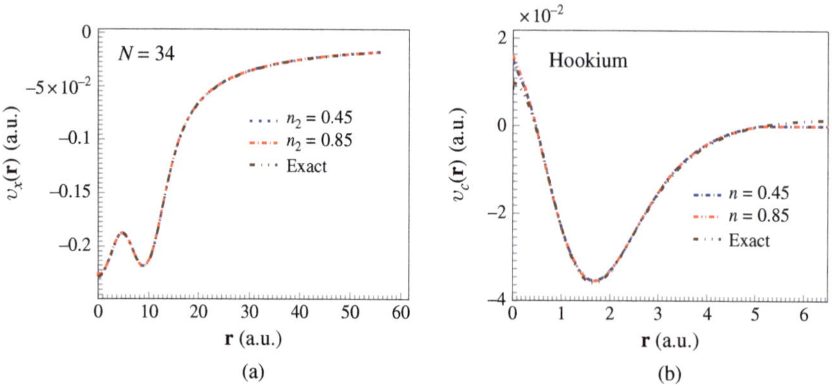

Figure 1.3 (a) Exchange potential of jellium sphere for $N = 34$ and (b) correlation potential of Hookium for hybrid functional $S_4[\rho]$. Here n_1 is fixed to a small value 0.05 and n_2 is taken to be 0.45 and 0.85.

1.2.5 Using Random Numbers to do Density-to-Potential Inversion

We now present a novel approach to do density-to-potential inversion using random numbers [89]. This is based on the observation that in any density-to-potential inversion method potential is updated as

$$v_{\mathrm{xc}}^{(i+1)}(\mathbf{r}) = v_{\mathrm{xc}}^{(i)}(\mathbf{r}) + \Delta v_{\mathrm{xc}}(\mathbf{r})$$

for the $(i+1)$th iteration by making $\Delta v_{\mathrm{xc}}(\mathbf{r})$ repulsive (attractive) in nature where the Kohn–Sham density $\rho^{(i)}(\mathbf{r})$ is larger (smaller) than the given ground-sate density $\rho_0(\mathbf{r})$. This can also be seen mathematically from the strong form of convergence condition Eq. (1.20). In Section 1.2.4 the correction in the potential $\Delta v_{\mathrm{xc}}(\mathbf{r}) = \left.\frac{\delta S[\rho]}{\delta\rho(\mathbf{r})}\right|_{\rho_0} - \left.\frac{\delta S[\rho]}{\delta\rho(\mathbf{r})}\right|_{\rho^{(i)}}$ for different inversion methods obtained by employing various generating functionals given in Tables 1.1 and 1.2. We now devise a purely numerical density-to-potential inversion method which updates $\Delta v_{\mathrm{xc}}(\mathbf{r})$ through random numbers. The method works as follows: Let the potential $v_{\mathrm{xc}}^{(i)}$ be exchange–correlation potential for the ith iteration. Then correction term in the potential, $\Delta v_{\mathrm{xc}}(\mathbf{r})$ for $(i+1)$th iteration is taken using the formula

$$\Delta v_{\mathrm{xc}}(\mathbf{r}) = \lambda f_{\mathrm{R}}(\mathbf{r})\mathrm{sgn}[\rho^{(i)}, \rho_0](\mathbf{r}) \tag{1.29}$$

where λ is a positive quantity which gives the magnitude of correction in the potential. It is chosen in such a way that it keeps on decreasing as the density $\rho^{(i)}(\mathbf{r})$ approaches the given ground-state density $\rho_0(\mathbf{r})$. $f_{\mathrm{R}}(\mathbf{r})$ is a function that takes random value between 0 and 1 at each point $\mathbf{r}$, and $\mathrm{sgn}[\rho^{(i)}, \rho_0](\mathbf{r})$ is as defined in Eq. (1.25). Observe that by using function $\mathrm{sgn}[\rho^{(i)}, \rho_0](\mathbf{r})$ we ensure that the convergence condition of Eq. (1.20) is satisfied. We have applied this algorithm to systems considered in Section 1.2.4. We have taken $\lambda = |\rho^{(i)^n}(\mathbf{r}) - \rho_0^n(\mathbf{r})|$ for the ith iteration, where $n = 1$ for atoms and $n = 0.01$ for the Hookium atom and jellium spheres. We start with mixing parameter ϵ, which is of the order of $10^{-4} - 10^{-3}$ and increase it with iterations as the quantity Δ_ρ starts decreasing. Other conditions are taken to be the same as earlier in Section 1.2.4. We found that the method achieved convergence of $\Delta_\rho < 10^{-6}$ for jellium spheres and Hookium, and $\Delta_\rho < 10^{-5}$ for atoms except for Na atom.

In Figure 1.4, we display the exchange potential for Be and Ar atom, $N = 58$ jellium sphere, and the correlation potential of Hookium. It is evident from the figure that the calculated potential is close to the corresponding exact results except for some fluctuations in the potential. The same is

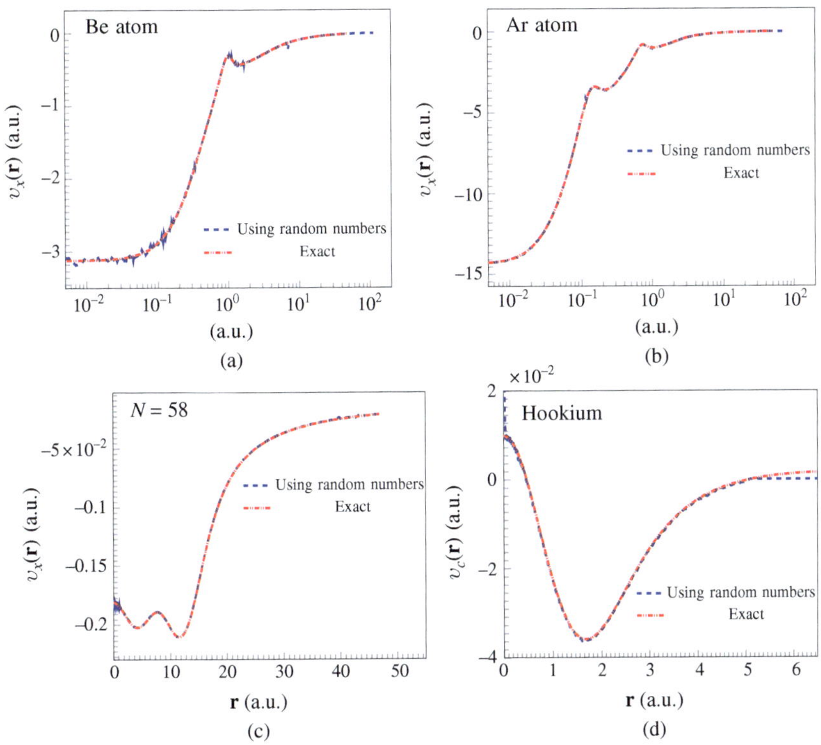

Figure 1.4 (a)–(c) Exchange potential of Be, Ar, and $N = 58$ jellium, and (d) correlation potential of Hookium calculated by using random numbers.

Figure 1.5 Exchange potential for Be atom after smoothing.

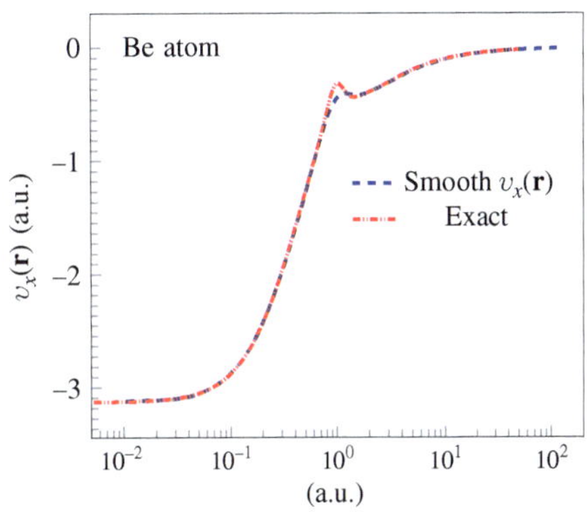

observed for other systems. The potential can be made smooth by removing fluctuation using any smoothing code. We smooth the Be atom's potential by fitting it with the Bezier curve in Gnuplot. The resulting potential after smoothing is shown in Figure 1.5. It can be seen that the exchange potential becomes closer to the exact one except in the inter shell region of the atom, where the bump in the potential softens as a result of smoothing.

We note that use of random numbers offers the following advantages: First, the use of functional $S[\rho]$ requires evaluation of its functional derivative for each iteration, which may be time-consuming making iterative process slow. Second, a fixed $S[\rho]$ treats the whole spatial region on equal- footing and sometimes does not work well for some regions of density where corresponding correction in potential $\Delta v_{xc}(\mathbf{r})$ is very small. This is seen for $S_2[\rho]$ when applied to jellium spheres. Algorithm based on the random numbers circumvents these two problems.

1.3 General Penalty Method for Density-to-Potential Inversion

One of the earliest KS inversion methods is the Zhao–Morrison–Parr (ZMP) method [35, 37, 40] whereby the KS orbitals are obtained by minimizing the noninteracting kinetic energy functional

$$T_S[\rho] = \sum_i \langle \phi_i | -\frac{1}{2}\nabla^2 | \phi_i \rangle \tag{1.30}$$

with the constraint that the orbitals lead to the target density $\rho_0(\mathbf{r})$ i.e.

$$\sum_i |\phi_i(\mathbf{r})|^2 - \rho_0(\mathbf{r}) = 0 \tag{1.31}$$

This is done by rewriting the above constraint as

$$\frac{1}{2}\iint \frac{[\rho(\mathbf{r}) - \rho_0(\mathbf{r})][\rho(\mathbf{r}') - \rho_0(\mathbf{r}')]}{|\mathbf{r}-\mathbf{r}'|} d\mathbf{r}d\mathbf{r}' = 0 \tag{1.32}$$

with $\rho(\mathbf{r}) = \sum_i |\phi_i(\mathbf{r})|^2$, and minimizing the functional

$$\sum_i \langle \phi_i | -\frac{1}{2}\nabla^2 | \phi_i \rangle + \frac{\lambda}{2}\iint \frac{[\rho(\mathbf{r}) - \rho_0(\mathbf{r})][\rho(\mathbf{r}') - \rho_0(\mathbf{r}')]}{|\mathbf{r}-\mathbf{r}'|} d\mathbf{r}d\mathbf{r}' \tag{1.33}$$

with respect to variation in the orbitals. This leads to the equation

$$\left[-\frac{1}{2}\nabla^2 + \lambda \int \frac{[\rho(\mathbf{r}') - \rho_0(\mathbf{r}')]}{|\mathbf{r}-\mathbf{r}'|} d\mathbf{r}'\right] \phi_i(\mathbf{r}) = \epsilon_i \phi_i(\mathbf{r}) \tag{1.34}$$

When solved in the limit $\lambda \to \infty$, it gives $\rho(\mathbf{r}) \to \rho_0(\mathbf{r})$ and the Kohn–Sham potential

$$v_{KS}(\mathbf{r}) = \lambda \int \frac{[\rho(\mathbf{r}') - \rho_0(\mathbf{r}')]}{|\mathbf{r}-\mathbf{r}'|} d\mathbf{r}' \tag{1.35}$$

In carrying out actual calculations, one separates the external potential $v_{ext}(\mathbf{r})$ and the exact Hartree potential $v_H[\rho_0](\mathbf{r}) = \int \frac{[\rho_0(\mathbf{r}')]}{|\mathbf{r}-\mathbf{r}'|} d\mathbf{r}'$ and writes

$$v_{KS}(\mathbf{r}) = v_{ext}(\mathbf{r}) + v_H[\rho_0](\mathbf{r}) + \lambda \int \frac{[\rho(\mathbf{r}') - \rho_0(\mathbf{r}')]}{|\mathbf{r}-\mathbf{r}'|} d\mathbf{r}' \tag{1.36}$$

in the limit $\lambda \to \infty$.

ZMP method falls under the category of the penalty method. We explain this below in the context of Kohn–Sham inversion. In penalty method, the functional $T_S[\rho] = T_S[\{\phi_i\}]$ written in terms of orbitals $\{\phi_i\}$ is minimized by adding to it a penalty functional $\lambda P[\rho, \rho_0]$, where $\rho(\mathbf{r}) = \sum_i |\phi_i(\mathbf{r})|^2$ and λ is the penalty parameter. The penalty functional has the following properties:

$$\begin{aligned} P[\rho, \rho_0] &> 0 \quad \text{if} \quad \rho(\mathbf{r}) \neq \rho_0(\mathbf{r}) \\ &= 0 \quad \text{if} \quad \rho(\mathbf{r}) = \rho_0(\mathbf{r}) \end{aligned} \tag{1.37}$$

The end result of the calculation is based on following theorem [90]:

Theorem 1.1 Let $\left\{\{\phi_i\}_k\right\}$, $k = 1, 2, \ldots, \infty$ be a convergent sequence of functions $\{\phi_i\}$ that minimize $T_P[\{\phi_i\}, \lambda_k] = T[\{\phi_i\}] + \lambda_k P[\rho, \rho_0]$ with $\lambda_{k+1} > \lambda_k > 0$. Then limit $\{\overline{\phi}_i\}$ of the set $\left\{\{\phi_i\}_k\right\}$ gives minimum of $T[\{\phi_i\}]$ satisfying the constraint $P[\rho, \rho_0] = 0$.

Evidently, the penalty functional in the ZMP method is expression for the Coulomb energy of the density difference $\rho(\mathbf{r}) - \rho_0(\mathbf{r})$ as given on the left-hand side of Eq. (1.32). Nevertheless, any functional of $\rho(\mathbf{r})$ that satisfies the criteria of Eq. (1.37) can be used as the penalty functional according to Theorem 1.1. What could this penalty functional be? The answer is quite straightforward and is provided by the condition of Eq. (1.19) satisfied by the generating functional $S[\rho]$. Thus, the functional $\int \left(\left.\frac{\delta S[\rho]}{\delta \rho(\mathbf{r})}\right|_{\rho} - \left.\frac{\delta S[\rho]}{\delta \rho(\mathbf{r})}\right|_{\rho_0} \right) \Delta\rho(\mathbf{r}) d\mathbf{r}$ for the generating functional employed in the Kohn–Sham inversion can be used as a penalty functional since it satisfies Eq. (1.37). This gives a general penalty method to carry out Kohn–Sham inversion based on Levy's constrained search approach. We now present the results of employing three such functionals. Before doing that, however, we note with satisfaction that the same generating functional can be used in the maximization scheme of Lieb and minimization of Levy, both of which lead to the universal functional for a given ground-state density.

We have applied general penalty method described to calculate exchange–correlation potential of spherical systems. In general penalty method, for a given functional $S[\rho]$ the Kohn–Sham potential is

$$v_{\mathrm{KS}}(\mathbf{r}) = v_{\mathrm{ext}}(\mathbf{r}) + v_{\mathrm{H}}[\rho_0](\mathbf{r}) + v^{\lambda}[\rho, \rho_0](\mathbf{r}) \tag{1.38}$$

where

$$v^{\lambda}[\rho, \rho_0](\mathbf{r}) = \lambda \left[\frac{\delta S[\rho]}{\delta \rho(\mathbf{r})} - \left.\frac{\delta S[\rho]}{\delta \rho(\mathbf{r})}\right|_{\rho_0(\mathbf{r})} + \int \frac{\delta^2 S[\rho]}{\delta \rho(\mathbf{r}) \delta \rho(\mathbf{r}')} (\rho(\mathbf{r}') - \rho_0(\mathbf{r}')) d\mathbf{r}' \right. \tag{1.39}$$

and exchange–correlation potential is obtained by

$$v_{\mathrm{xc}}(\mathbf{r}) = \lim_{\lambda \to \infty} v^{\lambda}[\rho, \rho_0](\mathbf{r}) \tag{1.40}$$

For our calculations, we have used functional $S_1[\rho] - S_3[\rho]$ given in Table 1.2. For the functional $S_2[\rho]$, n is taken to be 0.05. Potential $v^{\lambda}[\rho, \rho_0](\mathbf{r})$ corresponding to these functionals are given in Table 1.3. As is evident, the potential $v^{\lambda}[\rho, \rho_0](\mathbf{r})$ in general penalty method is much easier to calculate than the ZMP method because there is no integral to be performed.

In our calculations, we have iteratively updated the penalty parameter λ as suggested in Ref. [91]. We start with a small value λ say λ_j $(j = 1)$ and solve self-consistently the Kohn–Sham equation for the potential

$$v_{\mathrm{KS}}(\mathbf{r}) = v_{\mathrm{fixed}}^{\lambda_{j-1}}(\mathbf{r}) + v^{\lambda_j}[\rho_0, \rho](\mathbf{r}) \tag{1.41}$$

where $v_{\mathrm{fixed}}^{\lambda_{j-1}}(\mathbf{r})$ is the Kohn–Sham potential obtained for λ_{j-1}, which is known exactly and kept fixed and $v^{\lambda_j}[\rho_0, \rho](\mathbf{r})$ is updated during the self-consistent process. For $j = 0$ we have taken

Table 1.3 The potential $v^{\lambda}[\rho, \rho_0](\mathbf{r})$ of Eq. (1.39) for the functionals $S_1[\rho] - S_3[\rho]$ is given in Table 1.2.

Functionals $S[\rho]$	$v^{\lambda}[\rho, \rho_0](\mathbf{r}) = \lambda \left[\frac{\delta S[\rho]}{\delta \rho(\mathbf{r})} - \left.\frac{\delta S[\rho]}{\delta \rho(\mathbf{r})}\right|_{\rho_0(\mathbf{r})} + \int \frac{\delta^2 S[\rho]}{\delta \rho(\mathbf{r}) \delta \rho(\mathbf{r}')} (\rho(\mathbf{r}') - \rho_0(\mathbf{r}')) d\mathbf{r}' \right]$
$S_1[\rho]$	$\lambda \left[-\frac{1}{2} \frac{\nabla^2 \rho^{1/2}(\mathbf{r})}{\rho^{1/2}(\mathbf{r})} + \frac{1}{2} \frac{\nabla^2 \rho_0^{1/2}(\mathbf{r})}{\rho_0^{1/2}(\mathbf{r})} + \left(\frac{\rho(\mathbf{r}) - \rho_0(\mathbf{r})}{4\rho(\mathbf{r})} \right) \left(\left| \frac{\nabla \rho(\mathbf{r})}{\rho(\mathbf{r})} \right|^2 - \frac{\nabla^2 \rho(\mathbf{r})}{\rho(\mathbf{r})} \right) \right]$
$S_2[\rho]$	$\lambda \left[2\rho^n(\mathbf{r}) + \rho_0(\mathbf{r}) \rho^{n-1}(\mathbf{r}) - \rho_0^n(\mathbf{r}) \right]$
$S_3[\rho]$	$\lambda \left[\log \frac{\rho(\mathbf{r})}{\rho_0(\mathbf{r})} + \left(1 - \frac{\rho_0(\mathbf{r})}{\rho(\mathbf{r})} \right) \right]$

$v_{\text{fixed}}^{\lambda_0}(\mathbf{r}) = v_{\text{ext}}(\mathbf{r}) + (1 - 1/N)v_H[\rho_0](\mathbf{r})$, where N is the total numbers of electrons. Having obtained the self-consistent solution for λ_j, we check if $\Delta_\rho = \int |\rho_0(\mathbf{r}) - \rho(\mathbf{r})|d\mathbf{r}$ has become smaller than 1×10^{-5} for atoms and 1×10^{-6} for jellium spheres and Hookium atom. When this is true, then we stop. Otherwise we increase the value of λ from λ_j to λ_{j+1} and again solve the Kohn–Sham equation using Eq. (1.41).

In Figure 1.6, we give the exchange potential of atoms calculated by the general penalty method. It can be clearly seen that for all atoms, the potential calculated by each functional $S[\rho]$ matches with the corresponding exact result. The same is true for jellium spheres and Hookium as is evident from Figure 1.7.

A different approach to constrained minimization has been taken in Ref. [59]. This is based on partial differential equation constrained optimization [56]. In that work, the quantity $\int w(\mathbf{r})(\rho_0(\mathbf{r}) - \sum_i|\phi(\mathbf{r})|^2)^2 d\mathbf{r}$ is minimized with the constraint that $\{\phi_i(\mathbf{r})\}$ satisfy the Kohn–Sham equation, be normalized and simultaneously be orthogonal to auxiliary functions $\{p_i(\mathbf{r})\}$ that satisfy an inhomogeneous Kohn–Sham equation. Thus, the functional that is minimized is

$$L[v_{\text{xc}}(\mathbf{r}), \{\phi_i\}, \{\epsilon_i\}, \{p_i\}] = \int w(\mathbf{r})(\rho_0(\mathbf{r}) - \sum_i |\phi_i(\mathbf{r})|^2)^2 d\mathbf{r} + \sum_i \int p_i(\mathbf{r})(\hat{H} - \epsilon_i)\phi_i(\mathbf{r})d\mathbf{r} + \sum_i \mu_i \left(\int |\phi_i(\mathbf{r})|^2 d\mathbf{r} - 1 \right) \quad (1.42)$$

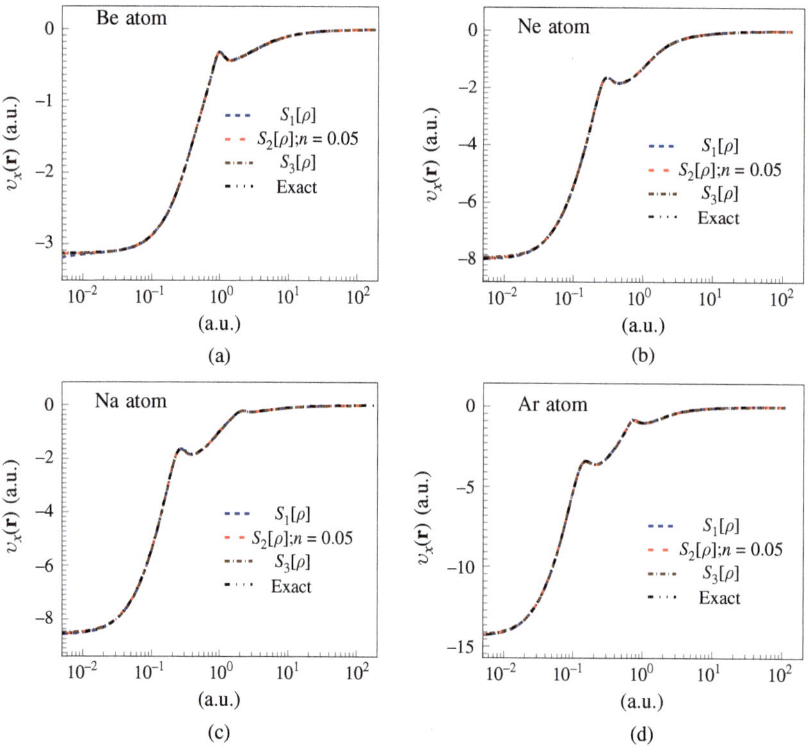

Figure 1.6 Exchange potential of Be (a), Ne (b), Na (c), and Ar (d) atoms calculated by the general penalty method using the functionals $S_1[\rho] - S_3[\rho]$.

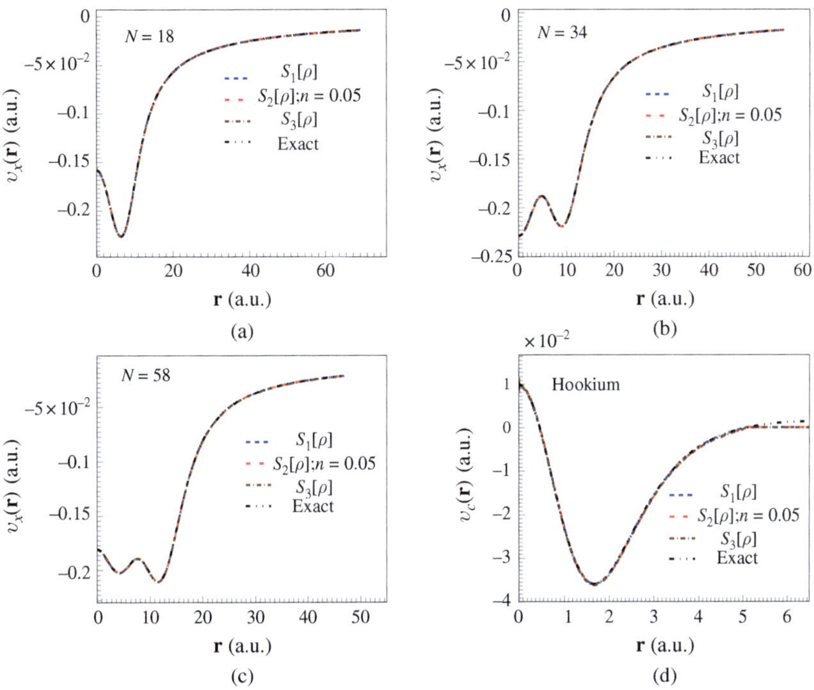

Figure 1.7 (a)–(c) Exchange potential for jellium spheres corresponding to $N = 18, 34, 58$ sodium atoms and (d) the correlation potential of Hookium atom obtained by the general penalty method using the functionals $S_1[\rho] - S_3[\rho]$.

where $\hat{H} = -\frac{1}{2}\nabla^2 + v_{\text{ext}}(\mathbf{r}) + v_{\text{H}}(\mathbf{r}) + v_{\text{xc}}(\mathbf{r})$ and minimization is done with respect to $\phi_i(\mathbf{r})$, $p_i(\mathbf{r})$, ϵ_i and μ_i. This leads to the equations

$$\hat{H}\phi_i(\mathbf{r}) = \epsilon_i\phi_i(\mathbf{r}) \tag{1.43}$$

$$\int |\phi_i(\mathbf{r})|^2 d\mathbf{r} = 1 \tag{1.44}$$

$$\int (\hat{H} - \epsilon_i)p_i(\mathbf{r}) = g_i(\mathbf{r}) \tag{1.45}$$

$$\int p_i(\mathbf{r})\phi_i(\mathbf{r})d\mathbf{r} = 0 \tag{1.46}$$

with

$$g_i(\mathbf{r}) = 8w(\mathbf{r})\left(\rho_0(\mathbf{r}) - \sum_i |\phi_i(\mathbf{r})|^2\right)\phi_i(\mathbf{r}) - 2\mu_i\phi_i(\mathbf{r}) \tag{1.47}$$

Exchange–correlation potential is updated using the derivative

$$\frac{\delta L}{\delta v_{\text{xc}}(\mathbf{r})} = \sum_i p_i(\mathbf{r})\phi_i(\mathbf{r})$$

via gradient-based optimization technique. The method has been applied to H_2, H_2O and C_6H_6 molecules.

Having discussed in detail the density-based inversion schemes, we now turn our attention to methods that employ the many-body wavefunctions directly to obtain the corresponding

Kohn–Sham potential. These methods have been developed and employed [49, 51–53, 60, 92–94] over the past decade and have an advantage over density-based methods when the latter is used with densities derived from wavefunction expanded in basis sets. For such densities, density-based methods generally lead to spurious features [95, 96] in the exchange–correlation potential in the form of large deviations from the exact potential. However, these features are not present when wavefunction-based methods are employed. In the next section, we explain using the LPS equation why density-based inversion leads to such features in the potential and why they disappear when wavefunction-based methods are employed.

1.4 Understanding Connection Between Density and Wavefunction-Based Inversion Methods Using LPS Equation [97]

For a ground-state density $\rho(\mathbf{r})$. The LPS equation gives

$$v_{\text{eff}}(\mathbf{r}) = \mu + \frac{1}{2}\frac{\nabla^2\sqrt{\rho(\mathbf{r})}}{\sqrt{\rho(\mathbf{r})}} - v_{\text{ext}}(\mathbf{r}) \tag{1.48}$$

and therefore

$$v_{\text{xc}}(\mathbf{r}) = \mu + \frac{1}{2}\frac{\nabla^2\sqrt{\rho(\mathbf{r})}}{\sqrt{\rho(\mathbf{r})}} - v_{\text{H}}(\mathbf{r}) - v_{\text{P}}(\mathbf{r}) - v_{\text{ext}}(\mathbf{r}) \tag{1.49}$$

Now if $\rho(\mathbf{r})$ is an approximate density that differs from the density corresponding to $v_{\text{ext}}(\mathbf{r})$, the true external potential $\overline{v}_{\text{ext}}(\mathbf{r})$ for $\rho(\mathbf{r})$ will be different [97] from $v_{\text{ext}}(\mathbf{r})$ by the Hohenberg–Kohn theorem. This difference $\Delta v_{\text{ext}}(\mathbf{r}) = \overline{v}_{\text{ext}}(\mathbf{r}) - v_{\text{ext}}(\mathbf{r})$ then becomes a part of the exchange–correlation potential obtained by inverting the density $\rho(\mathbf{r})$, as evident from Eq. (1.49). This is what gives rise to spurious oscillations in the exchange–correlation potential and makes it deviate from the true potential substantially. For example, if the density in far away regions of atoms decays like a Gaussian, the corresponding potential will increases [96] gradually with the distance, and this is reflected in the exchange–correlation potential. Similar large deviations from the exact $v_{\text{xc}}(\mathbf{r})$ occur near the nucleus. If one wishes to avoid these undesirable features, it is obvious that $\overline{v}_{\text{ext}}(\mathbf{r})$ should not be part of the expression for the exchange–correlation potential. That is accomplished by using LPS expression for $v_{\text{eff}}(\mathbf{r})$ in terms of wavefunctions.

For a many-body wavefunction Ψ, $v_{\text{eff}}[\Psi](\mathbf{r})$ is given as [67]

$$v_{\text{eff}}[\Psi](\mathbf{r}) = \int \frac{\rho_{N-1}(\mathbf{r};\mathbf{r}')}{|\mathbf{r}-\mathbf{r}'|}d\mathbf{r}' + \langle\Phi_{N-1}|H_{N-1} - E^0_{N-1}|\Phi_{N-1}\rangle + \frac{1}{2}\int |\nabla\Phi_{N-1}|^2 d\sigma d\mathbf{x}_{2-N} \tag{1.50}$$

where $\mathbf{x}$ represents spatial $\mathbf{r}$ and spin σ coordinates for a given electron, $\mathbf{x}_{i-j} = \mathbf{x}_i, \mathbf{x}_{i+1}, \dots, \mathbf{x}_j$. The function

$$\Phi_{N-1}(\mathbf{x}, \mathbf{x}_{2-N}) = \left(\frac{N}{\rho(\mathbf{r})}\right)^{1/2}\Psi(\mathbf{x}, \mathbf{x}_{2-N}) \tag{1.51}$$

is the conditional probability amplitude [98]. It is normalized for every value of $\mathbf{x}$. Furthermore, for a given electron at $\mathbf{x}$, $|\Phi_{N-1}(\mathbf{x}, \mathbf{x}_{2-N})|^2$ gives the probability of finding other electrons at $\mathbf{x}_{2-N}$. $\rho_{N-1}(\mathbf{r};\mathbf{r}')$ is the density of $N-1$ particles at $\mathbf{r}'$ associated with the function $\Phi_{N-1}(\mathbf{x}, \mathbf{x}_{2-N})$ and is given as

$$\rho_{N-1}(\mathbf{r};\mathbf{r}') = (N-1)\int |\Phi_{N-1}(\mathbf{x}, \mathbf{x}', \mathbf{x}_{3-N})|^2 d\sigma d\sigma' d\mathbf{x}_{3-N} \tag{1.52}$$

Next,

$$H_{N-1}(\mathbf{r}_{2-N}) = \sum_{i=2}^{N} -\frac{1}{2}\nabla_i^2 + \sum_{i=2}^{N} v_{\text{ext}}(\mathbf{r}_i) + \frac{1}{2}\sum_{\substack{i,j=2 \\ i\neq j}}^{N} \frac{1}{|\mathbf{r}_i - \mathbf{r}_j|} \tag{1.53}$$

is the Hamiltonian of $N-1$ interacting electrons moving in external potential $v_{\text{ext}}(\mathbf{r})$ and E^0_{N-1} is the corresponding ground-state energy. If the exact wavefunctions Ψ_0 and corresponding densities are used in Eqs. (1.50)–(1.52), we get the exact effective potential $v_{\text{eff}}[\Psi_0](\mathbf{r})$. We now show if an approximate wavefunction Ψ is used in the expressions, the resulting effective potential $v_{\text{eff}}[\Psi](\mathbf{r})$ does not differ much from the $v_{\text{eff}}[\Psi_0](\mathbf{r})$. For this expand the conditional probability amplitude from Ψ as

$$\Phi_{N-1}(\mathbf{x}, \mathbf{x}_{2-N}) = \sum_i f_i(\mathbf{x})\Psi^i_{N-1}(\mathbf{x}_{2-N}) \tag{1.54}$$

where because of Φ_{N-1} being normalized, $\sum_i |f_i(\mathbf{x})|^2 = 1$ so that $|f_i(\mathbf{x})| \leq 1$ for every value of $\mathbf{x}$; here Ψ^i_{N-1} are the eigenfunctions of H_{N_1} given in Eq. (1.53).

For well-behaved Ψ [99–101] and because $\rho(\mathbf{r})$ for the ground-state is nonzero except when $r \to \infty$, f_i and its gradient will be finite and smooth. Thus, all the terms in $v_{\text{eff}}[\Psi](\mathbf{r})$ viz.

$$\int \frac{\rho_{N-1}(\mathbf{r};\mathbf{r}')}{|\mathbf{r}-\mathbf{r}'|} d\mathbf{r}' = (N-1)\sum_{i,j} \int f_i^*(\mathbf{x}) f_j(\mathbf{x}) d\sigma \int \frac{\Psi^{i*}_{N-1}\Psi^j_{N-1}}{|\mathbf{r}-\mathbf{r}'|} d\mathbf{x}_{2-N}$$

$$\langle \Phi_{N-1}|H_{N-1} - E^0_{N-1}|\Phi_{N-1}\rangle = \sum_{i\neq 0}(E^i_{N-1} - E^0_{N-1}) \int |f_i(\mathbf{x})|^2 d\sigma$$

and

$$\frac{1}{2}\int |\nabla\Phi_{N-1}|^2 d\sigma d\mathbf{x}_{2-N} = \frac{1}{2}\sum_i \int |\nabla f_i(\mathbf{x})|^2 d\sigma$$

are also finite and do not become spuriously large. On the other hand, if in the expression there was a term having division by $|f|^2$ or $\rho(\mathbf{r})$, that term could have become large. Using the relations derived, we now obtain deviation of $v_{\text{eff}}[\Psi](\mathbf{r})$ from the exact one.

If $f_i^0(\mathbf{x})$ are the functions corresponding to the exact wavefunction Ψ_0, then for $f_i(\mathbf{x}) = f_i^0(\mathbf{x}) + \delta f_i(\mathbf{x})$ the effective potential

$$v_{\text{eff}}[\Psi](\mathbf{r}) = v_{\text{eff}}[\Psi_0](\mathbf{r}) + \delta v_{\text{LPS}}(\mathbf{r})$$

where

$$\begin{aligned}\delta v_{\text{LPS}}(\mathbf{r}) = & \sum_{i\neq 0}(E^i_{N-1} - E^0_{N-1}) \int \left(f_i^{0*}(\mathbf{x})\delta f_i(\mathbf{x}) + f_i^0(\mathbf{x})\delta f_i^*(\mathbf{x}) + |\delta f_i(\mathbf{x})|^2\right) d\sigma \\ & + (N-1)\sum_{i,j} \int \left(\delta f_i^*(\mathbf{x}) f_j^0(\mathbf{x}) + f_i^{0*}(\mathbf{x})\delta f_j(\mathbf{x}) + \delta f_i^*(\mathbf{x})\delta f_j(\mathbf{x})\right) d\sigma \int \frac{\Psi^{i*}_{N-1}\Psi^j_{N-1}}{|\mathbf{r}-\mathbf{r}'|} d\mathbf{x}_{2-N} \\ & + \frac{1}{2}\sum_i \int \left(\nabla f_i^{0*}(\mathbf{x})\cdot\nabla\delta f_i(\mathbf{x}) + \nabla f_i^0(\mathbf{x})\cdot\nabla\delta f_i^*(\mathbf{x}) + |\nabla\delta f_i(\mathbf{x})|^2\right) d\sigma\end{aligned} \tag{1.55}$$

As is apparent, $\delta v_{\text{LPS}}(\mathbf{r})$ has terms that do not grow large erroneously. Furthermore, for small $\{\delta f_i\}$, the difference is linear in $\{\delta f_i\}$. Thus, it can be safely concluded that $v_{\text{eff}}[\Psi](\mathbf{r})$ is close to $v_{\text{eff}}[\Psi_0](\mathbf{r})$.

Next, we observe the following. Since $v_{\text{eff}}[\Psi](\mathbf{r})$ is free from spuriously large deviations from $v_{\text{eff}}[\Psi_0](\mathbf{r})$, it is anticipated that exchange–correlation potential calculated from $v_{\text{eff}}[\Psi](\mathbf{r})$ will also not deviate arbitrarily from the exact potential. We now show this to be the case for the Kohn–Sham calculation.

For a given many-body wavefunction Ψ, the Kohn–Sham exchange–correlation potential $v_{xc}[\Psi](\mathbf{r})$ is obtained from $v_{eff}[\Psi](\mathbf{r})$ by subtracting from it $v_P[\{\phi_i\}](\mathbf{r})$, the Pauli potential for the Kohn–Sham orbitals. The convergence toward true $v_{xc}(\mathbf{r})$ is done by an iterative process. The resulting potential $v_{xc}[\Psi](\mathbf{r})$ will be smooth and will not have erroneously large deviation from the exact potential if the Pauli potential $v_P[\{\phi_i\}](\mathbf{r})$ is well behaved at each iteration for approximate $\{\phi_i\}$. Since Pauli potential [71, 97, 102] is the same as the LPS effective potential for the noninteracting electrons moving in Kohn–Sham external potential, it can be shown that the $v_P[\{\phi_i\}](\mathbf{r})$ is well behaved for each iteration.

For this, at the nth iteration, we write

$$\Phi_{N-1}^{KS}(\mathbf{x},\mathbf{x}_{2-N}) = \sum_i f_i^{KS(n)}(\mathbf{x})\Psi_{N-1}^{i,KS(n)}(\mathbf{x}_{2-N}) \tag{1.56}$$

where $\Psi_{N-1}^{i,KS(n)}(\mathbf{x}_{2-N})$ represents the determinant for the ith excited state of the KS Hamiltonian corresponding to the nth iteration and $\{f_i^{KS(n)}\}$ have the same properties as in the fully interacting case (see Eq. (1.54)). Then during each iteration

$$v_P^{(n)}[\{\phi_i\}](\mathbf{r}) = \sum_{i\neq 0}\left(E_{N-1}^{i,KS(n)} - E_{N-1}^{0,KS(n)}\right)\int |f_i^{KS(n)}(\mathbf{x})|^2 d\sigma + \frac{1}{2}\sum_i \int |\nabla f_i^{KS(n)}(\mathbf{x})|^2 d\sigma \tag{1.57}$$

Thus, if we start iterations with a reasonable approximation (say LDA) to v_{xc}, this potential is always going to be free of erroneous large terms and close to the exact Pauli potential for the nth iteration, thereby leading to a smooth exchange–correlation potential as iterations proceed. Furthermore, as shown by the Eq. (1.55), this exchange–correlation potential will be close to the true exchange–correlation potential after convergence.

The analysis above explains the well-behaved nature and accuracy of the exchange–correlation potential obtained in Refs. [49, 51, 93]. The expressions for the exchange–correlation potential used by these researchers have been shown [97] to be

$$v_{xc}^{RKS}(\mathbf{r}) = v_{eff}[\Psi](\mathbf{r}) + \frac{1}{8}\left|\frac{\nabla\rho(\mathbf{r})}{\rho(\mathbf{r})}\right|^2 - v_P[\{\phi_i^{KS}\}](\mathbf{r}) - \frac{1}{8}\left|\frac{\nabla\rho_{KS}(\mathbf{r})}{\rho_{KS}(\mathbf{r})}\right|^2 - v_H[\rho](\mathbf{r}) \tag{1.58}$$

where RKS stands for Ryabinkin et al. [51] and

$$v_{xc}^{mRKS}(\mathbf{r}) = v_{eff}[\Psi](\mathbf{r}) - v_P[\{\phi_i^{KS}\}](\mathbf{r}) - v_H[\rho](\mathbf{r}) \tag{1.59}$$

where mRKS stands for modified RKS [93]. Here, ρ is density calculated from many-body wavefunction and ρ_{KS} is KS density. Since we have shown that the LPS potential obtained from many-body wavefunction Ψ and Kohn–Sham orbitals $\{\phi_i^{KS}\}$ is well behaved, so the resulting exchange–correlation potential obtained from RKS method is also expected to show proper structure. However, Eq. (1.58) also contains density-dependent term $|\nabla\rho(\mathbf{r})/\rho(\mathbf{r})|^2 - |\nabla\rho_{KS}(\mathbf{r})/\rho_{KS}(\mathbf{r})|^2$ in it whose effect may appear in the resulting potential. For $\rho^{KS}(\mathbf{r}) \approx \rho(\mathbf{r})$, the contribution of density-dependent term is vanishingly small. However, for the finite-basis-set calculation $\rho^{KS}(\mathbf{r}) \neq \rho(\mathbf{r})$ and the quantity $|\nabla\rho(\mathbf{r})/\rho(\mathbf{r})|^2 - |\nabla\rho_{KS}(\mathbf{r})/\rho_{KS}(\mathbf{r})|^2$ may give significant contributions and the resulting potential could have pathological features. This is seen for Ar atom [93] where the exchange–correlation shows well-behaved nature only for a large basis-set calculation. On the other hand, $v_{xc}^{mRKS}(\mathbf{r})$ contains no such terms, and therefore, gives exchange–correlation potential that is free of wild oscillations irrespective of the size of the basis set used to obtain the wavefunction.

1.5 Concluding Remarks

Interest in density-to-potential inversion problem has various aspects. First and foremost, being able to obtain the exact Kohn–Sham potential for a system from its density and getting to learn about its structure is by itself intellectually quite stimulating. Once we get the exact potential, it can then be employed to analyze various exchange–correlation energy functionals and can lead to making them more accurate. Exact Kohn–Sham potential can also be employed to test theoretical predictions like ionization potential theorem [30], derivative discontinuity of the exchange–correlation [103], step structure of Kohn–Sham potential [104], and recently [105] has been applied to test the relationship between the fundamental band gap and Kohn–Sham gap in solids. In this chapter, we have discussed various methods of doing the density-to-potential inversion and shown them to be arising from a single algorithm based on the Euler–Lagrange equation for the density. For this, we first use the LPS equation to obtain an inversion method based on the von Wizscaker kinetic energy functional. We then use the insight that the Euler equation can assume various forms depending upon how the noninteracting kinetic energy functional is partitioned. Each of these forms gives a different method of density-to-potential inversion. Furthermore, the generating functional for these methods satisfies a mathematical condition. We have shown that using this condition, we can employ random numbers to carry out density-to-potential inversion. Finally, we have given a general penalty method for performing the inversion. An interesting aspect of our focus on the LPS equation is that it led to an explanation of the accuracy of wavefunction-based methods developed and employed in Refs. [49, 51, 60, 92, 93].

We have performed density-to-potential inversion computations by developing codes based on Herman–Skillman program [84]. Handling numerical instabilities with care is necessary when writing these codes. Interest in studying exact Kohn–Sham potential has led to some groups to make codes developed by them available to researchers. Two such codes are IDEA [106] and n2v [107].

Our focus in this chapter has been on obtaining the Kohn–Sham potential for a given ground-state density. An equally important research activity is to get the exact external potential for a given density where the electron–electron interaction strength is varied. This is done [108–111] to study adiabatic connection [112] in density-functional theory. In these references the inversion has been carried out by employing the Wu–Yang method [113, 114] or by using the ZMP method [35, 37, 40].

Another area of research where inversion methods are useful is the study of time-dependent Kohn–Sham potential. These studies have been carried out in Refs. [115–119] and in Refs. [103, 106, 120–122]. These have led to insights into dynamics of electrons in a time-dependent applied field. We would like to suggest and hopefully in future carry out ourselves a general method of inversion for time-dependent problems that are similar to the work reported in this chapter.

Acknowledgments

We thank Profs. P. K. Chattaraj and D. Chakraborty for invitation to present our work in this book.

References

1 Hohenberg, P. and Kohn, W. (1964). Inhomogeneous electron gas. *Phys. Rev.* 136: B864–B871.

2 Kohn, W. and Sham, L.J. (1965). Self-consistent equations including exchange and correlation effects. *Phys. Rev.* 140: A1133–A1138.

3 Parr, R.G. and Yang, W. (1995). *Density-Functional Theory of Atoms and Molecules*. Oxford Science Publications.

4 Dreizler, R.M. and Gross, E.K.U. (1990). *Density Functional Theory*. Berlin Heidelberg: Springer-Verlag.

5 Sahni, V. (2004). *Quantal Density Functional Theory*. Springer.

6 Engel, E. and Dreizler, R.M. (2011). *Density Functional Theory*. Berlin Heidelberg: Springer-Verlag.

7 Pribram-Jones, A., Gross, D.A., and Burke, K. (2015). DFT: A theory full of holes? *Annu. Rev. Phys. Chem.* 66: 283–304.

8 Jones, R.O. (2015). Density functional theory: its origins, rise to prominence, and future. *Rev. Mod. Phys.* 87: 897–923.

9 Singh, P. and Harbola, M.K. (2021). Density-functional theory of material design: fundamentals and applications-I. *Oxford Open Mater. Sci.* 1: itab018.

10 Ziman, J.M. (1975). *Elements of Advanced Quantum Theory*. Cambridge: Cambridge University Press.

11 Fetter, A.L. and Walecka, J.D. (2003). *Quantum Theory of Many-Particle Systems*. Courier Corporation.

12 Pariser, R. and Parr, R.G. (1953). A semi-empirical theory of the electronic spectra and electronic structure of complex unsaturated molecules. I. *J. Chem. Phys.* 21: 466–471.

13 Pariser, R. and Parr, R.G. (1953). A semi-empirical theory of the electronic spectra and electronic structure of complex unsaturated molecules. II. *J. Chem. Phys.* 21: 767–776.

14 Pople, J.A. (1953). Electron interaction in unsaturated hydrocarbons. *Trans. Faraday Soc.* 49: 1375–1385.

15 Hubbard, J. (1963). Electron correlations in narrow energy bands. *Proc. R. Soc. London A Math. Phys. Sci.* 276: 238–257.

16 Perdew, J.P. and Yue, W. (1986). Accurate and simple density functional for the electronic exchange energy: generalized gradient approximation. *Phys. Rev. B* 33: 8800–8802.

17 Becke, A.D. (1988). Density-functional exchange-energy approximation with correct asymptotic behavior. *Phys. Rev. A* 38: 3098–3100.

18 Lee, C., Yang, W., and Parr, R.G. (1988). Development of the Colle-Salvetti correlation-energy formula into a functional of the electron density. *Phys. Rev. B* 37: 785–789.

19 Perdew, J.P., Burke, K., and Ernzerhof, M. (1996). Generalized gradient approximation made simple. *Phys. Rev. Lett.* 77: 3865–3868.

20 Perdew, J.P., Kurth, S., Zupan, A., and Blaha, P. (1999). Accurate density functional with correct formal properties: a step beyond the generalized gradient approximation. *Phys. Rev. Lett.* 82: 2544–2547.

21 Tao, J., Perdew, J.P., Staroverov, V.N., and Scuseria, G.E. (2003). Climbing the density functional ladder: nonempirical meta–generalized gradient approximation designed for molecules and solids. *Phys. Rev. Lett.* 91: 146401.

22 Sun, J., Perdew, J.P., and Ruzsinszky, A. (2015). Semilocal density functional obeying a strongly tightened bound for exchange. *Proc. Natl. Acad. Sci. U.S.A.* 112: 685–689.

23 Sun, J., Ruzsinszky, A., and Perdew, J.P. (2015). Strongly constrained and appropriately normed semilocal density functional. *Phys. Rev. Lett.* 115: 036402.

24 Sun, J., Remsing, R.C., Zhang, Y. et al. (2016). Accurate first-principles structures and energies of diversely bonded systems from an efficient density functional. *Nat. Chem.* 8: 831–836.

25 Tao, J. and Mo, Y. (2016). Accurate semilocal density functional for condensed-matter physics and quantum chemistry. *Phys. Rev. Lett.* 117: 073001.

26 Patra, B., Jana, S., Constantin, L.A., and Samal, P. (2019). Efficient band gap prediction of semiconductors and insulators from a semilocal exchange-correlation functional. *Phys. Rev. B* 100: 045147.

27 Jana, S., Patra, A., Constantin, L.A. et al. (2019). Long-range screened hybrid-functional theory satisfying the local-density linear response. *Phys. Rev. A* 99: 042515.

28 Levy, M. and Perdew, J.P. (1985). Hellmann-Feynman, virial, and scaling requisites for the exact universal density functionals. Shape of the correlation potential and diamagnetic susceptibility for atoms. *Phys. Rev. A* 32: 2010–2021.

29 Gidopoulos, N.I. (2011). Progress at the interface of wave-function and density-functional theories. *Phys. Rev. A* 83: 040502.

30 Perdew, J.P., Parr, R.G., Levy, M., and Balduz, J.L. Jr. (1982). Density-functional theory for fractional particle number: derivative discontinuities of the energy. *Phys. Rev. Lett.* 49: 1691–1694.

31 Newton, R. (1970). Inverse problems in physics. *SIAM Rev.* 12: 346–356.

32 Carter, A.H. (2000). A class of inverse problems in physics. *Am. J. Phys.* 68: 698–703.

33 Sabatier, P.C. (2000). Past and future of inverse problems. *J. Math. Phys.* 41: 4082–4124.

34 Aryasetiawan, F. and Stott, M.J. (1988). Effective potentials in density-functional theory. *Phys. Rev. B* 38: 2974–2987.

35 Zhao, Q. and Parr, R.G. (1992). Quantities $T_s[n]$ and $T_c[n]$ in density-functional theory. *Phys. Rev. A* 46: 2337–2343.

36 Görling, A. (1992). Kohn-Sham potentials and wave functions from electron densities. *Phys. Rev. A* 46: 3753–3757.

37 Zhao, Q. and Parr, R.G. (1993). Constrained-search method to determine electronic wave functions from electronic densities. *J. Chem. Phys.* 98: 543–548.

38 Wang, Y. and Parr, R.G. (1993). Construction of exact Kohn-Sham orbitals from a given electron density. *Phys. Rev. A* 47: R1591–R1593.

39 van Leeuwen, R. and Baerends, E.J. (1994). Exchange-correlation potential with correct asymptotic behavior. *Phys. Rev. A* 49: 2421–2431.

40 Zhao, Q., Morrison, R.C., and Parr, R.G. (1994). From electron densities to Kohn-Sham kinetic energies, orbital energies, exchange-correlation potentials, and exchange-correlation energies. *Phys. Rev. A* 50: 2138–2142.

41 Tozer, D.J., Ingamells, V.E., and Handy, N.C. (1996). Exchange-correlation potentials. *J. Chem. Phys.* 105: 9200–9213.

42 Ingamells, V.E. and Handy, N.C. (1996). Towards accurate exchange-correlation potentials for molecules. *Chem. Phys. Lett.* 248: 373–378.

43 Jayatilaka, D. (1998). Wave function for beryllium from X-ray diffraction data. *Phys. Rev. Lett.* 80: 798–801.

44 Peirs, K., Van Neck, D., and Waroquier, M. (2003). Algorithm to derive exact exchange-correlation potentials from correlated densities in atoms. *Phys. Rev. A* 67: 012505.

45 Kadantsev, E.S. and Stott, M.J. (2004). Variational method for inverting the Kohn-Sham procedure. *Phys. Rev. A* 69: 012502.

46 Harbola, M.K. (2004). Exchange-correlation potentials in ground-and excited-state Kohn-Sham theory. *Phys. Rev. A* 69: 042512.

47 Helbig, N., Tokatly, I.V., and Rubio, A. (2009). Exact Kohn–Sham potential of strongly correlated finite systems. *J. Chem. Phys.* 131: 224105.

48 Ryabinkin, I.G. and Staroverov, V.N. (2012). Determination of Kohn–Sham effective potentials from electron densities using the differential virial theorem. *J. Chem. Phys.* 137: 164113.

49 Ryabinkin, I.G., Kananenka, A.A., and Staroverov, V.N. (2013). Accurate and efficient approximation to the optimized effective potential for exchange. *Phys. Rev. Lett.* 111: 013001.

50 Gould, T. and Toulouse, J. (2014). Kohn-Sham potentials in exact density-functional theory at noninteger electron numbers. *Phys. Rev. A* 90: 050502.

51 Ryabinkin, I.G., Kohut, S.V., and Staroverov, V.N. (2015). Reduction of electronic wave functions to Kohn-Sham effective potentials. *Phys. Rev. Lett.* 115: 083001.

52 Cuevas-Saavedra, R., Ayers, P.W., and Staroverov, V.N. (2015). Kohn–Sham exchange-correlation potentials from second-order reduced density matrices. *J. Chem. Phys.* 143: 244116.

53 Ryabinkin, I.G., Ospadov, E., and Staroverov, V.N. (2017). Exact exchange-correlation potentials of singlet two-electron systems. *J. Chem. Phys.* 147: 164117.

54 Hollins, T.W., Clark, S.J., Refson, K., and Gidopoulos, N.I. (2017). A local Fock-exchange potential in Kohn–Sham equations. *J. Phys.: Condens. Matter* 29: 04LT01.

55 Finzel, K., Ayers, P.W., and Bultinck, P. (2018). A simple algorithm for the Kohn–Sham inversion problem applicable to general target densities. *Theor. Chem. Acc.* 137: 30.

56 Jensen, D.S. and Wasserman, A. (2017). Numerical methods for the inverse problem of density functional theory. *Int. J. Quantum Chem.* 118 (1): e25425.

57 Kaiser, A. and Kümmel, S. (2018). Revealing the field-counteracting term in the exact Kohn-Sham correlation potential. *Phys. Rev. A* 98: 052505.

58 Zhang, X. and Carter, E.A. (2018). Kohn-Sham potentials from electron densities using a matrix representation within finite atomic orbital basis sets. *J. Chem. Phys.* 148: 034105.

59 Kanungo, B., Zimmerman, P.M., and Gavini, V. (2019). Exact exchange-correlation potentials from ground-state electron densities. *Nat. Commun.* 10: 4497. https://doi.org/10.1038/s41467-019-12467-0.

60 Ospadov, E., Tao, J., Staroverov, V.N., and Perdew, J.P. (2018). Visualizing atomic sizes and molecular shapes with the classical turning surface of the Kohn–Sham potential. *Proc. Natl. Acad. Sci. U.S.A.* 115: E11578–E11585.

61 Accorto, G., Brandolini, P., Marino, F. et al. (2020). First step in the nuclear inverse Kohn-Sham problem: from densities to potentials. *Phys. Rev. C* 101: 024315.

62 Accorto, G., Naito, T., Liang, H. et al. (2021). Nuclear energy density functionals from empirical ground-state densities. *Phys. Rev. C* 103: 044304.

63 Liardi, A., Marino, F., Colò, G. et al. (2022). Complete solution to the inverse Kohn-Sham problem: from the density to the energy. *Phys. Rev. C* 105: 034309.

64 Garrick, R., Natan, A., Gould, T., and Kronik, L. (2020). Exact generalized Kohn-Sham theory for hybrid functionals. *Phys. Rev. X* 10: 021040.

65 Shi, Y. and Wasserman, A. (2021). Inverse Kohn–Sham density functional theory: progress and challenges. *J. Phys. Chem. Lett.* 12: 5308–5318.

66 Bousiadi, S., Gidopoulos, N.I., and Lathiotakis, N.N. (2022). Density inversion method for local basis sets without potential auxiliary functions: inverting densities from RDMFT. *Phys. Chem. Chem. Phys.* 24: 19279–19286.

67 Levy, M., Perdew, J.P., and Sahni, V. (1984). Exact differential equation for the density and ionization energy of a many-particle system. *Phys. Rev. A* 30: 2745–2748.

68 Deb, B.M. and Chattaraj, P.K. (1989). Density-functional and hydrodynamical approach to ion-atom collisions through a new generalized nonlinear Schrödinger equation. *Phys. Rev. A* 39: 1696–1713.

69 Buijse, M.A., Baerends, E.J., and Snijders, J.G. (1989). Analysis of correlation in terms of exact local potentials: applications to two-electron systems. *Phys. Rev. A* 40: 4190–4202.

70 Kumar, A., Singh, R., and Harbola, M.K. (2019). Universal nature of different methods of obtaining the exact Kohn–Sham exchange-correlation potential for a given density. *J. Phys. B: At. Mol. Opt. Phys.* 52: 075007.

71 Levy, M. and Ou-Yang, H. (1988). Exact properties of the Pauli potential for the square root of the electron density and the kinetic energy functional. *Phys. Rev. A* 38: 625–629.

72 Lieb, E.H. (1983). Density functionals for Coulomb systems. *Int. J. Quantum Chem.* 24: 243–277.

73 Werden, S.H. and Davidson, E.R. (1984). On the calculation of potentials from densities. In: *Local Density Approximations in Quantum Chemistry and Solid State Physics* (ed. J.P. Dahl and J. Avery), 33–42. Boston, MA: Springer.

74 Gaudoin, R. and Burke, K. (2004). Lack of Hohenberg-Kohn theorem for excited states. *Phys. Rev. Lett.* 93: 173001.

75 Wagner, L.O., Baker, T.E., Stoudenmire, E.M. et al. (2014). Kohn-Sham calculations with the exact functional. *Phys. Rev. B* 90: 045109.

76 Harbola, M.K., Kumar, A., and Singh, R. (2018). Universal nature of different methods of obtaining the exact Kohn-Sham exchange-correlation potential for a given density. *APS March Meeting 2018/L60.325.*

77 Laufer, P.M. and Krieger, J.B. (1986). Test of density-functional approximations in an exactly soluble model. *Phys. Rev. A* 33: 1480–1491.

78 Kais, S., Herschbach, D.R., Handy, N.C. et al. (1993). Density functionals and dimensional renormalization for an exactly solvable model. *J. Chem. Phys.* 99: 417–425.

79 Brack, M. (1993). The physics of simple metal clusters: self-consistent jellium model and semi-classical approaches. *Rev. Mod. Phys.* 65: 677–732.

80 Bunge, C., Barrientos, J., and Bunge, A. (1993). Roothaan-Hartree-Fock ground-state atomic wave functions: slater-type orbital expansions and expectation values for Z = 2-54. *At. Data Nucl. Data Tables* 53: 113–162.

81 Harbola, M.K. and Sahni, V. (1989). Quantum-mechanical interpretation of the exchange-correlation potential of Kohn-Sham density-functional theory. *Phys. Rev. Lett.* 62: 489–492.

82 Harbola, M.K. (1992). Magic numbers for metallic clusters and the principle of maximum hardness. *Proc. Natl. Acad. Sci. U.S.A.* 89: 1036–1039.

83 Harbola, M.K. (2009). Exchange-correlation potential of Kohn–Sham theory: a physical perspective. In: *Chemical Reactivity Theory A Density Functional View* (ed. P.K. Chattaraj), 83–103. Singapore: Taylor & Francis.

84 Herman, F. and Skillman, S. (1963). *Atomic Structure Calculations.* Prentice-Hall Publication.

85 Almbladh, C.-O. and von Barth, U. (1985). *Phys. Rev. A* 31: 3231.

86 Sharp, R.T. and Horton, G.K. (1953). A variational approach to the unipotential many-electron problem. *Phys. Rev.* 90: 317–317.

87 Aashamar, K., Luke, T., and Talman, J. (1978). Optimized central potentials for atomic ground-state wavefunctions. *At. Data Nucl. Data Tables* 22: 443–472.

88 Engel, E. and Vosko, S.H. (1993). Accurate optimized-potential-model solutions for spherical spin-polarized atoms: evidence for limitations of the exchange-only local spin-density and generalized-gradient approximations. *Phys. Rev. A* 47: 2800–2811.

89 Kumar, A. and Harbola, M.K. (2021). Using random numbers to obtain Kohn-Sham potential for a given density. *Chem. Phys. Lett.* 779: 138851.

90 Kumar, A. and Harbola, M.K. (2020). A general penalty method for density-to-potential inversion. *Int. J. Quantum Chem.* 120: e26400.

91 Morrison, R.C. and Zhao, Q. (1995). Solution to the Kohn-Sham equations using reference densities from accurate, correlated wave functions for the neutral atoms helium through argon. *Phys. Rev. A* 51: 1980–1984.

92 Kohut, S.V., Polgar, A.M., and Staroverov, V.N. (2016). Origin of the step structure of molecular exchange-correlation potentials. *Phys. Chem. Chem. Phys.* 18: 20938–20944.

93 Ospadov, E., Ryabinkin, I.G., and Staroverov, V.N. (2017). Improved method for generating exchange-correlation potentials from electronic wave functions. *J. Chem. Phys.* 146: 084103.

94 Ospadov, E. and Staroverov, V.N. (2018). Construction of Fermi potentials from electronic wave functions. *J. Chem. Theory Comput.* 14: 4246–4253.

95 Mura, M.E., Knowles, P.J., and Reynolds, C.A. (1997). Accurate numerical determination of Kohn-Sham potentials from electronic densities: I. Two-electron systems. *J. Chem. Phys.* 106: 9659–9667.

96 Schipper, P.R.T., Gritsenko, O.V., and Baerends, E.J. (1997). Kohn-Sham potentials corresponding to Slater and Gaussian basis set densities. *Theor. Chem. Acc.* 98: 16–24.

97 Kumar, A., Singh, R., and Harbola, M.K. (2020). Accurate effective potential for density amplitude and the corresponding Kohn–Sham exchange–correlation potential calculated from approximate wavefunctions. *J. Phys. B: At. Mol. Opt. Phys.* 53: 165002.

98 Hunter, G. (1975). Conditional probability amplitudes in wave mechanics. *Int. J. Quantum Chem.* 9: 237–242.

99 Schiff, L. (1955). *Quantum Mechanics*. New York: McGraw-Hill.

100 Szabo, A. and Ostlund, N. (1996). *Modern Quantum Chemistry: Introduction to Advanced Electronic Structure Theory*. New York: Dover Publications.

101 Levine, I. (2009). *Quantum Chemistry*. London: Prentice Hall International.

102 March, N. (1986). The local potential determining the square root of the ground-state electron density of atoms and molecules from the Schrödinger equation. *Phys. Lett. A* 113: 476–478.

103 Hodgson, M.J.P., Ramsden, J.D., and Godby, R.W. (2016). Origin of static and dynamic steps in exact Kohn-Sham potentials. *Phys. Rev. B* 93: 155146.

104 Kraisler, E., Hodgson, M.J.P., and Gross, E.K.U. (2021). From Kohn–Sham to many-electron energies via step structures in the exchange-correlation potential. *J. Chem. Theory Comput.* 17: 1390–1407.

105 Aouina, A., Gatti, M., Chen, S. et al. (2022). Accurate Kohn-Sham auxiliary system from the ground state density of solids. arXiv preprint arXiv:2207.03919.

106 Hodgson, M.J.P., Ramsden, J.D., Chapman, J.B.J. et al. (2013). Exact time-dependent density-functional potentials for strongly correlated tunneling electrons. *Phys. Rev. B* 88: 241102.

107 Shi, Y., Chávez, V.H., and Wasserman, A. (2022). *n2v*: A density-to-potential inversion suite. A sandbox for creating, testing, and benchmarking density functional theory inversion methods. *WIREs Comput. Mol. Sci.* 12 (6): e1617.

108 Peach, M.J.G., Miller, A.M., Teale, A.M., and Tozer, D.J. (2008). Adiabatic connection forms in density functional theory: H_2 and the He isoelectronic series. *J. Chem. Phys.* 129: 064105.

109 Teale, A.M., Coriani, S., and Helgaker, T. (2009). The calculation of adiabatic-connection curves from full configuration-interaction densities: two-electron systems. *J. Chem. Phys.* 130: 104111.

110 Chauhan, R.S. and Harbola, M.K. (2017). Study of adiabatic connection in density functional theory with an accurate wavefunction for two-electron spherical systems. *Int. J. Quantum Chem.* 117: e25344.

111 Singh, R. and Harbola, M.K. (2017). A study of accurate exchange-correlation functionals through adiabatic connection. *J. Chem. Phys.* 147: 144105.

112 Harris, J. and Jones, R.O. (1974). The surface energy of a bounded electron gas. *J. Phys. F: Met. Phys.* 4: 1170–1186.

113 Yang, W. and Wu, Q. (2002). Direct method for optimized effective potentials in density-functional theory. *Phys. Rev. Lett.* 89: 143002.

114 Wu, Q. and Yang, W. (2003). A direct optimization method for calculating density functionals and exchange-correlation potentials from electron densities. *J. Chem. Phys.* 118: 2498–2509.

115 Elliott, P. and Maitra, N.T. (2012). Propagation of initially excited states in time-dependent density-functional theory. *Phys. Rev. A* 85: 052510.

116 Elliott, P., Fuks, J.I., Rubio, A., and Maitra, N.T. (2012). Universal dynamical steps in the exact time-dependent exchange-correlation potential. *Phys. Rev. Lett.* 109: 266404.

117 Suzuki, Y., Lacombe, L., Watanabe, K., and Maitra, N.T. (2017). Exact time-dependent exchange-correlation potential in electron scattering processes. *Phys. Rev. Lett.* 119: 263401.

118 Dar, D., Lacombe, L., Feist, J., and Maitra, N.T. (2021). Exact time-dependent density-functional theory for nonperturbative dynamics of the helium atom. *Phys. Rev. A* 104: 032821.

119 Dar, D., Lacombe, L., and Maitra, N.T. (2022). The exact exchange–correlation potential in time-dependent density functional theory: choreographing electrons with steps and peaks. arXiv, 10.48550/arXiv.2205.03691, 2205.03691

120 Ramsden, J.D. and Godby, R.W. (2012). Exact density-functional potentials for time-dependent quasiparticles. *Phys. Rev. Lett.* 109: 036402.

121 Hodgson, M.J.P., Ramsden, J.D., Durrant, T.R., and Godby, R.W. (2014). Role of electron localization in density functionals. *Phys. Rev. B* 90: 241107.

122 Entwistle, M.T. and Godby, R.W. (2020). Exact nonadiabatic part of the Kohn-Sham potential and its fluidic approximation. *Phys. Rev. Mater.* 4: 035002.

2

Electron Density, Density Functional Theory, and Chemical Concepts

Swapan K. Ghosh

UM-DAE-Centre for Excellence in Basic Sciences, University of Mumbai, Mumbai, India

2.1 Introduction

This chapter attempts to provide a glimpse [1–3] of some of the aspects of the vast field of electronic density functional theory (DFT)-based concepts as applied to many-electron systems, concerning their structure, properties, and interaction as well as the various associated phenomena. As is well known, different systems, such as atoms, molecules, clusters, and solids, can be broadly viewed in a unified manner, as a set of nuclei, suitably placed in different geometrical configurations, and bound chemically by the electrons that essentially serve as glue, leading to the stability of the systems. The change of one spatial configuration to another, either within a molecule or when two or more atoms or molecules approach each other, leads to different physicochemical phenomena. Clearly, the process is associated with electron density redistribution (a cause or an effect!), which can serve as a vehicle for a journey to the vast land of chemistry for exploration of chemical binding, reactivity, and other conceptual domains, with a view to gaining insights into the available knowledge and thereby producing new design of novel molecules and materials with tailored properties and predicting new chemistry for the future. Thus, electron density rightly plays a vital role in chemistry; the theoretical framework that uses density as the basic variable is the so-called density functional theory (DFT) [4–6] and its subset version specifically dedicated to explore the conceptual aspects, by helping germination of new concepts or providing the foundation for the existing concepts, is popularly known as conceptual density functional theory (CDFT) [7]. A breakthrough in CDFT originated through the pioneering contributions [8] of Prof. Robert G. Parr in the mid-70s and successful follow-up by many leading scientists from all over the world. Thus, the seed planted by Prof. Parr has now grown into a big tree with many branches, full of flowers and fruits, radiating an aroma of pleasant surprises. The objective of this work is to present a brief overview of these developments.

2.2 Viewing Chemical Concepts Through a DFT Window

DFT (or to be more precise CDFT) has provided a sound foundation for many of the chemical concepts that have been introduced empirically from time to time for interpreting and rationalizing many of the experimental observations.

Electron Density: Concepts, Computation and DFT Applications, First Edition.
Edited by Pratim Kumar Chattaraj and Debdutta Chakraborty.

Consider an N-electron system characterized by an external potential $v(\mathbf{r})$. Its ground state electronic energy, within the framework of DFT, can be expressed as a functional $E_v[\rho]$ of the electron density $\rho(\mathbf{r})$ (evaluated as the expectation value of the density operator, viz. $\langle\Psi(r_1, r_2, \ldots r_N)|\Sigma_i\delta(r-r_i)|\Psi(r_1, r_2, \ldots r_N)\rangle$), as

$$E_v[\rho] = \int v(\mathbf{r})\rho(\mathbf{r})\, d\mathbf{r} + F[\rho] \tag{2.1}$$

Here, $F[\rho]$ represents a universal functional of density and consists of contributions from the kinetic energy and inter-electron repulsion energies. For fixed $v(\mathbf{r})$, the variational principle states that the energy functional $E_v[\rho]$ assumes a minimum value for the true density, obeying the condition $\delta[E_v[\rho] - \mu\{\int \rho(\mathbf{r})\, d\mathbf{r} - N\}] = 0$, with the normalization constraint $\int \rho(\mathbf{r})\, d\mathbf{r} = N$, imposed using the parameter μ as a Lagrange multiplier. This leads to the result

$$\mu = (\delta E_v[\rho]/\delta\rho(\mathbf{r})) = v(\mathbf{r}) + (\delta F[\rho]/\delta\rho(\mathbf{r})) \tag{2.2}$$

for the Lagrange multiplier μ, which represents the chemical potential of the electron cloud, and has been identified with the electronegativity parameter χ, defined as the first derivative of the energy E with respect to N, as [8]

$$\chi = -(\partial E/\partial N)_{v(\mathbf{r})} \tag{2.3}$$

$$= -\int (\delta E_v[\rho]/\delta\rho(\mathbf{r}))(\partial\rho(\mathbf{r})/\partial N)\, d\mathbf{r} = -\mu \int (\partial\rho(\mathbf{r})/\partial N)\, d\mathbf{r} = -\mu \tag{2.4}$$

thus equating χ with the negative of the chemical potential μ. Equations (2.2) and (2.4) provide a basis for the principle of electronegativity equalization during molecule formation.

Analogously, the second derivative, which represents the hardness parameter η, is given by [9]

$$\eta = (1/2)(\partial^2 E/\partial N^2)_{v(\mathbf{r})} = -(1/2)(\partial\chi/\partial N)_{v(\mathbf{r})} \tag{2.5}$$

The corresponding spin-polarized counterparts [10, 11], viz. χ_α and χ_β, representing, respectively, the up-spin and down-spin electronegativity, can be defined as $\chi_\alpha = (\partial E_v/\partial N_\alpha)_{v(\mathbf{r}),N_\beta}$ and $\chi_\beta = -(\partial E_v/\partial N_\beta)_{v(\mathbf{r}),N_\alpha}$, where N_α and N_β, are the spin-up and spin-down electron numbers, satisfying $N = N_\alpha + N_\beta$. One also has the identification of χ_α and χ_β with the respective spin-dependent chemical potentials μ_α and μ_β, as

$$\begin{aligned}\chi_\alpha &= -(\partial E_v/\partial N_\alpha)_{v(\mathbf{r}),N_\beta}\\ &= -\int (\delta E_v[\rho_\alpha, \rho_\beta]/\delta\rho_\alpha(\mathbf{r}))(\partial\rho_\alpha(\mathbf{r})/\partial N_\alpha)d\mathbf{r}\\ &= -\mu_\alpha \int (\partial\rho_\alpha(\mathbf{r})/\partial N_\alpha)d\mathbf{r} = -\mu_\alpha\end{aligned} \tag{2.6a}$$

$$\begin{aligned}\chi_\beta &= -(\partial E_v/\partial N_\beta)_{v(\mathbf{r}),N_\alpha}\\ &= -\int (\delta E_v[\rho_\alpha, \rho_\beta]/\delta\rho_\beta(\mathbf{r}))(\partial\rho_\beta(\mathbf{r})/\partial N_\beta)d\mathbf{r}\\ &= -\mu_\beta \int (\partial\rho_\beta(\mathbf{r})/\partial N_\beta)d\mathbf{r} = -\mu_\beta\end{aligned} \tag{2.6b}$$

where $\rho_\alpha(\mathbf{r})$ and $\rho_\beta(\mathbf{r})$ represent the up-spin and down-spin electron density counterparts. That satisfies $\rho(\mathbf{r}) = \rho_\alpha(\mathbf{r}) + \rho_\beta(\mathbf{r})$, and $\int \rho_\alpha(\mathbf{r})\, d\mathbf{r} = N_\alpha$ and $\int \rho_\beta(\mathbf{r})\, d\mathbf{r} = N_\beta$.

Similarly, the spin-dependent hardness parameters $\eta_{\alpha\alpha}$, $\eta_{\beta\beta}$, and $\eta_{\alpha\beta}$ are defined as [10, 11]

$$\eta_{\alpha\alpha} = (1/2)\left(\partial^2 E/\partial N_\alpha{}^2\right)_{v(\mathbf{r}),N_\beta} = -(1/2)(\partial\chi_\alpha/\partial N_\alpha)_{v(\mathbf{r}),N_\beta} \quad (2.7a)$$

$$\eta_{\beta\beta} = (1/2)\left(\partial^2 E/\partial N_\beta{}^2\right)_{v(\mathbf{r}),N_\alpha} = -(1/2)(\partial\chi_\beta/\partial N_\beta)_{v(\mathbf{r}),N_\alpha} \quad (2.7b)$$

$$\eta_{\alpha\beta} = \eta_{\beta\alpha} = (\partial^2 E/\partial N_\alpha \cdot \partial N_\beta)_{v(\mathbf{r})} = -(\partial\chi_\alpha/\partial N_\beta)_{v(\mathbf{r})} = -(\partial\chi_\beta/\partial N_\alpha)_{v(\mathbf{r})} \quad (2.7c)$$

Another class of reactivity parameters, known as Fukui function [12], can be defined as $f(\mathbf{r}) = (\partial\rho(\mathbf{r})/\partial N)$ and the spin polarized counterparts $f_{\alpha\alpha}(\mathbf{r}) = (\partial\rho_\alpha(\mathbf{r})/\partial N_\alpha)$, $f_{\alpha\beta}(\mathbf{r}) = (\partial\rho_\alpha(\mathbf{r})/\partial N_\beta)$, $f_{\beta\beta}(\mathbf{r}) = (\partial\rho_\beta(\mathbf{r})/\partial N_\beta)$, etc.

The χ and η parameters, defined here, can be used to develop a model for the chemical binding associated with molecule formation from the constituent atoms. From the relative magnitudes of the binding energy of molecules and the electronic energies of individual atoms, it is clear that consideration of the interatomic interaction as perturbation of the electron cloud can provide a good estimate of the binding energy of molecules. One can also introduce further simplification by introducing coarse-grained variables in terms of interatomic charge transfer as an approximation to the detailed electron density rearrangement.

Consideration of the overall electronegativity [8] and chemical hardness [9] of the atoms determines the direction and extent of interatomic charge flow and is suitable only for ionic binding. On the other hand, consideration of their spin-dependent counterparts can estimate the up-spin electron transfer from one atom to the other, with down-spin electron transfer in the opposite direction, driven by the up- and down-spin electronegativity parameters of the two atoms, which have different values. This can provide an estimate of covalent binding as well (in addition to the ionic binding). This model justifies the need for the presence of unpaired electrons in the atoms for covalent binding. For purely covalent binding, there may not be any resultant interatomic charge transfer, and it is the net spin transfer that can account for the covalent binding.

Thus, considering the energy as the functional $E[N_\alpha, N_\beta, v_\alpha(\mathbf{r}), v_\beta(\mathbf{r})]$, one can have a perturbative expansion in terms of change in the number of electrons and the potential due to the approach of the two atoms. The energy change ΔE associated with changes in these quantities, viz. δN_α, δN_β, $\delta v_\alpha(\mathbf{r})$, and $\delta v_\beta(\mathbf{r})$, can be expressed through a perturbation expansion in terms of them, retaining, for example, terms up to second order. A more detailed description can be provided by considering the dependence in terms of the densities, viz. $E[\rho_\alpha(\mathbf{r}), \rho_\beta(\mathbf{r}), v_\alpha(\mathbf{r}), v_\beta(\mathbf{r})]$. One can then derive the expressions for the chemical potentials corresponding to these energy expressions, equate them and calculate the charge transfer, which on substitution in the energy expression will result in the binding energy expression. Evaluation of these expressions using suitable values of the parameters has been found to yield a good estimate of the binding energies.

The other important factor for consideration for modeling of covalent binding is the accumulation of electron density in the bond region, which can be quantified by defining a chemical potential for the bonding region (bond midpoint, for example, for homonuclear molecules) [13]. The corresponding calculated results have been found promising [14–16].

For developing a lattice model for molecule formation, one can consider the total density change as a sum of the density changes at each atom, viz. $\delta\rho_\alpha(\mathbf{r}) = \Sigma_i \delta\rho_\alpha{}^{(i)}(r_i)$; $\delta\rho_\beta(\mathbf{r}) = \Sigma_i \delta\rho_\beta{}^{(i)}(r_i)$, and similarly the total potential change as sum over the atomic sites, i.e. $\delta v_\alpha(\mathbf{r}) = \Sigma_i \delta v_\alpha{}^{(i)}(r_i)$; $\delta v_\beta(\mathbf{r}) = \Sigma_i \delta v_\beta{}^{(i)}(r_i)$, where $\mathbf{r}_i = (\mathbf{r} - \mathbf{R}_i)$ denote the ith atom-centered coordinate for the electron. These substitutions lead to the energy change expressed as a perturbation expression in terms of atomic contributions. This approach has also been employed to obtain the response properties such as polarizability [17, 18].

The CDFT-based approach with this type of atom-centered coarse-grained electron density-based description has been immensely useful to describe not only chemical binding but also chemical reactivity [19]. Thus, the study of partial atomic charges [20], local hardness [21] and softness [22], Fukui function [12], various energy derivatives [23], response function parameters [24, 25], electrophilicity index [26, 27], etc., have been valuable for describing chemical reactivity and related issues.

2.3 Electron Fluid, Quantum Fluid Dynamics, Electronic Entropy, and a Local Thermodynamic Picture

The interpretive aspect of quantum mechanics is almost as old as quantum mechanics itself. Thus, soon after the Schrodinger equation was discovered, a fluid dynamical interpretation of a one-electron quantum system was proposed by Madelung [28]. Then, the description was further resurrected in the 50s by Bohm [29] and Takabayashi [30]. The main advantage of the quantum fluid dynamics (QFD) picture is the classical interpretation associated with it. Thus, a quantum system consisting of a quantum particle moving in an external potential is shown to be equivalent to a classical fluid, with the fluid elements obeying classical-like equation of motion, moving under the action of a force arising from not only the classical external potential but also an additional force of quantum origin, defined by a quantum potential (also known as Bohm potential). The derivation involves a simple substitution of the polar form of the wave function, $\Psi(r,t) = R(\mathbf{r},t)\exp[iS(\mathbf{r},t)]$, in the time-dependent (TD) Schrodinger equation $-(1/2)\nabla^2\Psi(\mathbf{r},t) + V(\mathbf{r},t)\Psi(r,t) = i(\partial\Psi(r,t)/\partial t)$, which leads to a continuity equation and an Euler type of fluid dynamical equation [2], given by

$$(\partial\rho(\mathbf{r},t)/\partial t) + \nabla\cdot(\rho\mathbf{v}) = 0 \tag{2.8a}$$

$$\rho(\mathbf{r},t)(d\mathbf{v}(\mathbf{r},t)/dt) = -\rho(\mathbf{r},t)\nabla(V(\mathbf{r},t) + V_{\mathrm{qu}}(\mathbf{r},t)) \tag{2.8b}$$

with

$$(d\mathbf{v}(\mathbf{r},t)/dt) = (\partial\mathbf{v}(\mathbf{r},t)/\partial t) + (\mathbf{v}\cdot\nabla)\mathbf{v} \tag{2.8c}$$

and

$$V_{\mathrm{qu}}(\mathbf{r},t) = -(1/2)(\nabla^2 R(\mathbf{r},t))/R(\mathbf{r},t) \tag{2.8d}$$

representing the total derivative and the quantum potential, respectively. It may be noted here that atomic units are used for simplicity.

Although QFD was first proposed for a one-particle system, it was later extended to many-electron systems within the framework of DFT [31], Hartree–Fock, natural orbital description [32], or other one-particle frameworks. The QFD equations corresponding to an electron fluid have been employed to calculate the trajectories [33], interpret quantum tunneling, describe collision phenomena, etc.

The time-dependent (TD) DFT has been developed by proposing TD Kohn–Sham equations initially for periodic time-dependence [31] and later for a general time-dependence [34]. These equations can easily be converted into the corresponding fluid dynamical equations, but they will have the character of a multicomponent fluid. Also, both TDDFT and QFD have been merged, and a newer so-called QFDFT formalism has been developed [35].

The fluid dynamical analogy and classical-like interpretation were further strengthened by introducing a local thermodynamic picture of the electron fluid [36] within the framework of DFT.

While the QFD and QFDFT were for TD or nonequilibrium situations, the thermodynamic picture has been formulated for the ground state DFT, which deals with equilibrium aspects. The concepts of local temperature, local entropy, local free energy density, etc., have been introduced for many-electron systems by invoking the concept of DFT, and a complete local thermodynamic picture was found to emerge [36].

The electronic entropy was defined [36] in terms of a phase space distribution function, and it was maximized by using the constraints of yielding the correct density on integration over the momentum variable and the correct kinetic energy density (KED) by defining the kinetic energy using the classical concept and integration over the momentum variables. The local KED is used to define a local temperature $T(\mathbf{r})$. All the thermodynamics equations were also derived [37].

Although the thermodynamic picture was developed originally for interpretive purposes mainly, a new route to calculate the momentum space property, such as the Compton profile, is proposed through a theoretical formulation [38] based on the phase space distribution corresponding to the local thermodynamic picture of the electronic fluid. Several subsequent extensions and generalizations have been formulated in the recent past, and further implications of the concept of local thermodynamics need more thoughts and insights.

2.4 Miscellaneous Offshoots from Electron Density Experience

From the brief survey presented here, it is clear that electron density is a rich source of information and insight into the chemical binding, reactivity, and interaction involving atomic and molecular systems. The density concept is, however, much more general and is applicable profitably in the mesoscopic length scale as well through the particle number density as the basic variable in the domain of statistical mechanics. While in quantum DFT, for the investigation of inhomogeneous electron density, one uses the Schrodinger-like Kohn–Sham equations. In the same spirit, in classical DFT [39], one uses Boltzmann-like distribution for the study of inhomogeneous particle number density. In both cases, one uses the mapping of the actual interacting system into a system of noninteracting particles moving under the influence of an effective potential to be determined by the density.

While both these densities are position-dependent, there is another class of density function that depends on a suitable parameter, leading to DFT in parameter space [40]. Here, instead of the position of the particles, one has a suitable phase space function (γ) such as energy E, on which the density depends. In this parameter space DFT, the basic variable is $\rho(\gamma)$ (for example, $\rho(E)$). It has been very useful in the study of folding/unfolding of chain molecules [41], dynamics of photoinduced processes in the condensed phase [40, 42–44], etc.

2.5 Concluding Remarks

The work presented here deals with the use of classical-like concepts to many-electron systems. The concepts of some of the descriptors for chemical binding through a CDFT-based view are presented. Concepts of the electron fluid, the description of QFD, and a local thermodynamic picture have been discussed. A practical application of the resulting phase space distribution function has also been indicated. Finally, several ideas of the electron density-based descriptions have been useful in other areas, leading to classical DFT or parameter space DFT, which have been pointed out. It must be mentioned that only a very brief introduction to the concepts within the framework of

DFT has been presented here, and a much more broader picture remains outside the coverage of this account.

Acknowledgments

The work and views reported here have been possible due to enlightening discussions that I had with many teachers, collaborators, and friends. In particular, I express my gratefulness to Prof. B.M. Deb and Prof. Robert G. Parr, from whom I have learned a lot over the years. I also thank my colleagues Dr. Alok Samanta, Dr. Tapan Ghanty, Dr. K.R.S. Chandrakumar, Dr. Amita Wadehra, and many others for helpful discussions and collaborations. Finally, I am thankful to Prof. Pratim Kumar Chattaraj for inviting me to contribute to this chapter and for showing enormous patience and encouragement during the delay of its completion.

References

1 Bamzai, A.S. and Deb, B.M. (1981). The role of single-particle density in chemistry. *Rev. Mod. Phys.* 53: 95–126.

2 Ghosh, S.K. and Deb, B.M. (1982). Densities, density functionals and electron fluids. *Phys. Rep.* 42: 1–44.

3 March, N.H. and Deb, B.M. (1987). *The Role of Single-Particle Density in Physics and Chemistry*. New York: Academic Press.

4 Hohenberg, P. and Kohn, W. (1964). Inhomogeneous electron gas. *Phys. Rev.* 136: B864.

5 Kohn, W. and Sham, L.J. (1965). Self-consistent equations including exchange and correlation effects. *Phys. Rev.* 140: A1133.

6 Parr, R.G. and Yang, W. (1989). *Density-Functional Theory of Atoms and Molecules*. New York/Oxford: Oxford University Press/Clarendon Press.

7 Geerlings, P., Chamorro, E., Chattaraj, P.K. et al. (2020). Conceptual density functional theory: status, prospects, issues. *Theor. Chem. Acc.* 139: 36.

8 Parr, R.G., Donnelly, R.A., Levy, M., and Palke, W.E. (1978). Electronegativity: the density functional viewpoint. *J. Chem. Phys.* 68: 3801–3807.

9 Parr, R.G. and Pearson, R.G. (1983). Absolute hardness: companion parameter to absolute electronegativity. *J. Am. Chem. Soc.* 105: 7512–7516.

10 Ghosh, S.K. (1994). Electronegativity, hardness, and a semiempirical density functional theory of chemical binding. *Int. J. Quantum Chem.* 49: 239–251.

11 Ghanty, T.K. and Ghosh, S.K. (1994). Spin-polarized generalization of the concepts of electronegativity and hardness and the description of chemical binding. *J. Am. Chem. Soc.* 116: 3943–3948.

12 Parr, R.G. and Yang, W. (1984). Density functional approach to the frontier-electron theory of chemical reactivity. *J. Am. Chem. Soc.* 106: 4049–4050.

13 Ghosh, S.K. and Parr, R.G. (1987). Toward a semiempirical density functional theory of chemical binding. *Theor. Chim. Acta* 72: 379–391.

14 Ghanty, T.K. and Ghosh, S.K. (1991). Electronegativity and covalent binding in homonuclear diatomic molecules. *J. Phys. Chem.* 95: 6512–6514.

15 Ghanty, T.K. and Ghosh, S.K. (1992). Electronegativity, hardness, and chemical binding in simple molecular systems. *Inorg. Chem.* 31: 1951–1955.

16 Ghanty, T.K. and Ghosh, S.K. (1994). A new simple density functional approach to chemical binding. *J. Phys. Chem.* 98: 1840–1843.

17 Wadehra, A. and Ghosh, S.K. (2005). A density functional theory-based chemical potential equalisation approach to molecular polarizability. *J. Chem. Sci.* 117: 401–409.

18 Ghosh, S.K. (2010). A coarse grained density functional theory, chemical potential equalization and electric response in molecular systems. *J. Mol. Struct. THEOCHEM* 943: 178–182.

19 Chattaraj, P.K. (ed.) (2009). *Chemical Reactivity Theory: A Density Functional View*. Boca Raton, FL: CRC Press.

20 Mortier, W.J., Ghosh, S.K., and Shankar, S. (1986). Electronegativity-equalization method for the calculation of atomic charges in molecules. *J. Am. Chem. Soc.* 108: 4315–4320.

21 Berkowitz, M., Ghosh, S.K., and Parr, R.G. (1985). On the concept of local hardness in chemistry. *J. Am. Chem. Soc.* 107: 6811–6814.

22 Berkowitz, M. and Parr, R.G. (1988). Molecular hardness and softness, local hardness and softness, hardness and softness kernels, and relations among these quantities. *J. Chem. Phys.* 88: 2554–2557.

23 Ghosh, S.K. (1990). Energy derivatives in density-functional theory. *Chem. Phys. Lett.* 172: 77–82.

24 Geerlings, P., Fias, S., Boisdenghien, Z., and De Proft, F. (2014). Conceptual DFT: chemistry from the linear response function. *Chem. Soc. Rev.* 43: 4989–5008.

25 Chakkamalayath, J., Chandrakumar, K.R.S., and Ghosh, S.K. (2020). Reactivity parameters and substitution effect in organic acids. *J. Phys. Chem. A* 124: 3770–3777.

26 Parr, R.G., Szentpály, L., and Liu, S. (1999). Electrophilicity index. *J. Am. Chem. Soc.* 121 (9): 1922–1924.

27 Chattaraj, P.K., Sarkar, U., and Roy, D.R. (2006). Electrophilicity index. *Chem. Rev.* 106: 2065–2091.

28 Madelung, E. (1927). Quantentheorie in hydrodynamischer Form. *Z. Phys.* 40: 322–326.

29 Bohm, D. (1952). A suggested interpretation of the quantum theory in terms of "hidden" variables. I & II. *Phys. Rev.* 85: 166, 18.

30 Takabayashi, T. (1952). On the formulation of quantum mechanics associated with classical picture. *Prog. Theor. Phys. Jpn.* 8: 143–182.

31 Deb, B.M. and Ghosh, S.K. (1982). Schrödinger fluid dynamics of many-electron systems in a time-dependent density-functional framework. *J. Chem. Phys.* 77: 342–348.

32 Ghosh, S.K. and Deb, B.M. (1982). Quantum fluid dynamics of many-electron systems in three-dimensional space. *Int. J. Quantum Chem.* 22: 871–888.

33 Hirschfelder, J.O. and Christoph, A.C. (1974). Quantum mechanical streamlines. I. Square potential barrier. *J. Chem. Phys.* 61: 5435.

34 Ghosh, S.K. and Dhara, A.K. (1988). Density-functional theory of many-electron systems subjected to time-dependent electric and magnetic fields. *Phys. Rev. A* 38: 1149.

35 Deb, B.M. and Chattaraj, P.K. (1988). Quantum fluid density functional theory of time-dependent phenomena: ion-atom collisions. *Chem. Phys. Lett.* 148: 550.

36 Ghosh, S.K., Berkowitz, M., and Parr, R.G. (1984). Transcription of ground-state density-functional theory into a local thermodynamics. *Proc. Natl. Acad. Sci. U.S.A.* 81: 8028–8031.

37 Ghosh, S.K. and Berkowitz, M. (1985). A classical fluid-like approach to the density-functional formalism of many-electron systems. *J. Chem. Phys.* 83 (6): 2976–2983.

38 Parr, R.G., Rupnik, K., and Ghosh, S.K. (1986). Phase-space approach to the density-functional calculation of Compton profiles of atoms and molecules. *Phys. Rev. Lett.* 56: 1555–1558.

39 Evans, R. (1979). The nature of the liquid–vapour interface and other topics in the statistical mechanics of non-uniform, classical fluids. *Adv. Phys.* 28: 143–200.

40 Patra, A., Samanta, A., and Ghosh, S.K. (2011). Mapping the reaction dynamics in Liouville space onto a reaction coordinate space: a first-principle-based theory. *Phys. Rev. E* 83: 026104.

41 Bandyopadhyay, T. and Ghosh, S.K. (2003). Diffusion assisted end–to–end relaxation of a flexible Rouse polymer chain: fluorescence quenching through a model energy transfer. *J. Chem. Phys.* 119: 572–584.

42 Samanta, A. and Ghosh, S.K. (2008). A one-dimensional energy diffusion approach to multidimensional dynamical processes in the condensed phase. *J. Phys. Chem.* 112: 752–757.

43 Gupta, K., Patra, A., Dhole, K. et al. (2017). Optically controlled electron-transfer reaction kinetics and solvation dynamics: effect of Franck–Condon states. *J. Phys. Chem. Lett.* 8 (18): 4545–4549.

44 Das, M., Samanta, A., and Ghosh, S.K. (2020). Concept of reaction coordinate for dynamics of optically controlled non-equilibrium processes in condensed phase. *Theor. Chem. Acc.* 139 (2): 22.

3

Local and Nonlocal Descriptors of the Site and Bond Chemical Reactivity of Molecules

José L. Gázquez[1], Paulino Zerón[1], Maurizio A. Pantoja-Hernández[1] and Marco Franco-Pérez[2]

[1]*Departamento de Química, Universidad Autónoma Metropolitana-Iztapalapa, Ciudad de Mexico, Mexico*
[2]*Facultad de Química, Universidad Nacional Autónoma de México, Cd. Universitaria, Ciudad de Mexico, Mexico*

3.1 Introduction

The distribution of electrons around the nuclei in a molecule contains fundamental information about the behavior it will show in a chemical interaction. A simple picture of this electronic density consists of atomic-like densities that decay from a maximum value at the nucleus in their core region and overlap in the bond regions with the atomic-like densities coming from the neighbor atoms. Thus, although this information is in itself basic for the description of chemical reactivity, its analysis may be enriched through response functions, that, even though they are evaluated at the situation when the molecule is isolated, they provide guidance about the way in which it will interact with different families of chemical species. This approach allows one to describe relevant aspects of the intrinsic chemical reactivity that molecules display because of their ground-state electronic structure.

The natural variables to study the changes in the energy and in the density due to the presence of another species are the number of electrons (N) and the external potential ($v(\mathbf{r})$). In its ground state, a chemical species A is characterized by the number of electrons N_A^0 and the external potential generated by the nuclei $v_A^0(\mathbf{r})$. However, when A is in the presence of another species, B, the number of electrons and the external potential change because of charge transfer among the two species and because the electrons of A are now in the presence of the nuclei of A, but also the nuclei and the electrons of B.

The response functions, also called reactivity descriptors, are then derivatives of the energy or the density with respect to the number of electrons and the external potential. Therefore, one may recognize three types of descriptors: global, local, and nonlocal. The first ones are independent of the position within the molecule and characterize it as a whole. The second ones depend on the position within the molecule so that through them, one can differentiate the reactivity of each site, and the third ones depend simultaneously on two or more points so that they can be used to distinguish some reactivity features of chemical bonds.

From the previous statements, one can see that density functional theory (DFT) provides a natural framework for the study of the intrinsic reactivity of a chemical species, since, in this approach, the energy (E) and the density ($\rho(\mathbf{r})$), for a given external potential and a number of electrons, are the basic variables used to determine the ground-state electronic structure [1–4].

Electron Density: Concepts, Computation and DFT Applications, First Edition.
Edited by Pratim Kumar Chattaraj and Debdutta Chakraborty.

Thus, this theoretical framework has been used to link intuitive concepts proposed over the years to describe many aspects of chemical interactions with some of the response functions, to get important insights about the general principles of chemical reactivity associated with them, and to develop new descriptors at the global, local and nonlocal levels, leading to what is known as conceptual density functional theory (CDFT) [4–14].

The starting point of CDFT came from the first and second derivatives of the energy with respect to the number of electrons at constant external potential, which led to two global reactivity descriptors, the chemical potential [15]

$$\mu = \left(\frac{\partial E}{\partial N}\right)_{v(\mathbf{r})} \tag{3.1}$$

and the hardness [16]

$$\eta = \left(\frac{\partial^2 E}{\partial N^2}\right)_{v(\mathbf{r})} = \left(\frac{\partial \mu}{\partial N}\right)_{v(\mathbf{r})} \tag{3.2}$$

The chemical potential defined in Eq. (3.1) is equal to the negative of the electronegativity proposed by Iczkowski and Margrave [17], which may be seen as a generalization of the expression suggested by Mulliken [18]. Thus, one can see that the electronegativity equalization principle of Sanderson [19, 20] can be seen as a chemical potential equalization. On the other hand, the hardness defined in Eq. (3.2) has been identified with the concept introduced by Pearson in the formulation of the hard and soft acids and bases principle [21–23].

In support of these two identifications, one may consider the second-order Taylor series expansion of the energy around a reference system with an integer number of electrons N_0, to make a smooth interpolation between the systems with $N_0 - 1$, N_0, and $N_0 + 1$ electrons, all evaluated at the external potential of the reference system. Using Eqs. (3.1) and (3.2) this smooth quadratic interpolation is found to be [16]

$$\Delta E = \mu \Delta N + \frac{1}{2}\eta\,(\Delta N)^2 \tag{3.3}$$

where $\Delta E = E - E_{N_0}$, $\Delta N = N - N_0$, while the chemical potential and the hardness of the reference system are equal to

$$\mu = -\frac{I + A}{2} \quad \text{and} \quad \eta = I - A \tag{3.4}$$

where $I = E_{N_0-1} - E_{N_0}$ and $A = E_{N_0} - E_{N_0+1}$ are the vertical ionization potential and electron affinity, respectively. One can see that the expression for μ is equal to the negative of the electronegativity of Mulliken, which is known to follow the same tendencies as the intuitive scale proposed by Pauling [24, 25] for this property, while the expression for the hardness also reproduces the tendencies of the qualitative scale developed by Pearson to compare the relative hardnesses of acids and bases. Accordingly, these results indicate that the relationship between the first and second derivatives of the energy with respect to N with the intuitive concepts of chemical potential (minus electronegativity) and hardness seems to be adequate.

The relevance of the chemical potential and the hardness as global descriptors may be highlighted through Eq. (3.3). This expression may be used to quantify the energy change associated with a charge transference in a chemical interaction, when the effects due to the change in the external potential are neglected, and as such, it has been the basis, through rather simple arguments with great chemical significance, to analyze several aspects of charge transfer processes [4, 16, 26–50] and to derive the chemical potential (electronegativity) equalization [16], the hard and soft acids and bases [51–54], the maximum hardness and the minimum electrophilicity principles, and the $|\Delta\mu|$ big is good rule [55–58].

It is important to mention that Parr, Von Szentpály, and Liu introduced the electrophilicity index [28] through Eq. (3.3), by determining that $\Delta N_{max} = -\mu/\eta > 0$ is the maximum amount of charge that a chemical species can accept when it is immersed in an idealized zero-temperature free electron sea of zero chemical potential. Since the energy change associated with this amount of charge transferred is $\Delta E = -\mu^2/2\eta < 0$, the electrophilicity index was defined as

$$\omega = \mu^2/2\eta \tag{3.5}$$

Recently, it has been argued [50] through CDFT that the chemical potential is closely related to the approach followed by Mulliken [18] to quantify the concept of electronegativity, while the electrophilicity index is closely related to the approach followed by Pauling [24] for this concept. The electrophilicity concept by itself has been used to explain many aspects of chemical reactivity [30–32, 34–36, 39–41].

The subsequent development of CDFT came from the first and second derivatives of the density with respect to the number of electrons at constant external potential, which led to two local reactivity descriptors, the Fukui function [59–61]

$$f(\mathbf{r}) = \left(\frac{\delta\mu}{\delta v(\mathbf{r})}\right)_N = \frac{\delta\partial E}{\delta v(\mathbf{r})\partial N} = \frac{\partial\delta E}{\partial N\delta v(\mathbf{r})} = \left(\frac{\partial\rho(\mathbf{r})}{\partial N}\right)_{v(\mathbf{r})} \tag{3.6}$$

and the dual descriptor [62, 63]

$$\Delta f(\mathbf{r}) = \left(\frac{\delta\eta}{\delta v(\mathbf{r})}\right)_N = \frac{\delta\partial^2 E}{\delta v(\mathbf{r})\partial N^2} = \frac{\partial^2\delta E}{\partial^2 N\delta v(\mathbf{r})} = \left(\frac{\partial^2\rho(\mathbf{r})}{\partial N^2}\right)_{v(\mathbf{r})} = \left(\frac{\partial f(\mathbf{r})}{\partial N}\right)_{v(\mathbf{r})} \tag{3.7}$$

where we have used Eq. (3.8)

$$\rho(\mathbf{r}) = \left(\frac{\delta E}{\delta v(\mathbf{r})}\right)_N \tag{3.8}$$

If one considers the smooth quadratic interpolation outlined before for the electron density, one finds that

$$\Delta\rho(\mathbf{r}) = f(\mathbf{r})\Delta N + \frac{1}{2}\Delta f(\mathbf{r})(\Delta N)^2 \tag{3.9}$$

where $\Delta\rho(\mathbf{r}) = \rho_N(\mathbf{r}) - \rho_{N_0}(\mathbf{r})$, and the Fukui function and the dual descriptor of the reference system are equal to

$$f(\mathbf{r}) = \frac{f^-(\mathbf{r}) + f^+(\mathbf{r})}{2} \quad \text{and} \quad \Delta f(\mathbf{r}) = f^+(\mathbf{r}) - f^-(\mathbf{r}) \tag{3.10}$$

where $f^-(\mathbf{r}) = \rho_{N_0}(\mathbf{r}) - \rho_{N_0-1}(\mathbf{r})$ and $f^+(\mathbf{r}) = \rho_{N_0+1}(\mathbf{r}) - \rho_{N_0}(\mathbf{r})$ are the directional Fukui functions [59]. The integral over the whole space of the Fukui function and the dual descriptor are equal to

$$\int f(\mathbf{r})d\mathbf{r} = 1 \quad \text{and} \quad \int \Delta f(\mathbf{r})d\mathbf{r} = 0 \tag{3.11}$$

A very relevant property of the Fukui function, as defined in Eq. (3.6), is that it is the function that minimizes the energy change that results from the remotion or addition of a given fraction of charge [64, 65]. This result indicates that $f^-(\mathbf{r})$ and $f^+(\mathbf{r})$ depict the best sites to donate or to accept charge, respectively. Another important aspect of the Fukui function comes from the fact that $f^-(\mathbf{r})$ and $f^+(\mathbf{r})$ are approximately equal to the charge densities of the highest occupied (H) and the lowest unoccupied (L) molecular orbitals, respectively, when one neglects the relaxation effects that result from the remotion or the addition of charge. Thus, this relationship has been used to provide additional support to the frontier orbital theory [66, 67] from the perspective of CDFT.

The importance of the Fukui function as a local descriptor was established originally by Parr and Yang [59] by postulating that the preferred direction of a chemical interaction is the one for which the $|\Delta\mu|$ initial is a maximum (known nowadays as the $|\Delta\mu|$ big is good rule), and from this statement, they concluded that the preferred direction is the one with the largest Fukui function at the reaction site. Additionally, they also established that $f^-(\mathbf{r})$, $f^+(\mathbf{r})$ and $f(\mathbf{r})$ (Eq. (3.10)) indicate the sites for electrophilic, nucleophilic, and radical attacks, respectively. The postulate, together with the connection mentioned before, of the directional Fukui functions with the charge densities of the highest occupied (H) and the lowest unoccupied (L) molecular orbitals and the approach developed by Berkowitz to frontier controlled reactions [68] highlight the importance of the Fukui functions as a site selectivity criterion.

In relation to the dual descriptor, one should note that according to Eq. (3.10), the regions where $\Delta f(\mathbf{r}) > 0$ indicate the places with a greater tendency to receive charge, while regions with $\Delta f(\mathbf{r}) < 0$ indicate the places with greater affinity to donate charge. This distinction is important because, in many chemical interactions, a molecule may exchange charge donating it from some places and accepting it in other regions. Also, the dual descriptor has shown to be able to explain the Woodward–Hoffmann rules, which have been typically discussed through the frontier orbitals, providing an additional indication of the connection between them and CDFT [69–74]. Furthermore, this concept can be extended through what is known as the state-specific dual descriptor [75–78], which involves excited states, so that this quantity can be used to analyze molecular excitations as charge transfer processes [79], or to incorporate the excited states in some chemical interactions.

In addition to the chemical potential and hardness, there have been developments oriented toward the formulation of the local and nonlocal (kernel) counterparts of these global indexes. Thus, the object of the present chapter is to give an integral perspective of a procedure to derive these descriptors and to show that while the global ones provide information on the reactivity of a molecule as a whole, the local and nonlocal counterparts lead to valuable information of site and bond reactivity, respectively.

3.2 Local and Nonlocal Reactivity Indexes

Early work in the direction just mentioned departed from the global hardness (Eq. (3.2)) and its inverse, the softness, to develop a local hardness [80] and a local softness [81], and later a hardness and a softness kernels [82] (nonlocal). However, through the procedure adopted, the local hardness and the two kernels were defined in terms of quantities that present some difficulties in the determination of numerical values required to apply these concepts to specific molecules. Additionally, it is important to mention that in this development, the global hardness and the global softness are defined as inverses of each other, so this characteristic was kept also for the corresponding local and nonlocal quantities. Nevertheless, the global hardness and softness are inverses of each other only at zero temperature [83, 84] so it does not seem to be necessary to propose new definitions of the local and nonlocal hardness with this inverse relationship with the local and nonlocal softness.

On the other hand, it is interesting to look at the properties of the softness. The global quantity at zero temperature is defined, as mentioned, as the inverse of the global hardness so that from the second equality of Eq. (3.2), one can see that [16, 81]

$$S = \left(\frac{\partial N}{\partial \mu}\right)_{v(\mathbf{r})} \tag{3.12}$$

while the local softness is defined as [81, 82]

$$s(\mathbf{r}) = \left(\frac{\partial \rho(\mathbf{r})}{\partial \mu}\right)_{v(\mathbf{r})} = \left(\frac{\partial N}{\partial \mu}\right)_{v(\mathbf{r})} \left(\frac{\partial \rho(\mathbf{r})}{\partial N}\right)_{v(\mathbf{r})} = Sf(\mathbf{r}) \tag{3.13}$$

and the softness kernel is expressed in terms of the intrinsic potential, $u(\mathbf{r}) = v(\mathbf{r}) - \mu$ as

$$s(\mathbf{r}, \mathbf{r}') = -\frac{\delta \rho(\mathbf{r})}{\delta u(\mathbf{r})} \tag{3.14}$$

Now, from these definitions, one can show that the global, the local and the kernel quantities are related through the expressions [82]

$$s(\mathbf{r}) = \int s(\mathbf{r}, \mathbf{r}')d\mathbf{r}' \tag{3.15}$$

and

$$S = \int s(\mathbf{r})d\mathbf{r} = \int\int s(\mathbf{r}, \mathbf{r}')d\mathbf{r}d\mathbf{r}' \tag{3.16}$$

One can see that, according to the first equality in Eq. (3.16), the local softness can be interpreted as the function that leads to the distribution of the global property S all over the molecule, while according to Eq. (3.15), the softness kernel leads to the distribution of the local property $s(\mathbf{r})$ at point $\mathbf{r}$ all over the molecule. Thus, it seems that for any global property, one could try to develop a procedure to find the local and the nonlocal counterparts that satisfy the corresponding Eqs. (3.15) and (3.16).

The development of the local counterpart of the chemical potential and the hardness, together with the development of the nonlocal counterpart of the Fukui function and the dual descriptor, may be done through two different procedures. One is based on a temperature-dependent approach combined with a grand canonical ensemble composed of the systems with $N_0 - 1$, N_0, and $N_0 + 1$ electrons in their ground state [84–86]. However, there is another procedure [86–89], which will be the one described in this chapter. The first approach leads to the same results as the second approach in the zero-temperature limit.

In order to achieve this goal, consider first the chain rule for the derivative of the energy with respect to the number of electrons in the following form [87]

$$\left(\frac{\partial E}{\partial N}\right)_{v(\mathbf{r})} = \int \left(\frac{\delta E}{\delta \rho(\mathbf{r})}\right)_{v(\mathbf{r})} \left(\frac{\partial \rho(\mathbf{r})}{\partial N}\right)_{v(\mathbf{r})} d\mathbf{r} \tag{3.17}$$

Since in the variational procedure of DFT, one must minimize the total energy with respect to the electron density at constant external potential, subject to the constraint that the latter must integrate to the total number of electrons in the system, the corresponding Euler–Lagrange equation is

$$\left(\frac{\delta E}{\delta \rho(\mathbf{r})}\right)_{v(\mathbf{r})} = \mu \tag{3.18}$$

in which the chemical potential is identified with the undetermined Lagrange multiplier that is constant throughout the whole space. Thus, substituting Eqs. (3.6) and (3.18) in Eq. (3.17), one finds that

$$\left(\frac{\partial E}{\partial N}\right)_{v(\mathbf{r})} = \int \mu f(\mathbf{r})d\mathbf{r} = \mu \tag{3.19}$$

where Eq. (3.11) is used to establish the second equality. This result reproduces Eq. (3.1). However, additionally, it allows to define a local chemical potential from the integrand in the first equality of Eq. (3.19) given by

$$\mu(\mathbf{r}) = \mu f(\mathbf{r}) \tag{3.20}$$

that describes how the chemical potential is distributed in the molecule. On the other hand, one can derive Eq. (3.19) with respect to the number of electrons at constant external potential so that, considering Eqs. (3.2) and (3.7), one obtains that [87]

$$\left(\frac{\partial^2 E}{\partial N^2}\right)_{v(\mathbf{r})} = \int (\eta f(\mathbf{r}) + \mu \Delta f(\mathbf{r}))d\mathbf{r} = \eta \tag{3.21}$$

where Eq. (3.11) is used to obtain the second equality, and from the integrand of this expression, one can define a local hardness as

$$\eta(\mathbf{r}) = \eta f(\mathbf{r}) + \mu \Delta f(\mathbf{r}) \tag{3.22}$$

that describes how the hardness is distributed in the molecule.

In the case of electrophilicity, there have been several proposals to derive a local function for this global reactivity descriptor [90–92]. However, through the approach outlined in this work, one can see [88] that if one multiply Eq. (3.19) by the factor $\mu/2\eta$ one finds that

$$\omega = \frac{\mu^2}{2\eta} = \int \frac{\mu^2}{2\eta} f(\mathbf{r})d\mathbf{r} = \int \omega f(\mathbf{r})d\mathbf{r} \tag{3.23}$$

so that from the integrand of this expression, one can define the local electrophilicity as

$$\omega(\mathbf{r}) = \omega f(\mathbf{r}) \tag{3.24}$$

Thus, through this procedure, one can obtain local counterparts of the global chemical potential, hardness, and electrophilicity that satisfy a similar property to the one described in the first equality of Eq. (3.16) for the softness, in the sense that the local property integrated over the whole space leads to the global property.

Now, one can proceed in a similar way to construct the kernels associated with these two properties. In this case, one needs first to take the derivative with respect to the external potential at a constant number of electrons on both sides of Eq. (3.19) to find that [86]

$$\left(\frac{\delta \mu}{\delta v(\mathbf{r}')}\right)_N = \int f(\mathbf{r})\left(\frac{\delta \mu}{\delta v(\mathbf{r}')}\right)_N d\mathbf{r} + \int \mu \left(\frac{\delta f(\mathbf{r})}{\delta v(\mathbf{r}')}\right)_N d\mathbf{r} \tag{3.25}$$

Using Eq. (3.6) and taking the derivative with respect to N out in the second integral, this expression can be written as

$$f(\mathbf{r}') = \int f(\mathbf{r})f(\mathbf{r}')d\mathbf{r} + \mu \frac{\partial}{\partial N} \int \left(\frac{\delta \rho(\mathbf{r})}{\delta v(\mathbf{r}')}\right)_N d\mathbf{r} = \int f(\mathbf{r})f(\mathbf{r}')d\mathbf{r} \tag{3.26}$$

where in the last equality, it has been taken into account that the integral of the linear response function, which is the function in the integrand of the second integral after the first equality, is equal to zero. From the integrand of the second equality, one can define the Fukui kernel [86]

$$f(\mathbf{r}, \mathbf{r}') = f(\mathbf{r})f(\mathbf{r}') \tag{3.27}$$

Additionally, if one derives Eq. (3.26) with respect to the number of electrons at constant external potential, taking into account Eq. (3.7), one obtains that

$$\Delta f(\mathbf{r}') = \int (f(\mathbf{r})\Delta f(\mathbf{r}') + f(\mathbf{r}')\Delta f(\mathbf{r}))d\mathbf{r} \tag{3.28}$$

that may be used to define, from the integrand, a dual descriptor kernel as

$$\Delta f(\mathbf{r},\mathbf{r}') = f(\mathbf{r})\Delta f(\mathbf{r}') + f(\mathbf{r}')\Delta f(\mathbf{r}) \tag{3.29}$$

Clearly, from Eq. (3.11), one can see that the integral over the two variables of the kernels of the Fukui function and the dual descriptor are equal to

$$\iint f(\mathbf{r},\mathbf{r}')d\mathbf{r}d\mathbf{r}' = 1 \quad \text{and} \quad \iint \Delta f(\mathbf{r},\mathbf{r}')d\mathbf{r}d\mathbf{r}' = 0 \tag{3.30}$$

Now, using Eq. (3.26) to replace the Fukui function in Eq. (3.19), one finds that,

$$\mu = \int \mu f(\mathbf{r})d\mathbf{r} = \iint \mu f(\mathbf{r})f(\mathbf{r}')d\mathbf{r}d\mathbf{r}' \tag{3.31}$$

so that from the integrand in the second equality, one can define the chemical potential kernel,

$$\mu(\mathbf{r},\mathbf{r}') = \mu f(\mathbf{r},\mathbf{r}') = \mu f(\mathbf{r})f(\mathbf{r}') \tag{3.32}$$

Proceeding in a similar way in the case of hardness, one first rewrites Eq. (3.21) in the form

$$\eta = \int (\eta f(\mathbf{r}) + \mu\Delta f(\mathbf{r}))d\mathbf{r} = \int \left(\eta f(\mathbf{r}) + \mu\left(\frac{\partial f(\mathbf{r})}{\partial N}\right)_{v(\mathbf{r})}\right)d\mathbf{r} \tag{3.33}$$

and then substitutes the Fukui function through Eq. (3.26) to obtain

$$\eta = \iint \left(\eta f(\mathbf{r})f(\mathbf{r}') + \mu\left(\frac{\partial (f(\mathbf{r})f(\mathbf{r}'))}{\partial N}\right)_{v(\mathbf{r})}\right)d\mathbf{r}d\mathbf{r}' \tag{3.34}$$

By performing the derivative with respect to the number of electrons and using Eq. (3.7), one is led to

$$\eta = \iint [\eta f(\mathbf{r})f(\mathbf{r}') + \mu(f(\mathbf{r})\Delta f(\mathbf{r}') + f(\mathbf{r}')\Delta f(\mathbf{r}))]d\mathbf{r}d\mathbf{r}' \tag{3.35}$$

and from the integrand, one can see that the hardness kernel may be expressed as

$$\begin{aligned}\eta(\mathbf{r},\mathbf{r}') &= \eta f(\mathbf{r},\mathbf{r}') + \mu\Delta f(\mathbf{r},\mathbf{r}') \\ &= \eta f(\mathbf{r})f(\mathbf{r}') + \mu(f(\mathbf{r})\Delta f(\mathbf{r}') + f(\mathbf{r}')\Delta f(\mathbf{r}))\end{aligned} \tag{3.36}$$

where we have used Eqs. (3.27) and (3.29).

In order to derive the kernel associated with the electrophilicity concept, one can proceed in the same way as it was done for the chemical potential kernel, leading to the expression

$$\omega(\mathbf{r},\mathbf{r}') = \omega f(\mathbf{r},\mathbf{r}') = \omega f(\mathbf{r})f(\mathbf{r}') \tag{3.37}$$

Thus, one can see that also, in this case, the nonlocal counterparts of the local chemical potential, hardness, and electrophilicity satisfy a similar property to the one described in Eq. (3.15) and in the second equality of Eq. (3.16) for the softness, in the sense that the nonlocal property integrated over the whole space for one of the variables leads to the local property.

A relevant aspect of the local and nonlocal reactivity descriptors described above is that they are composed by the product of the corresponding global property and the local Fukui function or the kernels of the Fukui function and the dual descriptor. This product indicates that analogous sites or bonds in different molecules that may have directional Fukui functions with similar values, will be multiplied by a different factor given by the global part. This situation implies then that one can perform the study of the behavior of analogous sites or bonds in different molecules that imply different chemical environments.

3.3 Site and Bond Reactivities

In order to calculate numerical values for these quantities, one needs to establish the behavior of the energy as a function of the number of electrons. If one considers the grand canonical ensemble composed by the systems with $N_0 - 1$, N_0, and $N_0 + 1$ electrons in their ground state, one finds [83, 93–95] that the energy and the density as functions of N are given by a series of straight lines connecting the integer values of N. Then, for this behavior, the first derivative of the energy and of the density will be different if one approaches N_0 from the left or the right, and the second derivative, in both cases, will be ill-defined. Thus, in order to have well-defined first and second derivatives of the energy and the density, one can make use of the smooth quadratic interpolation expressed in Eqs. (3.3) and (3.9), so that the global chemical potential and hardness are given by Eq. (3.4), and the Fukui function and the dual descriptor are given by Eq. (3.10).

Additionally, one can make use of the frontier orbitals to determine the global and local properties that are present in the definitions of the global, local, and nonlocal (kernel) chemical potential, hardness, and electrophilicity. That is, for the global reactivity indexes, Eq. (3.4) becomes

$$\mu = \frac{\varepsilon_{\mathrm{H}} + \varepsilon_{\mathrm{L}}}{2} \quad \text{and} \quad \eta = \varepsilon_{\mathrm{L}} - \varepsilon_{\mathrm{H}} \tag{3.38}$$

where ε_{H} and ε_{L} are the eigenvalues of the highest (H) occupied and lowest (L) unoccupied molecular orbitals.

On the other hand, in order to study site and bond reactivities, it is more convenient to replace the directional Fukui functions by the frontier orbitals densities and since these take values at every point of space $\mathbf{r}$, one can replace them with the condensed to atom Fukui functions [96, 97]. These can be determined directly through a Hirshfeld [98] analysis of the density of the HOMO for f_k^- and the density of the LUMO for f_k^+, where the subindex k indicates the kth atom in the molecule. Therefore, in terms of the condensed Fukui functions, Eq. (3.10) adopts the form

$$f_k = \frac{f_k^- + f_k^+}{2} \quad \text{and} \quad \Delta f_k = f_k^+ - f_k^- \tag{3.39}$$

Therefore, through the local reactivity indexes, since they are functions evaluated at one point, one can establish the characteristics of each atomic site with the condensed Fukui function of each atom in the molecule multiplied by the corresponding global descriptor, while through the kernel, since it is given by a function that must be evaluated simultaneously at two points, one can select the condensed Fukui functions of the two atoms associated with a particular bond multiplied by the corresponding global descriptor.

As a first example, the study of the electrophilic addition of hydrogen halides (HX) to several substituted ethenes will be presented. In Figure 3.1, one can see the general scheme of this type of reaction, where Markovnikov's rule [99] establishes that the hydrogen addition occurs at the carbon atom of the double bond that has more H atoms bonded, C1 in Figure 3.1.

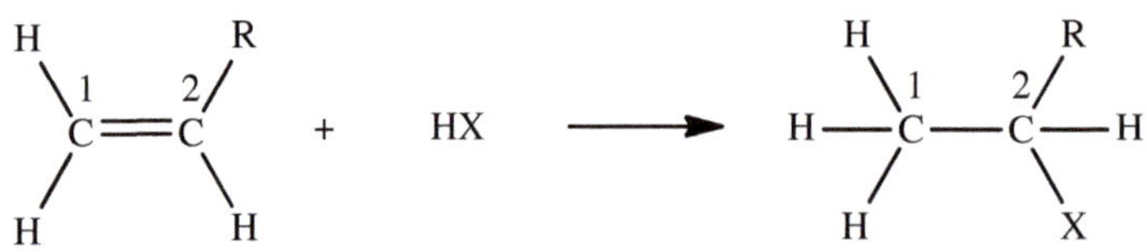

Figure 3.1 General scheme of an electrophilic addition reaction with Markovnikov orientation of hydrogen halides (HX) to substituted ethenes.

The substituents considered are NH_2, OH, CH_3, OCH_3, $NHCH_3$, NHOH, $N(CH_3)_2$, $NHNH_2$, CH_2CH_3, and F. The analysis presented is focused on the activation energy calculated for these cases [100], to know if there is a correlation between this kinetic property and the local and nonlocal descriptors. In this case, one can look at the local indexes at C1 because the H atom bonds at this site. However, it is also important to look at the nonlocal indexes associated with the double bond, since the two carbon atoms participate in the addition.

The electronic structure calculations required to determine the values of the descriptors were done with the PBE0 exchange–correlation energy functional [101–103] and the 6-311G** basis set [104, 105], with a modified version of the NWChem program [106].

The activation energy as a function of the local chemical potential at C1 is presented on the left panel of Figure 3.2, where one can see that there is a very good correlation between these two quantities. Also, since it is an electrophilic addition in which C1 donates charge to the H of HX, one can expect that the greater the chemical potential at C1, the easier it will donate and, therefore, the lower the activation energy. Thus, the negative slope observed agrees with the trend. A relevant aspect is related to the fact that the values of the Fukui function at C1 are rather similar for all the substituents, so that the differentiation between the different molecules comes from the multiplication by the global chemical potential. This situation confirms the previous statement about the importance of the product global local, to compare a specific site, in this case, C1, but in different chemical environments. In the right panel of Figure 3.2, one can see that there is also a very good correlation between the activation energies and the chemical potential kernel. However, in this case, the values refer to the bond between the two carbon atoms, but the argumentation of the slope and the importance of the product global local is the same one, since in this case, it is also found that the values of the Fukui function kernel are rather similar for all substituents.

A similar situation occurs with the local hardness and the hardness kernel, Figure 3.3. One can see that there is also a good correlation. However, in these cases, the slopes are positive because the greater the hardness, the greater the activation energy.

Finally, in Figure 3.4, the local electrophilicity and the electrophilicity kernel also show a good correlation and the correct tendency, since greater electrophilicity implies greater capacity to accept

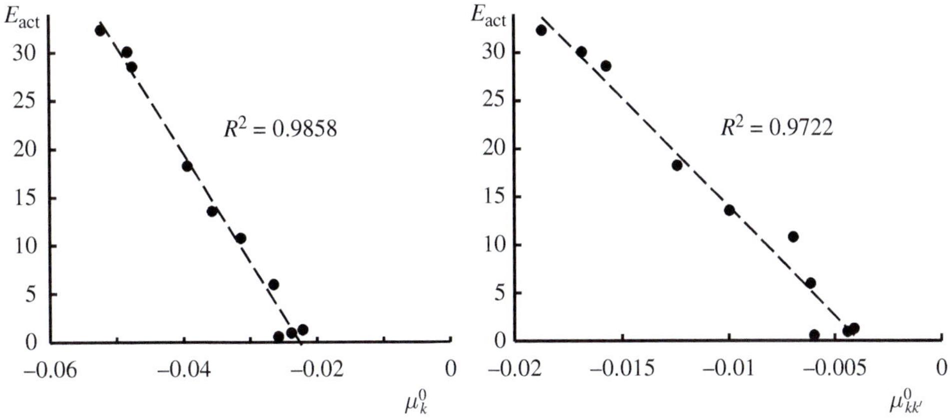

Figure 3.2 Correlation profiles for the activation energy in kcal/mol with the local chemical potential (a) and with the chemical potential kernel (b).

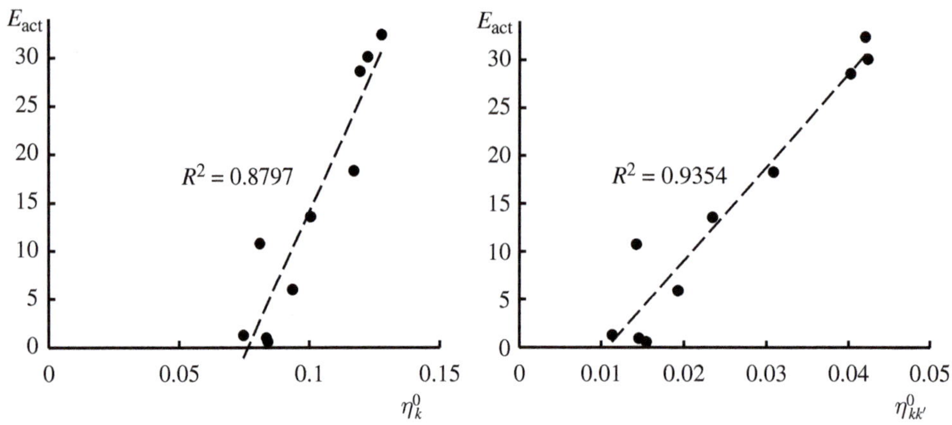

Figure 3.3 Correlation profiles for the activation energy in kcal/mol with the local hardness (a) and with the hardness kernel (b).

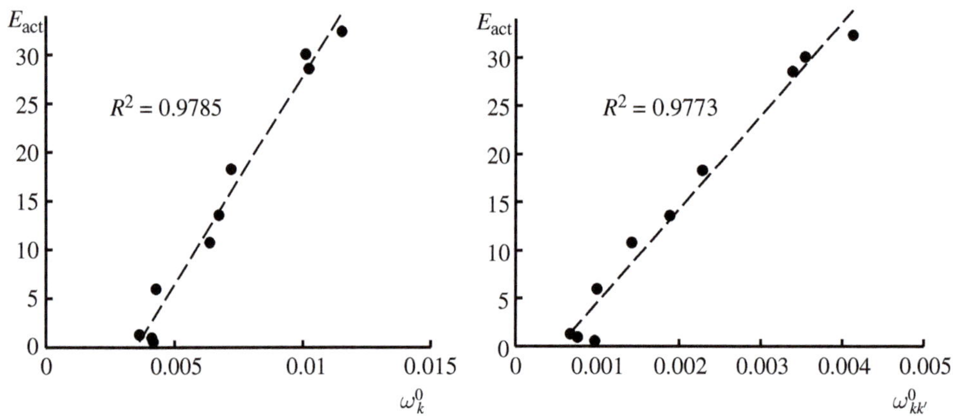

Figure 3.4 Correlation profiles for the activation energy in kcal/mol with the local electrophilicity (a) and with the electrophilicity kernel (b).

charge, so that being an electrophilic addition, the molecules with lower electrophilicity will be the ones with lower activation energy.

These electrophilic additions in ethenes are dominated by the characteristics of C1 and the double bond. Therefore, the results obtained indicate that the three local indexes provide significative information about this site, while the results obtained with the three kernels provide similar information, but in this case focused on the double bond.

As a second example of the usefulness of these reactivity descriptors, one can consider bond enthalpies, which are values obtained in a chemical reaction where there is a homolytic bond break. This property represents a good test to study the kernels proposed, since they can be calculated precisely for the bond that is going to break.

In Table 3.1, one can see the systems studied, that are classified in two groups, because they show opposite tendencies with respect to the different indexes. The experimental values were taken from Ref. [107]. The name of each molecule is assigned according to the fragments that form part of the products after the bond breaks.

Table 3.1 Molecules conforming to the two groups studied and bond enthalpies in kcal/mol.

Group 1	Bond enthalpy	Group 2	Bond enthalpy
Vinyl-I	63.5	Vinyl-CN	133.0
Vinyl-Br	80.8	Vinyl-vinyl	116.0
Vinyl-Cl	91.2	Vinyl-tertbutyl	97.8
Vinyl-F	123.3	Vinyl-ethyl	100.0
Methoxy-H	104.6	Vinyl-methyl	101.4
		Vinyl-isopropyl	99.2
		H-allyl	88.8
		H-benzyl	89.7
		H-CH_2-allyl	85.0
		Phenyl-methoxy	101.0
		Phenyl-NH_2	104.2

The electronic structure calculations to determine the indexes were also done with the PBE0 exchange–correlation energy functional [101–103], and the Def2-TZVP basis set [108], in a modified version of the NWChem program.

In these cases, one should not expect to have good correlations with the local indexes since the homolytic bond breaking clearly involves the two atoms forming the bond. This situation was numerically confirmed, since all the values for the local chemical potential, hardness, and electrophilicity show a rather larger dispersion, indicating that, indeed, there is no correlation with the bond enthalpies.

However, in the case of the kernels, the situation is different, given that it involves the two atoms forming the bond that is going to be broken. In Figure 3.5, one can see that the chemical potential kernel has a good correlation with the bond enthalpies. One can observe that in the first group, this property increases when the kernel increases, implying that the larger the accepting charge

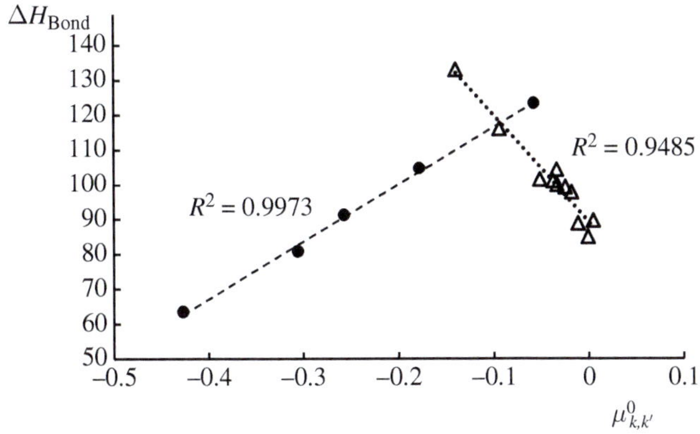

Figure 3.5 Correlation profile for the bond enthalpy in kcal/mol with the chemical potential kernel.

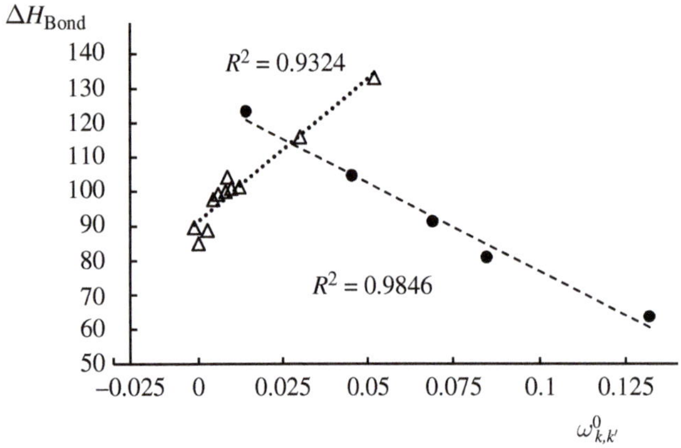

Figure 3.6 Correlation profile for the bond enthalpy in kcal/mol with the electrophilicity kernel.

character (more negative values of this kernel), the lower the bond enthalpy, which implies a more reactive molecule because it is easier to break the bond. This situation is reversed in the second group indicating that the larger the donating charge character (less negative values of this kernel), the lower the bond enthalpy, implying a more reactive molecule.

In the case of the electrophilicity kernel, there is a good correlation with the tendencies correctly described since, for this kernel, the group that shows accepting charge character implies that the larger the electrophilicity, the lower the bond enthalpy, while for the donating charge character group the lower the electrophilicity, the lower the bond enthalpy. The results for this case are presented in Figure 3.6.

Finally, in the case of the hardness kernel, the correlation is rather bad, indicating that this property does not have an important influence on the bond enthalpy and, in light of the previous results, that it will governed by the donating or accepting charge character of the bond.

Thus, one can see that the local counterpart of a global descriptor contains valuable information of site reactivity, while the nonlocal (kernel) counterpart may be used to rationalize aspects of bond reactivity.

3.4 Concluding Remarks

CDFT has become an important approach that allows one to study chemical reactivity through a set of indexes whose computation requires first doing an electronic structure calculation. In this context, one can look at CDFT as a methodology that transforms the information contained in the wavefunction into chemically meaningful quantities that can be used to understand which are the most relevant aspects of the electronic structure that are present in a chemical interaction.

Therefore, the development of new indexes offers the possibility of having a more complete vision within CDFT that allows one to obtain a deeper insight into the chemical reactivity of a group of molecules. In the present chapter, we have presented precisely a procedure that can be used to establish the local and nonlocal counterparts of a global descriptor, and that these new indexes can be used to study site and bond reactivities.

A relevant aspect of this approach relies on the fact that it leads to reactivity descriptors that are given by the product of the Fukui function or the Fukui function kernel for the local and nonlocal parts and the global index. Therefore, with these descriptors, one can study a particular site or a bond in different chemical environments.

Also, the present development highlights the role of the Fukui function to perform the distribution of a property within the molecule, so that one can establish which are the sites or the bonds (through the kernels) with greater or lower values of important properties like the chemical potential, the hardness or the electrophilicity. Since the Fukui function is the first derivative of the density with respect to the number of electrons, it indirectly highlights the relevance of the electronic density in describing chemical reactivity.

In summary, the general perspective presented in this chapter shows that by incorporating the local and nonlocal information of a global descriptor, one is led to a more complete and integrated view of chemical reactivity through CDFT.

Acknowledgment

We would like to thank Conacyt for grant Sinergia 1561802.

References

1 Hohenberg, P. and Kohn, W. (1964). Inhomogeneous electron gas. *Phys. Rev. B* 136: B864–B871.

2 Kohn, W. and Sham, L.J. (1965). Self-consistent equations including exchange and correlation effects. *Phys. Rev.* 140: A1133–A1138.

3 Parr, R.G. (1983). Density functional theory. *Annu. Rev. Phys. Chem.* 34: 631–656.

4 Parr, R.G. and Yang, W.T. (1989). *Density-Functional Theory of Atoms and Molecules*. New York: Oxford University Press.

5 Parr, R.G. and Yang, W.T. (1995). Density-functional theory of the electronic-structure of molecules. *Annu. Rev. Phys. Chem.* 46: 701–728.

6 Chermette, H. (1999). Chemical reactivity indexes in density functional theory. *J. Comput. Chem.* 20: 129–154.

7 Geerlings, P., De Proft, F., and Langenaeker, W. (2003). Conceptual density functional theory. *Chem. Rev.* 103: 1793–1873.

8 Ayers, P.W., Anderson, J.S.M., and Bartolotti, L.J. (2005). Perturbative perspectives on the chemical reaction prediction problem. *Int. J. Quantum Chem.* 101: 520–534.

9 Gázquez, J.L. (2008). Perspectives on the density functional theory of chemical reactivity. *J. Mex. Chem. Soc.* 52: 3–10.

10 Chattaraj, P.K. (2009). *Chemical Reactivity Theory: A Density Functional View*. Boca Raton, FL: CRC Press.

11 Liu, S.B. (2009). Conceptual density functional theory and some recent developments. *Acta Phys. Chim. Sin.* 25: 590–600.

12 Johnson, P.A., Bartolotti, L., Ayers, P.W. et al. (2012). Charge density and chemical reactions: a unified view from conceptual DFT. In: *Modern Charge-Density Analysis* (ed. C. Gatti and P. Macchi), 715–764. Dordrecht: Springer.

13 Fuentealba, P. and Cardenas, C. (2015). Density functional theory of chemical reactivity. In: *Chemical Modelling: A Specialist Periodical Report*, The Royal Society of Chemistry, 11: 151–174.

14 Geerlings, P., Chamorro, E., Chattaraj, P.K. et al. (2020). Conceptual density functional theory: status, prospects, issues. *Theor. Chem. Acc.* 139: 36.

15 Parr, R.G., Donnelly, R.A., Levy, M., and Palke, W.E. (1978). Electronegativity – density functional viewpoint. *J. Chem. Phys.* 68: 3801–3807.

16 Parr, R.G. and Pearson, R.G. (1983). Absolute hardness – companion parameter to absolute electronegativity. *JACS* 105: 7512–7516.

17 Iczkowski, R. and Margrave, J.L. (1961). Electronegativity. *JACS* 83: 3547–3551.

18 Mulliken, R.S. (1934). A new electroaffinity scale; together with data on valence states and on valence ionization potentials and electron affinities. *J. Chem. Phys.* 2: 782–793.

19 Sanderson, R.T. (1951). An interpretation of bond lengths and a classification of bonds. *Science* 114: 670–672.

20 Sanderson, R.T. (1971). *Chemical Bonds and Bond Energy*. New York: Academic Press.

21 Pearson, R.G. (1963). Hard and soft acids and bases. *JACS* 85: 3533–3539.

22 Pearson, R.G. (1966). Acids and bases. *Science* 151: 172–177.

23 Pearson, R.G. (1997). *Chemical Hardness*. Weinheim, Germany: Wiley-VCH.

24 Pauling, L. (1932). The nature of the chemical bond. IV. The energy of single bonds and the relative electronegativity of atoms. *JACS* 54: 3570–3582.

25 Pauling, L. (1960). *The Nature of the Chemical Bond*, 3e. New York: Cornell University Press.

26 Pearson, R.G. (1988). Absolute electronegativity and hardness: application to inorganic chemistry. *Inorg. Chem.* 27: 734–740.

27 Pearson, R.G. (1989). Absolute electronegativity and hardness: application to organic chemistry. *J. Org. Chem.* 54: 1423–1430.

28 Parr, R.G., Von Szentpaly, L., and Liu, S.B. (1999). Electrophilicity index. *JACS* 121: 1922–1924.

29 Ayers, P.W., Anderson, J.S.M., Rodriguez, J.I., and Jawed, Z. (2005). Indices for predicting the quality of leaving groups. *Phys. Chem. Chem. Phys.* 7: 1918–1925.

30 Parthasarathi, R., Elango, M., Subramanian, V., and Chattaraj, P.K. (2005). Variation of electrophilicity during molecular vibrations and internal rotations. *Theor. Chem. Acc.* 113: 257–266.

31 Chattaraj, P.K., Sarkar, U., and Roy, D.R. (2006). Electrophilicity index. *Chem. Rev.* 106: 2065–2091.

32 Chattaraj, P.K. and Roy, D.R. (2007). Update 1 of: electrophilicity index. *Chem. Rev.* 107: PR46–PR74.

33 Gázquez, J.L., Cedillo, A., and Vela, A. (2007). Electrodonating and electroaccepting powers. *J. Phys. Chem. A* 111: 1966–1970.

34 Chattaraj, P.K. (2007). A minimum electrophilicity perspective of the HSAB principle. *Indian J. Phys.* 81: 871–879.

35 Padmanabhan, J., Parthasarathi, R., Subramanian, V., and Chattaraj, P.K. (2007). Electrophilicity-based charge transfer descriptor. *J. Phys. Chem. A* 111: 1358–1361.

36 Chattaraj, P.K., Chakraborty, A., and Giri, S. (2009). Net electrophilicity. *J. Phys. Chem. A* 113: 10068–10074.

37 Bagaria, P., Saha, S., Murru, S. et al. (2009). A comprehensive decomposition analysis of stabilization energy (CDASE) and its application in locating the rate-determining step of multi-step reactions. *Phys. Chem. Chem. Phys.* 11: 8306–8315.

38 Saha, S., Roy, R.K., and Pal, S. (2010). CDASE – a reliable scheme to explain the reactivity sequence between Diels–Alder pairs. *Phys. Chem. Chem. Phys.* 12: 9328–9338.

39 Chattaraj, P.K., Giri, S., and Duley, S. (2010). Electrophilicity equalization principle. *J. Phys. Chem. Lett.* 1: 1064–1067.

40 Chattaraj, P.K., Giri, S., and Duley, S. (2011). Update 2 of: electrophilicity index. *Chem. Rev.* 111: PR43–PR75.

41 Chakraborty, A., Das, R., Giri, S., and Chattaraj, P.K. (2011). Net reactivity index (Delta omega(+/−)(R)). *J. Phys. Org. Chem.* 24: 854–864.

42 Sarmah, A. and Roy, R.K. (2013). Understanding the interaction of nucleobases with chiral semiconducting single-walled carbon nanotubes: an alternative theoretical approach based on density functional reactivity theory. *J. Phys. Chem. C* 117: 21539–21550.

43 Sarmah, A. and Roy, R.K. (2015). Interaction between small gold clusters and nucleobases: a density functional reactivity theory based study. *J. Phys. Chem. C* 119: 17940–17953.

44 Hamid, A., Anand, A., and Roy, R.K. (2017). The charge transfer limit of a chemical adduct: the role of perturbation on external potential. *Phys. Chem. Chem. Phys.* 19: 10905–10912.

45 Orozco-Valencia, A.U., Gázquez, J.L., and Vela, A. (2017). Global and local partitioning of the charge transferred in the Parr–Pearson model. *J. Phys. Chem. A* 121: 4019–4029.

46 Orozco-Valencia, U., Gázquez, J.L., and Vela, A. (2017). Donation and back-donation analyzed through a charge transfer model based on density functional theory. *J. Mol. Model.* 23: 207.

47 Orozco-Valencia, U., Gázquez, J.L., and Vela, A. (2018). Role of reaction conditions in the global and local two parabolas charge transfer model. *J. Phys. Chem. A* 122: 1796–1806.

48 Orozco-Valencia, U., Gázquez, J.L., and Vela, A. (2018). Reactivity of indoles through the eyes of a charge-transfer partitioning analysis. *Acta Phys. Chim. Sin.* 34: 692–698.

49 Orozco-Valencia, U., Gázquez, J.L., and Vela, A. (2018). Global and local charge transfer in electron donor-acceptor complexes. *J. Mol. Model.* 24.

50 Franco-Pérez, M. and Gázquez, J.L. (2019). Electronegativities of Pauling and Mulliken in density functional theory. *J. Phys. Chem. A* 123: 10065–10071.

51 Ayers, P.W. (2005). An elementary derivation of the hard/soft-acid/base principle. *J. Chem. Phys.* 122: 141102.

52 Chattaraj, P.K. and Ayers, P.W. (2005). The maximum hardness principle implies the hard/soft acid/base rule. *J. Chem. Phys.* 123: 086101.

53 Ayers, P.W., Parr, R.G., and Pearson, R.G. (2006). Elucidating the hard/soft acid/base principle: a perspective based on half-reactions. *J. Chem. Phys.* 124: 194107.

54 Chattaraj, P.K., Ayers, P.W., and Melin, J. (2007). Further links between the maximum hardness principle and the hard/soft acid/base principle: insights from hard/soft exchange reactions. *Phys. Chem. Chem. Phys.* 9: 3853–3856.

55 Miranda-Quintana, R.A. (2017). Note: The minimum electrophilicity and the hard/soft acid/base principles. *J. Chem. Phys.* 146: 046101.

56 Miranda-Quintana, R.A. and Ayers, P.W. (2018). Note: Maximum hardness and minimum electrophilicity principles. *J. Chem. Phys.* 148: 196101.

57 Miranda-Quintana, R.A., Heidar-Zadeh, F., and Ayers, P.W. (2018). Elementary derivation of the "Delta mu Big Is Good" rule. *J. Phys. Chem. Lett.* 9: 4344–4348.

58 Miranda-Quintana, R.A. and Ayers, P.W. (2019). The "Delta mu big is good" rule, the maximum hardness, and minimum electrophilicity principles. *Theor. Chem. Acc.* 138: 44.

59 Parr, R.G. and Yang, W.T. (1984). Density functional approach to the frontier-electron theory of chemical reactivity. *JACS* 106: 4049–4050.

60 Yang, W.T., Parr, R.G., and Pucci, R. (1984). Electron-density, Kohn–Sham frontier orbitals, and Fukui functions. *J. Chem. Phys.* 81: 2862–2863.

61 Ayers, P.W. and Levy, M. (2000). Perspective on "Density functional approach to the frontier-electron theory of chemical reactivity" by Parr RG, Yang W (1984). *Theor. Chem. Acc.* 103: 353–360.

62 Morell, C., Grand, A., and Toro-Labbé, A. (2005). New dual descriptor for chemical reactivity. *J. Phys. Chem. A* 109: 205–212.

63 Morell, C., Grand, A., and Toro-Labbé, A. (2006). Theoretical support for using the $\Delta f(r)$ descriptor. *Chem. Phys. Lett.* 425: 342–346.

64 Ayers, P.W. and Parr, R.G. (2000). Variational principles for describing chemical reactions: the Fukui function and chemical hardness revisited. *JACS* 122: 2010–2018.

65 Chattaraj, P.K., Cedillo, A., and Parr, R.G. (1995). Variational method for determining the Fukui function and chemical hardness of an electronic system. *J. Chem. Phys.* 103: 7645–7646.

66 Fukui, K., Yonezawa, T., and Shingu, H. (1952). A molecular orbital theory of reactivity in aromatic hydrocarbons. *J. Chem. Phys.* 20: 722–725.

67 Fukui, K. (1982). Role of frontier orbitals in chemical-reactions. *Science* 218: 747–754.

68 Berkowitz, M. (1987). Density functional-approach to frontier controlled reactions. *JACS* 109: 4823–4825.

69 Chattaraj, P.K., Fuentealba, P., Gomez, B., and Contreras, R. (2000). Woodward-Hoffmann rule in the light of the principles of maximum hardness and minimum polarizability: DFT and ab initio SCF studies. *JACS* 122: 348–351.

70 De Proft, F., Ayers, P.W., Fias, S., and Geerlings, P. (2006). Woodward-Hoffmann rules in density functional theory: initial hardness response. *J. Chem. Phys.* 125: 214101.

71 Ayers, P.W., Morell, C., De Proft, F., and Geerlings, P. (2007). Understanding the Woodward-Hoffmann rules by using changes in electron density. *Chem. Eur. J.* 13: 8240–8247.

72 Sablon, N., de Proft, F., and Geerlings, P. (2009). Reformulating the Woodward-Hoffmann rules in a conceptual density functional theory context: the case of sigmatropic reactions. *Croat. Chem. Acta* 82: 157–164.

73 Jaque, P., Correa, J.V., De Proft, F. et al. (2010). Regaining the Woodward-Hoffmann rules for chelotropic reactions via conceptual DFT. *Can. J. Chem.* 88: 858–865.

74 Geerlings, P., Ayers, P.W., Toro-Labbe, A. et al. (2012). The Woodward-Hoffmann rules reinterpreted by conceptual density functional theory. *Acc. Chem. Res.* 45: 683–695.

75 Tognetti, V., Morell, C., Ayers, P.W. et al. (2013). A proposal for an extended dual descriptor: a possible solution when Frontier Molecular Orbital Theory fails. *Phys. Chem. Chem. Phys.* 15: 14465–14475.

76 De Proft, F., Forquet, V., Ourri, B. et al. (2015). Investigation of electron density changes at the onset of a chemical reaction using the state-specific dual descriptor from conceptual density functional theory. *Phys. Chem. Chem. Phys.* 17: 9359–9368.

77 Merzoud, L., Saal, A., Morell, C., and Chermette, H. (2019). Substituent effect on the Himbert intramolecular arene/allene Diets–Alder reaction: NBO analysis and state specific dual descriptors. *J. Phys. Chem. A* 123: 10730–10738.

78 Guegan, F., Pigeon, T., De Proft, F. et al. (2020). Understanding chemical selectivity through well selected excited states. *J. Phys. Chem. A* 124: 633–641.

79 Carmona-Espíndola, J. and Gázquez, J.L. (2022). Charge transfer excitations and constrained density functional theory. *Theor. Chem. Acc.* 141.

80 Berkowitz, M., Ghosh, S.K., and Parr, R.G. (1985). On the concept of local hardness in chemistry. *JACS* 107: 6811–6814.

81 Yang, W.T. and Parr, R.G. (1985). Hardness, softness, and the Fukui function in the electronic theory of metals and catalysis. *PNAS* 82: 6723–6726.

82 Berkowitz, M. and Parr, R.G. (1988). Molecular hardness and softness, local hardness and softness, hardness and softness kernels, and relations among these quantities. *J. Chem. Phys.* 88: 2554–2557.

83 Franco-Pérez, M., Gázquez, J.L., Ayers, P.W., and Vela, A. (2015). Revisiting the definition of the electronic chemical potential, chemical hardness, and softness at finite temperatures. *J. Chem. Phys.* 143: 154103.

84 Gázquez, J.L., Franco-Pérez, M., Ayers, P.W., and Vela, A. (2019). Temperature-dependent approach to chemical reactivity concepts in density functional theory. *Int. J. Quantum Chem.* 119: e25797.

85 Franco-Pérez, M., Ayers, P.W., Gázquez, J.L., and Vela, A. (2017). Local chemical potential, local hardness, and dual descriptors in temperature dependent chemical reactivity theory. *Phys. Chem. Chem. Phys.* 19: 13687–13695.

86 Franco-Pérez, M., Polanco-Ramirez, C.A., Ayers, P.W. et al. (2017). New Fukui, dual and hyper-dual kernels as bond reactivity descriptors. *Phys. Chem. Chem. Phys.* 19: 16095–16104.

87 Polanco-Ramírez, C.A., Franco-Pérez, M., Carmona-Espíndola, J. et al. (2017). Revisiting the definition of local hardness and hardness kernel. *Phys. Chem. Chem. Phys.* 19: 12355–12364.

88 Robles, A., Franco-Pérez, M., Gázquez, J.L. et al. (2018). Local electrophilicity. *J. Mol. Model.* 24: 245.

89 Franco-Pérez, M., Polanco-Ramírez, C.A., Gázquez, J.L., and Ayers, P.W. (2018). Local and nonlocal counterparts of global descriptors: the cases of chemical softness and hardness. *J. Mol. Model.* 24: 285.

90 Chattaraj, P.K., Maiti, B., and Sarkar, U. (2003). Philicity: a unified treatment of chemical reactivity and selectivity. *J. Phys. Chem. A* 107: 4973–4975.

91 Cedillo, A. and Contreras, R. (2012). A local extension of the electrophilicity index concept. *J. Mex. Chem. Soc.* 56: 257–260.

92 Morell, C., Gázquez, J.L., Vela, A. et al. (2014). Revisiting electroaccepting and electrodonating powers: proposals for local electrophilicity and local nucleophilicity descriptors. *Phys. Chem. Chem. Phys.* 16: 26832–26842.

93 Perdew, J.P., Parr, R.G., Levy, M., and Balduz, J.L. (1982). Density-functional theory for fractional particle number – derivative discontinuities of the energy. *Phys. Rev. Lett.* 49: 1691–1694.

94 Yang, W.T., Zhang, Y.K., and Ayers, P.W. (2000). Degenerate ground states and a fractional number of electrons in density and reduced density matrix functional theory. *Phys. Rev. Lett.* 84: 5172–5175.

95 Ayers, P.W. (2008). The dependence on and continuity of the energy and other molecular properties with respect to the number of electrons. *J. Math. Chem.* 43: 285–303.

96 Yang, W.T. and Mortier, W.J. (1986). The use of global and local molecular-parameters for the analysis of the gas-phase basicity of amines. *JACS* 108: 5708–5711.

97 Bultinck, P., Fias, S., Van Alsenoy, C. et al. (2007). Critical thoughts on computing atom condensed Fukui functions. *J. Chem. Phys.* 127: 034102.

98 Hirshfeld, F.L. (1977). Bonded-atom fragments for describing molecular charge-densities. *Theor. Chim. Acta* 44: 129–138.

99 Markovnikov, V.I. (1870). Ueber die Abhängigkeit der verschiedenen Vertretbarkeit des Radicalwasserstoffs in den isomeren Buttersäuren. *Liebigs Ann. Chem.* 153: 228–259.

100 Meneses, L., Araya, A., Pilaquinga, F. et al. (2007). Local hardness: an application to electrophilic additions. *Chem. Phys. Lett.* 446: 170–175.

101 Adamo, C. and Barone, V. (1999). Toward reliable density functional methods without adjustable parameters: the PBE0 model. *J. Chem. Phys.* 110: 6158–6170.

102 Ernzerhof, M. and Scuseria, G.E. (1999). Assessment of the Perdew-Burke-Ernzerhof exchange-correlation functional. *J. Chem. Phys.* 110: 5029–5036.

103 Adamo, C., Scuseria, G.E., and Barone, V. (1999). Accurate excitation energies from time-dependent density functional theory: assessing the PBE0 model. *J. Chem. Phys.* 111: 2889–2899.

104 Krishnan, R., Binkley, J.S., Seeger, R., and Pople, J.A. (1980). Self-consistent molecular orbital methods. XX. Basis set for correlated wavefunctions. *J. Chem. Phys.* 72: 650–654.

105 McLean, A.D. and Chandler, G.S. (1980). Contracted Gaussian-basis set for molecular calculations. 1. 2nd row atoms $Z = 11$–18. *J. Chem. Phys.* 72: 5639–5648.

106 Valiev, M., Bylaska, E.J., Govind, N. et al. (2010). NWChem: a comprehensive and scalable open-source solution for large scale molecular simulations. *Comput. Phys. Commun.* 181: 1477–1489.

107 Blanksby, S.J. and Ellison, G.B. (2003). Bond dissociation energies of organic molecules. *Acc. Chem. Res.* 36: 255–263.

108 Weigend, F. and Ahlrichs, R. (2005). Balanced basis sets of split valence, triple zeta valence and quadruple zeta valence quality for H to Rn: design and assessment of accuracy. *Phys. Chem. Chem. Phys.* 7: 3297–3305.

4

Relativistic Treatment of Many-Electron Systems Through DFT in CCG

Shamik Chanda and Amlan K. Roy

Department of Chemical Sciences, Indian Institute of Science Education and Research Kolkata, Mohanpur, Nadia, West Bengal, India

4.1 Introduction

The mathematical description of physical systems has gone beyond the standard classical mechanics after the invention of Einstein's special theory of relativity. According to this, a physical system should obey two basic postulates of special relativity [1] instead of Galilei relativity (nonrelativistic mechanics); i.e. (i) speed of light should be finite (rather than infinite in Galilean relativity) and constant for all inertial frame of reference, and (ii) the fundamental equation of motion (EOM) should take the same form, in all inertial reference frame. In other words, they must be Lorentz co-variant. For many physical systems, a nonrelativistic description gives fairly accurate results, and they may not differ significantly from a correct relativistic description. For example, for bigger objects, Newtonian mechanics is enough to describe their motions. But many physical situations exist where the non-relativistic theory differs considerably from experiment. In such cases, corrections occurring from special relativity need to be incorporated into their mathematical formulation. The importance of relativity in molecular science was first realized in the 1970s [2–6], which came as a surprise as it was unexpected that for low-energy molecular physics, we need prescription beyond Schrödinger quantum mechanics. Since average speed of valence electrons is low, it was originally thought (even Dirac believed [7]) that relativity was not important, until its impact on chemistry of systems containing heavy elements came out with the celebrated consequences like yellow color, trivalency, and nobility of gold as well as large effects on bond lengths, spectroscopic properties of molecules containing heavy elements like lanthanides and actinides. Even liquidicity of mercury at normal temperature is a consequence of relativity, as it is responsible for its low melting point. One recent communication [8], found that, relativistic effects are responsible for 1.7–1.8 V in a standard 2-V lead-acid battery. In the authors' words we say, "cars start due to the relativity."

The notion of relativistic effect implies mainly the difference between (four-component, two-component or scalar) relativistic and nonrelativistic expectation values. Although this definition is very straightforward, but the difficulties actually lie in the details. So, the relativistic effects are not in general observable in the case of usual observables of quantum mechanics such as energy, position, and momentum. But, the relativistic effects are clearly manifest in the chemistry of heavy elements, as already mentioned. When an atomic/molecular property, like spin–orbit

Electron Density: Concepts, Computation and DFT Applications, First Edition.
Edited by Pratim Kumar Chattaraj and Debdutta Chakraborty.

(SO) coupling or electronic g-shifts (defined in terms of deviations from the free-electron g value), vanishes in the nonrelativistic limit, then the consequences of relativistic effects are directly noticeable. However, one of the primary objectives of molecular quantum chemistry is to examine the impacts of relativity on chemistry beyond the celebrated examples of relativity, such as the yellow color of solid gold or liquidity of mercury at room temperature, as well as new methodological developments for relativistic quantum chemical calculations with increasingly accurate results, especially in the realm of molecular spectroscopic properties.

We can understand the essence of relativity by considering a simple problem of a free electron. In the nonrelativistic case, the energy of the particle is $E_{\text{non-rel}} = \frac{p^2}{2m} = \frac{mv^2}{2}$ where m, $p = mv$, and v are the rest-mass, linear momentum, and velocity of the electron. In relativistic theory, $E_{\text{rel}} = \sqrt{p^2c^2 + m_0^2c^4} = m_0c^2 + \frac{p^2}{2m_0} - \frac{p^4}{8m_0^3c^2} + \cdots$, where p signifies relativistic momentum, while m_0c^2 is the rest-mass energy, which shifts the origin of energy scale. The mass of the electron m is increased with velocity when $v \approx c$ through $m = \gamma m_0 = \frac{m_0}{\sqrt{1-\frac{v^2}{c^2}}}$. The kinetic energy gives a leading order in c^{-2}, after the inclusion of relativistic correction, and varies with p^2 of the particle. If we consider an electron in an atom of nuclear charge Ze, the nonrelativistic kinetic energy becomes proportional to the expectation value of p^2; that also means, proportional to Z^2. Hence, the leading order relativistic correction for electron energies, therefore, varies as $\frac{Z^4}{c^2}$. Here, in atomic unit, $c \approx 137$. So, it is very clear that the relativistic energies of one-electron atoms follow an approximate Z^4 trend. Moreover, an increase in mass leads to a smaller Bohr radius for one-electron atom as $a_0 = \frac{\hbar^2}{me^2}$. This leads to a relativistic contraction and stabilization of all s and most of the p orbitals in a given many-electron atom. Consequently, the stronger screening of nuclear attraction by the contracted s and p orbitals, gives rise to the relativistic expansion and destabilization in case of d and f orbitals. These effects contribute significantly for the differences in the chemical properties between periods V (Rb through Xe) and VI (Cs through Rn) of the periodic table. These direct and indirect effects, as well as the SO splitting increase in case of valence shells down a given column approximately as Z^2. Note that, here, Z is full nuclear charge. In case of H-like systems, one gets the Z^4 trend. Relativistic effects on the electronic structure of the system increase not the screened, but the full charges of the nuclei of a system.

In most of the cases, relativistic calculations are not much expensive when compared to their nonrelativistic counterparts. Sometimes relativistic calculations are mandatory, like for the chemical reactions in which educts and products have electronic ground states with different spins. This indicates that the spin symmetry is changed on the reaction path (mediated by SO coupling). These kinds of events are quite common in the field of transition metal chemistry. Moreover, there exist a number of molecular properties, that may not be calculable at all within a nonrelativistic framework. Like for example, the electronic g-tensor of ESR spectroscopy may not be looked at without considering SO coupling, while for calculation of ESR isotropic hyper-fine coupling constant a purely scalar-relativistic framework is seen to be accurate well enough.

The most accurate approach to do relativistic calculation for many-electron system is to perform the fully relativistic four-component calculations (see [9] for review), where the basis-set expansion is taken in the form of generally contracted kinetically balanced Gaussian-type spinors. These are followed by several groups [10]. The most popular program for performing routine relativistic *ab initio* calculation is DIRAC [11]. However, four-component methods are computationally quite demanding and limited to small systems. Over the years, various methods were developed to do relativistic calculations efficiently, by reducing computational burden in an approximate manner. As molecular quantum mechanical studies need quite extensive computational resources, the most popular route to incorporate relativistic effects involves the relativistic effective core potentials. The replacement of core electrons of a heavy atom by some suitable model potential, often augmented with an additional core-polarization potential, permits a direct application of standard non-relativistic quantum chemical calculations in such compounds. The restriction of a given calculation to valence and sub-valence electrons leads to a cost-efficient procedure that also permits the application of relatively more demanding electron correlation methods.

Now, in chemistry, the behavior of electrons is of vital importance. So, the electronic states are our central focus of attention. As an alternative to rigorous four-component method, two-component quasi-relativistic approaches are proposed, centering on how to retrieve the effects lost by neglecting the positronic part of the solutions. Several approaches based on a transformation of the Hamiltonians facilitate considerable progress with respect to accurate *ab initio* relativistic methods. Besides, applications to large chemical systems became accessible. In particular, few important ones pertain to the unitary transformations, which result in two-component electron-only Hamiltonians after decoupling the electronic and positronic parts of the solutions. Thus, several two-component methods were developed over the years, e.g. zeroth-order regular approximation (ZORA) [12, 13], direct perturbation theory (DPT) [14], Breit-Pauli Hamiltonian, Dyall's modified Dirac [15], and Douglas–Kroll–Hess (DKH) [16–18] method. The latter method, up to second order, was extensively studied by Hess and coworkers in a series of papers [19–23]. This has become the most familiar method for quasi-relativistic calculation as it can capture most of the scalar relativistic effects. Later DKH Hamiltonian up to 3rd order was derived [24]. A generalized DKH transformation up to fifth order was also proposed [22] and subsequently extended to obtaining infinite-order DKH Hamiltonian [25–27]. However, it can be shown that DKH method up to second or third order gives electronic properties sufficiently converged [27–30]. Transformation of the two-body matrix elements was also obtained [21, 31]. One of the advantages is that scalar or spin-free DKH Hamiltonian directly fits into any standard non-relativistic quantum chemistry program. Appropriate basis sets for DKH calculations were also developed [32–35] and they have been brought into Hartree–Fock (HF), CCSD and various other levels of theory.

Now, in the last few decades, density functional theory (DFT) [36] has been brought into atomic or molecular electronic structure calculations over a wide range of fields, from solid-state physics to organic chemistry, as it permits for a quite accurate quantum mechanical description at a relatively lesser computational cost [37–40]. Practical applications of DFT, to a large extent, completely depend on the solution of Kohn–Sham (KS) equation [41] or generalized (G)KS equation [42]. Within this framework, the interacting many-electron potential is mapped onto an effective single-particle one by means of a local one-body potential, called KS potential, $v_{KS}(\mathbf{r})$, keeping the ground-state density fixed. *In principle*, the theory is *exact*, and has the ability to incorporate the many-body effects fully, satisfactorily. But all the non-classical interactions are taken as exchange-correlation (XC) potential $v_{xc}(\mathbf{r})$, defined as the functional derivative of XC energy, $E_{xc}[\rho(\mathbf{r})]$ expressed as, $\frac{\delta E_{xc}[\rho(\mathbf{r})]}{\delta \rho(\mathbf{r})}$. The XC energy is approximated under the so-called density functional approximation (DFA). To obtain it accurately, remains one of the fundamental interests in the development of modern DFT. Then DFT has been extended to relativistic regime by taking Dirac equation (DEQ) as a starting point [43, 44] as a generalization of so called Hohenberg–Kohn–Sham theory [36, 41] for in-homogeneous relativistic electron gas. Moreover, four-component Dirac–Kohn–Sham (DKS) equation was also deduced from QED [43]. Several two-components methods are also augmented in density functional calculations mainly due to their cost efficiency. We have witnessed the extension of DKH method in the framework of linear combination of Gaussian type orbitals (GTO) approach to DFT [45–48].

In the past few years, a KS-DFT scheme for atoms and molecules in Cartesian Coordinate Grid (CCG) has been found to be very successful [49–60] for calculation of properties in ground state, and also extended to some excited states via a virial theorem and exciton model. By employing an accurate representation of Coulomb interaction kernel (CIK) through Fourier convolution technique (FCT) in CCG, as well as through LCAO ansatz of KS MOs, relevant quantities such as electron density, basis function, as well as classical Hartree and non-classical XC potentials are directly calculated on a 3D real CCG. In this chapter, we will discuss our recent extension of spin-free $1e^-$ DKH Hamiltonian in CCG-DFT. The two-body terms are kept non relativistic in this case, approximating the relativistic effects only through one-body terms. A short review of relativistic quantum mechanics, namely Dirac equation, relativistic DFT as well as DKH methodology and KS DFT in CCG is presented in Section 4.2. Various aspects of computational and technical details are discussed in Section 4.3. The validity of our implementation is checked by showing results for one-electron systems in Section 4.1. The accuracy of our results to give total energy of few representative atoms, molecules, ions in case of LDA, BLYP, B3LYP and PBE0 XC functionals is offered in Section 4.2. Finally Section 4.3 provides an application of the proposed method to estimate energy and ionization potential (IP) of some highly charged ions of He- and Li-isoelectronic ions for ionic charge Z falling in between 20 and 100. Additionally we employ some range-separated hybrid (RSH) functionals like LC-PBE, CAM-B3LYP, LC-PBE, CAM-PBE0, LRCωPBEh* to check their performance for these highly charged systems. The trends of energy and IP with varying Z for such ions are studied graphically. Lastly, a few conclusions and future direction is presented in Section 4.5.

4.2 Theoretical Framework

4.2.1 Dirac Equation

The dynamics of a particle in quantum mechanics is governed by the time-dependent Schrödinger equation (TDSE),

$$i\hbar\frac{\partial\psi(\vec{r},t)}{\partial t} = H\psi \tag{4.1}$$

Following Dirac [7, 61, 62], the relativistic Hamiltonian is given by,

$$H = c\vec{\alpha}.\vec{p} + \beta mc^2 \tag{4.2}$$

where the constant c refers to the speed of light $\vec{p}(= -i\vec{\nabla})$ refers to the momentum operator. The Dirac matrices α and β obey following relations:

$$\begin{aligned}\alpha_i\alpha_j + \alpha_j\alpha_i &= 2\delta_{ij}I \\ \alpha_i\beta + \beta\alpha_i &= 0 \\ \alpha_i^2 + \alpha_j^2 &= I\end{aligned} \tag{4.3}$$

It can be shown that the explicit form of α and β which will satisfy all the above relationships are given by the following matrices,

$$\alpha = \begin{pmatrix} O_2 & \vec{\sigma} \\ \vec{\sigma} & O_2 \end{pmatrix}, \qquad \beta = \begin{pmatrix} I_2 & O_2 \\ O_2 & -I_2 \end{pmatrix} \tag{4.4}$$

where $\vec{\sigma}$ is 2×2 Pauli spin matrix, while O_2 and I_2 are 2×2 null and identity matrices. Thus consistency of the proposed Eq. (4.2) requires that ψ must be a 4×1 column matrix, known as Dirac spinor wave function. Now from Eqs. (4.1) and (4.2), we have,

$$i\hbar\frac{\partial\psi}{\partial t} = -i\hbar c\alpha_i\frac{\partial\psi}{\partial x^i} + \beta mc^2\psi \tag{4.5}$$

Taking complex conjugate and right multiplying with ψ and left multiplying Eq. (4.5) with $\psi^\dagger$ and subtracting we will get,

$$\frac{\partial}{\partial t}(\psi^\dagger\psi) + \frac{\partial}{\partial x^i}(c\psi^\dagger\alpha_i\psi) = 0 \tag{4.6}$$

Hence,

$$\frac{\partial\rho}{\partial t} + \vec{\nabla}.\vec{j} = 0 \tag{4.7}$$

This is the familiar equation of continuity in relativistic quantum mechanics where, $\rho = \psi^\dagger\psi$ and $\vec{j} = c\psi^\dagger\alpha_i\psi$ are probability density and probability current density respectively. Here ρ is always a positive definite number. In four-vector notation, the co-variant form of Dirac equation can be written as,

$$\gamma^\mu\frac{\partial\psi}{\partial x^\mu} + \frac{imc}{\hbar}\psi = 0 \tag{4.8}$$

where $\gamma^\mu = (\gamma^0, \gamma^i)$ ($\mu = 0, 1, 2, 3$ and $i = 1, 2, 3$) given by,

$$\gamma^0 = \beta, \qquad \gamma^i = \beta\alpha^i \tag{4.9}$$

These γ matrices satisfy the relation,

$$\gamma^\mu\gamma^\nu + \gamma^\nu\gamma^\mu = 2g^{\mu\nu} \tag{4.10}$$

where $g^{\mu\nu}$ is the metric tensor given by,

$$g^{\mu\nu} = \begin{pmatrix} 1 & 0 & 0 & 0 \\ 0 & -1 & 0 & 0 \\ 0 & 0 & -1 & 0 \\ 0 & 0 & 0 & -1 \end{pmatrix} \tag{4.11}$$

Now,

$$\begin{aligned} \rho &= \psi^\dagger\psi = \psi^\dagger\gamma^0\gamma^0\psi = \tilde{\psi}\gamma^0\psi \\ \vec{j} &= \psi^\dagger\vec{\alpha}\psi = \psi^\dagger\gamma^0\gamma^0\vec{\alpha}\psi = \psi^\dagger\gamma^0\overline{\gamma}\psi = \tilde{\psi}\overline{\gamma}\psi \end{aligned} \tag{4.12}$$

where $(\gamma^0)^2 = 1$ and $\tilde{\psi} = \psi^\dagger\gamma^0$. Then, the Dirac four-current density is expressed as,

$$j^\mu = (\rho, \bar{j}) = \tilde{\psi}\gamma^\mu\psi, \qquad (\mu = 0, 1, 2, 3) \tag{4.13}$$

Now plugging the wave function in Schrödinger picture, $\phi(r, t) = \psi(r)e^{-\frac{iEt}{\hbar}}$ in Eq. (4.1) then,

$$H\psi = E\psi \tag{4.14}$$

Expanding above equation in terms of Eq. (4.2), we get,

$$\begin{pmatrix} mc^2 & -c(\vec{\sigma}.\vec{p}) \\ -c(\vec{\sigma}.\vec{p}) & -mc^2 \end{pmatrix} \begin{pmatrix} \psi_i^{2l} \\ \psi_i^{2s} \end{pmatrix} = E\begin{pmatrix} \psi_i^{2l} \\ \psi_i^{2s} \end{pmatrix} \tag{4.15}$$

where $\psi_i = \begin{pmatrix} \psi_i^{2l} \\ \psi_i^{2s} \end{pmatrix}$ is the four-component spinor solution. The two-component vectors ψ_i^{2l} and ψ_i^{2s} are the large- and small-components corresponding to the (+)ve-energy (electronic) and (−)ve-energy (positronic) states.

The free-particle solution of the above equation is (actually these two components are always coupled by the odd operator), given by,

$$\psi^{2s} = \frac{c(\vec{\sigma}.\vec{p})}{E + mc^2}\psi^{2l} \tag{4.16}$$

The energy eigen value corresponding to the above equation is given by,

$$E = \pm\sqrt{p^2c^2 + m^2c^4} \tag{4.17}$$

That means, the energy spectrum of Dirac equation corresponds to two branches. The (+)ve branch spans at $E = mc^2$ to $+\infty$ and (−)ve branch spans from $E = -mc^2$ to $-\infty$. There is a forbidden gap of $2mc^2$ between these two. But still, (−)ve energy solutions are admitted, that means Dirac equation cannot get rid out of the (−)ve energy solutions. Till now we have two positive energy solutions and two negative energy solutions. This is illustrated in Figure 4.1. This corresponds to the spin-$\frac{1}{2}$ particle so that, there are two independent spin orientations for a given momentum and energy.

Now, how to reconcile the negative energy solutions? We cannot discard them by saying they are nonphysical, because they are necessary to make the vector space complete. But a particle with (+)ve energy will fall down through the (−)ve energy states, which will make (+)ve energy states unstable. For the validity of Dirac equation, the (+)ve energy states must be stable. Here Dirac comes with a masterstroke! Accordingly, he postulated that all the (−)ve energy states are completely occupied! Then vacuum is actually the filled-up, but unobservable sea of negative energy levels, called the Dirac sea. If we accept this, then the stability problem is solved by the Pauli exclusion principle, which prevents (+)ve energy particle to cascade down in the (−)ve energy levels. Let us assume a spin-$\frac{1}{2}$ particle–electron. So, the vacuum is the state where all (−)ve energy levels are occupied by electrons; that means it has infinitely (−)ve charge and energy. Since all observations lead to finite fluctuations of charge and energy with respect to the vacuum, we can re-scale the vacuum to be without any charge and energy (charge, energy, and spin of the vacuum are considered to be 0). Let us assume that, an electron with energy $-E$ and spin up is withdrawn from the Dirac sea; then we have a positive energy electron with spin up, but also a "hole" is created in the Dirac sea. As the occupied Dirac sea is unobservable, the hole is observable. It represents the increase in charge by $+e$ and increase in energy by $+E$, if $-E$ was the energy ejected from the sea. Thus a hole with charge $+e$ and energy $+E$ and spin down (opposite to the ejected electron) is created along with the electron. Thus the absence of a (−)ve energy electron with spin up can be thought of, as the presence of a (+)ve energy and positively charged "hole" with spin down. The "hole" represents the antiparticle of the electron (i.e, positron). So the unoccupied (−)ve energy states according to Dirac, represent (+)ve energy antiparticles. Thus in order to give stability to the (+)ve energy levels, Dirac predicted the existence of positron. Discovery of positron after a few years of Dirac proposition of "hole," was a triumph of the hole theory. The whole effects coming due to the relativistic corrections are actually the contribution coming from (−)ve energy (positronic) part of the solutions.

4.2.2 Relativistic Density Functional Theory: Dirac–Kohn–Sham Method

In a central potential $V(\vec{r})$, the one-electron Dirac equation may be written as,

$$H_{\mathrm{D}}\psi = [c\vec{\alpha}.\vec{p} + (\beta - 1)mc^2 + V(\vec{r})]\psi = E\psi \tag{4.18}$$

We have shifted the energy scale by $-mc^2$ to make energies comparable with non-relativistic cases. To extend the Dirac Hamiltonian to chemically interesting many-electron systems, it is

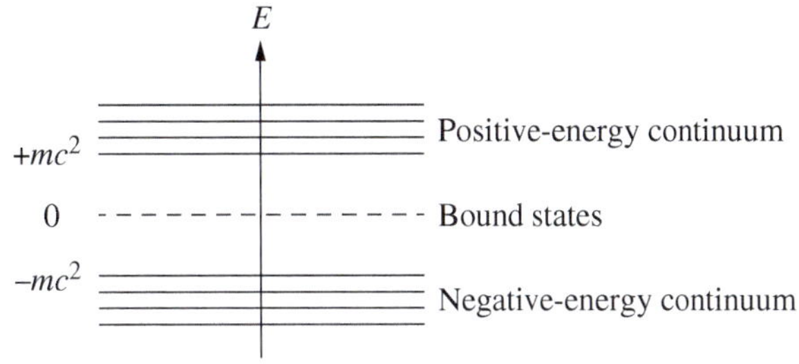

Figure 4.1 Eigen-spectra of the Dirac Hamiltonian (schematic).

augmented with Coulomb or Breit operator as a two-electron term to yield the Dirac Coulomb (DC) or Dirac–Coulomb–Breit (DCB) Hamiltonian, as follows,

$$H_{\mathrm{DC(B)}} = \sum_{i} H_{\mathrm{D}}(r_i) + \sum_{i>j} g_{ij} \tag{4.19}$$

where $g_{ij} = \frac{1}{r_{ij}}$ and $\frac{1}{r_{ij}} - \frac{1}{2}\left(\frac{\vec{\alpha}_i.\vec{\alpha}_j}{r_{ij}} + \frac{(\vec{\alpha}_i.\vec{r}_{ij})(\vec{\alpha}_j.\vec{r}_{ij})}{r_{ij}^3}\right)$ for DC and DCB Hamiltonians.

The first term represents the $1e^-$ Dirac operator, which from Eq. (4.18), is written as,

$$H_{\mathrm{D}} = \begin{pmatrix} V(\vec{r}) & c\vec{\sigma}.\vec{p} \\ c\vec{\sigma}.\vec{p} & V(\vec{r}) - 2mc^2 \end{pmatrix} \tag{4.20}$$

As already mentioned, DFT has been popularized as an attractive alternative to include correlation effects in quantum chemical calculations of molecular systems. Relativistic DFT was formulated [43] as a generalization of so-called Hohenberg–Kohn–Sham theory of inhomogeneous electron gas. Also the corresponding four-component DKS [44] equation has been deduced from the QED Hamiltonian. Like non-relativistic DFT scheme, a fictitious, non-interacting system built from a set of KS orbitals is introduced [41], in order to compute major part of the kinetic energy accurately. The remaining fairly small portion is nothing but the non-classical contribution of electron-electron repulsion. So electrons will now move in an effective single-particle KS potential, while the mapped auxiliary system will have same ground-state energy as real interacting system.

The DKS one-electron effective Hamiltonian derived from QED, is given by,

$$h_{\mathrm{DKS}} = h_{\mathrm{D}} + \upsilon_h(\vec{r}) + \upsilon_{\mathrm{xc}}(\vec{r}) \tag{4.21}$$

where h_{D} refers to the one electron Dirac Hamiltonian and the $\upsilon_h(\vec{r})$ refers to the Hartree potential, given by,

$$\upsilon_h(\vec{r}) = \int \frac{\rho(\vec{r}')}{|\vec{r} - \vec{r}'|} d\vec{r}' \tag{4.22}$$

Here $\rho(\vec{r})$ refers to the charge density, expressed in terms of a set of auxiliary occupied one electron spinors as,

$$\rho(\vec{r}) = \sum_{i}^{N^{\mathrm{elec}}} |\psi_i(\vec{r})|^2 \tag{4.23}$$

and $\upsilon_{\mathrm{xc}}(\vec{r})$ signifies the all important XC potential. Now, using the no-sea approximation positronic states are ignored. The DKS equation deduces to a matrix eigenvalue equation when basis set expansion of the wave function is introduced,

$$F_{\mathrm{DKS}}C = \mathrm{SCE}. \tag{4.24}$$

In the above equation, F_{DKS} is known as Dirac-Fock (DF) matrix and S is the overlap matrix. For every matrices, each component is a 4×4 block. Here, the external field is the potential $V(\vec{r})$ due to the nuclei. This four-component DKS formalism has been carried forward successfully to a large number of atomic, molecular, and solid-state systems.

4.2.3 Decoupling of Dirac Hamiltonian: DKH Methodology

In chemistry, the primary interest is not the (−)ve-energy solutions of four-component Dirac Hamiltonian. Also, the four-component calculations are computationally very demanding and hard to apply even for medium systems. Therefore, one of the primary motives of molecular scientists is to develop and implement approximate quantum chemical methodologies which explicitly treat only the electronic part of the solution, referred to as two- and one-component relativistic methods. This analysis is also true for the formulation of relativistic DFT approach and the solutions of corresponding DKS equation.

Now, the (+)ve and (−)ve energy solutions of a $1e^-$ Dirac equation are represented by states where either electronic or positronic contributions of four-component spinors are dominating respectively and these two are called as "large" and "small" components. However, we cannot easily ignore "small" components, because they also contribute in a fully relativistic quantum chemical description of a system. Thus, a projection step, in which positronic parts of the solution are discarded, can only be applied after a suitable decoupling of large- and small-component solutions. That means, the effects of small-component part of the solutions are implicitly retrieved by some mathematical strategy, simply called the transformation techniques.

The basic idea of these techniques is to block-diagonalize the Dirac Hamiltonian H_{D} having only the block diagonal even-terms, while off-diagonal odd (coupling) terms are annihilated in this strategy,

$$H_{\mathrm{bd}} = UH_{\mathrm{D}}U^{+} = \begin{pmatrix} h^{+} & 0 \\ 0 & h^{-} \end{pmatrix}. \tag{4.25}$$

Here, h^+ and h^- refer to the decoupled electronic and positronic part of the Hamiltonian. The Foldy–Wouthuysen (FW) transformation [63] is shown to give a decoupling of Dirac Hamiltonian, which is, *in principle*, exact, but leads to a singularity in $\frac{1}{c^2}$ in case of a Coulombic potential. Later Douglas and Kroll (DK) [16] proposed an alternative decoupling method, which is based on a sequential approximate transformations, first step being a free-particle FW transformation, followed by the transformations where external potential $V(\vec{r})$ is taken as the expansion parameter.

For a $1e^-$ DKS Hamiltonian in presence of an external potential $V(\vec{r})$, the free-particle FW transformation U_0 leads to coupling ("odd") terms between large- and small-components of the solution, which are first order in V. Eigenfunctions of the corresponding (two-component) Hamiltonian possessed a large contamination due to (−)ve energy states with eigenvalues which lie lower than those of the corresponding DKS equation. So, it is recommended to continue the decoupling process one step further, so that the coupling term becomes second order in $V(\vec{r})$. Thus the first-order DK transformation can be expressed by a unitary matrix, as follows:

$$U_1 = (1 + W_1^2)^{1/2} + W_1 \tag{4.26}$$

with an anti-Hermitian generator $W_1 = -W_1^{\dagger}$, which depends on V in first order. If applied to the $1e^-$ DKS Hamiltonian h_{DKS}, along with the free-particle FW transformation U_0, $U = U_1U_0$, we will obviously get the second-order DK Hamiltonian, which approximately has block-diagonal (decoupled) form, as given below,

$$H_{\text{DKH2}} = UH_{\text{D}}U^{+} = \begin{pmatrix} h^{+}_{\text{DKH2}} & O_2 \\ O_2 & h^{-}_{\text{DKH2}} \end{pmatrix} \tag{4.27}$$

The coupling (odd) term O_2 is of the second order in V. Now, the electronic part of the Hamiltonian h^{+}_{DKH2} is sufficient to serve as a fully relativistic electron-only operator. The second order DK transformation can be thought of as a mapping of four-component operator h_{DKS} to the two-component operator h_{DKH2}.

Some recent studies on higher-order DK Hamiltonian is based on the exponential type parametrization of unitary transformation, written as,

$$U_n = \exp(W_n) \tag{4.28}$$

In general, an anti-Hermitian generator W_n of the order n in V produces DK Hamiltonian up to $2n+1$ order, which simplifies its formulation significantly.

Now, theoretically all the coupling or "odd" terms can be annihilated by repeating further unitary transformations, which leads to the more accurate higher order DKH Hamiltonians [25, 64], or leads to the so-called exact decoupling, expressed as,

$$H_{\text{bd}} = \text{UH}_{\text{D}}U^{+} = \cdots U_3U_2U_1U_0H_{\text{D}}U_0^{+}U_1^{+}U_2^{+}U_3^{+}\cdots = \sum_k \varepsilon_k = \sum_k \begin{pmatrix} \varepsilon_k^{+} & 0 \\ 0 & \varepsilon_k^{-} \end{pmatrix} \tag{4.29}$$

The matrices U_i are chosen in such a way, that the off-diagonal (odd) terms diminishes. The index k here denotes the order of expansion parameter, which is external potential $V(\vec{r})$, in this case. Hence ε_k will contain all block-diagonal (even) terms of power k in $V(\vec{r})$. All the even terms summed up to order k gives the DKH k Hamiltonian.

Now, the operators resulting from DK Hamiltonian can be directly expressed in an "approximate momentum space representations," employing the eigenvectors of the operator p^2 in basis-set representation. In a practical calculation, for the construction of DKH Hamiltonian we need three operator p^2, V, $\vec{p}V\vec{p} + i\vec{\sigma}\cdot(\vec{p}V \times \vec{p})$. The relativistic DKH calculations can be split into scalar and SO terms. If the latter terms (particularly related to $i\vec{\sigma}\cdot(\vec{p}V \times \vec{p})$) are ignored, then the two-component methodology is now converted into an effective one-component treatment, having a trivial spin-dependence on the wave functions. These methods are called the scalar or spin-free relativistic approaches. They very easily fit into the existing non-relativistic quantum chemistry program [64]. The detailed mathematical description can be found in the references [22, 25–27, 64, 65].

Thus on the basis of above discussions, we obtain the even operators of spin-free DKH Hamiltonian up to fourth order, given by,

$$\begin{aligned}
\varepsilon_0^{+}(i) &= E_i - mc^2 \\
\varepsilon_1^{+}(i,j) &= A_i(V_{ij} + P_iV_{ij}P_j)A_j \\
\varepsilon_2^{\dagger}(i,j,k) &= \frac{1}{2}\Bigg[-A_iP_iV'_{ij}P_jA_jA_jV_{jk}A_k + A_iP_iV'_{ij}P_jA_j\frac{1}{P_j^2}A_jP_jV_{jk}P_kA_k + A_iV'_{ij}A_jP_j^2A_jV_{jk}A_k \\
&\quad - A_iV'_{ij}A_jA_jP_jV_{jk}P_kA_k - A_iP_iV_{ij}P_jA_jA_jV'_{jk}A_k + A_iP_iV_{ij}P_jA_j\frac{1}{P_j^2}A_jP_jV'_{jk}P_kA_k \\
&\quad + A_iV_{ij}A_jP_j^2A_jV'_{jk}A_k - A_iV_{ij}A_jA_jP_jV'_{jk}P_kA_k\Bigg]
\end{aligned} \tag{4.30}$$

where $E_i = \sqrt{p_i^2c^2 + m^2c^4}$, $A_i = \sqrt{\frac{E_i+mc^2}{2E_i}}$, $P_i = \frac{c\vec{p}_i}{E_i+mc^2}$, $V'_{ij} = \frac{V_{ij}}{E_i+E_j} = \frac{V(p_i,p_j)}{E_i+E_j}$. The expressions of $\varepsilon_3^{\dagger}$, $\varepsilon_4^{\dagger}$ are quite involved and are not given in lowest level expressions in literature. They can be obtained from the expressions of operators given in [64]. The relevant matrix elements including *pVp* integrals, $\langle \chi_i | pVp | \chi_j \rangle = \langle \vec{\nabla}\chi_i | V | \vec{\nabla}\chi_j \rangle$, in Gaussian basis, are incorporated in our InDFT [66] code using established recursion relations [67].

4.2.4 DFT in Cartesian Grid

4.2.4.1 Basic Methodology

The single-particle KS equation [41], for a particular many-electron system, is given by,

$$\left[-\frac{1}{2}\nabla^2 + v_{\text{ext}}(\vec{r}) + v_h(\vec{r}) + v_{\text{xc}}[\rho(\vec{r})]\right] \psi_i^\sigma(\vec{r}) = \varepsilon_i \psi_i^\sigma(\vec{r}) \tag{4.31}$$

where $v_{\text{ext}}(\vec{r})$ designates el-nu attraction potential, while $v_h[\rho(\mathbf{r})] = \int \frac{\rho(\mathbf{r}')}{|\mathbf{r}-\mathbf{r}'|} d\mathbf{r}'$ describes the usual classical electrostatic (Hartree) potential amongst electrons, and $v_{\text{xc}}[\rho(\mathbf{r})]$ represents the non-classical part of many-electron Hamiltonian, $\mathbf{H}$ as $\frac{\delta E_{\text{xc}}[\rho(\mathbf{r})]}{\delta \rho(\mathbf{r})}$, the functional derivative of XC energy. Then, we can write the single-particle charge density as,

$$\rho(\mathbf{r}) = \rho^\alpha(\mathbf{r}) + \rho^\beta(\mathbf{r}) = \sum_i f_i^\alpha |\phi_i^\alpha(\mathbf{r})|^2 + \sum_i f_i^\beta |\phi_i^\beta(\mathbf{r})|^2 \tag{4.32}$$

where $\{\phi_i^\sigma, \sigma = \alpha \text{ or } \beta\}$ refers to a set of N occupied orthonormal spin MO and f_i^σ's indicate occupation numbers of ith spin-MO. Integral of this electron density gives the total number of electrons, which is N.

For the realistic solution of Eq. (4.31), we follow the convenient basis-set method through LCAO-MO ansatz. The KS MO's can be expanded in terms of K appropriately picked, known basis functions $\{\chi_\mu(\mathbf{r}); \mu = 1, 2, 3, \ldots, K\}$, called as atomic orbitals (AO), in a practice homologous to that in Roothaan-HF method, such as,

$$\phi_i(\mathbf{r}) = \sum_{\mu=i}^{K} C_{\mu i} \chi_\mu(\mathbf{r}), \quad i = 1, 2, 3, \ldots K \tag{4.33}$$

Then the electron density can be expressed in case of basis set representation, as given by,

$$\rho(\mathbf{r}) = \sum_{i=1}^{N_v} \sum_{\mu=1}^{K} \sum_{\nu=1}^{K} C_{\mu i} C_{\nu i} \chi_\mu(\mathbf{r}) \chi_\nu(\mathbf{r}) \tag{4.34}$$

In principle, we need a *complete* basis set ($K = \infty$) to get exact expansion of MOs, but practically it is not possible. Subsequently, appropriate truncation is required for real computational purposes; hence it suffices to work with a finite basis set.

Now inserting Eq. (4.33) in Eq. (4.31), multiplying left side of obtained equation with $\chi_\nu^\star(\mathbf{r})$, then by doing some mathematical manipulations, we get the following KS matrix eigenvalue equation, in a similar manner to HF case,

$$\mathbf{F}^{\text{KS}}\mathbf{C} = \mathbf{S}\mathbf{C}\epsilon \tag{4.35}$$

where **F** and **S** denote the $K \times K$ real, symmetric KS and overlap matrices respectively. The eigenvector matrix **C** will contain all AO expansion coefficients $\mathbf{C}_{\mu i}$ and diagonal matrix ϵ contains the orbital energies ϵ_i of the MO's. This is generally solved by standard numerical techniques of linear algebra. The individual components of the above KS matrix are given by,

$$\begin{aligned} F_{\mu\nu}^{\text{KS}} &= \int \chi_\mu(\mathbf{r}) \left[h^{\text{core}} + v_{\text{hxc}}(\mathbf{r})\right] \chi_\nu(\mathbf{r}) d\mathbf{r} \\ &= H_{\mu\nu}^{\text{core}} + \langle \chi_\mu(\mathbf{r}) | v_{\text{hxc}}(\mathbf{r}) | \chi_\nu(\mathbf{r}) \rangle = H_{\mu\nu}^{\text{core}} + J_{\mu\nu} + V_{\mu\nu}^{\text{xc}} \\ v_{\text{hxc}}(\mathbf{r}) &= v_{\text{h}}(\mathbf{r}) + v_{\text{x}}(\mathbf{r}) + v_{\text{c}}(\mathbf{r}) \end{aligned} \tag{4.36}$$

where the first term, $H_{\mu\nu}^{\text{core}}$ contains the kinetic energy and nuclear-electron attraction integrals, which represents 1e$^-$ contributions. These are evaluated analytically with the assistance of well-established recursion relations [67].

$$H_{\mu\nu}^{\text{core}} = \langle\mu|T|\nu\rangle + \langle\mu|V|\nu\rangle \tag{4.37}$$

These two terms i.e. non relativistic kinetic and potential energy integrals are transformed into DKH Hamiltonian. The two-electron terms remain unchanged from the non-relativistic case. The second term $\upsilon_{\text{hxc}}(\mathbf{r})$ includes all the two-electron interaction terms containing classical Coulomb repulsion and non-classical XC potential. The $J_{\mu\nu}$ term represents matrix element of $\upsilon_{\text{h}}(\mathbf{r})$, while $V_{\mu\nu}^{\text{xc}}$ represents matrix element related to the XC contributions of 2e$^-$ interactions. The development of this, remains one of the fundamentally interesting steps in KS-DFT. Now, in absence of any analytical method, the latter terms can be calculated numerically or be fitted by an auxiliary set of Gaussian functions, proposed in [68–70]. The present method does not involve any fitting procedure. In case of gradient-corrected functionals, the non-local XC potential matrix is implemented through an expansion in form of finite-orbital basis-sets, without evaluating the density Hessians. Thus, for such cases, XC contribution of the KS matrix is given in a convenient working form [71],

$$F_{\mu\nu}^{\text{xc}\alpha} = \int \left[\frac{\partial f}{\partial\rho_\alpha}\chi_\mu\chi_\nu + \left(2\frac{\partial f}{\partial\gamma_{\alpha\alpha}}\nabla\rho_\alpha + \frac{\partial f}{\partial\gamma_{\alpha\beta}}\nabla\rho_\beta\right)\cdot\nabla(\chi_\mu\chi_\nu)\right] d\mathbf{r} \tag{4.38}$$

where $\gamma_{\alpha\alpha} = |\nabla\rho_\alpha|^2$, $\gamma_{\alpha\beta} = \nabla\rho_\alpha\cdot\nabla\rho_\beta$, $\gamma_{\beta\beta} = |\nabla\rho_\beta|^2$ and f is a function of only local quantities ρ_α, ρ_β and their gradients.

To pursue further, all the related quantities, like basis function, electron density, MO as well as all 2e$^-$ contribution of the potential are directly set up on a real 3D CCG,

$$r_i = r_0 + (i-1)h_r, \quad i = 1, 2, 3, \ldots, N_r; \quad \text{for} \quad r \in \{x, y, z\}, \tag{4.39}$$

where h_r,N_r indicate grid spacing and total number of grid points respectively ($r_0 = -\frac{N_r h_r}{2}$). As needs be, the single-particle charge density in active grid can be given as,

$$\rho(\mathbf{r}_g) = \sum_{\mu\nu} P_{\mu\nu}\chi_\mu(\mathbf{r}_g)\chi_\nu(\mathbf{r}_g) \tag{4.40}$$

where $\mathbf{r}_g$ is the gth grid point in the simulation box. In a similar manner, other multi-centre integration engaged in KS-DFT, like classical Hartree energy, non-classical XC energy, and so forth, *in principle*, can be straightforwardly set up in 3D real-space grid utilizing a small 3D unit volume,

$$I = \int F(\mathbf{r})d\mathbf{r} \approx \sum_i F(\mathbf{r}_i)h_x h_y h_z \tag{4.41}$$

The 2e$^-$ matrix elements can be obtained through direct numerical integration in CCG,

$$\langle\chi_\mu(\mathbf{r})|\upsilon_{hxc}(\mathbf{r})|\chi_\nu(\mathbf{r})\rangle \approx h_x h_y h_z \sum_{\text{grid}} \chi_\mu(\mathbf{r})\upsilon_{hxc}(\mathbf{r})\chi_\nu(\mathbf{r}) \tag{4.42}$$

4.2.4.2 Hartree Potential in CCG

A major issue in grid-based DFT approaches is the accurate computation of Hartree potential $\upsilon_h(\mathbf{r})$. For small to medium systems, the easiest route to calculate this is by doing direct numerical integration on the grid. However, the most popular approach involves solution of the corresponding Poisson equation,

$$\nabla^2\upsilon_h(\mathbf{r}) = -4\pi\rho(\mathbf{r}) \tag{4.43}$$

The popular standard method to tackle this is by conjugate gradients [72] or through multi-grid solvers [73]. As an alternative option, we choose a conventional FCT, originally proposed in [74–76], and later also extended in the context of molecular modelling [77, 78]. To give a brief idea of this theorem, let us consider two functions $f(\mathbf{r})$ and $F(\mathbf{k})$ in $\mathbf{r}$ and $\mathbf{k}$ spaces, which are Fourier transforms (FT) (denoted by $\mathcal{F}$) of one another, as given below,

$$f(\mathbf{r}) = \mathcal{F}(\mathbf{k}) = \int_{-\infty}^{\infty} F(\mathbf{k})e^{2\pi i \mathbf{r}\mathbf{k}} d\mathbf{k}$$
$$F(\mathbf{k}) = \mathcal{F}^{-1}(\mathbf{r}) = \int_{-\infty}^{\infty} f(\mathbf{r})e^{-2\pi i \mathbf{r}\mathbf{k}} d\mathbf{r} \tag{4.44}$$

With two functions $f(\mathbf{r})$, $g(\mathbf{r})$, one can write the convolution, characterized by,

$$f \star g = \int_{-\infty}^{\infty} f(\mathbf{r}')g(\mathbf{r} - \mathbf{r}')d\mathbf{r}' \tag{4.45}$$

The FCT directs that FT of convolution is nothing but the product of individual FTs,

$$\mathcal{F}(f \bigstar g) = F(\mathbf{k}).G(\mathbf{k}), \quad f \bigstar g = \mathcal{F}^{-1}(F(\mathbf{k})\cdot G(\mathbf{k})) \tag{4.46}$$

The above theorem is utilized to construct $v_{\mathrm{h}}(\mathbf{r})$ in an efficient manner, as follows,

$$v_h(\mathbf{r}) = \mathrm{FFT}^{-1}\{v_{\mathrm{h}}^c(\mathbf{k})\rho(\mathbf{k})\} \quad \text{and} \quad \rho(\mathbf{k}) = \mathrm{FFT}\{\rho(\mathbf{r})\} \tag{4.47}$$

where $\rho(\mathbf{k})$ and $v_{\mathrm{h}}^c(\mathbf{k})$ denote Fourier integrals of density and CIK respectively. The quantity $\rho(\mathbf{k})$ can be calculated by doing discrete Fourier transform to its real-space value. One major concern lies in the calculation of CIK which contains singularity in real space. In this scenario, we use an Ewald summation-type approach [78], which involves expansion of the Hartree kernel into long-range (LR) and short-range (SR) parts,

$$v_h^c(\mathbf{r}) = \frac{\mathrm{erf}(\gamma r)}{r} + \frac{\mathrm{erfc}(\gamma r)}{r} \equiv v_{h_{\mathrm{lr}}}^c(\mathbf{r}) + v_{h_{\mathrm{sr}}}^c(\mathbf{r}) \tag{4.48}$$

where erf(x) and erfc(x) represent error function and its complementary respectively. The FT of SR segment is evaluated analytically, while the LR part is calculated directly from FFT of its real-space values. We have to use a convergence parameter γ to adjust the range of $v_{h_{\mathrm{sr}}}^c(\mathbf{r})$, so that the error gets minimized. Following the speculation suggested in [74], we employ $\gamma \times L = 7$ (L denotes smallest side of the box), which offers quite precise outcomes [52–54, 60]. A few other courses which evaluate LR part proficiently are fast multi-pole [79], multi-level summation [80], fast Fourier-Poisson [81] method, etc.

In the next step, we follow a simple grid optimization strategy, given by,

$$\{N_x, N_y, N_z\}_i \iff \rho_i^{\mathrm{SCF}} \iff E_{\mathrm{tot},i}^{\mathrm{SCF}} \tag{4.49}$$

for a particular grid spacing h_r. Here "i" denotes the ith combination of N_x, N_y, N_z for the active box. Given this grid parameter, the corresponding self-consistent field (SCF) density and total energy can be labelled as ρ_i^{SCF} and $E_{\mathrm{tot},i}^{\mathrm{SCF}}$ respectively. Then, one can systematically increase $\{N_x, N_y, N_z\}$ starting from an appropriate beginning, which eventually leads to an optimal value of $\{N_x, N_y, N_z\}$ such that,

$$\{N_x, N_y, N_z\}_i = \{N_x, N_y, N_z\}_{\mathrm{opt}}, \quad \text{when} \quad \Delta E = (E_{\mathrm{tot},i}^{\mathrm{SCF}} - E_{\mathrm{tot},i-1}^{\mathrm{SCF}}) < \text{thresh} \tag{4.50}$$

Here *thresh* is a particular grid accuracy for the convergence of total energy i.e. the energy difference between two repetitive SCF calculations with different grid parameters. A detailed demonstration of this simple optimization strategy has been well documented earlier [52, 53, 82]. Recently, we have extended our program InDFT [66] to relativistic domain as discussed earlier.

The various aspects of scalar relativistic DKH strategy to solve DKS problem in Cartesian-grid will be discussed in Section 4.3.

4.2.4.3 Hartree Fock Exchange Through FCT in CCG

For the development of better density functionals, we need exact exchange (EEX) to combine empirically or non-empirically with semi-local functionals, to decrease self-interaction error (SIE) and improve asymptotic nature of the functionals. Following this, the global hybrid functionals, like B3LYP were developed.

If the MO's are expanded in form of GTOs, the HF exchange energy and corresponding matrix elements can be estimated analytically through four-center electron repulsion integrals (ERI) analytically. This gives the following contribution to KS-Fock matrix,

$$F_{\mu\nu}^{X\sigma} = \sum_{\lambda\eta} P_{\lambda\eta}^{\sigma}(\mu\lambda|\eta\nu) = \sum_{\lambda\eta} P_{\lambda\eta}^{\sigma} \iint \frac{\chi_{\mu,\sigma}(\mathbf{r})\chi_{\lambda,\sigma}(\mathbf{r})\chi_{\nu,\sigma}(\mathbf{r}')\chi_{\eta,\sigma}(\mathbf{r}')}{|\mathbf{r}-\mathbf{r}'|} d\mathbf{r}d\mathbf{r}' \tag{4.51}$$

Here $(\mu\lambda|\eta\nu)$ is the ERI with μ, ν, λ, η denoting the contracted AO basis, and $P_{\lambda\eta}^{\sigma}$ corresponds to a matrix-element of a single-particle spin density matrix, P^{σ} with spin σ.

The corresponding HF exchange energy density may be expressed as,

$$\begin{aligned} e_{\sigma}^{X}(\mathbf{r}) &= -\sum_{i}^{\text{occ}}\sum_{j}^{\text{occ}} \int \frac{\phi_{i,\sigma}^{\star}(\mathbf{r})\phi_{j,\sigma}^{\star}(\mathbf{r})\phi_{i,\sigma}(\mathbf{r}')\phi_{j,\sigma}(\mathbf{r}')}{|\mathbf{r}-\mathbf{r}'|} d\mathbf{r}' \\ &= -\sum_{\mu\nu}\sum_{\lambda\eta} P_{\mu\nu}^{\sigma} P_{\lambda\eta}^{\sigma} \int \frac{\chi_{\mu,\sigma}(\mathbf{r})\chi_{\lambda,\sigma}(\mathbf{r})\chi_{\nu,\sigma}(\mathbf{r}')\chi_{\eta,\sigma}(\mathbf{r}')}{|\mathbf{r}-\mathbf{r}'|} d\mathbf{r}'. \end{aligned} \tag{4.52}$$

The first and second expressions are written in form of occupied MO (ϕ_{σ}) and AO (χ_{σ}). Here, the complex conjugate sign is withdrawn, as, the density matrix and basis functions both are in real form. It appears computationally more expensive than normal exchange energy calculation since it needs to be evaluated *at every grid points* with four AO indices. But computational cost can be reduced via a pair-atomic resolution of identity [83–85] or a semi-numerical (SNR) [86–89] approximation.

Here, we present an alternative strategy to calculate HF exchange energy density and relevant matrix elements in CCG. These are essential components for some XC functionals (especially the orbital-dependent XC functionals). This is inspired from [89], where a crucial step is the evaluation of *two-centered electrostatic potential* (ESP), as given below,

$$v_{\nu\eta}(\mathbf{r}) = \int \frac{\chi_{\nu}(\mathbf{r}')\chi_{\eta}(\mathbf{r}')}{|\mathbf{r}-\mathbf{r}'|} d\mathbf{r}' = \sum_{p}\sum_{q} \int \frac{\varphi_{\nu}^{p}(\mathbf{r}')\varphi_{\eta}^{q}(\mathbf{r}')}{|\mathbf{r}-\mathbf{r}'|} d\mathbf{r}' \tag{4.53}$$

The two expressions are in terms of AO basis and primitive basis functions, φ_{ν}^{p} for a given χ_{ν} (spin indices ignored for simplicity). In present formalism, a direct, efficient *numerical* (NR) strategy for accurate, reliable estimation of ESP integrals is offered. This is based on FCT, which involves a RS technique. It eventually splits the CIK into a LR and SR interaction. The framing of and optimal RS parameter for an accurate mapping of CIK in CCG from the *first principles*, can be obtained through a proper grid optimization strategy [52, 53, 60] with respect to total energy in CCG, given an well-defined constraint.

4.2.4.4 Orbital-Dependent Hybrid Functionals via RS-FCT

The EEX is combined with LYP or PBE correlation to form mainly three global hybrid XC functionals, which are named as, B3LYP, PBE0 and BHLYP, each having a variable amount of EEX, and conventional XC functional. The corresponding XC energy to each functional can be expressed as,

$$\begin{aligned} E_{\rm xc}^{\rm B3LYP} &= (1-a_0)E_{\rm LSDA}^{\rm x} + a_0 E_{\rm HF}^{\rm x} + a_{\rm x}E_{\rm B88}^{\rm x} + a_{\rm c}E_{\rm LYP}^{\rm c} + (1-a_{\rm c})E_{\rm VWN}^{\rm c}, \\ E_{\rm xc}^{\rm PBE0} &= b_0 E_{\rm HF}^{\rm x} + (1-b_0)E_{\rm PBE}^{\rm x} + E_{\rm PBE}^{\rm c}, \\ E_{\rm xc}^{\rm BHLYP} &= (1-c_0)E_{\rm LSDA}^{\rm x} + c_0 E_{\rm HF}^{\rm x} + E_{\rm LYP}^{\rm c} \end{aligned} \tag{4.54}$$

Following [90], $a_0, a_{\rm x}, a_{\rm c}$ will have values 0.2, 0.72, 0.81 respectively in case of B3LYP, while for PBE0 [91], we have $b_0 = 0.25$. It should be noted that, in PBE0, the amount of HF exchange contribution is slightly higher than B3LYP, but a larger amount ($c_0 = 0.5$) is attributed in BHLYP.

Now the XC functionals have piece-wise linearity (PWL) problem, as well as suffers from SIE and improper asymptotic nature. *In principle*, the range-separated hybrid (RSH) functionals, are with correct asymptotic behavior at LR part. These functionals [92] are based on expanding the CIK into SR and LR segments, through an RS operator, $g(\gamma, \mathbf{r})$, and an RS parameter, γ, in the following form,

$$\frac{1}{\mathbf{r}} = \frac{\tilde{g}(\gamma, \mathbf{r})}{\mathbf{r}} + \frac{g(\gamma, \mathbf{r})}{\mathbf{r}} \tag{4.55}$$

where, $\tilde{g}(\gamma, \mathbf{r})$ denotes the complementary RS operator. In this case, γ parameter is utilized to adjust the contribution of EEX between LR and SR region, for a given $g(\gamma, \mathbf{r})$. In practice, the SR part is described through a "modified inter-electronic distance-dependent local or semi-local DFA," while the LR part is represented using EEX with the $g(\gamma, \mathbf{r})/\mathbf{r}$ correction. So, the XC energy, is then defined as,

$$\begin{aligned} E_{\rm xc} &= a_{\rm eex}^{\rm sr} E_{\rm eex}^{\rm sr}(\gamma) + (1 - a_{\rm eex}^{\rm sr})E_{\rm dfa}^{\rm x,sr}(\gamma) + b_{\rm eex}^{\rm lr} E_{\rm eex}^{\rm lr}(\gamma) \\ &\quad + (1 - b_{\rm eex}^{\rm lr})E_{\rm dfa}^{\rm x,lr}(\gamma) + E_{\rm dfa}^{\rm c}, \end{aligned} \tag{4.56}$$

where $E_{\rm eex}^{\rm sr}, E_{\rm eex}^{\rm lr}$ signify EEX contribution while $E_{\rm dfa}^{\rm x,sr}, E_{\rm dfa}^{\rm x,lr}$ refer to the respective DFAs for exchange, at SR and LR region. A precise set of ($a_{\rm eex}^{\rm sr}, b_{\rm eex}^{\rm lr}$) constitutes a specific mode of partitioning for a given $g(\gamma, \mathbf{r})$. It is known that, while Koopmans' theorem underestimates the ionization energy from standard LDA, GGA and hybrid XC functionals, better results may be obtained for RSH functionals. RSH functional with $b_{\rm eex}^{\rm lr} = 1$ offers an asymptotically correct XC potential at LR segment. Moreover, an appropriate balance between EEX and dynamical correlation is achieved by choosing an optimal value of $a_{\rm eex}^{\rm sr}$. As a consequence, these XC functionals are not completely SIE free, and also, for the desired system, they may not follow PWL criteria unless some optimal tuning is done. This occurs because of the default γ, which is generally obtained semi-empirically through some fitting procedure from the reference data [93–96].

4.3 Computational Details

For the implementation of DKH method in our program InDFT [66], we have followed the approach suggested in [17, 64]. We first evaluate the matrix elements with momentum operator in an "approximate momentum space representation." Then, the transformation matrix between position and momentum space, and an approximate p^2 eigenvalues are evaluated by diagonalizing non-relativistic kinetic energy matrix, expanding in form of uncontracted AO basis. The p^2 eigenvalues are used to get E_i, A_i, P_i terms. In case of the term $E_i - mc^2$ in Eq. (4.30), we need subtraction of two large numbers, which is avoided by resorting to terms such as $c^2p^2/(E_i + mc^2)$. The evaluation of the full matrix of 1e$^-$ DK integrals are generally performed by a sequence of matrix multiplication. At first, V', $p.V'p$, and $p \times V'p$ integrals are transformed to momentum space. Then using the kinematic factors (E_i, A_i, P_i), these integrals are scaled (row and/or column wise). At last, these are transformed back to position space. The computation of $V', p.V'p, p \times V.p$

terms in Eq. (4.30) needs an element-wise scaling by $1/(E_i + E_j)$ of $V, p.Vp, p \times Vp$ matrices after they are transformed to momentum space.

Here, we have incorporated the DKH Hamiltonian up to 4th order through Eq. (4.30) in our in-house InDFT [66] program. The atomic basis sets are used in an uncontracted manner to obtain the momentum-space representation. The DKH integrals, and therefore the calculated relativistic effects, generally depend on the value of velocity of light c. The one used here is 137.0359895 a.u., consistent with the fully relativistic codes such as MOLFDIR [98] and DIRAC [11].

The Jorge-DZP-DKH basis set [35] in an uncontracted form are taken from EMSL basis set library [99]. The $1e^-$ non-relativistic integrals including kinetic energy, nuclear-electron attraction and pVp matrix elements are generated by standard recurrence relation scheme [67]. The relevant LDA- and GGA-type of XC functionals corresponding to B3LYP, and PBE0 are given as follows: (i) Vosko-Wilk-Nusair (VWN)–with homogeneous electron gas correlation suggested in parametrization formula V [100] (ii) B88–incorporating Becke [101] semi-local exchange (iii) Lee-Yang-Parr (LYP) [102] semi-local correlation (iv) PBE [103] in case of semi-local exchange and correlation. Also, the range-separated functionals LC-BLYP, CAM-B3LYP, LC-PBE, CAM-PBE0 and LRC-ωPBEh* are employed in accordance with the procedure adopted in [58]. Some of the correlation functionals are directly adopted from density functional repository program [104]. During an SCF process, changes in following quantities are looked after, *viz.*, (i) orbital energy deviation between two successive SCF iterations and (ii) absolute difference in a density matrix element. Both of these are needed to remain below a precisely defined threshold set to 10^{-8} a.u.; this implies that total energy should generally maintain at least this much convergence. In order to perform discrete Fourier transform, standard FFTW3 package [105] is utilized. The resulting generalized matrix-eigenvalue equation is solved accurately in an efficient manner, through standard LAPACK routine [106]. Some molecular calculations are performed taking their experimental geometries from NIST computational chemistry database [107]. For molecules not having any experimental geometry (like for H_3^+), we have taken the full CCSD or full MP3 level of theoretical geometry from NIST computational chemistry database [107]. Other mathematical details including scaling properties may be found in references [49–60, 82, 108, 109].

4.4 Results and Discussion

4.4.1 One-Electron Atoms

At first, we present some HF results for H-like ions in Table 4.1 for Z ranging from 20 to 120 at intervals of 20. The respective *exact* analytical energies from Dirac equation (DEQ) are known [29]; these are quoted in column 6. The DKH energies are tabulated for orders 2–4, in columns 3 to 5 along with the non-relativistic HF energies in column 2. We have engaged highly accurate relativistic universal even-tempered Gaussian basis [110] set consisting of 32 s functions, augmented up to 50 s functions for higher Z (> 90) for better accuracy. The HF calculations are done with special strategy incorporating EEX discussed in Section 2.4.3. From column 2, one notices that the non-relativistic HF results are very close to *exact* non-relativistic energies. The small deviation in larger Z ($\geq$ 70) may be due to an incompleteness of basis set. As expected, DKH2 gives substantial improvement in energy compared to column-2 results. For larger Z, it is observed that higher-order DKH (DKH3, DKH4) show closer agreement to *exact* Dirac eigenvalue in last column. As it is revealed, the DKH2 recovers almost $\approx$99.6% of DEQ energy. All these above variations of energy with Z are graphically depicted in Figure 4.2. Obviously, for lower Z, all the plots including non-relativistic and DEQ, tend

Table 4.1 DKH-HF ground-state energies (in a.u.) of $1e^-$ atoms, with varying Z.

Z	NR	DKH2	DKH3	DKH4	DEQ [29]
20	−199.9999	− 201.0725	− 201.0766	− 201.0763	− 201.0725
40	−799.9999	− 817.6157	− 817.8197	− 817.7935	− 817.8074
60	−1799.9999	−1893.8961	−1895.8368	−1895.5030	−1895.6823
80	−3199.9990	−3523.2816	−3533.0170	−3531.2481	−3532.1921
100	−4999.9997	−5905.2611	−5940.9219	−5936.2320	−5939.1954
120	−7199.9993	−9594.0959	−9712.9311	−9725.0823	−9710.7840

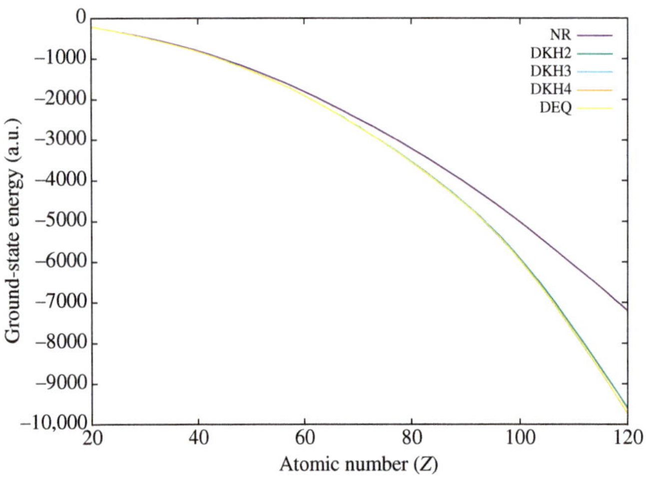

Figure 4.2 Changes in DKH-HF energies (in a.u.) of H-like ions with Z along with DEQ.

to coincide with each other. But as Z grows, the non-relativistic energies deviate significantly from both DEQ and DKH energies. However, all the DKH energies (orders 2–4) and DEQ plots also form a family, indicating their closeness to each other. But as already evident from the table, a closer examination of the plots also shows the differences of various DKH orders from each other, for $Z > 80$. Some of the deficiency may be accounted for due to an incomplete basis set. Furthermore, it may be possible to improve the results for such species by engaging an even higher order DKH operator. However this is beyond the objective of this communication and are not undertaken here.

4.4.2 Many-Electron Systems

4.4.2.1 Grid Optimization

Besides one-electron systems we have performed all-electron DKH-DFT calculation for some atoms and small molecules. Before proceeding for main results, let us first discuss the grid optimization for the convergence of total energy, E_{tot}. To illustrate, we consider two XC functionals LDA and BLYP with DKH2 Hamiltonian. At first, the upper segment of Table 4.2 shows this for DKH2-LDA for HeH^+ molecule taken at its experimental geometry (bond length 0.7899Å). The total energy is given in non-uniform grid with respect to *sparsity* (regulated by N_x, N_y, N_z) for a fixed grid spacing

Table 4.2 Convergence of DKH2 energy (in a.u.) of $HeH^+$and He in the grid ($h_r = 0.1$) for LDA and BLYP functionals.

Set I				Set II			
N_x	N_y	N_z	⟨E⟩	N_x	N_y	N_z	⟨E⟩
			HeH^+ (LDA)				
64	64	64	−2.890867	64	64	96	−2.891744
—	—	68	−2.891302	64	68	—	−2.891861
—	—	72	−2.891522	64	72	—	−2.891919
—	—	76	−2.891633	64	76	—	−2.891947
—	—	80	−2.891690	64	80	—	−2.891961
—	—	84	−2.891718	64	84	—	−2.891967
-	—	88	−2.891732	64	88	—	−2.891970
—	—	92	−2.891740	64	92	—	−2.891970
—	—	96	−2.891744	68	88	—	−2.892085
—	—	100	−2.891744	72	—	—	−2.892143
—	—	—	—	76	—	—	−2.892170
—	—	—	—	80	—	—	−2.892186
—	—	—	—	84	—	—	−2.892193
—	—	—	—	88	—	—	−2.892202
—	—	—	—	92	—	—	−2.892202
			He (BLYP)				
64	64	64	−2.897770	64	64	92	−2.898204
—	—	68	−2.897996	64	68	—	−2.898429
—	—	72	−2.898107	64	72	—	−2.898540
—	—	76	−2.898160	64	76	—	−2.898593
—	—	80	−2.898185	64	80	—	−2.898618
—	—	84	−2.898196	64	84	—	−2.898629
—	—	88	−2.898201	64	88	—	−2.898635
—	—	92	−2.898204	64	92	—	−2.898637
—	—	96	−2.898204	64	96	—	−2.898637
—	—	—	—	68	92	—	−2.898862
—	—	—	—	72	—	—	−2.898973
—	—	—	—	76	—	—	−2.899027
—	—	—	—	80	—	—	−2.899051
—	—	—	—	84	—	—	−2.899063
—	—	—	—	88	—	—	−2.899070
—	—	—	—	92	—	—	−2.899076
—	—	—	—	96	—	—	−2.899076

The respective energies from GAMESS package are: −2.892202, −2.899075 a.u.
Source: Adapted from [97]/John Wiley & Sons.

(given by h_r, here it is 0.1). Following our previous works [52–54, 58] we firstly increase N_z, the number of grid points along the inter-nuclear axis, keeping them static at certain reasonable value, say $N_x = N_y = 64$ along *xy* plane. As shown in Set I in left hand side, N_z is eventually varied from 64 to 100 with an increment of 4, and a smooth convergence in energy is obtained at around $N_z = 96$, when the energy difference between two successive steps is set to about 5×10^{-6} a.u. We say it as grid accuracy. Then in Set II in right-hand side, we increase firstly N_y and then N_x, along *xy* plane keeping the N_z fixed at its previously determined value of 96. We can see that the convergence in energy takes place at $N_x = 88$ and $N_y = 88$ with same grid accuracy of Set I. To complete the illustration, the respective energy is also given from GAMESS [97] in caption. In the lower segment of Table 4.2, we show the similar convergence of DKH2-BLYP energy using the same procedure mentioned above. This time, we consider He atom as a specimen case. Here also the respective energy value is given from GAMESS in caption. The performance of our calculated energies from Table 4.2 is seen to be fairly accurate with the reference results for a given grid accuracy.

4.4.2.2 Ground-State Energy of Atoms and Molecules

To broaden the scope of applicability, we now focus on the ground-state energy for some sample atoms and molecules, following the grid optimization strategy of Table 4.2 presented in Section 4.2.1. Maintaining the same accuracy of Table 4.2, DKH2 binding energies for LDA, BLYP, B3LYP and PBE0 XC functionals are compared with results from *ab initio* program GAMESS [97], in Table 4.3. Solving with DKH Hamiltonian means that the upper block of block-diagonalized Dirac Hamiltonian is taken, yielding the eigenvalue of electronic states, which represents fully relativistic states. Thus one gets a relativistically corrected eigenspectrum. Here, both positive

Table 4.3 LDA, BLYP, B3LYP and PBE0 energies of selected atoms and molecules for DKH2 Hamiltonian.

	DKH2-LDA		DKH2-BLYP		DKH2-B3LYP		DKH2-PBE0	
System	**PR**	**Literature**	**PR**	**Literature**	**PR**	**Literature**	**PR**	**Literature**
H^-	−0.475668	−0.475667	−0.492543	−0.492547	−0.496778	−0.496771	−0.493172	−0.493172
He	−2.827315	−2.827315	−2.899076	−2.899075	−2.900672	−2.900674	−2.888108	−2.888110
Li^+	−7.143122	−7.143125	−7.279789	−7.279790	−7.278994	−7.278994	−7.262824	−7.262820
Be^{2+}	−13.445141	−13.445141	−13.650010	−13.649926	−13.646795	−13.646795	−13.629104	−13.629110
Li	−7.344241	−7.344256	−7.483021	−7.483012	−7.482974	−7.482942	−7.467523	−7.467534
Be^+	−14.115425	−14.115436	−14.330892	−14.330902	−14.327358	−14.327396	−14.307018	−14.307011
Be	−14.444476	−14.444515	−14.658782	−14.658824	−14.656913	−14.656965	−14.634786	−14.634799
H_2	−1.135084	−1.135085	−1.167683	−1.167683	−1.171313	−1.171313	−1.166669	−1.166669
H_3^+	−1.221812	−1.221813	−1.247934	−1.247935	−1.246512	−1.246512	−1.243508	−1.243508
HeH^+	−2.892202	−2.892202	−2.963433	−2.963434	−2.962983	−2.962983	−2.951664	−2.951664
HeH	−3.303632	−3.303633	−3.394434	−3.394434	−3.397547	−3.397547	−3.387394	−3.387397
He_2^+	−4.897870	−4.897872	−5.018537	−5.018534	−5.011365	−5.011363	−4.989273	−4.989284
LiH^+	−7.627195	−7.627191	−7.782268	−7.782264	−7.782350	−7.782354	−7.768041	−7.768042
LiH	−7.918880	−7.918897	−8.042473	−8.042474	−8.074068	−8.074095	−8.051917	−8.051931
BeH^+	−14.713817	−14.713844	−14.940296	−14.940310	−14.939148	−14.939156	−14.914776	−14.914782

Reference results are taken from [97]. PR implies Present Result.
Source: Adapted from [97]/John Wiley /& Sons.

and negative ions are considered in addition to neutral atoms, and also some molecular cations besides neutral species are taken. For all species excepting Be, Be^+, Jorge-DKH-DZP basis set was engaged. For $H_3{}^+$, we take the full CCSD level theoretical geometry and for HeH full MP3 level geometry is taken; for all other molecules, experimental geometry is adopted from NIST computational chemistry database [107]. It is very clear from the table, that current results are very close to GAMESS results, for all the cases. In some cases, two energies completely coincide.

4.4.3 Application to Highly Charged Ions: He- and Li-Isoelectronic Series

In order to extend the method further, it is now applied to generate DKH2, DKH3 and DKH4 energies in some highly charged ions where the relativistic effects are usually more pronounced. It has been observed in Table 4.2 that in small- to medium-Z systems, the differences amongst DKH2, DKH3 and DKH4 are not markedly significant. But for highly charged systems, various DKH methods considerably differ from each other and thus it may be worthwhile to examine the performance of these methods in such systems. It has been reported in literature [29, 112] that, for higher order DKH Hamiltonian, the one-electron energy is converged. But, as such, there is no report of convergence of DKH energies for a many-electron system. In chemistry, DKH2 Hamiltonian is considered to be the most variationally stable and generally provides sufficiently converged results [27–30]. That is why, DKH2 Hamiltonian is quite popular for chemical systems and implemented in most of the quantum chemistry and DFT programs, like GAMESS [97], Gaussian [113], NWChem [114] etc. As a representative case, we now consider highly charged He- and Li-like Ag and Au cations. In Table 4.4, DKH2, DKH3 and DKH4 energies of these ions are compared with NIST theoretical value, employing the LDA functional, and relativistic universal even-tempered Gaussian basis set [110]. To check the validity of our calculation, it also offers similar results engaging same level of calculation from GAMESS suite of program [97] for DKH2 and DKH3; DKH4 result can not be produced as it is not available in GAMESS. One can see that, DKH3 usually overestimates and DKH4 energies lie slightly lower than DKH3. We can expect convergence for even higher order; but DKH2 can be considered to be the best compromise by delivering accurate results while maintaining affordable computational overhead. Of course a certain portion of the deviation may be caused due to the incompleteness of involved basis sets. Also it is possible to get improved results with better functionals, as it will be evident from the following discussion.

In order to illustrate the impact of functional, DKH2 binding energies of He-isoelectronic series corresponding to Z ranging from 20 to 100, are reported in Table 4.5. For such calculations, we have chosen a rather finer grid spacing of $h_r = 0.01$ in order to capture the delicate electron density distribution. Note that these calculations have employed basis sets delineated in Section 4.1.

Table 4.4 LDA energies for He- and Li-like Ag and Au ions for different order of DKH.

	NR		DKH2		DKH3		DKH4		
System	**PR**	**Ref.**	**PR**	**Ref.**	**PR**	**Ref.**	**PR**	**Ref.**	**NIST** [111]
Ag^{45+}	−2175.88	−2175.88	−2241.81	− 2241.81	−2242.80	− 2242.80	− 2242.18	—	−2244.02
Ag^{44+}	− 2433.35	−2433.35	−2508.94	−2508.94	−2509.98	−2509.98	− 2509.85	—	− 2511.24
Au^{77+}	−6185.07	−6185.07	−6784.28	−6784.28	−6802.21	−6802.21	−6798.94	—	−6790.18
Au^{76+}	−6933.62	−6933.62	−7624.58	−7624.58	−7643.65	−7643.65	−7641.15	—	−7630.57

Table 4.5 DKH2 binding energies (a.u.) for He-like ions for a few XC functionals.

Z	LDA	BLYP	B3LYP	LC-BLYP	CAM-B3LYP	PBE0	LC-PBE	CAM-PBE0	LRCωPBEh*	Ref. [111].
20	−388.0908	−389.5687	−389.4867	−389.4212	−389.5047	−389.4730	−389.3103	−389.3866	−387.8934	−389.7957
30	−889.3286	−891.7651	−891.5359	−891.5359	−891.6541	−891.5286	−891.3442	−891.4234	−889.1067	−891.3132
40	−1605.2552	−1608.6354	−1608.3116	−1608.3989	−1608.4780	−1608.3107	−1608.1277	−1608.1671	−1605.0142	−1607.8785
50	−2547.8937	−2552.2036	−2551.8172	−2552.0415	−2552.0173	−2551.8160	−2552.0415	−2551.6051	−2547.6371	−2552.2235
60	−3741.3482	−3746.8227	−3746.1231	−3746.6404	−3746.3496	−3746.0937	−3746.2108	−3745.7783	−3741.0772	−3743.1085
70	−5205.7438	−5212.4980	−5211.2090	−5212.2327	−5211.4467	−5211.1074	−5211.7105	−5210.6507	−5205.4608	−5207.3
80	−6977.7147	−6986.3846	−6985.2177	−6986.1355	−6985.7073	−6985.2299	−6985.4836	−6984.9291	−6977.4073	−6983.195
90	−9118.2980	−9128.8010	−9127.0239	−9128.5388	−9127.5874	−9126.9964	−9127.7838	−9126.5595	−9117.9885	−9126.32
100	−11713.8899	−11726.4737	−11723.6401	−11726.1977	−11724.2723	−11723.4707	−11725.3313	−11722.8435	−11713.5692	−11721.45
ME	−3.89	2.20	1.31	2.06	1.60	1.27	1.64	0.97	−4.16	
MAE	3.89	2.34	1.47	2.19	1.71	1.43	1.79	1.20	4.16	

With an increase in Z, the electron density contracts; this helps lowering the size of effective simulation box. This is reflected in the fact that, energy gets converged in a comparatively smaller grid size for $Z > 50$. Leaving aside the LDA case, XC functionals are categorized into two (B3LYP and PBE0) blocks such that functionals having B88 exchange and LYP correlation (BLYP, B3LYP, LC-BLYP, CAM-B3LYP) belong to columns 3–6, while the ones including PBE (PBE0, LC-PBE, CAM-PBE0, LRCωPBEh*) are produced in columns 7–10. The mean absolute error (MAE) and mean error (ME) from statistical analysis have been also given at the bottom to facilitate a detailed comparison with available *ab initio* theoretical results [111]. For lower Z, the DKH2 energy is quite compatible with literature values, while for higher Z, the discrepancies start developing. For $Z > 60$, DKH2 energy actually overestimates, possibly due to the missing non-dynamic correlation. Moreover, orbitals of ions of higher Z are more contracted to the nucleus causing these differences. The growth in deviations for very high Z also indicates that contributions coming from Breit interaction term and QED correction needs to be incorporated in the theory. A closer examination reveals that the four best functionals may be ordered in terms of their performance as: CAM-PBE0, B3LYP, CAM-B3LYP, PBE0.

Now we proceed for a consideration of ionization potential, as $-\epsilon_{\text{HOMO}}$. Here, we note that, the "KS analogue of Koopmans' theorem in HF theory" [115–117] advocates,

$$\text{IE}(M) = -\epsilon_{\text{HOMO}} \tag{4.57}$$

where IE(M) refers to the first ionization energy of a specified M-electron system. In the context of LDA or GGA-type functionals, strictly speaking, Eq. (4.57) is no longer a valid statement; the HOMO energy is usually underestimated. Moreover, this will not work for other functionals outside of KS regime. The RSH functionals, *in principle*, have correct asymptotic behavior at LR region, but the essence of HOMO and its energy is possible only through a generalized (G)KS version of Koopmans' theorem. While Koopmans' theorem underestimates the ionization energy from standard LDA, GGA and hybrid XC functionals, better results may be obtained for RSH functionals. The calculated ionization energies for He-like ions are given in Table 4.6 along with the theoretical estimates from NIST [111]. For very high Z (90, 100), only relativistic corrections are not sufficient, but also QED corrections need to be incorporated. And because of the assumption that, no relativistic effects are incorporated in two-body interaction, this deviation may arise. So one needs to include DKH transformation of the electron-electron interaction terms for better accuracy.

The trends of ground-state energy and ionization energy for He-like ions, produced in previous two tables are displayed as functions of Z, in Figure 4.3a,b. These are generated for several XC functionals along with NIST results [111]. Both these curves reflect quadratic nature. It is well known [2, 5, 6] that, for many-electron systems, the relativistic energy varies as Z^2 due to effective shielding of nuclear potential caused by a contraction of s and p orbitals. That is why we consider a relation of the form $f(Z) = aZ^2 + bZ + c$ for these two quantities. One can estimate the values of a, b, c by fitting each of these curves, using standard interpolation techniques. These are listed in left and right sides of Table 4.7 for energies and ionization energies. It appears that a remains nearly constant for all the XC functionals in both sides. This term represents the kinetic energy of a system, which should remain nearly constant for all XC functionals and it has a large value causing the relativistic effects. Our estimation for coefficient a is exact in both cases, in the sense that it coincides with NIST value. The presence of linear term b corresponding to the coulomb interaction, which differs from functional to functional in both cases. Moreover it shows opposite signs in the two panels. The b value differs significantly from the reference, indicating that a major source of error perhaps lies in the two-electron part. Hence, relativistic correction to the functionals in the form of DKH transformation may improve the result, especially for highly charged ions ($Z > 80$).

Table 4.6 Ionization energies, $-\epsilon_{HOMO}$ for selected He-like ions in a.u.

Z	B3LYP	LC-BLYP	CAM-B3LYP	PBE0	LC-PBE	CAM-PBE0	LRC-ωPBEh*	Ref. [111].
20	184.925	184.178	186.344	185.141	184.118	185.237	183.349	188.482
30	430.739	429.550	430.803	431.080	429.445	431.159	428.328	436.028
40	783.721	782.068	783.765	784.192	781.932	784.235	780.445	790.715
50	1249.919	1247.806	1249.923	1250.520	1247.637	1250.498	1245.757	1258.926
60	1841.707	1839.198	1841.655	1842.423	1839.013	1842.300	1836.682	1849.948
70	2569.352	2566.577	2569.221	2570.155	2566.360	2569.896	2563.571	2576.974
80	3445.478	3441.700	3445.467	3446.569	3441.406	3446.462	3437.855	3459.019
90	4508.675	4504.423	4508.604	4509.931	4504.085	4509.692	4499.529	4523.345
100	5799.594	5795.056	5799.434	5800.968	5794.671	5800.543	5789.615	5811.995
ME	9.04	11.65	8.91	8.27	11.86	8.38	14.48	
MAE	9.04	11.65	8.91	8.27	11.86	8.38	14.48	

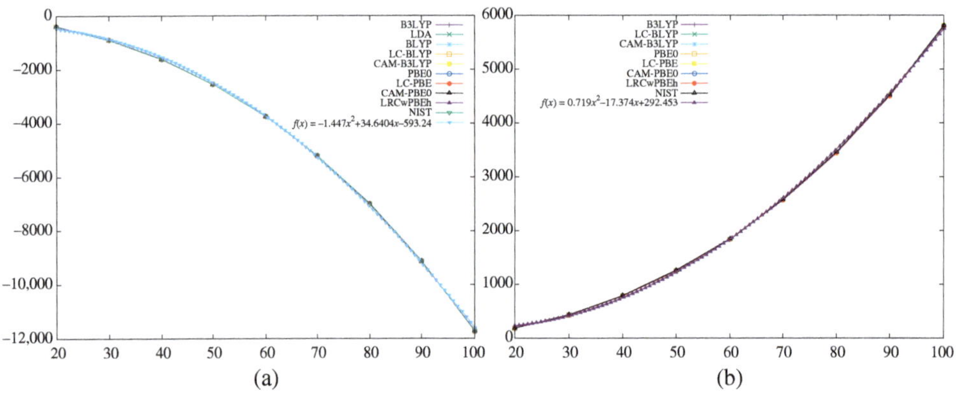

Figure 4.3 Variation of ground-state energy and ionization energy of He-like ions with ionic charge Z in (a) and (b) panels respectively. More details are in the text.

Also some discrepancy may occur due to an incompleteness in basis set. The c values also vary in each curves. Like b, here also the signs change in energy and ionization energy plots. An analogous analysis for Li-like ions was also carried out. The calculated binding energies and ionization energies are graphically shown in Figure 4.4a,b, in a similar manner as in He-isoelectronic counterpart. The findings are in harmoniums to those in Figure 4.3.

4.5 Future and Outlook

For an accurate reliable solution of DKS-DFT in CCG, we have incorporated scalar relativistic DKH Hamiltonian in our program. Lot of complications of four-component approaches, as well as their computational costs, can be reduced by adopting a simple, effective one-component strategy. Its feasibility and reliability has been investigated by applying it to a host of neutral atoms,

Table 4.7 Estimated parameters a, b, c for energy and ionization energy for He-like ions.

	Co-efficients for energy			Co-efficients for ionization energy		
	a	b	c	a	b	c
LDA	−1.446	34.526	−589.692	—	—	—
BLYP	−1.447	34.587	−590.557	—	—	—
B3LYP	−1.447	34.498	−590.484	0.720	−17.532	291.742
LC-BLYP	−1.447	34.507	−590.389	0.720	−17.568	291.750
CAM-B3LYP	−1.447	34.495	−590.402	0.721	−17.593	293.772
PBE0	−1.447	34.492	−590.315	0.720	−17.529	291.782
LC-PBE	−1.447	34.503	−590.218	0.720	−17.570	291.716
CAM-PBE0	−1.447	34.484	−590.036	0.720	−17.527	291.950
LRCωPBEh*	−1.446	34.529	−589.538	0.720	−17.577	291.203
NIST [111]	−1.447	34.620	−593.240	0.719	−17.374	292.453

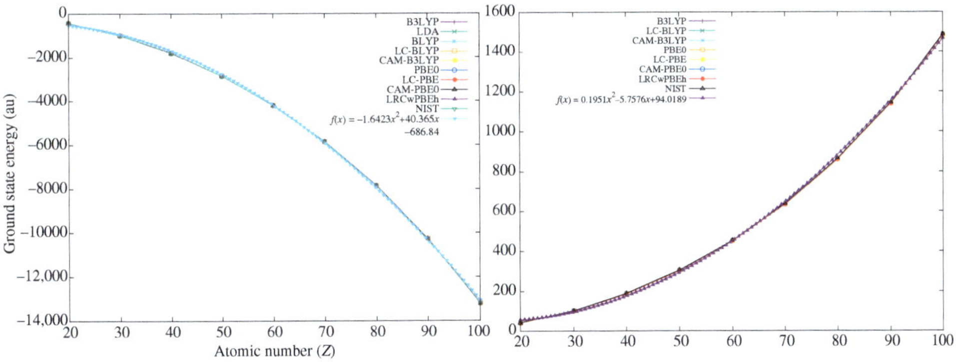

(a) (b)

Figure 4.4 Variation of ground-state energy and ionization energy of Li-like ions with ionic charge Z in (a) and (b) panel respectively. More details can be found in the text.

molecules as well as ions, in terms of total energies and orbital energies. Some results are presented for highly charged ions, where the effects are prominent. The role of several XC functionals (including some RSH ones) is analyzed, which seems to capture the correlation of such systems quite well. Obtained results are compared with the standard quantum chemistry package, as well as other highly accurate and sophisticated theoretical methods. A careful observation of the plots reveals that the energies can be fitted to *quadratic polynomial* forms. One simple conclusion can be drawn that, the RSH functionals generally offer better ionization potentials, at least for lower Z, but for too high Z (> 90), QED corrections are necessary to reach the reference values. The results may further improve if optimally tuned RSH functionals [58] are used, which could be done in future. The method can be extended to get the molecular properties exploiting the famous "picture change" effects in chemistry [64]. Also transformation of the two-body matrix elements, including SO coupling, can be pursued later.

Acknowledgement

We are very much grateful to the editors for their kind invitation for this prestigious book chapter, and extending the deadline for submission. SC thanks IISER Kolkata for a Junior Research Fellowship. AKR acknowledges funding support from DST SERB, New Delhi, India (sanction order: CRG/2019/000293). Partial financial support from MATRICS, SERB (sanction order: MTR/2019/000012) has been very helpful. It is a pleasure to thank Dr. Sangita Majumdar for valuable discussions.

References

1 Einstein, A. (1905). Zur elektrodynamik bewegter körper. *Ann. Phys. (Leipzig)* 17: 891.

2 Pyykkö, P. and Desclaux, J.P. (1979). Relativity and the periodic system of elements. *Acc. Chem. Res.* 12: 276.

3 Pitzer, K. (1979). Relativistic effects on chemical properties. *Acc. Chem. Res.* 12: 271.

4 Pyykkö, P. (1988). Relativistic effects in structural chemistry. *Chem. Rev.* 88: 563.

5 Pyykkö, P. (2012). The physics behind chemistry and the periodic table. *Chem. Rev.* 112: 371.

6 Pyykkö, P. (2012). Relativistic effects in chemistry: more common than you thought. *Annu. Rev. Phys. Chem.* 63: 45.

7 Dirac, P.A.M. (1929). Quantum mechanics of many-electron systems. *Proc. R. Soc. London, Ser. A* 123: 714.

8 Ahuja, R., Blomqvist, A., Larsson, P., and Pyykkö, P. (2011). Relativity and the lead-acid battery. *Phys. Rev. Lett.* 106: 018301.

9 Reiher, M. and Hinze, J. (2002). Four-component *ab-initio* methods for electronic structure calculations of atoms, molecules, and solids. In: *Relativistic Effects in Heavy-Element Chemistry and Physics* (ed. B.A. Hess), 61–88. Chichester: Wiley.

10 Saue, T. and Helgaker, T. (2002). Four-component relativistic Kohn–Sham theory. *J. Comput. Chem.* 23: 814.

11 Jensen, H.J.A., Saue, T., and Visscher, L. (2004). *DIRAC, a relativistic ab-initio electronic structure program, Release DIRAC04.0.*

12 van Lenthe, E., Baerends, E.J., and Snijders, J.G. (1993). Relativistic regular two-component Hamiltonians. *J. Chem. Phys.* 99: 4597.

13 van Lenthe, E., Baerends, E.J., and Snijders, J.G. (1994). Relativistic total energy using regular approximations. *J. Chem. Phys.* 101: 9783.

14 Rutkowski, A. (1986). Relativistic perturbation theory. I. A new perturbation approach to the Dirac equation. *J. Phys. B* 19: 149.

15 Dyall, K.G. (1998). Interfacing relativistic and nonrelativistic methods. II. Investigation of a low-order approximation. *J. Chem. Phys.* 109: 4201.

16 Douglas, M. and Kroll, N.M. (1974). Quantum electrodynamical corrections to the fine structure of helium. *Ann. Phys. (N.Y.)* 82: 89.

17 Hess, B.A. (1985). Applicability of the no-pair equation with free-particle projection operators to atomic and molecular structure calculations. *Phys. Rev. A* 32: 756.

18 Hess, B.A. (1986). Relativistic electronic-structure calculations employing a two-component no-pair formalism with external-field projection operators. *Phys. Rev. A* 33: 3742.

19 Jansen, G. and Hess, B.A. (1989). Revision of the Douglas-Kroll transformation. *Phys. Rev. A* 39: 6016.

20 Samzow, R. and Hess, B.A. (1991). Spin–orbit effects in the Br atom in the framework of the no-pair theory. *Chem. Phys. Lett.* 184: 491.

21 Samzow, R., Hess, B.A., and Jansen, G. (1992). The two-electron terms of the no-pair Hamiltonian. *J. Chem. Phys.* 96: 1227.

22 Wolf, A., Reiher, M., and Hess, B.A. (2002). The generalized Douglas–Kroll transformation. *J. Chem. Phys.* 117: 9215.

23 Hess, B.A. (ed.) (2003). *Relativistic Effects in Heavy Elements Chemistry*. Wiley.

24 Nakajima, T. and Hirao, K. (2000). The higher-order Douglas–Kroll transformation. *J. Chem. Phys.* 113: 7786.

25 Reiher, M. and Wolf, A. (2004). Exact decoupling of the Dirac Hamiltonian. I. General theory. *J. Chem. Phys.* 121: 2037.

26 Reiher, M. and Wolf, A. (2004). Exact decoupling of the Dirac Hamiltonian. II. The generalized Douglas–Kroll–Hess transformation up to arbitrary order. *J. Chem. Phys.* 121: 10945.

27 Wolf, A., Reiher, M., and Hess, B.A. (2004). Correlated *ab initio* calculations of spectroscopic parameters of SnO within the framework of the higher-order generalized Douglas–Kroll transformation. *J. Chem. Phys.* 120: 8624.

28 Neese, F., Wolf, A., Fleig, T. et al. (2005). Calculation of electric-field gradients based on higher-order generalized Douglas–Kroll transformations. *J. Chem. Phys.* 122: 204107.

29 Wolf, A. and Reiher, M. (2006). Exact decoupling of the Dirac Hamiltonian. IV. Automated evaluation of molecular properties within the Douglas-Kroll-Hess theory up to arbitrary order. *J. Chem. Phys.* 124: 064103.

30 Masterlerz, R., Barone, G., Lindh, R., and Reiher, M. (2007). Analytic high-order Douglas-Kroll-Hess electric field gradients. *J. Chem. Phys.* 127: 074105.

31 van Wullen, C. (2004). Relation between different variants of the generalized Douglas–Kroll transformation through sixth order. *J. Chem. Phys.* 120: 7307.

32 Neto, A.C. and Jorge, F.E. (2013). All-electron double zeta basis sets for the most fifth-row atoms: application in DFT spectroscopic constant calculations. *Chem. Phys. Lett.* 582: 158.

33 Fantin, P.A., Barbieri, P.L., Neto, A.C., and Jorge, F.E. (2007). Augmented Gaussian basis sets of triple and quadruple zeta valence quality for the atoms H and from Li to Ar: applications in HF, MP2, and DFT calculations of molecular dipole moment and dipole (hyper)polarizability. *J. Mol. Struct. THEOCHEM* 810: 103.

34 Camiletti, G.G., Machado, S.F., and Jorge, F.E. (2008). Gaussian basis set of double zeta quality for atoms K through Kr: application in DFT calculations of molecular properties. *J. Comput. Chem.* 29: 2434.

35 Jorge, F.E., Neto, A.C., Camiletti, G.G., and Machado, S.F. (2009). Contracted Gaussian basis sets for Douglas–Kroll–Hess calculations: estimating scalar relativistic effects of some atomic and molecular properties. *J. Chem. Phys.* 130: 064108.

36 Hohenberg, P. and Kohn, W. (1964). Inhomogeneous electron gas. *Phys. Rev.* 136: B864.

37 Cohen, A.J., Mori-Sánchez, P., and Yang, W. (2011). Challenges for density functional theory. *Chem. Rev.* 112: 289–320.

38 Burke, K. (2012). Perspective on density functional theory. *J. Chem. Phys.* 136: 150901.

39 Becke, A.D. (2014). Perspective: fifty years of density-functional theory in chemical physics. *J. Chem. Phys.* 140: 18A301.

40 Jones, R.O. (2015). Density functional theory: its origins, rise to prominence, and future. *Rev. Mod. Phys.* 87: 897.

41 Kohn, W. and Sham, L.J. (1965). Self-consistent equations including exchange and correlation effects. *Phys. Rev.* 140: A1133.

42 Seidl, A., Görling, A., Vogl, P. et al. (1996). Generalized Kohn-Sham schemes and the band-gap problem. *Phys. Rev. B* 53: 3764.
43 Rajagopal, A.K. and Callaway, J. (1973). Inhomogeneous electron gas. *Phys. Rev. B* 7: 1912.
44 Rajagopal, A.K. (1978). Inhomogeneous relativistic electron gas. *J. Phys. C* 11: L943.
45 Mayer, M., Krugger, S., and Rosch, N. (2001). A two-component variant of the Douglas–Kroll relativistic linear combination of Gaussian-type orbitals density-functional method: spin–orbit effects in atoms and diatomics. *J. Chem. Phys.* 115: 4411.
46 Matveev, A. and Rosch, N. (2003). The electron–electron interaction in the Douglas–Kroll–Hess approach to the Dirac–Kohn–Sham problem. *J. Chem. Phys.* 118: 3997.
47 Majumder, S., Matveev, A.V., and Rosch, N. (2003). Spin–orbit interaction in the Douglas–Kroll approach to relativistic density functional theory: the screened nuclear potential approximation for molecules. *Chem. Phys. Lett.* 382: 186.
48 Nasluzov, V.A. and Rosch, N. (1996). Density functional based structure optimization for molecules containing heavy elements: analytical energy gradients for the Douglas-Kroll-Hess scalar relativistic approach to the LCGTO-DF method. *Chem. Phys.* 210: 413.
49 Roy, A.K. (2008). Grid-based density functional calculations of many-electron systems. *Int. J. Quantum Chem.* 108: 837.
50 Roy, A.K. (2008). Pseudopotential density functional treatment of atoms and molecules in cartesian coordinate grid. *Chem. Phys. Lett.* 461: 142.
51 Roy, A.K. (2011). Density functional calculation of many-electron systems in cartesian coordinate grid. *J. Math. Chem.* 49: 1687.
52 Ghosal, A. and Roy, A.K. (2016). DFT calculations of atoms and molecules in cartesian grids. In: *Specialist Periodical Reports: Chemical Modelling, Applications and Theory*, vol. 13 (ed. M. Springborg and J.-O. Joswig), 221–260. London: Royal Society of Chemistry.
53 Ghosal, A., Mandal, T., and Roy, A.K. (2018). Density functional electric response properties of molecules in cartesian grid. *Int. J. Quantum Chem.* 118: e25708.
54 Ghosal, A., Mandal, T., and Roy, A.K. (2019). Efficient HF exchange evaluation through Fourier convolution in cartesian grid for orbital-dependent density functionals. *J. Chem. Phys.* 150: 064104.
55 Ghosal, A., Gupta, T., Mahato, K., and Roy, A.K. (2021). Excitation energies through Becke's exciton model within a cartesian-grid KS DFT. *Theor. Chem. Acc.* 140: 2.
56 Roy, R., Ghosal, A., and Roy, A.K. (2021). A simple effective ΔSCF method for computing optical gaps in organic chromophores. *Chem. Asian J.* 16: 1.
57 Roy, R., Ghosal, A., and Roy, A.K. (2022). Charge-transfer excitation within a hybrid-(G)KS framework through cartesian grid DFT. *J. Phys. Chem. A* 126: 1448.
58 Ghosal, A. and Roy, A.K. (2022). A self-consistent systematic optimisation of range-separated hybrid functionals from first principles. *Mol. Phys.* 120: e1983056.
59 Ghosal, A. and Roy, A.K. (2022). A real-time TDDFT scheme for strong-field interaction in cartesian coordinate grid. *Chem. Phys. Lett.* 796: 139562.
60 Majumdar, S. and Roy, A.K. (2022). Recent advances in Cartesian gird DFT in atoms and molecules. *Front. Chem.* 10: 926916.
61 Dirac, P.A.M. (1928). The quantum theory of the electron. *Proc. R. Soc. London, Ser. A* 117: 610.
62 Dirac, P.A.M. (1928). The quantum theory of the electron. Part II. *Proc. R. Soc. London, Ser. A* 118: 351.
63 Foldy, L.L. and Wouthuysen, S.A. (1950). On the Dirac theory of spin-1/2 particles and its non-relativistic limit. *Phys. Rev.* 78: 29.

64 Wolf, A., Reiher, M., and Hess, B.A. (2002). Two-component methods and the generalised Douglas–Kroll transformation. In: *Relativistic Electronic Structure Theory, Part I* (ed. P. Schwerdtfeger), 622. Berlin: Elsevier.

65 Wolf, A., Reiher, M., and Hess, B.A. (2004). The generalised Douglas–Kroll transformation. In: *Recent Advances in Relativistic Effects in Chemistry* (ed. K. Hirao and Y. Ishikawa), 137. Singapore: World Scientific.

66 Roy, A.K., Ghosal, A., Mandal, T., and Chanda, S. (2022). *InDFT: A DFT Program for Atoms and Molecules in CCG. IISER Kolkata, India: Theoretical Chemistry Laboratory*. This is based on the extension of an initial version of the code, established by A.K. Roy in 2008, whose results were published in Refs. [49–60, 82, 108, 109].

67 Obara, S. and Saika, A. (1986). Efficient recursive computation of molecular integrals over cartesian Gaussian functions. *J. Chem. Phys.* 84: 3963–3974.

68 Sambe, H. and Felton, R.H. (1975). A new computational approach to Slater's SCF–Xα equation. *J. Chem. Phys.* 62: 1122.

69 Dunlap, B.I., Connolly, J.W.D., and Sabin, J.R. (1979). On some approximations in applications of $X\ \alpha$ theory. *J. Chem. Phys.* 71: 3396.

70 Dunlap, B.I., Connolly, J.W.D., and Sabin, J.R. (1979). On first-row diatomic molecules and local density models. *J. Chem. Phys.* 71: 4993.

71 Pople, J.A., Gill, P.M.W., and Johnson, B.G. (1992). Kohn-Sham density-functional theory within a finite basis set. *Chem. Phys. Lett.* 199: 557.

72 Saad, Y. (2003). *Iterative Methods for Sparse Linear Systems*. SIAM.

73 Brandt, A. (1977). Multi-level adaptive solutions to boundary-value problems. *Math. Comput.* 31: 333.

74 Martyna, G.J. and Tuckerman, M.E. (1999). A reciprocal space based method for treating long range interactions in *ab initio* and force-field-based calculations in clusters. *J. Chem. Phys.* 110: 2810–2821.

75 Minary, P., Tuckerman, M.E., Pihakari, K.A., and Martyna, G.J. (2002). A new reciprocal space based treatment of long range forces on surfaces. *J. Chem. Phys.* 116: 5351–5351.

76 Skylaris, C.-K., Mostofi, A.A., Haynes, P.D. et al. (2002). Nonorthogonal generalized Wannier function pseudopotential plane-wave method. *Phys. Rev. B* 66: 035119.

77 Hine, N.D.M., Dziedzic, J., Haynes, P.D., and Skylaris, C.-K. (2011). Electrostatic interactions in finite systems treated with periodic boundary conditions: application to linear-scaling density functional theory. *J. Chem. Phys.* 135: 204103.

78 Chang, C.-M., Shao, Y., and Kong, J. (2012). Ewald mesh method for quantum mechanical calculations. *J. Chem. Phys.* 136: 114112.

79 Beatson, R. and Greengard, L. (1997). A short course on fast multipole methods. *Wavelets, Multilevel Methods and Elliptic PDEs* 1: 1.

80 Skeel, R.D., Tezcan, I., and Hardy, D.J. (2002). Multiple grid methods for classical molecular dynamics. *J. Comput. Chem.* 23: 673.

81 York, D. and Yang, W. (1994). The fast Fourier Poisson method for calculating Ewald sums. *J. Chem. Phys.* 101: 3298.

82 Mandal, T., Ghosal, A., and Roy, A.K. (2019). Static polarizability and hyperpolarizability in atoms and molecules through a cartesian-grid DFT. *Theor. Chem. Acc.* 138: 10.

83 Proynov, E., Shao, Y., and Kong, J. (2010). Efficient self-consistent DFT calculation of nondynamic correlation based on the B05 method. *Chem. Phys. Lett.* 493: 381–385.

84 Proynov, E., Liu, F., Shao, Y., and Kong, J. (2012). Improved self-consistent and resolution-of-identity approximated Becke'05 density functional model of nondynamic electron correlation. *J. Chem. Phys.* 136: 034102.

85 Liu, F., Proynov, E., Yu, J.-G. et al. (2012). Comparison of the performance of exact-exchange-based density functional methods. *J. Chem. Phys.* 137: 114104.

86 Kossmann, S. and Neese, F. (2009). Comparison of two efficient approximate Hartree–Fock approaches. *Chem. Phys. Lett.* 481: 240–243.

87 Bahmann, H. and Kaupp, M. (2015). Efficient self-consistent implementation of local hybrid functionals. *J. Chem. Theory Comput.* 11: 1540–1548.

88 Laqua, H., Kussmann, J., and Ochsenfled, C. (2018). Efficient and linear-scaling seminumerical method for local hybrid density functionals. *J. Chem. Theory Comput.* 14: 3451–3458.

89 Liu, F. and Kong, J. (2017). Efficient computation of exchange energy density with Gaussian basis functions. *J. Chem. Theory Comput.* 13: 2571–2580.

90 Stephens, P.J., Devlin, F.J., Chabalowski, C.F., and Frisch, M.J. (1994). *Ab Initio* calculation of vibrational absorption and circular dichroism spectra using density functional force fields. *J. Phys. Chem.* 98: 11623.

91 Perdew, J.P., Ernzerhof, M., and Burke, K. (1996). Rationale for mixing exact exchange with density functional approximations. *J. Chem. Phys.* 105: 9982–9985.

92 Savin, A. and Flad, H.J. (1995). Density functionals for the Yukawa electron-electron interaction. *Int. J. Quantum Chem.* 56: 327.

93 Iikura, H., Tsuneda, T., Yanai, T., and Hirao, K. (2001). A long-range correction scheme for generalized-gradient-approximation exchange functionals. *J. Chem. Phys.* 115: 3540.

94 Tawada, Y., Tsuneda, T., Yanagisawa, S. et al. (2004). A long-range-corrected time-dependent density functional theory. *J. Chem. Phys.* 120: 8425.

95 Yanai, T., Tew, D.P., and Handy, N.C. (2004). A new hybrid exchange-correlation functional using the Coulomb-attenuating method (CAM-B3LYP). *Chem. Phys. Lett.* 393: 51.

96 Lange, A.W., Rohrdanz, M.A., and Herbert, J.M. (2008). Charge-transfer excited states in a π-stacked adenine dimer, as predicted using long-range-corrected time-dependent density functional theory. *J. Phys. Chem. B* 112 (20): 6304.

97 Schmidt, M.W., Baldridge, K.K., Boatz, J.A. et al. (1993). General atomic and molecular electronic structure system. *J. Comput. Chem.* 14: 1347.

98 Visscher, L., Aerts, O.V.P.J.C., Merenga, H., and Nieuwpoor, W.C. (1994). Relativistic quantum chemistry - the MOLFDIR program package. *Comput. Phys. Commun.* 81: 120.

99 Feller, D. (1996). The role of databases in support of computational chemistry calculations. *J. Comput. Chem.* 17: 1571–1586.

100 Vosko, S.H., Wilk, L., and Nusair, M. (1980). Accurate spin-dependent electron liquid correlation energies for local spin density calculations: a critical analysis. *Can. J. Phys.* 58: 1200–1211.

101 Becke, A.D. (1988). Density-functional exchange-energy approximation with correct asymptotic behavior. *Phys. Rev. A* 38: 3098.

102 Lee, C., Yang, W., and Parr, R.G. (1988). Development of the Colle-Salvetti correlation-energy formula into a functional of the electron density. *Phys. Rev. B* 37: 785.

103 Perdew, J.P., Burke, K., and Ernzerhof, M. (1996). Generalized gradient approximation made simple. *Phys. Rev. Lett.* 77: 3865.

104 Quantum Chemistry Group (2001). *Density Functional Repository*. Daresbury, Cheshire, UK: CCLRC Daresbury Laboratory.

105 Frigo, M. and Johnson, S.G. (2005). The design and implementation of FFTW3. *Proc. IEEE* 93: 216–231.

106 Anderson, E., Bai, Z., Bischof, C. et al. (1999). *LAPACK Users' Guide*, vol. 9. SIAM.

107 Johnson, R.D. III, (ed.) (2016). *NIST Computational Chemistry Comparisons and Benchmark Database, NIST Standard Reference Database, Number, Release 18*. Gaithersburg, MD: NIST.

108 Roy, A.K. (2009). A new density functional method for electronic structure calculation of atoms and molecules. In: *Handbook of Computational Chemistry Research* (ed. C.T. Collett and C.D. Robson), 409–434. New York: Nova Publishers.

109 Roy, A.K. (2010). A new DFT method for atoms and molecules in cartesian grid. *Trends Phys. Chem.* 14: 27.

110 Malli, G.L., Silva, A.B.F.D., and Ishikawa, Y. (1993). Universal Gaussian basis set for accurate *ab-initio* relativistic Dirac-Fock calculations. *Phys. Rev. A* 47: 143.

111 Kramida, A., Ralchenko, Y., Reader, J., and NIST ASD Team (2021). *NIST Atomic Spectra Database (Version 5.9)*. Gaithersburg, MD: National Institute of Standards and Technology. https://physics.nist.gov/asd (accessed 31 January 2016).

112 Hirao, K. and Ishikawa, Y. (2004). *Recent Advances in Relativistic Molecular Theory*. World Scientific.

113 Frisch, M., Trucks, G., Schlegel, H. et al. (2004). *Gaussian 03, Revision C. 02*. Wallingford, CT: Gaussian, Inc.

114 Valiev, M., Bylaska, E., Govind, N. et al. (2010). NWChem: A comprehensive and scalable open-source solution for large scale molecular simulations. *Comput. Phys. Commun.* 181 (9): 1477–1489.

115 Perdew, J.P., Parr, R.G., Levy, M., and Balduz, J.L. Jr. (1982). Density-functional theory for fractional particle number: derivative discontinuities of the energy. *Phys. Rev. Lett.* 49: 1691.

116 Levy, M., Perdew, J.P., and Sahni, V. (1984). Exact differential equation for the density and ionization energy of a many-particle system. *Phys. Rev. A* 30: 2745.

117 Perdew, J.P. and Levy, M. (1997). Comment on "significance of the highest occupied Kohn-Sham eigenvalue". *Phys. Rev. B* 56: 16021.

5

Relativistic Reduced Density Matrices: Properties and Applications

Somesh Chamoli[1], *Malaya K. Nayak*[2,3] *and Achintya Kumar Dutta*[1]

[1] *Department of Chemistry, Indian Institute of Technology Bombay, Powai, Mumbai, India*
[2] *Theoretical Chemistry Section, Bhabha Atomic Research Centre, Trombay, Powai, Mumbai, India*
[3] *Homi Bhabha National Institute, BARC Training School Complex, Anushakti Nagar, Mumbai, India*

5.1 Introduction

Density matrices [1–4] play an important role in many-body quantum mechanics. They can have many meaningful insights into the electronic structures of atoms and molecules [5, 6]. Husimi [7] has introduced the concept of reduced density matrices in nonrelativistic many-body quantum mechanics. The one-body reduced density matrices (1-RDM) and their application to nonrelativistic quantum mechanics have been extensively studied in the literature [8–14]. The extension of 1-RDMs into the relativistic domain requires the solution of the Hamiltonian that takes relativistic effects into account. The Schrödinger equation is not invariant to Lorentzian transformation and cannot be used in the relativistic domain. One needs to solve the Dirac equation to include the relativistic effect in quantum mechanical calculations. Since relativistic effects in a molecule expand in proportion to the atom's nuclear charge, relativistic effects are more important in the case of heavy elements where the speed of the electrons approaches the speed of light.

Density functional theory (DFT) [15] is often the method of choice for systems including heavier elements due to its favorable scaling and reasonable accuracy. Despite being the most used method, DFT occasionally fails to produce results that can be trusted, and there is no systematic way to improve the results, which limits the scope of its application [16–19]. The results in the relativistic wave function-based methods can always be improved, albeit at the expense of enhanced computational cost. Among all the relativistic wave function-based methods, the relativistic coupled-cluster methods [20, 21] is considered the most accurate and systematically improvable method for the electronic structure of systems containing heavy elements. Additionally, the ability of coupled-cluster methods, in its multi-reference variants, to effectively address quasi-degenerate electronic states [22–25] makes them the preferred method for relativistic correlation calculation. However, due to the high computation scaling of coupled-cluster methods, their extensive use is limited to atoms. 1-RDMs provide an alternate way to directly calculate the correlation energy using the relativistic reduced density matrix functional theory (Rel-RDMFT) [26, 27]. One can also use 1-RDMs to get a compact orbital space for already well-established wave function-based methods [28, 29]. At the same time, they can be used to extract physically meaningful quantities out of the quantum mechanically calculated wave function. In the following section, we will discuss the properties of the relativistic 1-RDM and its various applications.

Electron Density: Concepts, Computation and DFT Applications, First Edition.
Edited by Pratim Kumar Chattaraj and Debdutta Chakraborty.

5.2 Relativistic One-Body Reduced Density Matrix

The full 1-RDM can be obtained as the expectation value of the 1-RDM operator in a given state by

$$D(r : r') = \langle \Psi | \hat{D}(r : r') | \Psi \rangle \tag{5.1}$$

where the 1-RDM operator is defined as

$$\hat{D}(r : r') = N[\overline{\psi}^{\dagger}(r') \otimes \overline{\psi}(r)] \tag{5.2}$$

N denotes the normal ordering with respect to the mean field vacuum $|0\rangle$ and $[\overline{\psi}^{\dagger}(r') \otimes \overline{\psi}(r)]$ is the normal ordered tensor product. In Eq. (5.2) $\overline{\psi}(r)$ is the quantized Dirac field given by

$$\overline{\psi}(r) = \sum_{I} \hat{b}_I \psi_I(r) + \sum_{R} \hat{d}_R^{\dagger} \psi_R(r) \tag{5.3}$$

with $\hat{b}_I$ is the electron annihilation operator for I belonging to the positive energy region and $\hat{d}_R^{\dagger}$ as positron creation operator for R in the negative energy spectrum. In Eq. (5.3), $\psi_q(r)$ for $q \in [I, R]$are general four-spinor wave function denoted as

$$\psi_q(r) = \begin{bmatrix} \Phi_1^q(r) \\ \Phi_2^q(r) \\ \Phi_3^q(r) \\ \Phi_4^q(r) \end{bmatrix} \tag{5.4}$$

And its conjugate transpose is given by

$$\psi_q^{\dagger}(r) = \begin{bmatrix} \Phi_1^{q*}(r) & \Phi_2^{q*}(r) & \Phi_3^{q*}(r) & \Phi_4^{q*}(r) \end{bmatrix} \tag{5.5}$$

Substituting Eq. (5.3) in Eq. (5.2), the 1-RDM operator can be written as

$$\hat{D}(r : r') = N\left[\left[\sum_{I} \hat{b}_I \psi_I(r') + \sum_{R} \hat{d}_R^{\dagger} \psi_R(r')\right]^{\dagger} \otimes \left[\sum_{I} \hat{b}_I \psi_I(r) + \sum_{R} \hat{d}_R^{\dagger} \psi_R(r)\right]\right] \tag{5.6}$$

which, on expanding, becomes

$$\begin{aligned} \hat{D}(r : r') = {} & \sum_{I\,J} \hat{b}_I^{\dagger} \hat{b}_J \left[\psi_I^{\dagger}(r') \otimes \psi_J(r)\right] + \sum_{R} \sum_{J} \hat{d}_R \hat{b}_J \left[\psi_R^{\dagger}(r') \otimes \psi_J(r)\right] \\ & + \sum_{I} \sum_{S} \hat{b}_I^{\dagger} \hat{d}_S^{\dagger} \left[\psi_I^{\dagger}(r') \otimes \psi_S(r)\right] + \sum_{R\,S} \hat{d}_R \hat{d}_S^{\dagger} \left[\psi_R^{\dagger}(r') \otimes \psi_S(r)\right] \end{aligned} \tag{5.7}$$

In Eq. (5.7), the indexing I and J correspond to positive energy spinors, whereas R and S correspond to negative energy spinors. The full 1-RDM may now be represented using the expression for the 1-RDM operator.

$$D(r : r') = \left\langle \Psi \left| \begin{array}{l} \sum_{I\,J} \hat{b}_I^{\dagger} \hat{b}_J \left[\psi_I^{\dagger}(r') \otimes \psi_J(r)\right] + \sum_{R} \sum_{J} \hat{d}_R \hat{b}_J \left[\psi_R^{\dagger}(r') \otimes \psi_J(r)\right] \\ + \sum_{I} \sum_{S} \hat{b}_I^{\dagger} \hat{d}_S^{\dagger} \left[\psi_I^{\dagger}(r') \otimes \psi_S(r)\right] + \sum_{R\,S} \hat{d}_R \hat{d}_S^{\dagger} \left[\psi_R^{\dagger}(r') \otimes \psi_S(r)\right] \end{array} \right| \Psi \right\rangle \tag{5.8}$$

$$\begin{aligned} D(r : r') = {} & \sum_{I\,J} W_{IJ} \left[\psi_I^{\dagger}(r') \otimes \psi_J(r)\right] + \sum_{R} \sum_{J} W_{RJ} \left[\psi_R^{\dagger}(r') \otimes \psi_J(r)\right] \\ & + \sum_{I} \sum_{S} W_{IS} \left[\psi_I^{\dagger}(r') \otimes \psi_S(r)\right] + \sum_{R\,S} W_{RS} \left[\psi_R^{\dagger}(r') \otimes \psi_S(r)\right] \end{aligned} \tag{5.9}$$

where $W_{IJ} = \left\langle \Psi \left| \hat{b}_I^\dagger \hat{b}_J \right| \Psi \right\rangle$, $W_{IS} = \left\langle \Psi \left| \hat{b}_I^\dagger \hat{d}^\dagger{}_S \right| \Psi \right\rangle$, $W_{RJ} = \langle \Psi | \hat{d}_R \hat{b}_J | \Psi \rangle$ and $W_{RS} = -\left\langle \Psi \left| \hat{d}^\dagger{}_S \hat{d}_R \right| \Psi \right\rangle$ are 1-RDM coefficients.

The expression $\left[\psi_I^\dagger(r') \otimes \psi_J(r)\right]$ in Eq. (5.9) can be simplified as

$$\left[\psi_I^\dagger(r') \otimes \psi_J(r)\right] = \left[\Phi_1^{I*}(r') \quad \Phi_2^{I*}(r') \quad \Phi_3^{I*}(r') \quad \Phi_4^{I*}(r')\right] \otimes \begin{bmatrix} \Phi_1^J(r) \\ \Phi_2^J(r) \\ \Phi_3^J(r) \\ \Phi_4^J(r) \end{bmatrix}$$

$$= \begin{bmatrix} \Phi_1^{I*}(r')\Phi_1^J(r) & \Phi_2^{I*}(r')\Phi_1^J(r) & \Phi_3^{I*}(r')\Phi_1^J(r) & \Phi_4^{I*}(r')\Phi_1^J(r) \\ \Phi_1^{I*}(r')\Phi_2^J(r) & \Phi_2^{I*}(r')\Phi_2^J(r) & \Phi_3^{I*}(r')\Phi_2^J(r) & \Phi_4^{I*}(r')\Phi_2^J(r) \\ \Phi_1^{I*}(r')\Phi_3^J(r) & \Phi_2^{I*}(r')\Phi_3^J(r) & \Phi_3^{I*}(r')\Phi_3^J(r) & \Phi_4^{I*}(r')\Phi_3^J(r) \\ \Phi_1^{I*}(r')\Phi_4^J(r) & \Phi_2^{I*}(r')\Phi_4^J(r) & \Phi_3^{I*}(r')\Phi_4^J(r) & \Phi_4^{I*}(r')\Phi_4^J(r) \end{bmatrix} \tag{5.10}$$

Similarly, we have a 4×4 matrix form for the $\left[\psi_R^\dagger(r') \otimes \psi_J(r)\right]$, $\left[\psi_I^\dagger(r') \otimes \psi_S(r)\right]$ and $\left[\psi_R^\dagger(r') \otimes \psi_S(r)\right]$. So, the full 1-RDM takes the form as

$$D(r : r') = \sum_{IJ} W_{IJ} \begin{bmatrix} \Phi_1^{I*}(r')\Phi_1^J(r) & \Phi_2^{I*}(r')\Phi_1^J(r) & \Phi_3^{I*}(r')\Phi_1^J(r) & \Phi_4^{I*}(r')\Phi_1^J(r) \\ \Phi_1^{I*}(r')\Phi_2^J(r) & \Phi_2^{I*}(r')\Phi_2^J(r) & \Phi_3^{I*}(r')\Phi_2^J(r) & \Phi_4^{I*}(r')\Phi_2^J(r) \\ \Phi_1^{I*}(r')\Phi_3^J(r) & \Phi_2^{I*}(r')\Phi_3^J(r) & \Phi_3^{I*}(r')\Phi_3^J(r) & \Phi_4^{I*}(r')\Phi_3^J(r) \\ \Phi_1^{I*}(r')\Phi_4^J(r) & \Phi_2^{I*}(r')\Phi_4^J(r) & \Phi_3^{I*}(r')\Phi_4^J(r) & \Phi_4^{I*}(r')\Phi_4^J(r) \end{bmatrix}$$

$$+ \sum_R \sum_J W_{RJ} \begin{bmatrix} \Phi_1^{R*}(r')\Phi_1^J(r) & \Phi_2^{R*}(r')\Phi_1^J(r) & \Phi_3^{R*}(r')\Phi_1^J(r) & \Phi_4^{R*}(r')\Phi_1^J(r) \\ \Phi_1^{R*}(r')\Phi_2^J(r) & \Phi_2^{R*}(r')\Phi_2^J(r) & \Phi_3^{R*}(r')\Phi_2^J(r) & \Phi_4^{R*}(r')\Phi_2^J(r) \\ \Phi_1^{R*}(r')\Phi_3^J(r) & \Phi_2^{R*}(r')\Phi_3^J(r) & \Phi_3^{R*}(r')\Phi_3^J(r) & \Phi_4^{R*}(r')\Phi_3^J(r) \\ \Phi_1^{R*}(r')\Phi_4^J(r) & \Phi_2^{R*}(r')\Phi_4^J(r) & \Phi_3^{R*}(r')\Phi_4^J(r) & \Phi_4^{R*}(r')\Phi_4^J(r) \end{bmatrix}$$

$$+ \sum_I \sum_S W_{IS} \begin{bmatrix} \Phi_1^{I*}(r')\Phi_1^S(r) & \Phi_2^{I*}(r')\Phi_1^S(r) & \Phi_3^{I*}(r')\Phi_1^S(r) & \Phi_4^{I*}(r')\Phi_1^S(r) \\ \Phi_1^{I*}(r')\Phi_2^S(r) & \Phi_2^{I*}(r')\Phi_2^S(r) & \Phi_3^{I*}(r')\Phi_2^S(r) & \Phi_4^{I*}(r')\Phi_2^S(r) \\ \Phi_1^{I*}(r')\Phi_3^S(r) & \Phi_2^{I*}(r')\Phi_3^S(r) & \Phi_3^{I*}(r')\Phi_3^S(r) & \Phi_4^{I*}(r')\Phi_3^S(r) \\ \Phi_1^{I*}(r')\Phi_4^S(r) & \Phi_2^{I*}(r')\Phi_4^S(r) & \Phi_3^{I*}(r')\Phi_4^S(r) & \Phi_4^{I*}(r')\Phi_4^S(r) \end{bmatrix}$$

$$+ \sum_{RS} W_{RS} \begin{bmatrix} \Phi_1^{R*}(r')\Phi_1^S(r) & \Phi_2^{R*}(r')\Phi_1^S(r) & \Phi_3^{R*}(r')\Phi_1^S(r) & \Phi_4^{R*}(r')\Phi_1^S(r) \\ \Phi_1^{R*}(r')\Phi_2^S(r) & \Phi_2^{R*}(r')\Phi_2^S(r) & \Phi_3^{R*}(r')\Phi_2^S(r) & \Phi_4^{R*}(r')\Phi_2^S(r) \\ \Phi_1^{R*}(r')\Phi_3^S(r) & \Phi_2^{R*}(r')\Phi_3^S(r) & \Phi_3^{R*}(r')\Phi_3^S(r) & \Phi_4^{R*}(r')\Phi_3^S(r) \\ \Phi_1^{R*}(r')\Phi_4^S(r) & \Phi_2^{R*}(r')\Phi_4^S(r) & \Phi_3^{R*}(r')\Phi_4^S(r) & \Phi_4^{R*}(r')\Phi_4^S(r) \end{bmatrix} \tag{5.11}$$

5.3 Properties of Relativistic 1-RDM

The 1-RDM is of prime importance for the calculation of the average value of any local or nonlocal one-electron operator. The average value expression is given by

$$\langle O \rangle = \mathrm{Trace}[O_1 D(r : r')] \tag{5.12}$$

where O_1 is one electron operator. The trace of the 1-RDM is equal to the total charge Q.

$$\sum_q W_{qq} = Q \quad q \in \{I, R\} \tag{5.13}$$

Since only the positive spectrum of the energy contributes to the no-pair approximation, the sole term that remains in Eq. (5.11) will be

$$D(r:r') = \sum_{I\,J} \ddot{W}_{IJ} \begin{bmatrix} \Phi_1^{I*}(r')\Phi_1^J(r) & \Phi_2^{I*}(r')\Phi_1^J(r) & \Phi_3^{I*}(r')\Phi_1^J(r) & \Phi_4^{I*}(r')\Phi_1^J(r) \\ \Phi_1^{I*}(r')\Phi_2^J(r) & \Phi_2^{I*}(r')\Phi_2^J(r) & \Phi_3^{I*}(r')\Phi_2^J(r) & \Phi_4^{I*}(r')\Phi_2^J(r) \\ \Phi_1^{I*}(r')\Phi_3^J(r) & \Phi_2^{I*}(r')\Phi_3^J(r) & \Phi_3^{I*}(r')\Phi_3^J(r) & \Phi_4^{I*}(r')\Phi_3^J(r) \\ \Phi_1^{I*}(r')\Phi_4^J(r) & \Phi_2^{I*}(r')\Phi_4^J(r) & \Phi_3^{I*}(r')\Phi_4^J(r) & \Phi_4^{I*}(r')\Phi_4^J(r) \end{bmatrix} \tag{5.14}$$

with the modified matrix coefficient due to the projection onto positive energy electron space only. As a result, the trace of 1-RDM in no-pair approximation exactly corresponds to the total number of electrons mirroring the nonrelativistic results.

$$\int D(r:r)dr = \sum_{I} \ddot{W}_{II} = N \tag{5.15}$$

To treat post Dirac-Hartree-Fock (DHF) wave function-based methods in the relativistic framework, the use of no-pair approximation [30] is recommended. It makes the formulation consistent with the corresponding post-Hartree–Fock electron correlation methods in the nonrelativistic picture. Similarly, for all practical applications in the relativistic realm, 1-RDM is used in its no-pair approximation form. One can directly include the correlation effect into the relativistic calculations by using the Rel-RDMFT. Formulation of Rel-RDMFT based on the local [26] and nonlocal [27] characteristics of the relativistic 1-RDM has already been described in the literature and is a very active field of research.

The majority of information on the chemical bonding and electron distribution in a molecule is contained in the 1-RDM, also known as the charge-and-bond-order matrix [31]. Since the 1-RDM's diagonal elements reflect a spinor's charge order, which indicates how its electrons are distributed, and its non-diagonal elements depict the bond order of two spinors, which gives information about their chemical bonding. Another way to think about the charge order is the average number of particles in the spinor.

Due to the Hermitian nature of the charge-and-bond-order matrix, diagonalization of this matrix is possible via a unitary transformation.

$$DV = Vn \tag{5.16}$$

The eigenvectors V obtained after diagonalization are known as natural spinors (relativistic counterpart of natural orbitals [8]), which are eigenfunctions of the 1-RDM of a spin–orbit coupled wave function derived from a relativistic electron correlation theory. The charge orders are the eigenvalues n of the matrix, which are also known as occupation numbers of the corresponding natural spinors. The occupation number of a natural spinor takes a value in the range of

$$0 \le n \le 1 \tag{5.17}$$

1-RDM calculated from an uncorrelated wave function (DHF) gives occupation 1 for all the occupied spinors and 0 for all the virtual spinors. On the other hand, for 1-RDM constructed by a correlated wave function, occupation of natural spinors comes in a range of $0 \le n < 1$, indicating less average number of particles in a spinor due to the explicit interaction between them. So, on the basis of occupation info, natural spinors can efficiently describe electron correlation in the many-electron system.

5.3.1 Natural Spinors: An Efficient Framework for Low-cost Calculations

Wave function-based computational methods that include electron correlation for the assessments of energy and properties scale unfavorably with the basis set size. As an example, the popular coupled-cluster methods, with the commonly used singles–doubles truncation of the cluster operator, has an unfavorable scaling of iterative O(N^6) power of the basis set. Noniterative treatment of triples (i.e. CCSD(T)) scales even worse and has a formal scaling of O(N^7) power of the basis set. The computation costs further increased in the case of relativistic coupled-cluster calculations due to the complex algebra involved and the larger dimension of the basis set since a separate contraction scheme is required for spin–orbit split orbitals and also for the large and small components of a spinor. This confines the utilization of the relativistic coupled-cluster approach to atoms or diatomic molecules.

Mass computational savings can be made by freezing a portion of the virtual spinors. Merely freezing virtual spinors on the basis of an energy-based threshold scheme leads to considerable errors in correlation energy, as the energy, in particular, basis converges slowly with virtual spinor space size. Natural spinors generated out of the approximate correlation method offer a more compact description of the spinor space than the corresponding canonical spinors obtained from Dirac–Hartree–Fock (DHF) calculations and show a rapid convergence behavior with respect to the size of the virtual space. Natural spinors can be used to truncate the dimension of the basis set by using the spinor's occupation number info, resulting in the reduced computational cost of relativistic correlated electronic structure calculations.

For the nonrelativistic correlation calculations, there is extensive literature on the methods based on natural orbitals for the reduction of computational cost. For example, MP2 natural orbitals [32], frozen natural orbitals [33–38], natural transition orbitals [39–41], and pair natural orbitals [42–46]-based methods have been applied to speed up ground and excited state coupled-cluster approaches. Among all-natural orbital-based methods, natural orbitals obtained through second-order Moller–Plesset theory (MP2) are most commonly used due to their ability to retrieve most of the correlation effect and its favorable scaling. Therefore, the MP2 method is generally used to generate natural spinors. What follows is an outline of how to obtain MP2 natural spinors.

Initially, a DHF calculation is done, followed by a partial integral transformation in the no-pair approximation. The MP2 approach is then used to generate the virtual–virtual block of the 1-RDM matrix as

$$D_{ab} = \sum_{cij} \frac{\langle ac \| ij \rangle \langle ij \| bc \rangle}{\varepsilon_{ij}^{ac} \varepsilon_{ij}^{bc}} \tag{5.18}$$

where

$$\varepsilon_{ij}^{ac} = \varepsilon_i + \varepsilon_j - \varepsilon_a - \varepsilon_c \tag{5.19}$$

$$\varepsilon_{ij}^{bc} = \varepsilon_i + \varepsilon_j - \varepsilon_b - \varepsilon_c \tag{5.20}$$

with ε_i, ε_j, ε_a, ε_b, and ε_c being the DHF molecular spinor energies and $\langle ac\|ij\rangle$and $\langle ij\|bc\rangle$ being antisymmetrized two-electron integrals. Next, the 1-RDM's virtual–virtual block is diagonalized

$$D_{ab}V = Vn \tag{5.21}$$

In Eq. (5.21), V are virtual natural spinors, and their respective occupation numbers are denoted by n. Using the occupation numbers, we may group these virtual spinors according to how important they are to the total correlation energy. The virtual space can be reduced by keeping only the virtual natural spinors whose occupancy numbers are higher than a set cutoff or threshold. Then, the virtual–virtual block of the Fock matrix is subsequently converted into the truncated natural spinor basis

$$F_{vv}^{NS} = \tilde{V}^{\dagger} F_{vv} \tilde{V} \tag{5.22}$$

F_{vv} in Eq. (5.22) denotes the virtual–virtual block of the canonical Fock matrix. $\tilde{V}$ are natural spinors in the truncated basis and F_{vv}^{NS} represents the virtual–virtual block of the Fock matrix in the truncated natural spinor basis. This Fock matrix is subsequently diagonalized in the natural spinor basis to get the semi-canonical virtual natural spinors and the associated orbital energies.

$$F_{vv}^{NS} \tilde{Z} = \tilde{Z} \grave{o} \tag{5.23}$$

Here $\tilde{Z}$ indicates, semi-canonical virtual natural spinors, and associated orbital energies are indicated by $\grave{o}$. Using $\tilde{Z}$ and $\tilde{V}$, a transformation matrix U can be formed with which canonical DHF virtual spinor space can be transformed into the semi-canonical natural virtual spinor space.

$$U = \tilde{Z} \tilde{V} \tag{5.24}$$

We have used the natural spinors from relativistic MP2 calculations to approximate the virtual space only, and the occupied region is kept the same as obtained from DHF calculations, so the procedure is also termed "frozen natural spinor (FNS) method." In the FNS basis, higher-order correlation calculations can be carried out at a reduced computational expense and preferably without substantial errors. In the nonrelativistic domain, the accuracy of coupled-cluster methods based on natural orbitals was significantly improved by the perturbative corrections [33, 47]. By approximating the $\Delta E_{\text{CCSD/CCSD(T)}}$ with the ΔE_{MP2}, one can undertake a perturbative correction for the truncated virtual space.

$$\Delta E_{\text{CCSD/CCSD(T)}} = E_{\text{CCSD/CCSD(T)}}^{\text{Canonical}} - E_{\text{CCSD/CCSD(T)}}^{\text{FNS}} \tag{5.25}$$

$$\Delta E_{\text{CCSD/CCSD(T)}} \approx \Delta E_{\text{MP2}} \tag{5.26}$$

where

$$\Delta E_{\text{MP2}} = E_{\text{MP2}}^{\text{Canonical}} - E_{\text{MP2}}^{\text{FNS}} \tag{5.27}$$

The development version of BAGH [48] features both the FNS-CCSD and FNS-CCSD(T) methods. The one- and-two-electron integrals, as well as the converged four-component DHF spinors, are calculated using the PYSCF [49] interface of BAGH. We have used the example of the fluorine halide series (X–F, X = F, Cl, Br, I) to assess the performance of the FNS-CCSD and FNS-CCSD(T) methods. For F and Cl, the aug-cc-pVTZ basis set is used, whereas for Br and I, the dyall.ae3z basis set is employed. For all calculations, the basis sets were left uncontracted, and the frozen core approximation was applied (except for F_2). For all the molecules, experimental geometry has been adopted. Along with CCSD and CCSD(T) energies, molecular properties such as bond length and frequencies are calculated. Numerical derivative techniques were used to get the properties.

5.3.1.1 Correlation Energy

The selection of appropriate criteria for truncating the natural spinor virtual space is crucial. The truncation framework must be uniform among various atoms, and molecules and be equally applicable to energies and different properties. It has been well established in the literature [28, 29, 35, 47] that the occupation of the natural orbital/spinor is a more legitimate criterion to order the natural orbitals/spinors according to their significance toward total correlation energy. Therefore, we have also employed the occupation number as the virtual space truncation criterion.

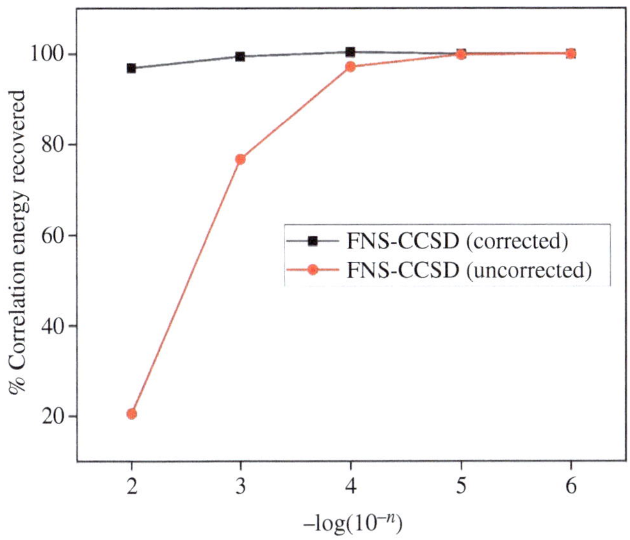

Figure 5.1 The convergence of the correlation energies was calculated using FNS-CCSD method with respect to the FNS truncation threshold (10^{-n}) for ClF.

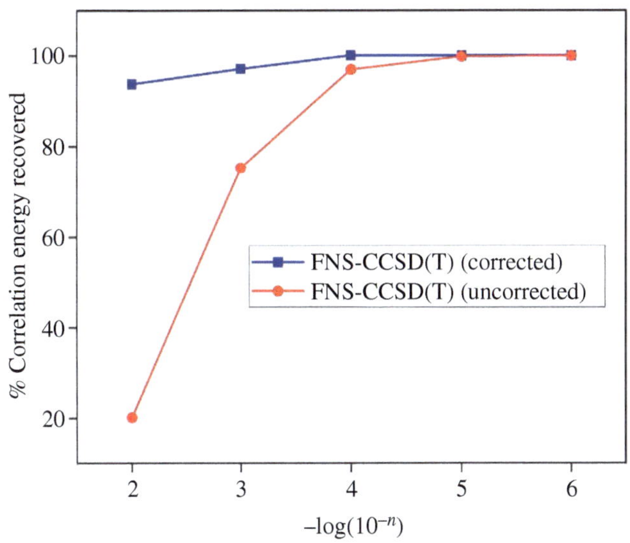

Figure 5.2 The convergence of the correlation energies was calculated using FNS-CCSD(T) method with respect to the FNS truncation threshold (10^{-n}) for ClF.

The convergence of the FNS-CCSD correlation energy for ClF is shown in Figure 5.1 with regard to the FNS truncation threshold (10^{-n}) or occupation number. It is clear from the figure that in the uncorrected FNS-CCSD method, it is important to keep virtual natural spinors with occupancy numbers up to 10^{-5} in order to recover more than 99.9% of the correlation energy. The convergence with respect to the truncation threshold can be significantly accelerated by the addition of the MP2 perturbative correction. However, a truncation threshold of 10^{-5} appears to provide comparable results in both uncorrected and corrected conditions. The FNS-CCSD(T) method shows a similar pattern (see Figure 5.2). The same threshold must apply to all molecules for the FNS-CCSD and

Table 5.1 The percentage of the correlation energy recovered in FNS-CCSD and FNS-CCSD(T) methods with an FNS truncation threshold of 10^{-5} with respect to the full canonical basis.

Fluorine halide						
			% Correlation energy recovered			
Molecule	**Total virtual spinors**	**Active virtual spinors**	**FNS-CCSD (corrected)**	**FNS-CCSD (uncorrected)**	**FNS-CCSD(T) (corrected)**	**FNS-CCSD(T) (uncorrected)**
F_2	214	168	100.03	98.70	99.99	98.70
ClF	240	160	100.03	99.82	100.01	99.80
BrF	402	228	100.07	99.34	100.04	99.33
IF	520	228	100.07	99.37	100.02	99.35

FNS-CCSD(T) methods to be considered a black box. A benchmarking of the energy and property calculations for different molecules needs to be done to check that.

Table 5.1 presents the correlation energy in the FNS-CCSD and FNS-CCSD(T) methods for the fluorine halide series at a truncation threshold of 10^{-5}. It can be seen that both methods can recover more than 99.5% of the canonical correlation energy when the MP2 corrections are considered.

5.3.1.2 Bond Length and Harmonic Vibrational Frequency

Using the FNS-CCSD and FNS-CCSD(T) methods, we have determined the bond length and harmonic vibrational frequency as molecular properties by numerically differentiating the total energy with the DIRAC19's [50] TWOFIT utility scheme. A polynomial of fifth-degree was utilized, and the convergence criteria for CCSD energies were set up to 10^{-12} a.u. Figure 5.3 depicts the convergence of FNS-CCSD bond lengths with respect to the FNS threshold for ClF. The convergence behavior is not as systematic as that observed in the case of correlation energy. It can be smoothed by the use of the MP2 perturbative correction. Similar trends are observed for the FNS-CCSD(T)

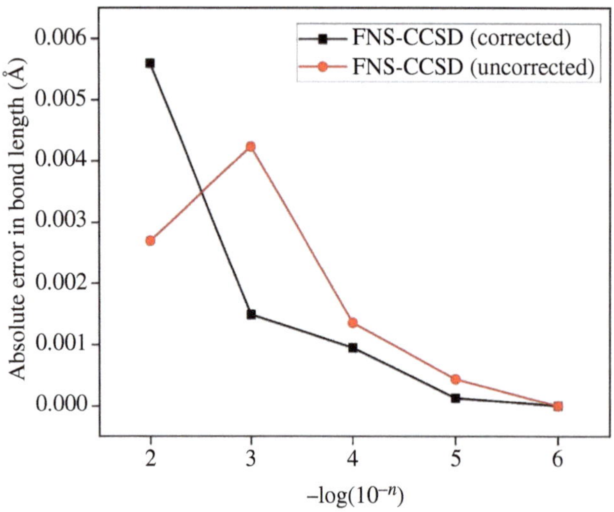

Figure 5.3 The convergence of the bond length was calculated using FNS-CCSD method with respect to the FNS truncation threshold (10^{-n}) for ClF.

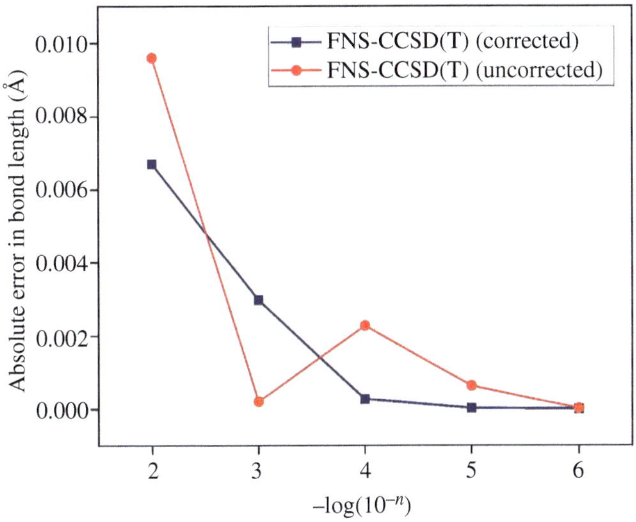

Figure 5.4 The convergence of the bond length was calculated using FNS-CCSD(T) method with respect to the FNS truncation threshold (10^{-n}) for ClF.

(see Figure 5.4). In both cases, a threshold of 10^{-5} can give a converged bond length with respect to the truncation threshold. Table 5.2 lists the error in the bond length of fluorine halides with an FNS threshold of 10^{-5}. The experimental bond length has been used as a reference. The uncorrected FNS-CCSD method's highest error relative to its experimental value is less than 0.021 Å. The MP2 correction further reduces the inaccuracy. In uncorrected FNS-CCSD(T), the triples correction reduces the error to a maximum of 0.0178 Å. The FNS-CCSD(T) results in this instance are within two picometers of the experimental values.

Figure 5.5 gives the convergence of the harmonic vibrational frequencies in the FNS-CCSD method with respect to the FNS truncation threshold for ClF. In the uncorrected FNS-CCSD approach, the convergence behavior is not systematic. However, the inclusion of perturbative correction makes convergence smoother. The FNS-CCSD(T) method follows a similar trend (see Figure 5.6). From Figures 5.5 and 5.6, it is clear that harmonic vibrational frequency convergence

Table 5.2 The error in bond length (Å) was calculated using FNS-CCSD and FNS-CCSD(T) methods with an FNS truncation threshold of 10^{-5} with respect to the experimental value.

Fluorine halide					
	Error in bond length				
Molecule	FNS-CCSD (corrected)	FNS-CCSD (uncorrected)	FNS-CCSD(T) (corrected)	FNS-CCSD(T) (uncorrected)	Experimental value
F_2	−0.0168	−0.0209	0.0061	0.0040	1.4119 [51]
ClF	0.0057	0.0051	0.0184	0.0177	1.6283 [51]
BrF	−0.0037	−0.0045	0.0091	0.0082	1.7589 [51]
IF	−0.0013	−0.0023	0.0106	0.0095	1.9097 [51]

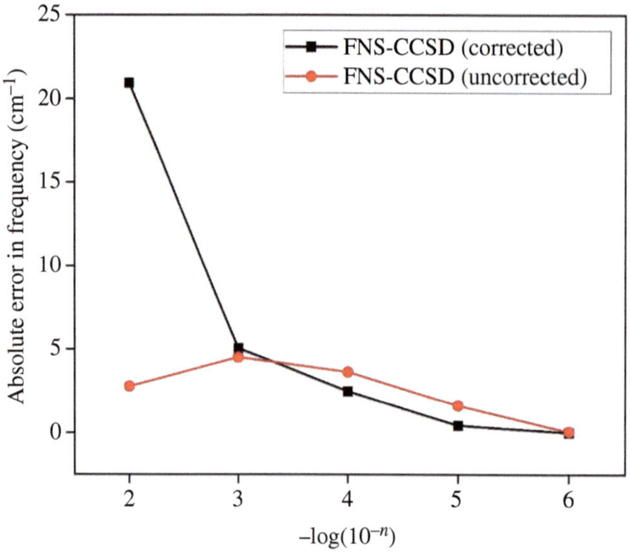

Figure 5.5 The convergence of the harmonic vibrational frequency was calculated using FNS-CCSD method with respect to the FNS truncation threshold (10^{-n}) for ClF.

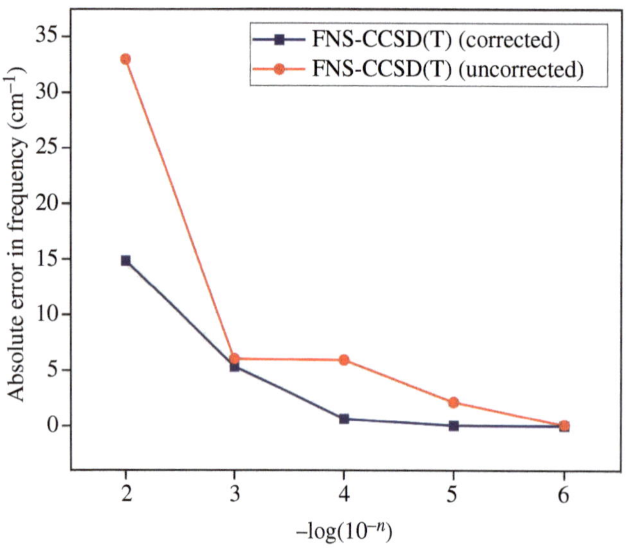

Figure 5.6 The convergence of the harmonic vibrational frequency was calculated using FNS-CCSD(T) method with respect to the FNS truncation threshold (10^{-n}) for ClF.

in the FNS-CCSD and FNS-CCSD(T) methods can be attained with a threshold of 10^{-5} for the ClF molecule when the MP2 perturbative correction is taken into account. Table 5.3 presents the error in the harmonic vibrational frequency of fluorine halides with respect to the experimental result at the FNS threshold of 10^{-5}. The predicted results for harmonic vibrational frequencies from the FNS-CCSD(T) method are in close agreement with the available experimental results, with a maximum error within 13 cm^{-1}.

Table 5.3 The error in the harmonic vibrational frequency (cm^{-1}) was calculated using FNS-CCSD and FNS-CCSD(T) methods with an FNS truncation threshold of 10^{-5} with respect to the experimental value.

Fluorine halide					
	Error in frequency				
Molecule	FNS-CCSD (corrected)	FNS-CCSD (uncorrected)	FNS-CCSD(T) (corrected)	FNS-CCSD(T) (uncorrected)	Experimental value
F_2	97.58	60.88	−1.65	−30.48	917 [52]
ClF	22.33	24.40	−12.19	−9.95	783 [52]
BrF	20.05	20.09	−8.55	−8.27	670 [51]
IF	12.52	13.99	−10.18	−8.66	610 [51]

5.3.2 Natural Spinors as an Interpretive Tool

Natural spinors are not only effective cost-reducing techniques but also a very robust tool for the interpretation of prominent spin-dependent phenomena in the relativistic domain. The natural spinors produced from two-step spin-orbit coupled wave functions were demonstrated to exhibit identical symmetry features to the canonical spinors [53], when the 1-RDM is correctly averaged. In Ref. [53], the natural spinors ($6p_{1/2}$ and $6p_{3/2}$ of the Tl and At atoms and $5d_{3/2}$ and $5d_{5/2}$ of the Lu atom) were compared to their equivalent canonical spinors in terms of their radial function, and it was discovered that natural spinors could quantitatively imitate the spin–orbit splittings of the sub-shells. Additionally, natural spinors can be used to explain the significance of spin–orbit coupling in covalent bonding and the Jahn–Teller phenomena. They are utilized to describe the antibonding spin–orbit effects [54] on TlH and Tl_2 molecules and also show how spin–orbit coupling can both weakens [54] and strengthens [55] the Jahn–teller effect in WF_5 and WF_6^- anion.

5.4 Concluding Remarks

The purpose of this chapter is mainly to introduce the aspect of reduced density matrices in the relativistic realm. After an overview of the theory and fundamental concepts underlying the relativistic 1-RDM, various properties of the relativistic 1-RDM have been discussed. The importance of the relativistic 1-RDM to obtain a compact set of orbitals, which can reduce the computational cost of relativistic wave function calculation, is discussed. Natural spinors obtained after diagonalization of relativistic 1-RDM offer a more compact basis set for relativistic coupled-cluster calculations of energy and properties without any significant loss in accuracy. The effect of relativistic spin–orbit coupling on general phenomena of chemistry, such as structure and bonding, can be well described with natural spinors.

Acknowledgments

The authors express their gratitude to ISRO, the Prime Minister's Research Fellowship, IIT Bombay, CRG and Matrix Project of DST-SERB, CSIR-India, and DST-Inspire Faculty Fellowship for their financial support, and C-DAC supercomputing resources (Param Bramha, Param Smriti) for computational time.

References

1 von Neumann, J. (1927). Nachr. Akad. Wiss. Göttingen, Math. Physik. kl. IIa Math. Physik. Chem. Abt. 245.

2 Dirac, P.A.M. (1929). The basis of statistical quantum mechanics. *Math. Proc. Cambridge Philos. Soc.* 25 (1): 62–66.

3 Dirac, P.A.M. (1930). Note on exchange phenomena in the Thomas atom. *Math. Proc. Cambridge Philos. Soc.* 26 (3): 376–385.

4 Dirac, P.A.M. (1931). Note on the interpretation of the density matrix in the many-electron problem. *Math. Proc. Cambridge Philos. Soc.* 27 (2): 240–243.

5 Mulliken, R.S. (1955). Electronic population analysis on LCAO–MO molecular wave functions. I. *J. Chem. Phys.* 23 (10): 1833–1840.

6 Ruedenberg, K. (1962). The physical nature of the chemical bond. *Rev. Mod. Phys.* 34 (2): 326–376.

7 Husimi, K. (1940). Some formal properties of the density matrix. *Proc. Phys. Math. Soc. Jpn.* 22 (4): 264–314.

8 Löwdin, P.O. (1955). Quantum theory of many-particle aystems. I. Physical interpretations by means of density matrices, natural spin-orbitals, and convergence problems in the method of configurational interaction. *Phys. Rev.* 97 (6): 1474–1489.

9 Shull, H. and Löwdin, P.O. (1955). Natural spin orbitals for helium. *J. Chem. Phys.* 23 (8): 1565–1565.

10 Löwdin, P.O. and Shull, H. (1956). Natural orbitals in the quantum theory of two-electron systems. *Phys. Rev.* 101 (6): 1730–1739.

11 Shull, H. and Löwdin, P.O. (1959). Superposition of configurations and natural spin orbitals. applications to the He problem. *J. Chem. Phys.* 30 (3): 617–626.

12 Shull, H. (1959). Natural spin orbital analysis of hydrogen molecule wave functions. *J. Chem. Phys.* 30 (6): 1405–1413.

13 Hurley, A.C. (1960). Electronic structure and binding energy of carbon monoxide. *Rev. Mod. Phys.* 32 (2): 400–411.

14 McWeeny, R. (1960). Some recent advances in density matrix theory. *Rev. Mod. Phys.* 32 (2): 335–369.

15 Burke, K. (2012). Perspective on density functional theory. *J. Chem. Phys.* 136 (15): 150901.

16 Kervazo, S., Réal, F., Virot, F. et al. (2019). Accurate predictions of volatile plutonium thermodynamic properties. *Inorg. Chem.* 58 (21): 14507–14521.

17 Aebersold, L.E. and Wilson, A.K. (2021). Considering density functional approaches for actinide species: the An66 molecule set. *J. Phys. Chem. A* 125 (32): 7029–7037.

18 Aquino, F., Govind, N., and Autschbach, J. (2010). Electric field gradients calculated from two-component hybrid density functional theory including spin–orbit coupling. *J. Chem. Theory Comput.* 6 (9): 2669–2686.

19 Sunaga, A. and Saue, T. (2021). Towards highly accurate calculations of parity violation in chiral molecules: relativistic coupled-cluster theory including QED effects. *Mol. Phys.* 119: e1974592.

20 Kaldor, U., Eliav, E., and Landau, A. (2002). *Reviews of Modern Quantum Chemistry*, 260–292. World Scientific Publishing Co Pte Ltd.

21 Liu, J. and Cheng, L. (2021). Relativistic coupled-cluster and equation-of-motion coupled-cluster methods. *Wiley Interdiscip. Rev.: Comput. Mol. Sci.* 11: e1536.

22 Ghosh, A., Chaudhuri, R.K., and Chattopadhyay, S. (2016). Relativistic state-specific multireference coupled cluster theory description for bond-breaking energy surfaces. *J. Chem. Phys.* 145 (12): 124303.

23 Eliav, E. and Kaldor, U. (2010). Relativistic four-component multireference coupled cluster methods: towards a covariant approach. In: *Recent Progress in Coupled Cluster Methods. Challenges and Advances in Computational Chemistry and Physics*, vol. 11 (ed. P. Cársky, J. Paldus, and J. Pittner), 113–144. Dordrecht: Springer.

24 Fleig, T., Sørensen, L.K., and Olsen, J. (2007). A relativistic 4-component general-order multi-reference coupled cluster method: initial implementation and application to HBr. *Theor. Chem. Acc.* 118 (2): 347–356.

25 Landau, A., Eliav, E., Ishikawa, Y., and Kaldor, U. (2000). Intermediate Hamiltonian Fock-space coupled-cluster method: excitation energies of barium and radium. *J. Chem. Phys.* 113 (22): 9905–9910.

26 Ohsaku, T., Yamanaka, S., Yamaki, D., and Yamaguchi, K. (2002). Quantum electrodynamical density-matrix functional theory and group theoretical consideration of its solution. *Int. J. Quantum Chem.* 90 (1): 273–281.

27 Rodríguez-Mayorga, M., Giesbertz, K.J.H., and Visscher, L. (2022). Relativistic reduced density matrix functional theory. *SciPost Chem.* 1 (2): 004.

28 Chamoli, S., Surjuse, K., Jangid, B. et al. (2022). A reduced cost four-component relativistic coupled cluster method based on natural spinors. *J. Chem. Phys.* 156 (20): 204120.

29 Yuan, X., Visscher, L., and Gomes, A.S.P. (2022). Assessing MP2 frozen natural orbitals in relativistic correlated electronic structure calculations. *J. Chem. Phys.* 156 (22): 224108.

30 Sucher, J. (1980). Foundations of the relativistic theory of many-electron atoms. *Phys. Rev. A* 22 (2): 348–362.

31 Davidson, E.R. (1972). Natural orbitals. *Adv. Quantum Chem.* 6: 235–266.

32 Jensen, H.J.A., Jørgensen, P., Ågren, H., and Olsen, J. (1988). Second-order Møller–Plesset perturbation theory as a configuration and orbital generator in multiconfiguration self-consistent field calculations. *J. Chem. Phys.* 88 (6): 3834–3839.

33 Taube, A.G. and Bartlett, R.J. (2005). Frozen natural orbitals: systematic basis set truncation for coupled-cluster theory. *Collect. Czech. Chem. Commun.* 70 (6): 837–850.

34 Barr, T.L. and Davidson, E.R. (1970). Nature of the configuration-interaction method in ab initio calculations. I. Ne ground state. *Phys. Rev. A* 1 (3): 644–658.

35 Landau, A., Khistyaev, K., Dolgikh, S., and Krylov, A.I. (2010). Frozen natural orbitals for ionized states within equation-of-motion coupled-cluster formalism. *J. Chem. Phys.* 132 (1): 014109.

36 Pokhilko, P., Izmodenov, D., and Krylov, A.I. (2020). Extension of frozen natural orbital approximation to open-shell references: theory, implementation, and application to single-molecule magnets. *J. Chem. Phys.* 152 (3): 034105.

37 Gyevi-Nagy, L., Kállay, M., and Nagy, P.R. (2021). Accurate reduced-cost CCSD(T) energies: parallel implementation, benchmarks, and large-scale applications. *J. Chem. Theory Comput.* 17 (2): 860–878.

38 Kumar, A. and Crawford, T.D. (2017). Frozen virtual natural orbitals for coupled-cluster linear-response theory. *J. Phys. Chem. A* 121 (3): 708–716.

39 Mata, R.A. and Stoll, H. (2011). An incremental correlation approach to excited state energies based on natural transition/localized orbitals. *J. Chem. Phys.* 134 (3): 034122.

40 Mester, D., Nagy, P.R., and Kállay, M. (2017). Reduced-cost linear-response CC2 method based on natural orbitals and natural auxiliary functions. *J. Chem. Phys.* 146 (19): 194102.

41 Folkestad, S.D. and Koch, H. (2020). Multilevel CC2 and CCSD methods with correlated natural transition orbitals. *J. Chem. Theory Comput.* 16 (1): 179–189.

42 Edmiston, C. and Krauss, M. (1965). Configuration-interaction calculation of H_3 and H_2. *J. Chem. Phys.* 42 (3): 1119–1120.

43 Ahlrichs, R. and Kutzelnigg, W. (1968). Direct calculation of approximate natural orbitals and natural expansion coefficients of atomic and molecular electronic wavefunctions. II. Decoupling of the pair equations and calculation of the pair correlation energies for the Be and LiH ground states. *J. Chem. Phys.* 48 (4): 1819–1832.

44 Neese, F., Wennmohs, F., and Hansen, A. (2009). Efficient and accurate local approximations to coupled-electron pair approaches: an attempt to revive the pair natural orbital method. *J. Chem. Phys.* 130 (11): 114108.

45 Nagy, P.R., Samu, G., and Kallay, M. (2018). Optimization of the linear-scaling local natural orbital CCSD(T) method: improved algorithm and benchmark applications. *J. Chem. Theory Comput.* 14 (8): 4193–4215.

46 Schwilk, M., Ma, Q., Koppl, C., and Werner, H.J. (2017). Scalable electron correlation methods. 3. Efficient and accurate parallel local coupled cluster with pair natural orbitals (PNO-LCCSD). *J. Chem. Theory Comput.* 13 (8): 3650–3675.

47 Neese, F., Hansen, A., and Liakos, D.G. (2009). Efficient and accurate approximations to the local coupled cluster singles doubles method using a truncated pair natural orbital basis. *J. Chem. Phys.* 131 (6): 064103.

48 Dutta, A.K., Manna, A., Jangid, B., et al. (2021). BAGH: a quantum chemistry software package.

49 Sun, Q.M., Berkelbach, T.C., Blunt, N.S. et al. (2018). PYSCF: the Python-based simulations of chemistry framework. *Wiley Interdiscip. Rev.: Comput. Mol. Sci.* 8 (1): ee1340.

50 Bast, R., Gomes, A.S.P., Saue, T. et al. (2021). DIRAC, a relativistic ab initio electronic structure program, Release DIRAC21. http://dx.doi.org/10.5281/zenodo.4836496 and http://www.diracprogram.org (accessed 20 November 2021).

51 Huber, K.P. and Herzberg, G. (1979). *Constants of Diatomic Molecules*. New York: Van Nostrand-Reinhold.

52 Irikura, K.K. (2007). Experimental vibrational zero-point energies: diatomic molecules. *J. Phys. Chem. Ref. Data* 36 (2): 389–397.

53 Zeng, T., Fedorov, D.G., Schmidt, M.W., and Klobukowski, M. (2011). Two-component natural spinors from twostep spin-orbit coupled wave functions. *J. Chem. Phys.* 134 (21): 214107.

54 Zeng, T., Fedorov, D.G., Schmidt, M.W., and Klobukowski, M. (2011). Effects of spin-orbit coupling on covalent bonding and the Jahn-Teller effect are revealed with the natural language of spinors. *J. Chem. Theory Comput.* 7 (9): 2864–2875.

55 Zeng, T., Fedorov, D.G., Schmidt, M.W., and Klobukowski, M. (2012). Natural spinors reveal how the spin–orbit coupling affects the Jahn–Teller distortions in the hexafluorotungstate(V) anion. *J. Chem. Theory Comput.* 8 (9): 3061–3071.

6

Many-Body Multi-Configurational Calculation Using Coulomb Green's Function

Bharti Kapil[1], Shivalika Sharma[2], Priyanka Aggarwal[2], Harsimran Kaur[3], Sunny Singh[1] and Ram Kuntal Hazra[1]

[1] *Department of Chemistry, Physical Division, University of Delhi, Delhi, India*
[2] *Department of Theoretical Physics, Asia Pacific Center for Theoretical Physics, POSTECH, Pohang, Korea*
[3] *Department of Chemistry, Indian Institute of Technology Bombay, Powai, Mumbai, Maharashtra, India*

6.1 Introduction

The Schrödinger equation (SE) of a many-body quantum system has been one of the most interesting problems for a long time due to difficulty in solving Coulomb interaction along with inevitable divergence issues [1–12]. The electronic, optical, magnetic, and other transport properties of multi-electron and multi-exciton quantum systems can be studied by monitoring gate voltage, transverse magnetic field, dielectric constant, and mass of carriers. The study of these properties is also important for a comprehensive knowledge of fundamental physical phenomena such as the Meissner effect, Josephson effect, fractional quantum hall effect, Kondo effect and Coulomb blockade [13–15]. Nuclei/holes, lateral electrical, and transverse magnetic confinements define exactly solved unperturbed models of 2D and 3D H-like, harmonic oscillator and particle in a box, respectively [16–29]. Salient features of electronic structure, chemical potential, capacitive energy, magnetization, etc., could be explored by monitoring either linear and nonlinear parameters of wavefunctions for electron–electron (e–e), electron–hole (e–h), and electron–nucleus (e–n) interactions or their densities in zero and non-zero magnetic field [30–36]. In 2D and 3D quantum systems, Coulomb correlation in the presence of magnetic field has been investigated by various numerical and a few analytical calculations [16–24, 37–58]. Since graphene, transition metal dichalcogenides (TMDCs), carbon nanotube (CNT) and boron nitride (BN) have multilayered generations, 3D N–e model became necessary to elucidate their electronic structure. In 1996, Fujito improvised an unrestricted Hartree–Fock method to study many-electron ground states of 3D anisotropic parabolic quantum dot (QD) at zero magnetic field [37]. Synergistic effect of interaction and anisotropy on electronic spectra, statistical properties and shell structure of 3D 2-e and 3-e axial dots revealed quantum effects on level clustering/dynamical symmetries in response to magnetic field that had been improvised by Hamaguchi et al. [16–18, 38–44]. In light of 2D QDs, certain milestone literature was found expounding conjectures and numerical methods for analytical treatment to strongly correlated systems [19–21, 44–58]. Laughlin extemporized a unique set of basis functions to illustrate anomalous quantum hall effect in 2D electrons in a strong magnetic field [45, 46]. Later, Bryant and others illustrated the role of Coulomb interaction in QDs with zero and non-zero cyclotron frequency [47]. Chakraborty et al., Hamaguchi et al., Taut and others made valuable contributions in solving the 2D multi-electron systems [19–21, 44–51].

Electron Density: Concepts, Computation and DFT Applications, First Edition.
Edited by Pratim Kumar Chattaraj and Debdutta Chakraborty.

An exciton being realized as quantum of excitation from fermi sea was improvised by Frenkel [59]. Later on, investigations established its resemblance to H-like systems and also accounted for novel phenomena of excitons, bi-excitons and tri-excitons in terms of hidden symmetry, exciton–phonon condensation, exciton–polariton condensation, optoelectronics, composite materials, electric batteries, field emitters, solid-state gas sensors and excitonic systems in strong magnetic field. Interestingly, in weaker confinement the ground state exhibited higher degeneracy of states whereas in stronger cyclotron field the trend of forming Landau bunching could be realized by invoking multipole expansion [22–24, 60–66]. The occurrence of neutral and charged exciton complexes further shed deeper insight at mesoscopic scale [66–72]. Tunable direct band-gap, high binding energy, photosensitivity, etc., made these excitonic systems highly desirable for a wide range of technological applications. Exciton-complexes of TMDCs (MX_2; M = Mo, W; X = S, Se, Te) and QDs could be characterized by the presence of strong Coulomb interaction [22–24, 70–72]. In the realm of atoms, molecules and ions, He-isoelectronic series became the benchmark. In this context, Hylleraas, Kinoshita, Pekeris, Scherr, Clementi, Roetti, Morgan, Sharma, Kapil and others paved the way in determining the ground state energy (GSE) [25–29, 73–100]. He-isoelectronic ions attracted volumes of works with GSE −2.90363 a.u. as experimental evidence as well as numerical solutions by modifying trial wavefunction as reciprocal of nuclear charge or hybrid perturbation-variational calculations. Recently, Kapil et al. examined Period-II elements along with He-isoelectronic ions for $N = 2, 3, 4, 5, 6, 7, 8, 9, 10$, respectively at nonrelativistic quantum domain using the analytical multiconfiguration method to include Pauli exclusion principle [28]. In this chapter, we have presented a modest analytical approach to Coulomb (exchange) interactions using 3D and 2D Green's function furnishing simple, finitely summed and terminable integrals in terms of Lauricella function. The electronic structure of extensive variety of materials from mesoscopic to atomic systems as a function of magnetic field has been investigated [16–29]. The development of current density for 2-e systems has also been discussed, which can be extended to many electron system [27, 28].

6.2 Theoretical Development

The SE of a small microcluster with lateral electrical and transverse magnetic confinements of electrons ($\omega_e, \vec{A}(r_i)$) and nuclei ($\omega_n, \vec{A}(R_I)$) reads [16–29]:

$$H = \sum_{I=1}^{W} \left\{ \frac{1}{2M_I} \left(\vec{p}_I - \frac{Z_I e}{c} \vec{A}(R_I) \right)^2 + \frac{1}{2} M_I \omega_n^2 R_I^2 \right\} + \sum_{i=1}^{N} \left\{ \frac{1}{2m_e} \left(\vec{p}_i + \frac{e}{c} \vec{A}(r_i) \right)^2 + \frac{1}{2} m_e \omega_e^2 r_i^2 \right\}$$
$$- \frac{1}{4\pi\epsilon_0\epsilon} \sum_{I=1}^{W} \sum_{i=1}^{N} \frac{Z_I e^2}{|\vec{R}_I - \vec{r}_i|} + \frac{1}{4\pi\epsilon_0\epsilon} \frac{1}{2} \sum_{i=1}^{N} \sum_{j\neq i}^{N} \frac{e^2}{r_{ij}} + \frac{1}{4\pi\epsilon_0\epsilon} \frac{1}{2} \sum_{I=1}^{W} \sum_{J\neq I}^{W} \frac{Z_I Z_J e^2}{R_{IJ}} \tag{6.1}$$

where Ze, M, m_e, W, N and ϵ are the charge of nucleus/hole, mass of nucleus/hole, mass of electron, number of nuclei, number of electrons, and dielectric constant of the medium, respectively. The above equation can be written in a.u. ($e = 1, c = 1, \hbar = 1$ and $\frac{1}{4\pi\epsilon_0} = 1$) as:

$$H = \sum_{I=1}^{W} \left\{ \frac{1}{2M_I} \left(\vec{p}_I - Z_I \vec{A}(R_I) \right)^2 + \frac{1}{2} M_I \omega_n^2 R_I^2 \right\} + \sum_{i=1}^{N} \left\{ \frac{1}{2m_e} \left(\vec{p}_i + \vec{A}(r_i) \right)^2 + \frac{1}{2} m_e \omega_e^2 r_i^2 \right\}$$
$$- \frac{1}{\epsilon} \sum_{I=1}^{W} \sum_{i=1}^{N} \frac{Z_I}{|\vec{R}_I - \vec{r}_i|} + \frac{1}{2\epsilon} \sum_{i=1}^{N} \sum_{j\neq i}^{N} \frac{1}{r_{ij}} + \frac{1}{2\epsilon} \sum_{I=1}^{W} \sum_{J\neq I}^{W} \frac{Z_I Z_J}{R_{IJ}} \tag{6.2}$$

In the Sections 6.2.1 and 6.2.2, we have addressed the Schrödinger equation of various 3D and 2D systems both in presence and absence of magnetic field.

6.2.1 Presence of Magnetic Field

In presence of magnetic field, interacting electron gas, and exciton complex models can be represented by the Hamiltonian as (Eq. (6.2)) [16–24, 101, 102]:

$$H = \sum_{I=1}^{W} \left\{ \frac{1}{2M_I}\left(\vec{p}_I - \vec{A}(R_I)\right)^2 + \frac{1}{2}M_I\omega_n^2 R_I^2 \right\} + \sum_{i=1}^{N} \left\{ \frac{1}{2m_e}\left(\vec{p}_i + \vec{A}(r_i)\right)^2 + \frac{1}{2}m_e\omega_e^2 r_i^2 \right\}$$
$$-\frac{1}{\epsilon}\sum_{I=1}^{W}\sum_{i=1}^{N}\frac{1}{|\vec{R}_I - \vec{r}_i|} + \frac{1}{2\epsilon}\sum_{i=1}^{N}\sum_{j\neq i}^{N}\frac{1}{r_{ij}} + \frac{1}{2\epsilon}\sum_{I=1}^{W}\sum_{J\neq I}^{W}\frac{1}{R_{IJ}} \tag{6.3}$$

6.2.1.1 3D Electron Gas Model

Hamiltonian for a 3D electron gas with N electrons can be elucidated as ($W = 0$) (Eq. (6.3)):

$$H = \sum_{i=1}^{N} \left\{ \frac{1}{2m_e}\left(\vec{p}_i + \vec{A}(r_i)\right)^2 + \frac{1}{2}m_e\omega_e^2 r_i^2 \right\} + \frac{1}{2\epsilon}\sum_{i=1}^{N}\sum_{j\neq i}^{N}\frac{1}{r_{ij}} \tag{6.4}$$

For 3D N–e anisotropic system ($\omega_{x_i} = \omega_{y_i} = \omega_0$ and $\omega_{z_i} = \omega_1$), the Hamiltonian can be represented as [18]:

$$H = \underbrace{\sum_{i=1}^{N} \left\{ \frac{1}{2m_e}\left(\vec{p}_i + \vec{A}(r_i)\right)^2 + \frac{1}{2}m_e\omega_0^2 r_i^2 + \frac{1}{2}m_e\left(\omega_1^2 - \omega_0^2\right)z_i^2 \right\}}_{H_i} + \underbrace{\frac{1}{2\epsilon}\sum_{i=1}^{N}\sum_{j\neq i}^{N}\frac{1}{r_{ij}}}_{H'} \tag{6.5}$$

In transverse magnetic field (symmetric gauge) $\vec{A}(r_i) = \frac{1}{2}(\vec{B}\times\vec{r}_i) = \frac{1}{2}B(-y_i, x_i, 0)$ where $w_c = \frac{eB}{c}$, the Hamiltonian H_i can be resolved for eigensolutions as:

$$H_i\xi_{n,l,m}(r_i,\theta_i,\phi_i) = E_{i(n,l,m)}\xi_{n,l,m}(r_i,\theta_i,\phi_i)$$
$$\left\{ -\frac{1}{2}\left(r_i^{-1}\frac{d^2}{dr_i^2}r - \frac{L^2}{r_i^2} + i\omega_c\frac{\partial}{\partial\phi_i} \right) + \frac{r_i^2}{2}\left(m_e^2\omega_o^2 + \frac{\omega_c^2}{4} \right) + \frac{1}{2}\left(m_e^2\left(\omega_1^2 - \omega_o^2\right) - \frac{\omega_c^2}{4} \right) r_i^2\cos^2\theta_i \right\}$$
$$\xi_{n,l,m}(r_i,\theta_i,\phi_i) = m_i E_{i(n,l,m)}\xi_{n,l,m}(r_i,\theta_i,\phi_i) \tag{6.6}$$

where $H_i = H_i^0 + H_i'$, and anisotropy, $H_i' = \frac{1}{2}\left(m_e^2\left(\omega_1^2 - \omega_o^2\right) - \frac{\omega_c^2}{4}\right) r_i^2\cos^2\theta_i$. Now, writing the eigenvalue equation for a particular ith coordinate $H^0\xi_{n,l,m} = E^0_{n,l,m}\xi_{n,l,m}$ in spherical polar coordinates with ansatz wavefunction $\xi_{n,l,m}(r_i,\theta_i,\phi_i) = Y_l^{\,m}(\theta_i,\phi_i)\frac{u(r_i)}{r_i}$, $m = 0, \pm1, \pm2, \ldots$ where $u(r_i)$ and $Y_l^{\,m}(\theta_i,\phi_i) = \sqrt{\frac{(2l+1)(l-|m|)!}{4\pi(l+|m|)!}}P_l^{|m|}(\theta)e^{im\phi_i}$ are radial and angular parts, respectively, results in following radial equation [16–18]:

$$\left[\frac{d^2}{dr^2} - \frac{l(l+1)}{r^2} - m\omega_c - \Omega^2 r^2 + 2m_e E^0_{n,l,m} \right] u(r) = 0 \tag{6.7}$$

where $\Omega = \sqrt{\left(m_e^2\omega_0^2 + \frac{\omega_c^2}{4}\right)}$. On substituting $z = \Omega_r r^2$ with $u(r) \Longrightarrow U(z)$ and by making the transformation $U(z) = z^{-\frac{1}{4}}M(z)$, it yields:

$$M''(z) + \left\{ \frac{\left(\frac{3}{16} - \frac{l(l+1)}{4}\right)}{z^2} - \frac{1}{4} + \frac{\lambda'}{z} \right\} M(z) = 0 \tag{6.8}$$

where $M(z)$ is Whittaker-M function and $\lambda' = \frac{\left(2m_e E^0_{n,l,m} \pm m\omega_c\right)}{4\Omega}$. Comparing with Whittaker-M equation yields the corresponding eigenvalue and eigenfunction for N–e system as $m_e E^0_{(n,l,m)} = (2n + l + \frac{3}{2})\Omega + \frac{m}{2}\omega_c$ and $\xi_{n,l,m}(z, \theta, \phi) = N_{n,l} z^{-\left(\frac{3}{4}\right)} M_{\kappa,\nu}(z) Y_l^m(\theta, \phi)$, respectively, where $N_{n,l} = \sqrt{\left(\frac{2\Omega^{\frac{3}{2}}(n+l+\frac{1}{2})!}{n!(l+\frac{1}{2})!^2}\right)}$, $\kappa = \nu + n + \frac{1}{2}$ and $l + \frac{1}{2} = 2\nu$ [16–18, 103].

6.2.1.1.1 Anisotropy (H_i')

Integrals of H_i' can be represented by [18]:

$$I = \underbrace{N_{n_1,l_1} N_{n_2,l_2} C \left\langle z^{-\frac{3}{4}} M_{\kappa_1,\nu_1}(z) \left| \frac{z^{\frac{3}{2}}}{\Omega^{\frac{5}{2}}} \right| z^{-\frac{3}{4}} M_{\kappa_2,\nu_2}(z) \right\rangle}_{I_1} \times \underbrace{\left\langle Y_{l_1}^{m_1} |\cos^2\theta| Y_{l_2}^{m_2} \right\rangle}_{I_2} \tag{6.9}$$

where $C = \frac{1}{2}\left(m_e^2\left(\omega_1^2 - \omega_0^2\right) - \frac{\omega_c^2}{4}\right) r_i^2 \cos^2\theta_i$. The above integrals are analyzed into diagonal and off-diagonal terms using standard integral (Eq. (6.A.3)) with $b = 0$, $a_1 = a_2 = 1$, $A = \frac{1}{2}(a_1 + a_2) = 1$, $\gamma_1 - \frac{1}{2} = \frac{l_1}{2} + \frac{1}{4}$, $\gamma_2 - \frac{1}{2} = \frac{l_2}{2} + \frac{1}{4}$, $M = \gamma_1 + \gamma_2 = \frac{l_1}{2} + \frac{l_2}{2} + \frac{3}{2}$ and $\varrho = 1$ as [18]:

Diagonal integral ($n_1 = n_2 = n; l_1 = l_2 = l; m_1 = m_2 = m$):

$$I_1 = \frac{C}{\Omega} \frac{(n + l + \frac{1}{2})!}{n!(l + \frac{1}{2})!^2} \Gamma\left(l + \frac{5}{2}\right) \sum_{i=0}^{n} \sum_{j=0}^{n} \frac{(l + \frac{5}{2})_{i+j} (-n)_i (-n)_j}{(l + \frac{3}{2})_i (l + \frac{3}{2})_j i! j!}$$

$$I_2 = \left(\frac{l^2 - |m^2|}{4l^2 - 1} + \frac{(l+1)^2 - |m^2|}{(2l+1)(2l+3)}\right)$$

Off-diagonal integral ($l_1 = l_2 = l; m_1 = m_2 = m$):

$$I_1 = -\frac{C}{\Omega} \sqrt{\left(\frac{(n_1 + l_1 + \frac{1}{2})!}{n_1!(l_1 + \frac{1}{2})!^2} \frac{(n_2 + l_2 + \frac{1}{2})!}{n_2!(l_2 + \frac{1}{2})!^2}\right)} \Gamma\left(\frac{l_1}{2} + \frac{l_2}{2} + \frac{5}{2}\right)$$
$$\times \sum_{i=0}^{n_1} \sum_{j=0}^{n_2} \frac{(\frac{l_1}{2} + \frac{l_2}{2} + \frac{5}{2})_{i+j} (-n_1)_i (-n_2)_j}{(l_1 + \frac{3}{2})_i (l_2 + \frac{3}{2})_j i! j!}$$

$$I_2 = \frac{1}{2l_2 - 1} \sqrt{\frac{(l_2^2 - |m_2|^2)((l_2 - 1)^2 - |m_2|^2)}{(2l_2 + 1)(2l_2 - 3)}} \delta_{l_1, l_{2-2}} \delta_{m_1, m_2}$$
$$+ \frac{1}{2l_2 + 3} \sqrt{\frac{((l_2 + 2)^2 - |m_2|^2)((l_2 + 1)^2 - |m_2|^2)}{(2l_2 + 1)(2l_2 + 5)}} \delta_{l_1, l_{2+2}} \delta_{m_1, m_2}$$

where for any number "n," Pochhammer symbol $(n)_0 = 1$.

6.2.1.1.2 Coulomb Interaction: Monopole and Dipole Factors

The individual interaction can be unfolded using multipole expansion as [16–24, 101]:

$$V = \frac{1}{\epsilon} \sum_{\mathcal{P}=0}^{\infty} V^{(\mathcal{P})} \tag{6.10}$$

where V is the sum of Coulomb interactions between all ith and jth charge carriers, with mono-, di-, quadru and octupole terms and so on as:

$$V_m^{(0)} = \frac{1}{2}\sum_{i=1}^{N}\sum_{j\neq i}^{N}\frac{e_i e_j}{|\vec{r}_i|} \tag{6.10a}$$

$$V_d^{(1)} = \frac{1}{2}\sum_{i=1}^{N}\sum_{j\neq i}^{N}\frac{e_i e_j \vec{r}_i.\vec{r}_j}{|\vec{r}_i|^3} \tag{6.10b}$$

$$V_q^{(2)} = \frac{1}{2}\frac{1}{2}\sum_{i=1}^{N}\sum_{j\neq i}^{N} e_i e_j \left\{ \frac{3(\vec{r}_i.\vec{r}_j)^2}{|\vec{r}_i|^5} - \frac{\vec{r}_j^{\,2}}{|\vec{r}_i|^3} \right\}$$

$$\cdots = \cdots \tag{6.10c}$$

Monopole Factor Monopole contribution of the interaction term for ith coordinate (Eq. (6.10a)) can be expressed as $I_m = \left\langle \Xi(Z,\Theta,\Phi) \left| \frac{1}{|\vec{r}_i|} \right| \Xi(Z,\Theta,\Phi) \right\rangle$ where $\Xi(Z,\Theta,\Phi) = \prod_{i=1}^{N} \xi(z_i,\theta_i,\phi_i)$ [18]. The integral is represented as:

$$I_m = \frac{N_{n_1,l_1} N_{n_2,l_2}}{2\Omega} \left\langle z_i^{-\frac{3}{2}} M_{\kappa_1,\nu_1}(z_i) M_{\kappa_2,\nu_2}(z_i) \right\rangle \tag{6.11}$$

where quantum numbers of all $j \neq i$ coordinates, "l" and "m" of ith coordinate are conserved (Eq. (6.10a)). Diagonal and off-diagonal integrals are evaluated using standard integral (Eq. (6.A.3)). For the above integral, $b = 0$, $a_1 = a_2 = 1$, $A = \frac{1}{2}(a_1 + a_2) = 1$, $\gamma_1 - \frac{1}{2} = \frac{l_1}{2} + \frac{1}{4}$, $\gamma_2 - \frac{1}{2} = \frac{l_2}{2} + \frac{1}{4}$, $M = \gamma_1 + \gamma_2 = \frac{l_1}{2} + \frac{l_2}{2} + \frac{3}{2}$, $\varrho = -\frac{1}{2}$, $2\nu = 0, 1, 2, \ldots$; $\kappa - \nu - 1/2 = 0, 1, 2, \ldots$. The off-diagonal and diagonal integrals can be expressed in single summed form using F_2 to ${}_2F_1$ relation and Chu–Vandermonde identity as (Eqs. (6.A.5), (6.A.6)) [18]:

Off-diagonal integral ($l_1 = l_2 = l$; $m_1 = m_2 = m$):

$$I_m = \sqrt{\left(\Omega\frac{(n_1 + l + \frac{1}{2})!}{n_1!(l + \frac{1}{2}!)^2}\frac{(n_2 + l + \frac{1}{2})!}{n_2!(l + \frac{1}{2}!)^2}\right)} \times \Gamma(l+1)\sum_{k=0}^{n_1}\frac{(l+1)_k(-n_1)_k(-k+\frac{1}{2})_{n_2}}{(l+\frac{3}{2})_k(l+\frac{3}{2})_{n_2}k!}$$

Diagonal integral ($n_1 = n_2 = n$):

$$I_m = \Omega^{\frac{1}{2}}\frac{(n + l + \frac{1}{2})!}{n!(l + \frac{3}{2})}\Gamma(l+1)\sum_{k=0}^{n}\frac{(l+1)_k(-n)_k(-k+\frac{1}{2})_n}{(l+\frac{3}{2})_k(l+\frac{3}{2})_n k!}$$

Dipole Factor The second-order dipole factor for any ith and jth coordinate (Eq. (6.10b)) is represented as [18]:

$$I_d = \left\langle \Xi(Z,\Theta,\Phi) \left| \frac{\vec{r}_j.\vec{r}_i}{|\vec{r}_i|^3} \right| \Xi(Z,\Theta,\Phi) \right\rangle$$

Introducing discretized projection over a complete basis having the form $\sum_{l=0} |\Xi_l(Z,\Phi)\rangle\langle\Xi_l(Z,\Phi)|$, the integral can be written as [18]:

$$I_d = \sum_{l=0} \left\langle \Xi(Z,\Theta,\Phi) \left| \frac{r_j \chi}{r_i} \right| \Xi_l(\Xi,\Theta,\Phi) \right\rangle \left\langle \Xi_l(Z,\Theta,\Phi) \left| \frac{1}{r_i} \right| \Xi(Z,\Theta,\Phi) \right\rangle$$

where $\chi = [\sin\theta_i \sin\theta_j \cos(\phi_i - \phi_j)) + \cos\theta_i \cos\theta_j]$. Decoupling I_d into ith, jth, and qth spaces reveals [18]:

$$I_d = \sum_{q=0} \left\{ I_{d_1} I_{d_2} I_{d_3} \right\} \times I_{d_4} \tag{6.12}$$

where,

$$I_{d_1} = \left\langle N_{n_{1,j}\cdots} z_j^{-\frac{3}{4}} M_{\kappa_{1,j};\nu_{1,j}} \left| r_j \right| N_{n_{2,q_j}\cdots} z_{q_j}^{-\frac{3}{4}} M_{\kappa_{2,q_j};\nu_{2,q_j}} \right\rangle, I_{d_2} = \left\langle N_{n_{1,i}\cdots} z_i^{-\frac{3}{4}} M_{\kappa_{1,i};\nu_{1,i}} \left| \frac{1}{r_i} \right| N_{n_{2,q_i}\cdots} z_{q_i}^{-\frac{3}{4}} M_{\kappa_{2,q_i};\nu_{2,q_i}} \right\rangle,$$

$$I_{d_3} = \left\langle Y_{l_1,i}^{m_1,i} Y_{l_1,j}^{m_1,j} \left| \chi \right| Y_{l_2,q_i}^{m_2,q_i} Y_{l_2,q_j}^{m_2,q_j} \right\rangle, I_{d_4} = \left\langle \xi^0_{n_{1,q_i}\cdots} \left| \left(\frac{\Omega}{z_i}\right)^{\frac{1}{2}} \right| \xi_{n_{2,i}\cdots} \right\rangle \delta_{(l_{1,q_i};l_{2,i})} \delta_{(m_{1,q_i};m_{2,i})}$$

I_d is composed of Coulomb-type (I_{d_2}, I_{d_4}), dipole-type (I_{d_1}), and angular momentum-type integrals (I_{d_3}) (Eq. (6.12)). The Coulomb-type integrals (I_{d_2}, I_{d_4}) are evaluated similar to monopole factor (Section 6.2.1.1.2). $I_{d_1}(n_1, l_1; n_2, l_2)$ for every z_j coordinate is interpreted using the standard integral with $b = 0, a_1 = a_2 = 1, A = \frac{1}{2}(a_1 + a_2) = 1, \gamma_1 - \frac{1}{2} = \frac{l_1}{2} + \frac{1}{4}, \gamma_2 - \frac{1}{2} = \frac{l_2}{2} + \frac{1}{4}, M = \gamma_1 + \gamma_2 = \frac{l_1}{2} + \frac{l_2}{2} + \frac{3}{2}, \varrho = \frac{1}{2}, 2\nu = 0, 1, 2, \ldots; \kappa - \nu - 1/2 = 0, 1, 2, \ldots$ The integral I_{d_1} can thus be written in single summation form using F_2 to ${}_2F_1$ relation and Chu–Vandermonde identity as (Eqs. (6.A.5), (6.A.6)) [18]:

$$I_{d_1} = \frac{1}{\Omega^{\frac{1}{2}}} \sqrt{\left(\frac{(n_1 + l_1 + \frac{1}{2})!}{n_1!(l_1 + \frac{1}{2})!^2} \frac{(n_2 + l_2 + \frac{1}{2})!}{n_2!(l_2 + \frac{1}{2})!^2} \right)} \Gamma\left(\frac{l_1}{2} + \frac{l_2}{2} + 2\right)$$
$$\times \sum_{k=0}^{n_1} \frac{\left(\frac{l_1}{2} + \frac{l_2}{2} + 2\right)_k (-n_1)_k \left(\frac{-l_1}{2} + \frac{l_2}{2} - k - \frac{1}{2}\right)_{n_2}}{(l_1 + \frac{3}{2})_k (l_2 + \frac{3}{2})_{n_2} k!}$$

Unwinding I_{d_3} into ith, jth, and qth spaces reframe [18]:

$$I_{d_3} = \frac{I_{d_{3a}} + I_{d_{3b}}}{2} + I_{d_{3c}}$$

where

$$I_{d_{3a}} = \left\langle Y_{l_1,i}^{m_1,i} Y_{l_1 j}^{m_1 j} \left| \sin\theta_i \sin\theta_j e^{i\phi_i} e^{-i\phi_j} \right| Y_{l_2,l_i}^{m_2,l_i} Y_{l_2,l_j}^{m_2,l_j} \right\rangle$$
$$I_{d_{3b}} = \left\langle Y_{l_1,i}^{m_1,i} Y_{l_1 j}^{m_1 j} \left| \sin\theta_i \sin\theta_j e^{-i\phi_i} e^{i\phi_j} \right| Y_{l_2,l_i}^{m_2,l_i} Y_{l_2,l_j}^{m_2,l_j} \right\rangle$$
$$I_{d_{3c}} = \left\langle Y_{l_1,i}^{m_1,i} Y_{l_1 j}^{m_1 j} \left| \cos\theta_i \cos\theta_j \right| Y_{l_2,l_i}^{m_2,l_i} Y_{l_2,l_j}^{m_2,l_j} \right\rangle$$

$I_{d_{3a}}$ is evaluated using the recurrence relations [18, 103]:

$$I_{d_{3a}} = \left[-\left\{ \frac{(l_2 + |m_2| + 1)(l_2 + |m_2| + 2)}{(2l_2 + 1)(2l_2 + 3)} \right\}^{\frac{1}{2}} \times \delta_{m_1, m_{2+1}} \delta_{l_1, l_{2+1}} \right.$$
$$\left. + \left\{ \frac{(l_2 - |m_2|)(l_2 - |m_2| - 1)}{(2l_2 - 1)(2l_2 + 1)} \right\}^{\frac{1}{2}} \times \delta_{m_1, m_{2+1}} \delta_{l_1, l_{2-1}} \right]$$
$$\times \left[\left\{ \frac{(l_2' + |m_2'| + 1))(l_2' + |m_2'| + 2)}{(2l_2' + 1)(2l_2' + 3)} \right\}^{\frac{1}{2}} \times \delta_{m_1', m_{2-1}'} \delta_{l_1', l_{2+1}'} \right.$$
$$\left. - \left\{ \frac{(l_2' + |m_2'|)(l_2' + |m_2'| - 1)}{(2l_2' - 1)(2l_2' + 1)} \right\}^{\frac{1}{2}} \times \delta_{m_1', m_{2-1}'} \delta_{l_1', l_{2-1}'} \right]$$

Similarly, $I_{d_{3b}}$ can also be calculated using recurrence relations. $I_{d_{3c}}$ is analyzed as $(m_1 = m_2 = m)$ [18, 103]:

$$I_{d_{3c}} = \left[\left\{\frac{(l_2)^2 - |m|^2}{(4l_2^2 - 1)}\right\}^{\frac{1}{2}} \delta_{l_1,l_{2-1}} + \left\{\frac{(l_2+1)^2 - |m|^2}{(2l_2+1)(2l_2+3)}\right\}^{\frac{1}{2}} \times \delta_{l_1,l_{2+1}}\right]$$
$$\left[\left\{\frac{(l'_2+1)^2 - |m|^2}{(2l'_2+1)(2l'_2+3)}\right\}^{\frac{1}{2}} \delta_{l'_1,l'_{2+1}} + \left\{\frac{l_2'^2 - |m|^2}{(4l_2'^2 - 1)}\right\}^{\frac{1}{2}} \times \delta_{l'_1,l'_{2-1}}\right]$$

Substituting I_{d_1}, I_{d_2}, I_{d_3}, and I_{d_4} into Eq. (6.12) obtains dipole factor for "N"–e using Eq. (6.10b).

6.2.1.2 2D Electron Gas Model

The multi-electron Hamiltonian for 2D anisotropic system ($\omega_{x_i} = \omega_0$ and $\omega_{y_i} = \omega_1$) can be expressed as [21]:

$$H = \underbrace{\sum_{i=1}^{N}\left\{\frac{1}{2m_e}\left(\vec{p}_i + \vec{A}(r_i)\right)^2 + \frac{1}{2}m_e\omega_0^2 r_i^2 + \frac{1}{2}m_e(\omega_y^2 - \omega_0^2)y_i^2\right\}}_{H_i} + \underbrace{\frac{1}{2\epsilon}\sum_{i=1}^{N}\sum_{j\neq i}^{N}\frac{1}{r_{ij}}}_{H'} \tag{6.13}$$

with $H_i = H_i^0 + H_i'$. Schrödinger equation $H_i\xi_{n,m}^0(r_i,\phi_i) = E_{i(n,m)}\xi_{n,m}^0(r_i,\phi_i)$ for 2D systems in magnetic field (symmetric gauge, $A_i = \frac{1}{2}(\vec{B}\times\vec{r}_i) = \frac{1}{2}B(-y_i, x_i, 0)$) can be cast as:

$$\left\{-\frac{1}{2}\left[r_i^{-\frac{1}{2}}\frac{d^2}{dr_i^2}r_i^{\frac{1}{2}} + \frac{1}{r_i^2}\left(\frac{\partial^2}{\partial\phi_i^2} + \frac{1}{4}\right) + i\omega_c\frac{\partial}{\partial\phi_i}\right] + \frac{1}{2}\left(m_e^2\omega_0^2 + \frac{1}{4}\omega_c^2\right)r_i^2\right\}\xi_{n,m}(r_i,\phi_i) = m_e E_{i(n,m)}\xi_{n,m}^0(r_i,\phi_i) \tag{6.14}$$

with $\xi_{n,m}^0(r_i,\phi_i) = \frac{e^{im\phi_i}}{\sqrt{2\pi}}\frac{u(r_i)}{\sqrt{r_i}}$, $m = 0, \pm1, \pm2, \ldots$ where $u(r_i)$ is the radial part of the wavefunction. Schrödinger equation for every ith coordinate ($H^0\xi_{n,m}^0 = E_{n,m}^0\xi_{n,m}^0$) can be written as [19–21]:

$$\left\{\frac{d^2}{dr^2} + \left(\frac{1}{4} - m^2\right)\frac{1}{r^2} - m\omega_c - \Omega^2 r^2 + 2m_e E_{n,m}^0\right\}u(r) = 0 \tag{6.15}$$

Transformations $u(r) \Longrightarrow U(z)$ with $z = \Omega r^2$, $\Omega = \sqrt{m_e^2\omega_0^2 + \frac{\omega_c^2}{4}}$ and then $U(z) = z^{-\frac{1}{4}}M(z)$ in Eq. (6.15) lead to:

$$M''(z) + \left\{\frac{\left(\frac{1}{4} - \frac{m^2}{4}\right)}{z^2} - \frac{1}{4} + \frac{\lambda'}{z}\right\}M(z) = 0 \tag{6.16}$$

where $\lambda' = \left(2m_e E_{n,m}^0 - m\omega_c\right)/4\Omega$ and $M(z)$ is Whittaker-M function. Mapping Eq. (6.16) with Whittaker-M equation gives $m_e E_{n,m}^0 = (2n + |m| + 1)\Omega + \frac{1}{2}m\omega_c$ and $\xi_{n,m}^0(z,\phi) = N_{n,|m|}z^{-\left(\frac{1}{2}\right)} M_{\kappa,\nu}(z)\frac{e^{im\phi}}{\sqrt{2\pi}}$ as the ith energy eigenvalue and corresponding eigenfunction, respectively, where $N_{n,|m|} = \sqrt{\left(\frac{2\Omega(n+|m|)!}{n!|m|!^2}\right)}$, $\kappa = \nu + n + \frac{1}{2}$ and $2\nu = |m|$ [19–21, 103].

6.2.1.2.1 Anisotropy (H_i')

The anisotropic term (H_i') is evaluated as [21]:

$$I' = \left\langle \xi_{n_1,m_1}(z,\phi)\left|H_i'\right|\xi_{n_2,m_2}(z,\phi)\right\rangle \tag{6.17}$$

Isolating the radial and the angular part:

$$I' = \frac{1}{2}m_i(\omega_y^2 - \omega_0^2)\underbrace{\left\langle N_{n_1,m_1}z^{-1/2}M_{k_1,v_1}(z)\left|\frac{z}{\Omega}\right|N_{n_2,m_2}z^{-1/2}M_{k_2,v_2}(z)\right\rangle}_{I_1'}\underbrace{\left\langle\frac{e^{im_1\theta}}{\sqrt{2\pi}}\left|\sin^2\theta\right|\frac{e^{im_2\theta}}{\sqrt{2\pi}}\right\rangle}_{I_2'}$$

where $2v = 0, 1, 2, \ldots; k - v - 1/2 = 0, 1, 2, \ldots$. The solution of the above integral is obtained by using the standard integral (Eq. (6.A.3)). For the above integral (i.e. I_1'), $b = 0$, $a_1 = a_2 = 1$, $A = \frac{1}{2}(a_1 + a_2) = 1$, $\gamma_1 - \frac{1}{2} = \frac{|m_1|}{2}$, $\gamma_2 - \frac{1}{2} = \frac{|m_2|}{2}$, $M = \gamma_1 + \gamma_2 = \frac{|m_1|}{2} + \frac{|m_2|}{2} + 1$, $\varrho = 1$, $\varrho + M = \frac{|m_1|}{2} + \frac{|m_2|}{2} + 2$, $\gamma_1 - k_1 = -n_1$, $\gamma_2 - k_2 = -n_2$. Therefore, the solution to the radial part of integral arises as [21]:

$$I_1' = \frac{1}{\Omega}\sqrt{\frac{(n_1+|m_1|)!}{n_1!|m_1|!^2}}\sqrt{\frac{(n_2+|m_2|)!}{n_2!|m_2|!^2}}\Gamma\left(\frac{|m_1|}{2} + \frac{|m_2|}{2} + 2\right)$$
$$\times\sum_{i=0}^{n_1}\sum_{j=0}^{n_2}\frac{\left(\frac{|m_1|}{2} + \frac{|m_2|}{2} + 2\right)_{i+j}(-n_1)_i(-n_2)_j}{(|m_1|+1)_i(|m_2|+1)_j i!j!}$$

where for any number "n," Pochhammer symbol $(n)_0 = 1$. The diagonal term can be calculated by simply putting $n_1 = n_2 = n$, $m_1 = m_2 = m$ while for the off-diagonal term, $m_1 = m_2 = m$ needs to be substituted. Hence, the complete solution for the anisotropy can be represented as [21]:

$$I' = \frac{1}{2}\frac{m_i(\omega_y^2 - \omega_0^2)}{\Omega}\sqrt{\frac{(n_1+|m_1|)!}{n_1!|m_1|!^2}}\sqrt{\frac{(n_2+|m_2|)!}{n_2!|m_2|!^2}}\Gamma\left(\frac{|m_1|}{2} + \frac{|m_2|}{2} + 2\right)$$
$$\times\sum_{i=0}^{n_1}\sum_{j=0}^{n_2}\frac{\left(\frac{|m_1|}{2} + \frac{|m_2|}{2} + 2\right)_{i+j}(-n_1)_i(-n_2)_j}{(|m_1|+1)_i(|m_2|+1)_j i!j!}\left[\frac{1}{2}\delta_{m_1,m_2} - \frac{1}{4}\left(\delta_{m_1,m_{2+2}} + \delta_{m_{1+2},m_2}\right)\right]$$

6.2.1.2.2 Coulomb Interaction: Monopole, Dipole, and Quadrupole Factors

Monopole Factor The integral of ith monopole factor (Eq. (6.10a)) can be written as $I_m = \left\langle \Xi(Z,\Phi)\left|\frac{1}{|\vec{r_i}|}\right|\Xi(Z,\Phi)\right\rangle$ where $\Xi(Z,\Phi) = \prod_{i=1}^{N}\xi^0(z_i,\phi_i)$. Finally, I_m can be cast as:

$$I_m = \frac{N_{n_1,|m|}N_{n_2,|m|}}{2\sqrt{\Omega}}\left\langle z_i^{-\frac{3}{2}}M_{\kappa_1,v_1}(z_i)M_{\kappa_2,v_2}(z_i)\right\rangle \tag{6.18}$$

where quantum numbers of all $j \neq i$ coordinates and "m" are conserved (Eq. (6.10a)). The diagonal and off-diagonal integrals are then evaluated using the standard integral (Eq. (6.A.3)). In (Eq. (6.18)), $b = 0$, $a_1 = a_2 = 1$, $A = \frac{1}{2}(a_1 + a_2) = 1$, $\gamma_1 - \frac{1}{2} = \gamma_2 - \frac{1}{2} = \frac{|m|}{2}$, $M = \gamma_1 + \gamma_2 = |m| + 1$, $\varrho = -\frac{1}{2}$, $2\gamma = 0, 1, 2, \ldots; \kappa - \gamma - 1/2 = 0, 1, 2, \ldots$ [19–21].

Off-diagonal term:

$$I_m = \sqrt{\left(\Omega\frac{(n_1+|m|)!}{n_1!(|m|!)^2}\frac{(n_2+|m|)!}{n_2!(|m|!)^2}\right)}\Gamma\left(|m| + \frac{1}{2}\right)F_2\left(|m| + \frac{1}{2}; -n_1, -n_2; |m|+1, |m|+1; 1, 1\right)$$

The off-diagonal term can be expressed into single summed form by exploiting Eqs. (6.A.5), (6.A.6):

$$I_m = \sqrt{\left(\Omega\frac{(n_1+|m|)!}{n_1!(|m|!)^2}\frac{(n_2+|m|)!}{n_2!(|m|!)^2}\right)}\Gamma\left(|m| + \frac{1}{2}\right)\sum_{k=0}^{n_1}\frac{(|m| + \frac{1}{2})_k(-n_1)_k(-k + \frac{1}{2})_{n_2}}{(|m|+1)_k(|m|+1)_{n_2}k!} \tag{6.19}$$

Diagonal term:

$$I_m = \Omega^{\frac{1}{2}} \frac{(n+|m|)!}{n!(|m|!)^2} \Gamma\left(|m| + \frac{1}{2}\right) \sum_{k=0}^{n} \frac{(|m| + \frac{1}{2})_k (-n)_k (-k + \frac{1}{2})_n}{(|m|+1)_k (|m|+1)_n k!} \tag{6.20}$$

where for any number "n," Pochhammer symbol $(n)_0 = 1$. The monopole contribution of other coordinates can similarly be extended to N electrons using (Eq. (6.10a)). Alternatively, the same outcome (Eqs. (6.19), (6.20)) can be achieved using $\xi^0_{n,m}(z,\theta,\phi) = N'_{n,|m|} e^{-\frac{z}{2}} z^{\frac{|m|}{2}} L_n^{|m|}(z) \frac{e^{im\phi}}{\sqrt{2\pi}}$ where $N' = \sqrt{\frac{2\Omega n!}{(n+|m|)!}}$ and standard integral of the form (Eq. (6.A.7)).

Dipole Factor First order individual integral of dipole factor (Eq. (6.10b)) can be expressed as [20, 21]:

$$I_d = \frac{1}{2} \left\langle \Xi(Z,\Phi) \left| \frac{\vec{r_j} . \vec{r_i}}{|\vec{r_i}|^3} \right| \Xi(Z,\Phi) \right\rangle \tag{6.21}$$

The discretized closer relation over the complete orthonormal basis of the form $\sum_{l=0} |\Xi_l(Z,\Phi)\rangle\langle\Xi_l(Z,\Phi)|$ is inserted to evaluate the integral, leading to [20, 21, 104]:

$$I_d = \frac{1}{2} \sum_{l=0} \left\langle \Xi(Z,\Phi) \left| \frac{r_j \cos(\phi_{ij})}{r_i} \right| \Xi_l(Z,\Phi) \right\rangle \left\langle \Xi_l(Z,\Phi) \left| \frac{1}{r_i} \right| \Xi(Z,\Phi) \right\rangle$$

Unwinding I_d into individual ith, jth, and lth spaces as [20, 21]:

$$I_d = \frac{1}{2} \sum_{l=0} \left\{ I_{d_1} I_{d_2} I_{d_3} \right\} \times I_{d_4} \tag{6.22}$$

where,

$$I_{d_1} = \left\langle N_{n_{1,j} \dots} z_j^{-\frac{1}{2}} M_{\kappa_{1,j};\nu_{1,j}} \left| r_j \right| N_{n_{2,l_j} \dots} z_{l_j}^{-\frac{1}{2}} M_{\kappa_{2,l_j};\nu_{2,l_j}} \right\rangle, I_{d_2} = \left\langle N_{n_{1,i} \dots} z_i^{-\frac{1}{2}} M_{\kappa_{1,i};\nu_{1,i}} \left| \frac{1}{r_i} \right| N_{n_{2,l_i} \dots} z_{l_i}^{-\frac{1}{2}} M_{\kappa_{2,l_i};\nu_{2,l_i}} \right\rangle,$$

$$I_{d_3} = \left\langle \frac{e^{im_{1,i}\phi_i}}{\sqrt{2\pi}} \frac{e^{im_{1,j}\phi_j}}{\sqrt{2\pi}} \left| \cos(\phi_i - \phi_j) \right| \frac{e^{im_{2,l_i}\phi_{l_i}}}{\sqrt{2\pi}} \frac{e^{im_{2,l_j}\phi_{l_j}}}{\sqrt{2\pi}} \right\rangle I_{d_4} = \left\langle \xi^0_{n_{1,l_i};m_{1,l_i}} \left| \frac{1}{r_i} \right| \xi^0_{n_{2,i};m_{2,i}} \right\rangle \delta_{(m_{1,l_i};m_{2,i})}$$

I_d is composed of Coulomb-type (I_{d_2}, I_{d_4}), dipole-type (I_{d_1}) and angular momentum type (I_{d_3}) integrals. Coulomb-type integral (I_{d_2}, I_{d_4}) can be evaluated similarly as the monopole factor (Section 6.2.1.2.2). The dipole-type integral, $I_{d_1}(n_1, |m_1|; n_2, |m_2|)$ for every z_j coordinate can be cast as:

$$I_{d_1} = \frac{N_{n_1;|m_1|} N_{n_2;|m_2|}}{2\Omega^{\frac{3}{2}}} \left\langle z^{-\frac{1}{2}} M_{\kappa_1,\nu_1}(z) M_{\kappa_2,\nu_2}(z) \right\rangle$$

The above integral can be solved by using the standard integral (Eq. (6.A.3)) where $b = 0, a_1 = a_2 = 1, A = \frac{1}{2}(a_1 + a_2) = 1, \gamma_1 - \frac{1}{2} = \nu_1 = \frac{|m_1|}{2}, \gamma_2 - \frac{1}{2} = \nu_2 = \frac{|m_2|}{2}, M = \gamma_1 + \gamma_2 = \frac{|m_1|}{2} + \frac{|m_2|}{2} + 1, \varrho = \frac{1}{2}, 2\gamma = 0, 1, 2, \dots; \kappa - \gamma - 1/2 = 0, 1, 2, \dots$ and then exploiting F_2 to ${}_2F_1$ relation and Chu–Vandermonde identity as [Eqs. (6.A.5), (6.A.6)) [21]:

$$I_{d_1} = \sqrt{\left(\frac{1}{\Omega} \frac{(n_1 + |m_1|)!}{n_1!(|m_1|!)^2} \frac{(n_2 + |m_2|)!}{n_2!(|m_2|!)^2}\right)} \Gamma\left(\frac{|m_1| + |m_2| + 3}{2}\right)$$
$$\times \sum_{k=0}^{n_1} \frac{\left(\frac{|m_1|+|m_2|+3}{2}\right)_k (-n_1)_k \left(\frac{|m_2|-|m_1|-1}{2}\right)_{n_2}}{(|m_1|+1)_k (|m_2|+1)_{n_2} k!} \tag{6.23}$$

Now, evaluating angular momentum (I_{d_3}) type integral [21]:

$$I_{d_3} = \frac{1}{2}\left(\delta_{m_{(1,i)}+1,m_{(2,l_i)}} \times \delta_{m_{(1,j)},m_{(2,l_j)}+1} + \delta_{m_{(1,i)},m_{(2,l_i)}+1} \times \delta_{m_{(1,j)}+1,m_{(2,l_j)}}\right) \tag{6.24}$$

Substituting (Eqs. (6.23), (6.24)) and Coulomb-type integrals (I_{d_2}, I_{d_4}) in (Eq. (6.22)) to obtain dipole factor, which can be extended to N electrons using (Eq. (6.10b)).

Quadrupole Factor The integral representation of second-order quadrupole contribution (order of pole, $\mathcal{P}$=2) (Eq. (6.10c)) reads as [20, 21]:

$$I_q = \left\langle \Xi(Z,\Phi) \left| \frac{3(\vec{r}_j.\vec{r}_i)^2}{|\vec{r}_i|^5} - \frac{\vec{r}_j^{\,2}}{|\vec{r}_i|^3} \right| \Xi(Z,\Phi) \right\rangle = \left\langle \Xi(Z,\Phi) \left| \frac{3\vec{r}_j^{\,2}\cos^2(\phi_{ij})}{|\vec{r}_i|^3} - \frac{\vec{r}_j^{\,2}}{|\vec{r}_i|^3} \right| \Xi(Z,\Phi) \right\rangle$$

It further reduces to [20, 21]:

$$I_q = \left\langle \Xi(Z,\Phi) \left| \frac{r_j^2}{r_i^3}(3\cos^2(\phi_{ij}) - 1) \right| \Xi(Z,\Phi) \right\rangle$$

Alike dipole factor, the integral can be written by inserting discretized closer relation of the form $\sum_{l=0} |\Xi_l(Z,\Phi)\rangle\langle\Xi_l(Z,\Phi)|$ [104]. Splitting "I_q" to ith, jth, and lth coordinates leads to [20, 21]:

$$I_q = \sum_{l=0} \left\{ I_{q_1} I_{q_2} I_{q_3} \right\} \times I_{q_4} \tag{6.25}$$

where,

$$I_{q_1} = \left\langle N_{n_{1,j}}...z_j^{-\frac{1}{2}} M_{\kappa_{1,j};\nu_{1,j}} \left| r_j^2 \right| N_{n_{2,l_j}}...z_{l_j}^{-\frac{1}{2}} M_{\kappa_{2,l_j};\nu_{2,l_j}} \right\rangle, I_{q_2} = \left\langle N_{n_{1,i}}..z_i^{-\frac{1}{2}} M_{\kappa_{1,i};\nu_{1,i}} \left| \frac{1}{r_i} \right| N_{n_{2,l_i}}...z_{l_i}^{-\frac{1}{2}} M_{\kappa_{2,l_i};\nu_{2,l_i}} \right\rangle,$$

$$I_{q_3} = \left\langle \frac{e^{im_{1,i}\phi_i}}{\sqrt{2\pi}} \frac{e^{im_{1,j}\phi_j}}{\sqrt{2\pi}} \left| 3\cos^2(\phi_i - \phi_j) - 1 \right| \frac{e^{im_{2,l_i}\phi_{l_i}}}{\sqrt{2\pi}} \frac{e^{im_{2,l_j}\phi_{l_j}}}{\sqrt{2\pi}} \right\rangle, I_{q_4} = \left\langle \xi^0_{n_{1,l_i};m_{1,l_i}} \left| \frac{1}{r_i^2} \right| \xi^0_{n_{2,i};m_{2,i}} \right\rangle$$

Here, Eq. (6.25) is composed of angular momentum type (I_{q_3}), Coulomb-type (I_{q_2}, I_{q_4}) and square-position-type integrals (I_{q_1}). Coulomb-type integral can be calculated similar to monopole factor (Section 6.2.1.2.2). Uncoiling I_{q_4} and inserting discretized projection operator of the form $\sum_{k=0} |\Xi_k(Z,\Phi)\rangle\langle\Xi_k(Z,\Phi)|$ [20, 21]:

$$I_{q_4} = \sum_{k=0} \left\langle \xi^0_{n_{1,l_i};m_{1,l_i}} \left| \frac{1}{r_i} \right| \xi^0_{n_{2,k_i};m_{2,k_i}} \right\rangle \delta_{(m_{1,l_i};m_{2,k_i})} \left\langle \xi^0_{n_{1,k_i};m_{1,k_i}} \left| \frac{1}{r_i} \right| \xi^0_{n_{2,i};m_{2,i}} \right\rangle \delta_{(m_{1,k_i};m_{2,i})}$$

The square-position-type integral ($I_{q_1}(n_1, |m_1|; n_2, |m_2|)$) for every z_j coordinate can be represented as [20, 21]:

$$I_{q_1} = \frac{N_{n_1;|m_1|} N_{n_2;|m_2|}}{2\Omega^2} \left\langle z^0 M_{\kappa_1,\nu_1}(z) M_{\kappa_2,\nu_2}(z) \right\rangle$$

Using the standard integral (Eq. (6.A.3)) where $b = 0$, $a_1 = a_2 = 1$, $A = \frac{1}{2}(a_1 + a_2) = 1$, $\gamma_1 - \frac{1}{2} = \frac{|m_1|}{2}$, $\gamma_2 - \frac{1}{2} = \frac{|m_2|}{2}$, $M = \gamma_1 + \gamma_2 = \frac{|m_1|}{2} + \frac{|m_2|}{2} + 1$, $\varrho = 1$, $2\gamma = 0, 1, 2, ...$; $\kappa - \gamma - 1/2 = 0, 1, 2, ...$, and subsequently using F_2 to ${}_2F_1$ relation and Chu–Vandermonde identity (Eqs. (6.A.5), (6.A.6)), the integral leads to single summation form as [20, 21]:

$$I_{q_1} = \frac{1}{\Omega}\sqrt{\left(\frac{(n_1+|m_1|)!}{n_1!(|m_1|!)^2}\frac{(n_2+|m_2|)!}{n_2!(|m_2|!)^2}\right)}\Gamma\left(\frac{|m_1|+|m_2|+4}{2}\right)$$
$$\times\sum_{k=0}^{n_1}\frac{\left(\frac{|m_1|+|m_2|+3}{2}\right)_k(-n_1)_k\left(\frac{|m_2|-|m_1|-2}{2}\right)_{n_2}}{(|m_1|+1)_k(|m_2|+1)_{n_2}k!} \tag{6.26}$$

Now, analyzing angular momentum (I_{q_3}) type integral as [20, 21]:

$$I_{q_3} = \frac{3}{4}\left\{\delta_{m_{(1,i)},m_{(2,l_i)}-2}\times\delta_{m_{(1,j)},m_{(2,l_j)}+2}+\delta_{m_{(1,i)},m_{(2,l_i)}+2}\times\delta_{m_{(1,j)},m_{(2,l_j)}-2}\right.$$
$$\left.+2\left(\delta_{m_{(1,i)},m_{(2,l_i)}}\times\delta_{m_{(1,j)},m_{(2,l_j)}}\right)\right\}-\delta_{m_{(1,i)},m_{(2,l_i)}}\times\delta_{m_{(1,j)},m_{(2,l_j)}} \tag{6.27}$$

Substituting Eqs. (6.26), (6.27) and Coulomb-type integrals (I_{q_2}, I_{q_4}) in Eq. (6.25) to obtain quadrupole contribution factor which can be extended to N electrons using Eq. (6.10c).

6.2.1.3 3D Exciton Model

The Hamiltonian for e–h pair of 3D QD confined in harmonic potential (ω_o) with transverse magnetic field (ω_c) can be written as [22, 66]:

$$H = \frac{1}{2m_e}(p_e+A_e)^2+\frac{1}{2}m_e\omega_e^2r_e^2+\frac{1}{2m_h}(p_h-A_h)^2+\frac{1}{2}m_h\omega_h^2r_h^2-\frac{1}{\epsilon|\vec{r}_e-\vec{r}_h|} \tag{6.28}$$

Introduction of the centre-of-mass and relative coordinates $\vec{R}=\frac{(\vec{r}_e+\alpha\vec{r}_h)}{\alpha+1}$, $\vec{r}=\vec{r}_e-\vec{r}_h$ where $\alpha = m_h/m_e$ and ϵ is dielectric constant of the medium and applying symmetric gauge for both electron and hole as $A_e=-A_h=A_r=\frac{1}{2}\vec{B}\times\left(\vec{r}_e-\vec{r}_h\right)$ transform the Hamiltonian to:

$$\begin{aligned}H =&\frac{(1+\alpha)}{\alpha}\left[\frac{1}{2m_e}\left(p^2+\frac{A_r^2}{c^2}+\frac{(\alpha-1)}{(\alpha+1)c}(\vec{p}.\vec{A}_r+\vec{A}_r.\vec{p})\right)\frac{1}{2}m_e\omega_r^2r^2-\frac{\alpha}{1+\alpha}\frac{1}{\epsilon r}\right]\\&-\frac{2}{m_ec(1+\alpha)}P.A_r+\frac{1}{1+\alpha}\left(\frac{P^2}{2m_e}+\frac{1}{2}m_e\omega_R^2R^2\right)+\frac{\alpha m_e}{1+\alpha}\left(\omega_e^2-\omega_h^2\right)\vec{R}.\vec{r}\\=&\frac{(1+\alpha)}{\alpha}\left(H_r-\frac{\alpha}{1+\alpha}\frac{e^2}{\epsilon r}\right)+\frac{1}{1+\alpha}H_R^0+H_x\end{aligned}$$

where $H_x=\frac{\alpha m_e}{1+\alpha}\left(\omega_e^2-\omega_h^2\right)\vec{R}.\vec{r}-\frac{2e}{m_ec(1+\alpha)}\vec{P}.\vec{A}_r$, $\omega_r^2=\frac{\alpha^2}{(1+\alpha)^3}(\alpha\omega_e^2+\omega_h^2)$ and $\omega_R^2=(1+\alpha)(\alpha\omega_h^2+\omega_e^2)$.
The Hamiltonian $H_r=H_r^0+H_r'$ in spherical polar coordinates can be cast as:

$$\left\{-\frac{1}{2}\left(r^{-1}\frac{d^2}{dr^2}r-\frac{L^2}{r^2}+i\left(\frac{\alpha-1}{\alpha+1}\right)\omega_{c\_r}\frac{\partial}{\partial\phi}\right)+\frac{r^2}{2}\left(m_e^2\omega_r^2+\frac{\omega_{c\_r}^2}{4}\right)-\frac{1}{8}\omega_{c\_r}^2r^2\cos^2\theta\right\}\xi_{n,l,m}(r,\theta,\phi)$$
$$= m_eE_{n,l,m}\xi_{n,l,m}(r,\theta,\phi) \tag{6.29}$$

where $w_{c\_r}=\frac{eB}{c}$, $\Omega_r=\sqrt{m_e^2\omega_r^2+\frac{\omega_{c\_r}^2}{4}}$ and anisotropy, $H_r'=-\frac{1}{8}\omega_{c\_r}^2r^2\cos^2\theta$. The complete Hamiltonian can be separated into two parts namely unperturbed ($H_r^0+H_R^0$) and perturbed ($H'=-\frac{1}{\epsilon.r}+H_x+H_r'$). Introducing spherical harmonics, ansatz wavefunction becomes $\xi_{n,l,m}(r,\theta,\phi)=\overline{Y}_l^m(\theta,\phi)\frac{u(r)}{r}$, $m=0,\ \pm1,\pm\ 2,\ldots$ where $u(r)$ is a radial part and $\overline{Y}_l^m(\theta,\phi)=\sqrt{\frac{(2n+1)(l-|m|)!}{4\pi(l+|m|)!}}P_l^{|m|}(\theta)e^{im\phi}$ is angular part. Eigenvalues of H_r^0 and H_R^0 are $E_{n,l,m}^0$ and $E_{\mathcal{N},\mathcal{L},\mathcal{M}}^0$, respectively. Now, solving $H_r^0\xi_{n,l,m}=E_{n,l,m}^0\xi_{n,l,m}$ results in the following radial equation [22]:

$$\left[\frac{d^2}{dr^2}-\frac{l(l+1)}{r^2}-m\left(\frac{\alpha-1}{\alpha+1}\right)\omega_{c\_r}-\Omega^2r^2+2m_eE_{n,l,m}^0\right]u(r)=0 \tag{6.30}$$

Substitution $x = \Omega_r r^2$ with $u(r) \Longrightarrow U(x)$ transforms the above equation to:

$$\left[4x\frac{d^2}{dx^2} + 2\frac{d}{dx} - \frac{l(l+1)}{x} - m\left(\frac{\alpha-1}{\alpha+1}\right)\frac{m\omega_{c\_r}}{\Omega_r} - x + \frac{2m_e E^0_{n,l,m}}{\Omega_r}\right] U(x) = 0 \tag{6.31}$$

and finally making transformation $U(x) = x^{-\frac{1}{4}} M(x)$, it attributes to:

$$M''(x) + \left\{\frac{\left(\frac{3}{16} - \frac{l(l+1)}{4}\right)}{x^2} - \frac{1}{4} + \frac{\lambda'}{x}\right\} M(x) = 0 \tag{6.32}$$

where $M(x)$ is Whittaker-M function and $\lambda' = \frac{\left(2m_e E^0_{n,l,m} - m\left(\frac{\alpha-1}{\alpha+1}\right)\omega_{c\_r}\right)}{4\Omega_r}$. Comparing with Whittaker-M equation yields the corresponding eigenvalue as $m_e E^0_{n,l,m} = (2n + l + \frac{3}{2})\Omega_r + \frac{m}{2}\left(\frac{\alpha-1}{\alpha+1}\right)\omega_{c\_r}$, the normalized eigenfunction $\xi_{n,l,m}(x, \theta, \phi) = N_{n,l} x^{-\left(\frac{1}{4}\right)} M_{\kappa,\nu}(x) Y_l^m(\theta, \phi)$ where $N_{n,l} = \sqrt{\left(\frac{2\Omega_r^{\frac{3}{2}}(n+l+\frac{1}{2})!}{n!(l+\frac{1}{2})!^2}\right)}$, $\kappa = \nu + n + \frac{1}{2}$ and $l + \frac{1}{2} = 2\nu$ [22, 103]. The H_R^0 can be solved similarly with $E^0_{\mathcal{N},\mathcal{L},\mathcal{M}} = (2\mathcal{N} + \mathcal{L} + \frac{3}{2})\omega_R$.

Calculating integrals of $\langle n_1, l_1, m_1; \mathcal{N}_1, \mathcal{L}_1, \mathcal{M}_1 | H_x | n_2, l_2, m_2; \mathcal{N}_2, \mathcal{L}_2, \mathcal{M}_2 \rangle$ as [22]:

$$\begin{aligned}
I_1 &= \frac{\alpha m_e}{(1+\alpha)}\left(\omega_e^2 - \omega_h^2\right)\left\langle \overline{\Psi}_{n_1,l_1,m_1;\mathcal{N}_1,\mathcal{L}_1,\mathcal{M}_1} \left| \vec{R}.\vec{r} \right| \overline{\Psi}_{n_2,l_2,m_2;\mathcal{N}_2,\mathcal{L}_2,\mathcal{M}_2} \right\rangle \\
&= \frac{\alpha m_e}{(1+\alpha)}\left(\omega_e^2 - \omega_h^2\right) \langle n_1, l_1 \rangle | r | n_2, l_2 \ \langle \mathcal{N}_1, \mathcal{L}_1 | R | \mathcal{N}_2, \mathcal{L}_2 \ \rangle \\
&\quad \left\{ \left\langle \overline{Y}_{l_1}^{m_1} \overline{Y}_{\mathcal{L}_1}^{\mathcal{M}_1} \left| \frac{\sin\Theta \sin\theta e^{i(\Phi-\phi)}}{2} \right| \overline{Y}_{l_2}^{m_2} \overline{Y}_{\mathcal{L}_2}^{\mathcal{M}_2} \right\rangle \right. \\
&\quad \left. + \left\langle \overline{Y}_{l_1}^{m_1} \overline{Y}_{\mathcal{L}_1}^{\mathcal{M}_1} \left| \frac{\sin\Theta \sin\theta e^{-i(\Phi-\phi)}}{2} \right| \overline{Y}_{l_2}^{m_2} \overline{Y}_{\mathcal{L}_2}^{\mathcal{M}_2} \right\rangle + \left\langle \overline{Y}_{l_1}^{m_1} \overline{Y}_{\mathcal{L}_1}^{\mathcal{M}_1} \left| \cos\Theta \cos\theta \right| \overline{Y}_{l_2}^{m_2} \overline{Y}_{\mathcal{L}_2}^{\mathcal{M}_2} \right\rangle \right\}
\end{aligned}$$

I_1 can be evaluated using standard integral (Eq. (6.A.3)). For the above integral $b = 0$, $a_1 = a_2 = 1$, $A = \frac{1}{2}(a_1 + a_2) = 1$, $\gamma_1 - \frac{1}{2} = \frac{l_1}{2} + \frac{1}{4}$, $\gamma_2 - \frac{1}{2} = \frac{l_2}{2} + \frac{1}{4}$, $M = \gamma_1 + \gamma_2 = \frac{l_1}{2} + \frac{l_2}{2} + \frac{3}{2}$, $\varrho = \frac{1}{2}$ and $\varrho + M = \frac{l_1}{2} + \frac{l_2}{2} + 2$. Therefore, the radial factors of the integral are reshaped as [22]:

$$\begin{aligned}
\langle n_1, l_1 | r | n_2, l_2 \rangle &= \frac{1}{\Omega_r^{\frac{1}{2}}} \sqrt{\left(\frac{(n_1 + l_1 + \frac{1}{2})!}{n_1!(l_1 + \frac{1}{2})!^2} \frac{(n_2 + l_2 + \frac{1}{2})!}{n_2!(l_2 + \frac{1}{2})!^2}\right)} \Gamma\left(\frac{l_1}{2} + \frac{l_2}{2} + 2\right) \\
&\quad \times \sum_{i=0}^{n_1} \sum_{j=0}^{n_2} \frac{(\frac{l_1}{2} + \frac{l_2}{2} + 2)_{i+j}(-n_1)_i(-n_2)_j}{(l_1 + \frac{3}{2})_i (l_2 + \frac{3}{2})_j i! j!}
\end{aligned}$$

where $(n)_0 = 1$ and for angular part we make use of the recurrence relations [22]:

$$\begin{aligned}
\left\langle \overline{Y}_{\mathcal{L}_1}^{\mathcal{M}_1} \left| sin\Theta e^{i\Phi} \right| \overline{Y}_{\mathcal{L}_2}^{\mathcal{M}_2} \right\rangle &= -\left[\frac{(\mathcal{L}_2 + |\mathcal{M}_2| + 1)(\mathcal{L}_2 + |\mathcal{M}_2| + 2)}{(2\mathcal{L}_2 + 1)(2\mathcal{L}_2 + 3)}\right]^{\frac{1}{2}} \delta_{\mathcal{M}_1,\mathcal{M}_2+1} \delta_{\mathcal{L}_1,\mathcal{L}_2+1} \\
&\quad + \left[\frac{(\mathcal{L}_2 - |\mathcal{M}_2|)(\mathcal{L}_2 - |\mathcal{M}_2| - 1)}{(2\mathcal{L}_2 + 1)(2\mathcal{L}_2 + 3)}\right]^{\frac{1}{2}} \delta_{\mathcal{M}_1,\mathcal{M}_2+1} \delta_{\mathcal{L}_1,\mathcal{L}_2-1}
\end{aligned}$$

$$\begin{aligned}
\left\langle \overline{Y}_{\mathcal{L}_1}^{\mathcal{M}_1} \left| \sin\Theta e^{-i\Phi} \right| \overline{Y}_{\mathcal{L}_2}^{\mathcal{M}_2} \right\rangle &= \left[\frac{(\mathcal{L}_2 - |\mathcal{M}_2| + 1)(\mathcal{L}_2 - |\mathcal{M}_2| + 2)}{(2\mathcal{L}_2 + 1)(2\mathcal{L}_2 + 3)}\right]^{\frac{1}{2}} \delta_{\mathcal{M}_1,\mathcal{M}_2-1} \delta_{\mathcal{L}_1,\mathcal{L}_2+1} \\
&\quad - \left[\frac{(\mathcal{L}_2 + |\mathcal{M}_2|)(\mathcal{L}_2 + |\mathcal{M}_2| - 1)}{(2\mathcal{L}_2 + 1)(2\mathcal{L}_2 + 3)}\right]^{\frac{1}{2}} \delta_{\mathcal{M}_1,\mathcal{M}_2-1} \delta_{\mathcal{L}_1,\mathcal{L}_2-1}
\end{aligned}$$

$$\left\langle \overline{Y}_{\mathcal{L}_1}^{\mathcal{M}_1} \left| \cos\Theta \right| \overline{Y}_{\mathcal{L}_2}^{\mathcal{M}_2} \right\rangle = -\left[\frac{(\mathcal{L}_2 - |\mathcal{M}_2| + 1)(\mathcal{L}_2 + |\mathcal{M}_2| + 1)}{(2\mathcal{L}_2 + 1)(2\mathcal{L}_2 + 3)}\right]^{\frac{1}{2}} \delta_{\mathcal{M}_1,\mathcal{M}_2}\delta_{\mathcal{L}_1,\mathcal{L}_2+1} + \left[\frac{(\mathcal{L}_2 - |\mathcal{M}_2|)(\mathcal{L}_2 + |\mathcal{M}_2|)}{(2\mathcal{L}_2 - 1)(2\mathcal{L}_2 + 1)}\right]^{\frac{1}{2}} \delta_{\mathcal{M}_1,\mathcal{M}_2}\delta_{\mathcal{L}_1,\mathcal{L}_2-1}$$

Center-of-mass coordinates have identical eigen solutions with those of a one particle system.

6.2.1.3.1 *Coulomb Interaction: Monopole Factor*

Unwinding the Coulomb interaction for e–h problem, the integral can be cast as ($\mathcal{N}$, $\mathcal{L}$ and $\mathcal{M}$ conserved) (Eq. (6.10a)) [22]:

$$I = N_{n_1,l_1,|m|_1} N_{n_2,|m|_2,l_2} \left\langle M_{\kappa_1,\nu_1} \left| \frac{x^{-\frac{3}{2}}}{2\Omega_r \epsilon} \right| M_{\kappa_2,\nu_2} \right\rangle \delta_{l_1,l_2}\delta_{m_1,m_2} \tag{6.33}$$

Diagonal and off-diagonal matrix elements are elucidated using standard integral (6.A.3) where $b = 0$, $a_1 = a_2 = 1$, $A = \frac{1}{2}(a_1 + a_2) = 1$, $\gamma_1 - \frac{1}{2} = \frac{l_1}{2} + \frac{1}{4}$, $\gamma_2 - \frac{1}{2} = \frac{l_2}{2} + \frac{1}{4}$, $M = \gamma_1 + \gamma_2 = \frac{l_1}{2} + \frac{l_2}{2} + \frac{3}{2}$, $\varrho = -\frac{1}{2}$ and $\varrho + M = \frac{l_1}{2} + \frac{l_2}{2} + 1$ [22].

Diagonal ($n_1 = n_2 = n; l_1 = l_2 = l; m_1 = m_2 = m$):

$$I = \frac{\Omega_r^{\frac{1}{2}}}{\epsilon} \frac{(n + l + \frac{1}{2})!}{n!(l + \frac{1}{2})!^2} \Gamma(l + 1) \sum_{i=0}^{n}\sum_{j=0}^{n} \frac{(l + 1)_{i+j}(-n)_i(-n)_j}{(l + \frac{3}{2})_i (l + \frac{3}{2})_j i! j!} \tag{6.34}$$

Off-diagonal ($l_1 = l_2 = l; m_1 = m_2 = m$):

$$I = \frac{1}{\epsilon}\sqrt{\left(\frac{\Omega_r(n_1 + l + \frac{1}{2})!}{n_1!(l + \frac{1}{2})!^2} \frac{(n_2 + l + \frac{1}{2})!}{n_2!(l + \frac{1}{2})!^2}\right)} \Gamma(l + 1) \sum_{i=0}^{n_1}\sum_{j=0}^{n_2} \frac{(l + 1)_{i+j}(-n_1)_i(-n_2)_j}{(l + \frac{3}{2})_i (l + \frac{3}{2})_j i! j!} \tag{6.35}$$

6.2.1.4 2D Exciton Model

For an anisotropic 2D excitonic QD system confined by harmonic potential with transverse magnetic field (ω_c) and dielectric constant (ϵ) (in a.u.), the spin-free model Hamiltonian can be written [24]:

$$H = \sum_{i=1}^{N} \left\{ \frac{1}{2m_i}(p_i - A(r_i))^2 + \frac{1}{2} m_i \omega_{x_i}^2 x_i^2 + \frac{1}{2} m_i \omega_{y_i}^2 y_i^2 \right\} + \frac{1}{\epsilon}\left(\frac{1}{2}\sum_{i=1}^{N}\sum_{j\neq i}^{N} \frac{1}{r_{ij}}\right) \tag{6.36}$$

where m_i is the mass (a.u.) on the ith charge carrier in the excitonic system. The model Hamiltonian for a 2D axial QD, under symmetric gauge, $A(r_i) = \frac{1}{2}(\vec{B} \times \vec{r_i})$, and anisotropic confinement conditions, $\omega_{x_i} = \omega_o$ and $\omega_{y_i} = \omega_1$ (where anisotropy factor, $\alpha = \frac{\omega_0}{\omega_1}$) is transformed (in polar coordinates) into:

$$H = \sum_{i=1}^{N} \left\{ \frac{1}{2m_i}\left(p_i - \frac{1}{c}A(r_i)\right)^2 + \frac{1}{2} m_i \omega_o^2 r_i^2 + \frac{1}{2} m_i \left(\omega_1^2 - \omega_o^2\right) y_i^2 \right\} + \frac{1}{\epsilon}\left(\frac{1}{2}\sum_{i=1}^{N}\sum_{j\neq i}^{N} \frac{1}{r_{ij}}\right) \tag{6.37}$$

The Hamiltonian gets factorized into the following form:

$$H = \sum_{i=1}^{N} H_i^0 + \sum_{i=1}^{N} H_i' + \sum_{i=1}^{N}\sum_{j\neq i}^{N} H_{ij} \tag{6.38}$$

H_i' accounts for the anisotropy while H_i^0 describes an isotropic 2D excitonic system. The most vital term H_{ij}, represents the Coulomb interactions amongst the charge carriers (e–h, e–e, h–h) in the excitonic system. The eigensolutions for the unperturbed Hamiltonian (H_i^0) are obtained as [23, 24, 103]:

$$H_i^0 \xi_{n,m}(r_i,\theta_i) = \mathrm{E}^0_{i(n,m)} \xi_{n,m}(r_i,\theta_i)$$

$$\left[-\frac{1}{2}\left\{r_i^{-\frac{1}{2}}\frac{d^2}{dr_i^2}r_i^{\frac{1}{2}} + \frac{1}{r_i^2}\left(\frac{d^2}{d\theta_i^2}+\frac{1}{4}\right)\right\} + \frac{i\omega_c}{2}\frac{\partial}{\partial\theta_i} + \frac{r_i^2}{2}\left(m_i^2\omega_r^2 + \frac{\omega_c^2}{4}\right)\right]\xi_{n,m}(r_i,\theta_i) = m_i E^0_{n,m}\xi_{n,m}(r_i,\theta_i) \tag{6.39}$$

where ω_c is defined as $\frac{eB}{c}$. Determining the eigensolution for a particular ith charge carrier using the ansatz wavefunction, $\xi_{n,m}(r,\theta) = \frac{u(r)}{\sqrt{r}}\frac{e^{im\theta}}{\sqrt{2\pi}}$, where $u(r)$ is the radial part and $e^{im\theta}$ is the angular part having $m = 0, \pm1, \pm2$ and $\Omega = \sqrt{\left(m_i^2\omega_r^2 + \frac{\omega_c^2}{4}\right)}$ leads to [23, 24]:

$$\left[\frac{d^2}{dr^2} + \frac{1}{r^2}\left(\frac{1}{4} - m^2\right) \pm m\omega_c - \Omega^2 r^2 + 2m_i E^0_{n,m}\right] u(r) = 0 \tag{6.40}$$

where the sign preceding the term $m\omega_c$ is positive (+) when the ith charge carrier is a hole and negative (−) in case of an electron. Substituting $z = \Omega r^2$, $u(r) \Longrightarrow U(z)$ reforms the above equation as [23, 24]:

$$\left[z\frac{d^2}{dz^2} + \frac{1}{2}\frac{d}{dz} + \frac{1}{4z}\left(\frac{1}{4} - m^2\right) \pm \frac{m\omega_c}{4\Omega} - \frac{z}{4} + \frac{m_i \mathrm{E}^0_{n,m}}{2\Omega}\right] U(z) = 0 \tag{6.41}$$

The above differential equation is then transformed in terms of Whittaker-M function, $M(z)$, using $U(z) = z^{-1/4}M(z)$ [23, 24]:

$$M''(z) + \left\{\frac{\left(\frac{1}{4} - \frac{m^2}{4}\right)}{z^2} - \frac{1}{4} + \frac{\lambda}{z}\right\} M(z) = 0 \tag{6.42}$$

where λ denotes $\frac{2m_i E^0_{n,m} \pm m\omega_c}{4\Omega}$. The quantization condition for ith charge carrier emerges as [23, 24]:

$$\mathrm{E}^0_{n,m} = (2n + |m| + 1)\frac{\Omega}{m_e} + \frac{m\omega_c}{2m_e} \quad \text{(for electron)} \tag{6.43}$$

$$\mathrm{E}^0_{n,m} = (2n + |m| + 1)\frac{\Omega}{m_h} - \frac{m\omega_c}{2m_h} \quad \text{(for hole)} \tag{6.44}$$

The corresponding normalized eigenfunction has the form $\xi_{n,m}(z,\theta) = N_{n,|m|} z^{-\left(\frac{1}{2}\right)} M_{\kappa,\nu}(z)\frac{e^{im\theta}}{\sqrt{2\pi}}$, where $N_{n,|m|} = \sqrt{\left(\frac{2\Omega(n+|m|)!}{n!|m|!^2}\right)}$, $\kappa = \nu + n + \frac{1}{2}$ and $2\nu = |m|$ [23, 24].

6.2.1.4.1 Anisotropy (H_i')

The variational term, H_i', is evaluated as [24]:

$$I' = \left\langle \xi_{n_1,m_1}(z,\theta) | H_i' | \xi_{n_2,m_2}(z,\theta) \right\rangle \tag{6.45}$$

Segregating the radial and angular parts [24]:

$$I' = \frac{1}{2}m_i(\omega_1^2 - \omega_0^2)\underbrace{\left\langle N_{n_1,m_1} z^{-1/2} M_{k_1,\nu_1}(z) \left| \frac{z}{\Omega} \right| N_{n_2,m_2} z^{-1/2} M_{k_2,\nu_2}(z)\right\rangle}_{I_1'} \underbrace{\left\langle \frac{e^{im_1\theta}}{\sqrt{2\pi}} \left| \sin^2\theta \right| \frac{e^{im_2\theta}}{\sqrt{2\pi}}\right\rangle}_{I_2'} \tag{6.46}$$

where $2\nu = 0, 1, 2, \ldots; k - \nu - 1/2 = 0, 1, 2, \ldots$ For the present integral (i.e.$I_1^{'}$), $b = 0$, $a_1 = a_2 = 1$, $A = \frac{1}{2}(a_1 + a_2) = 1$, $\gamma_1 - \frac{1}{2} = \frac{|m_1|}{2}$, $\gamma_2 - \frac{1}{2} = \frac{|m_2|}{2}$, $M = \gamma_1 + \gamma_2 = \frac{|m_1|}{2} + \frac{|m_2|}{2} + 1$, $\rho = 1$, $\rho + M = \frac{|m_1|}{2} + \frac{|m_2|}{2} + 2$, $\gamma_1 - k_1 = -n_1$ and $\gamma_2 - k_2 = -n_2$ using standard integral (Eq. (6.A.3)). The solution to the radial part of integral can be written as [24]:

$$I_1' = \frac{1}{\Omega}\sqrt{\frac{(n_1 + |m_1|)!}{n_1!|m_1|^2!}}\sqrt{\frac{(n_2 + |m_2|)!}{n_2!|m_2|^2!}}\Gamma\left(\frac{|m_1|}{2} + \frac{|m_2|}{2} + 2\right) \times \sum_{i=0}^{n_1}\sum_{j=0}^{n_2}\frac{\left(\frac{|m_1|}{2} + \frac{|m_2|}{2} + 2\right)_{i+j}(-n_1)_i(-n_2)_j}{(|m_1| + 1)_i(|m_2| + 1)_j i!j!} \tag{6.47}$$

where for any number "n," Pochhammer symbol $(n)_0 = 1$. The diagonal term can be evaluated by simply putting $n_1 = n_2 = n$, $m_1 = m_2 = m$ while for the off-diagonal term a substitution of only $m_1 = m_2 = m$ is required. The complete solution corresponding to the variational part due to anisotropy is given by [24]:

$$I' = \frac{1}{2}\frac{m_i(\omega_1^2 - \omega_0^2)}{\Omega}\sqrt{\frac{(n_1 + |m_1|)!}{n_1!|m_1|^2!}}\sqrt{\frac{(n_2 + |m_2|)!}{n_2!|m_2|^2!}}\Gamma\left(\frac{|m_1|}{2} + \frac{|m_2|}{2} + 2\right) \times \sum_{i=0}^{n_1}\sum_{j=0}^{n_2}\frac{\left(\frac{|m_1|}{2} + \frac{|m_2|}{2} + 2\right)_{i+j}(-n_1)_i(-n_2)_j}{(|m_1| + 1)_i(|m_2| + 1)_j i!j!}\left[\frac{1}{2}\delta_{m_1,m_2} - \frac{1}{4}\left(\delta_{m_1,m_{2+2}} + \delta_{m_{1+2},m_2}\right)\right] \tag{6.48}$$

The Coulomb interactions can be formulated using Eqs. (6.10a)–(6.10c) as the mono-, di-, quadru-pole terms, respectively.

6.2.1.4.2 Coulomb Interaction: Monopole, Dipole, and Quadrupole Factors

Monopole Factor Integral for monopole factor can be expanded over the product of excitonic eigen-states given by $\Xi(Z, \Theta) = \prod_{i=1}^{N} \xi(z_i, \theta_i)$ is (Eq. (6.10a)):

$$I_m = \left\langle \Xi(Z, \Theta) \left| \frac{1}{|\vec{r_i}|} \right| \Xi(Z, \Theta) \right\rangle \tag{6.49}$$

Considering the ith term of the monopole factor with quantum numbers "m" for all coordinates $j \neq i$ being conserved:

$$I_m = \frac{N_{n_1,|m_1|}N_{n_2,|m_2|}}{2\sqrt{\Omega}}\left\langle z_i^{-3/2}M_{k_1,\nu_1}(z_i)M_{k_2,\nu_2}(z_i)\right\rangle \delta_{m_1,m_2} \tag{6.50}$$

The integral for I_m is resolved by employing the standard integral (Eq. (6.A.3)). For I_m on comparison with standard integral it is found that $b = 0$, $a_1 = a_2 = 1$, $A = \frac{1}{2}(a_1 + a_2) = 1$, $\gamma_1 - \frac{1}{2} = \frac{|m_1|}{2}$, $\gamma_2 - \frac{1}{2} = \frac{|m_2|}{2}$, $M = \gamma_1 + \gamma_2 = \frac{|m_1|}{2} + \frac{|m_2|}{2} + 1$ and $\rho = -\frac{1}{2}$, $2\nu = 0, 1, 2, \ldots$ and $k - \nu - 1/2 = 0, 1, 2, \ldots$ which leads to [23, 24]:

$$I_m = \Omega^{\frac{1}{2}}\sqrt{\frac{(n_1 + |m_1|)!}{n_1!|m_1|^2!}}\sqrt{\frac{(n_2 + |m_2|)!}{n_2!|m_2|^2!}}\Gamma\left(\frac{|m_1|}{2} + \frac{|m_2|}{2} + \frac{1}{2}\right) \times F_2\left(\frac{|m_1|}{2} + \frac{|m_2|}{2} + \frac{1}{2}; -n_1, n_2; |m_1| + 1, |m_2| + 1; 1, 1\right)\delta_{m_1,m_2}$$

The off-diagonal and diagonal terms are written in a single summation form using F_2 to ${}_2F_1$ relation and Chu–Vandermonde identity (Eqs. (6.A.5), (6.A.6)) given by [23, 24]:

Off-diagonal integral ($m_1 = m_2 = m$):

$$I_m = \Omega^{\frac{1}{2}} \sqrt{\frac{(n_1+|m|)!}{n_1!|m|^2!}} \sqrt{\frac{(n_2+|m|)!}{n_2!|m|^2!}} \Gamma(|m|+\frac{1}{2}) \sum_{k=0}^{n_1} \frac{(|m|+\frac{1}{2})_k (-n_1)_k (-k+\frac{1}{2})_{n_2}}{(|m|+1)_k k! (|m|+1)_{n_2}} \tag{6.51}$$

Diagonal integral ($n_1 = n_2 = n$):

$$I_m = \Omega^{\frac{1}{2}} \frac{(n+|m|)!}{n!|m|^2!} \Gamma(|m|+\frac{1}{2}) \sum_{k=0}^{n} \frac{(|m|+\frac{1}{2})_k (-n)_k (-k+\frac{1}{2})_n}{(|m|+1)_k k! (|m|+1)_n} \tag{6.52}$$

Dipole Factor The dipole factor after the insertion of discretized projection states in terms of closure relations over the complete orthonormal basis defined by $\sum_{l=0} |\Xi_l(Z,\Theta)\rangle\langle\Xi_l(Z,\Theta)|$ as [24]:

$$I_d = \left\langle \Xi(Z,\Theta) \left| \frac{\vec{r}_j . \vec{r}_i}{|\vec{r}_i|^3} \right| \Xi(Z,\Theta) \right\rangle \tag{6.53}$$

The insertion of discretized projection operators modifies the dipole integral as [24]:

$$I_d = \sum_{l=0} \left\langle \Xi(Z,\Theta) \left| \frac{r_j \cos\theta_{ij}}{r_i} \right| \Xi_l(\Xi,\Theta) \right\rangle \left\langle \Xi_l(Z,\Theta) \left| \frac{1}{r_i} \right| \Xi(Z,\Theta) \right\rangle \tag{6.54}$$

Sorting out I_d into radial and angular components yields [24]:

$$I_d = \sum_{l=0} \left\{ I_{d_1} I_{d_2} I_{d_\theta} \right\} \times I_{d_3} \tag{6.55}$$

The final integral is a product of three radial and one angular coordinates dependent integral, which can be evaluated independently. The individual integrals denoted by $I_{d_1}, I_{d_2}, I_{d_3}$ and I_{d_θ} are expressed as [24]:

$$I_{d_1} = \left\langle N_{n_{1,j}\ldots} z_j^{-\frac{1}{2}} M_{\kappa_{1,j};\nu_{1,j}} \left| r_j \right| N_{n_{2,l_j}\ldots} z_{l_j}^{-\frac{1}{2}} M_{\kappa_{2,l_j};\nu_{2,l_j}} \right\rangle, I_{d_2} = \left\langle N_{n_{1,i}\ldots} z_i^{-\frac{1}{2}} M_{\kappa_{1,i};\nu_{1,i}} \left| \frac{1}{r_i} \right| N_{n_{2,l_i}\ldots} z_{l_i}^{-\frac{1}{2}} M_{\kappa_{2,l_i};\nu_{2,l_i}} \right\rangle$$

$$I_{d_3} = \left\langle \xi_{n_{1,l_i};m_{1,l_i}} \left| \frac{1}{r_i} \right| \xi_{n_{2,i};m_{2,i}} \right\rangle, I_{d_\theta} = \left\langle \frac{e^{im_{1,i}\theta_i}}{\sqrt{2\pi}} \frac{e^{im_{1,j}\theta_j}}{\sqrt{2\pi}} \left| \cos\theta_{ij} \right| \frac{e^{im_{2,l_i}\theta_{l_i}}}{\sqrt{2\pi}} \frac{e^{im_{2,l_j}\theta_{l_j}}}{\sqrt{2\pi}} \right\rangle$$

I_{d_2} and I_{d_3} can be evaluated using the form of Coulomb-type integrals [Section 6.2.1.4.2]. Here I_{d_1} is the dipole-type integral which can be simplified for any z_j coordinate (quantum numbers for all $i \neq j$ remain conserved) as follows [24]:

$$I_{d_1} = N_{n_1,m_1} N_{n_2,m_2} \left\langle z_j^{-\frac{1}{2}} M_{\kappa_1,\nu_1}(z_j) \left| \frac{z_j^{1/2}}{2\Omega^{3/2}} \right| z_j^{-\frac{3}{4}} M_{\kappa_2,\nu_2}(z_j) \right\rangle \tag{6.56}$$

comparing the above integral with the standard integral (Eq. (6.A.3)), $b = 0$, $a_1 = a_2 = 1$, $A = \frac{1}{2}(a_1 + a_2) = 1$, $\gamma_1 - \frac{1}{2} = \frac{|m_1|}{2}$, $\gamma_2 - \frac{1}{2} = \frac{|m_2|}{2}$, $M = \gamma_1 + \gamma_2 = \frac{|m_1|}{2} + \frac{|m_2|}{2} + 1$, $\varrho = \frac{1}{2}$, $2\nu = 0, 1, 2, \ldots$; $\kappa - \nu - 1/2 = 0, 1, 2, \ldots$ and applying F_2 to ${}_2F_1$ relation and Chu–Vandermonde identity (Eqs. (6.A.5), (6.A.6)) leads to [24]:

$$I_{d_1} = \frac{1}{\sqrt{\Omega}} \sqrt{\frac{(n_1+|m_1|)!}{n_1!|m_1|^2!}} \sqrt{\frac{(n_2+|m_2|)!}{n_2!|m_2|^2!}} \Gamma\left(\frac{|m_1|}{2} + \frac{|m_2|}{2} + \frac{3}{2}\right)$$
$$\times \sum_{k=0}^{n_1} \frac{\left(\frac{|m_1|}{2} + \frac{|m_2|}{2} + \frac{3}{2}\right)_k (-n_1)_k \left(\frac{-|m_1|}{2} + \frac{|m_2|}{2} - k - \frac{1}{2}\right)_{n_2}}{(|m_1|+1)_k k! (|m_2|+1)_{n_2}}$$

I_{d_θ} is resolved in the following manner [24]:

$$I_{d_\theta} = \left\langle \frac{e^{im_{1,i}\theta_i}}{\sqrt{2\pi}} \frac{e^{im_{1,j}\theta_j}}{\sqrt{2\pi}} \left| \frac{e^{i\theta_{ij}} + e^{i\theta_{ij}}}{2} \right| \frac{e^{im_{2,l_i}\theta_{l_i}}}{\sqrt{2\pi}} \frac{e^{im_{2,l_j}\theta_{l_j}}}{\sqrt{2\pi}} \right\rangle$$
$$= \frac{1}{2}\left[\delta_{m_{1,i},m_{2,l_i+1}} \delta_{m_{1,j+1},m_{2,l_j}} + \delta_{m_{1,i+1},m_{2,l_i}} \delta_{m_{1,j},m_{2,l_j+1}}\right]$$

Quadrupole Factor The quadrupole term is evaluated by adopting a similar approach as the dipole term. The discretized projector states is inserted twice i.e. $\sum_{l=0}\sum_{m=0} |\Xi_l(Z,\Theta)\rangle\langle\Xi_l(Z,\Theta)|\Xi_m(Z,\Theta)\rangle$ $\langle\Xi_m(Z,\Theta)|$ to decouple the quadrupole integral [24]:

$$I_q = \left\langle \Xi(Z,\Theta) \left| \frac{3(\vec{r_i}.\vec{r_j})^2}{|\vec{r_i}|^5} - \frac{\vec{r_j}^2}{|\vec{r_i}|^3} \right| \Xi(Z,\Theta) \right\rangle$$

Upon inserting the discretized projection operators, I_q transforms as [24]:

$$I_q = \sum_{l=0} \left\langle \Xi(Z,\Theta) \left| \frac{r_j(3\cos^2\theta_{ij} - 1)}{r_i} \right| \Xi_l(\Xi,\Theta) \right\rangle$$
$$\times \sum_{l=0} \left\langle \Xi_l(Z,\Theta) \left| \frac{r_j}{r_i} \right| \Xi_m(\Xi,\Theta) \right\rangle \left\langle \Xi_m(Z,\Theta) \left| \frac{1}{r_i} \right| \Xi(Z,\Theta) \right\rangle$$

Thus, the above integral is separated into the radial and angular integrals as [24]:

$$I_q = \sum_{l=0} \left\{ I_{q_1} I_{q_2} I_{q_{\theta_1}} \right\} \times \sum_{m=0} \left\{ I_{q_3} I_{q_4} I_{q_{\theta_2}} \right\} \times I_{q_5}$$

where I_{q_1} and I_{q_3} are dipole-type integrals and I_{q_2} and I_{q_4} are Coulomb-type integrals having the form [24]:

$$I_{q_1} = \left\langle N_{n_{1,j}\ldots} z_j^{-\frac{1}{2}} M_{\kappa_{1,j};\nu_{1,j}} \left| r_j \right| N_{n_{2,l_j}\ldots} z_{l_j}^{-\frac{1}{2}} M_{\kappa_{2,l_j};\nu_{2,l_j}} \right\rangle, I_{q_3} = \left\langle N_{n_{1,l_j}\ldots} z_{l_j}^{-\frac{1}{2}} M_{\kappa_{1,l_j};\nu_{1,l_j}} \left| r_j \right| N_{n_{2,m_j}\ldots} z_{m_j}^{-\frac{1}{2}} M_{\kappa_{2,m_j};\nu_{2,m_j}} \right\rangle$$
$$I_{q_2} = \left\langle N_{n_{1,i}\ldots} z_i^{-\frac{1}{2}} M_{\kappa_{1,i};\nu_{1,i}} \left| \frac{1}{r_i} \right| N_{n_{2,l_i}\ldots} z_{l_i}^{-\frac{1}{2}} M_{\kappa_{2,l_i};\nu_{2,l_i}} \right\rangle, I_{q_4} = \left\langle N_{n_{1,l_i}\ldots} z_{l_i}^{-\frac{1}{2}} M_{\kappa_{1,l_i};\nu_{1,l_i}} \left| \frac{1}{r_i} \right| N_{n_{2,m_i}\ldots} z_{m_i}^{-\frac{1}{2}} M_{\kappa_{2,m_i};\nu_{2,m_i}} \right\rangle$$

I_{q_5} is identical to I_{d_3}. Using the approach developed in Section 6.2.1.4.2, I_{q_1}, I_{q_2}, I_{q_3} and I_{q_4} can be easily calculated. The angular integrals $I_{q_{\theta_1}}$ and $I_{q_{\theta_2}}$ can be evaluated as [24]:

$$I_{\theta_1} = \left\langle \frac{e^{im_{1,i}\theta_i}}{\sqrt{2\pi}} \frac{e^{im_{1,j}\theta_j}}{\sqrt{2\pi}} \left| \frac{3\cos 2\theta_{ij} + 1}{2} \right| \frac{e^{im_{2,l_i}\theta_{l_i}}}{\sqrt{2\pi}} \frac{e^{im_{2,l_j}\theta_{l_j}}}{\sqrt{2\pi}} \right\rangle$$
$$= \frac{3}{4}\left[\delta_{m_{1,i},m_{2,l_i+2}} \delta_{m_{1,j+2},m_{2,l_j}} + \delta_{m_{1,i+2},m_{2,l_i}} \delta_{m_{1,j},m_{2,l_j+2}}\right] + \frac{1}{2}(\delta_{m_{1,i},m_{2,l_i}} \delta_{m_{1,j},m_{2,l_j}}) \tag{6.57}$$

$$I_{\theta_2} = \left\langle \frac{e^{im_{1,i}\theta_i}}{\sqrt{2\pi}} \frac{e^{im_{1,j}\theta_j}}{\sqrt{2\pi}} \frac{e^{im_{2,l_i}\theta_{l_i}}}{\sqrt{2\pi}} \frac{e^{im_{2,l_j}\theta_{l_j}}}{\sqrt{2\pi}} \right\rangle = (\delta_{m_{1,i},m_{2,l_i}} \delta_{m_{1,j},m_{2,l_j}}) \tag{6.58}$$

the quadrupole correction is found to be exceptionally small to cause any aberration in the results [24].

6.2.2 Absence of Magnetic Field

In absence of magnetic field for single nucleus ($W = 1$) and zero confinements ($\omega_e = \omega_n = 0.0$ a.u.), the Hamiltonian can be recast for atoms within nonrelativistic limit as (Eq. (6.2)) [26–28, 102, 105]:

$$H = \frac{P^2}{2M} + \sum_{i=1}^{N} \left\{ \frac{p_i^2}{2m_e} - \frac{Z}{\epsilon|\vec{R}-\vec{r}_i|} + \frac{1}{2\epsilon}\sum_{j\neq i}^{N}\frac{1}{r_{ij}} \right\} \tag{6.59}$$

Relative mass and the center-of-mass coordinates constitute the following relations:

$$\vec{R}_{\mathrm{CM}} = \frac{M\vec{R} + m_e\sum_{i=1}^{N}\vec{r}_i}{M + Nm_e} \quad \text{and} \quad |\vec{R} - \vec{r}_i| = \vec{\mathfrak{r}}_{\mathrm{rel.},i}$$

Implementation of Born-Oppenheimer (BO) approximation (i.e. $M \ggg m_e$) $\Rightarrow \vec{R}_{\mathrm{CM}} = \vec{R}$ and translation of origin to center-of-mass $\Rightarrow \vec{\mathfrak{r}}_{\mathrm{rel.},i} = \vec{r}_i$. Therefore, atoms exhibiting Abelian topology incur the absence of nonadiabatic coupling terms, conical intersections, and consequent complexity of potential energy surfaces [106–111]. Thus, the Hamiltonian for He-isoelectronic ions ($\epsilon = 1.0$ for vacuum) can be cast as [27, 28, 94, 112]:

$$H = \underbrace{\sum_{i=1}^{2}\overbrace{\left\{\frac{p_i^2}{2} - \frac{Z}{r_i}\right\}}^{H_i^0}}_{H^0} + \underbrace{\frac{1}{2}\sum_{i=1}^{2}\sum_{j\neq i}^{2}\overbrace{\frac{1}{r_{ij}}}^{H'_{ij}}}_{H'} \tag{6.60}$$

6.2.2.1 3D He-Isoelectronic Ions

In spherical polar coordinates, $H^0\xi(r,\phi) = E^0\xi(r,\phi)$ single particle 3D H-like system reads as (Eq. (6.60)):

$$\left\{-\frac{1}{2}\left[r^{-2}\left(\frac{\partial}{\partial r}r^2\frac{\partial}{\partial r}\right) - \frac{L^2}{r^2}\right] - \frac{Z}{r}\right\}\xi(r,\theta,\phi) = \mathcal{E}^0\xi(r,\theta,\phi) \tag{6.61}$$

where wavefunction, $\xi^0(r,\theta,\phi) = u(r)Y_l^m(\theta,\phi)$, is product of radial and angular part, respectively, where $m = 0, \pm1, \pm2, \ldots, \pm l$. Using $z = \alpha r$, $u(r) \Longrightarrow U(z)$, $\alpha^2 = -8\mathcal{E}^0$, $\beta = 2Z/\alpha$ and subsequently $U(z) = z^{-1}M(z)$ for bound H-like orbitals ($\mathcal{E}^0 < 0$) transforms the radial SE as [25–27]:

$$M''(z) + \left\{\frac{\beta}{z} - \frac{l(l+1)}{z^2} - \frac{1}{4}\right\}M(z) = 0 \tag{6.62}$$

where $M(z)$ is Whittaker-M function and $\beta = 2Z\sqrt{-\frac{1}{8\mathcal{E}^0}}$. Solving the radial SE in Whittaker-M function yields the energy and the corresponding eigenfunction of bound state as $\mathcal{E}^0 = -\frac{Z^2}{2n^2}$ and $\xi^0(r,\theta,\phi) = \mathcal{N}(\alpha r)^{-1}M_{\kappa,\nu}(\alpha r)Y_l^m(\theta,\phi)$ respectively where n is an integer, $\mathcal{N} = \sqrt{\alpha^3\left(\frac{(n+l)!}{2n(n-l-1)!(2l+1)!^2}\right)}$, $\alpha = \frac{2Z}{n}$, $\kappa = n$ and $l + 1/2 = \nu$. Relating Eq. (6.62) with associated Laguerre differential equation gives eigenfunction, $\xi^0(r,\theta,\phi) = \mathcal{B}e^{-\frac{\alpha r}{2}}(\alpha r)^l L_{\mathcal{N}}^{2l+1}(\alpha r)Y_l^m(\theta,\phi)$, where $\mathcal{N} = n\text{-}l\text{-}1$, n is principal quantum number and $\mathcal{B} = \sqrt{\alpha^3\left(\frac{(n-l-1)!}{2n(n+l)!}\right)}$ [25–27, 103]. Spin has been incorporated by taking the spin-orbital wavefunction as the product of the spatial wavefunction $\xi^0(\tau)$ with $\tau \Rightarrow (r,\theta,\phi)$ and spin wavefunction, $\omega(\sigma_\mu) = |\alpha\rangle$ or $|\beta\rangle$.

6.2.2.1.1 Coulomb Interaction: Monopole ($l = 0$) and Dipole ($l = 1$) Factors

Coulomb Green's function can be utilized to expand $H'_{ij}(1/r_{ij})$ in multipoles as [101, 113]:

$$
\begin{aligned}
G(\vec{r}_i, \vec{r}_j) &= \sum_{l=0}^{\infty}\sum_{m=-l}^{l} \frac{4\pi}{2l+1}\frac{r_<^l}{r_>^{l+1}} Y_l^m(\theta_i, \phi_i) Y_l^{m*}(\theta_j, \phi_j) \\
&= 4\pi \left[\frac{1}{r_>} Y_0^0(\theta_i, \phi_i) Y_0^{0*}(\theta_j, \phi_j) + \frac{r_<}{3r_>^2} \left\{ Y_1^1(\theta_i, \phi_i) Y_1^{1*}(\theta_j, \phi_j) + Y_1^0(\theta_i, \phi_i) Y_1^{0*}(\theta_j, \phi_j) \right.\right. \\
&\quad \left.\left. + Y_1^{-1}(\theta_i, \phi_i) Y_1^{-1*}(\theta_j, \phi_j) \right\} + \sum_{l=2}^{\infty}\sum_{m=-l}^{+l} \cdots \right]
\end{aligned}
\tag{6.63}
$$

where first, second, and third factors denote monopole (H'_{M}), dipole (H'_{D}), quadrupole (H'_{Q}) terms, respectively.

Monopole Factor ($l = 0$) For monopole expansion ($l = 0$ and $m = 0$) (Eq. (6.63)), the integral for the total wavefunctions can be written as:

$$
\begin{aligned}
\mathcal{H}'_{\mathrm{M}} &= 4\pi \left\langle \xi_I^0(\tau_i)\xi_J^0(\tau_j) \left| \frac{1}{r_>} \right| \xi_K^0(\tau_i)\xi_L^0(\tau_j) \right\rangle \times \langle \omega_I(\sigma_i)|\omega_K(\sigma_i)\rangle \langle \omega_J(\sigma_j)|\omega_L(\sigma_j)\rangle \\
&= \underbrace{\left\langle \xi_I^0(\alpha_1 r_i)\xi_J^0(\alpha_3 r_j) \left| \frac{1}{r_>} \right| \xi_K^0(\alpha_2 r_i)\xi_L^0(\alpha_4 r_j) \right\rangle}_{I_{\mathrm{M}}} \\
&\quad \times 4\pi \left\langle Y_{l_1,i}^{m_1,i}\, Y_{l_3,j}^{m_3,j} \left| Y_{0,i}^{0,i}\, Y_{0,j}^{0,j} \right| Y_{l_2,i}^{m_2,i}\, Y_{l_4,j}^{m_4,j} \right\rangle \delta_{I,K}(\sigma_i)\delta_{J,L}(\sigma_j) \\
&\text{where } \delta_{I,K}(\sigma_i) = 1 \text{ if } I = K \\
&\qquad\qquad\quad = 0 \text{ if } I \neq K
\end{aligned}
\tag{6.64}
$$

and scaling factors for bra-space coordinates as α_1=2Z/n_1, $\alpha_3 = 2Z/n_3$ and for ket-space coordinates as α_2=2Z/n_2, α_4=2Z/n_4, subject to singly excited ($n_2 = 1, n_4 \neq 1$) or doubly excited states ($n_2, n_4 \neq 1$) for the radial integral (I_{M}) [25–27]. Evaluation of the angular part results in the cancellation of the factor 4π using Eq. (6.A.4). Solving the radial part by splitting the limits of the integral as:

$$
\begin{aligned}
I_{\mathrm{M}} &= \underbrace{\int_0^{\infty} \xi(\alpha_1 r_i)\xi(\alpha_2 r_i) \left[\int_0^{r_i} \frac{\xi(\alpha_3 r_j)\xi(\alpha_4 r_j) r_j^2 dr_j}{r_i} \right] r_i^2 dr_i}_{I_{\mathrm{M}_1}} \\
&\quad + \underbrace{\int_0^{\infty} \xi(\alpha_1 r_i)\xi(\alpha_2 r_i) \left[\int_{r_i}^{\infty} \frac{\xi(\alpha_3 r_j)\xi(\alpha_4 r_j) r_j^2 dr_j}{r_j} \right] r_i^2 dr_i}_{I_{\mathrm{M}_2}}
\end{aligned}
\tag{6.65}
$$

Solving the above integrals by using Eqs. (6.A.1)–(6.A.3), the final forms read as [25–27]:

$$
\begin{aligned}
I_{\mathrm{M}_1} &= \frac{\mathscr{D}}{\alpha_3} \sum_{k_1=0}^{\mathcal{N}_3}\sum_{k_2=0}^{\mathcal{N}_4} (-1)^{k_1+k_2} \frac{w^{l_4+k_2} g^{\mathscr{A}+3} (\mathscr{A}+2)!}{k_1! k_2!} \binom{n_3 + l_3}{2l_3 + k_1 + 1} \binom{n_4 + l_4}{2l_4 + k_2 + 1} \\
&\quad \times \left[\alpha_1^{l_1+1} \alpha_2^{l_2+1} \left(\frac{\alpha_1 + \alpha_2}{2} \right)^{-(l_1+l_2+2)} \Gamma(l_1 + l_2 + 2) \right.
\end{aligned}
$$

$$\times \sum_{p=0}^{\mathcal{N}_1}\sum_{q=0}^{\mathcal{N}_2} \frac{(l_1+l_2+2)_{(p+q)}(-\mathcal{N}_1)_p(-\mathcal{N}_2)_q}{(2l_1+2)_p(2l_2+2)_q p!q!} \left(\frac{2\alpha_1}{\alpha_1+\alpha_2}\right)^p \left(\frac{2\alpha_1}{\alpha_1+\alpha_2}\right)^q - \sum_{s=0}^{\mathcal{A}+2} \frac{1}{s!}\left(\frac{\alpha_3}{g}\right)^s$$
$$\times \alpha_1^{l_1+1}\alpha_2^{l_2+1}\left(\frac{(w+1)\alpha_3+\alpha_1+\alpha_2}{2}\right)^{-(s+l_1+l_2+2)} \Gamma(s+l_1+l_2+2)$$
$$\times \sum_{p=0}^{\mathcal{N}_1}\sum_{q=0}^{\mathcal{N}_2} \frac{(s+l_1+l_2+2)_{(p+q)}(-\mathcal{N}_1)_p(-\mathcal{N}_2)_q}{(2l_1+2)_p(2l_2+2)_q p!q!}$$
$$\times \left(\frac{2\alpha_1}{(w+1)\alpha_3+\alpha_1+\alpha_2}\right)^p \left(\frac{2\alpha_2}{(w+1)\alpha_3+\alpha_1+\alpha_2}\right)^q \Bigg]$$
$$I_{M_2} = \mathcal{D}\sum_{k_1=0}^{\mathcal{N}_3}\sum_{k_2=0}^{\mathcal{N}_4} (-1)^{k_1+k_2} \frac{w^{l_4+k_2} g^{+2}(\mathcal{A}+1)!}{k_1!k_2!} \binom{n_3+l_3}{2l_3+k_1+1}\binom{n_4+l_4}{2l_4+k_2+1}$$
$$\times \left\{ \sum_{s=0}^{\mathcal{A}+2} \frac{1}{s!}\left(\frac{\alpha_3}{g}\right)^s \alpha_1^{l_1+1}\alpha_2^{l_2+1}\left(\frac{(w+1)\alpha_3+\alpha_1+\alpha_2}{2}\right)^{-(s+l_1+l_2+3)} \Gamma(s+l_1+l_2+3)\right.$$
$$\times \sum_{p=0}^{\mathcal{N}_1}\sum_{q=0}^{\mathcal{N}_2} \frac{(s+l_1+l_2+3)_{(p+q)}(-\mathcal{N}_1)_p(-\mathcal{N}_2)_q}{(2l_1+2)_p(2l_2+2)_q p!q!}$$
$$\left. \times \left(\frac{2\alpha_1}{(w+1)\alpha_3+\alpha_1+\alpha_2}\right)^p \left(\frac{2\alpha_2}{(w+1)\alpha_3+\alpha_1+\alpha_2}\right)^q \right\} \tag{6.66}$$

where $\mathcal{A} = l_3+l_4+k_1+k_2$ and $\mathcal{D} = \frac{\mathcal{N}_1\mathcal{N}_2 B_3 B_4}{\alpha_1\alpha_2\alpha_3^2}$. Equation (6.66) gives the exact value of the integral of the monopole factor (Eq. (6.64)).

Dipole Factor (l = 1) The dipole moment has significant role for transition integral among bound states [112]. Similarly, the integral of dipole factor (Eq. (6.63)) can be evaluated as:

$$\mathcal{H}'_{\mathrm{D}} = \underbrace{\left\langle \xi_I^0(\tau_i)\xi_J^0(\tau_j) \left| \frac{r_<}{r_>^2} \right| \xi_K^0(\tau_i)\xi_L^0(\tau_j) \right\rangle}_{I_{\mathrm{D}}}$$
$$\underbrace{\times \frac{4\pi}{3}\sum_{m=-1}^{1} \left\langle Y_{l_1,i}^{m_1,i}\, Y_{l_3,j}^{m_3,j} \middle| Y_{1,i}^{m,i}\, Y_{l,j}^{m,j} \middle| Y_{l_2,i}^{m_2,i}\, Y_{l_4,j}^{m_4,j} \right\rangle}_{I_{\mathrm{DA}}} \times \delta_{I,K}(\sigma_i)\,\delta_{J,L}(\sigma_j)$$
$$\text{where } \delta_{I,K}(\sigma_i) = 1 \text{ if } I = K$$
$$= 0 \text{ if } I \neq K \tag{6.67}$$

First solving the radial part (I_{D}) by segregating the integral for each r_i considering integral over r_j and consequently integrating over r_i:

$$I_{\mathrm{D}} = \underbrace{\int_0^\infty \xi(\alpha_1 r_i)\xi(\alpha_2 r_i)\left[\int_0^{r_i} \frac{\xi(\alpha_3 r_j)\xi(\alpha_4 r_j) r_j^3 dr_j}{r_i^2}\right] r_i^2 dr_i}_{I_{\mathrm{D}_1}} + \underbrace{\int_0^\infty \xi(\alpha_1 r_i)\xi(\alpha_2 r_i)\left[\int_{r_i}^{\infty} \frac{\xi(\alpha_3 r_j)\xi(\alpha_4 r_j) r_i r_j^2 dr_j}{r_j^2}\right] r_i^2 dr_i}_{I_{\mathrm{D}_2}} \tag{6.68}$$

Solving the above integrals by using (Eqs. (6.A.1)–(6.A.3)), the final forms read as [27]:

$$
\begin{aligned}
I_{D_1} &= \frac{\mathscr{D}}{\alpha_3}\sum_{k_1=0}^{\mathcal{N}_3}\sum_{k_2=0}^{\mathcal{N}_4}(-1)^{k_1+k_2}\frac{w^{l_4+k_2}g^{\mathscr{A}+4}(\mathscr{A}+3)!}{k_1!k_2!}\binom{n_3+l_3}{2l_3+k_1+1}\binom{n_4+l_4}{2l_4+k_2+1}\\
&\times\left[\alpha_1^{l_1+1}\alpha_2^{l_2+1}\left(\frac{\alpha_1+\alpha_2}{2}\right)^{-(l_1+l_2+1)}\Gamma(l_1+l_2+1)\sum_{p=0}^{\mathcal{N}_1}\sum_{q=0}^{\mathcal{N}_2}\frac{(l_1+l_2+1)_{(p+q)}(-\mathcal{N}_1)_p(-\mathcal{N}_2)_q}{(2l_1+2)_p(2l_2+2)_q p!q!}\right.\\
&\times\left(\frac{2\alpha_1}{\alpha_1+\alpha_2}\right)^p\left(\frac{2\alpha_1}{\alpha_1+\alpha_2}\right)^q-\sum_{s=0}^{\mathscr{A}+3}\frac{1}{s!}\left(\frac{\alpha_3}{g}\right)^s\\
&\times\alpha_1^{l_1+1}\alpha_2^{l_2+1}\left(\frac{(w+1)\alpha_3+\alpha_1+\alpha_2}{2}\right)^{-(s+l_1+l_2+1)}\Gamma(s+l_1+l_2+1)\\
&\times\sum_{p=0}^{\mathcal{N}_1}\sum_{q=0}^{\mathcal{N}_2}\frac{(s+l_1+l_2+1)_{(p+q)}(-\mathcal{N}_1)_p(-\mathcal{N}_2)_q}{(2l_1+2)_p(2l_2+2)_q p!q!}\\
&\left.\times\left(\frac{2\alpha_1}{(w+1)\alpha_3+\alpha_1+\alpha_2}\right)^p\left(\frac{2\alpha_2}{(w+1)\alpha_3+\alpha_1+\alpha_2}\right)^q\right]\\
I_{D_2} &= \mathscr{D}\sum_{k_1=0}^{\mathcal{N}_3}\sum_{k_2=0}^{\mathcal{N}_4}(-1)^{k_1+k_2}\frac{w^{l_4+k_2}g^{\mathscr{A}+1}\mathscr{A}!}{k_1!k_2!}\binom{n_3+l_3}{2l_3+k_1+1}\binom{n_4+l_4}{2l_4+k_2+1}\sum_{s=0}^{\mathscr{A}}\frac{1}{s!}\left(\frac{\alpha_3}{g}\right)^s\alpha_1^{l_1+1}\alpha_2^{l_2+1}\\
&\times\left(\frac{(w+1)\alpha_3+\alpha_1+\alpha_2}{2}\right)^{-(s+l_1+l_2+4)}\Gamma(s+l_1+l_2+4)\\
&\times\sum_{p=0}^{\mathcal{N}_1}\sum_{q=0}^{\mathcal{N}_2}\frac{(s+l_1+l_2+4)_{(p+q)}(-\mathcal{N}_1)_p(-\mathcal{N}_2)_q}{(2l_1+2)_p(2l_2+2)_q p!q!}\mathscr{A}\\
&\times\left(\frac{2\alpha_1}{(w+1)\alpha_3+\alpha_1+\alpha_2}\right)^p\left(\frac{2\alpha_2}{(w+1)\alpha_3+\alpha_1+\alpha_2}\right)^q
\end{aligned}
\tag{6.69}
$$

where $\mathscr{A}=l_3+l_4+k_1+k_2$ and $\mathscr{D}_1=\frac{\mathcal{N}_1\mathcal{N}_2B_3B_4}{\alpha_1\alpha_2\alpha_3}$. Summing I_{D_1} and I_{D_2} (Eq. (6.69)) gives the exact value of the integral for the radial part of the dipole factor (Eq. (6.68)).

Now, evaluating the angular part:

$$
\begin{aligned}
I_{DA} &= \frac{4\pi}{3}\sum_{m=-1}^{1}\left\langle Y_{l_1,i}^{m_1,i}\,Y_{l_3j}^{m_3j}\middle|Y_{1_i}^{m_i}\,Y_{1_j}^{m_j*}\middle|Y_{l_2,i}^{m_2,i}\,Y_{l_4j}^{m_4j}\right\rangle\\
&= \frac{4\pi}{3}\left\langle Y_{l_1,i}^{m_1,i}Y_{l_3j}^{m_3j}\left|Y_1^{-1}Y_1^{-1*}+Y_1^{0*}Y_1^{0}+Y_1^{1*}Y_1^{1}\right|Y_{l_2,i}^{m_2,i}Y_{l_4j}^{m_4j}\right\rangle\\
&= \frac{4\pi}{3}\left\{\left\langle Y_{l_1,i}^{m_1,i}Y_{l_3j}^{m_3j}\left|Y_1^{-1}\,Y_1^{-1*}\right|Y_{l_2,i}^{m_2,i}Y_{l_4j}^{m_4j}\right\rangle+\left\langle Y_{l_1,i}^{m_1,i}Y_{l_3j}^{m_3j}\left|Y_1^{0}\,Y_1^{0*}\right|Y_{l_2,i}^{m_2,i}Y_{l_4j}^{m_4j}\right\rangle\right.\\
&\quad\left.+\left\langle Y_{l_1,i}^{m_1,i}Y_{l_3j}^{m_3j}\left|Y_1^{1}\,Y_1^{1*}\right|Y_{l_2,i}^{m_2,i}Y_{l_4j}^{m_4j}\right\rangle\right\}
\end{aligned}
\tag{6.70}
$$

The above integrals are evaluated by making use of the Recurrence relations (Eq. (6.A.4)).

6.2.2.1.2 Multiconfigurations for He and Period-II Elements

In Hartree–Fock SCF method, SC closed-shell Slater determinantal variation method has been manipulated without analytical Coulomb (exchange) integrals for H-like SOs leading to pseudo-eigenvalue Fock equation [102]. Later on, post Hartree-Fock SCF method succeeded with open-shell and restricted closed-shell calculations [102, 114]. However, here we have applied multiconfiguration (MC) variation principle with four possibilities for determinants having H-like

SO differences of zero, one, two, and more between Hartree products of bra- and ket-spaces respectively. The MCs encircle closed-, open-, restricted closed-, and other shells. We have considered SOs- 1s, $\overline{1s}$, 2s, $\overline{2s}$, 3s, $\overline{3s}$, … and 2p, $\overline{2p}$, 3p, $\overline{3p}$, 4p, $\overline{4p}$, etc., to incorporate monopole and dipole factors. Moreover, we have incurred the idea of exchanging electrons among SOs of innermost to outermost orbits, to say, core- and valence- shells. Full anti-symmetrization is implemented to address Pauli exclusion principle for $N = 2$–10 electrons corresponding to He, Li, Be, B, C, N, O, F, and Ne, respectively. Some of the MCs (Slater determinantal wavefunction, D_q) utilized have been presented here.

He ($1s^2$): $\left|1s\ \overline{1s}\right|$, $|1s\ 2s|$, $\left|1s\ \overline{2s}\right|$, …, $\left|\overline{11s}\ \overline{12s}\right|$, $\left|12s\ \overline{12s}\right|$, $|1s\ 2p_{-1}|$, $|1s\ 2p_0|$, $|1s\ 2p_{+1}|$, $\left|2p_0\ \overline{2p_0}\right|$, and so on.

Li ($1s^22s^1$): $\left|1s\ \overline{1s}\ 2s\right|$, $\left|1s\ \overline{1s}\ \overline{2s}\right|$, $\left|1s\ \overline{1s}\ 3s\right|$, $\left|1s\ \overline{1s}\ \overline{3s}\right|$, $\left|\overline{1s}\ 2s\ \overline{2s}\right|$, $\left|\overline{1s}\ 2s\ 3s\right|$, and so on.

Be ($1s^22s^2$): $\left|1s\ \overline{1s}\ 2s\ \overline{2s}\right|$, $\left|1s\ \overline{1s}\ 2s\ 3s\right|$, $\left|1s\ \overline{1s}\ 2s\ \overline{3s}\right|$, $\left|\overline{1s}\ 2s\ \overline{2s}\ 3s\right|$, $\left|\overline{1s}\ 2s\ \overline{2s}\ \overline{3s}\right|$, and so on.

B ($1s^22s^22p^1$): $\left|1s\ \overline{1s}\ 2s\ \overline{2s}\ 2p_{-1}\right|$, $\left|1s\ \overline{1s}\ 2s\ \overline{2s}\ 2p_0\right|$, $\left|1s\ \overline{1s}\ 2s\ \overline{2s}\ 2p_{+1}\right|$, $\left|1s\ \overline{1s}\ 2s\ \overline{2s}\ \overline{2p_{-1}}\right|$, and so on.

C ($1s^22s^22p^2$): $\left|1s\ \overline{1s}\ 2s\ \overline{2s}\ 2p_{-1}\ 2p_0\right|$, $\left|1s\ \overline{1s}\ 2s\ \overline{2s}\ 2p_{-1}\ 2p_{+1}\right|$, $\left|1s\ \overline{1s}\ 2s\ \overline{2s}\ 2p_0\ \overline{2p_{-1}}\right|$, and so on.

N ($1s^22s^22p^3$): $\left|1s\ \overline{1s}\ 2s\ \overline{2s}\ 2p_{-1}\ 2p_0\ 2p_{+1}\right|$, $\left|1s\ \overline{1s}\ 2s\ \overline{2s}\ 2p_{-1}\ 2p_0\ \overline{2p_{-1}}\right|$, $\left|1s\ \overline{1s}\ 2s\ \overline{2s}\ 2p_{-1}\ 2p_0\ \overline{2p_0}\right|$, and so on.

O ($1s^22s^22p^4$): $\left|1s\ \overline{1s}\ 2s\ \overline{2s}\ 2p_{-1}\ 2p_0\ 2p_{+1}\ \overline{2p_{-1}}\right|$, $\left|1s\ \overline{1s}\ 2s\ \overline{2s}\ 2p_{-1}\ 2p_0\ \overline{2p_{-1}}\ \overline{2p_0}\right|$, $\left|1s\ \overline{1s}\ 2s\ \overline{2s}\ 2p_{-1}\ 2p_0\ \overline{2p_0}\ \overline{2p_{+1}}\right|$, and so on.

F ($1s^22s^22p^5$): $\left|1s\ \overline{1s}\ 2s\ \overline{2s}\ 2p_{-1}\ 2p_0\ 2p_{+1}\ \overline{2p_{-1}}\ \overline{2p_0}\right|$, $\left|1s\ \overline{1s}\ 2s\ \overline{2s}\ 2p_{-1}\ 2p_0\ \overline{2p_{-1}}\ \overline{2p_0}\ \overline{2p_{+1}}\right|$, $\left|1s\ \overline{1s}\ 2s\ \overline{2s}\ 2p_{-1}\ 2p_0\ \overline{2p_{-1}}\ \overline{2p_0}\ 3p_{-1}\right|$, and so on.

Ne ($1s^22s^22p^6$): $\left|1s\ \overline{1s}\ 2s\ \overline{2s}\ 2p_{-1}\ 2p_0\ 2p_{+1}\ \overline{2p_{-1}}\ \overline{2p_0}\ \overline{2p_{+1}}\right|$, $\left|1s\ \overline{1s}\ 2s\ \overline{2s}\ 2p_{-1}\ 2p_0\ 2p_{+1}\ \overline{2p_{-1}},\ \overline{2p_0}\ 3p_{-1}\right|$ $\left|1s\ \overline{1s}\ 2s\ \overline{2s}\ 2p_{-1}\ 2p_0\ 2p_{+1}\ \overline{2p_{-1}}\ \overline{2p_0}\ 3p_0\right|$, and so on.

6.2.2.1.3 Multiconfigurational Calculation of *H′*

The algebra of total Hamiltonian consisting of mono-electronic (H_i^0) and two-electron Coulomb (exchange) operators (H'_{ij}), respectively, can be written as [102]:

Case-1: ($D_p \sim D_q = 0$ SO, i.e. diagonal terms)

$$\langle D_p|H^0|D_q\rangle = \sum_{I=1}^{N} \left(\langle S_I(\mu)|H_I^0|S_I(\mu)\rangle\right) = \sum_{I=1}^{N} \mathcal{E}_I^0$$

$$\langle D_p|H'|D_q\rangle = \sum_{I=1}^{N}\sum_{J>I}^{N} \left(\left\langle S_I(\mu)S_J(\nu)\middle|H'_{\mu\nu}\middle|S_I(\mu)S_J(\nu)\right\rangle - \left\langle S_I(\mu)S_J(\nu)\middle|H'_{\mu\nu}\middle|S_J(\mu)S_I(\nu)\right\rangle\right)$$

Case-2: ($D_p \sim D_q = 1$ SO)

$$\langle D_p|H^0|D_q\rangle = 0$$

$$\langle D_p|H'|D_q\rangle = \sum_{J\neq I\neq L}^{N} \left(\left\langle S_I(\mu)S_J(\nu)\middle|H'_{\mu\nu}\middle|S_L(\mu)S_J(\nu)\right\rangle - \left\langle S_I(\mu)S_J(\nu)\middle|H'_{\mu\nu}\middle|S_J(\mu)S_L(\nu)\right\rangle\right)$$

Case-3: ($D_p \sim D_q = 2$ SOs)

$$\langle D_p|H^0|D_q\rangle = 0$$

$$\langle D_p|H'|D_q\rangle = \left(\left\langle S_I(\mu)S_J(\nu)\middle|H'_{\mu\nu}\middle|S_K(\mu)S_L(\nu)\right\rangle - \left\langle S_I(\mu)S_J(\nu)\middle|H'_{\mu\nu}\middle|S_L(\mu)S_K(\nu)\right\rangle\right)$$

Case-4: ($D_p \sim D_q = 3$ or more SOs)

$$\langle D_p | H^0 | D_q \rangle = 0$$
$$\langle D_p | H' | D_q \rangle = 0$$

where $D = \left| \prod_{I,\mu}^{N} S_I(\mu) \right|$. It is evident from above equations that for exact description of Coulomb interaction, calculation of two-electron coordinate integral, $\mathcal{H}'_{\text{PQ}} = \left\langle S_I(\mu) S_J(\nu) \left| \frac{1}{r_{\mu\nu}} \right| S_K(\mu) S_L(\nu) \right\rangle$ becomes vital. $1/r_{ij}$ can be written in multipoles using free electrostatic Green's function expansion as [101, 113]. The total contribution of multipoles can be summarized as:

$$\begin{aligned}
\mathcal{H}'_{\text{PQ}} &= \left\langle S_I(\mu) S_J(\nu) \left| H'_{\text{M}} + H'_{\text{D}} + H'_{\text{Q}} + \cdots \right| S_K(\mu) S_L(\nu) \right\rangle \\
&= \left\{ \left\langle \xi_I^0(\tau_i) \xi_J^0(\tau_j) \left| H'_{\text{M}} \right| \xi_K^0(\tau_i) \xi_L^0(\tau_j) \right\rangle + \left\langle \xi_I^0(\tau_i) \xi_J^0(\tau_j) \left| H'_{\text{D}} \right| \xi_K^0(\tau_i) \xi_L^0(\tau_j) \right\rangle \right. \\
&\quad \left. + \left\langle \xi_I^0(\tau_i) \xi_J^0(\tau_j) \left| H'_{\text{Q}} + \cdots \right| \xi_K^0(\tau_i) \xi_L^0(\tau_j) \right\rangle \right\} \times \langle \omega_{\text{P}}(\sigma_i) | \omega_{\text{Q}}(\sigma_i) \rangle \langle \omega_{\text{P}}(\sigma_j) | \omega_{\text{Q}}(\sigma_j) \rangle \\
\mathcal{H}'_{\text{PQ}} &= \mathcal{H}'_{\text{M}} + \mathcal{H}'_{\text{D}} + \mathcal{H}'_{\text{Q}} + \cdots
\end{aligned} \tag{6.71}$$

6.2.2.2 2D He-Isoelectronic Ions

Schrödinger equation of the unperturbed Hamiltonian for the ith particle of planar H-like species can be represented as (Eq. (6.60)):

$$\begin{aligned}
&\text{H}^0 \xi(r,\phi) = \text{E}^0 \xi(r,\phi) \\
&\left[-\frac{1}{2} \left(\frac{\partial^2}{\partial r^2} + \frac{1}{r}\frac{\partial}{\partial r} + \frac{1}{r^2}\frac{\partial^2}{\partial \phi^2} \right) - \frac{Z}{r} \right] \xi(r,\phi) = \text{E}^0 \xi(r,\phi)
\end{aligned} \tag{6.72}$$

With ansatz wavefunction $\xi^0(r,\phi) = \frac{e^{im\phi}}{\sqrt{2\pi}} u(r)$, $m = 0, \pm 1, \pm 2, \ldots$, where $u(r)$ is the radial part, the above equation reads:

$$\left\{ \left(\frac{d^2}{dr^2} + \frac{1}{r}\frac{d}{dr} \right) - \frac{m^2}{r^2} + \frac{2Z}{r} - \frac{\alpha^2}{4} \right\} u(r) = 0 \tag{6.73}$$

where $\alpha^2 = -8\text{E}^0$, restricting only to bound states ($\text{E}^0 < 0$). Transformations $z = \alpha r$, $u(r) \Longrightarrow U(z)$ with $\beta = 2Z/\alpha$ and then $U(z) = z^{-\frac{1}{2}} M(z)$ simplify the above equation to [28]:

$$M''(z) + \left\{ \frac{1}{z^2}\left(\frac{1}{4} - m^2 \right) + \frac{\beta}{z} - \frac{1}{4} \right\} M(z) = 0 \tag{6.74}$$

where $M(z)$ is Whittaker-M function and $\beta = 2Z\sqrt{-\frac{1}{8\text{E}^0}}$. Frobenius series solution of self-adjoint differential equation of Whittaker-M function gives the energy and the corresponding bound state wavefunction respectively as [28, 103]:

$$\text{E}^0 = -\frac{Z^2}{2(n-\frac{1}{2})^2}; \qquad \xi^0(r,\phi) = N(\alpha r)^{-\frac{1}{2}} M_{\kappa,\nu}(\alpha r) \frac{e^{im\phi}}{\sqrt{2\pi}} \tag{6.75}$$

where n is an integer, $N = \sqrt{\alpha^2 \left(\frac{(n+|m|-1)!}{2(n-\frac{1}{2})(n-|m|-1)!(2|m|)!^2} \right)}$, $\alpha = \frac{2Z}{(n-\frac{1}{2})}$, $\kappa = n - \frac{1}{2}$ and $\nu = |m|$. Also, Eq. (6.74) can be compared with associated Laguerre differential equation to yield the eigenfunction [28, 103]:

$$\xi^0(r,\phi) = \mathfrak{B} e^{-\frac{\alpha r}{2}} (\alpha r)^{|m|} L_{\mathcal{N}}^{2|m|}(\alpha r) \frac{e^{im\phi}}{\sqrt{2\pi}} \tag{6.76}$$

where $\mathcal{N} = n - |m| - 1$, n is the principal quantum number and $\mathfrak{B} = \sqrt{\alpha^2 \left(\frac{(n-|m|-1)!}{2(n-\frac{1}{2})(n+|m|-1)!} \right)}$.

6.2.2.2.1 Coulomb Interaction: Monopole ($l = 0$) and Dipole ($l = 1$) Factors

The interaction term (H'_{ij}) can be evaluated by employing 2D Green's function multipole expansion [28, 101, 113]:

$$\frac{1}{r_{ij}} = \sum_{l=0}^{\infty} \frac{r_<^l}{r_>^{l+1}} P_l(\cos(\phi_i - \phi_j)) \tag{6.77}$$

Monopole Factor ($l = 0$) The integral involving the monopole factor ($l = 0$) (Eq. (6.77)) can be written as:

$$H_{\mathrm{PQ}}^{(1)} = \left\langle \Xi(R, \Phi)_{\mathrm{P}} \left| \frac{1}{r_>} \right| \Xi(R, \Phi)_{\mathrm{Q}} \right\rangle \tag{6.78}$$

For ground-state planar He-isoelectronic ions with two electrons corresponding to individual ith and jth spaces, the above integral can be recast as:

$$\begin{aligned} H_{\mathrm{PQ}}^{(1)} &= \left\langle \xi(r_i, \phi_i)\xi(r_j, \phi_j) \left| \frac{1}{r_>} \right| \xi(r'_i, \phi'_i)\xi(r'_j, \phi'_j) \right\rangle \\ &= \underbrace{\left\langle \xi(\alpha_1 r_i)\xi(\alpha_2 r_j) \left| \frac{1}{r_>} \right| \xi(\alpha'_1 r_i)\xi(\alpha'_2 r_j) \right\rangle}_{I_{\mathrm{M}}} \left\langle \frac{e^{im_1\phi_i}}{\sqrt{2\pi}} \frac{e^{im_2\phi_j}}{\sqrt{2\pi}} \mid \frac{e^{im'_1\phi_i}}{\sqrt{2\pi}} \frac{e^{im'_2\phi_j}}{\sqrt{2\pi}} \right\rangle \end{aligned} \tag{6.79}$$

Angular part gives unity value with $m_1 = m'_1$ and $m_2 = m'_2$ and the above integral reduces to the following radial integral:

$$I_{\mathrm{M}} = \left\langle \xi(\alpha_1 r_i)\xi(\alpha_2 r_j) \left| \frac{1}{r_>} \right| \xi(\alpha'_1 r_i)\xi(\alpha'_2 r_j) \right\rangle \tag{6.80}$$

Splitting limits of the integral over r_j at the point r_i and consequently integrating both the integrals over r_i:

$$\begin{aligned} I_{\mathrm{M}} &= I_{\mathrm{M}_1} + I_{\mathrm{M}_2} \\ I_{\mathrm{M}} &= \int_0^{\infty} \xi(\alpha_1 r_i)\xi(\alpha_2 r_i) \left[\int_0^{r_i} \frac{\xi(\alpha'_1 r_j)\xi(\alpha'_2 r_j) r_j dr_j}{r_i} \right] r_i dr_i \\ &+ \int_0^{\infty} \xi(\alpha_1 r_i)\xi(\alpha_2 r_i) \left[\int_{r_i}^{\infty} \frac{\xi(\alpha'_1 r_j)\xi(\alpha'_2 r_j) r_j dr_j}{r_j} \right] r_i dr_i \end{aligned} \tag{6.81}$$

With normalized associated Laguerre wavefunction (Eq. (6.76)), $z_j = \alpha_2 r_j$, $w = \frac{\alpha'_2}{\alpha_2}$, using Eqs. (6.A.1) and (6.A.2) and subsequently employing standard integral (Eq. (6.A.3)), the integrals I_{M_1} and I_{M_2} assume the form (for bound states with circular symmetry, $|m_2| = |m'_2| = 0$) [28]:

$$\begin{aligned} I_{\mathrm{M}_1} &= \frac{N_1 N'_1 \mathfrak{B}_2 \mathfrak{B}'_2}{\alpha_2^2} \sum_{k_1=0}^{(n_2-1)} \sum_{k_2=0}^{(n'_2-1)} \frac{(-1)^{(k_1+k_2)} w^{k_2}}{k_1!\, k_2!} g^{(k_1+k_2+2)} (k_1 + k_2 + 1)! \binom{n_2 - 1}{k_1} \binom{n'_2 - 1}{k_2} \\ &\times \left\{ \left(\frac{\alpha_1 + \alpha'_1}{2} \right)^{-1} \sum_{p=0}^{\mathcal{N}_1} \sum_{q=0}^{\mathcal{N}'_1} \frac{(1)_{p+q} (-\mathcal{N}_1)_p (-\mathcal{N}'_1)_q}{(1)_p (1)_q\, p!\, q!} \left(\frac{2\alpha_1}{\alpha_1 + \alpha'_1} \right)^p \left(\frac{2\alpha'_1}{\alpha_1 + \alpha'_1} \right)^q \right. \\ &- \sum_{s=0}^{k_1+k_2+1} \frac{1}{s!} \left(\frac{\alpha_2}{g} \right)^s \left(\frac{(w+1)\alpha_2 + \alpha_1 + \alpha'_1}{2} \right)^{-(s+1)} \Gamma(s+1) \sum_{p=0}^{\mathcal{N}_1} \sum_{q=0}^{\mathcal{N}'_1} \frac{(s+1)_{p+q} (-\mathcal{N}_1)_p (-\mathcal{N}'_1)_q}{(1)_p (1)_q\, p!\, q!} \\ &\left. \times \left(\frac{2\alpha_1}{(w+1)\alpha_2 + \alpha_1 + \alpha'_1} \right)^p \left(\frac{2\alpha'_1}{(w+1)\alpha_2 + \alpha_1 + \alpha'_1} \right)^q \right\} \end{aligned}$$

$$I_{\mathrm{M}_2} = \frac{N_1 N_1' \mathfrak{B}_2 \mathfrak{B}_2'}{\alpha_2} \sum_{k_1=0}^{(n_2-1)} \sum_{k_2=0}^{(n_2'-1)} \frac{(-1)^{(k_1+k_2)}\, w^{k_2}}{k_1!\, k_2!} g^{(k_1+k_2+1)} (k_1+k_2)! \binom{n_2-1}{k_1} \binom{n_2'-1}{k_2}$$
$$\times \left\{ \sum_{s=0}^{k_1+k_2} \frac{1}{s!} \left(\frac{\alpha_2}{g}\right)^s \left(\frac{(w+1)\alpha_2+\alpha_1+\alpha_1'}{2}\right)^{-(s+2)} \Gamma(s+2) \sum_{p=0}^{\mathcal{N}_1} \sum_{q=0}^{\mathcal{N}_1'} \frac{(s+2)_{p+q}(-\mathcal{N}_1)_p(-\mathcal{N}_1')_q}{(1)_p(1)_q\, p!\, q!} \right.$$
$$\left. \times \left(\frac{2\alpha_1}{(w+1)\alpha_2+\alpha_1+\alpha_1'}\right)^p \left(\frac{2\alpha_1'}{(w+1)\alpha_2+\alpha_1+\alpha_1'}\right)^q \right\} \tag{6.82}$$

The exact value of the integral for monopole factor of 2D He-isoelectronic series can be obtained by Eq. (6.82).

Dipole Factor (l = 1) The integral of dipole factor ($m = \pm 1$) (Eq. (6.77)) involving p-orbitals can be evaluated as:

$$H_{\mathrm{PQ}}^{(1)} = \left\langle \Xi(R,\Phi)_{\mathrm{P}} \left| \left(\frac{r_<}{r_>^2}\right) \cos(\phi_i - \phi_j) \right| \Xi(R,\Phi)_{\mathrm{Q}} \right\rangle \tag{6.83}$$

The wavefunction for individual particles can be written as product of radial and angular parts for ground-state of planar He-isoelectronic ions for individual ith and jth spaces:

$$H_{\mathrm{PQ}}^{(1)} = \underbrace{\left\langle \xi(\alpha_1 r_i)\xi(\alpha_2 r_j) \left| \frac{r_<}{r_>^2} \right| \xi(\alpha_1' r_i)\xi(\alpha_2' r_j) \right\rangle}_{I_{\mathrm{D}}} \underbrace{\left\langle \frac{e^{im_1\phi}}{\sqrt{2\pi}} \frac{e^{im_2\phi}}{\sqrt{2\pi}} \left| \cos(\phi_i - \phi_j) \right| \frac{e^{im_1'\phi}}{\sqrt{2\pi}} \frac{e^{im_2'\phi}}{\sqrt{2\pi}} \right\rangle}_{I_{\mathrm{DA}}} \tag{6.84}$$

First evaluating the radial part by segregating the integral for each r_i considering integral over r_j and consequently integrating over r_i:

$$I_{\mathrm{D}} = I_{\mathrm{D}_1} + I_{\mathrm{D}_2}$$
$$I_{\mathrm{D}} = \int_0^\infty \frac{\xi(\alpha_1 r_i)\xi(\alpha_1' r_i)}{r_i} \left[\int_0^{r_i} r_j^2 \xi(\alpha_2 r_j)\xi(\alpha_2' r_j) dr_j \right] dr_i$$
$$+ \int_0^\infty r_i^2 \xi(\alpha_1 r_i)\xi(\alpha_2 r_i) \left[\int_{r_i}^\infty \frac{\xi(\alpha_2 r_j)\xi(\alpha_2' r_j) dr_j}{r_j} \right] r_i^2 dr_i \tag{6.85}$$

Letting $z_j = \alpha_2 r_j$, $w = \frac{\alpha_2'}{\alpha_2}$ and employing normalized associated Laguerre wavefunction and Eqs. (6.A.1), (6.A.2) and then (6.A.3) shape the integrals I_{D_1} and I_{D_2} to [28]:

$$I_{\mathrm{D}_1} = \frac{BB'NN'\alpha_1^{(|m_1|)}(\alpha_1')^{(|m_1'|)}}{\alpha_2^3} \sum_{k_1=0}^{\mathcal{N}_2} \sum_{k_2=0}^{\mathcal{N}_2'} \frac{(-1)^{k_1+k_2}}{k_1!\, k_2!} (D+2)!\, w^{|m_2'|+k_1} g^{(D+3)}$$
$$\times \binom{n_2+|m_2|-1}{2|m_2|+k_1} \binom{n_2'+|m_2'|-1}{2|m_2'|+k_2}$$
$$\times \left[\left(\frac{\alpha_1+\alpha_2}{2}\right)^{-(|m_1|+|m_1'|)} \Gamma\left(|m_1|+|m_1'|\right) \sum_{p=0}^{\mathcal{N}_1} \sum_{q=0}^{\mathcal{N}_1'} \frac{(|m_1|+|m_1'|)_{(p+q)}(-\mathcal{N}_1)_p(-\mathcal{N}_1')_q}{(2|m_1|+1)_p\,(2|m_1'|+1)_q\, p!\, q!} \right.$$
$$\times \left(\frac{2\alpha_1}{\alpha_1+\alpha_1'}\right)^p \left(\frac{2\alpha_1'}{\alpha_1+\alpha_1'}\right)^q \sum_{s=0}^{(A-1)} \left(\frac{\alpha_1+\alpha_2}{2}\right)^{-(s+|m_1|+|m_1'|)} \Gamma\left(s+|m_1|+|m_1'|\right)$$

$$\times \sum_{p=0}^{\mathcal{N}_1} \sum_{q=0}^{\mathcal{N}_1'} \frac{(s+|m_1|+|m_1'|)_{(p+q)}(-\mathcal{N}_1)_p(-\mathcal{N}_1')_q}{(2|m_1|+1)_p\,(2|m_1'|+1)_q\; p!\; q!} \left(\frac{2\alpha_1}{(w+1)\alpha_2+\alpha_1+\alpha_1'}\right)^p \left(\frac{2\alpha_1'}{(w+1)\alpha_2+\alpha_1+\alpha_1'}\right)^q \Bigg]$$

$$I_{\mathrm{D}_2} = NN'BB'\alpha_1^{|m_1|}\alpha_1^{|m_1'|} \sum_{k_1=0}^{\mathcal{N}_2} \sum_{k_1=0}^{\mathcal{N}_2'} \frac{(-1)^{k_1+k_2}}{k_1!\,k_2!} w^{(|m_2|+k_2)} g^{(D)} \binom{n_2+|m_2|-1}{2|m_2|+k_1} \binom{n_2+|m_2|-1}{2|m_2'|+k_2}$$

$$\times \left[\sum_{s=0}^{(D-1)} \frac{1}{s!}\left(\frac{\alpha_2}{g}\right)^s \Gamma\left(s+|m_2|+|m_2'|+3\right) \left(\frac{(w+1)\alpha_3+\alpha_1+\alpha_2}{2}\right)^{-(s+|m_2|+|m_2'|+3)}\right.$$

$$\times \sum_{p=0}^{\mathcal{N}_1} \sum_{q=0}^{\mathcal{N}_2} \frac{(s+|m_2|+|m_2'|+3)_{p+q}(-\mathcal{N}_1)_p(-\mathcal{N}_2)_q}{(2|m_1|+1)_p(2|m_1'|+1)_q\; p!\; q!}$$

$$\times \left(\frac{2\alpha_1}{(w+1)\alpha_2+\alpha_1+\alpha_1'}\right)^p \left(\frac{2\alpha_1'}{(w+1)\alpha_2+\alpha_1+\alpha_1'}\right)^q \Bigg] \tag{6.86}$$

Now, evaluating the angular part of the integral (Eq. (6.84)):

$$\begin{aligned} I_{\mathrm{DA}} &= \left\langle \frac{e^{im_1\phi_i}}{\sqrt{2\pi}} \frac{e^{im_2\phi_j}}{\sqrt{2\pi}} \middle| \cos(\phi_i-\phi_j) \middle| \frac{e^{im_1'\phi_i}}{\sqrt{2\pi}} \frac{e^{im_2'\phi_j}}{\sqrt{2\pi}} \right\rangle \\ &= \left\langle \frac{e^{im_1\phi_i}}{\sqrt{2\pi}} \frac{e^{im_2\phi_j}}{\sqrt{2\pi}} \middle| \frac{e^{i(\phi_i-\phi_j)}+e^{-i(\phi_i-\phi_j)}}{2} \middle| \frac{e^{im_1'\phi_i}}{\sqrt{2\pi}} \frac{e^{im_2'\phi_j}}{\sqrt{2\pi}} \right\rangle \\ &= \frac{1}{2}\left(\delta_{(m_{1,i};m_{1,i}'+1)}\delta_{(m_{1,j};m_{1,j}'-1)} + \delta_{(m_{1,i};m_{1,i}'-1)}\delta_{(m_{1,j};m_{1,j}'+1)}\right) \end{aligned} \tag{6.87}$$

6.2.2.3 Energy Calculation Through Perturbation

The higher order perturbation calculation can be represented as [25, 115, 116]:

$$\begin{aligned} \mathrm{E_P} &= \mathrm{E_P^{(0)}} + \mathrm{E_P^{(1)}} + \mathrm{E_P^{(2)}} + \mathrm{E_P^{(3)}} + \cdots + \cdots \\ \Psi_\mathrm{P} &= \Psi_\mathrm{P}^{(0)} + \Psi_\mathrm{P}^{(1)} + \Psi_\mathrm{P}^{(2)} + \Psi_\mathrm{P}^{(3)} + \cdots + \cdots \end{aligned} \tag{6.88}$$

where

$$\mathrm{E_P^0} = \left\langle \Xi_\mathrm{P}^0 \left| H_1^0 + H_2^0 \right| \Xi_\mathrm{P}^0 \right\rangle, \tag{6.88a}$$

$$\mathrm{E_P^{(1)}} = H_\mathrm{PP}^{(1)} = \left\langle \Xi_\mathrm{P}^0 \left| H' \right| \Xi_\mathrm{P}^0 \right\rangle, \tag{6.88b}$$

$$\begin{aligned} \Psi_\mathrm{P}^{(1)} &= \sum_{\mathrm{Q}\neq\mathrm{P}} \frac{H_\mathrm{QP}^{(1)}}{\mathrm{E_P^0}-\mathrm{E_Q^0}} \Psi_\mathrm{Q}^{(0)}, \\ \mathrm{E_P^{(2)}} &= \sum_{\mathrm{Q}\neq\mathrm{P}} \frac{H_\mathrm{PQ}^{(1)} H_\mathrm{QP}^{(1)}}{\mathrm{E_P^0}-\mathrm{E_Q^0}} = \sum_{\mathrm{Q}\neq\mathrm{P}} \frac{\left\langle \Xi_\mathrm{P}^0 \left| H' \right| \Xi_\mathrm{Q}^0 \right\rangle \left\langle \Xi_\mathrm{Q}^0 \left| H' \right| \Xi_\mathrm{P}^0 \right\rangle}{\mathrm{E_P^0}-\mathrm{E_Q^0}}, \end{aligned} \tag{6.88c}$$

$$\begin{aligned} \Psi_\mathrm{P}^{(2)} &= \sum_{\mathrm{Q}\neq\mathrm{P}} \sum_{\mathrm{M}\neq\mathrm{P}} \frac{H_\mathrm{QP}^{(1)} H_\mathrm{MQ}^{(1)}}{(\mathrm{E_P^0}-\mathrm{E_Q^0})(\mathrm{E_P^0}-\mathrm{E_M^0})} \Psi_\mathrm{M}^{(0)} - H_\mathrm{PP}^{(1)} \sum_{\mathrm{M}\neq\mathrm{P}} \frac{H_\mathrm{MP}^{(1)}}{(\mathrm{E_P^0}-\mathrm{E_M^0})^2} \Psi_\mathrm{M}^{(0)}, \\ \mathrm{E_P^{(3)}} &= \sum_{\mathrm{Q}\neq\mathrm{P}} \sum_{\mathrm{M}\neq\mathrm{P}} \frac{H_\mathrm{PQ}^{(1)} H_\mathrm{QM}^{(1)} H_\mathrm{MP}^{(1)}}{(\mathrm{E_P^0}-\mathrm{E_Q^0})(\mathrm{E_P^0}-\mathrm{E_M^0})} H_\mathrm{PP}^{(1)} \sum_{\mathrm{Q}\neq\mathrm{P}} \frac{H_\mathrm{PQ}^{(1)} H_\mathrm{QP}^{(1)}}{(\mathrm{E_P^0}-\mathrm{E_Q^0})^2}, \\ \cdots &= \cdots \end{aligned} \tag{6.88d}$$

where Eqs. (6.88a)–(6.88d) represent the zeroth (unperturbed), first, second, third-order energy corrections (ECs) respectively and Ξ is the product of wavefunction of the individual particles.

6.2.2.4 Current Density of 2-e System

Obeying laws of conservation of mass, charge and probabilities on continuity equation with 2-e wavefunction ($\Psi(1,2) = \sum_p c_p \phi_i(1)\phi_j(2) \Longrightarrow \Psi^*(1,2) = \sum_p c_q \phi_k^*(1)\phi_l^*(2)$) for 2D and 3D He-isoelectronic ions, we unprecedentedly developed current densities (**J**s) of two electrons, respectively, as [27, 28, 104, 117, 118]:

$$\mathbf{J}_1 = \left(\frac{\hbar}{2m_e i}\right)\left\{\sum_p \sum_q c_p^* c_q (\phi_i^* \vec{\nabla}\phi_k - \phi_k \vec{\nabla}\phi_i^*)\right\}$$
$$\mathbf{J}_2 = \left(\frac{\hbar}{2m_e i}\right)\left\{\sum_p \sum_q c_p^* c_q (\phi_j^* \vec{\nabla}\phi_l - \phi_l \vec{\nabla}\phi_j^*)\right\} \tag{6.89}$$

6.3 Results and Discussion

In the present work, we investigated the electronic structure of different isotropic and anisotropic 2D and 3D systems as a function of magnetic field (ω_c). The anisotropy was introduced in the geometry of the dot through the gate voltage as deformation factor, η. For 3D dot, $\eta = \omega_z/\omega_0$ with ω_0 as the in-plane and ω_z vertical confinement frequencies respectively. In case of 2D dot, $\eta = \omega_z/\omega_0$ where ω_z and ω_0 are confinements along x- and y-axes respectively. For the brevity of the chapter, we have explained the electronic structure of only a few systems.

6.3.1 3D Interacting Electron Gas

Diagonalization was accomplished with 729 basis set to capture the spectra of e–e interaction in 3D isotropic 2-e QD at confinement frequency, $\omega_0 = 0.002$ and 0.00002 a.u., $m_e = 0.1$ a.u., and ϵ=8.0 a.u. as a function of ω_c. (Figure 6.1) [16]. The ground state prevailed spherical symmetry with ω_c for noninteracting system [16]. The states with positive and negative "m" manifested in Landau bunching at intermediate ω_c. The anisotropy due to different vertical and lateral confinements removed Landau bunching at high magnetic field (Figure 6.1) [16]. The complexity in the spectra was increased upon inclusion of the e–e interaction which led to the change in ground state symmetry along with multiple degeneracies (Figure 6.1) [16]. The Coulomb integral for higher angular momentum (l) in 3D converged to:

$$\left\langle \xi_{0,l,m}(r) \left| \frac{1}{r} \right| \xi_{0,l,m}(r) \right\rangle \simeq \frac{1}{\epsilon}\sqrt{\frac{\Omega_r}{l+1}} \tag{6.90}$$

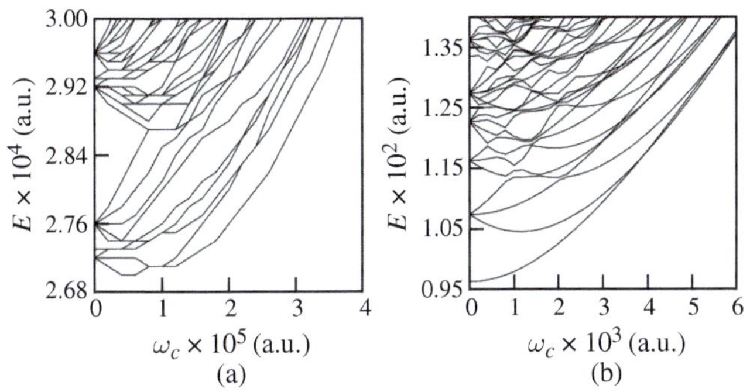

Figure 6.1 Energy level diagram of 3D isotropic ($\eta = 1$) QD as a function of ω_c with $m_e = 1.0$ a.u., $\epsilon = 8.0$ a.u., $N = 2$, and $\omega_0 = 0.002$ and 0.00002 a.u. (a, b), respectively.

The Coulomb interaction being inversely related to l caused huge mixing among bound states [16].

Energy spectra of 3D anisotropic ($\eta = 2.0$) atom-like QD ($\epsilon = 1.0$, $m_e = 1.0$ a.u.) for $N = 2$ and 4 interacting electrons were studied with the basis span of 1000–2000 as function of ω_c at $\omega_0 = 0.004$ a.u. and $\omega_z = 0.008$ a.u. (Figure 6.2) [17]. Convergence was achieved by quadrupole factor (Figure 6.2) [17]. For $N = 2$, anisotropy ($\eta = 2$) removed degeneracies among lowest bound states with subsequently increased energy gap between ground and first excited states in absence of the magnetic field (Figure 6.2) [17]. In the presence of magnetic field, evolution of states became dormant and the crossing of ground state shifted to higher ω_c (~0.0031 a.u.) in comparison to isotropic case ($\eta = 1.0$) (Figure 6.2) [17]. For $N = 4$, Coulomb interaction was found to dominate over gate voltage, magnetic field, and anisotropy. It incurred the loss of degeneracy at $\omega_c = 0.0$ a.u. and fast evolution of states without any clustering among levels (Figure 6.1, Figure 6.1b). Despite the anisotropy ($\eta = 2.0$) slowing down the evolution of the states, impression of level clustering had not been exhibited as Coulomb interactions dominated (Figure 6.2) [17].

The study of 3D anisotropic ($\eta = 2.0$) ZnO QD having higher m_e (=0.24 a.u.) and low ϵ (=8.5) for N=2 and 4 at $\omega_0 = 0.004$ a.u. was carried over 100–300 MCs in contrast to single configuration Hartree–Fock SCF method (Figure 6.3) [18]. The energy spectra showed some peculiar patterns and formation of shell structure due to the synergistic effect of Coulomb interaction, full-scale fermionic exchange symmetry and anisotropy [18, 40]. For isotropic ($\eta = 1.0$) dot, a mixing among lower bound states was observed due to the interplay of Coulomb interactions and magnetic field. Increasing particle number emerged as the rapid evolution of states with ω_c along with earlier crossing/anticrossing of ground state. It was reflected in the form of pronounced level bunching/accidental degeneracies among lower bound states for N =6–8 [18]. Such level

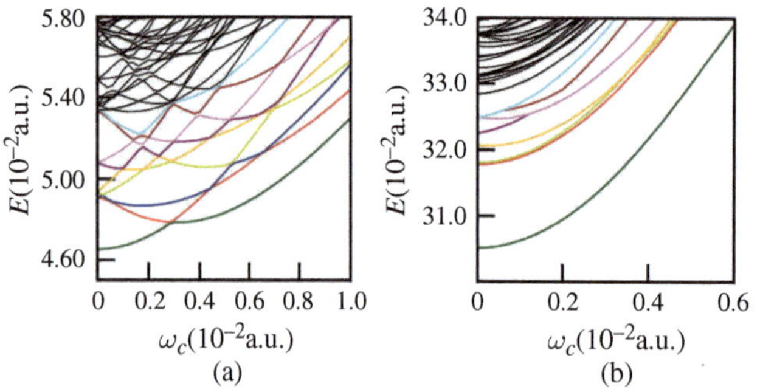

Figure 6.2 Energy level diagram of 3D anisotropic ($\eta = 2$) atom-like QD as a function of ω_c with $m_e = 1.0$ a.u., $\epsilon = 1.0$ a.u. and $\omega_0 = 0.004$ a.u. for $N = 2$ and 4 (a, b), respectively.

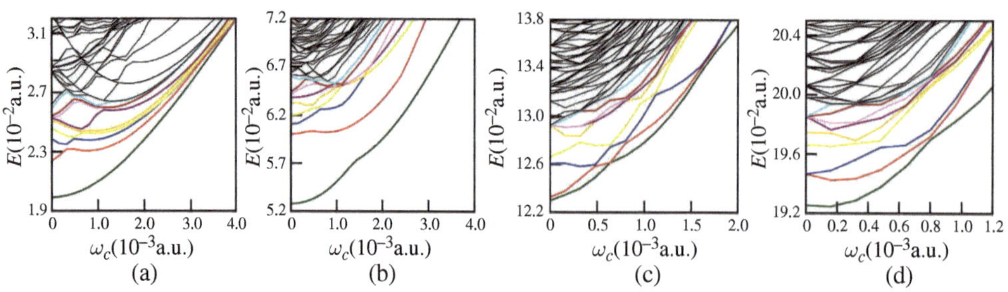

Figure 6.3 Energy level diagram of 3D anisotropic ($\eta = 2$) ZnO QD as a function of ω_c with $m_e = 1.0$ a.u., $\epsilon = 1.0$ a.u. and $\omega_0 = 0.004$ a.u. for $N = 2$, 4, 6 and 8 (a–d), respectively.

clusterings implied a high degree of symmetry in the system at specific magnetic fields [18, 119]. The anisotropy ($\eta = 2.0$) resulted in slower evolution of states as a function of magnetic field in comparison to isotropic case along with reduction of level crossings/anticrossings and loss of degeneracies (Figure 6.3) [18]. For $N > 4$, interplay between anisotropy and Coulomb interactions caused earlier crossing among lowest bound states to form level clustering at ω_c ~0.0025 a.u. [18]. Accidental degeneracies/level clusterings represented restoring of dynamic spherical symmetry in the system at certain magnetic fields (Figure 6.3) [18, 40].

6.3.2 2D Interacting Electron Gas

We utilized 756 bases set for diagonalization to examine the energy level diagram as function of m_e, ω_0, ω_c, ϵ and e–e interaction (upto monopole factor) on the energy level diagram. Integral of Coulomb interaction for monopole factor could be approximated for states with higher angular momentum (m) to,

$$\left\langle \xi_{0,m}(r) \left| \frac{1}{r} \right| \xi_{0,m}(r) \right\rangle \simeq \frac{1}{\epsilon} \sqrt{\frac{\Omega_r}{|m| + \frac{1}{2}}} \tag{6.91}$$

virtually merging to Laughlin's results at high ω_c [45]. The energy spectra as a function of ω_c with $\omega_0 = 0.002$, 0.00002 a.u., $m_e = 0.01$ a.u. and $\epsilon = 8.0$ was studied to capture *CdS* system (Figure 6.4) [19]. Since Coulomb interaction occurred in factor of $\sqrt{m_e}$ and $\frac{1}{\epsilon}$, unusual features emerged in electronic structures as crossing/anti-crossings among states, change in degeneracy, level clustering formation of same or different sets of "magic number" for shell closure [37–39, 41, 120–122]. Therefore, the degeneracy pattern of energy levels was changed and finally at strong magnetic field Landau bunching appeared. The ground state remained node-less throughout ω_c for a non-interacting system [19]. At the onset of Coulomb interaction, the states with $m = -1, -2, -3 \ldots$ became the ground state at a certain value of $m_e\omega_0/\omega_c$ (0.10 for $m_e = 0.01$) (Figure 6.4) [19]. At high confinement ($\omega_0 = 0.002$ a.u.) and dielectric medium ($\epsilon = 8$ a.u.), the energy spectrum closely resembled the non-interacting systems (Figure 6.4) [19]. A 100 times decrease in confinement frequency reduced the noninteracting part of energy in same fraction whereas interaction energy was decreased roughly by 10 times (Eq. (6.91)). Consequently, an early crossing of ground state was observed at certain threshold ω_c=7×10^{-5} and 1.6×10^{-5} a.u. for $\omega_0 = 0.002$ and 0.00002 a.u. respectively [19].

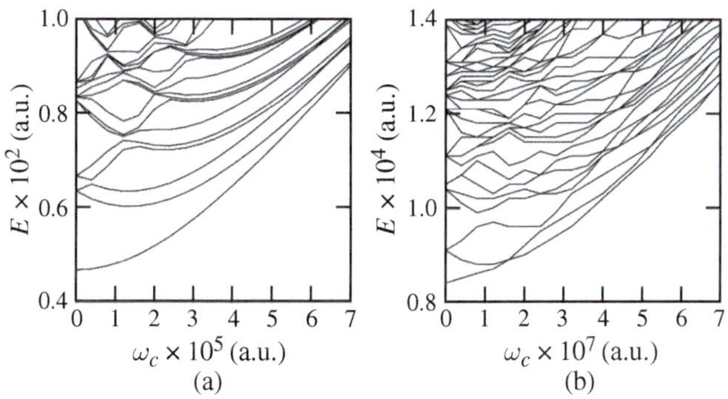

Figure 6.4 Energy level diagram of 2D isotropic ($\eta = 1$) QD as a function of ω_c with $m_e = 0.01$ a.u., $\epsilon = 8.0$ a.u., $N = 2$ and ω_0= 0.002 and 0.00002 a.u. (a, b), respectively.

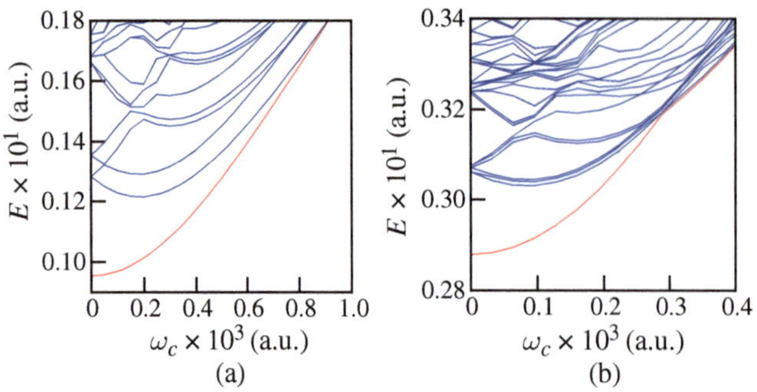

Figure 6.5 Energy level diagram of 2D isotropic ($\eta = 1$) GaAs QD as a function of ω_c with $m_e = 0.07$ a.u., $\epsilon = 12.35$ a.u. and $\omega_0 = 0.004$ a.u. for $N = 2$ and 4 (a, b), respectively.

Monopole and dipole integrals had significant contributions to Coulomb (exchange) correlations in comparison to quadrupole factor. Energy levels were raised more with the carrier density ($N = 2$, 3, 4 and 5) for atomic systems (He, Li, Be) than other systems (GaAs and BN) because of low dielectric constant. The electronic structure of GaAs QD ($m_e = 0.07$ a.u. and $\epsilon = 12.35$) at $\omega_0 = 0.004$ a.u. for $N = 2$ and 4 registered an unusual pattern on energy spectra (Figure 6.5) [20]. In systems with high confinement ($\omega_0 = 0.004$ a.u.), weak correlations showed a little effect on the energy spectrum (Figure 6.5) [20]. The early crossing of ground state shifts monotonically to lower field with increase in the carrier density (ω_c ~0.0009 a.u. and 0.0003 a.u. for $N = 2$ and 4 respectively) (Figure 6.5) [20]. Lowering of ω_0 by 10 times, a three times decrease of Coulomb integrals was observed in comparison to non-interacting energy (Eq. (6.91)) [20].

Combined interplay among anisotropy (η), Coulomb interactions and Fermionic exchange symmetry (FES) as function of ω_c was exhibited through parametric dependence on function of $N = 3, 4, 5, 6, \ldots 20$. Incorporation of the multi configuration Slater determinant effectively addressed Pauli's exclusion principle [10, 123]. The effect of Coulomb and exchange interactions with increasing particle number for N–e isotropic ($\eta = 1$) and anisotropic ($\eta = 2$) GaAs QD for $N = 2$, 5, 8, and 14 was examined. For isotropic dot with $N = 2$, crossing of ground state shifted to lower ω_c ~0.0006 a.u. with incorporation of FES [21]. For $N = 5$, a doubly degenerate ground state occurred due to enhanced exchange Coulomb correlations and FES [21]. For $N = 8$, degeneracy was removed at $\omega_c = 0$ and the ground state was observed to participate in crossing and anti-crossing for non-zero magnetic field [21]. It was revealed that $N = 14$ to 19, the ground state preserved degeneracy throughout the magnetic field suppressing Zeeman effect. This unusual characteristic exhibited the effect of high Coulomb correlation, FES and occupation number of energy levels due to effect of Fermi-hole and Fermi-heap phenomena [21]. The anisotropy ($\eta = 2$) withdrew the degeneracy at zero magnetic field till $N = 8$ for $\eta = 2$ but increasing N suppressed the anisotropy and preserved the degeneracy (Figure 6.6) [21]. A ground state with double degeneracy and little stabilization was seen for $\eta = 2$ in comparison to the spectra of $\eta = 1$ (Figure 6.6) [21]. Moreover, crossing and anti-crossing of the ground state of 2-e dot was observed at ω_c ~0.0015–0.0025 a.u. in comparison to isotropic dot (Figure 6.6). Similar aspect of degeneracy was observed for $N = 14$ upto ω_c ~0.0003 a.u. for η=2 (Figure 6.6) [21].

6.3.3 3D Exciton Complexes

A detailed examination of the interplay among $\omega_e = \omega_h = \omega_0$, ω_c, ϵ and e–h/e–e/h–h interactions, m_e and m_h on the energy spectra of BN ($m_e = 0.7$ a.u., $\epsilon = 7.1$) exciton complex was carried out.

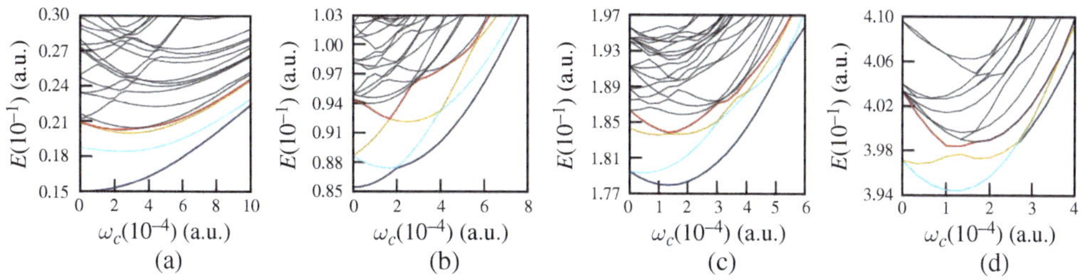

Figure 6.6 Energy level diagram of 2D anisotropic ($\eta = 2$) GaAs QD as a function of ω_c with $m_e = 0.07$ a.u., $\epsilon = 12.35$ a.u. and $\omega_0 = 0.004$ a.u. for $N = 2, 5, 8$ and 14 (a – d), respectively.

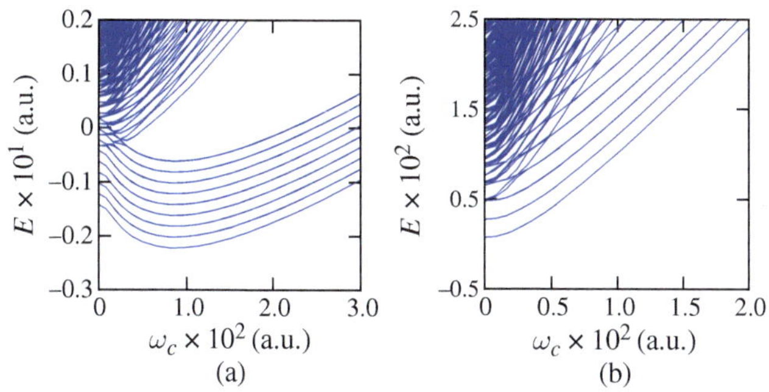

Figure 6.7 Energy level diagram of 3D isotropic ($\eta = 1$) BN QD as a function of ω_c with $m_e = 0.7$ a.u., $\epsilon = 7.1$ a.u. at $\omega_0 = 0.0002$ and 0.002 a.u. (a, b) respectively.

The diagonalization was accomplished over 1296 product basis set. Incorporation of $\left(\frac{\alpha-1}{\alpha}\right)\frac{m\omega_c}{2m_e}$ and perturbation term including Coulomb interaction produced more crossings and anti-crossing among bound states. Therefore, the degeneracy pattern was reformed under strong magnetic field ($\omega_c \gg \omega_0$) realizing Landau levels. The excess stabilization of lowest states giving rise to less crossings and anti-crossings was governed by enhanced Coulomb (exchange) interactions dominating over ω_o and ω_c (Figure 6.7) [22]. At moderate ω_c, Coulomb correlation became more stabilizing factor than exchange interaction and the value read as:

$$\left\langle \xi_{0,l,m}(r) \left| \frac{1}{r} \right| \xi_{0,l,m}(r) \right\rangle \simeq \frac{1}{\epsilon}\sqrt{\frac{\Omega_r}{l+1}} \tag{6.92}$$

The ground state was highly degenerate at $\omega_0 = 0.0$ a.u. but the degeneracy was eliminated with the onset of ω_c (Figure 6.7) [22]. This feature could be assumed analogous to the well-known Zeeman effect (Figure 6.7) [22]. For non-interacting exciton complex of BN QD, the energy of the ground state increased along with enhanced crossings and anti-crossings as function of ω_c [22]. At low $\omega_o = 0.0002$ a.u., the energy gap between ground and higher excited states was reduced for interacting excitons (Figure 6.7) [22]. At high $\omega_o = 0.002$ a.u., crossings/anti-crossings were intensified with increasing ω_c and α (1.0 to 10.0) (Figure 6.7) [22]. However, features of the energy spectra remained similar with increasing α from 10.0 to 100.0 [22].

6.3.4 2D Exciton Complexes

Diagonalization was carried out over 1980 basis set to examine the electronic structure of 2D isotropic BN QD. A systemic investigation of the interplay among $\omega_e = \omega_h = \omega_o$ (isotropic

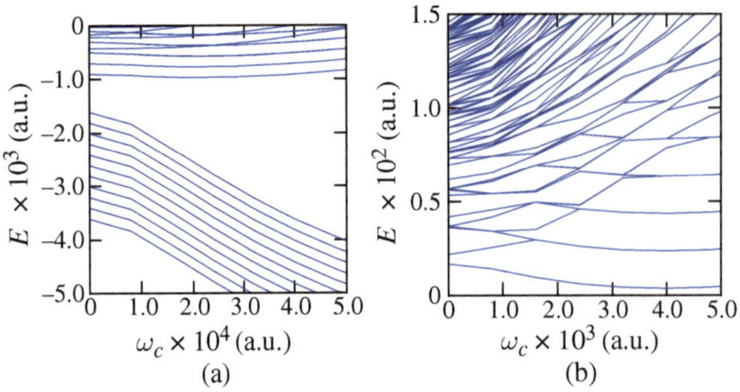

Figure 6.8 Energy level diagram of 2D isotropic ($\eta = 1$) BN QD as a function of ω_c with $m_e = 0.7$ a.u., $\epsilon = 7.1$ a.u. at $\omega_0 = 0.0002$ and 0.002 a.u. (a, b) respectively.

environment), ω_c, ϵ, e–h/e–e/h–h interactions, m_e and m_h on the energy levels was accomplished. Incorporation of the factor $\left(\frac{\alpha-1}{\alpha+1}\right)\frac{m\omega_{c\_r}}{2m_e}$ showed no remarkable change in pattern of crossings and anti-crossings among lower bound states. The lower bound states did not show much of mixing with the onset of magnetic field as the Coulomb (exchange) integrals stabilized the system (Figure 6.8) [23]. The extent of stabilization was monitored by both ω_0 and ω_c. At low ω_c, Coulomb interaction dominated whereas for high ω_c, Landau levels appeared in spite of the stabilization due to exchange interaction (Figure 6.8) [23]. This trend was further enhanced at higher ω_0 (Figure 6.8) [23]. In moderate magnetic field, Coulomb interaction seemed to stabilize more than its exchange counterpart. Ground state energy of non-interacting BN exciton complex was observed to increase rapidly with increase in ω_c (Figure 6.8) [23]. The energy gap between ground and higher excited states was further reduced for interacting systems at low $\omega_0 = 0.0002$ a.u., but at high $\omega_0 = 0.002$ a.u., shuffling among the hydrogenic and the gaussian states enhanced crossings/anti-crossings with increasing ω_c and α from 1.0 to 10.0 (Figure 6.8) [23]. At $\omega_0 = 0.0$ a.u., few lowest bound states were stabilized. The ground state exhibited unusually high degeneracy at $\omega_0 = 0.0$ a.u. in the presence of Coulomb interaction, which was removed immediately in presence of magnetic field, reminiscent of Zeeman-effect [23].

The electronic structure of exciton complexes of 2D isotropic and anisotropic MoS_2 ($m_e = 0.60$, $\epsilon = 4.26$) QD with N upto 3 was investigated with interplay among ω_0, ω_c and many-body Coulomb potentials (Figure 6.9) [24]. ω_c was seen to enhance Coulomb (exchange) interactions stabilizing the bound states. Signature of anisotropy and many-body Coulomb interactions indicated significant difference between isotropic and anisotropic environment. Anisotropy was noticed to destroy degeneracy of the system. The ground state was stabilized more with increasing number of electrons and hole (Figure 6.9) [24]. The stabilization of lowest bound states was enhanced with increasing ω_0 and their energy became lowest at a threshold $\omega_c = 0.0015$ a.u. due to anisotropy and Coulomb (exchange) interactions (Figure 6.9) [24]. For weaker confinement, similar but steeper trend of stabilization was also noticed (Figure 6.9) [24].

6.3.5 3D He-Isoelectronic Species

GSEs of He-isoelectronic species were obtained by analytical perturbation calculation upto third order EC incorporating monopole and dipole factors of Coulomb Green's function (Table 6.1) [25–27]. It was observed that the poles higher than dipole factor contributed negligibly to the GSE [25–27].

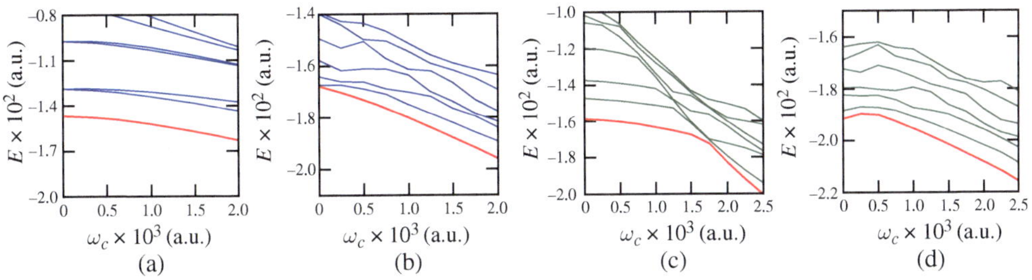

Figure 6.9 Energy level diagram of exciton complexes in 2D MoS_2 QD as a function of ω_c with $m_e = 0.60$ a.u., $\epsilon = 4.26$ a.u., $\omega_0 = 0.0002$ a.u. and $\eta = 1$ for exciton and charged exciton/trion (a, b), respectively, and $\eta = 2$ for exciton and charged exciton/trion (c, d), respectively.

Table 6.1 Comparison of GSEs (E_g) of H^- to Ne^{8+} in a.u. obtained by our analytical (perturbation and variational) treatment with reported values [81, 124]. Contribution of first order ($E_0^{(1)}$), second order ($E_0^{(2)}$), and third order ($E_0^{(3)}$) EC to ground-state energies for different Z is shown.

	Perturbation calculation					MC variational Calculation	
Z	**$E_0^{(0)}$ (a.u.)**	**$E_0^{(1)}$ (a.u.)**	**$E_0^{(2)}$ (a.u.)**	**$E_0^{(3)}$ (a.u.)**	**E_g(a.u.)**	**E_g(a.u.)**	**E_{exp}(a.u.)**
1 (H^-)	−1.0	0.625	−0.060882	−0.055491	−0.491373	−0.516737	−0.52776
2 (He)	−4.0	1.250	−0.060882	−0.027746	−2.838627	−2.849910	−2.90378
3 (Li^+)	−9.0	1.875	−0.060882	−0.018216	−7.204098	−7.209930	−7.28041
4 (Be^{2+})	−16.0	2.500	−0.060882	−0.013873	−13.574754	−13.578288	−13.65744
5 (B^{3+})	−25.0	3.125	−0.060882	−0.011052	−21.946933	−21.949615	−22.03603
6 (C^{4+})	−36.0	3.750	−0.060882	−0.009108	−32.319990	−32.322297	−32.41733
7 (N^{5+})	−49.0	4.375	−0.060882	−0.007271	−44.693153	−44.695704	−44.80351
8 (O^{6+})	−64.0	5.000	−0.060882	−0.006936	−59.067818	−59.069542	−59.19580
9 (F^{7+})	−81.0	5.625	−0.060882	−0.006131	−75.442013	−75.443659	−75.59658
10 (Ne^{8+})	−100.0	6.250	−0.060882	−0.005526	−93.816408	−93.817964	−94.00835

6.3.5.1 Analysis of $E_0^{(2)}$ of He-Isoelectronic Ions

Monopole and dipole factors were reported to contribute significantly to $E_0^{(2)}$ (Eq. (6.88c)) (Table 6.1). Singly and doubly excited states associated with spherical ($l = 0$) and dumbbell ($l = 1$) symmetries enhanced GSEs of H^-, He, Li^+, Be^{2+}, B^{3+}, C^{4+}, N^{5+}, O^{6+}, F^{7+} and Ne^{8+} by −0.06088169 a.u. but in absence of the dipole factor $E_0^{(2)}$ appeared to be −0.06005083 a.u. and independent of Z (Table 6.1) [25–27].

6.3.5.2 Analysis of $E_0^{(3)}$ of He-Isoelectronic Ions

The range of $E_0^{(3)}$ was found to be −0.055491 a.u. to −0.005523 a.u. for He-isoelectronic ions. Cumulative EC due to $E_0^{(2)}$ and $E_0^{(3)}$ was realized to be exactly 57.63% (100×0.08862731/0.15378) for He [25–27]. Inclusion of dipole factor caused a change of −0.00608594 a.u. to −0.00028042 a.u. in $E_0^{(3)}$ for H^- to Ne^{8+} due to interdependence between monopole and dipole factors [25–27]. Alike $E_0^{(2)}$, dipole factor escalated the contribution due to monopole factor for third-order EC. However,

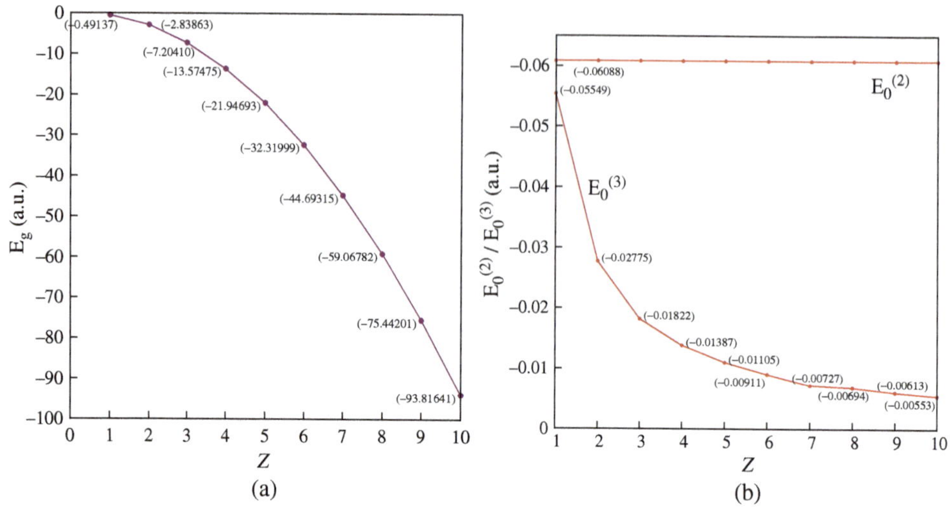

Figure 6.10 (a) GSEs (through perturbation approach) of different Z. (b) Total contribution of $E_0^{(2)}$ and $E_0^{(3)}$ with increasing Z (ref. Table I).

reflection symmetry of $E_0^{(3)}$ plot against principal quantum numbers (n_2, n_4) of He including dipole factor also proved the reciprocity of Green's theorem and indistinguishability of electrons [25–27].

We also employed MCs to examine the GSEs of He-isoelectronic ions with electron configuration $(1s...2s...)^t\ (2p...3p...)^{N-t}$ with $t = 0, 1, \ldots, N$ variationally and analogized with the literature [29, 81, 124] (Table 6.1, Figure 6.10). 1128 MCs seemed sufficient for the convergence of the GSEs of He-isoelectronic sequence. The runtime for diagonalization was ~4 minutes for He-isoelectronic series in IBM System (x3400 M2/x3300 M4 with Quad-Core Intel Xeon processor) [29]. GSEs were noted to lie in far better agreement than the literature [27, 29]. It was noticed that MCs other than so-called electron configuration for He-isoelectronic series had realizable contribution. The GSE of H^- shows interestingly better result from −0.491373 a.u. to −0.516737 a.u. differing only 2.18% from the reported value (Table 6.1) [29]. The first excited states of He-isoelectronic series exhibited triple degeneracy [29].

The perturbed ground state (Eqs. (6.88), (6.88a)–(6.88d)) expanded over singly and doubly excited spherically symmetric hydrogenic orbitals ($l = 0$) emerged zero current density within the ions. Monopole and dipole factros could be anticipated to generate non-zero current density within He-isoelectronic ions by introducing orbitals of both spherical ($l = 0$) and dumbell ($l = 1$) symmetries [27].

6.3.6 2D He-Isoelectronic Species

The GSEs of 2D He-isoelectronic series (H^-, He, Li^+, Be^{2+}, B^{3+}, C^{4+}, N^{5+}, O^{6+}, F^{7+} and Ne^{8+}) were examined analytically utilizing monopole ($m = 0$) and dipole ($m = 1$) factors of 2D Coulomb Green's function for bound states with circular and dumbell symmetries [28]. The $E_0^{(1)}$ of GSE for 2D He-isoelectronic ions was estimated to be $2Z$ [28]. The $E_0^{(2)}$ stabilized planar He-isoelectronic ions by −0.06754743 a.u. [28]. Alike 3D He-isoelectronic ions, $E_0^{(2)}$ was found to be independent of Z whereas $E_0^{(3)}$ showed inverse proportionality to Z. The GSE of 2D He atom seemed to be −12.09614727 a.u., which varied only 1.82% from the literature [28, 125].

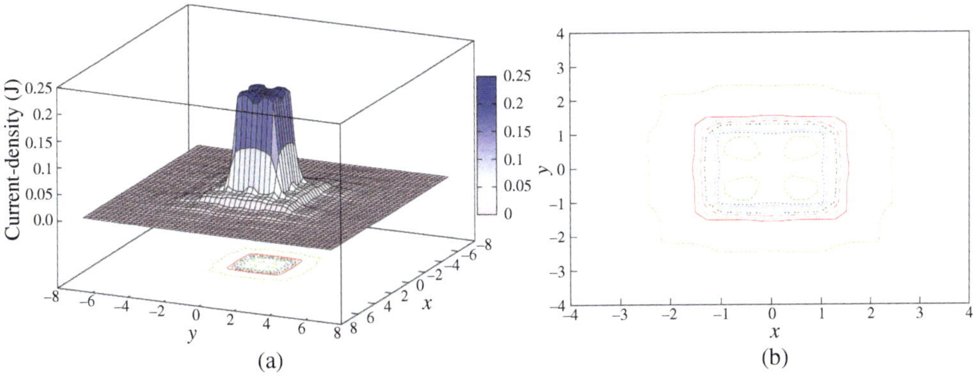

Figure 6.11 (a) Current density along x- and y-axes for heavy-hole trions of monolayer TMDCs and (b) Contour plot for current density along x- and y-axes for heavy-hole trions of monolayer TMDCs.

A steady local current was exhibited around the heavy-hole of trions (Figure 6.11). As the surface plots majorly showed the current density along x- and y-axes in absence of circular symmetry. It could be concluded that alone monolpole factor had no contribution to current density. The combined effect of dipole and monopole factors orginated as non-zero current density around the heavy-hole due to both s- and p-orbitals (Figure 6.11) [28].

6.4 Concluding Remarks

Analytical, terminable, simple closed forms of integrals for Coulomb Green's function for 2D/3D H-like and oscillator systems under perpendicular magnetic field were improvised to cover wide domain of materials from atomic to mesoscopic scales within non-relativistic limit. Such integrals did not visit any divergence problem. As the trial wavefunctions were constructed of electrons' generic coordinates, implementation of Pauli exclusion principle became very simple and convenient for large number of electrons. Therefore, the theory did not necessitate any pseudo-potentials to mimic bare Coulomb interaction and furnished drastically reduced computation cost even for MC calculations. The analytical MC calculation could throw light on the unexplored corner beyond the closed-, restricted- and open-shell calculations. It was able to produce key results of Fermi liquid model, exciton complexes, He-isoelectronic ions and Period-II elements. Among other state-of-art theories, this method could be a modest alternate in electronic structure theory. The combined effect of monopole and dipole factors gave rise to non-zero current density of electrons of 2D heavy-hole trions and 3D He-isoelectronic ions.

Acknowledgments

We express our sincere indebtedness to Professor Shankar Prasad Bhattacharyya for his untiring inspiration and support. Our sincere thanks go to CSIR (SRF scheme) and FRP Grant under Institution of Eminence, University of Delhi (Ref. No./IoE/2021/12/FRP) for their financial support.

6.A Standard Equations and Integrals

Associated Laguerre Polynomial [103]

$$L_k^{\mu}(x) = \sum_{m=0}^{k} (-1)^m \frac{(\mu+k)!}{(k-m)!(\mu+m)!m!} x^m \quad \text{where} \quad \mu > -1 \tag{6.A.1}$$

Lower and Upper Incomplete Gamma function [103]

$$\gamma(a,x) = \int_0^x e^{-t} t^{a-1} dt = (a-1)! \left(1 - e^{-x} \sum_{s=0}^{a-1} \frac{x^s}{s!}\right)$$

$$\Gamma(a,x) = \int_x^{\infty} e^{-t} t^{a-1} dt = (a-1)! \left(e^{-x} \sum_{s=0}^{a-1} \frac{x^s}{s!}\right) \quad \text{where} \quad \mathrm{Re}(a) \geq 0 \tag{6.A.2}$$

Standard Integral-I (Erdélyi's Integral) [126]

$$\int_0^{\infty} x^{(\varrho-1)} e^{-cx} M_{\kappa_1,\gamma_1-\frac{1}{2}}(a_1 x) M_{\kappa_2,\gamma_2-\frac{1}{2}}(a_2 x) dx = a_1^{\gamma_1} a_2^{\gamma_2} (c+A)^{-\varrho-M} \Gamma(\varrho+M)$$
$$\times F_2 \left(\varrho+M; \gamma_1-\kappa_1, \gamma_2-\kappa_2; 2\gamma_1, 2\gamma_2; \frac{a_1}{c+A}, \frac{a_2}{c+A}\right)$$
$$\text{where } \mathrm{Re}(\varrho+M) > 0, Re\left(c \pm \frac{1}{2}a_1 \pm \frac{1}{2}a_2\right) > 0 \text{ and } M = \gamma_1 + \gamma_2 \tag{6.A.3}$$

Recurrence Relations [103]

$$(a) \int Y_l^{m*} Y_0^0 Y_l^m d\Omega = \sqrt{\frac{1}{4\pi}}$$

$$(b) \int Y_{l+1}^{m*} Y_1^0 Y_l^m d\Omega = \sqrt{\frac{3}{4\pi}} \sqrt{\frac{(l-m+1)(l+m+1)}{(2l+1)(2l+3)}}$$

$$(c) \int Y_{l-1}^{m*} Y_1^0 Y_l^m d\Omega = \sqrt{\frac{3}{4\pi}} \sqrt{\frac{(l-m)(l+m)}{(2l+1)(2l-1)}}$$

$$(d) \int Y_{l+1}^{m+1*} Y_1^1 Y_l^m d\Omega = \sqrt{\frac{3}{8\pi}} \sqrt{\frac{(l+m+1)(l+m+2)}{(2l+1)(2l+3)}}$$

$$(e) \int Y_{l-1}^{m+1*} Y_1^1 Y_l^m d\Omega = -\sqrt{\frac{3}{8\pi}} \sqrt{\frac{(l-m)(l-m-1)}{(2l+1)(2l-1)}}$$

$$(f) \int Y_{l+1}^{m-1*} Y_1^{-1} Y_l^m d\Omega = \sqrt{\frac{3}{8\pi}} \sqrt{\frac{(l-m+1)(l-m+2)}{(2l+1)(2l+3)}}$$

$$(g) \int Y_{l-1}^{m-1*} Y_1^{-1} Y_l^m d\Omega = -\sqrt{\frac{3}{8\pi}} \sqrt{\frac{(l+m)(l+m-1)}{(2l+1)(2l-1)}} \tag{6.A.4}$$

F_2 ***to*** ${}_2F_1$ ***relation*** [103]

$$F_2|m| + 1/2, -n_1, -n_2|m| + 1, |m| + 11, 1 = \sum_{k=0}^{n_1} \frac{(|m|+\frac{1}{2})_k (-n_1)_k}{(|m|+1)_k k!}$$
$$\times {}_2F^1 - n_2, |m| + 1/2 + k|m| + 11 \tag{6.A.5}$$

Chu-Vandermonde identity [127]

$${}_2F^1 - b', a + kc'd' = \frac{(c' - (a+k))_{b'}}{(c')_{b'}} \qquad \text{where} Re(c') \geq 0 \text{and } d' = 1 \tag{6.A.6}$$

Standard Integral-II (Erdélyi's Integral) [128]

$$\int_0^\infty x^{(\sigma-1)}\, e^{-bx}\, L_{k_1}^{\mu_1}(a_1x)\, L_{k_2}^{\mu_2}(a_2x)\, dx = (-1)^{(k_1+k_2)} \binom{-(\mu_1+1)}{k_1} \binom{-(\mu_2+1)}{k_2}$$
$$\times\, b^{-\sigma}\, \Gamma(\sigma) F_2\left(\sigma; -k_1, -k_2; \mu_1+1, \mu_2+1; \frac{a_1}{b}, \frac{a_2}{b}\right) \tag{6.A.7}$$

where $\text{Re}(\sigma) > 0$, $\text{Re}(b) > 0$.

References

1 Reed, M.A. (1993). *Sci. Am.* 268: 118.
2 Ashoori, R.C. (1996). *Nature* 379: 413.
3 Schmid, G. (2004). *Nanoparticles: From Theory to Application*. Wiley.
4 Van der Wiel, W.G., De Franceschi, J.M., Elzerman, S. et al. (2002). *Rev. Mod. Phys.* 75: 1.
5 Sargent, E.H. (2005). *Adv. Mater.* 17: 515.
6 Nielsen, M.A. and Chuang, I.L. (2000). *Quantum Computation and Quantum Information*. Cambridge University Press.
7 Murray, C.B., Sun, S., Gaschler, W. et al. (2001). *IBM J. Res. Dev.* 45: 47.
8 Chakraborty, T. (1999). *Quantum Dots: A Survey of the Properties of Artificial Atoms*. Elsevier Science.
9 Spinborg, M. (2000). *Methods of Electronic-Structure Calculations: From Molecules to Solids*. Wiley.
10 Springborg, M. and Joswig, J.-O. (ed.) (2018). *Chemical Modelling, SPR - Chemical Modelling*, vol. 14, P001–P254. The Royal Society of Chemistry.
11 Chakraborty, D. and Chattaraj, P.K. (2019). *J. Phys. Chem. A* 123: 4513.
12 Chakraborty, D. and Chattaraj, P.K. (2021). *Chem. Sci.* 12: 6264.
13 Meissner, W. and Ochsenfeld, R. (1933). *Naturwissenschaften* 21: 787.
14 Josephson, B. (1962). *Phys. Lett.* 1: 251.
15 Schwarzschild, B. (1998). *Phys. Today* 51: 17.
16 Aggarwal, P., Sharma, S., Kaur, H. et al. (2017). *Physica E* 85: 56.
17 Sharma, S., Aggarwal, P., Kaur, H. et al. (2018). *Physica E* 104: 206.
18 Sharma, S., Aggarwal, P., Kaur, H., and Hazra, R.K. (2018). *AIP Adv.* 8: 095116.
19 Aggarwal, P., Sharma, S., Singh, S. et al. (2017). *Physica E* 88: 26.
20 Aggarwal, P., Sharma, S., Kaur, H. et al. (2019). *J. Phys. Commun.* 3: 035011.
21 Aggarwal, P., Sharma, S., and Hazra, R.K. (2020). *Mol. Phys.* 118: e1761472.
22 Singh, S., Kaur, H., Sharma, S. et al. (2017). *Physica E* 88: 289.
23 Kaur, H., Singh, S., Aggarwal, P. et al. (2017). *ACS Omega* 2: 7410.
24 Kaur, H., Sharma, S., Aggarwal, P., and Hazra, R.K. (2019). *Physica E* 108: 347.
25 Sharma, S., Aggarwal, P., Kaur, H., and Hazra, R.K. (2019). *J. Indian Chem. Soc.* 96: 775.
26 Sharma, S., Aggarwal, P., and Hazra, R.K. (2020). *Mol. Phys.* 118: e1770881.
27 Sharma, S., Kapil, B., Aggarwal, P., and Hazra, R.K. (2022). *Phys. Open* 11: 100107.
28 Kapil, B., Sharma, S., Aggarwal, P., and Hazra, R.K. (2022). *Eur. Phys. J. Plus* 137: 809.
29 Kapil, B., Sharma, S., Aggarwal, P., and Hazra, R.K. (2023). *Theor. Chem. Acc.* 142: 90.
30 Macucci, M., Hess, K., and Iafrate, G.J. (1995). *J. Appl. Phys.* 77: 3267. https://doi.org/10.1063/1.358680.
31 Sheng, W. and Xu, H. (1998). *Physica B* 256–258: 152.
32 Ashoori, R.C., Stormer, H.L., Weiner, J.S. et al. (1993). *Phys. Rev. Lett.* 71: 613.

33 Toulouse, J., Colonna, F., and Savin, A. (2004). *Phys. Rev. A* 70: 062505.
34 Cremer, D. (2001). *Mol. Phys.* 99: 1899.
35 Pak, M.V., Chakraborty, A., and Hammes-Schiffer, S. (2007). *J. Phys. Chem. A* 111: 4522.
36 Morrison, R.C. (2002). *J. Chem. Phys.* 117: 10506.
37 Fujito, M., Natori, A., and Yasunaga, H. (1996). *Phys. Rev. B* 53: 9952.
38 Ezaki, T., Mori, N., and Hamaguchi, C. (1997). *Phys. Rev. B* 56: 6428.
39 Drouvelis, P.S., Schmelcher, P., and Diakonos, F.K. (2004). *Phys. Rev. B* 69: 035333.
40 Nazmitdinov, R.G. (2009). *Phys. Part. Nucl.* 40: 71.
41 Li, Y., Yannouleas, C., and Landman, U. (2007). *Phys. Rev. B* 76: 245310.
42 Kagan, M.Y., Val'kov, V.V., and Aksenov, S.V. (2017). *Phys. Rev. B* 95: 035411. https://doi.org/10.1103/physrevb.95.035411.
43 Rontani, M., Rossi, F., Manghi, F., and Molinari, E. (1999). *Phys. Rev. B* 59: 10165.
44 Hamaguchi, C. (2009). *Basic Semiconductor Physics*. Berlin Heidelberg: Springer-Verlag.
45 Laughlin, R.B. (1983). *Phys. Rev. B* 27: 3383.
46 Laughlin, R.B. (1983). *Phys. Rev. Lett.* 50: 1395.
47 Bryant, G.W. (1987). *Phys. Rev. Lett.* 59: 1140.
48 Kamilla, R.K. and Jain, J.K. (1995). *Phys. Rev. B* 52: 2798.
49 Maksym, P. and Chakraborty, T. (1990). *Phys. Rev. Lett.* 65: 108.
50 Taut, M. (1994). *J. Phys. A: Math. Gen.* 27: 1045.
51 Taut, M. (2000). *Physica E* 6: 479.
52 Ralko, A. and Truong, T.T. (2002). *J. Phys. A: Math. Gen.* 35: 9671.
53 Wagner, M., Merkt, U., and Chaplik, A.V. (1992). *Phys. Rev. B* 45: 1951.
54 Merkt, U., Huser, J., and Wagner, M. (1991). *Phys. Rev. B* 43: 7320.
55 Dineykhan, M. and Nazmitdinov, R.G. (1997). *Phys. Rev. B* 55: 13707.
56 Ciftja, O. and Kumar, A.A. (2004). *Phys. Rev. B* 70: 205326.
57 Ciftja, O. and Faruk, M.G. (2005). *Phys. Rev. B* 72: 205334.
58 Mathew, A. and Nandy, M.K. (2013). *Physica B* 421: 127.
59 Frenkel, J. (1931). *Phys. Rev.* 37: 17.
60 Bryant, G.W. (1988). *Phys. Rev. B* 37: 8763.
61 Bayer, M., Stern, O., Hawrylak, P. et al. (2000). *Nature* 405: 923.
62 Yoshida, M., Popert, A., and Kato, Y.K. (2016). *Phys. Rev. B* 93: 041402.
63 Raymond, S., Hawrylak, P., Gould, C. et al. (1997). *Solid State Commun.* 101: 883.
64 Borri, P., Langbein, W., Woggon, U. et al. (2003). *Phys. Rev. Lett.* 91: 267401.
65 Taut, M. (2000). *J. Phys.: Condens. Matter* 12: 3689.
66 Halonen, V., Chakraborty, T., and Pietiläinen, P. (1992). *Phys. Rev. B* 45: 5980.
67 Patton, B., Langbein, W., and Woggon, U. (2003). *Phys. Rev. B* 68: 125316.
68 Ramasubramaniam, A. (2012). *Phys. Rev. B* 86: 115409.
69 Korolev, A. and Liberman, M. (1994). *Phys. Rev. Lett.* 72: 270.
70 Berkelbach, T.C., Hybertsen, M.S., and Reichman, D.R. (2013). *Phys. Rev. B* 88: 045318.
71 Kylänpää, I. and Komsa, H.-P. (2015). *Phys. Rev. B* 92: 205418.
72 Mayers, M.Z., Berkelbach, T.C., Hybertsen, M.S., and Reichman, D.R. (2015). *Phys. Rev. B* 92: 161404.
73 Kellner, G.W. (1927). *Z. Angew. Phys.* 44: 91.
74 Hylleraas, E.A. (1930). *Z. Angew. Phys.* 65: 209.
75 Hylleraas, E.A. (1929). *Z. Angew. Phys.* 54: 347.
76 Kinoshita, T. (1957). *Phys. Rev.* 105: 1490.
77 Pekeris, C.L. (1959). *Phys. Rev.* 115: 1216.

78 Frankowski, K. and Pekeris, C.L. (1966). *Phys. Rev.* 146: 46.
79 Scherr, C.W. (1960). *J. Chem. Phys.* 33: 317. https://doi.org/10.1063/1.1731129.
80 Scherr, C.W. and Knight, R.E. (1963). *Rev. Mod. Phys.* 35: 436.
81 Nakashima, H. and Nakatsuji, H. (2007). *J. Chem. Phys.* 127: 224104. https://doi.org/10.1063/1.2801981.
82 Bartlett, J.H. (1937). *Phys. Rev.* 51: 661.
83 Bhattacharyya, S., Bhattacharyya, A., Talukdar, B., and Deb, N.C. (1996). *J. Phys. B: At. Mol. Opt. Phys.* 29: L147.
84 Thakkar, A.J. and Koga, T. (2003). *Theor. Chem. Acc.* 109: 36.
85 Freund, D.E., Huxtable, B.D., and Morgan, J.D. (1984). *Phys. Rev. A* 29: 980.
86 Umrigar, C.J., Wilson, K.G., and Wilkins, J.W. (1988). *Phys. Rev. Lett.* 60: 1719.
87 Korobov, V.I. (2002). *Phys. Rev. A* 66: 024501.
88 Roothaan, C.C.J. (1960). *Rev. Mod. Phys.* 32: 179.
89 Thakkar, A.J. and Koga, T. (1994). *Phys. Rev. A* 50: 854.
90 Rodriguez, K.V., Gasaneo, G., and Mitnik, D.M. (2007). *J. Phys. B: At. Mol. Opt. Phys.* 40: 3923.
91 Sahni, V. and Krieger, J.B. (1971). *Int. J. Quantum Chem.* 5: 47.
92 Combescot, R. (2017). *Phys. Rev. X* 7: 041035.
93 Fock, V. (1954). *Izv. Akad. Nauk SSSR Ser. Fiz.* 18: 161.
94 Kato, T. (1957). *Commun. Pure Appl. Math.* 10: 151.
95 Poszwa, A. (2020). *Physica E* 124: 114247.
96 Clementi, E. and Roetti, C. (1974). *At. Data Nucl. Data Tables* 14 (3–4): 177–478.
97 Schmidt, K.E. and Moskowitz, J.W. (1990). *J. Chem. Phys.* 93: 4172.
98 Moskowitz, J.W. and Schmidt, K.E. (1992). *J. Chem. Phys.* 97: 3382.
99 Alexander, S.A. and Coldwell, R.L. (1995). *J. Chem. Phys.* 103: 2572.
100 Lin, X., Zhang, H., and Rappe, A.M. (2000). *J. Chem. Phys.* 112: 2650.
101 Griffiths, D. (1999). *Introduction to Electrodynamics*. Prentice Hall.
102 Pilar, F.L. (1968). *Elementary Quantum Chemistry*, Quantum Mechanics. McGraw-Hill.
103 Arfken, G. and Weber, H. (2005). *Mathematical Methods for Physicists*. Elsevier.
104 Cohen-Tannoudji, C., Diu, B., and Lalöe, F. (1977). *Quantum Mechanics*. Wiley.
105 McQuarrie, D.A. (2008). *Quantum Chemistry*. University Science Books.
106 Dutta, J., Mukherjee, S., Naskar, K. et al. (2020). *Phys. Chem. Chem. Phys.* 22: 27496.
107 Baer, M. (2002). *Phys. Rep.* 358: 75.
108 Mukherjee, B., Naskar, K., Mukherjee, S. et al. (2019). *Int. Rev. Phys. Chem.* 38: 287.
109 Feynman, R.P. (1939). *Phys. Rev.* 56: 340.
110 Lichten, W. (1967). *Phys. Rev.* 164: 131.
111 Smith, F.T. (1969). *Phys. Rev.* 179: 111.
112 Messiah, A. (1999). *Quantum Mechanics*, Dover Books on Physics. Dover Publications.
113 Jackson, J. (1999). *Classical Electrodynamics*. Wiley.
114 Merzbacher, E. (1970). *Quantum Mechanics*. New York: Wiley.
115 Eyring, H., Walter, J., and Kimball, G. (1961). *Quantum Chemistry*. Wiley.
116 Landau, L. and Lifshitz, E. (1981). *Quantum Mechanics: Non-Relativistic Theory*, Course of Theoretical Physics. Elsevier Science.
117 Shankar, R. (1977). *Principles of Quantum Mechanics*. Springer.
118 Feller, W. (1968). *An Introduction to Probability Theory and Its Applications*. Wiley.
119 Reimann, S.M. and Manninen, M. (2002). *Rev. Mod. Phys.* 74: 1283.
120 Tarucha, S., Honda, T., Austing, D. et al. (1998). *Physica E* 3: 112.
121 Madhav, A.V. and Chakraborty, T. (1994). *Phys. Rev. B* 49: 8163.

122 Maksym, P. (1998). *Physica B* 249–251: 233.
123 Tennie, F., Vedral, V., and Schilling, C. (2017). *Phys. Rev. A* 95: 022336. https://doi.org/10.1103/physreva.95.022336.
124 Friedrich, H. (1991). *Theoretical Atomic Physics*, 115. Springer-Verlag.
125 Patil, S.H. (2008). *Eur. J. Phys.* 29: 517.
126 Erdélyi, A. (1937). *Monatsh. Math. Phys.* 46: 1.
127 Cerquetti, A. (2010). ArXiv e-prints, arXiv:1012.1243 [math.PR].
128 Buchholz, H. (1969). *The Confluent Hypergeometric Function with Special Emphasis on its Applications*, 15. Springer-Verlag.

7

Excited State Electronic Structure – Effect of Environment

Supriyo Santra and Debashree Ghosh

School of Chemical Sciences, Indian Association for the Cultivation of Science, Kolkata, India

7.1 Introduction

Electronic structure of ground state species and their reactions have long been studied using accurate electronic structure tools. Methods in this field are well-developed and have led to many pioneering predictions and interpretations. However, study of excited states is a relatively less studied area. One of the reasons for that is the breakdown of mean-field approximation, which is the bed-rock starting point of most quantum chemical calculations. Breakdown of mean-field approximation can be alternatively viewed as the inadequacy of a single electronic configuration (obtained from the molecular orbitals calculated via mean-field approximation) to depict the correct qualitative picture of the state. This can also be referred to as multi-reference or multi-configurational nature of the state [1]. Excited states are typically quite multi-reference in nature, and therefore, single-reference methods such as most of the perturbative and coupled-cluster-based post Hartree–Fock methods are inadequate to understand such systems [2].

However, excited states chemistry is crucial to the understanding of many biological processes, such as vision, photosynthesis, DNA damage and so on [3–8]. It is also important in electronic materials and photo-catalytic effects. Furthermore, these real-life systems are large and complicated, which makes them computationally unaffordable with most quantum chemical techniques. Therefore, these large systems have been mainly treated via embedding methods or hybrid quantum mechanical/molecular mechanical (QM/MM) methods [9–13]. The main idea of these methods is to identify the most important part of the system where there are significant changes in electron density or the location of bond breaking and formation. These important parts are treated quantum mechanically with high-level methods. The rest of the system, i.e. the less important part can be thought of as providing a field to change the energetics of the QM (important) part to some extent and this less important part (environment) is treated with less accurate methods, which are predominantly classical in origin (classical force fields or MM methods). The details of the different methods that are most commonly used will be described in Section 7.2 (see Figure 7.1). In this chapter, we will discuss the difficulties associated with understanding excited states in condensed phases and some interesting results that have been obtained with state-of-the-art methods that have been developed to circumvent these difficulties.

Electron Density: Concepts, Computation and DFT Applications, First Edition.
Edited by Pratim Kumar Chattaraj and Debdutta Chakraborty.

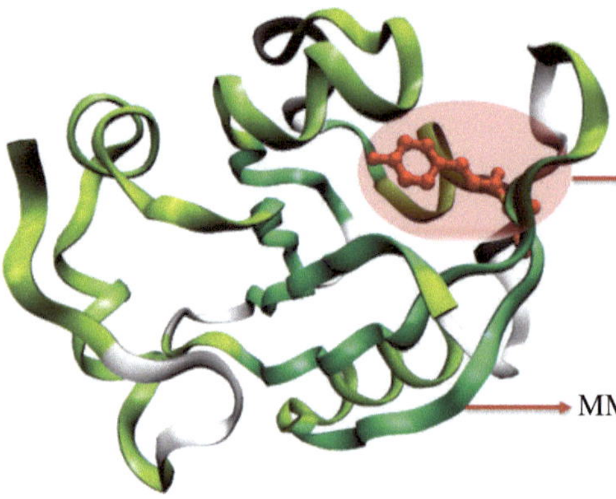

Figure 7.1 QM and MM regions in a protein (PDB id: 6P5G).

7.2 Methodology

As mentioned above, the most efficient approach to understanding excited state processes in condensed phases has been various flavours of hybrid QM/MM method. To achieve the required accuracy and speed, one requires QM methods that can sufficiently deal with the multi-reference nature of the excited states and MM methods that are economical.

7.2.1 Quantum Mechanical Methods

7.2.1.1 Time-Dependent Density Functional Theory

The most commonly used and computationally efficient method for calculating excited states is time-dependent density functional theory (TD-DFT) [14–16]. Just as DFT forms the work-horse of much of ground state calculations, TD-DFT forms the most affordable counterpart in excited state chemistry. However, it is plagued by two major disadvantages – (i) it can be severely affected by the that is used for the calculation and that reduces its predictive property, and (ii) it suffers from self-interaction error, which manifests in underestimation of excitation energies of charge–transfer excited states. Since, charge–transfer excited states are crucial to the low-lying excited state manifold of most large, real-life, heterogeneous systems, this makes TD-DFT quite inadequate in most of these cases. It should be noted that to circumvent such situations, constrained DFT techniques have been developed, which have been successfully applied to understand excited state phenomena such as singlet fission [17].

Further complications can also arise in systems where the ground state is also multi-reference in nature. As an example, the ligand field splittings and excited states of transition metal complexes are quite difficult to ascertain accurately from TD-DFT approaches [18].

7.2.1.2 Active Space-Based Methods

In order to understand multireference states, active space (shown in Figure 7.2) based approaches are used most often, i.e. the multiple references or electronic configurations are generated within the valence space (often denoted as active space, shown in Figure 7.2). Methods such as complete active space self-consistent field (CASSCF) [19, 20] and restricted active space self-consistent field

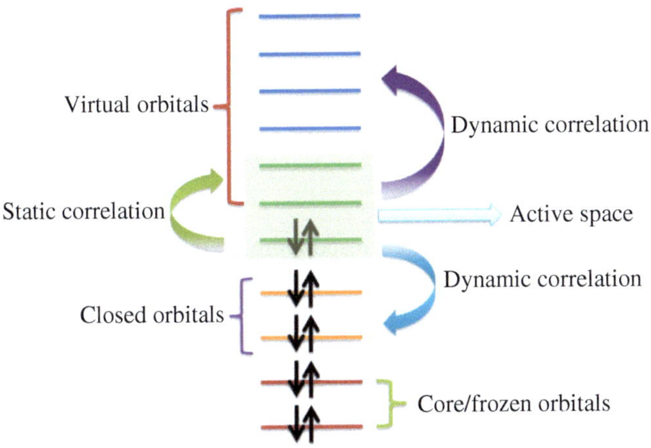

Figure 7.2 Separation of molecular orbital space and electronic correlation considered in active space-based methods.

(RASSCF) [21] have been developed depending on the different criterion for creating the electronic configurations. The CASSCF or RASSCF wavefunction can be written as,

$$\Psi_{\text{MCSCF}} = \sum_i c_i \phi_i \tag{7.1}$$

where ϕ_i denotes ith configuration (or Slater determinant) and c_i is its coefficient or relative importance. Since, several configurations are included, it can treat both ground and excited state multi-reference states and are therefore, ideal for such systems and one obtains qualitatively correct wavefunctions. However, at this point, the correlation between the internal (active) and external (core and virtual) spaces (referred to as dynamic correlation) is still completely absent and therefore, one cannot hope for quantitative accuracy, which is crucial for chemical predictions. In order to introduce dynamic correlation, perturbative or configuration interaction-based approaches as used on top of CASSCF or RASSCF reference wavefunctions. These are then referred to as CASPT2, NEVPT2 or MRCI group of methods [22–28]. While these are extremely accurate approaches, they are also computationally quite expensive. Furthermore, they scale exponentially with the active space, which means that they cannot be used for large active spaces. Anything higher than (16e,16o) active space is virtually impossible to calculate due to storage and memory requirements. Methods such as density matrix renormalization group (DMRG) can circumvent the exponential scaling with the active space and therefore, allow one to use larger active spaces where the problems demand [29].

7.2.1.3 Configuration Interaction-Based Approaches

Starting from a Hartree–Fock reference, the simplest approach to include correlation is by including the singles, doubles, or higher excited determinants/configurations. This forms the basis of the so-called configuration interaction class of methods [30–32]. Since the total number of determinants/configurations scales exponentially with system size (i.e. no of atoms and basis functions etc.), one requires to judiciously truncate. This truncation has traditionally been on the basis of degree of excitation, i.e. inclusion of only single excited determinants leads to configuration interaction singles (CIS), singles and doubles determinants give rise to configuration interaction singles doubles (CISD), etc.,

$$|\Psi_{\text{CIS}}\rangle = |\psi_{\text{HF}}\rangle + \sum_{i,a} c_i^a a_a^\dagger a_i$$
$$|\Psi_{\text{CISD}}\rangle = |\psi_{\text{HF}}\rangle + \sum_{i,a} c_i^a a_a^\dagger a_i + \sum_{i,j,a,b} c_{ij}^{ab} a_a^\dagger a_b^\dagger a_i a_j \tag{7.2}$$

where the terms c_i^a and c_{ij}^{ab} refers to the CI coefficients corresponding to the specific one or two-electron excited determinants. Due to the truncation at a specific excited determinant level, the CI class of methods in this traditional form lack size extensivity and size consistency.

The CI methods have long been used successfully to calculate the excited state energies and phenomena of many systems, especially when the ground state of the system is relatively single reference in nature. Of late variants of CISD such as SOS-CISD have been extremely successful in excited state properties and energies [33, 34].

7.2.1.4 Equation of Motion Coupled Cluster

Equation of motion coupled cluster (EOM-CC) [35–40] is an extremely accurate and powerful approach for the calculation of various properties. It starts from a coupled cluster (CC) reference wavefunction denoted by,

$$|\Psi_{\text{CC}}\rangle = e^T|\Psi_{\text{HF}}\rangle \tag{7.3}$$

where the T denotes the excitation operators (also called cluster operators), which can be $T_1 + T_2 + T_3 + \cdots$, i.e. one, two, three…etc. body excitations. Most commonly singles doubles excitations are used, i.e. it is truncated as $T_1 + T_2$. EOM-CC refers to application of a EOM operator on the CC reference function. It can be written as,

$$R_1 = \sum_{i,j} a_i^\dagger a_j$$
$$R_2 = \sum_{i,j,k,l} = a_i^\dagger a_j^\dagger a_k a_l \tag{7.4}$$

where a and $a^\dagger$ refer to annihilation and creation operators and R_1 and R_2 refer to the one and two-body EOM operators of 1h1p and 2h2p nature. EOM-CC suite of methods are black box in nature and can treat a wide range of excited states as long as one can obtain a reference state that is suitably single reference in nature. However, they are often prohibitively expensive especially when dynamical studies are required.

7.2.2 Molecular Mechanical Methods

The less important part of the system is often treated with molecular mechanical methods, which only contain the atomic degrees of freedom and not the electronic degrees of freedom. This reduces the computational complexity many-fold. The main effects of the environment are: (i) electrostatics, (ii) polarization, (iii) dispersion and (iv) exchange repulsion [41]. Electrostatic interactions are the interactions between permanent charges, dipole etc. This can also sometimes contain charge–transfer-type interactions. Polarization denotes the effect of a permanent charge/dipole on a electron cloud, which can thus induce a dipole on that. This charge-induced dipole interaction is often called polarization is a second-order effect. In polar and charged system, polarization can be quite important. Both these terms have long-range effect, i.e. the system can feel this effect even

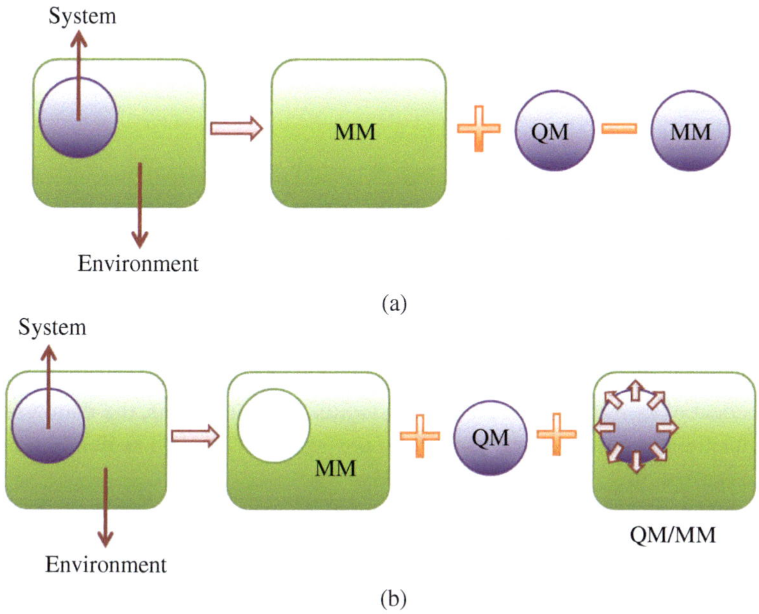

Figure 7.3 Two approaches of performing a QM/MM calculation (a) subtractive scheme and (b) additive scheme.

when the charge or dipole is relatively far away, due to the rather slow decay of Coulombic $1/r$ term. Dispersion refers to the induced dipole–induced dipole interaction that are most important in non-polar systems, e.g. between benzene dimers. Exchange repulsion is the repulsion the atoms feel when they are too close to each other, which essentially originates from the electronic repulsion when they are two close to each other. It can be shown to be dependent on the overlap integral of orbitals.

The MM methods can be divided into subtractive and additive schemes (see Figure 7.3).

7.2.2.1 ONIOM

ONIOM [42, 43] is an example of subtractive scheme (shown in Figure 7.3a). Here, separate calculations are performed on the system with QM and MM methods, and system + environment with MM method. The total energy of the full system (system + environment) can then be put together as,

$$E = E_{\text{system}}^{\text{QM}} + E_{\text{system+env}}^{\text{MM}} - E_{\text{system}}^{\text{MM}} \tag{7.5}$$

such that there is no double counting of interactions. However, this also means that the interaction between QM and MM regions are then calculated purely at a MM level and effects of higher-level interactions and electronic effects will be missed out. While this is the simplest and oldest approach, it is seldom used for excited states.

7.2.2.2 Mechanical Embedding

Additive QM/MM scheme calculates the energy as,

$$E = E_{\text{system}}^{\text{QM}} + E_{\text{env}}^{\text{MM}} + E_{\text{system+env}}^{\text{QM-MM}} \tag{7.6}$$

where the first two terms are calculated with purely QM and MM levels of theory (shown in Figure 7.3b). The third interaction term between the system and environment is calculated at different levels when it pertains to mechanical or electronic embedding schemes [44]. In the mechanical embedding, the interaction term is given by,

$$E^{\text{QM-MM}}_{\text{system+env}} = \sum_i \int \frac{\rho_{\text{QM}}(r)q_i}{|r - r_i|} dr \tag{7.7}$$

where $\rho_{\text{QM}}(r)$ denotes the electron density of the QM region and q_i are the charges on the atoms in the MM region, r and r_i denote the position of the QM density to be integrated over and the position of ith MM charge, respectively. This interaction energy can therefore, be calculated as a post-facto correction.

The mechanical embedding approach in additive QM/MM method calculates the interaction between the system and environment separately and therefore, the energy components coming from the different regions can be separated. However, the interaction energy is calculated post-facto and therefore, there is no effect of the MM region on the QM region. In other words, the manifestation of the MM region on the Hamiltonian and therefore, properties of the QM region is not explicitly considered in this embedding.

7.2.2.3 Electronic Embedding

To address the issues of the mechanical embedding, an electronic embedding can be envisaged [43]. In this approach, the Hamiltonian of the QM region is modified perturbatively due to the effect of the environment or MM region,

$$\hat{H} = \hat{H}^{\text{QM}}_{\text{system}} + \hat{H}^{\text{MM}}_{\text{env}} + \hat{H}'_{\text{QM-MM}} \tag{7.8}$$

where $\hat{H}'_{\text{QM-MM}}$ denotes the perturbative term that is added to the quantum mechanical Hamiltonian $\hat{H}^{\text{QM}}_{\text{system}}$. Therefore, at this level when the properties of such a system is calculated, it automatically contains the effect of the environment, albeit perturbatively. The electronic effect of the MM can be included as point charges as well as distributed multipole moments.

However, as the electron density of the ground and excited states may be significantly different, those effects need to be considered in the environmental effect. Such interactions are not included in the electronic embedding.

7.2.2.4 Polarizable Embedding

Polarizable embedding (PE) considers not only the effect of the MM (environment) on the QM (system/chromophore) region, but also the effect of the QM region on the MM region. In other words, it includes the self-consistent effect between the QM and MM regions [45]. It can be included as fluctuating charges or induced dipoles on atoms. Effective fragment potential (EFP) [46–48] or polarizable embedding (PE) [49, 50] induced dipole moments to include polarization given by,

$$\vec{\mu}_{\text{ind}} = \alpha \vec{F} \tag{7.9}$$

where α is the polarizability tensor and the induced dipole is formed due to the action of the field $\vec{F}$.

Therefore, the electrostatics and polarization interaction terms for the QM and EFP regions can be written as,

$$\begin{aligned} E^{\text{elec}}_{\text{QM/EFP}} &= \sum_k \int_{\text{QM}} \left\{ \frac{q_k}{R_k} + \sum_a \mu^k_a \frac{F^k_a}{R^3_k} \sum_{a,b} \Theta^k_{ab} \frac{F^k_{ab}}{3R^5_k} + \sum_{a,b,c} \Omega^k_{abc} \frac{F^k_{abc}}{5R^7_k} \right\} \rho_{\text{QM}}(\vec{r}) dr \\ E^{\text{pol}}_{\text{QM/EFP}} &= \frac{1}{2} \sum_p \sum_a^{x,y,z} \vec{\mu}^p_{\text{conj},a} F^{p,(ai)}_a - \vec{\mu}^p_a F^{p,(\text{mult+nuc})}_a \end{aligned} \tag{7.10}$$

k and p denote the multipole and polarizability points. The induced dipole moments and the conjugated induced dipole moments act on the fields due to the *ab initio* or QM part, and the MM part (multipoles and nuclear charges), respectively.

7.3 Representative Examples

We will consider a few examples of excited state processes and phenomena where QM/MM with or without polarizable embedding have been used. The changes in the electronic structure and electron density due to the environment at different levels of embedding will be discussed.

7.3.1 Photo-Isomerization of Rhodopsin

Rhodopsins form an important class of proteins involved in vision process in vertebrates and in ion transport in microbes [51, 52]. Excited state quantum mechanical and hybrid QM/MM methods have been crucial in elucidating these processes, especially in the close confinement of the protein environment. Retinal and more specifically retinal protonated Schiff base (rpSb) is the main chromophore in these proteins. Photoisomerization (11-cis to trans) of rpSb can be envisaged via different mechanisms (shown in Figure 7.4a) – one bond flip, bicycle pedal, hula twist and folding table. The photoisomerization process starts with excitation by a single photon, which forms the optically active excited state predominantly on the chromophore, i.e. rpSb [53–55]. The excited state involves significant redistribution of charge density, i.e. electron density moves into the protonated Schiff base end, thus activating the double bonds in that end. This in turn facilitates cis–trans isomerization, which in the excited state completely downhill (shown in Figure 7.4b).

Gas phase calculation on the rpSb moiety and its photoisomerization when considering a minimal model shows that isomerization can occur via 1-bond rotation [56]. These are CASSCF and CASPT2 levels of theory predictions [55]. However, it is easy to notice that the 1-bond rotation is the least volume conserving rotation and will therefore, be severely hindered in protein due to steric effects [57, 58]. Warshel had an earlier prediction of the isomerization via bicycle pedal rotation [59], and a similar two-bond rotation called 'wring-a-towel-mode' was noticed by Ishida and coworkers from QM/MM calculations [60]. Crankshaft mode was noticed from QM/MM calculations by Olivucci and coworkers [61]. This clearly shows that while QM prediction only on the chromophore can show one mechanism, considering the environment can significantly change the preferred mechanism of isomerization.

7.3.2 DNA-Base Excited States in Solution

Excitations, ionization and electron attachment on DNA bases have long interested researchers due to their implication in DNA damage. Earlier works have used TDDFT to address this problem. However, these systems have low-lying charge transfer states, which are severely underestimated with TDDFT level of theory. Therefore, EOM-CCSD and SOS-CISD methods have been applied extensively to solve this problem.

The lowest ionization energies of the nucleobases follow the patterns: $G < A < C < T$. Furthermore, when the geometric effects subsequent to ionization is noticed, they follow the trend: $G > C > A > T$ [62]. Many of these results have been corroborated by photoelectron spectroscopic measurements. Significant effects and cooperativity of π-stacking and hydrogen bonding interactions on the ionization energies have been found [63]. It is the interplay of these interactions

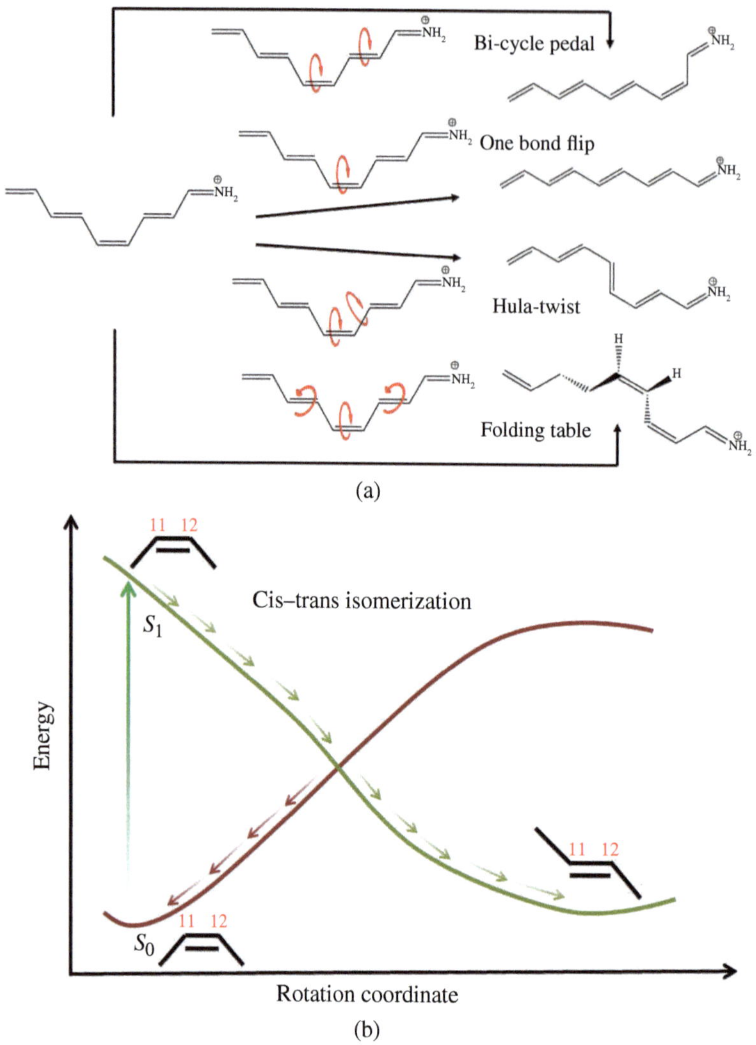

Figure 7.4 (a) Different rotation mechanisms possible in rhodopsin. (b) Schematic potential energy diagram of cis–trans isomerization.

that leads to the ionization thresholds of the nucleic acid bases in DNA. Since the calculations of ionization energies of nucleic acid bases in DNA require the consideration of much larger systems, they have been studied with the conjunction of pure QM methods and hybrid QM/MM approaches. Furthermore, the effects of solvation on the ionization energies need to be evaluated. Significant effects ($\approx$1 eV shifts) have been noticed on the ionization energies due to bulk solvation and here it has been noticed that both electrostatic and polarization interactions are crucial to capture the full extent of the effects. Therefore, polarizable QM/MM methods (such as QM/EFP) have been imperative to understand these effects [64].

Electron attachment problem is more difficult to accurately estimate and studies show that EOM-CCSD methods are the method of choice. For this problem, it has been noticed that different nucleobases have either valence-bound or dipole-bound anions after electron attachment. For uracil and thymine, the lowest electron attached state is valence bound while for all other nucleobases, i.e. cytosine, adenine and guanine have dipole-bound states [65] When effect of solvents is studied for the electron attachment problem using hybrid QM/MM methods, it is noticed that there can be surface-bound and bulk bound electrons, as well as electrons that are bound on nucleobases versus water [66, 67].

7.3.3 Green Fluorescent Proteins

Fluorescent proteins form a class of proteins that are interesting for its spectacular photo processes. It finds its uses in bio-imaging and probes for molecular environments in vivo. The first studies on these systems were towards understanding the spectra of the fluorescent proteins. It was evident quite quickly, that both the effects of changes to the chromophore as well as the neighbouring protein environment could have significant effects on the spectra (shown in Figure 7.5). In fact in many cases, contrary to expectations, it was the long-range electrostatics that was crucial to the spectral shifts. To understand these, high-level QM methods had to be used along with hybrid QM/MM approaches that accurately estimated electrostatic and polarizable environment effects [68, 69].

Furthermore, among the low-lying excited states, there are both locally excited and charge–transfer states, which are sometimes quasi-degenerate. Therefore, the interplay between these states is known to have significant effect on the functioning of the fluorescent proteins. For example, in the fluorescent protein Dreikling, the charge transfer and locally excited states cause the switching mechanism. When light shines on the resting state, it absorbs the photon and transforms into the locally excited state, which is quasidegenerate with the charge transfer state. Once the molecule moves into the charge transfer state, it re-organizes and forms the dark state of the protein [70]. This, therefore, forms the photoswitching mechanism of the protein. To elucidate these mechanisms, which involve large electronic rearrangement, the density functional approaches have proved to be quite inadequate.

Further difficulties in modelling this class of systems arise when one tries to understand the brightness and photostability of the proteins. To understand the brightness, quantum yields can be calculated. However, to answer the question of photostability, the excited state potential energy surface needs to be modelled and ideally, one would like to couple a high-level quantum

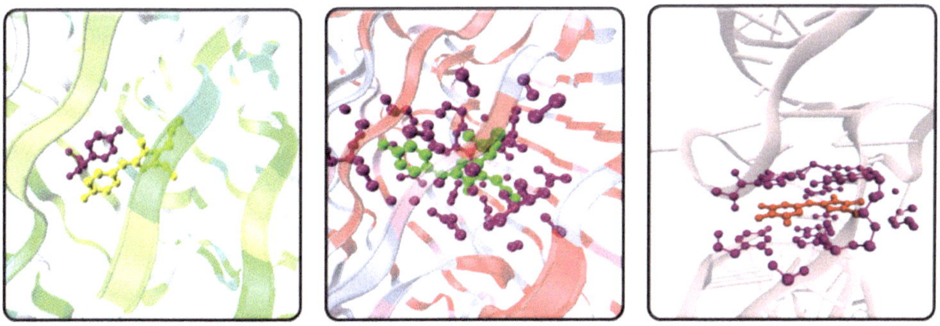

(a) (b) (c)

Figure 7.5 (a) Yellow fluorescent protein chromophore (shown in yellow) with a nearby residue (shown in purple) (PDB id: 1MYW). There is π-interaction between chromophore and nearby tyrosine. (b) Green fluorescent protein choromophore (shown in green) and the nearby residues (shown in purple). (PDB id: 2AWM). (c) For RNA spinach aptamer (PDB id: 4KZD), fluorescent chromophore (in red) and nearby residues (in purple) are shown. Here, long range electrostatic interaction is important. In all the three cases the chromophore is HBDI or a substituted HBDI. However the spectra changes due to neighboring group interactions.

chemical method with the subsequent quantum dynamics. This approach remains computationally unfeasible till this date. Therefore, the affordable solution is to perform a classical molecular dynamics simulation on the excited state surface or a quasi-classical approach such as surface hopping dynamics in conjunction with multireference quantum mechanical methods. Many such difficulties and the existing solutions are explained quite succinctly in Ref. [71].

7.4 Conclusion

In this chapter, we have described some of the computational approaches that are currently used to study excited state electronic structure of real systems, especially bio-systems. We have shown with the help of three representative examples what can be achieved within the existing frameworks that are available in commercially available softwares. We have also highlighted the difficulties one faces in such studies and where the current limitations of our understanding and methodologies stand.

Acknowledgement

We acknowledge SERB-POWER (SPF/2021/000194) fellowship for funding.

References

1 Lischka, H., Nachtigallova, D., Aquino, A.J. et al. (2018). *Chem. Rev.* 118: 7293–7361.
2 Balkov, A. and Bartlett, R.J. (1994). *J. Chem. Phys.* 101: 8972–8987.
3 Herek, J.L., Wohlleben, W., Cogdell, R.J. et al. (2002). *Nature* 417: 533–535.
4 Scholes, G.D., Fleming, G.R., Olaya-Castro, A., and Van Grondelle, R. (2011). *Nat. Chem.* 3: 763–774.
5 Krause, G.H. and Weis, E. (1991). *Annu. Rev. Plant Biol.* 42: 313–349.
6 Brixner, T., Stenger, J., Vaswani, H.M. et al. (2005). *Nature* 434: 625–628.
7 Warshel, A. and Barboy, N. (1982). *J. Am. Chem. Soc.* 104: 1469–1476.
8 Fox, M.A. and Dulay, M.T. (1993). *Chem. Rev.* 93: 341–357.
9 Levitt, M. (1976). *J. Mol. Biol.* 104: 59–107.
10 Warshel, A. and Levitt, M. (1976). *J. Mol. Biol.* 103: 227–249.
11 Lin, H. and Truhlar, D.G. (2007). *Theor. Chem. Acc.* 117: 185–199.
12 Senn, H.M. and Thiel, W. (2009). *Angew. Chem. Int. Ed.* 48: 1198–1229.
13 Sneskov, K., Schwabe, T., Christiansen, O., and Kongsted, J. (2011). *Phys. Chem. Chem. Phys.* 13: 18551–18560.
14 Marques, M.A. and Gross, E.K. (2004). *Annu. Rev. Phys. Chem.* 55: 427–455.
15 Burke, K., Werschnik, J., and Gross, E. (2005). *J. Chem. Phys.* 123: 062206.
16 Ullrich, C.A. (2011). *Time-Dependent Density-Functional Theory: Concepts and Applications.* Oxford: OUP.
17 Yost, S.R., Lee, J., Wilson, M.W. et al. (2014). *Nat. Chem.* 6: 492–497.
18 Seth, K. and Ghosh, D. (2019). *J. Indian Chem. Soc.* 96: 801–807.
19 Olsen, J. (2011). *Int. J. Quantum Chem.* 111: 3267–3272.
20 Malmqvist, P.-Å. and Roos, B.O. (1989). *Chem. Phys. Lett.* 155: 189–194.
21 Malmqvist, P., Rendell, A., and Roos, B.O. (1990). *J. Phys. Chem.* 94: 5477–5482.

22 Andersson, K., Malmqvist, P.-Å., and Roos, B.O. (1992). *J. Chem. Phys.* 96: 1218–1226.
23 Dyall, K.G. (1995). *J. Chem. Phys.* 102: 4909–4918.
24 Celani, P. and Werner, H.-J. (2000). *J. Chem. Phys.* 112: 5546–5557.
25 Angeli, C., Evangelisti, S., Cimiraglia, R., and Maynau, D. (2002). *J. Chem. Phys.* 117: 10525–10533.
26 Angeli, C., Calzado, C.J., Cimiraglia, R., and Malrieu, J.-P. (2006). *J. Chem. Phys.* 124: 234109.
27 Lischka, H., Shepard, R., Brown, F.B., and Shavitt, I. (1981). *Int. J. Quantum Chem.* 20: 91–100.
28 Shamasundar, K., Knizia, G., and Werner, H.-J. (2011). *J. Chem. Phys.* 135: 054101.
29 Ghosh, D., Hachmann, J., Yanai, T., and Chan, G.K.-L. (2008). *J. Chem. Phys.* 128: 144117.
30 Hegarty, D. and Robb, M.A. (1979). *Mol. Phys.* 38: 1795–1812.
31 Pople, J.A., Head-Gordon, M., and Raghavachari, K. (1987). *J. Chem. Phys.* 87: 5968–5975.
32 Daudey, J.-P., Heully, J.-L., and Malrieu, J.-P. (1993). *J. Chem. Phys.* 99: 1240–1254.
33 Casanova, D., Rhee, Y.M., and Head-Gordon, M. (2008). *J. Chem. Phys.* 128: 164106.
34 Grimme, S., Goerigk, L., and Fink, R.F. (2012). *Wiley Interdiscip. Rev.: Comput. Mol. Sci.* 2: 886–906.
35 Stanton, J.F. and Bartlett, R.J. (1993). *J. Chem. Phys.* 98: 7029–7039.
36 Geertsen, J., Rittby, M., and Bartlett, R.J. (1989). *Chem. Phys. Lett.* 164: 57–62.
37 Krylov, A.I. (2008). *Annu. Rev. Phys. Chem.* 59: 433–462.
38 Adnan, S., Bhattacharya, S., and Mukherjee, D. (1980). *Mol. Phys.* 39: 519–524.
39 Nooijen, M. and Bartlett, R.J. (1997). *J. Chem. Phys.* 107: 6812–6830.
40 Hirata, S. (2004). *J. Chem. Phys.* 121: 51–59.
41 Gordon, M.S., Slipchenko, L., Li, H., and Jensen, J.H. (2007). *Annu. Rep. Comput. Chem.* 3: 177–193.
42 Vreven, T. and Morokuma, K. (2006). *Annu. Rep. Comput. Chem.* 2: 35–51.
43 Chung, L.W., Sameera, W., Ramozzi, R. et al. (2015). *Chem. Rev.* 115: 5678–5796.
44 Ghosh, D. (2017). *J. Phys. Chem. A* 121: 741–752.
45 Olsen, J.M., Aidas, K., and Kongsted, J. (2010). *J. Chem. Theory Comput.* 6: 3721–3734.
46 Gordon, M.S., Freitag, M.A., Bandyopadhyay, P. et al. (2001). *J. Phys. Chem. A* 105: 293–307.
47 Ghosh, D., Kosenkov, D., Vanovschi, V. et al. (2010). *J. Phys. Chem. A* 114: 12739–12754.
48 Gordon, M.S. (2017). *Fragmentation: Toward Accurate Calculations on Complex Molecular Systems*. Wiley.
49 Reinholdt, P., Kongsted, J., and Lipparini, F. (2021). *J. Chem. Theory Comput.* 18: 344–356.
50 Bondanza, M., Nottoli, M., Cupellini, L. et al. (2020). *Phys. Chem. Chem. Phys.* 22: 14433–14448.
51 Patel, A.B., Crocker, E., Eilers, M. et al. (2004). *Proc. Natl. Acad. Sci. U.S.A.* 101: 10048–10053.
52 Prokhorenko, V.I., Nagy, A.M., Waschuk, S.A. et al. (2006). *Science* 313: 1257–1261.
53 Sen, S., Kar, R.K., Borin, V.A., and Schapiro, I. (2022). *Wiley Interdiscip. Rev.: Comput. Mol. Sci.* 12: e1562.
54 Send, R. and Sundholm, D. (2007). *J. Phys. Chem. A* 111: 8766–8773.
55 Fantacci, S., Migani, A., and Olivucci, M. (2004). *J. Phys. Chem. A* 108: 1208–1213.
56 Garavelli, M., Vreven, T., Celani, P. et al. (1998). *J. Am. Chem. Soc.* 120: 1285–1288.
57 Marsili, E., Olivucci, M., Lauvergnat, D., and Agostini, F. (2020). *J. Chem. Theory Comput.* 16: 6032–6048.
58 Rinaldi, S., Melaccio, F., Gozem, S. et al. (2014). *Proc. Natl. Acad. Sci. U.S.A.* 111: 1714–1719.
59 Warshel, A. (1976). *Nature* 260: 679–683.
60 Chung, W.C., Nanbu, S., and Ishida, T. (2012). *J. Phys. Chem. B* 116: 8009–8023.
61 Frutos, L.M., Andruniw, T., Santoro, F. et al. (2007). *Proc. Natl. Acad. Sci. U.S.A.* 104: 7764–7769.

62 Bravaya, K.B., Kostko, O., Dolgikh, S. et al. (2010). *J. Phys. Chem. A* 114: 12305–12317.

63 Bravaya, K.B., Epifanovsky, E., and Krylov, A.I. (2012). *J. Phys. Chem. Lett.* 3: 2726–2732.

64 Ghosh, D., Isayev, O., Slipchenko, L.V., and Krylov, A.I. (2011). *J. Phys. Chem. A* 115: 6028–6038.

65 Dutta, A.K., Sengupta, T., Vaval, N., and Pal, S. (2015). *Int. J. Quantum Chem.* 115: 753–764.

66 Verma, P., Ghosh, D., and Dutta, A.K. (2021). *J. Phys. Chem. A* 125: 4683–4694.

67 Mukherjee, M., Tripathi, D., and Dutta, A.K. (2020). *J. Chem. Phys.* 153: 044305.

68 Bose, S., Chakrabarty, S., and Ghosh, D. (2016). *J. Phys. Chem. B* 120: 4410–4420.

69 Bose, S., Chakrabarty, S., and Ghosh, D. (2017). *J. Phys. Chem. B* 121: 4790–4798.

70 Sen, T., Ma, Y., Polyakov, I.V. et al. (2021). *J. Phys. Chem. B* 125: 757–770.

71 Mroginski, M.-A., Adam, S., Amoyal, G.S. et al. (2021). *Photochem. Photobiol.* 97: 243–269.

8

Electron Density in the Multiscale Treatment of Biomolecules

Soumyajit Karmakar, Sunita Muduli*, Atanuka Paul*, and Sabyashachi Mishra*

Department of Chemistry, Indian Institute of Technology Kharagpur, Kharagpur, India

8.1 Introduction

Chemical bonding and reactivity constitute the central narrative of chemistry. With the rise of quantum mechanics, the bonding and reactivity of molecules are now placed on a firm footing via the valence bond theory and molecular orbital theory [1–4]. The growth in computational resources has brought quantum chemistry closer to traditional synthetic chemistry. The solution of Schrödinger's equation results in the energy and wavefunction of the quantum mechanical (QM) system. Molecular properties can then be evaluated as the expectation value of the concerned operator with the wavefunction defining the state of the system. The wavefunction, a function of $3N$ spatial and N spin coordinates becomes increasingly complex with a larger number of electrons (N). In an alternative approach, molecular quantum mechanics is developed around the central theme of electron density based on the two fundamental theorems of Hohenberg and Kohn [5, 6]. Subsequently, the density functional theory (DFT) progresses by using the electron density distribution of the ground state to determine the physical and chemical properties of molecules, while the change in electron density was used for determining molecular reactivity. Electron density is defined as the measure of the relative probability of finding an electron in an infinitesimal volume surrounding a given point in the molecular space. For an N electronic system, the electron density ($\rho(r)$) is represented by,

$$\rho(r) = N \int |\psi(r_1, r_2, r_3, \ldots, r_N)|^2 dr_2 \cdots dr_N \tag{8.1}$$

Electron density and its topological analysis offer effective tools for the investigation of molecular structure, bonding, and reactivity [7]. These concepts are routinely used in molecules and small clusters [8].

The complexity, however, increases by many folds when the system of interest is a large biomolecule. Although just like small molecules, biomolecules are composed of electrons and nuclei, and their dynamic behavior is regulated by combined nuclear-and-electronic correlations, their large size makes a full quantum mechanical treatment impractical. The size, complexity, and timescale associated with large biomolecules are best described by classical force-field-based MD simulations of the nuclear degrees of freedom [9], although these methods cannot be used

* Equal contribution.

Electron Density: Concepts, Computation and DFT Applications, First Edition.
Edited by Pratim Kumar Chattaraj and Debdutta Chakraborty.

to describe the bond formation and bond breaking. For a system containing N atoms, the cost of classical molecular dynamics (MD) simulation scales as $\mathcal{O}(N^2)$ in the simplest cases. Although, with efficient algorithms the cost is reduced to somewhere in between $\mathcal{O}(N)$ and $\mathcal{O}(N^2)$ [10]. Compared to this, a QM treatment scales as $\sim \mathcal{O}(M^3)$, where M is the number of electronic basis functions. In a typical QM calculation, the number of basis functions is a few times an order of magnitude greater than the number of electrons. Therefore, *ab initio* electronic structure calculations are not realistic for large bio-molecular systems [11].

The best way to deal with this conundrum is the hybrid multi-scale approach, where the speed of MD simulation is coupled with the accuracy of QM treatment. To describe the reaction mechanism of large and complex chemical systems, in 1976, Warshel and Levitt proposed the hybrid quantum mechanics-molecular mechanics (QM/MM) method [12]. This approach was later popularized with a wide range of applications in the 1990s, with Karplus and coworkers at the forefront [13]. In recognition of this, the 2013 Chemistry Nobel Prize was awarded to Warshel, Levitt, and Karplus for their significant contribution to the development of multiscale models for complex chemical systems [14]. In the hybrid QM/MM approach, the system is separated into two subsystems. For biomolecular systems like enzymes, the part of the enzyme where chemical reactions take place (the active site) is treated with quantum chemical theory, while the rest of the enzyme is described using classical molecular mechanics force field, thus allowing the study of a diverse class of reactions in the condensed phase. In the last 15–20 years, several review articles and book chapters have documented the progress in different aspects of the hybrid QM/MM approach, its applications to biologically important molecules, current methodological developments, the combination with the quantum nature of atomic motion, and free energy pathways of crucial enzymatic reactions [12, 15–25]. Recent applications of the hybrid QM/MM approach are not restricted to enzymatic reactions, they have found applications in organic, inorganic, and solid-state reactions in different solvents [26–32].

In the present chapter, we do not provide a detailed account of QM/MM methodology. Instead, we focus on those aspects of QM/MM methodology, where the electron density contributes to improving the accuracy of the QM/MM calculation. We also discuss some applications of QM/MM calculations where the electron density is used for property evaluation. In the following section, we provide a brief account of the standard QM calculation and MD simulation, followed by a concise discussion on the combination of the two. In the subsequent sections, we discuss the use of polarizable density embedding schemes [33] and multiscale methods with localized molecular orbitals [34] and multiple active zones [35]. We then discuss the usage of reactivity descriptors for QM/MM modeling and treatment of hydrogen bonding with QM/MM methods [36]. Finally, we discuss quantum refinement of X-ray crystal structure with QM/MM models [37].

8.2 Theoretical Background

The hybrid QM/MM approach offers the best of two worlds: the accuracy of the QM methods and the speed of the MM method. Successful implementation of the QM/MM strategy needs an easy-to-compute, reliable, and transferable force field for the MM calculation and an accurate QM method of choice.

Biomolecular systems like proteins have a large number of atoms. Systems with such complex architecture require simulations of the nanosecond to microsecond timescale to show their crucial dynamic behavior responsible for their physiological functions. A force-field-based molecular mechanics model, with time evolution according to the classical Newtonian equations of motion, i.e. MD simulation, is appropriate for such systems. A typical force field for a system with N atoms is given by [9],

$$\begin{aligned}
V_{\mathrm{MM}} &= V_{\mathrm{bonded}} + V_{\mathrm{non\text{-}bonded}} \\
&= \left(V_{\mathrm{bond}} + V_{\mathrm{angle}} + V_{\mathrm{torsion}}\right) + \left(V_{\mathrm{electrostatic}} + V_{\mathrm{vdW}}\right). \\
&= \sum_{i=1}^{N_{\mathrm{bond}}} k_i (R_i - R_0)^2 + \sum_{j=1}^{N_{\mathrm{angle}}} k_j (\theta_j - \theta_0)^2 + \sum_{\mathrm{dihedral}} k_{\phi,n} \left(\cos(n\phi + \delta n) + 1\right) \\
&\quad + \sum_{i>j}^{N} \frac{Q_i Q_j e^2}{4\pi\epsilon_0 |R_i - R_j|} + \sum_{i>j}^{N} 4\epsilon \left[\left(\frac{\sigma}{R_{ij}}\right)^{12} - \left(\frac{\sigma}{R_{ij}}\right)^{6}\right]
\end{aligned} \tag{8.2}$$

The bonded interactions, such as V_{bond} and V_{angle} are described by harmonic potentials with force constants k_i and k_j defined around their equilibrium values R_0 and θ_0, respectively. V_{torsion} is described by the periodic potential function. The nonbonded interactions are given by the sum of the electrostatic and van der Waals (vdW) interactions. The electrostatic interactions are defined by Coulomb's interaction between nuclei (with nuclear charges Q_i), and the vdW interactions are given by the Lennard–Jones (LJ) potential with ϵ and σ as the well-known LJ parameters. The first term in the LJ potential accounts for the short-range Pauli repulsion, while the second term allows the long-range London dispersion attraction. In a typical MD simulation algorithm, the forces and acceleration of individual atoms are calculated at each time step by taking the gradient of the potential energy function described by the MM force field in Eq. (8.2). Thus, the time evolution of a classical system is obtained by neglecting all direct electronic contributions. Hence, MD simulation cannot model chemical reactions where electronic rearrangement and electron correlation are the driving forces. For such systems, quantum mechanics is inevitable.

Quantum mechanical treatment of a molecular system requires the solution of the Schrödinger equation with the Hamiltonian,

$$\hat{H}_{\mathrm{QM}} = \hat{T}_{\mathrm{Nu}} + \hat{T}_{\mathrm{e}} + \hat{V}_{\mathrm{Nu\text{-}e}} + \hat{V}_{\mathrm{e\text{-}e}} + \hat{V}_{\mathrm{Nu\text{-}Nu}} \tag{8.3}$$

where, the first two terms represent the nuclear and electronic kinetic energy operators, respectively, while the latter three terms describe the electron-nucleus, inter-electron, and inter-nuclear interaction energy. Under the Born–Oppenheimer approximation ($\hat{T}_{\mathrm{Nu}} = 0$ and $\hat{V}_{\mathrm{Nu\text{-}Nu}}$ is a constant), the total Hamiltonian reduces to the electronic Hamiltonian ($\hat{H}_{\mathrm{el}}$), whose solution is given by,

$$\hat{H}_{\mathrm{el}} \psi_{\mathrm{el}}(\vec{r};\vec{R}) = E_{\mathrm{el}}(R) \psi_{\mathrm{el}}(\vec{r};\vec{R}) \tag{8.4}$$

where the wavefunction shows an explicit dependence on the electronic coordinates ($\vec{r}$) and a parametric dependence on the nuclear coordinates ($\vec{R}$), while $E_{\mathrm{el}}(R)$ provides the multi-dimensional potential-energy surface of the system. An exact solution of $\hat{H}_{\mathrm{el}}$ is not possible for systems with two or more electrons. In such cases approximate solutions are obtained, where explicit two-electron interaction terms ($\hat{V}_{\mathrm{e-e}}$) are approximated as the sum of several effective one-electron operators. These effective one-electron Fock operators describe the interaction of *an* electron with the *field* created by the rest of the electrons and the nuclei. The Fock operator can be written as,

$$\hat{F}(i) = \hat{h}(i) + \sum_{j=1}^{n/2} [2\hat{J}_j(i) - \hat{K}_j(i)] \tag{8.5}$$

where $\hat{h}(i)$ denotes the one-electron Hamiltonian operator for the ith electron and its interaction with the rest of the system is given by the (classical) Coulomb integral ($\hat{J}_j(i)$) and (quantum) exchange integral ($\hat{K}_j(i)$). The final solution, obtained through the self-consistent (iterative) method provides the so-called Hartree–Fock (HF) energy and the wavefunction. This method, known as the Hartree–Fock self-consistent field (HF-SCF) method, is the starting point of most other advanced (more accurate) electronic structure methods. In quantum chemistry, the HF-SCF equations are solved in matrix form by expanding the wavefunction through a basis set $\{\chi_\mu\}$.

8.2.1 Hybrid Quantum Mechanics–Molecular Mechanics Approach

Biomolecular systems, such as enzymes, have several thousands of atoms whose quantum mechanical treatment is extremely expensive, even with modern-day supercomputers. During enzyme catalysis, the chemical reactions take place at the active site, while the rest of the enzyme provides an environment for the catalysis [9]. Hence, hybrid QM/MM methods have been developed that segment the system into a smaller part (the active site) treated with QM methods and a larger part (the rest of the system) treated with MM methods (Figure 8.1). In a hybrid QM/MM simulation, the potential energy of the whole system is a combination of the potential energy due to the interaction between the QM atoms, the potential energy due to the interaction between the MM atoms, and the potential energy for the interaction of the QM atoms with the MM atoms. To compute the interaction between QM–MM atoms, two different schemes, the subtractive and additive coupling schemes, are generally employed.

In the subtractive coupling scheme introduced by Morokuma and Maseras in 1995 [38], the energy of the QM/MM system is calculated as a collection of three energy terms, i.e. (a) the energy of the QM region using a QM method E^{QM}(model), (b) the energy of the whole system using the MM force field (E^{MM}(full)), and (c) the energy of the QM region using MM force field (E^{MM}(model)). The final QM/MM energy is estimated as,

$$E(QM/MM) = E^{QM}(model) + E^{MM}(full) - E^{MM}(model) \tag{8.6}$$

The third term is subtracted from the sum of the first two energy terms to correct for the double counting of the QM region, hence the name, subtractive QM/MM scheme (Figure 8.1). This simple two-layer scheme can also be extended to multiple layers, where different QM levels (such as

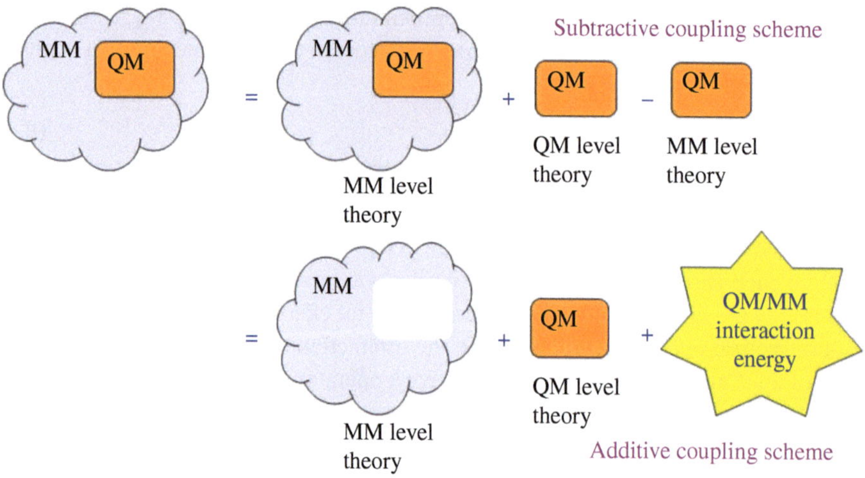

Figure 8.1 Subtractive and additive coupling schemes for QM/MM calculation.

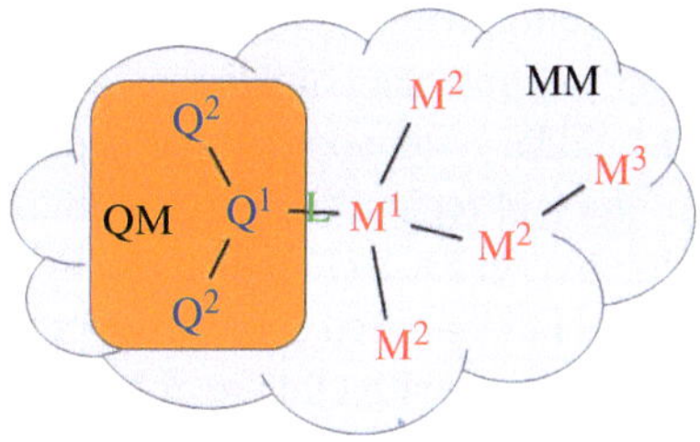

Figure 8.2 Separation of atoms between the QM and MM regions. The link-atom (L) is placed at the QM/MM boundary to saturate the valency of the QM atom Q^1 which loses the covalent bond with the MM atom M^1.

semiempirical, DFT, and coupled cluster) can be used to increase accuracy at a reduced cost. The scheme, popularly known as the ONIOM method, has been widely employed as a two-layer or three-layer ONIOM scheme for a wide variety of biological systems. In the additive coupling scheme, the energy of the QM/MM system is calculated as the summation of the energy of the model subsystem treated with the QM level of theory (E^{QM}(model)), the energy of the remaining part of the system E^{MM}(remnant), and explicit interaction energy between the QM and MM regions $E^{QM\text{-}MM}$(model–remnant) (Figure 8.1), [9, 16], i.e.,

$$E(QM/MM) = E^{QM}(\text{model}) + E^{MM}(\text{remnant}) + E^{QM\text{-}MM}(\text{model–remnant}) \tag{8.7}$$

An important element of QM/MM calculation is the QM–MM boundary. If the QM and MM region are connected by one or more covalent bonds, care should be taken to consider a nonpolar bond as the QM–MM boundary. The boundary in the close vicinity of charged MM atoms or electronegative atoms can introduce an erroneous artificial net charge near the QM electron density [16]. Hence, the choice of boundary is very crucial. There are three schemes for QM/MM boundary treatment: the link atom, the boundary atoms, and the frozen-localized orbital schemes.

In the link-atom scheme when covalent bonds are present between the QM region and MM region, an extra atom is introduced between the two regions, which is not a part of the real system (Figure 8.2). The link atom and the QM charge density are very close to the MM atom in the boundary, which can overpolarize the QM electron density [16]. This problem becomes more crucial when the MM charges are closer to the QM density and the QM electron density is spatially flexible [16]. However, there are different successful approaches introduced for the reduction of this overpolarization in the link-atom approach [39–45]. In the boundary atoms method, an atom of Janus-like character is introduced across the boundary that participates in the MM calculations and also has a character like a QM atom [16]. In the frozen-localized orbital scheme, the capping is done by placing localized orbitals at the boundary while some of them remain frozen and do not participate in SCF calculations. Some popular methods under this scheme are local self-consistent field [46], frozen orbitals [47], generalized hybrid orbitals [48], and effective fragment potentials [49].

The evaluation of the interaction energy (between QM and MM layers) can be obtained with different levels of embedding schemes with varying costs and accuracy. The most basic way of calculating the explicit QM/MM interaction term is the mechanical embedding scheme, where the QM region is mechanically placed in the MM region, with their interaction treated with MM model. If two atoms in two different layers are connected via internal coordinates, their interaction is treated with harmonic potential energy (Eq. (8.2)). The nonbonded atoms in two different layers interact with each other via electrostatic interaction (with force field charge on the MM atom and partial charge on the QM atom obtained from the fragmentation of the QM charge density) and vdW interaction with the force-field LJ parameters [9, 16]. While the mechanical embedding scheme is easy to implement and less expensive, it suffers from several limitations: (i) the QM calculation of the model system does not include the polarization effect by the MM region on the electron density of the QM region, (ii) the LJ-parameters used in calculations are not updated, although the reaction may cause a change in the hybridization of the atoms involved, (iii) the MM force field gives an overall balanced description of the MM charges, which also implies that the MM point charges that interact with QM are not real charge distributions of the MM region [9, 16].

In the electrostatic embedding scheme, the polarization of the QM charge density by the MM point charges is introduced by a new one-electron operator [9] in the QM Hamiltonian (Eq. (8.5)). The new one-electron operator includes the interaction of the QM electron density with N number of MM point charges (Q_i) given by,

$$\hat{h}'(i) = \hat{h}(i) - \sum_{j=1}^{N} \frac{Q_j e^2}{4\pi\epsilon_0 |r_i - R_j|} \tag{8.8}$$

where r_i and R_j represent the electronic and nuclear coordinates, respectively. The calculation of the extra one-electron integral in this scheme increases the computational cost of the model and modifies the QM/MM algorithm [9]. Field et al. first used this scheme in the interface of Mopac quantum chemistry package with CHARMM molecular mechanics package [13, 50, 51]. While the electrostatic embedding model takes care of the underpolarization of the mechanical embedding model, the presence of MM point charges near the QM atoms can lead to overpolarization of the QM region, in particular, when the QM–MM boundary crosses through a covalent bond and when a large, diffused basis set is used for the QM atoms. Near the interface of two subsystems, the MM point charges interact (attract or repel) with the QM electron density so strongly that electron density can spill into the MM region. To avoid such errors in calculation, the smeared-out charges for the MM atoms are considered (instead of the point charges) using a Gaussian distribution centered at the MM atoms [9] given by,

$$\Omega_j^{\mathrm{MM}}(r) = \sqrt{\frac{Q_j^{\mathrm{MM}}}{\pi\alpha^3}} \exp\left[-\frac{|(r - R_j|^2}{2\alpha^2}\right] \tag{8.9}$$

where $|\Omega_j^{\mathrm{MM}}(r)|^2$ is the charge density at r due to the MM atom J with charge (Q_j), and α is the width of the distribution that needs to be calibrated. These new charges modify the one-electron operator due to the interaction between the QM electrons (in molecular orbital ϕ) and the MM atoms via [9],

$$h_{ij}(r_1) = Q_j \int \phi_i^*(r_1) \frac{\mathrm{erf}(|r_1 - R_j|/\alpha)}{|r_1 - R_j|} \phi_j(r_1) d(r_1) \tag{8.10}$$

Although the electrostatic embedding schemes provide accuracy, the associated computational cost makes them expensive methods. On the other hand, the more affordable mechanical embedding scheme suffers from a lack of polarization of the QM region by the MM charges. One middle ground is the so-called self-consistent mechanical embedding scheme [52–54]. Here, the QM/MM optimization is carried out first with a mechanical embedding model. The electrostatic potential of the QM region of the optimized QM/MM geometry is then calculated, and partial charges on the QM atoms are obtained from an RESP charge fitting scheme. The new charges on the QM atoms are used in the previously optimized QM/MM geometry, and another round of geometry optimization is done (with the updated QM charges). This procedure (Figure 8.3) is repeated until convergence of geometry, energy, and QM atom charges are converged to a desired level of accuracy [52–54].

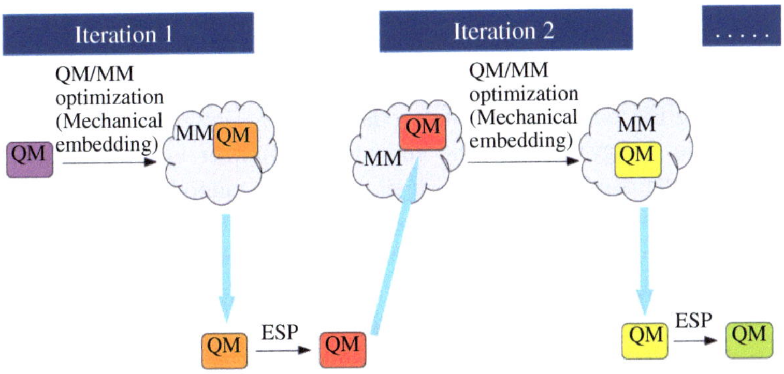

Figure 8.3 Self-consistent mechanical embedding scheme.

8.3 Polarizable Density Embedding

Just like the MM point charges polarize the QM electron density, the QM electron density can similarly polarize the MM charges. This reverse polarization is absent in the electrostatic embedding scheme [9, 16], where the MM atoms are still described by the force field charges that were parameterized to describe the complete MM region. Hence, reverse polarization of the MM charges (by QM density) can be incorporated by adopting a flexible MM charge model. In such a polarization embedding scheme, further back-and-forth polarization between the two layers is ignored [16]. Although classical polarizable force fields are recently being used for MD simulations [55], their versatility is yet to be demonstrated. On the other hand, the QM/MM calculations with a polarization embedding scheme are extremely expensive. A possible way out of the conundrum is to treat a portion of the MM region which is in close vicinity to the QM region with a polarizable embedding scheme while the outer MM region is treated as in the ordinary MM model [56].

One such scheme has been proposed by Olsen et al., where the system is divided into three subsystems: the core region deals with QM level, and the surrounding is divided into inner and outer layers. The atoms of the inner layer are treated with their exact QM-calculated density, while the outer region is treated with atom-centered multipole moments [33]. The electric fields of each fragment can induce a charge redistribution described by an induced dipole moment. The polarized density embedding (PDE) potential of the inner and outer regions ($\hat{v}^{\text{PDE}}$) is then added to the SCF Fock operator ($\hat{F}$) to get the new effective Fock operator $\hat{F}^{\text{eff}}$. The PDE potential ($\hat{v}^{\text{PDE}}$) can have a contribution from the electrostatic charge distribution ($\hat{v}^{\text{es}}$), induced charge distribution ($\hat{v}^{\text{ind}}$), and a repulsive potential added to treat the non-electrostatic repulsion term ($\hat{v}^{\text{rep}}$).

The interaction of the electron density of the QM region with the fragment density of the inner region (denoted by a that is composed of M_a nuclei of charge Z_m and coordinate R_m) in an atomic orbital basis ($\{\mu\nu\}$ for core electrons and $\{\gamma\delta\}$ for electrons in inner core) is given by,

$$\sum_{a=1}^{N_{\text{in}}}\left(-\sum_{m\in a}^{M_a} Z_m v_{\mu\nu}(R_m) + \sum_{\gamma\delta\in a} D_{\gamma\delta} v_{\mu\nu,\gamma\delta}\right)$$

where $v_{\mu\nu}$ and $v_{\mu\nu,\gamma\delta}$ are two and four-centred, one and two-electron integrals, respectively, and D is the density matrix. The first term represents the interaction of the QM electron density with the nuclei in the inner region and the second term accounts for the interaction of the QM electron density with the electron density of the inner region [33]. Similarly, the induction potential (due to charge polarization), as well as the repulsion potential, are expressed in terms of the atomic orbital integrals, density matrices, and overlap integrals. The HF-SCF equations are then solved with the new effective Fock operator.

Olsen et al. chose three dimeric systems: formamide, water, and benzene, for interaction energy calculations by employing the polarized density model [33]. The interaction energy is obtained as the difference in the energy between the embedded monomer and the isolated monomer where the embedded monomer was the system in which one monomer belonged to the core region while the other monomer belonged to either the inner or the outer region. Since the system size is small, a pure QM model (HF/aug-cc-pVDZ) was considered as the reference. Six different models, namely, polarizable embedding (PE) model, polarizable density embedding model without nonelectrostatic repulsion (PDE-R), PDE model (PDE), PE model with non-scaled nonelectrostatic repulsion (PE+R), PDE model with nonelectrostatic repulsion scaled by a factor of 0.8 (PDE*), and a relaxed embedding potential density model (DE(it)), were compared with the reference QM method. While the PDE-R and PE models deviated significantly from the QM reference value of the interaction energy for formamide dimer, the PDE model overestimated the nonelectrostatic repulsion at a shorter distance. The PDE* model provided the best comparison with the QM-reference values (Figure 8.4).

Table 8.1 provides a comparison of the formamide dimer interaction energy from PDE and PDE* models (with respect to the pure QM results) for three different basis sets. While for all basis sets

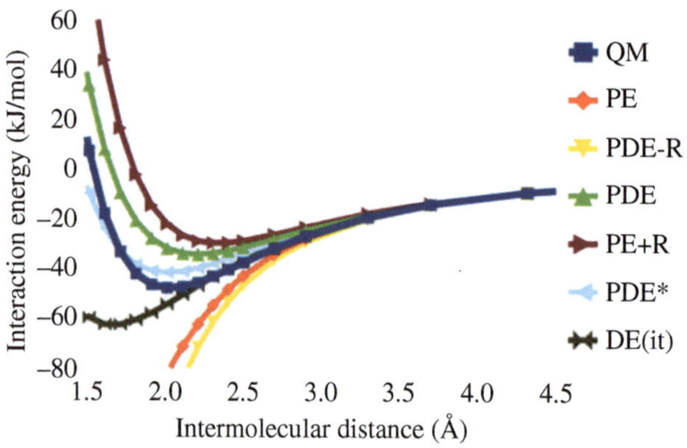

Figure 8.4 Interaction energy of formamide dimer with six-different QM/MM models compared with pure QM model using aug-cc-pVDZ basis. Source: With permission from [33]/American Chemical Society.

Table 8.1 Basis set effects on the formamide dimer interaction energy differences (in kJ/mol) in PDE and PDE* models.

		E_{int}^{PDE-QM}			$E_{int}^{PDE^*-QM}$	
***R* (Å)**	**6-31G**	**6-31G***	**6-31+G***	**6-31G**	**6-31G***	**6-31+G***
1.8	14.90	15.47	18.28	−1.64	−0.92	0.60
1.9	11.68	12.55	15.98	0.20	1.16	2.75
2.0	9.26	10.11	13.57	1.40	2.29	3.70
2.1	7.44	8.16	11.32	2.10	2.84	3.95
2.2	5.96	6.64	9.32	2.36	3.04	3.82

Source: Adapted from [33].

(with and without polarization functions and diffused functions), PDE* performs better than PDE model, the latter shows very large errors when polarized, and diffused functions are used. The larger basis sets lead to an increase in the fragment density of the inner region extended too far into the core region [33]. This affects the non-electrostatic repulsive interactions (Pauli-repulsion or exchange repulsion), and hence the PDE* model that uses a scaling factor for the nonelectrostatic repulsion provides better results. This is an excellent example of unintended error and mitigation in an improvised QM/MM model.

8.4 Multi-Scale QM/MM with Extremely Localized Molecular Orbitals

Due to the poor scaling of QM methods with increasing system size, the QM/MM methods for modeling bio-molecular reactions have remained a challenge, despite a rapid increase in computational resources. Reducing the size of the QM region can compromise the accuracy of results [57]. One way to address this issue is proposed by Genoni and coworker via a novel three-layer multi-scale embedding approach, the so-called QM/ELMO/MM (quantum mechanical/extremely localized molecular orbital/molecular mechanics) technique [34] and its two-layer QM/ELMO formulation [58–60]. The ELMOs are molecular orbitals strictly localized on small molecular fragments, such as atoms, bonds, and functional groups [34]. They are transferable and can be regarded as electronic building blocks. Therefore, to immediately compute approximation wavefunctions and electron densities of large biological systems, libraries of extremely localized molecular orbitals containing all possible elementary components of the 20 natural amino acids have been developed [61].

In the QM/ELMO/MM method, the system is partitioned into three subsystems, with the middle layer represented by the ELMOs which link the QM region with the MM region (Figure 8.5a). The inner QM region is treated with conventional QM methods, while the atoms of the ELMO region are described by frozen ELMOs either transformed from certain ELMO libraries or suitable model molecules [34]. The outer region is treated with a MM model. The total energy of the system can be represented by,

$$E = E_{QM/ELMO} + E_{MM} + E_{QM/ELMO/MM} \tag{8.11}$$

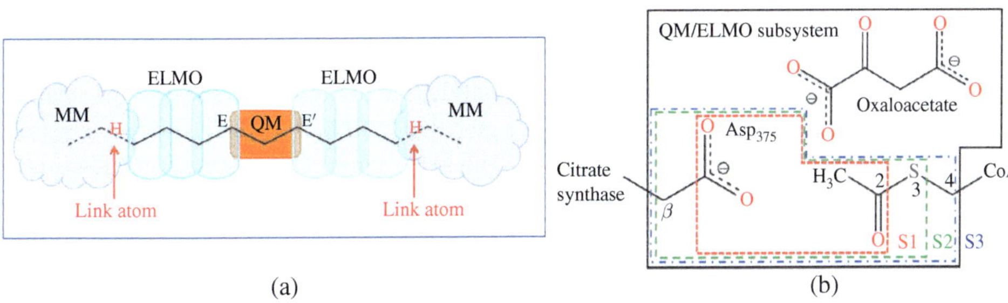

Figure 8.5 (a) Schematic representation of the QM, ELMO, and MM regions in the QM/ELMO/MM approach, where E and E′ indicate frontier atoms between the QM and ELMO sub-units. (b) An example of the partitioning schemes employed in the QM/ELMO/MM calculations. The black solid frame covers the whole QM/ELMO subsystem; red, green, and blue (also for the QM/MM calculation with scheme S3-MM) dashed frames contain the QM region for the partitioning schemes S1, S2, and S3, respectively. Source: These figures have been adapted from Reference [34].

where $E_{QM/ELMO}$ is the QM energy of the QM and the ELMO subsystems and E_{MM} is the standard MM energy. The last term $E_{QM/ELMO/MM}$ refers to the interactions of the MM region with the QM and ELMO regions. This interaction is treated classically (i.e., via electrostatic and vdW interactions; and harmonic potential if a covalent boundary is present between the ELMO and MM subsystems). Here, the most important term is the quantum contribution term ($E_{QM/ELMO}$) [58, 59, 61]. The SCF procedure for the QM/ELMO calculation is carried out in the atomic orbital basis $\{\chi\}$ of the QM and ELMO regions. The Fock matrix is constructed in this "supermolecular basis set" for the entire QM/ELMO region [58, 59]. In such a case, the Fock matrix is the summation of the original one-electron (two-center) integrals ($h^{\text{core}}_{\mu\nu}$), the two-electron Coulomb and exchange integrals involving the QM basis ($F^{\text{QM}}_{\mu\nu}$) and ELMO basis ($F^{\text{ELMO}}_{\mu\nu}$), and finally, the one-electron electrostatic interaction between the MM point charges (q_K) and the QM/ELMO electrons ($F^{\text{MM}}_{\mu\nu}$), i.e.

$$\begin{aligned} F_{\mu\nu} &= h^{\text{core}}_{\mu\nu} + F^{\text{QM}}_{\mu\nu} + F^{\text{ELMO}}_{\mu\nu} + F^{\text{MM}}_{\mu\nu} \\ &= \left\langle \chi_\mu \middle| \hat{h}^{\text{core}} \middle| \chi_\nu \right\rangle + \sum_{\lambda,\sigma=1}^{M} P^{\text{QM}}_{\lambda\sigma} \left(\chi_\mu\chi_\nu \| \chi_\sigma\chi_\lambda\right) \\ &\quad + \sum_{\lambda,\sigma\in\text{ELMO}} P^{\text{ELMO}}_{\lambda\sigma} \left(\chi_\mu\chi_\nu \| \chi_\sigma\chi_\lambda\right) + \sum_{K\in\text{MM}} \left\langle \chi_\mu \middle| \frac{q_K}{R_{iK}} \middle| \chi_\nu \right\rangle \end{aligned} \tag{8.12}$$

Here $\hat{h}^{\text{core}}$ is the one-electron Hamiltonian operator, $\mathbf{P}^{\text{QM/ELMO}}$ are the QM or ELMO one-electron reduced density matrices. $(\chi_\alpha\chi_\beta \,||\, \chi_\gamma\chi_\delta)$ represents the Coulomb-exchange integral. This approach can be easily extended to the post-HF method by introducing an exchange–correlation matrix into the above equation [34, 58–62].

The QM/ELMO/MM technique has been applied to investigate the deprotonation of acetyl coenzyme A (Ac-CoA) by aspartate in citrate synthase enzyme, which plays a vital role in the Krebs cycle. The reaction is initiated by the deprotonation of the α-carbon of Ac-CoA by Asp375, resulting in the formation of an enolate intermediate that shows a nucleophilic attack on the carbonyl carbon of oxaloacetate (OAA) to produce the citryl-CoA intermediate. Citrate and coenzyme A are ultimately produced as a result of the hydrolysis of citryl-CoA [63]. The initial geometry was obtained from QM/MM (B3LYP/6-31+G(d)//CHARMM27) optimization [64]. The reaction coordinate was chosen to be (RC = $d(C_{\text{Ac-CoA}} - H) - d(O_{\text{Asp375}} - H)$), which represents the reaction path accurately [63]. The QM region was treated at RHF, B3LYP, M06-2X, MP2, CCSD, and CCSD(T) levels of theory, the ELMO subunit was described by the properly transferred ELMOs, and the MM region was treated with the ff14SB/GAFF force field [65, 66].

Four alternative partitioning schemes (S1 to S4) were used to investigate how the dimensions of the QM and ELMO regions influence the QM/ELMO/MM calculations (Figure 8.5b). From scheme S1 to S3, the size of the QM region gradually increases while that of the ELMO subsystems steadily shrinks in a squeezebox fashion. Thus, the size of the global QM/ELMO region remained constant across partitioning schemes, as did the dimension of the MM region. This allowed an assessment of the significance of the full QM region and decided its appropriate size for the enzyme reaction under investigation. The references for all QM/ELMO/MM calculations were the corresponding full-QM/MM calculations (scheme S3, Figure 8.5b). The partitioning scheme S1 (with the smallest QM subsystem) shows the least accurate results as compared to the full-QM/MM reference values. This improves with the intermediate S2 scheme, and a significant improvement in the accuracy is observed in the S3 scheme for all correlated methods (Table 8.2). To assess the significance of treating the entire OAA molecule at the ELMO level, additional QM/MM calculations were also performed with partitioning scheme S3-MM, where the QM subsystem corresponds to the scheme S3 in the QM/ELMO/MM computations, but with OAA included in the MM region.

Table 8.2 Activation energy (in kcal/mol) barriers (ΔE_{act}) obtained at the reference full-QM/MM levels (ΔEact(QM/MM)) and absolute discrepancies from these values ($\Delta\Delta E_{act}$), as resulting from QM/ELMO/MM and QM/MM calculations with the cc-pVDZ basis set and by using different partitioning schemes.

QM method	$\Delta E_{act}^{(QM/MM)}$	$\Delta\Delta E_{act}^{(S1)}$	$\Delta\Delta E_{act}^{(S2)}$	$\Delta\Delta E_{act}^{(S3)}$	$\Delta\Delta E_{act}^{(S3-MM)}$
RHF	19.01	2.67	0.09	0.23	0.91
B3LYP	7.36	3.09	0.52	0.19	0.46
M06-2X	6.40	3.18	0.70	0.23	0.59
MP2	8.52	1.59	0.29	0.03	0.46
CCSD	12.07	2.37	0.19	0.07	0.46
CCSD(T)	10.55	2.12	0.27	0.05	0.42

Source: The table has been reproduced from Reference [34] with permission.

This approximation always produced poorer results than the analogous S3 and S2 schemes. This suggests that the oxaloacetate molecule should be described at least at an approximation quantum mechanical level, such as the one provided by the transferred and frozen extremely localized molecular orbitals.

8.5 Multiple Active Zones in QM/MM Modelling

Most of the QM/MM calculations are restricted to systems where a spatially localized region is treated with QM, and the rest of the system is treated with the MM method. In other words, there is one active zone (AZ). However, there are a few systems where multiple disjointed active sites contribute to the global behavior of the whole system. This requires a QM/MM set up with multiple active zones (MAZ) [24, 67–69]. Torras has recently proposed a scheme for multi-center disjointed MAZ-QM/MM calculation [70]. Here, the system is divided into multiple AZ. For each AZ, the concerned region is the inner region (I) to be treated with the QM method, while the remainder of the system is considered as the outer part (O), treated in the MM region. The entire system (S) is the sum of all inner regions and the outer region (Figure 8.6).

The total energy within an additive QM/MM scheme for such a set-up is given by the sum of the QM energies, the MM energy and the interaction energy between the QM regions and the MM region, i.e.

$$E = E_{QM}(I) + E_{MM}(O) + E_{QM\text{-}MM}(I, O) \tag{8.13}$$

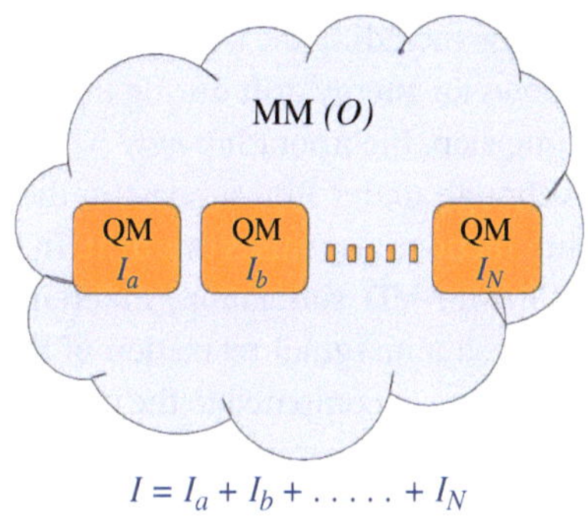

Figure 8.6 Partition scheme of the system (S) in an outer region (O) and multiple active zones (I_a, I_b, I_c, …).

The last term explains the interaction between the QM and MM regions treated classically (via bonded and nonbonded interactions). Since the AZs are (by definition) not connected to each other [71], there is no bonded interaction between one inner zone with the other. The overall system energy is, therefore, the sum of the N individual QM regions and their corresponding QM/QM and QM/MM coupling terms, i.e.

$$\mathrm{E} = \mathrm{E}_{\mathrm{MM}}(O) + \sum_{a}^{N} \left(\mathrm{E}_{\mathrm{QM}}(I_a) + \mathrm{E}_{\mathrm{QM\text{-}MM}}(I_a, O)\right) \tag{8.14}$$

The interaction of one inner zone (I_a) with another (I_b) can be considered as the sum of their vdW interaction and the electrostatic interaction. The latter between two inner zones is the interaction between the electron density of one of the regions (I_a) and the grid of charges (obtained from any charge decomposition scheme, say Mulliken) of the other (I_b), i.e.

$$\begin{aligned}\mathrm{E}_{\mathrm{QM\text{-}QM}}(I_a, I_b) &= \mathrm{E}^{\mathrm{vdW}}_{\mathrm{QM\text{-}QM}}(I_a, I_b) \\ &+ \frac{1}{2}\sum_{a>b}^{N} \left(\mathrm{E}^{\mathrm{el}}_{\mathrm{QM\text{-}QM}}(I_a : \rho_a, I_b : Q_b) + \mathrm{E}^{\mathrm{el}}_{\mathrm{QM\text{-}QM}}(I_b : \rho_b, I_a : Q_a)\right)\end{aligned} \tag{8.15}$$

The basic requirement of the multiple AZ method is that the density of two inner regions does not overlap. In such cases, the long-range electrostatic interactions need to be considered. The total electron density (ρ) can be partitioned into the density of individual inner zones (ρ_a) by several decomposition schemes. The long-range electrostatic interaction energy can then be described as,

$$\mathrm{E}(\rho + q, \rho + q) \cong \sum_{a}^{N} \left(\mathrm{E}(\rho_a, \rho_a) + \mathrm{E}(\rho_a, q)\right) + \frac{1}{2}\sum_{a \neq b}^{N} \mathrm{E}(\rho_a, Q_b) + \mathrm{E}(q, q) \tag{8.16}$$

The first two terms denote interactions among the QM atoms of an inner region and the MM atoms (charge q) within a certain cut-off distance. The following term represents the interaction between two inner zones (a and b), and the last term accounts for the interaction due to the MM point charges (q). In a typical quantum-chemical calculation, the long-range interactions are calculated in a straight-forward real-space cut-off-based approach. Correction terms to the above equations are added when these interactions are calculated under the periodic boundary conditions [72].

Ferritin is a protein with a cage-like structure with a 24-mer Cu-ferritin cage dependent on the divalent Cu-binding (PDB ID: 4DYX). Each monomer has three Ca(II) and four Cu(II) metal ions (Figure 8.7). The MAZ-QM/MM calculations were performed on one monomer, where five AZs contained a total of 92 atoms in the QM region, as shown in Table 8.3. The enzyme was solvated in a water box, and QM/MM-MD simulations were carried out at 298 K temperature, 1 atm pressure and by including the periodic boundary conditions (PBC) [35]. The MAZ-QM/MM-MD with PBC shows an energy drift during the MD simulation. The fluctuation in energy comes from the finite time step. Incorporating new AZ does not change the stability of the QM/MM MD without PBC, although under PBC a boost in the energy drift is seen. The addition of long-range electrostatic interactions also raises this drift. In Cu-ferritin protein, the metal ions stabilize the protein. During QM/MM-MD simulation, different AZs showed fluctuation to a different extent. While some showed a marginal relaxation of the metal-ligand coordination, the calcium ions showed ligand exchange as compared to the crystal structure. Furthermore, the reduced electron density gradient shows a large interaction between Cu203 and the Glu27, Glu62, and His65 residues, whereas Cu204 exhibits two strong interactions with the Glu62 and Glu107 residues [35].

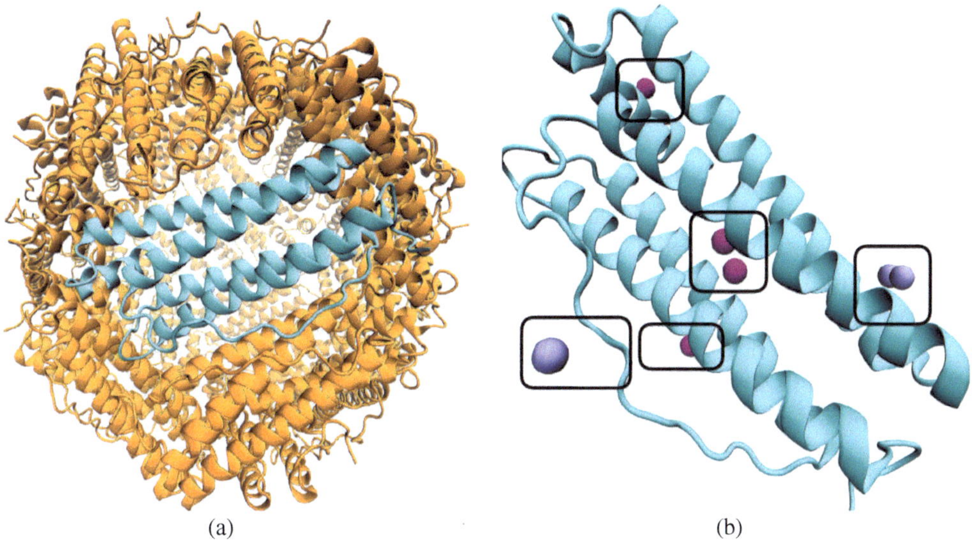

Figure 8.7 (a) Cu-ferritin cage with one building block monomer shown in cyan. (b) The active zones (marked by boxes) in a monomer of the Cu-ferritin cage. The figure is prepared from PDB ID 4DYX.

Table 8.3 The residues in the five active zones [35].

AZ	Residues	QM atoms
QM1	E27 E62 E107 H65 Cu203 Cu204	31
QM2	H63 H67 Cu201	23
QM3	H173 Cu202	12
QM4	Q86 Ca205	12
QM5	D131 E134 Ca206 Ca207	14

8.6 Reactivity Descriptors with QM/MM Modeling

The active site reactivity is largely controlled by the enzyme environment that allows efficient enzyme catalysis. Several reactivity descriptors have been proposed to explain the stability, bonding, and reactivity of molecular systems. The Fukui function, widely used in conceptual density functional theory as a reactivity indicator, is given in terms of the change of electron density [73, 74],

$$f(r) = \frac{\delta\rho(r)}{\delta N} \tag{8.17}$$

whereas the condensed Fukui function of a single atom is defined as the partial charge difference upon addition or removal of an electron [74, 75]. This helps to define reactivity at the specific site of a molecule (or an enzyme). In enzyme catalysis, amino-acid residues are the basic building blocks. Hence, the residue-wise condensed Fukui functions can be defined as [76, 77],

$$f^{+}_{\text{RES}} = q^{\text{RES}}(N+1) - q^{\text{RES}}(N) \tag{8.18}$$

$$f^{-}_{\text{RES}} = q^{\text{RES}}(N) - q^{\text{RES}}(N-1) \tag{8.19}$$

for the addition of electrons and removal of electrons, respectively. Here, $q^{RES}(N)$ is obtained by the summation of all $q_i(N)$ with i representing atoms of a given residue.

The Fukui shift analysis (FSA) technique [76], proposed by Kulik and others, measures the reactivity of a chemical reaction by analyzing how the frontier states of an active site are changed by the introduction of an additional QM residue. This offers insight into the requirement of QM treatment of a residue. The Fukui function achieves the maximum value in the most reactive site of a molecule. This can be used to rationally design the QM region of a QM/MM calculation. The strategy followed in such an analysis starts by calculating the Fukui functions with the minimal active site described by the QM method while treating the remaining protein residues at the MM level. All residues within a defined cut-off value are then added one by one to the QM region, and the condensed Fukui functions are recalculated. The relative importance of an active site residue (ASR) is estimated by the root-sum-squared (RSS) difference in all residue-wise condensed Fukui functions on the active site with respect to the median reference ($\tilde{f}$) [77], i.e.

$$\mathrm{RSS} = \sqrt{\sum_{j}^{\mathrm{ASR}} \left(\left(f_j^+ - \tilde{f}_j^+ \right)^2 + \left(f_j^- - \tilde{f}_j^- \right)^2 \right)} \tag{8.20}$$

Two amino-acid pair models have been studied for Fukui shift analysis. These models, where one residue was Asp, included Asp–Arg, Asp–Asn, Asp–His1, Asp–His2, and Asp–Ser pairs (see Figure 8.8), where the second residue provides a changing environment for Asp [77]. In order to obtain the Fukui function in the presence of the second residue, an electron from the overall residue pair was removed, and the partial charges were estimated for the Asp residue in both the N and $N-1$ electron systems. The shift in the Fukui function (ωPBEh/6-31G* and Mulliken charges) in amino-acid pairs reflects various noncovalent interactions. This shift in the condensed Fukui function is treated as a tool for constructing the QM region in QM/MM simulation [77].

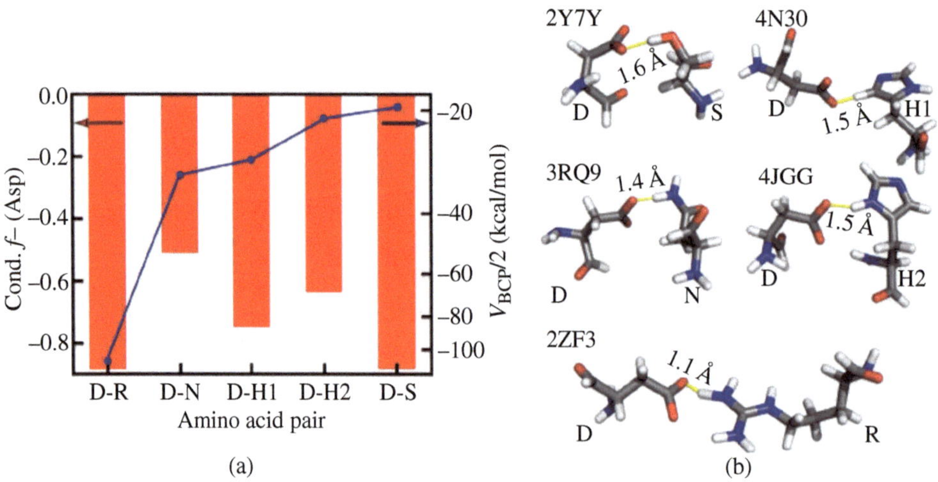

Figure 8.8 (a) Condensed Fukui function on an Asp residue (left axis, red bars) as a function of the paired residue with potential energy at the bond critical point, $V_{BCP/2}$ overlaid (right axis, blue points) for five representative Asp-residue pairs and (b) Five representative Asp-res residue pairs extracted from crystal structures and labeled by their corresponding PDB ID, where each residue is shown by the single letter amino acid code. The shortest distance between the two residues is indicated with a yellow line. Source: With permission from [77]/Massachusetts Institute of Technology.

Figure 8.8 shows the shift in the Fukui function of Asp in different pairs of residues. No correlation was observed between the shift in the Fukui function of Asp residue with the strength of the hydrogen bonding. However, the chemical identity plays a vital role in the deviation of the condensed Fukui function. The presence of basic residue Arg does not shift the Fukui function of Asp, as it is difficult to ionize due to the salt-bridge interaction (Figure 8.8). On the other hand, the neutral residues show greater Fukui function shifts. The Asp–Asn pair exhibits the highest electrophilic condensed Fukui function of all five residue pairs due to strong hydrogen bonding between Asp and the N–H hydrogen bond donor of Asn sidechain. Furthermore, the N–H and C–H hydrogen bond donors of His1 and His2 have relatively lower condensed Fukui functions of −0.7 and −0.8, respectively, although they have comparable bond distances. This is due to the lower Pauling electronegativity of carbon than nitrogen, causing the His pair (D-H2) to hold onto less electron density and hence lose less when an electron is withdrawn. The summed condensed Fukui function [76] of an Asp–Ser pairs with a strong O–H hydrogen bond donor, on the other hand, deviates from the ideal value by a relatively modest amount. Thus, the orientation (e.g. H1 versus H2 Asp–His pairs) and identity of adjacent residues (e.g. Asp–Asn versus Asp–Arg) influence the shift of a condensed Fukui function summed over a central residue [77]. It can be deduced that the presence of proximal residues in a short hydrogen bonding configuration causes some Fukui shifts.

8.7 Treatment of Hydrogen Bonding with QM/MM

Hydrogen bonds play a significant role in the structural stability of enzymes as well as in their physiological functions. The formation of a hydrogen bond is associated with a net change in the total electron density caused by polarization effects [12, 13, 36]. The change in electron density due to the formation of hydrogen bond can be estimated by subtracting the sum of total electron densities of the monomers (donor D or acceptor A) constituting the hydrogen-bonded complex from the total electron density of the hydrogen-bonded complex (DA), i.e.

$$\Delta\rho_{QM} = \rho_{DA,QM} - (\rho_{D,QM} + \rho_{A,QM}) \tag{8.21}$$

In a QM/MM set-up, the change in electron density due to hydrogen bonding ($\Delta\rho_{QM/MM}$) can be obtained from the total density of the hydrogen-bonded complex at QM/MM level after subtracting the sum of the densities obtained in two separate QM/MM calculations where each monomer (D or A) is treated at the QM region while the other in the MM region [36]. Hence,

$$\Delta\rho_{QM/MM} = (\rho_{D^*A,QM/MM} + \rho_{DA^*,QM/MM}) - (\rho_{D,QM} + \rho_{A,QM}) \tag{8.22}$$

where the asterisk indicates that the corresponding monomer is treated at the QM level in a QM/MM calculation. A series of hydrogen-bonded pairs, e.g. water-N-methylacetamide, water-formaldehyde, water-imidazole, water dimer, imidazole-acetate, water-acetate, phenol-water, phenol-imidazole, and guanidinium-acetate, were considered for QM/MM calculations to understand the effect of polarization on the density changes of the complexes using different levels of theory [36]. Two separate QM/MM calculations were performed on the hydrogen-bonded complexes. In the first case, the H-bond donor was treated with the QM method, while the acceptor was treated at the MM level. It was reversed in the second case. These results were compared with the full QM models.

The change in the electron density upon the hydrogen bond formation in the dimeric donor–acceptor complexes, calculated at the (i) QM ($\Delta\rho_{QM}$) and (ii) QM/MM ($\Delta\rho_{QM/MM}$) levels, are shown in Figure 8.9, where the red and blue regions indicate the decrease and increase

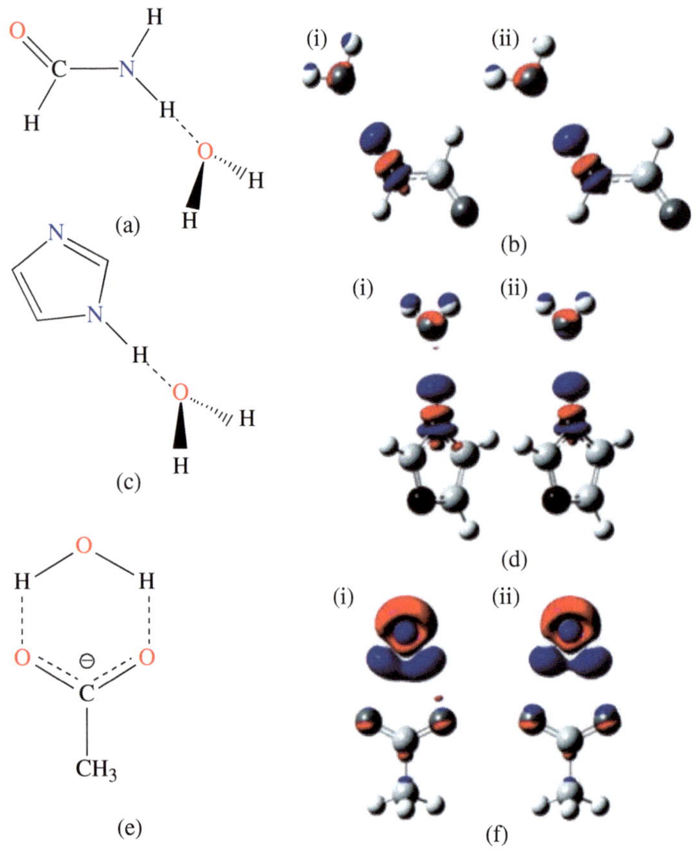

Figure 8.9 Schematic representation of the hydrogen-bond donor–acceptor complexes, namely, (a) formamide-water, (c) imidazole-water, and (e) acetate-water as optimized by QM/MM methods. Change in electron density of the corresponding complexes ((b), (d), and (f), respectively) due to the formation of hydrogen bond at the (i) QM ($\Delta\rho_{QM}$) and (ii) QM/MM ($\Delta\rho_{QM/MM}$) levels. Source: With permission from [36]/Royal Society Publishing.

in the electron density owing to the formation of the hydrogen bond. The formation of a hydrogen-bonded complex results in polarization of the electron density within the hydrogen bond acceptor toward the region containing its lone pair of electrons. In addition, polarization occurs at the hydrogen bond donor site, which subsequently increases the positive charge on the hydrogen or decreases the electron density on the hydrogen atom. But, the difference in the electron densities calculated at both fully QM and hybrid QM/MM methods provide remarkably similar results for the hydrogen-bonded dimers investigated, which further implies that the QM/MM approach gives a good description of the polarization effects. However, for the charged monomer, the electronic polarization effects are relatively larger, and thus, small differences in the densities are observed between the QM and QM/MM levels of description.

8.8 Quantum Refinement of Crystal Structure with QM/MM

The three-dimensional structures of biomolecules control their physiological functions. X-ray diffraction is the most popular method for the structure determination of biomolecules. To get a reasonable chemical structure, standard crystallographic refinement combines classical force field and experimental X-ray diffraction data. The standard crystallographic refinement for biomolecules with transition metal ions is challenging due to the unavailability of a well-described force field [78, 79]. Levitt and coworker have proposed the combination of crystallographic data and computational chemistry data to optimize the biomolecule structure [80]. Ryde and others have used QM/MM quantum refinement approach to do so [81].

Recently, Yan and others have proposed a new strategy called multiscale quantum refinement scheme (MQRS), which combines the X-ray quantum refinement method and flexible multiscale ONIOM schemes [37]. MQRS contains a contribution from computational chemistry and experimental data to obtain the total energy,

$$E_{total} = E_{comput} + \omega_{\alpha} E_{X\text{-}ray} \tag{8.23}$$

The computational energy (E_{comput}) and its gradients are calculated using either MM, semi-empirical models (SE), QM, fragmentation QM (or SE), hybrid QM/MM, and multiscale ONIOM schemes. $E_{X\text{-}ray}$ accounts for the difference between the structure factors of the observed structure and calculated X-ray structures. This difference is obtained from the crystallography and NMR system (CNS) program. The CNS program calculates the crystallographic penalty function and its derivatives from a given initial structure and the X-ray structure. ω_{α} is the weighing factor used to control the relative contribution of the computational energy to the overall energy. Based on both components' energy and gradients, the total energy is minimized in an iterative fashion until convergence is achieved. The final minimized structure is a refined structure that accounts for the QM effects in the X-ray structure determination.

The method was used to obtain a refined structure of monocopper(+2) containing rusticyanin (RC) from thiobacillus ferrooxidans (PDB ID:1RCY, 1.90 Å resolution) [82]. In total eight different models were considered (Table 8.4). The refined geometries from models M1–M4 give a similar range of metal-ligand bond lengths and bond angles. This concludes that different DFT functionals give similar results. The use of electrostatic embedding shows slight elongation in the bond lengths by 0.01 to 0.02 Å as compared to mechanical embedding. Hence, the latter was used due to its low

Table 8.4 QM/MM models employed for quantum refinement of X-ray crystal structure.

Model	Method
M1	ONIOM2(B3LYP/BS1:MM)
M2	ONIOM2(B3LYP-D3/BS1:MM)
M3	ONIOM2(M06-L/BS1:MM)
M4	ONIOM2(B3LYP/BS2:MM)
M5	ONIOM3(B3LYP/BS1:SE:MM)
M6	ONIOM2(B3LYP/BS1:SE)
M7	ONIOM2(CCSD-in-B3LYP/BS3:MM)
M8	ONIOM3(CCSD-in-B3LYP/BS3:SE:MM)

Grimme's tight-binding GFN2-xTB is the SE model. BS1, BS2, BS3 are three different basis sets: 6-31G(d); SDD for Cu(II) and 6-31G(d) for the rest; and def2-SVP for the WF-in-DFT (Zn(II) site) BS1 for the rest.
Source: With permission from [37]/American Chemical Society.

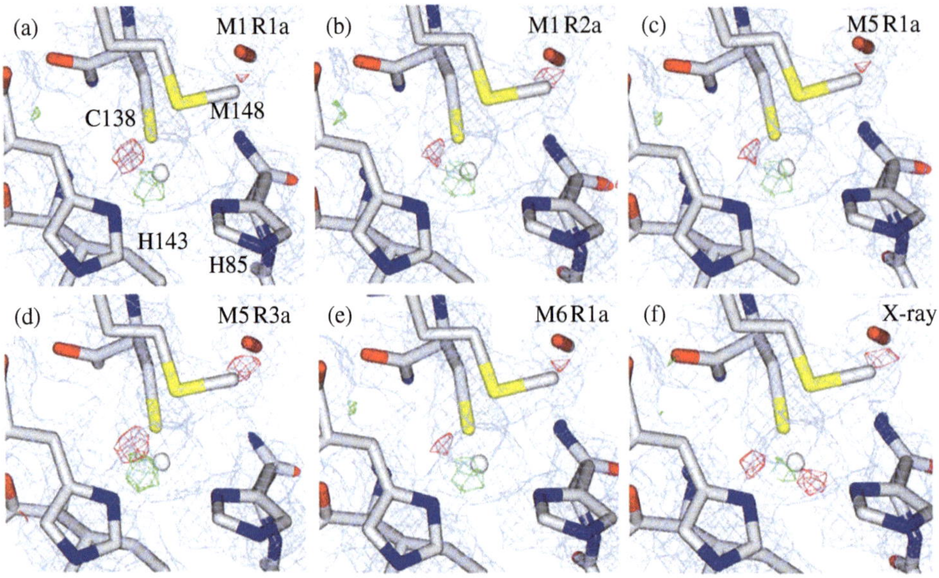

Figure 8.10 The Cu(II) binding site in the rusticyanin protein from various quantum refinement schemes ($\omega_\alpha = 0.606$) using different QM/MM models, including electron density map (2mFo-DFc maps, contoured at 1.0 σ (blue), mFo-DFc maps, contoured at +3.0 σ (green), and mFo-DFc maps, contoured at −3.0 σ (red)). (a)–(e) show data from different QM/MM models employed and (f) shows the X-ray crystal structure data. See Reference [37] for details. Source: With permission from [37]/American Chemical Society.

computational cost. The novel wavefunction in DFT approach improved the structure prediction capability. Compared to two-layer models, the three-layer ONIOM schemes (models 5 and 8) gave more accurate results at a reduced computational cost. The electron density mapping from different quantum refinement schemes is shown in Figure 8.10, where the negative difference density (red contours) is absent in all QM schemes compared to the crystal structure. This comes from the elongated Cu – N_{HIE85} (0.07–0.13 Å) distance. Furthermore, the real-space difference density *Z* score (RSZD) for Cu and the residues are less in the quantum refined structures, particularly in the structures from the three-layer ONIOM scheme.

8.9 Concluding Remarks

Electron density and its topological analysis have been widely used to describe the molecular structure, bonding, and reactivity in small molecules and clusters. Their use in multi-scale simulation methods, such as hybrid QM/MM models, is rather less common. However, with increasing computational resources and the development of novel computational strategies, it is now not only possible to use electron density for more efficient QM/MM models, but also to use them as a descriptor of bonding and reactivity in biomolecules. In this chapter, we have described some recent developments in the usage of electron density in multi-scale modeling of biomolecules. With parameterized polarizable electron density modeling, the hybrid QM/MM models have been shown to reproduce the full QM results. The usage of localized molecular orbitals from a library has been shown to be an efficient way to treat extended QM regions. The multiple active zone QM/MM method has been shown as an efficient way to treat those systems where the active site cannot be confined to a single region of space in the biomolecule. Using reactivity descriptors, such as the condensed Fukui functions, a systematic approach to the rational construction of the QM region is discussed. The treatment of hydrogen bonding and refinement of X-ray electron density by QM/MM models is discussed to show the wide range of applications of multi-scale simulation methods. In the coming years, the popularity of multi-scale simulations is expected to rise even further, and consequent growth of the role of electron density is expected as a central theme for novel QM/MM methodology and as a descriptor.

Acknowledgments

SK acknowledges the PMRF, Ministry of Education, Government of India, S. Muduli acknowledges UGC, New Delhi, India, and AP acknowledges IIT Kharagpur for research fellowship.

References

1 Mulliken, R.S. (1967). Spectroscopy, molecular orbitals, and chemical bonding. *Science* 157: 13–24.
2 Pauling, L. (1948). The metallic state. *Nature* 161: 1019–1020.
3 Slater, J.C. (1931). Molecular energy levels and valence bonds. *Phys. Rev.* 38: 1109–1144.
4 Slater, J.C. (1931). Directed valence in polyatomic molecules. *Phys. Rev.* 37: 481–489.
5 Hohenberg, P. and Kohn, W. (1964). Inhomogeneous electron gas. *Phys. Rev.* 136: B864–B871.
6 Kohn, W. and Sham, L.J. (1965). Self-consistent equations including exchange and correlation effects. *Phys. Rev.* 140: A1133–A1138.
7 Domingo, L.R. (2016). Molecular electron density theory: a modern view of reactivity in organic chemistry. *Molecules* 21: 1319.
8 Chakraborty, D. and Chattaraj, P.K. (2019). Bonding, reactivity, and dynamics in confined systems. *J. Phys. Chem. A* 123: 4513–4531.
9 Groenhof, G. (2013). Introduction to QM/MM simulations. In: *Biomolecular Simulations*, Methods in Molecular Biology, vol. 924 (ed. L. Monticelli and E. Salonen), 43–66. Totowa, NJ: Humana Press.
10 Hollingsworth, S.A. and Dror, R.O. (2018). Molecular dynamics simulation for all. *Neuron* 99: 1129–1143.
11 Tuckerman, M.E., Ungar, P.J., Von Rosenvinge, T., and Klein, M.L. (1996). *Ab initio* molecular dynamics simulations. *J. Phys. Chem.* 100: 12878–12887.
12 Warshel, A. and Levitt, M. (1976). Theoretical studies of enzymic reactions: dielectric, electrostatic and steric stabilization of the carbonium ion in the reaction of lysozyme. *J. Mol. Biol.* 103: 227–249.
13 Field, M.J., Bash, P.A., and Karplus, M. (1990). A combined quantum mechanical and molecular mechanical potential for molecular dynamics simulations. *J. Comput. Chem.* 11: 700–733.
14 André, J.-M. (2014). The Nobel Prize in Chemistry 2013. *Chem. Int.* 36: 2–7.
15 Lin, H. and Truhlar, D.G. (2007). QM/MM: what have we learned, where are we, and where do we go from here? *Theor. Chim. Acta* 117: 185–199.
16 Senn, H.M. and Thiel, W. (2009). QM/MM methods for biomolecular systems. *Angew. Chem. Int. Ed.* 48: 1198–1229.
17 Field, M.J. (2002). Simulating enzyme reactions: challenges and perspectives. *J. Comput. Chem.* 23: 48–58.

18 Ridder, L. and Mulholland, A.J. (2003). Modeling biotransformation reactions by combined quantum mechanical/molecular mechanical approaches: from structure to activity. *Curr. Top. Med. Chem.* 3: 1241–1256.

19 Senn, H.M. and Thiel, W. (2007). QM/MM studies of enzymes. *Curr. Opin. Chem. Biol.* 11: 182–187.

20 Garcia-Viloca, M., Gao, J., Karplus, M., and Truhlar, D.G. (2004). How enzymes work: analysis by modern rate theory and computer simulations. *Science* 303: 186–195.

21 Gao, J., Ma, S., Major, D.T. et al. (2006). Mechanisms and free energies of enzymatic reactions. *Chem. Rev.* 106: 3188–3209.

22 Wang, B. and Merz, K.M. (2006). A fast QM/MM (quantum mechanical/molecular mechanical) approach to calculate nuclear magnetic resonance chemical shifts for macromolecules. *J. Chem. Theory Comput.* 2: 209–215.

23 Li, H., Li, W., Li, S., and Ma, J. (2008). Fragmentation-based QM/MM simulations: length dependence of chain dynamics and hydrogen bonding of polyethylene oxide and polyethylene in aqueous solutions. *J. Phys. Chem.* 112: 7061–7070.

24 Okamoto, T., Ishikawa, T., Koyano, Y. et al. (2013). A minimal implementation of the AMBER-PAICS interface for *ab initio* FMO-QM/MM-MD simulation. *Bull. Chem. Soc. Jpn.* 86: 210–222.

25 Ghysels, A., Woodcock, H.L. III, Larkin, J.D. et al. (2011). Efficient calculation of QM/MM frequencies with the mobile block hessian. *J. Chem. Theory Comput.* 7: 496–514.

26 Gordon, M.S., Fedorov, D.G., Pruitt, S.R., and Slipchenko, L.V. (2012). Fragmentation methods: a route to accurate calculations on large systems. *Chem. Rev.* 112: 632–672.

27 Leszczynski, J. (2003). *Computational Chemistry: Reviews of Current Trends*, vol. 8. World Scientific.

28 Magistrato, A., Togni, A., and Rothlisberger, U. (2006). Enantioselective palladium-catalyzed hydrosilylation of styrene: influence of electronic and steric effects on enantioselectivity and catalyst design via hybrid QM/MM molecular dynamics simulations. *Organometallics* 25: 1151–1157.

29 Bühl, M., Grigoleit, S., Kabrede, H., and Mauschick, F.T. (2006). Simulation of 59Co NMR chemical shifts in aqueous solution. *Chem. Eur. J.* 12: 477–488.

30 Chatwin, S.L., Davidson, M.G., Doherty, C. et al. (2006). H—X bond activation via hydrogen transfer to hydride in ruthenium N-heterocyclic carbene complexes: density functional and synthetic studies. *Organometallics* 25: 99–110.

31 French, S., Sokol, A., Bromley, S. et al. (2003). Identification and characterization of active sites and their catalytic processes—the Cu/ZnO methanol catalyst. *Top. Catal.* 24: 161–172.

32 Catlow, C., French, S., Sokol, A., and Thomas, J. (2005). Computational approaches to the determination of active site structures and reaction mechanisms in heterogeneous catalysts. *Philos. Trans. R. Soc. London, Ser. A* 363: 913–936.

33 Olsen, J.M.H., Steinmann, C., Ruud, K., and Kongsted, J. (2015). Polarizable density embedding: a new QM/QM/MM-based computational strategy. *J. Phys. Chem. A* 119: 5344–5355.

34 Macetti, G. and Genoni, A. (2021). Three-layer multiscale approach based on extremely localized molecular orbitals to investigate enzyme reactions. *J. Phys. Chem. A* 125: 6013–6027.

35 Torras, J. (2015). Multiple active zones in hybrid QM/MM molecular dynamics simulations for large biomolecular systems. *Phys. Chem. Chem. Phys.* 17: 9959–9972.

36 Senthilkumar, K., Mujika, J.I., Ranaghan, K.E. et al. (2008). Analysis of polarization in QM/MM modelling of biologically relevant hydrogen bonds. *J. R. Soc. Interface* 5: S207–S2165.

37 Yan, Z., Li, X., and Chung, L.W. (2021). Multiscale quantum refinement approaches for metalloproteins. *J. Chem. Theory Comput.* 17: 3783–3796.

38 Maseras, F. and Morokuma, K. (1995). IMOMM: a new integrated *ab initio* + molecular mechanics geometry optimization scheme of equilibrium structures and transition states. *J. Comput. Chem.* 16: 1170–1179.

39 Field, M.J., Albe, M., Bret, C. et al. (2000). The dynamo library for molecular simulations using hybrid quantum mechanical and molecular mechanical potentials. *J. Comput. Chem.* 21: 1088–1100.

40 Reuter, N., Dejaegere, A., Maigret, B., and Karplus, M. (2000). Frontier bonds in QM/MM methods: a comparison of different approaches. *J. Phys. Chem. A* 104: 1720–1735.

41 Lin, H. and Truhlar, D.G. (2005). Redistributed charge and dipole schemes for combined quantum mechanical and molecular mechanical calculations. *J. Phys. Chem. A* 109: 3991–4004.

42 Waszkowycz, B., Hillier, I.H., Gensmantel, N., and Payling, D.W. (1991). A quantum mechanical/molecular mechanical model of inhibition of the enzyme phospholipase A2. *J. Chem. Soc., Perkin Trans.* 2: 1819–1832.

43 Waszkowycz, B., Hillier, I.H., Gensmantel, N., and Payling, D.W. (1991). A combined quantum mechanical/molecular mechanical model of the potential energy surface of ester hydrolysis by the enzyme phospholipase A2. *J. Chem. Soc., Perkin Trans.* 2: 225–231.

44 De Vries, A.H., Sherwood, P., Collins, S.J. et al. (1999). Zeolite structure and reactivity by combined quantum-chemical-classical calculations. *J. Phys. Chem. B* 103: 6133–6141.

45 Sherwood, P., de Vries, A.H., Guest, M.F. et al. (2003). Others QUASI: a general purpose implementation of the QM/MM approach and its application to problems in catalysis. *J. Mol. Struct. THEOCHEM* 632: 1–28.

46 Théry, V., Rinaldi, D., Rivail, J.-L. et al. (1994). Quantum mechanical computations on very large molecular systems: the local self-consistent field method. *J. Comput. Chem.* 15: 269–282.

47 Philipp, D.M. and Friesner, R.A. (1999). Mixed *ab initio* QM/MM modeling using frozen orbitals and tests with alanine dipeptide and tetrapeptide. *J. Comput. Chem.* 20: 1468–1494.

48 Pu, J., Gao, J., and Truhlar, D.G. (2004). Combining self-consistent-charge density-functional tight-binding (SCC-DFTB) with molecular mechanics by the generalized hybrid orbital (GHO) method. *J. Phys. Chem. A* 108: 5454–5463.

49 Day, P.N., Jensen, J.H., Gordon, M.S. et al. (1996). An effective fragment method for modeling solvent effects in quantum mechanical calculations. *J. Chem. Phys.* 105: 1968–1986.

50 Brooks, B. and Karplus, M. (1983). Harmonic dynamics of proteins: normal modes and fluctuations in bovine pancreatic trypsin inhibitor. *Proc. Natl. Acad. Sci. U.S.A.* 80: 6571–6575.

51 Dewar, M.J. (1983). Development and status of MINDO/3 and MNDO. *J. Mol. Struct.* 100: 41–50.

52 Dutta, D. and Mishra, S. (2014). The structural and energetic aspects of substrate binding and the mechanism of action of the DapE-encoded *N*-succinyl-L,L-diaminopimelic acid desuccinylase (DapE) investigated using a hybrid QM/MM method. *Phys. Chem. Chem. Phys.* 16: 26348–26358.

53 Dutta, D. and Mishra, S. (2016). Loss of catalytic activity in the E134D, H67A, and H349A mutants of DapE: mechanistic analysis with QM/MM investigation. *J. Phys. Chem. B* 120: 11654–11664.

54 Paul, A. and Mishra, S. (2021). Metal–ion promiscuity of microbial enzyme DapE at its second metal-binding site. *J. Biol. Inorg. Chem.* 26: 569–582.

55 Jing, Z., Liu, C., Cheng, S.Y. et al. (2019). Polarizable force fields for biomolecular simulations: recent advances and applications. *Annu. Rev. Biophys.* 48: 371.

56 Zhang, Y., Lin, H., and Truhlar, D.G. (2007). Self-consistent polarization of the boundary in the redistributed charge and dipole scheme for combined quantum-mechanical and molecular-mechanical calculations. *J. Chem. Theory Comput.* 3: 1378–1398.

57 Jindal, G. and Warshel, A. (2021). Exploring the dependence of QM/MM calculations of enzyme catalysis on the size of the QM region. *J. Phys. Chem. B* 120: 6013–6027.

58 Meyer, B., Guillot, B., Ruiz-Lopez, M.F., and Genoni, A. (2016). Libraries of extremely localized molecular orbitals. 1. Model molecules approximation and molecular orbitals transferability. *J. Chem. Theory Comput.* 15: 1052–1067.

59 Meyer, B., Guillot, B., Ruiz-Lopez, M.F. et al. (2016). Libraries of extremely localized molecular orbitals. 2. Comparison with the pseudoatoms transferability. *J. Chem. Theory Comput.* 12: 1068–1081.

60 Macetti, G. and Genoni, A. (2019). Quantum mechanics/extremely localized molecular orbital method: a fully quantum mechanical embedding approach for macromolecules. *J. Phys. Chem. A* 123: 9420–9428.

61 Meyer, B. and Genoni, A. (2018). Libraries of extremely localized molecular orbitals. 3. Construction and preliminary assessment of the new databanks. *J. Phys. Chem. A* 122: 8965–8981.

62 Macetti, G., Wieduwilt, E.K., Assfeld, X., and Genoni, A. (2020). Localized molecular orbital-based embedding scheme for correlated methods. *J. Chem. Theory Comput.* 16: 3578–3596.

63 Mulholland, A.J. and Richards, W.G. (1997). Acetyl-CoA enolization in citrate synthase: a quantum mechanical/molecular mechanical (QM/MM) study. *Proteins* 27: 9–25.

64 van der Kamp, M.W., Żurek, J., Manby, F.R. et al. (2010). Testing high-level QM/MM methods for modeling enzyme reactions: acetyl-CoA deprotonation in citrate synthase. *J. Phys. Chem. B* 114: 11303–11314.

65 Maier, J.A., Martinez, C., Kasavajhala, K. et al. (2015). ff14SB: improving the accuracy of protein side chain and backbone parameters from ff99SB. *J. Chem. Theory Comput.* 11: 3696–3713.

66 Wang, J., Wolf, R.M., Caldwell, J.W. et al. (2004). Development and testing of a general amber force field. *J. Comput. Chem.* 25: 1157–1174.

67 Hopkins, B.W. and Tschumper, G.S. (2003). A multicentered approach to integrated QM/QM calculations. Applications to multiply hydrogen bonded systems. *J. Comput. Chem.* 24: 1563–1568.

68 Kiyota, Y., Hasegawa, J.-Y., Fujimoto, K. et al. (2009). A multicore QM/MM approach for the geometry optimization of chromophore aggregate in protein. *J. Comput. Chem.* 30: 1351–1359.

69 Asada, N., Fedorov, D.G., Kitaura, K. et al. (2012). An efficient method to evaluate intermolecular interaction energies in large systems using overlapping multicenter ONIOM and the fragment molecular orbital method. *J. Phys. Chem. Lett.* 3: 2604–2610.

70 Torras, J., Deumens, E., and Trickey, S.B. (2006). Software integration in multi-scale simulations: the PUPIL system. *J. Comput.-Aided Mater. Des.* 13: 201–212.

71 Bakowies, D. and Thiel, W. (1996). Hybrid models for combined quantum mechanical and molecular mechanical approaches. *J. Phys. Chem.* 100: 10580–10594.

72 Nam, K., Gao, J., and York, D.M. (2004). An efficient linear-scaling Ewald method for long-range electrostatic interactions in combined QM/MM calculations. *J. Chem. Theory Comput.* 1: 2–13.

73 Geerlings, P., De Proft, F., and Langenaeker, W. (2003). Conceptual density functional theory. *Chem. Rev.* 103: 1793–1874.

74 Parr, R.G. and Yang, W. (1984). Density functional approach to the frontier-electron theory of chemical reactivity. *J. Am. Chem. Soc.* 106: 4049–4050.

75 Yang, W. and Mortier, W.J. (1986). The use of global and local molecular parameters for the analysis of the gas-phase basicity of amines. *J. Am. Chem. Soc.* 108: 5708–5711.

76 Karelina, M. and Kulik, H.J. (2017). Systematic quantum mechanical region determination in QM/MM simulation. *J. Chem. Theory Comput.* 13: 563–576.

77 Qi, H.W., Karelina, M., and Kulik, H.J. (2017). Quantifying electronic effects in QM and QM/MM biomolecular modeling with the Fukui function. *Acta Phys. -Chem. Sin.* 33: 81–91.

78 Kleywegt, G.J. (2007). Crystallographic refinement of ligand complexes. *Acta Crystallogr. D* 63: 94–100.

79 Hu, L. and Ryde, U. (2011). Comparison of methods to obtain force-field parameters for metal sites. *J. Chem. Theory Comput.* 7: 2452–2463.

80 Jack, A. and Levitt, M. (1978). Refinement of large structures by simultaneous minimization of energy and *R* factor. *Acta Crystallogr. A* 34: 931–935.

81 Cao, L., Caldararu, O., Rosenziweig, A.C., and Ryde, U. (2018). Quantum refinement does not support dinuclear copper sites in crystal structures of particulate methane monooxygenase. *Angew. Chem. Int. Ed.* 57: 162–166.

82 Walter, R.L., Ealick, S.E., Friedman, A.M. et al. (1996). Multiple wavelength anomalous diffraction (MAD) crystal structure of rusticyanin: a highly oxidizing cupredoxin with extreme acid stability. *J. Mol. Biol.* 263: 730–751.

9

Subsystem Communications and Electron Correlation

Roman F. Nalewajski

Department of Theoretical Chemistry, Jagiellonian University, Cracow, Poland

9.1 Introduction

Quantum mechanics (QM) and information theory (IT) provide a solid basis for determining the electronic structure of molecules and understanding their reactivity trends. The former establishes the system's electronic states and all their physical attributes, while the latter provides an additional interpretative tool in an entropic, chemical outlook on molecular structural organization and dynamics, explaining general trends in the substrate chemical behavior, e.g. [1–5]. The most important conceptual framework for the quantum interpretations in chemistry is generated by the *one*-determinantal representation of the ground-state electron configuration in the self-consistent field (SCF) molecular orbital (MO) approaches in the familiar Hartree–Fock (HF) wave function theory and Kohn–Sham (KS) density-functional theory (DFT). The Slater determinant is then defined by the occupied MO's, which determine the ground-state pattern of chemical bonds.

In the analytical linear combinations of atomic orbitals (LCAO) MO approaches, the configuration MO's are conventionally expanded as LC of AO. In orbital communication theory (OCT) [1–4, 6–17], this choice of analytical functions also determines a discrete set of elementary electronic "events," in the "input" and "output" of the associated molecular channel. It is defined by the probability-scattering network generated by the state charge-and-bond-order (CBO) matrix. In the local treatment, the CBO matrix similarly establishes the canonical first-order density matrix, which ultimately determines the conditional probabilities of the *output*-location events, given the *input*-positions of an electron in the bond system of a molecule. This bond-propagation approach also generates the "chemical" descriptors of both the overall bond multiplicity and its covalent/ionic composition [1–5, 14, 15].

The classical IT [18–25] of Fisher [18, 19] and Shannon [20, 21] has been successfully applied to interpret the electronic structure of molecules [26–31] and the polarization (P) and charge–transfer (CT) displacements in reactive complexes [1–5, 32–35]. It allows one to explore the information/entropy content of electronic probability distributions in terms of the complementary measures of the Fisher information for locality events, a "compactness" descriptor of electronic distribution, or the global Shannon entropy reflecting its "spread." The information principles have been explored [36–38], atomic pieces from the IT division of molecular electron density [1–5, 26, 39–44] have provided a solid basis for the intuitive (stockholder) partition of Hirshfeld [45], and the Hammond [46] postulate of reactivity theory has been tackled [1–5, 38]. Moreover, patterns of the chemical bond orders have been extracted in OCT [14, 15], and the *bridge*-bonds

Electron Density: Concepts, Computation and DFT Applications, First Edition.
Edited by Pratim Kumar Chattaraj and Debdutta Chakraborty.

originating from cascade propagations via intermediate orbitals have been identified [47–52]. The nonadditive Fisher information [53] has been related to the electron localization function (ELF) [54–56] of DFT, facilitating the development of an effective Contragradience (CG) probe [57] for localizations of electrons and chemical bonds in molecules.

In a holistic IT description [4, 5, 58–70], one ultimately requires the *resultant* extensions of the *classical* measures of the state information/entropy content, appropriate for complex amplitudes of molecular QM. In these overall measures, the wavefunction phase or its gradient determining the effective velocity of probability "fluid" give rise to *nonclassical* supplements of the classical (probability/modulus) descriptors. In reactivity theory, this extension allows one to distinguish between the bonded (entangled) states of the mutually *open* molecular fragments [71, 72], from their nonbonded (disentangled), *closed*-system analogs exhibiting the same set of electron densities. The resultant gradient information in the specified quantum state is then proportional to the expectation value of the kinetic energy of the system electrons. This relation allows one to reinterpret the *variational principle* for electronic energy as an equivalent *information rule* and to use the molecular virial theorem (VT) [73] in an IT probe of chemical reactivity [5, 32, 35, 72, 74, 75].

In the present analysis, we shall further explore the molecular information systems in the discrete (orbital) and continuous (local) representations. In these resolutions the probability scatterings are, respectively, related to elements of the CBO matrix and the density-matrix kernel of LCAO MO theory. Their roles in shaping the direct molecular communications via the system chemical bonds, as well as their entropy (bond covalency) and information (bond iconicity) descriptors will be reexamined. The previous *one*-electron development, based upon the superposition principle (SP) of QM [76], will be compared with the *two*-electron conditional probabilities defining the correlation holes. It will be argued that electronic communications are related to the dependent, correlation-hole part of *two*-electron distributions, and the information conservation in the transmission of signals in molecular channels will be discussed. The *partial* density matrices of subsystems will be introduced and used to determine communications between interacting substrates. An illustrative application of this formalism to CT complexes [4, 5, 77–85] will also be presented.

9.2 Discrete and Local Probability Networks in Molecular Bond Systems

A central concept in the communication theory of the chemical bond (CTCB) [1–5] is the molecular information network. A selection of electronic events, which can differ in the network *input* (Source, S) and *output* (Receiver, R), can involve both the discrete and continuous sets, appropriate for the molecular descriptors in question. In the specified electronic state $\Psi(N)$ of the N-electron system, the communication system is determined by *conditional* probabilities of observing the output events in R, given the input events in S. For example, a discrete set of the AO (output) events $\boldsymbol{b} = \{b_l\}$ will be observed with probabilities $\mathbf{P}(\boldsymbol{b}|\boldsymbol{a})$, conditional on the previously recorded AO (input) events $\boldsymbol{a} = \{a_k\}$:

$$\mathbf{P}(\boldsymbol{b}|\boldsymbol{a}) = \{P(b_l|a_k) = P(a_k, b_l)/P(a_k) \equiv P(l|k) \equiv P(k \to l)\} \equiv \mathbf{P}(\boldsymbol{a} \to \mathbf{b}) \tag{9.1}$$

Here, $P(a_k, b_l) \equiv P(k, l)$ stands for the *joint* probability of the network input and output basis states. The initial distribution $\boldsymbol{P}(\boldsymbol{a}) = \{P(a_k)\}$, determining the *input*-signal and hence also the actual network exploitation, ultimately shapes the channel *output*-signal:

$$\boldsymbol{Q}(\boldsymbol{b}) = \boldsymbol{P}(\boldsymbol{a})\mathbf{P}(\boldsymbol{b}|\boldsymbol{a}) \text{ or } \mathbf{P}(\boldsymbol{b}|\boldsymbol{a}) = \partial \boldsymbol{Q}(\mathbf{b})/\partial \boldsymbol{P}(\boldsymbol{a}) \tag{9.2}$$

The adopted AO basis χ of SCF LCAO MO calculations also determines a discrete variety of both the *input*, $\mathbf{a} = \chi = \{\chi_k\}$ and *output*, $\mathbf{b} = \chi' = \{\chi_l\}$, events of the molecular communication network in OCT.

Consider the electron configuration defined by Slater determinant

$$\Psi(N) = \det \boldsymbol{\psi} = |\psi_1 \psi_2 \dots \psi_N|, \{\psi_s(\mathbf{q}) = \varphi_{s\sigma}(\mathbf{r})\zeta_s(\sigma) \equiv \varphi_s(\mathbf{r})\zeta_s(\sigma)\} \tag{9.3}$$

where the *spin* states of an electron,

$$\zeta_s(\sigma) \in \{\alpha(\sigma), \text{spin} - \textit{up} \text{ state}; \beta(\sigma), \text{spin} - \textit{down} \text{ state}\}$$

explore its two admissible orientations: $\sigma \in (\uparrow, \downarrow)$. In SCF LCAO MO calculations using the orthonormal AO basis set $\chi(\mathbf{r}) = \{\chi_k(\mathbf{r})\} \equiv \langle \mathbf{r}|\chi\rangle\}$ of analytical functions contributed by the system constituent atoms, the *spatial*-MO

$$\begin{aligned} \boldsymbol{\varphi}(\mathbf{r}) &\equiv \langle \mathbf{r}|\boldsymbol{\varphi}\rangle = [\varphi_1(\mathbf{r}), \varphi_2(\mathbf{r}), \dots, \varphi_N(\mathbf{r})] = \{\varphi_s(\mathbf{r}) \equiv \langle \mathbf{r}|\varphi_s\rangle\} = \chi(\mathbf{r})\mathbf{C} \\ \mathbf{C} &= \langle \chi|\boldsymbol{\varphi}\rangle, \quad \mathbf{C}^\dagger\mathbf{C} = \langle \boldsymbol{\varphi}|\chi\rangle\langle \chi|\boldsymbol{\varphi}\rangle = \langle \boldsymbol{\varphi}|\boldsymbol{\varphi}\rangle = \mathbf{I} \end{aligned} \tag{9.4}$$

determine the molecular *bond system*. It is defined by the MO projection onto the configuration-occupied subspace:

$$P_\psi = |\boldsymbol{\psi}\rangle\langle\boldsymbol{\psi}| = \sum_s |\psi_s\rangle\langle\psi_s| \tag{9.5}$$

These (*singly* occupied) MO can be subsequently grouped into the *spin* subsets, $\boldsymbol{\varphi}(\mathbf{r}) = \{\boldsymbol{\varphi}^\sigma(\mathbf{r})\}$, each combining *one*-electron states of N^σ electrons exhibiting the same spin orientation σ,

$$\boldsymbol{\varphi}^\sigma(\mathbf{r}) = \left[\varphi_1^{\ \sigma}(\mathbf{r}), \varphi_2^{\ \sigma}(\mathbf{r}), \dots, \varphi_N^{\ \sigma}(\mathbf{r})\right] \equiv \chi(\mathbf{r})\mathbf{C}^\sigma \tag{9.6}$$

The *spin*-MO's are defined by their selections of expansion coefficients $\mathbf{C} = \{\mathbf{C}^\sigma = \langle \chi|\boldsymbol{\varphi}^\sigma\rangle\}$.

In this SCF LCAO MO framework, the AO representation of MO projection $|\boldsymbol{\varphi}\rangle\langle\boldsymbol{\varphi}|$ onto the bond system of a molecule ultimately defines the CBO matrix of quantum chemistry:

$$\begin{aligned} \boldsymbol{\gamma} &= \langle \chi|\boldsymbol{\varphi}\rangle\langle\boldsymbol{\varphi}|\chi\rangle = \mathbf{C}\mathbf{C}^\dagger \\ &= \sum_\sigma \langle \chi|\boldsymbol{\varphi}_\sigma\rangle\langle\boldsymbol{\varphi}_\sigma|\chi\rangle = \sum_\sigma \mathbf{C}_\sigma\mathbf{C}_\sigma^{\ \dagger} \equiv \sum_\sigma \boldsymbol{\gamma}_\sigma \end{aligned} \tag{9.7}$$

This AO representation of the configuration *bond*-projector,

$$|\boldsymbol{\varphi}\rangle\langle\boldsymbol{\varphi}| = \sum_\sigma |\boldsymbol{\varphi}_\sigma\rangle\langle\boldsymbol{\varphi}_\sigma|$$

satisfies the following idempotency relations:

$$\begin{aligned} \boldsymbol{\gamma}_\sigma\boldsymbol{\gamma}_{\sigma'} &= \langle \chi|\boldsymbol{\varphi}_\sigma\rangle\langle\boldsymbol{\varphi}_\sigma|\chi\rangle \left\langle \chi|\boldsymbol{\varphi}_{\sigma'}\right\rangle \left\langle \boldsymbol{\varphi}_{\sigma'}|\chi\right\rangle \\ &= \langle \chi|\boldsymbol{\varphi}_\sigma\rangle \left\langle \boldsymbol{\varphi}_\sigma|\boldsymbol{\varphi}_{\sigma'}\right\rangle \left\langle \boldsymbol{\varphi}_{\sigma'}|\chi\right\rangle = \langle \chi|\boldsymbol{\varphi}_\sigma\rangle\mathbf{I}_\sigma\delta_{\sigma,\sigma'} \left\langle \boldsymbol{\varphi}_{\sigma'}|\chi\right\rangle = \boldsymbol{\gamma}_\sigma\delta_{\sigma,\sigma'} \quad \text{and} \\ \boldsymbol{\gamma}^2 &= \sum_\sigma\sum_{\sigma'} \boldsymbol{\gamma}_\sigma\boldsymbol{\gamma}_{\sigma'} = \sum_\sigma \boldsymbol{\gamma}_\sigma = \boldsymbol{\gamma} \end{aligned} \tag{9.8}$$

It ultimately shapes the probability distributions in the molecular network. Indeed, the CBO diagonal elements define the configuration's (normalized) *input*-signal in terms of AO average occupancies $\{N_k = \gamma_{k,k}\}$,

$$\mathbf{P}(\chi) = \{P_k = N_k/N\}, \quad \sum_k P_k = 1 \tag{9.9}$$

which establishes the *molecular* exploitation of the channel. The amplitudes $\mathbf{A}(\chi, \chi') = \{A_{k,l}\}$ and probabilities $\mathbf{P}(\chi, \chi') = \{P_{k,l}\}$ of the *joint* AO input (k) and output (l) events in configuration $\Psi(N)$ similarly read:

$$A_{k,l} = \gamma_{k,l}/N^{1/2} \text{ and } P_{k,l} = |A_{k,l}|^2 = \gamma_{k,l}\gamma_{l,k}/N = |\gamma_{k,l}|^2/N$$
$$\sum_l P_{k,l} = P_k \tag{9.10}$$

They eventually determine channel's *one*-electron communications between AO $\mathbf{P}_{\text{com.}}(\chi'|\chi) = \{P_{\text{com.}}(l|k) \equiv P_{k\to l}\} \equiv \mathbf{P}(\chi \to \chi')$, reflected by the associated conditional probabilities of the output AO events in the configuration bond system. These propagations are generated by the squared moduli of the associated amplitudes $\{A_{\text{com.}}(l|k) \equiv A_{k\to l}\} \equiv \mathbf{A}(\chi \to \chi')$ reflecting the Hilbert-space projections contained in the CBO matrix elements of Eq. (9.7):

$$\mathbf{P}_{\text{com.}}(\chi'|\chi) = \left\{P_{\text{com.}}(l|k) = P_{k,l}/P_k = \gamma_{k,l}\gamma_{l,k}/\gamma_{k,k} \equiv P_{k\to l} \equiv |A_{\text{com.}}(l|k)|^2 \equiv |A_{k\to l}|^2\right\}$$
$$\sum_l P_{\text{com.}}(l|k) = 1 \tag{9.11}$$

For the stationary, *molecular* exploitation $\Psi(N)$ of this AO channel,

$$\boldsymbol{P}(\Psi) \to \chi \to \mathbf{P}_{\text{com.}}(\chi'|\chi) \to \chi' \to \boldsymbol{Q}(\Psi) \tag{9.12}$$

the input (source) probabilities $\boldsymbol{P}(\chi) = \boldsymbol{P}(\Psi)$ of Eq. (9.9) determine the same output distribution in the network receiver:

$$\boldsymbol{Q}(\Psi) \equiv \boldsymbol{Q}(\chi') = \boldsymbol{P}(\chi)\mathbf{P}_{\text{com.}}(\chi'|\chi) = \left\{Q_l = \sum_k P_k P_{\text{com.}}(l|k) = \sum_k P_{k,l} = P_l\right\}$$
$$= \boldsymbol{P}(\chi') \equiv \boldsymbol{P}(\Psi) \tag{9.13}$$

It should be realized, however, that communication channels can be probed by any admissible source signal. Therefore, one can also envisage alternative ways of using the molecular communication system, in extracting the bond descriptor of interest.

The continuous (complete) set of elementary events $\{|\boldsymbol{r}\rangle\}$ of the precise electronic locations, eigenstates determining the position representation of QM,

$$\mathbf{r}|\boldsymbol{r}\rangle = \boldsymbol{r}|\boldsymbol{r}\rangle \tag{9.14}$$

similarly defines the *local* electronic propagations in molecular channels:

$$\boldsymbol{p}(\boldsymbol{r}) \to \boldsymbol{r} \to P_{\text{com.}}(\boldsymbol{r}'|\boldsymbol{r}) \to \boldsymbol{r}' \to \boldsymbol{q}(\boldsymbol{r}') \tag{9.15}$$

The relevant *joint* probabilities in specified molecular state $\Psi(N)$ are then determined by the local kernel

$$\gamma(\boldsymbol{r}, \boldsymbol{r}') = \langle \boldsymbol{r}|\boldsymbol{\varphi}\rangle\langle\boldsymbol{\varphi}|\boldsymbol{r}'\rangle = \sum_s \varphi_s(\boldsymbol{r})\varphi_s^*(\boldsymbol{r}') = \langle\boldsymbol{r}|\chi\rangle[\langle\chi|\boldsymbol{\varphi}\rangle\langle\boldsymbol{\varphi}|\chi\rangle]\langle\chi|\boldsymbol{r}'\rangle = \chi(\boldsymbol{r})\boldsymbol{\gamma}\chi^\dagger(\boldsymbol{r}')$$
$$= \sum_\sigma \langle\boldsymbol{r}|\boldsymbol{\varphi}_\sigma\rangle\langle\boldsymbol{\varphi}_\sigma|\boldsymbol{r}'\rangle = \langle\boldsymbol{r}|\chi\rangle\sum_\sigma[\langle\chi|\boldsymbol{\varphi}_\sigma\rangle\langle\boldsymbol{\varphi}_\sigma|\chi\rangle]\langle\chi|\boldsymbol{r}'\rangle$$
$$= \sum_\sigma \chi(\boldsymbol{r})\boldsymbol{\gamma}_\sigma\chi^\dagger(\boldsymbol{r}') \equiv \sum_\sigma \gamma_\sigma(\boldsymbol{r}, \boldsymbol{r}') \tag{9.16}$$

satisfying all relevant idempotency conditions for the density matrix (see Eq. (9.8)):

$$\int \gamma_\sigma(\boldsymbol{r}, \boldsymbol{r}')\gamma_{\sigma'}(\boldsymbol{r}', \boldsymbol{r}'')d\boldsymbol{r}' = \chi(\boldsymbol{r})\boldsymbol{\gamma}_\sigma\left[\int \chi^\dagger(\boldsymbol{r}')\chi(\boldsymbol{r}')d\boldsymbol{r}'\right]\boldsymbol{\gamma}_{\sigma'}\chi^\dagger(\boldsymbol{r}'') = \chi(\boldsymbol{r})\left[\boldsymbol{\gamma}_\sigma{}^2\mathbf{I}_\sigma\right]\chi^\dagger(\boldsymbol{r}'')\delta_{\sigma,\sigma'}$$
$$= \chi(\boldsymbol{r})\boldsymbol{\gamma}_\sigma\chi^\dagger(\boldsymbol{r}'')\delta_{\sigma,\sigma'} = \gamma_\sigma(\boldsymbol{r}, \boldsymbol{r}'')\delta_{\sigma,\sigma'}$$

$$\int \gamma(\boldsymbol{r},\boldsymbol{r}')\gamma(\boldsymbol{r}',\boldsymbol{r}'')d\boldsymbol{r}' = \sum_{\sigma}\sum_{\sigma'}\int \gamma_{\sigma}(\boldsymbol{r},\boldsymbol{r}')\gamma_{\sigma'}(\boldsymbol{r}',\boldsymbol{r}'')d\boldsymbol{r}'$$
$$= \chi(\boldsymbol{r})\left\{\sum_{\sigma}\sum_{\sigma'}\boldsymbol{\Upsilon}_{\sigma}\left[\int \chi^{\dagger}(\boldsymbol{r}')\chi(\boldsymbol{r}')d\boldsymbol{r}'\right]\boldsymbol{\Upsilon}_{\sigma'}\right\}\chi^{\dagger}(\boldsymbol{r}'')$$
$$= \chi(\boldsymbol{r})\left[\sum_{\sigma}\sum_{\sigma'}\boldsymbol{\Upsilon}_{\sigma}\boldsymbol{\Upsilon}_{\sigma'}\right]\chi^{\dagger}(\boldsymbol{r}'') = \sum_{\sigma}\boldsymbol{\Upsilon}_{\sigma}(\boldsymbol{r},\boldsymbol{r}'') = \gamma(\boldsymbol{r},\boldsymbol{r}'') \tag{9.17}$$

Its diagonal element determines the local electron density,

$$\rho(\boldsymbol{r}) = \gamma(\boldsymbol{r},\boldsymbol{r}) = \sum_{s}|\varphi_s(\boldsymbol{r})|^2, \quad \int \rho(\boldsymbol{r})d\boldsymbol{r} = N \tag{9.18}$$

and hence also the state probability distribution $p(\boldsymbol{r}) = \rho(\boldsymbol{r})/N$. The joint amplitudes $A(\boldsymbol{r}, \boldsymbol{r}') = \gamma(\boldsymbol{r}, \boldsymbol{r}')/N^{1/2}$ and probabilities $P(\boldsymbol{r}, \boldsymbol{r}') = |A(\boldsymbol{r}, \boldsymbol{r}')|^2$ of simultaneously observing an electron in the molecular bond system $\boldsymbol{\varphi}$ at the input $(\boldsymbol{r})$ and output $(\boldsymbol{r}')$ locations,

$$P(\boldsymbol{r},\boldsymbol{r}') = \gamma(\boldsymbol{r},\boldsymbol{r}')\gamma(\boldsymbol{r}',\boldsymbol{r})/N = |\gamma(\boldsymbol{r},\boldsymbol{r}')|^2/N$$
$$\int P(\boldsymbol{r},\boldsymbol{r}')d\boldsymbol{r}' = \gamma(\boldsymbol{r},\boldsymbol{r})/N = p(\boldsymbol{r}) \tag{9.19}$$

ultimately determine the state local communications in the configuration bond system, the conditional probabilities $\{P_{\text{com.}}(\boldsymbol{r}'|\boldsymbol{r}) \equiv P(\boldsymbol{r} \to \boldsymbol{r}')\}$ in the local channel of Eq. (9.15):

$$P_{\text{com.}}(\boldsymbol{r}'|\boldsymbol{r}) = P(\boldsymbol{r},\boldsymbol{r}')/p(\boldsymbol{r}) = |\gamma(\boldsymbol{r},\boldsymbol{r}')|^2/\rho(\boldsymbol{r}), \quad \int P_{\text{com.}}(\boldsymbol{r}'|\boldsymbol{r})\,d\boldsymbol{r}' = 1 \tag{9.20}$$

The molecular input signal $p(\boldsymbol{r})$ in the source of this network thus produces the same probability distribution at its output:

$$q(\boldsymbol{r}') = \int p(\boldsymbol{r})P_{\text{com.}}(\boldsymbol{r}'|\boldsymbol{r})\,d\boldsymbol{r} = \int P(\boldsymbol{r},\boldsymbol{r}')\,d\boldsymbol{r} = p(\boldsymbol{r}') \tag{9.21}$$

This probability transformation also implies that communication links can be regarded as functional derivatives of the output distribution (compare Eq. (9.2)):

$$P_{\text{com.}}(\boldsymbol{r}'|\boldsymbol{r}) = \delta q(\boldsymbol{r}')/\delta p(\boldsymbol{r}) \tag{9.22}$$

This "sensitivity" interpretation implies that the probability scattering in the bond system of a molecule, in fact, reflects a *dependence* of the electron probability at the *output* location, with respect to the probability at the *input* position of an electron. These molecular communications are thus related to the interdependence of the state probabilities at alternative spatial locations, as epitomized by the familiar *correlation* effects between electrons [5].

This (*one*-electron) Hilbert-space approach to molecular communications should be thus linked to the *two*-electron densities and correlation holes of the electronic structure theories. It is thus of great interest to explore this *Communications* ↔ *Correlation* relation in more detail. Before addressing this intriguing problem, let us briefly summarize the bond descriptors of molecular information channels, which cover both the IT covalent/ionic composition and an overall bond-multiplicity measure of electronic wavefunctions [1–4, 6–17].

9.3 Bond Descriptors of Molecular Communication Channels

The *conditional-entropy* of the AO channel measures an average "noise" in electronic communications [23, 25]. This indeterminacy index reflects the IT "covalency" of the system chemical bonds [1–17], a degree of uncertainty in electronic propagations. For the AO and local resolutions, respectively, this descriptor reads:

$$\Delta S(\boldsymbol{Q}|\boldsymbol{P}) = -\sum_k \sum_l P_{k,l} \ln P_{\text{com.}}(l|k)$$

or

$$\Delta S[q|p] = -\iint P(\boldsymbol{r}, \boldsymbol{r}') \ln P_{\text{com.}}(\boldsymbol{r}'|\boldsymbol{r}) d\boldsymbol{r} d\boldsymbol{r}' \quad (9.23)$$

It represents the scattered part of the initial (molecular) information content, which has been dissipated in transmission of electronic signals as the propagation "noise."

This loss is explicitly seen in the channel complementary descriptor of the IT bond "iconicity" [1–17]. It is reflected by the *mutual-information* in system's inputs and outputs [23, 25], which measures the resultant amount of the information flowing in this scattering network. It indeed represents the molecular information diminished by the "noise" contribution:

$$\Delta I(\boldsymbol{P}:\boldsymbol{Q}) = \sum_k \sum_l P_{k,l} \ln[P_{\text{com.}}(l|k)/Q_l] = S(\boldsymbol{Q}) - \Delta S(\boldsymbol{Q}|\boldsymbol{P})$$

or

$$\Delta I[p:q] = \iint P(\boldsymbol{r}, \boldsymbol{r}') \ln[P_{\text{com.}}(\boldsymbol{r}'|\boldsymbol{r})/q(\boldsymbol{r}')] d\boldsymbol{r} d\boldsymbol{r}' = S[q] - \Delta S[q|p] \quad (9.24)$$

The bond-character indices of Eqs. (9.23) and (9.24) ultimately give rise to the corresponding *resultant* molecular descriptor of IT bond-multiplicity [1–17], reflecting the conserved Shannon's entropy in the channel input and output (see Eqs. (9.13) and (9.21)):

$$\Delta M(\boldsymbol{P}; \boldsymbol{Q}) = \Delta S(\boldsymbol{Q}|\boldsymbol{P}) + \Delta I(\boldsymbol{P}:\boldsymbol{Q}) = -\sum_l Q_l \ln Q_l \equiv S(\boldsymbol{Q})$$
$$= -\sum_k P_k \ln P_k \equiv S(\boldsymbol{P})$$

or

$$\Delta M[p; q] = \Delta S[q|p] + \Delta I[p:q] = -\int q(\boldsymbol{r}') \ln q(\boldsymbol{r}') d\boldsymbol{r}' \equiv S[q]$$
$$= -\int p(\boldsymbol{r}) \ln p(\boldsymbol{r}) d\boldsymbol{r} \equiv S[p] \quad (9.25)$$

To summarize, in the molecularly exploited channel, the *input* entropies $S(\boldsymbol{P})$ or $S[p]$ reflect the system's *initial* information content. This amount is preserved in the network *output*: $S(\boldsymbol{P}) = S(\boldsymbol{Q})$ or $S[p] = S[q]$.

The conservation of an overall information, reflected by the net-descriptor ΔM of the preceding equation, is thus due to a cancellation between the information lost ΔS of Eq. (9.23), dissipated as communication noise, and the scattered contribution in the *information* preserved ΔI in Eq. (9.24). This information conservation is schematically summarized in the following flow chart:

Source (S) ⟶ **Information Scattering S→R** ⟶ **Receiver (R)**

$$\textit{Input Information } S \rightarrow \left\{ \begin{matrix} \textit{Dissipated Information } \Delta S \\ \\ \textit{Flowing Information } \quad \Delta I \end{matrix} \right\} \rightarrow \textit{Output Information } \Delta M = S \quad (9.26)$$

It illustrates the IT partition of the molecular entropy content into its chemical components: the bond-covalency, due to electron-sharing between bonded atoms, and bond-iconicity, reflecting a deterministic aspect of electron-allocations to atoms-in-molecules.

9.4 Hartree–Fock Communications and Fermi Correlation

Consider first the probability distribution of two electrons, one at the "input" position $\boldsymbol{r}$ and another at the "output" location $\boldsymbol{r}'$, in the Hartree (H) theory, not corrected for the self-interaction, when the variational wavefunction is given by the product of occupied MO,

$$\Psi^{\mathrm{H}}(N) = \varphi_1{}^{\mathrm{H}}(1)\varphi_2{}^{\mathrm{H}}(2)\ldots\varphi_N{}^{\mathrm{H}}(N) \equiv \Pi\boldsymbol{\varphi}^{\mathrm{H}}$$

In this independent reference approximation, the probability density combines MO contributions,

$$p^{\mathrm{H}}(\boldsymbol{r}) = N^{-1}\sum_s \left|\varphi_s{}^{\mathrm{H}}(\boldsymbol{r})\right|^2 = \rho^{\mathrm{H}}(\boldsymbol{r})/N \equiv \gamma^{\mathrm{H}}(\boldsymbol{r},\boldsymbol{r})/N, \quad \int p^{\mathrm{H}}(\boldsymbol{r})\,d\boldsymbol{r} = 1 \tag{9.27}$$

and the *two*-electron probability is given by the product of local probabilities at these two monitoring positions:

$$\pi^{\mathrm{H}}(\boldsymbol{r},\boldsymbol{r}') = p^{\mathrm{H}}(\boldsymbol{r})q^{\mathrm{H}}(\boldsymbol{r}') \equiv \pi_{\mathrm{ind.}}(\boldsymbol{r},\boldsymbol{r}'), \quad \iint \pi^{\mathrm{H}}(\boldsymbol{r},\boldsymbol{r}')\,d\boldsymbol{r}\,d\boldsymbol{r}' = 1 \tag{9.28}$$

The associated *two*-electron density $\rho_2{}^{\mathrm{H}}(\boldsymbol{r},\boldsymbol{r}')$ of such exactly *independent* electrons is determined by the product distribution,

$$\rho_2{}^{\mathrm{H}}(\boldsymbol{r},\boldsymbol{r}') = \rho^{\mathrm{H}}(\boldsymbol{r})\rho^{\mathrm{H}}(\boldsymbol{r}') = N^2\pi^{\mathrm{H}}(\boldsymbol{r},\boldsymbol{r}'), \quad \iint \rho_2{}^{\mathrm{H}}(\boldsymbol{r},\boldsymbol{r}')\,d\boldsymbol{r}\,d\boldsymbol{r}' = N^2 \tag{9.29}$$

which is seen to exhibit incorrect normalization.

This self-interaction problem can be remedied by including in $\rho_2(\boldsymbol{r},\boldsymbol{r}')$ the *dependent* term due to the Fermi correlation,

$$\begin{aligned}&\rho_{\mathrm{dep.}}{}^{\mathrm{HF}}(\boldsymbol{r},\boldsymbol{r}') = \rho_x(\boldsymbol{r},\boldsymbol{r}') = -\gamma^{\mathrm{HF}}(\boldsymbol{r},\boldsymbol{r}')\gamma^{\mathrm{HF}}(\boldsymbol{r}',\boldsymbol{r}) \equiv \rho^{\mathrm{HF}}(\boldsymbol{r})h_x(\boldsymbol{r}'|\boldsymbol{r}), \quad \int h_x(\boldsymbol{r}'|\boldsymbol{r})d\boldsymbol{r}' = -1\\ &\iint \rho_x(\boldsymbol{r},\boldsymbol{r}')d\boldsymbol{r}d\boldsymbol{r}' = -\int \rho^{\mathrm{HF}}(\boldsymbol{r})d\boldsymbol{r} = -N\end{aligned} \tag{9.30}$$

reflecting the *exchange* (x) hole $h_x(\boldsymbol{r}'|\boldsymbol{r})$ in HF theory. The latter uses the Slater determinant of Eq. (9.3) as the method variational wave function,

$$\Psi^{\mathrm{HF}}(N) = |\varphi_1{}^{\mathrm{HF}}\varphi_2{}^{\mathrm{HF}}\ldots\varphi_N{}^{\mathrm{HF}}| = \det\boldsymbol{\varphi}^{\mathrm{HF}}$$

The x-corrected *two*-electron density,

$$\begin{aligned}\rho_2{}^{\mathrm{HF}}(\boldsymbol{r},\boldsymbol{r}') &= \rho^{\mathrm{HF}}(\boldsymbol{r})\rho^{\mathrm{HF}}(\boldsymbol{r}') - \gamma^{\mathrm{HF}}(\boldsymbol{r},\boldsymbol{r}')\gamma^{\mathrm{HF}}(\boldsymbol{r}',\boldsymbol{r}) \equiv \rho_{\mathrm{ind.}}{}^{\mathrm{HF}}(\boldsymbol{r},\boldsymbol{r}') + \rho_{\mathrm{dep.}}{}^{\mathrm{HF}}(\boldsymbol{r},\boldsymbol{r}')\\ &\equiv N(N-1)\left\{\pi_{\mathrm{ind.}}{}^{\mathrm{HF}}(\boldsymbol{r},\boldsymbol{r}') - \pi_{\mathrm{dep.}}{}^{\mathrm{HF}}\left(\boldsymbol{r},\boldsymbol{r}'\right)\right\} \equiv N(N-1)\pi^{\mathrm{HF}}(\boldsymbol{r},\boldsymbol{r}')\end{aligned} \tag{9.31}$$

then satisfies the correct *two*-electron condition

$$\iint \rho_2{}^{\mathrm{HF}}(\boldsymbol{r},\boldsymbol{r}')\,d\boldsymbol{r}\,d\boldsymbol{r}' = N(N-1)$$

It follows from Eqs. (9.30) and (9.31) that the dependent (exchange) probability factor

$$\begin{aligned}\pi_{\mathrm{dep.}}{}^{\mathrm{HF}}(\boldsymbol{r},\boldsymbol{r}') &= -[N(N-1)]^{-1}\rho_{\mathrm{dep.}}{}^{\mathrm{HF}}(\boldsymbol{r},\boldsymbol{r}') = [N(N-1)]^{-1}\gamma^{\mathrm{HF}}(\boldsymbol{r},\boldsymbol{r}')\gamma^{\mathrm{HF}}(\boldsymbol{r}',\boldsymbol{r})\\ &= -(N-1)^{-1}p^{\mathrm{HF}}(\boldsymbol{r})h_x(\boldsymbol{r}'|\boldsymbol{r})\end{aligned} \tag{9.32}$$

exhibits the following normalization:

$$\int\left[\int \pi_{\mathrm{dep.}}{}^{\mathrm{HF}}(\boldsymbol{r},\boldsymbol{r}')\,d\boldsymbol{r}'\right]d\boldsymbol{r} = (N-1)^{-1}\int p^{\mathrm{HF}}(\boldsymbol{r})\,d\boldsymbol{r} = (N-1)^{-1} \tag{9.33}$$

The product probabilities of Eq. (9.28) predict the associated conditional probabilities for the *independent* electrons in H-theory:

$$\pi^{\mathrm{H}}(\boldsymbol{r}'|\boldsymbol{r}) = \pi^{\mathrm{H}}(\boldsymbol{r},\boldsymbol{r}')/p^{\mathrm{H}}(\boldsymbol{r}) = q^{\mathrm{H}}(\boldsymbol{r}') \tag{9.34}$$

and hence: $\Delta I^{\mathrm{H}} = 0$, $\Delta S^{\mathrm{H}} = \Delta M^{\mathrm{H}} = -\int q^{\mathrm{H}}(\boldsymbol{r}')\ln q^{\mathrm{H}}(\boldsymbol{r}')\,d\boldsymbol{r}' = S[q^{\mathrm{H}}] = S[p^{\mathrm{H}}]$. This overall *two*-electron treatment of then exactly independent local events in the network "input" and "output" thus implies the vanishing flow of information measured by the bond "ionicity" descriptor in H-channel. The overall bond index and overall covalency measure then reflect the entropy content $S[p^{\mathrm{H}}]$ in the method's *one*-electron probability. The exact independence of the local events in the Hartree channel is thus marked by the network vanishing mutual information. The system (*a priori*) input information content $S[p^{\mathrm{H}}]$ is then exactly equal to its (*a posteriori*) scattered amount $S[q^{\mathrm{H}}]$ reflecting the average communication noise in the channel, so that there is no net information flow in such a reference description.

However, in the Hilbert-space approach of Eq. (9.20), the method (*one*-electron) *communications* between the two locations via the system chemical bonds, differ from the *separable* conditional probability of the preceding equation:

$$P_{\mathrm{com.}}{}^{\mathrm{H}}(\boldsymbol{r}'|\boldsymbol{r}) = \gamma^{\mathrm{H}}(\boldsymbol{r},\boldsymbol{r}')\gamma^{\mathrm{H}}(\boldsymbol{r}',\boldsymbol{r})/\rho^{\mathrm{H}}(\boldsymbol{r}) \tag{9.35}$$

These *one*-electron propagations involve AO projections into the molecular bond space, spanned by the method occupied MO, $\boldsymbol{\varphi}^{\mathrm{H}}(\boldsymbol{r}) = \chi(\boldsymbol{r})\mathbf{C}^{\mathrm{H}} = \{\varphi_s{}^{\mathrm{H}}(\boldsymbol{r})\}$, which define the method CBO matrix $\boldsymbol{\gamma}^{\mathrm{H}} = \langle \chi|\boldsymbol{\varphi}^{\mathrm{H}}\rangle\langle\boldsymbol{\varphi}^{\mathrm{H}}|\chi\rangle$ and the associated density-matrix kernel

$$\gamma^{\mathrm{H}}(\boldsymbol{r},\boldsymbol{r}') = \sum_s \varphi_s{}^{\mathrm{H}}(\boldsymbol{r})\varphi_s{}^{\mathrm{H}*}(\boldsymbol{r}') = \chi(\boldsymbol{r})[\mathbf{C}^{\mathrm{H}}\mathbf{C}^{\mathrm{H}\dagger}]\chi^{\dagger}(\boldsymbol{r}') \equiv \chi(\boldsymbol{r})\boldsymbol{\gamma}^{\mathrm{H}}\chi^{\dagger}(\boldsymbol{r}') \tag{9.36}$$

They are seen to be related to the *dependent* probability content $\pi_{\mathrm{dep.}}{}^{\mathrm{HF}}(\boldsymbol{r},\,\boldsymbol{r}')$ of Eq. (9.32), determining the following *two*-electron propagations $\pi_{\mathrm{com.}}{}^{\mathrm{HF}}(\boldsymbol{r}'|\boldsymbol{r}) = \pi^{\mathrm{HF}}(\boldsymbol{r}\to\boldsymbol{r}')$ in HF theory:

$$\begin{aligned}\pi_{\mathrm{com.}}{}^{\mathrm{HF}}(\boldsymbol{r}'|\boldsymbol{r}) &= \pi_{\mathrm{dep.}}{}^{\mathrm{HF}}(\boldsymbol{r},\boldsymbol{r}')/p^{\mathrm{HF}}(\boldsymbol{r}) = (N-1)^{-1}[\gamma^{\mathrm{HF}}(\boldsymbol{r},\boldsymbol{r}')\gamma^{\mathrm{HF}}(\boldsymbol{r}',\boldsymbol{r})/\rho^{\mathrm{HF}}(\boldsymbol{r})]\\ &\equiv (N-1)^{-1}P_{\mathrm{com.}}{}^{\mathrm{HF}}(\boldsymbol{r}'|\boldsymbol{r})\end{aligned} \tag{9.37}$$

For the Slater determinant in the HF method, which exactly accounts for the Fermi (exchange) correlation between electrons, one thus finds the following expression for the *two*-electron density of simultaneously locating the *reference*-electron at $\boldsymbol{r}$ and another *dependent*-electron of the remaining $N-1$ particles, at $\boldsymbol{r}'$:

$$\begin{aligned}\rho_2{}^{\mathrm{HF}}(\boldsymbol{r},\boldsymbol{r}') &= \rho^{\mathrm{HF}}(\boldsymbol{r})\rho^{\mathrm{HF}}(\boldsymbol{r}') - |\gamma^{\mathrm{HF}}(\boldsymbol{r},\boldsymbol{r}')|^2 \equiv \rho_{\mathrm{ind.}}{}^{\mathrm{HF}}(\boldsymbol{r},\boldsymbol{r}') + \rho_x{}^{\mathrm{HF}}(\boldsymbol{r},\boldsymbol{r}')\\ &\equiv N(N-1)\left[\pi_{\mathrm{ind.}}{}^{\mathrm{HF}}(\boldsymbol{r},\boldsymbol{r}') + \pi_{\mathrm{dep.}}{}^{\mathrm{HF}}(\boldsymbol{r},\boldsymbol{r}')\right] \equiv N(N-1)\pi^{\mathrm{HF}}(\boldsymbol{r},\boldsymbol{r}'),\\ \iint \pi^{\mathrm{HF}}(\boldsymbol{r},\boldsymbol{r}')d\boldsymbol{r}d\boldsymbol{r}' &= \int p^{\mathrm{HF}}(\boldsymbol{r})d\boldsymbol{r} = 1\end{aligned} \tag{9.38}$$

It contains the independent, H-like part $\rho_{\mathrm{ind.}}{}^{\mathrm{HF}}(\boldsymbol{r},\,\boldsymbol{r}')$ given by the product of electron densities at the "input" ($\boldsymbol{r}$) and "output" ($\boldsymbol{r}'$) locations, and the exchange (x) correction

$$\rho_x{}^{\mathrm{HF}}(\boldsymbol{r},\boldsymbol{r}') = -\gamma^{\mathrm{HF}}(\boldsymbol{r},\boldsymbol{r}')\gamma^{\mathrm{HF}}(\boldsymbol{r}',\boldsymbol{r}) \equiv -N(N-1)\pi_x{}^{\mathrm{HF}}(\boldsymbol{r},\boldsymbol{r}') \tag{9.39}$$

Its probability factor,

$$\pi_x{}^{\mathrm{HF}}(\boldsymbol{r},\boldsymbol{r}') = -\rho_x{}^{\mathrm{HF}}(\boldsymbol{r},\boldsymbol{r})/[N(N-1)] = [N(N-1)]^{-1}|\gamma^{\mathrm{HF}}(\boldsymbol{r},\boldsymbol{r}')|^2 \tag{9.40}$$

determines the associated *two*-electron conditional probability:

$$\pi_x{}^{\mathrm{HF}}(\boldsymbol{r}'|\boldsymbol{r}) = \pi_x{}^{\mathrm{HF}}(\boldsymbol{r},\boldsymbol{r}')/p^{\mathrm{HF}}(\boldsymbol{r}) = (N-1)^{-1}|\gamma^{\mathrm{HF}}\left(\boldsymbol{r},\boldsymbol{r}'\right)|^2/\rho^{\mathrm{HF}}(\boldsymbol{r}) \tag{9.41}$$

It is seen to be related to the method *one*-electron kernel of local communications, $P_{com.}{}^{HF}(\boldsymbol{r}'|\boldsymbol{r}) = P^{HF}(\boldsymbol{r} \rightarrow \boldsymbol{r}')$ in Eq. (9.37):

$$\pi_x{}^{HF}(\boldsymbol{r}'|\boldsymbol{r}) = (N-1)^{-1} P_{com.}{}^{HF}(\boldsymbol{r}'|\boldsymbol{r}) = -(N-1)^{-1} h_x(\boldsymbol{r}'|\boldsymbol{r})$$

or

$$P_{com.}{}^{HF}(\boldsymbol{r}'|\boldsymbol{r}) = -h_x(\boldsymbol{r}'|\boldsymbol{r}) \tag{9.42}$$

It can be thus concluded that in HF theory, the local network, determined by the defining set of *one*-electron conditional probabilities, can be linked only to the (renormalized) "correlation" part, defined by x-hole, of the approximation *two*-electron distribution. The "independent" component of the latter approach does not participate in the theory AO communications, which define molecular information network in OCT:

$$\text{HF } \textit{communications} \Leftrightarrow \textit{Exchange correlation} \tag{9.43}$$

9.5 Communication Partitioning of *Two*-Electron Probabilities

The *Fermi*, x-correlation, effective within groups of $\{N_\sigma\}$ electrons exhibiting identical spin orientation, is exactly accounted for in HF approximation, which totally neglects the *Coulomb*, c-correlation between electrons, due to their mutual repulsion. The former effect is embodied in the system's overall x-hole,

$$\begin{aligned} h_x(\boldsymbol{r}'|\boldsymbol{r}) &= -\rho^{HF}(\boldsymbol{r})^{-1} \sum_\sigma \left\{ \sum_n \left[\varphi_{n\sigma}{}^{HF}(\boldsymbol{r})\varphi_{n\sigma}{}^{HF*}(\boldsymbol{r}')\right] \right\} \left\{ \sum_m \left[\varphi_{m\sigma}{}^{HF}(\boldsymbol{r}')\varphi_{m\sigma}{}^{HF*}(\boldsymbol{r})\right] \right\} \\ &= -\rho^{HF}(\boldsymbol{r})^{-1} \sum_\sigma \left[\gamma_\sigma{}^{HF}(\boldsymbol{r},\boldsymbol{r}')\gamma_\sigma \mathrm{HF}\left(\boldsymbol{r}',\boldsymbol{r}\right)\right] \\ &= -\gamma^{HF}(\boldsymbol{r},\boldsymbol{r}')\gamma^{HF}(\boldsymbol{r}',\boldsymbol{r})/\rho^{HF}(\boldsymbol{r}) \end{aligned}$$

$$\int h_x(\boldsymbol{r}'|\boldsymbol{r})d\boldsymbol{r}' = -1 \tag{9.44}$$

defining the local electron communication $P_{com.}{}^{HF}(\boldsymbol{r}'|\boldsymbol{r}) = P_{com.}{}^{HF}(\boldsymbol{r} \rightarrow \boldsymbol{r}')$, which satisfies the required conditional normalization:

$$\int P_{com.}{}^{HF}(\boldsymbol{r}'|\boldsymbol{r})\, d\boldsymbol{r}' = 1 \tag{9.45}$$

In the explicitly correlated approaches to molecular electronic structure, e.g. the configuration interaction (CI) method of quantum chemistry or density functional theory (DFT) [86–91], the remaining part of the mutual dependencies between electrons gives rise to the familiar c-hole $h_c(\boldsymbol{r}'|\boldsymbol{r})$. In DFT, which realistically accommodates both the exchange and Coulomb correlation effects, the resultant xc-hole

$$h_{xc}(\boldsymbol{r}'|\boldsymbol{r}) = h_x(\boldsymbol{r}'|\boldsymbol{r}) + h_c(\boldsymbol{r}'|\boldsymbol{r}), \int h_c(\boldsymbol{r}'|\boldsymbol{r})\, d\boldsymbol{r}' = 0 \tag{9.46}$$

ultimately determines the method density functional for an overall xc contribution to electronic energy:

$$E_{xc}[\rho] = 1/2 \iint \rho(\boldsymbol{r}) h_{xc}(\boldsymbol{r}'|\boldsymbol{r})\, d\boldsymbol{r}\, d\boldsymbol{r}' \tag{9.47}$$

In such theories, one naturally generalizes the association of Eqs. (9.42) and (9.43) into the *resultant* correlation effects:

$$\textit{Electronic communications} \Leftrightarrow \textit{Electron correlation} \tag{9.48}$$

or

$$P_{\text{com.}}(\boldsymbol{r}'|\boldsymbol{r}) = -h_{xc}(\boldsymbol{r}'|\boldsymbol{r}), \int P_{\text{com.}}(\boldsymbol{r}'|\boldsymbol{r})\, d\boldsymbol{r}' = 1 \tag{9.49}$$

The electronic communications in the bond system of a molecule represent the *one*-particle phenomenon involving the bond-subspace projections defining the CBO matrix and the associated density-matrix kernel, while the electron correlation refers to the *two*-particle scenario. It thus appears that the former is linked only to the dependent part $\rho_{\text{dep.}}(\boldsymbol{r}, \boldsymbol{r}')$ of the system (correlated) *two*-electron density:

$$\begin{aligned}\rho_2(\boldsymbol{r}, \boldsymbol{r}') &= \rho(\boldsymbol{r})[\rho(\boldsymbol{r}') + h_{\text{xc}}(\boldsymbol{r}'|\boldsymbol{r})] \equiv \rho_{\text{ind.}}(\boldsymbol{r}, \boldsymbol{r}') + \rho_{\text{dep.}}(\boldsymbol{r}, \boldsymbol{r}') \\ &\equiv N(N-1)[\pi_{\text{ind.}}(\boldsymbol{r}, \boldsymbol{r}') + \pi_{\text{dep.}}(\boldsymbol{r}, \boldsymbol{r}')] \equiv N(N-1)\pi(\boldsymbol{r}, \boldsymbol{r}') \\ \iint \pi(\boldsymbol{r}, \boldsymbol{r}') d\boldsymbol{r} d\boldsymbol{r}' &= \int p(\boldsymbol{r}) d\boldsymbol{r} = 1\end{aligned} \tag{9.50}$$

Since in OCT, the IT bond orders already provide quite an adequate representation of the accepted chemical intuition [2–4, 14, 15], one can conjecture that bond multiplicities of chemistry also belong to a subtle framework of the electron-correlation concepts:

$$\textit{Bond multiplicities} \Leftrightarrow \textit{Electron correlation} \tag{9.51}$$

The orbital propagations of electronic signals, transmitted via the system chemical bonds from the channel Source (S) to its Receiver (R), are thus determined solely by the *dependent* part $\mathbf{P}(\text{S}\wedge\text{R}) = \mathbf{P}_{\text{dep.}}(\text{S}, \text{R})$ of the device overall (*two*-electron) *joint*-probabilities. In the local resolution of Eq. (9.50), one finds:

$$\begin{aligned}\mathbf{P}(\text{S}, \text{R}) &= \mathbf{P}_{\text{ind.}}(\text{S}, \text{R}) + \mathbf{P}_{\text{dep.}}(\text{S}, \text{R}) = \{\pi(\boldsymbol{r}, \boldsymbol{r}')\} \\ \mathbf{P}_{\text{ind.}}(\text{S}, \text{R}) &= \{\pi_{\text{ind.}}(\boldsymbol{r}, \boldsymbol{r}')\} \text{ and } \mathbf{P}_{\text{dep.}}(\text{S}, \text{R}) = \{\pi_{\text{dep.}}(\boldsymbol{r}, \boldsymbol{r}')\}\end{aligned} \tag{9.52}$$

This partition of *two*-electron probability/density is schematically represented in a qualitative diagram of Figure 9.1. Consider the discrete orbital channel of molecular communications between AO events defining both the S "inputs" and R "outputs" of the molecular network. In the (*one*-electron) communication theory, the amplitude of a mutual (Hilbert-space) coupling between the *input* $\boldsymbol{a} = \{\chi_i\} \equiv \chi$ and *output* $\boldsymbol{b} = \{\chi_j\} \equiv \chi'$ AO "events" in the electron configuration of Eq. (9.3) is embodied in elements of the CBO matrix of Eq. (9.7), the squared moduli of which determine the condensed *joint* probabilities of AO in the configuration bond system:

$$\mathbf{P}_{\text{dep.}}(\text{S}, \text{R}) = \left\{P_{i,j} = N^{-1}\gamma_{i,j}\gamma_{j,i} = N^{-1}|\gamma_{i,j}|^2 \equiv |A_{i,j}|^2\right\}$$

or

$$\mathbf{A}_{\text{dep.}}(\text{S}, \text{R}) = \{A_{i,j} = N^{-1/2}\gamma_{i,j}\} \tag{9.53}$$

They ultimately determine the associated *conditional* probabilities $\mathbf{P}_{\text{com.}}(\chi'|\chi) = \{P_{\text{com.}}(j|i) = P(i \to j) \equiv |A_{i\to j}|^2\}$, defining the *direct* AO communications in the molecular channel,

$$P_{\text{com.}}(j|i) = P_{i,j}/P_i = \gamma_{i,j}\gamma_{j,i}/\gamma_{i,i} \tag{9.54}$$

$$\mathbf{P}(\mathrm{R},\mathrm{S}) = [\mathbf{P}(\mathrm{S}|\mathrm{R}) + \mathbf{P}(\mathrm{R}|\mathrm{S})] + \mathbf{P}(\mathrm{R}\wedge\mathrm{S}) \equiv \mathbf{P}_{\mathrm{ind.}}(\mathrm{S},\mathrm{R}) + \mathbf{P}_{\mathrm{dep.}}(\mathrm{S},\mathrm{R})$$

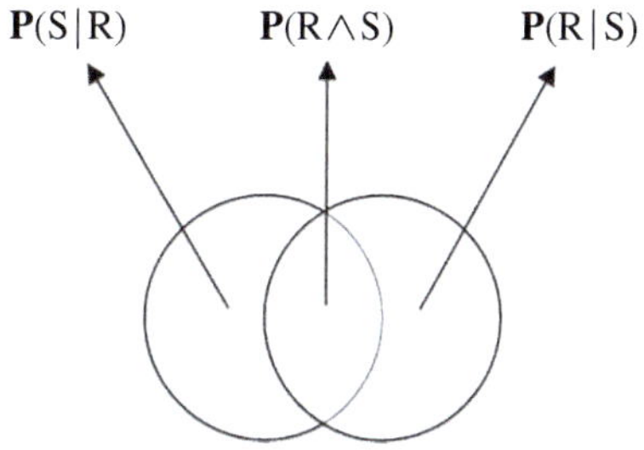

$$\mathbf{P}_{\mathrm{ind.}}(\mathrm{S}) = \mathbf{P}(\mathrm{S}|\mathrm{R}) \qquad \mathbf{P}_{\mathrm{ind.}}(\mathrm{R}) = \mathbf{P}(\mathrm{R}|\mathrm{S})$$

$$\mathbf{P}_{\mathrm{ind.}}(\mathrm{S},\mathrm{R}) = \mathbf{P}(\mathrm{S}|\mathrm{R}) + \mathbf{P}(\mathrm{R}|\mathrm{S})$$

$$\mathbf{P}_{\mathrm{dep.}}(\mathrm{S},\mathrm{R}) = \mathbf{P}(\mathrm{R}\wedge\mathrm{S})$$

Figure 9.1 *Two*-electron probability diagrams of the source (S) and receiver (R) in the molecular communication channel S → R. Two circles represent the (dependent) *joint*-probability distributions in the channel input and output, respectively: $\mathbf{P}(\mathrm{S}) = \mathbf{P}(\mathrm{S}|\mathrm{R}) + \mathbf{P}(\mathrm{R}\wedge\mathrm{S})$ and $\mathbf{P}(\mathrm{R}) = \mathbf{P}(\mathrm{R}|\mathrm{S}) + \mathbf{P}(\mathrm{R}\wedge\mathrm{S})$. The probability "overlap" area $\mathbf{P}(\mathrm{R}\wedge\mathrm{S})$ of the "input" (S) and "output" (R) circles represents the *dependent* part $\mathbf{P}_{\mathrm{dep.}}(\mathrm{S},\mathrm{R}) = \mathbf{P}(\mathrm{R}\wedge\mathrm{S})$ of the *two*-electron probability distribution $\mathbf{P}(\mathrm{R},\mathrm{S}) = [\mathbf{P}(\mathrm{S}|\mathrm{R}) + \mathbf{P}(\mathrm{R}|\mathrm{S})] + \mathbf{P}(\mathrm{R}\wedge\mathrm{S}) \equiv \mathbf{P}_{\mathrm{ind.}}(\mathrm{S},\mathrm{R}) + \mathbf{P}_{\mathrm{dep.}}(\mathrm{S},\mathrm{R})$, while the sum of the circle complements $\mathbf{P}(\mathrm{S}|\mathrm{R}) \equiv \mathbf{P}_{\mathrm{ind.}}(\mathrm{S})$ and $\mathbf{P}(\mathrm{R}|\mathrm{S}) \equiv \mathbf{P}_{\mathrm{ind.}}(\mathrm{R})$ defines its *independent* part.

where AO probability is determined by its effective occupation number $N_i(\Psi) = \gamma_{i,i}$ in $\Psi(N)$,

$$P_i(\Psi) = N_i(\Psi)/N = \sum_j P_{i,j} \tag{9.55}$$

This average electron occupancy represents the orbital expectation value of the configuration projection operator $\mathrm{P}_\Psi = |\Psi\rangle\langle\Psi| = |\boldsymbol{\varphi}\rangle\langle\boldsymbol{\varphi}| \equiv \mathrm{P}_{\boldsymbol{\varphi}}$. Indeed, the state orbital probability can be interpreted as the χ_i expectation value of the renormalized molecular projector $\mathrm{P}_{\boldsymbol{\varphi}}$ (or of MO density operator $\mathrm{d}_{\boldsymbol{\varphi}} = N^{-1}\mathrm{P}_{\boldsymbol{\varphi}}$),

$$\mathrm{P}_i(\Psi) = N^{-1}\langle i|\boldsymbol{\varphi}\rangle\langle\boldsymbol{\varphi}|i\rangle \equiv N^{-1}\langle i|\mathrm{P}_{\boldsymbol{\varphi}}|i\rangle \equiv \langle i|\mathrm{d}_{\boldsymbol{\varphi}}|i\rangle \tag{9.56}$$

In the *two*-electron scenario, the product of the state simultaneous occupations by one of N electrons on χ_i, $N_i(\Psi)$, and by another of remaining $(N-1)$ electrons on χ_j, $N_j(\Psi)$, reads:

$$N_i(\Psi)N_j(\Psi) = N(N-1)\pi_{i,j}(\Psi) \tag{9.57}$$

where the joint distribution of the *two*-electron *probability*

$$\pi_{i,j}(\Psi) = [N(N-1)]^{-1}[\gamma_{i,i}\gamma_{j,j} - \gamma_{i,j}\gamma_{j,i}] \equiv \pi_{i,j}{}^{\mathrm{ind.}}(\Psi) + \pi_{i,j}{}^{\mathrm{dep.}}(\Psi) \tag{9.58}$$

The *independent*, condensed AO events define the joint-probability contribution

$$\mathbf{P}^{\mathrm{ind.}}(\mathrm{S},\mathrm{R}) = \left\{\pi_{i,j}{}^{\mathrm{ind.}} = P_iP_j \equiv \left|A_{i,j}{}^{\mathrm{ind.}}\right|^2 = N^{-2}\gamma_{i,i}\gamma_{j,j}\right\}$$

or

$$\mathbf{A}^{\mathrm{ind.}}(\mathrm{S},\mathrm{R}) = \left\{A_{i,j}{}^{\mathrm{ind.}} = N^{-1}(\gamma_{i,i}\gamma_{j,j})^{1/2}\right\} \tag{9.59}$$

and the associated H-type conditional probabilities of χ_j given χ_i:

$$\mathbf{P}^{\mathrm{ind.}}(\mathrm{R}|\mathrm{S}) = \left\{\pi^{\mathrm{ind.}}(j|i) = \pi_{i,j}{}^{\mathrm{ind.}}/P_i = P_j\right\} \tag{9.60}$$

The "exchange" part of the condensed joint probabilities of AO in the molecular bond system, $\pi_{i,j}{}^{\mathrm{dep.}}(\Psi) = -[N(N-1)]^{-1}\gamma_{i,j}\gamma_{j,i}$, similarly predicts the *two*-electron conditional probability matrix

$\mathbf{P}^{\text{dep.}}(\text{R}|\text{S}) = \{\pi^{\text{dep.}}(j|i)\}$ related to (*one*-electron) probability propagations $\mathbf{P}_{\text{com.}}(\text{R}|\text{S}) = \{P_{\text{com.}}(j|i)\}$ (compare Eq. (9.42)):

$$\pi^{\text{dep.}}(j|i) \equiv -\pi_{i,j}{}^{\text{dep.}}/P_i = (\gamma_{i,j}\gamma_{j,i}/\gamma_{i,i})/(N-1) = -P_{\text{com.}}(j|i)/(N-1) \tag{9.61}$$

9.6 Communications in Interacting Subsystems

In describing the interacting subsystems it is useful to apply the *partial* density-matrix formalism of QM [92–94], see. e.g. [4, 5, 70, 71]. Consider the *closed* acceptor (acidic, A) and donor (basic, B) subsystems in the *polarized* reactive complex

$$\text{R}^+ = [\text{A}^+(\boldsymbol{x})|\text{B}^+(\boldsymbol{\xi})] \tag{9.62}$$

for the fixed *relative* coordinates of both substrates, which determine their "frozen" separation and mutual orientation. Here, $\boldsymbol{x}$ and $\boldsymbol{\xi}$ denote the fragment *internal* coordinates, with the subsystem Hamiltonian $\text{H}_\text{X}{}^+(N_\text{X}{}^+) = \text{H}_\text{X}(N_\text{X}{}^+;\, v)$ acting on the position variables of $N_\text{X}{}^+$ (integer) electrons originating from the polarized reactant X^+ and moving in the *molecular* external potential: $v = v_\text{A} + v_\text{B}$. Such closed reactants conserve the numbers of electrons in isolated subsystems $\{\text{X}^0\}$, described by Hamiltonians $\{\text{H}_\text{X}(N_\text{X}{}^0) = \text{H}_\text{X}(N_\text{X}{}^0;\, v_\text{X})\}$ and moving in the field of the *substrate* external potential v_X: $\{N_\text{X}{}^+ = N_\text{X}{}^0\}$.

The eigenstates of separate subsystems, corresponding to Hamiltonians $\text{H}_\text{A}(N_\text{A}{}^0) \equiv \text{H}_\text{A}(\boldsymbol{x})$ and $\text{H}_\text{B}(N_\text{B}{}^0) \equiv \text{H}_\text{B}(\boldsymbol{\xi})$, respectively,

$$\text{H}_\text{A}(\boldsymbol{x})\varphi_s{}^\text{A}(\boldsymbol{x}) = E_s{}^\text{A}\varphi_s{}^\text{A}(\boldsymbol{x}) \text{ and } \text{H}_\text{B}(\boldsymbol{\xi})\varphi_u{}^\text{B}(\boldsymbol{\xi}) = E_u{}^\text{B}\varphi_u{}^\text{B}(\boldsymbol{\xi}) \tag{9.63}$$

form the complete sets of states in isolated reactants, $\boldsymbol{\varphi}^\text{A}(\boldsymbol{x}) = \{\varphi_s{}^\text{A}(\boldsymbol{x}) \equiv \langle \boldsymbol{x}|s^\text{A}\rangle\}$ and $\boldsymbol{\varphi}^\text{B}(\boldsymbol{\xi}) = \{\varphi_u{}^\text{B}(\boldsymbol{\xi}) \equiv \langle \boldsymbol{\xi}|u^\text{B}\rangle\}$, both capable of representing the pure quantum state $\Psi(\boldsymbol{x}, \boldsymbol{\xi}) = \langle \boldsymbol{x}, \boldsymbol{\xi}|\Psi\rangle$ of the whole reactive complex:

$$\Psi(\boldsymbol{x}, \boldsymbol{\xi}) = \sum_s \varphi_s{}^\text{A}(\boldsymbol{x})\Phi_s{}^\text{A}(\boldsymbol{\xi}), \quad \Phi_s{}^\text{A}(\boldsymbol{\xi}) = \int \varphi_s{}^\text{A}(\boldsymbol{x})^*\Psi(\boldsymbol{x}, \boldsymbol{\xi})d\boldsymbol{x} \equiv \langle s^\text{A}|\Psi\rangle_\text{A} \quad \text{and}$$
$$\Psi(\boldsymbol{x}, \boldsymbol{\xi}) = \sum_u \varphi_u{}^\text{B}(\boldsymbol{\xi})\Phi_u{}^\text{B}(\boldsymbol{x}), \quad \Phi_u{}^\text{B}(\boldsymbol{x}) = \int \varphi_u{}^\text{B}(\boldsymbol{\xi})^*\Psi(\boldsymbol{x}, \boldsymbol{\xi})d\boldsymbol{\xi} \equiv \langle u^\text{B}|\Psi\rangle_\text{B} \tag{9.64}$$

The quantum operator of the physical property L(A) of subsystem A^+ acts only on the fragment internal variables $\boldsymbol{x}$: $\text{L(A)} = \text{L}(\boldsymbol{x})$. The expectation value of this observable in the "molecular" state $\Psi(\boldsymbol{x}, \boldsymbol{\xi})$ then reads:

$$\begin{aligned}\langle L\rangle_\Psi &= \iint \Psi(\boldsymbol{x}, \boldsymbol{\xi})^*\text{L}(\boldsymbol{x})\Psi(\boldsymbol{x}, \boldsymbol{\xi})d\boldsymbol{x}d\boldsymbol{\xi} \\ &= \sum_s\sum_{s'}\left[\int \Phi_s{}^\text{A}(\boldsymbol{\xi})\Phi_{s'}{}^\text{A}(\boldsymbol{\xi})^*d\boldsymbol{\xi}\right]\left[\int \varphi_{s'}{}^\text{A}(\boldsymbol{x})^*\text{L}(\boldsymbol{x})\varphi_s{}^\text{A}(\boldsymbol{x})d\boldsymbol{x}\right] \\ &\equiv \sum_s\sum_{s'}\rho_{s,s'}{}^\text{A}L_{s',s}{}^\text{A} \equiv \text{tr}_\text{A}[\boldsymbol{\rho}^\text{A}\mathbf{L}^\text{A}]\end{aligned} \tag{9.65}$$

Here $\mathbf{L}^\text{A} = \{L_{s',s}{}^\text{A} = \langle\varphi_{s'}{}^\text{A}|\text{L(A)}|\varphi_s{}^\text{A}\rangle_\text{A}\}$ represents L(A) in the basis set $\boldsymbol{\varphi}^\text{A}$ and

$$\boldsymbol{\rho}^\text{A} = \left\{\rho_{s,s'}{}^\text{A} = \int \Phi_s{}^\text{A}(\boldsymbol{\xi})\Phi_{s'}{}^\text{A}(\boldsymbol{\xi})^*d\boldsymbol{\xi} = \int \langle s^\text{A}|\Psi\rangle_\text{A}\langle\Psi|s'^\text{A}\rangle_\text{A}\, d\boldsymbol{\xi} \equiv \langle s^\text{A}|\Psi\rangle\langle\Psi|s'^\text{A}\rangle\right\} \tag{9.66}$$

denotes the *partial* density-matrix of subsystem $\text{A}^+(\boldsymbol{x})$, already integrated over coordinates of the complementary fragment $\text{B}^+(\boldsymbol{\xi})$. In the position representation, the matrix element $\rho_{s',s}{}^\text{A}$ can also be interpreted as a trace over the product of "molecular" and subsystem kernels:

$$\rho_{s,s'}{}^{\mathrm{A}} = \int \Phi_s{}^{\mathrm{A}}(\xi)\Phi_{s'}{}^{\mathrm{A}}(\xi)^* d\xi$$
$$= \iint \left[\int \Psi(\boldsymbol{x}, \xi)\Psi(\boldsymbol{x}', \xi)^* d\xi\right] \left[\varphi_{s'}{}^{\mathrm{A}}(\boldsymbol{x}')\varphi_s{}^{\mathrm{A}}(\boldsymbol{x})^*\right] d\boldsymbol{x}' d\boldsymbol{x}$$
$$\equiv \iint \rho_\Psi(\boldsymbol{x}, \boldsymbol{x}')\Omega_{s',s}{}^{\mathrm{A}}(\boldsymbol{x}', \boldsymbol{x}) d\boldsymbol{x}' d\boldsymbol{x} = \mathrm{tr}_{\mathrm{A}}\left[\int \rho_\Psi(\boldsymbol{x}, \boldsymbol{x}')\Omega_{s',s}{}^{\mathrm{A}}(\boldsymbol{x}', \boldsymbol{x})\, d\boldsymbol{x}'\right] \tag{9.67}$$

Writing Eq. (9.65) in the position representation,

$$\langle L\rangle_\Psi = \iint \rho^{\mathrm{A}}(\boldsymbol{x}, \boldsymbol{x}')\langle \boldsymbol{x}'|\mathrm{L}(\mathrm{A})|\boldsymbol{x}\rangle\, d\boldsymbol{x}'\, d\boldsymbol{x} \tag{9.68}$$

where $\langle \boldsymbol{x}'|\mathrm{L}(\mathrm{A})|\boldsymbol{x}\rangle = \delta(\boldsymbol{x}' - \boldsymbol{x})\mathrm{L}(\boldsymbol{x})$, one obtains the following expression for the subsystem density matrix:

$$\rho_{\mathrm{A}}(\boldsymbol{x}, \boldsymbol{x}') = \int \Psi(\boldsymbol{x}, \xi)\Psi(\boldsymbol{x}', \xi)^* d\xi = \sum_s \sum_{s'} \rho_{s,s'}{}^{\mathrm{A}}\Omega_{s',s}{}^{\mathrm{A}}(\boldsymbol{x}, \boldsymbol{x}') \equiv \mathrm{tr}_{\mathrm{A}}[\boldsymbol{\rho}^{\mathrm{A}}\boldsymbol{\Omega}^{\mathrm{A}}(\boldsymbol{x}, \boldsymbol{x}')] \tag{9.69}$$

where $\boldsymbol{\Omega}^{\mathrm{A}}(\boldsymbol{x}, \boldsymbol{x}') = \{\Omega_{s',s}{}^{\mathrm{A}}(\boldsymbol{x}, \boldsymbol{x}')\}$. The diagonal elements, for $\boldsymbol{x} = \boldsymbol{x}'$, then determine probabilities of the "local" events in A:

$$\rho_{\mathrm{A}}(\boldsymbol{x}, \boldsymbol{x}) = \int \Psi(\boldsymbol{x}, \xi)\Psi(\boldsymbol{x}, \xi)^* d\xi = \int P(\boldsymbol{x}, \xi)\, d\xi = p_{\mathrm{A}}(\boldsymbol{x}) \tag{9.70}$$

where $P(\boldsymbol{x}, \xi) = |\Psi(\boldsymbol{x}, \xi)|^2$ stands for the molecular probability of the joint local "events" on subsystems.

In accordance with SP of QM [76], the expansion coefficient $\Phi_s{}^{\mathrm{A}}(\xi)$ in Eq. (9.64) generates the quantum amplitude $a_{s(\mathrm{A})\to\Psi}(\xi)$ of the $|s^{\mathrm{A}}\rangle \to |\Psi\rangle$ scattering: $\Phi_s{}^{\mathrm{A}}(\xi) = \langle s^{\mathrm{A}}|\Psi\rangle_{\mathrm{A}} = a_{s(\mathrm{A})\to\Psi}(\xi)$. The associated conditional probability is then generated by its squared modulus integrated over coordinates of the complementary subsystem:

$$P(\Psi|s^{\mathrm{A}}) = \int |a_{s(\mathrm{A})\to\Psi}(\xi)|^2 d\xi = \int \left|\Phi_s{}^{\mathrm{A}}(\xi)\right|^2 d\xi = \rho_{s,s}{}^{\mathrm{A}}$$
$$= \int \langle s^{\mathrm{A}}|\Psi\rangle_{\mathrm{A}}\langle \Psi|s^{\mathrm{A}}\rangle_{\mathrm{A}} d\xi \equiv \langle s^{\mathrm{A}}|\Psi\rangle\langle \Psi|s^{\mathrm{A}}\rangle \equiv a_{s(\mathrm{A})\to\Psi\to s(\mathrm{A})} = \langle s^{\mathrm{A}}|\mathrm{P}_\Psi|s^{\mathrm{A}}\rangle \tag{9.71}$$

As also indicated above, this conditional probability determines the "bridge" amplitude $a_{s(\mathrm{A})\to\Psi\to s(\mathrm{A})}$ for the sequential, *cascade* communication $|s^{\mathrm{A}}\rangle \to |\Psi\rangle \to |s^{\mathrm{A}}\rangle$. Its square generates the diagonal bridge-probability:

$$P_{s(\mathrm{A})\to\Psi\to s(\mathrm{A})} = P(\Psi|s^{\mathrm{A}})P(s^{\mathrm{A}}|\Psi) = \left(\rho_{s,s}{}^{\mathrm{A}}\right)^2 \equiv P_\Psi(s^{\mathrm{A}} \to s^{\mathrm{A}})$$
$$= |a_{s(\mathrm{A})\to\Psi\to s(\mathrm{A})}|^2 = \langle s^{\mathrm{A}}|\mathrm{P}_\Psi|s^{\mathrm{A}}\rangle\langle s^{\mathrm{A}}|\mathrm{P}_\Psi|s^{\mathrm{A}}\rangle$$
$$\equiv \left\langle s^{\mathrm{A}}|\mathrm{P}_\Psi \mathrm{P}_s{}^{\mathrm{A}}\mathrm{P}_\Psi|s^{\mathrm{A}}\right\rangle \equiv \langle s^{\mathrm{A}}|\mathrm{P}_s(\Psi)|s^{\mathrm{A}}\rangle \tag{9.72}$$

The *off*-diagonal elements in the partial density matrix $\boldsymbol{\rho}^{\mathrm{A}}$ have an analogous interpretation as determining the *internal* bridge-communications in A^+:

$$\rho_{s,s'}{}^{\mathrm{A}} = \int \Phi_s{}^{\mathrm{A}}(\xi)\Phi_{s'}{}^{\mathrm{A}}(\xi)^* d\xi = \int \langle s^{\mathrm{A}}|\Psi\rangle_{\mathrm{A}}\langle \Psi|s'^{\mathrm{A}}\rangle_{\mathrm{A}} d\xi$$
$$\equiv \langle s^{\mathrm{A}}|\mathrm{P}_\Psi|s'^{\mathrm{A}}\rangle = a_{s(\mathrm{A})\to\Psi\to s'(\mathrm{A})} \tag{9.73}$$

$$P_{s(\mathrm{A})\to\Psi\to s'(\mathrm{A})} = \left|a_{s(\mathrm{A})\to\Psi\to s'(\mathrm{A})}\right|^2 = |\rho_{s,s'}{}^{\mathrm{A}}|^2 \equiv P_\Psi(s^{\mathrm{A}} \to s'^{\mathrm{A}})$$
$$= \langle s^{\mathrm{A}}|\mathrm{P}_\Psi|s'^{\mathrm{A}}\rangle\langle s'^{\mathrm{A}}|\mathrm{P}_\Psi|s^{\mathrm{A}}\rangle \equiv \left\langle s^{\mathrm{A}}|\mathrm{P}_{s'}(\Psi)|s^{\mathrm{A}}\right\rangle = \langle s'^{\mathrm{A}}|\mathrm{P}_s(\Psi)|s'^{\mathrm{A}}\rangle \tag{9.74}$$

Such internal probability propagations, within the complementary (interacting) molecular fragments, eventually generate the associated IT descriptors of the chemical bond patterns in the *polarized* subsystems. The conditional-entropy index $S[\boldsymbol{\varphi}'^{\mathrm{A}}|\boldsymbol{\varphi}^{\mathrm{A}}]$ of the fragment "output" states $\boldsymbol{\varphi}'^{\mathrm{A}}$ given the specified "input" states $\boldsymbol{\varphi}^{\mathrm{A}}$, reflects the average "noise" in internal communications $\boldsymbol{\varphi}^{\mathrm{A}} \to \boldsymbol{\varphi}'^{\mathrm{A}}$ within fragment A^+, which measures its overall entropic "covalency." The mutual-information descriptor $I[\boldsymbol{\varphi}^{\mathrm{A}}:\boldsymbol{\varphi}'^{\mathrm{A}}]$, measuring the network average information flow (capacity), then describes the subsystem information "iconicity."

The information system of probability scattering between the stationary states of reactants in the reactive system $R^+ = (\mathrm{A}^+|\mathrm{B}^+)$, $\mathbf{P}_\Psi(R^+)$, can be organized into the subsystem blocks of conditional probabilities:

$$\mathbf{P}_\Psi(R^+) = \{\mathbf{P}_\Psi(\mathrm{X} \to \mathrm{Y}) \equiv \mathbf{P}_\Psi(\boldsymbol{\varphi}^{\mathrm{X}} \to \boldsymbol{\varphi}'^{\mathrm{Y}});\quad \mathrm{X}, \mathrm{Y} \in (\mathrm{A}, \mathrm{B})\} \tag{9.75}$$

The *internal* (diagonal) and *external* (*off*-diagonal) blocks in this arrangement then define electronic communications *within* and *between* the internally polarized reactants, respectively. In OCT, the former give rise to multiplicities of the substrate's internal chemical bonds, reflecting the *intra*-reactant promotions relative to the promolecule, while the latter describe the external (*inter*-reactant) bond effects, which directly probe the reactivity behavior. The renormalized block $\mathbf{P}_\Psi(\mathrm{A} \to \mathrm{B})$ can be used as a separate channel focusing exclusively on the *inter*-reactant bond(s) in R.

The tensor product of subsystem spaces,

$$\begin{aligned}
\boldsymbol{\psi}(\boldsymbol{x}, \xi) &= \boldsymbol{\varphi}^{\mathrm{A}}(\boldsymbol{x}) \otimes \boldsymbol{\varphi}^{\mathrm{B}}(\xi) = \left\{\varphi_s{}^{\mathrm{A}}(\boldsymbol{x})\varphi_u{}^{\mathrm{B}}(\xi) \equiv \psi_w(\boldsymbol{x}, \xi)\right\} \\
\langle \psi_w | \psi_{w'} \rangle &= \left[\int \varphi_s{}^{\mathrm{A}}(\boldsymbol{x})^* \varphi_{s'}{}^{\mathrm{A}}(\boldsymbol{x})d\boldsymbol{x}\right]\left[\int \varphi_u{}^{\mathrm{B}}(\xi)^* \varphi_{u'}{}^{\mathrm{B}}(\xi)d\xi\right] \\
&= \langle \varphi_s{}^{\mathrm{A}} | \varphi_{s'}{}^{\mathrm{A}} \rangle_{\mathrm{A}} \langle \varphi_u{}^{\mathrm{B}} | \varphi_{u'}{}^{\mathrm{B}} \rangle_{\mathrm{B}} = \delta_{s,s'}\delta_{u,u'} = \delta_{w,w'}
\end{aligned} \tag{9.76}$$

constitutes an alternative (orthonormal and complete) basis, $\mathrm{P}_\psi = |\boldsymbol{\psi}\rangle\langle\boldsymbol{\psi}| = \sum_w P_w = 1$, for expanding the "molecular" state $\Psi(\boldsymbol{x}, \xi)$, of the whole reactive complex,

$$\begin{aligned}
\Psi(\boldsymbol{x}, \xi) &= \sum_w \psi_w(\boldsymbol{x}, \xi)C_w \equiv \psi(\boldsymbol{x}, \xi)\boldsymbol{C}, \quad \boldsymbol{C} = \langle \psi | \Psi \rangle = \{C_w = \langle \psi_w | \Psi \rangle\} \\
C_w &= \int \varphi_u{}^{\mathrm{B}}(\xi)^* \left[\int \varphi_s{}^{\mathrm{A}}(\boldsymbol{x})^* \Psi(\boldsymbol{x}, \xi)d\boldsymbol{x}\right] d\xi = \int \varphi_u{}^{\mathrm{B}}(\xi)^* \Phi_s(\xi)d\xi
\end{aligned} \tag{9.77}$$

The squared modulus of the expansion coefficient C_w now determines the conditional probability of observing in Ψ the product state ψ_w,

$$P(w|\Psi) = P(s^{\mathrm{A}}, u^{\mathrm{B}}|\Psi) = |C_w|^2 = \langle \Psi | \mathrm{P}_w | \Psi \rangle = \langle \psi_w | \mathrm{P}_\Psi | \psi_w \rangle, \ \sum_w P(w|\Psi) = 1 \tag{9.78}$$

These state probabilities also determine the associated density operator:

$$\mathrm{D}(\Psi) = \sum_w |\psi_w\rangle P(w|\Psi)\langle\psi_w| \equiv \sum_{s\in\mathrm{A}}\sum_{u\in\mathrm{B}} \left|\varphi_s{}^{\mathrm{A}}\varphi_u{}^{\mathrm{B}}\rangle P(s^{\mathrm{A}}, u^{\mathrm{B}}|\Psi)\langle\varphi_u{}^{\mathrm{B}}\varphi_s{}^{\mathrm{A}}\right|$$

Such ensembles of fragment states have been defined for the specified (pure) quantum state $\Psi(\boldsymbol{x}, \xi) = R(\boldsymbol{x}, \xi)\exp[i\phi(\boldsymbol{x}, \xi)]$ of the whole system, which represents their common "molecular" origin. Therefore, the modulus, $R(\boldsymbol{x}, \xi)$, and phase, $\phi(\boldsymbol{x}, \xi)$, components of this reference state also characterize the embedded (mutually open) subsystems in the *equilibrium* reactive complex $\mathrm{R}^* = (\mathrm{A}^* \vdots \mathrm{B}^*) \equiv \mathrm{R}$.

Let us qualitatively examine the *external* communications in R between the stationary states of the polarized acidic and basic reactants,

$$\mathrm{A} \to \mathrm{B} = \left\{\varphi_s{}^{\mathrm{A}} \to \varphi_u{}^{\mathrm{B}}\right\} \equiv \{|s^{\mathrm{A}}\rangle \to |u^{\mathrm{B}}\rangle\} \tag{9.79}$$

described by the corresponding block $\mathbf{P}_\Psi(\mathrm{A} \to \mathrm{B})$ in conditional probabilities $\mathbf{P}_\Psi(\mathrm{R}^+)$. This scattering between the stationary states of the closed subsystems is affected by the specified wavefunction $\Psi(\boldsymbol{x}, \boldsymbol{\xi}) = \langle \boldsymbol{x}, \boldsymbol{\xi} | \Psi \rangle$ of the whole reactive complex, which also establishes the channel molecular exploitation.

In accordance with SP of QM, the representative propagation $P(u^\mathrm{B}|s^\mathrm{A}) \equiv P_{s(\mathrm{A}) \to u(\mathrm{B})}$, of observing the *output* state $|\varphi_u{}^\mathrm{B}\rangle \equiv |u^\mathrm{B}\rangle$ in B, given the *input* state $|\varphi_s{}^\mathrm{A}\rangle \equiv$ in A, is determined by the squared modulus of the corresponding scattering amplitude, the product of projections of the fragment source ($|s^\mathrm{A}\rangle$) and receiver ($|u^\mathrm{B}\rangle$) states onto the "molecular" state $|\Psi\rangle$:

$$A_{s(\mathrm{A}) \to u(\mathrm{B})}(\Psi) \equiv A_\Psi(u^\mathrm{B}|s^\mathrm{A}) = \langle s^\mathrm{A}|\Psi\rangle\langle\Psi|u^\mathrm{B}\rangle \tag{9.80}$$

This external communication thus involves the intermediate (parameter) state $|\Psi\rangle$ in the *cascade*-communication $|s^\mathrm{A}\rangle \to |\Psi\rangle \to |u^\mathrm{B}\rangle$. Its amplitude is indeed determined by the product of *stage*-propagations in the preceding equation,

$$A_\Psi(u^\mathrm{B}|s^\mathrm{A}) = A_{s(\mathrm{A}) \to \Psi \to u(\mathrm{B})} = A_{s(\mathrm{A}) \to \Psi} A_{\Psi \to u(\mathrm{B})} = \langle s^\mathrm{A}|\Psi\rangle\langle\Psi|u^\mathrm{B}\rangle \equiv \langle s^\mathrm{A}|\mathrm{P}_\Psi|u^\mathrm{B}\rangle \tag{9.81}$$

where P_Ψ stands for the projection onto $|\Psi\rangle$. The associated conditional probability of this *bridge*-propagation $|s^\mathrm{A}\rangle \to |\Psi\rangle \to |u^\mathrm{B}\rangle$,

$$\begin{aligned} P_{s(\mathrm{A}) \to u(\mathrm{B})}(\Psi) &= |A_{s(\mathrm{A}) \to u(\mathrm{B})}(\Psi)|^2 = \langle s^\mathrm{A}|\Psi\rangle\langle\Psi|u^\mathrm{B}\rangle\langle u^\mathrm{B}|\Psi\rangle\langle\Psi|s^\mathrm{A}\rangle \\ &= \langle s^\mathrm{A}|\mathrm{P}_{u(\mathrm{B})}(\Psi)|s^\mathrm{A}\rangle = \langle u^\mathrm{B}|\mathrm{P}_{s(\mathrm{A})}(\Psi)|u^\mathrm{B}\rangle \equiv P_\Psi(u^\mathrm{B}|s^\mathrm{A}) \end{aligned} \tag{9.82}$$

then reads:

$$P_\Psi(u^\mathrm{B}|s^\mathrm{A}) = P_{s(\mathrm{A}) \to \Psi \to u(\mathrm{B})} = |A_{s(\mathrm{A}) \to \Psi \to u(\mathrm{B})}|^2 = P_{s(\mathrm{A}) \to \Psi} P_{\Psi \to u(\mathrm{B})} \tag{9.83}$$

It thus satisfies the following *bridge*-normalization condition:

$$\sum_u P_{s(\mathrm{A}) \to \Psi \to u(\mathrm{B})} = P_{s(\mathrm{A}) \to \Psi} \equiv P(\Psi|s^\mathrm{A})$$

The amplitudes of external communications $\mathrm{A} \to \mathrm{B}$ in the molecular state $\Psi(\boldsymbol{x}, \boldsymbol{\xi})$, between the stationary states $\{\varphi_s{}^\mathrm{A}(\boldsymbol{x}) = \langle \boldsymbol{x}|s^\mathrm{A}\rangle\}$ and $\{\varphi_u{}^\mathrm{B}(\boldsymbol{\xi}) = \langle \boldsymbol{\xi}|u^\mathrm{B}\rangle\}$ of polarized reactants, are thus determined by the *two* partial Hilbert-space projections:

$$A_{s(\mathrm{A}) \to \Psi}(\boldsymbol{\xi}) = \langle s^\mathrm{A}|\Psi\rangle_\mathrm{A} = \boldsymbol{\Phi}_s{}^\mathrm{A}(\boldsymbol{\xi}) \text{ and } A_{\Psi \to u(\mathrm{B})}(\boldsymbol{x}) = \langle\Psi|u^\mathrm{B}\rangle_\mathrm{B} = \boldsymbol{\Phi}_u{}^\mathrm{B}(\boldsymbol{x})^* \tag{9.84}$$

They generate the *local bridge* amplitude

$$A_{s(\mathrm{A}) \to \Psi \to u(\mathrm{B})}(\boldsymbol{x}, \boldsymbol{\xi}) = \Phi_s{}^\mathrm{A}(\boldsymbol{\xi})\Phi_u{}^\mathrm{B}(\boldsymbol{x})^* \tag{9.85}$$

and the associated *cascade*-probability density:

$$P_{s(\mathrm{A}) \to \Psi \to u(\mathrm{B})}(\boldsymbol{x}, \boldsymbol{\xi}) = \left|\Phi_s{}^\mathrm{A}(\boldsymbol{\xi})\right|^2 \left|\Phi_u{}^\mathrm{B}(\boldsymbol{x})\right|^2 = P_{s(\mathrm{A}) \to \Psi}(\boldsymbol{\xi}) P_{\Psi \to u(\mathrm{B})}(\boldsymbol{x}) \tag{9.86}$$

The related *condensed* communication link between these subsystem states then results from integrations over the subsystem coordinates:

$$\begin{aligned} P_{s(\mathrm{A}) \to \Psi \to u(\mathrm{B})} &= \iint P_{s(\mathrm{A}) \to \Psi \to u(\mathrm{B})}(\boldsymbol{x}, \boldsymbol{\xi}) d\boldsymbol{x} d\boldsymbol{\xi} \\ &= \left[\int \left|\Phi_s{}^\mathrm{A}(\boldsymbol{\xi})\right|^2 d\boldsymbol{\xi}\right] \left[\int \left|\Phi_u{}^\mathrm{B}(\boldsymbol{x})\right|^2 d\boldsymbol{x}\right] \equiv P(\Psi|s^\mathrm{A}) P(u^\mathrm{B}|\Psi) \end{aligned} \tag{9.87}$$

These *stage*-probabilities thus represent diagonal elements of the partial density matrices of subsystems (see Eqs. (9.64) and (9.66)):

$$P(\Psi|s^\mathrm{A}) \equiv P_{s(\mathrm{A}) \to \Psi} = \rho_{s,s}{}^\mathrm{A} \text{ and } P(u^\mathrm{B}|\Psi) \equiv P_{\Psi \to u(\mathrm{B})} = \rho_{u,u}{}^\mathrm{B} \tag{9.88}$$

The *inter*-reactant communication system, exploited in the specified molecular state Ψ, ultimately generates the entropic multiplicities (in bits) of chemical bonds between both substrates in R. The conditional entropy of the output (B) states, given the input (A) states, eventually defines the overall covalency of the external A—B bond(s), a measure of the channel *external*-communication noise, while the complementary descriptor of the overall bond ionicity reflects the mutual information in reactant states, a measure of the capacity of an information flow between reactants.

9.7 Illustrative Application to Reaction HSAB Principle

Let us now separate the geometrically accessible acidic (a) and basic (b) active sites of reactants $X \in (A, B)$, in the *donor* (basic) substrate $B = (a_B\vdots\ldots\vdots b_B) \equiv (a_B\vdots b_B)$ and its *acceptor* (acidic) companion $A = (a_A\vdots\ldots\vdots b_A) \equiv (a_A\vdots b_A)$, from their molecular remainders, which are immaterial for the mechanism of the bimolecular reaction

$$A + B \rightarrow [A\text{—}B]^{\ddagger} \rightarrow C + D$$

These crucial molecular fragments,

$$\lambda \in \{(a_A, b_A), (a_B, b_B)\} \tag{9.89}$$

thus define the substrate parts, which are directly active in the process of charge reconstruction via the elementary *internal* polarizations (P) and *external* transfer of electrons (CT).

There are two possible mutual arrangements of reactants in the reaction transition-state complex $[A\text{—}B]^{\ddagger}$ [95, 96]: the *complementary* (c) complex R_c, in which the a-site of one reactant faces the b-site of the other substrate,

$$R_c \equiv \begin{bmatrix} a_A - b_B \\ b_A - a_B \end{bmatrix} \tag{9.90}$$

and the regional hard/soft acids and bases (HSAB) [97–102], *parallel* (p) complex $R_{HSAB} = R_p$, when the coordinating sites in both reactants are of the *like* character:

$$R_{HSAB} \equiv \begin{bmatrix} a_A - a_B \\ b_A - b_B \end{bmatrix} \equiv R_p \tag{9.91}$$

The acidic (electron acceptor) site is relatively harder, i.e. less responsive to external perturbations, thus exhibiting lower values of the fragment Fukui Function (FF) [103] index or of the associated chemical softness descriptors [78, 79], while the basic (electron donor) fragment is more polarizable, as indeed reflected by its higher response descriptors. The acidic part of X thus exerts an electron-*accepting* (stabilizing) influence on the neighboring part of another reactant Y, while the basic part of X produces an electron-*donor* (destabilizing) effect on the coordinated fragment of Y in its vicinity. These simple electrostatic considerations might suggest that the "ionic" interactions in the complementary reactive complex of Figure 9.2 are dominant, while the "covalent" interactions of the familiar HSAB principle of chemistry [97, 98] point toward the parallel structure as the lowest-energy system. Numerical calculations [21] confirm, however, that *complementary* interactions of Figure 9.2, between the electron-*rich* (basic) fragment of one reactant and the electron-*deficient* (acidic) fragment of another substrate, indeed establish the most stable transition state.

This relative stability of R_c compared to R_p reflects the electrostatic dominance: a (repulsive) basic fragment of one reactant indeed prefers to face an (attractive) acidic part of the reaction partner.

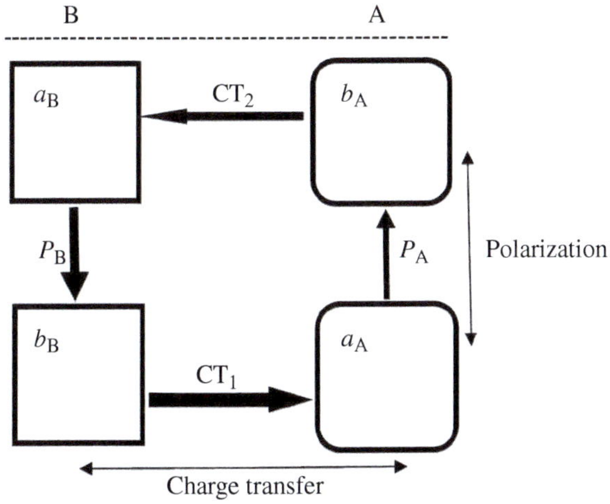

Figure 9.2 The CT $\{b_X \rightarrow a_Y\}$ and P(P_X) $\{a_X \rightarrow b_X\}$ electron flows involving acidic $A = (a_A \vdots \ldots \vdots b_A) \equiv (a_A \vdots b_A)$ and basic $B = (a_B \vdots \ldots \vdots b_B) \equiv (a_B \vdots b_B)$ reactants in the *complementary* arrangement R_c of their acidic (*a*) and basic (*b*) sites in the reactive complex $R_c = [A\text{—}B]$. These electronic flows are seen to produce a concerted ("circular") pattern of the P and CT fluxes of electronic probability in the final, equilibrium reactive system $R_c^* = (A^* \vdots B^*) = (a_A \vdots b_A \vdots a_B \vdots b_B)$, when all fragments are free to exchange electrons. Compared to the "promolecule" $R_c^0 = (A^0|B^0)$, it exhibits a "flow-through" pattern of electronic displacements on all active sites in both reactants, which precludes an exaggerated charge depletion or accumulation.

As shown in Figure 9.2, subsequent displacements in the reactant external potentials, due to the presence of the other substrate, trigger induced *polarizational* flows $\{P_X\}$, which restore the initially displaced *intra*-substrate equilibria. Finally, the *inter*-reactant CT takes place after the hypothetical opening of the initially polarized reactants. These P and CT displacements in molecular electron distribution, which can be directly inferred [83, 84, 95] from the electronegativity equalization (EE) principle [104, 105], give rise to different patterns of probability fluxes in these two reactive complexes:

$$R_c\!: \begin{array}{ccc} a_B & \longleftarrow & b_A \\ \downarrow & & \uparrow \\ b_B & \longrightarrow & a_A \end{array} \quad \text{and} \quad R_p\!: \begin{array}{ccc} a_B & \longrightarrow & a_A \\ \uparrow & & \uparrow \\ b_B & \longrightarrow & b_A \end{array},$$

concerted in R_c and disconcerted in R_p.

The maximum *site*-coherence of CT and P flows is thus predicted for the complementary complex. This concerted-flow pattern implies the least electron activation of both reactants in R_c, which also energetically favors the complementary structure relative to the parallel arrangement. The *disconcerted* flows indeed imply more exaggerated charge displacements on the crucial sites of both reactants (accentuated by bold symbols), which determine their chemical behavior: electron depletion on $\boldsymbol{b_B}$ and accumulation on $\boldsymbol{a_A}$. Only in R_c does one observe the *direct* $\boldsymbol{b_B} \rightarrow \boldsymbol{a_A}$ transfer of electrons, with the accompanying *double*-bridge CT, via the intermediates $\{a_B$ and $b_A\}$.

In this complementary coordination, the partial fluxes indeed show the "flow-through" pattern on all active molecular sites. The acidic parts of both reactants accept electrons as a result of CT and donate electrons due to substrate polarization. The opposite behavior of the basic sites is observed: they receive electrons due to P-mechanism and lose electrons as a result of CT. In the parallel complex, one observes a similar flow-through feature only on the a_B and b_A sites, $\boldsymbol{b_B}$ exclusively donates electrons both internally (P) and externally (CT), while $\boldsymbol{a}_{\mathrm{A}}$ accepts electrons from both its complementary fragment b_{A} (P) and from a_{B} part of the other reactant (CT).

To summarize: in $\mathrm{R_c}$, one observes the concerted flows of electrons involving all four active sites of both reactants, including the direct CT $\boldsymbol{b}_{\mathrm{B}} \rightarrow \boldsymbol{a}_{\mathrm{A}}$, with small net changes in electron populations on all active molecular fragments, while the charge reconstruction in $\mathrm{R_p}$ can be regarded as a transfer of electrons from $\boldsymbol{b}_{\mathrm{B}}$ to $\boldsymbol{a}_{\mathrm{A}}$ through the remaining (intermediate) sites a_{B} and b_{A}. The concerted flow pattern thus corresponds to the least-displaced electron populations on all reaction sites. This activation (promotion) perspective provides an additional physical explanation of the predicted complementary preference.

Finally, let us qualitatively examine the key electronic communications in R between active sites of Eq. (9.89) acting as whole, condensed units, which accompany these elementary charge reconstructions in reactive complexes. The alternative substrate coordinations are then defined by the conditional probabilities of observing in R the "output" ("receiver") sites $\lambda' = \{\lambda'\}$ (columns), for the specified "input" ("source") sites $\lambda = \{\lambda\}$ (rows):

$$\mathbf{P}(\lambda'|\lambda) = \{P(\lambda'|\lambda) = P(\lambda', \lambda)/P(\lambda) \equiv P_{\lambda\rightarrow\lambda'}\}, \sum_{\lambda'} P(\lambda'|\lambda) = 1 \tag{9.92}$$

Here, $P(\lambda)$ denotes the *site*-probability and $P(\lambda', \lambda)$ stands for the *joint*-probability of the occurrence of the specified *two*-site event in the bond system of R^+. The latter satisfies the usual normalization requirements:

$$\sum_{\lambda}\left[\sum_{\lambda'} P(\lambda', \lambda)\right] = \sum_{\lambda} P(\lambda) = 1 \tag{9.93}$$

Consider again the crucial CT sites of each reactant, again accentuated by bold symbols, which determine the substrate's overall donor/acceptor (DA) character in the reactive complex. They involve the *acceptor* (acidic) site $\boldsymbol{a}_{\mathrm{A}}$ of A and the *donor* (basic) site $\boldsymbol{b}_{\mathrm{B}}$ in B. In Figure 9.3, the dominating communications for both coordination, between the chemically bonded nearest neighbors, have been displayed. One recalls that in communication theory, each *intermediate* scattering involves the product of elementary direct propagations in the *bridge* in question. Therefore, the *direct* propagation $\boldsymbol{b}_{\mathrm{B}} \rightarrow \boldsymbol{a}_{\mathrm{A}}$, which is present only in $\mathrm{R_c}$, must dominate over all *indirect* (cascade) communications between these key reactant sites.

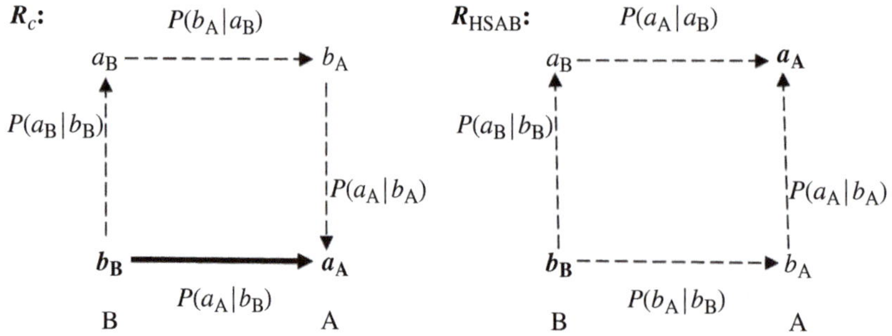

Figure 9.3 The dominating, nearest-neighbor communications between the chemically bonded sites of the Acid–Base complexes $\mathrm{R_c}$ and $\mathrm{R_p} = \mathrm{R_{HSAB}}$, respectively. The former involves a strong (direct) $\boldsymbol{b}_{\mathrm{B}} \rightarrow \boldsymbol{a}_{\mathrm{A}}$ propagation between the key sites $\boldsymbol{a}_{\mathrm{A}}$ and $\boldsymbol{b}_{\mathrm{B}}$, which determine the overall chemical character of reactants, and a weak intermediate (*double*-bridge) communication $\boldsymbol{b}_{\mathrm{B}} \rightarrow [a_{\mathrm{B}} \rightarrow b_{\mathrm{A}}] \rightarrow \boldsymbol{a}_{\mathrm{A}}$, while the latter exhibits two indirect (*single*-bridge) communications $\boldsymbol{b}_{\mathrm{B}} \rightarrow a_{\mathrm{B}} \rightarrow \boldsymbol{a}_{\mathrm{A}}$ and $\boldsymbol{b}_{\mathrm{B}} \rightarrow b_{\mathrm{A}} \rightarrow \boldsymbol{a}_{\mathrm{A}}$.

The R_c diagram in the figure shows that only the complementary arrangement exhibits a strong *direct* communication $\boldsymbol{b}_B \to \boldsymbol{a}_A$, between the crucial sites of both reactants, thus carrying the highest conditional probability $P(\boldsymbol{b}_B \to \boldsymbol{a}_A)$, besides a weak, *double*-cascade propagation $\boldsymbol{b}_B \to [a_B \to b_A] \to \boldsymbol{a}_A$ of relatively low probability:

$$P(\boldsymbol{b}_B \to [a_B \to b_A] \to \boldsymbol{a}_A) = P(\boldsymbol{b}_B \to a_B)P(a_B \to b_A)P(b_A \to \boldsymbol{a}_A) \ll P(\boldsymbol{b}_B \to \boldsymbol{a}_A) \tag{9.94}$$

The R_p coordination is seen to generate two *indirect*, *single*-cascade scatterings between these crucial sites of reactants:

$$P(\boldsymbol{b}_B \to a_B \to \boldsymbol{a}_A) = P(\boldsymbol{b}_B \to a_B)P(a_B \to \boldsymbol{a}_A) < P(\boldsymbol{b}_B \to \boldsymbol{a}_A) \quad \text{and}$$
$$P(\boldsymbol{b}_B \to b_A \to \boldsymbol{a}_A) = P(\boldsymbol{b}_B \to b_A)P(b_A \to \boldsymbol{a}_A) < P(\boldsymbol{b}_B \to \boldsymbol{a}_A) \tag{9.95}$$

This provides an additional communication rationale for the observed complementary preference in the acid-base interactions.

A softer (basic, diffused) fragment of reactant X is predicted to overlap more strongly with the neighboring sites of reactant Y, compared to its harder (acidic, more compact) fragment, thus also giving rise to stronger electron communications:

$$P(b_X \to \lambda') > P(a_X \to \lambda') \text{ and } P(\lambda \to b_Y) > P(\lambda \to a_Y) \tag{9.96}$$

The largest propagations can be thus expected for the most *covalently* bonded soft (basic) fragments, directly coordinating only in the HSAB complex, while the weakest probability scatterings are predicted for its two hard (acidic) sites, which bind *ionically*:

$$P(\boldsymbol{b}_B \to b_A) > P(a_B \to \boldsymbol{a}_A) \tag{9.97}$$

These qualitative arguments indeed confirm in R_p a *covalent* character of the chemical bond between the basic sites of reactants, and an *ionic* bond between their acidic fragments. In R_c, one similarly predicts a strong *coordination* bond between the crucial sites $\boldsymbol{a}_A$ and $\boldsymbol{b}_B$ and a predominantly *covalent* bond between their complements b_A and a_B.

9.8 Conclusion

Chemical reactions involve the interacting, externally and/or mutually open substrates, which cannot be characterized by the *pure* quantum states of subsystems or wavefunctions of the reactive complex as a whole. Each polarized fragment is, in fact, in an effective *mixed* state, also in the case of the pure *molecular* state of the whole system, which requires an ensemble description in terms of density operators and underlying (external) probabilities. The equilibrium states of *open* subsystems are also shaped by the applied external conditions, represented by ensemble probabilities depending on thermodynamic parameters. In this chapter, we have examined electronic communications involving open molecular fragments. Their entropy-*covalency* and information-*iconicity* descriptors reflect patterns of both the *internal* structure of the polarized substrates and *external* chemical bonds between substrates of chemical reactions.

In the SCF LCAO MO approach, the MO ensemble of the system ground-state configuration generates the CBO matrix in AO resolution and the associated kernel of the *first*-order reduced density matrix. In interacting fragments, described by the molecular state determining the common quantum origin of both reactants, a similar role is played by the *partial* density matrices of subsystems. These key concepts determine amplitudes of electron communications in the configuration *bond system*, ultimately establishing the discrete or local information channels for the molecular system in question. They also generate the IT multiplicity descriptors of chemical bonds at these two resolutions and their ionic/covalent composition.

The intriguing relation between descriptions of quantum communications via molecular bond system in the *one*-electron perspective and correlation holes of the *two*-electron approach, respectively, has been explored using the illustrative case of the AO and local channels in HF approximation, when the probability scatterings reflect the exchange (Fermi) correlation part of the *two*-electron density. We have also examined a hierarchy of dominating communications in the DA systems, the CT complexes involving the basic and acidic reactants. The alternative transition-state structures have been examined and the preferred *concerted* pattern of electronic flows in the "complementary" arrangement of reactants has been established using EE arguments. The *disconcerted* scheme of electronic fluxes in the alternative, "parallel" (regional HSAB) structure was also predicted, differences in communications between the substrate crucial active sites have been commented upon, and their implications for the character of chemical bonds between reactants have been qualitatively examined.

Entropic theories of molecular electronic structure call for the *resultant* generalization of the classical (probability) measures of the information/entropy content in molecular states [4, 5, 58–70], combining contributions from all structure manifestations. Such overall descriptors unite the familiar classical (probability) terms, due to wavefunction modulus, and their nonclassical (current) complements due to the state phase determining the "convection" velocity of probability flux. Indeed, the electron density and current distributions, both "encoded" in the system wavefunction, must ultimately contribute to the overall ("resultant") amount of information in complex electronic states. The classical term probes only the entropic content of the *incoherent* (disentangled) local "events," while the nonclassical supplement explores their mutual *coherence* (entanglement). Such generalized descriptors are also required in an information distinction between the bonded (entangled) and nonbonded (disentangled) states of the *open* and *closed* molecular subsystems, respectively, e.g. substrates in reaction complexes. Indeed, such hypothetical stages are addressed in modern reactivity theory [1–5].

Another promising development in the IT description of molecular electronic structure deals with *property networking* in quantum electronic states [106, 107]. It introduces the molecular information systems for specific physical observables, for example, the virial components of electronic energy, with the physical entity itself shaping the physical information channel between the basis functions of molecular calculations and the system quantum state determining its molecular exploitation. This approach ultimately explores the information content of different physical properties and examines their transport in the specified molecular state. It also allows one to design the information cascade consisting of the *potential* and *kinetic* stages for the *bound* → *intrinsic* information transfer of the extreme physical information (EPI) principle [19].

References

1 Nalewajski, R.F. (2006). *Information Theory of Molecular Systems*. Amsterdam: Elsevier.

2 Nalewajski, R.F. (2010). *Information Origins of the Chemical Bond*. New York: Nova Science Publishers.

3 Nalewajski, R.F. (2012). *Perspectives in Electronic Structure Theory*. Heidelberg: Springer.

4 Nalewajski, R.F. (2016). *Quantum Information Theory of Molecular States*. New York: Nova Science Publishers.

5 Nalewajski, R.F. (2023). *Chemical Reactivity in Quantum Mechanics and Information Theory*. New York: Elsevier.
6 Nalewajski, R.F. (2000). Entropic measures of bond multiplicity from the information theory. *J. Phys. Chem. A* 104: 11940–11951.
7 Nalewajski, R.F. (2004). Entropy descriptors of the chemical bond in information theory: I. Basic concepts and relations. *Mol. Phys.* 102: 531–546.
8 Nalewajski, R.F. (2004). Entropy descriptors of the chemical bond in information theory: II. Application to simple orbital models. *Mol. Phys.* 102: 547–566.
9 Nalewajski, R.F. (2004). Entropic and difference bond multiplicities from the two-electron probabilities in orbital resolution. *Chem. Phys. Lett.* 386: 265–271.
10 Nalewajski, R.F. (2005). Reduced communication channels of molecular fragments and their entropy/information bond indices. *Theor. Chem. Acc.* 114: 4–18.
11 Nalewajski, R.F. (2011). Entropy/information descriptors of the chemical bond revisited. *J. Math. Chem.* 49: 2308–2329.
12 Nalewajski, R.F. (2014). Quantum information descriptors and communications in molecules. *J. Math. Chem.* 52: 1292–1323.
13 Nalewajski, R.F. (2009). Multiple, localized and delocalized/conjugated bonds in the orbital-communication theory of molecular systems. *Adv. Quantum Chem.* 56: 217–250.
14 Nalewajski, R.F., Szczepanik, D., and Mrozek, J. (2011). Bond differentiation and orbital decoupling in the orbital communication theory of the chemical bond. *Adv. Quantum Chem.* 61: 1–48.
15 Nalewajski, R.F. (2012). Information theory insights into molecular electronic structure and reactivity. In: *Applications of Density Functional Theory to Chemical Reactivity*, Structure and Bonding, vol. 149 (ed. M.V. Putz and D.M.P. Mingos), 51–94. Berlin, Heidelberg: Springer.
16 Nalewajski, R.F., Szczepanik, D., and Mrozek, J. (2012). Basis set dependence of molecular information channels and their entropic bond descriptors. *J. Math. Chem.* 50: 1437–1457.
17 Nalewajski, R.F. (2017). Electron communications and chemical bonds. In: *Frontiers of Quantum Chemistry* (ed. M. Wójcik, H. Nakatsuji, B. Kirtman, and Y. Ozaki), 315–351. Singapore: Springer.
18 Fisher, R.A. (1925). Theory of statistical estimation. *Proc. Cambridge Philos. Soc.* 22: 700–725.
19 Frieden, B.R. (2004). *Physics from the Fisher Information – A Unification*. Cambridge: Cambridge University Press.
20 Shannon, C.E. (1948). The mathematical theory of communication. *Bell Syst. Tech. J.* 7: 379–493, 623–656.
21 Shannon, C.E. and Weaver, W. (1949). *The Mathematical Theory of Communication*. Urbana: University of Illinois Press.
22 Kullback, S. and Leibler, R.A. (1951). On information and sufficiency. *Ann. Math. Stat.* 22: 79–86.
23 Kullback, S. (1959). *Information Theory and Statistics*. New York: Wiley.
24 Abramson, N. (1963). *Information Theory and Coding*. New York: McGraw-Hill.
25 Pfeifer, P.E. (1978). *Concepts of Probability Theory*. New York: Dover.
26 Nalewajski, R.F. and Parr, R.G. (2000). Information theory, atoms-in-molecules and molecular similarity. *PNAS* 97: 8879–8882.
27 Nalewajski, R.F., Świtka, E., and Michalak, A. (2002). Information distance analysis of molecular electron densities. *Int. J. Quantum Chem.* 87: 198–213.
28 Nalewajski, R.F. (2008). Use of Fisher information in quantum chemistry. *Int. J. Quantum Chem.* (Jankowski, K. issue) 108: 2230–2252.

29 López-Rosa, S. (2010). Information theoretic measures of atomic and molecular systems. PhD thesis. University of Granada: Granada.

30 López-Rosa, S., Esquivel, R.O., Angulo, J.C. et al. (2010). Fisher information study in position and momentum spaces for elementary chemical reactions. *J. Chem. Theory Comput.* 6: 145–154.

31 Esquivel, R.O., Liu, S., Angulo, J.C. et al. (2011). Fisher information and steric effect: study of internal rotation barrier in ethane. *J. Phys. Chem. A* 115: 4406–4415.

32 Nalewajski, R.F. (2019). On entropy/information description of reactivity phenomena. In: *Advances in Mathematics Research*, vol. 26 (ed. A.R. Baswell), 97–157. New York: Nova Science Publishers.

33 Nalewajski, R.F. (2019). Role of electronic kinetic energy (resultant gradient information) in chemical reactivity. *J. Mol. Model.* (Latajka, Z. issue) 25: 259–278.

34 Nalewajski, R.F. (2020). Resultant information approach to donor-acceptor systems. In: *An Introduction to Electronic Structure Theory* (ed. N.T. Paulsen), 1–58. New York: Nova Science Publishers.

35 Nalewajski, R.F. (2021). Quantum information perspective on chemical reactivity. In: *Mathematics Applied to Engineering in Action: Advanced Theories, Methods, and Models* (ed. N. Islam, S. Bir-Singh, P. Ranjan, and A.K. Haghi), 1–40. Palm Bay: Apple Academic Press.

36 Nalewajski, R.F. (2003). Information principles in the theory of electronic structure. *Chem. Phys. Lett.* 272: 28–34.

37 Nalewajski, R.F. (2003). Information principles in the Loge Theory. *Chem. Phys. Lett.* 375: 196–203.

38 Nalewajski, R.F. and Broniatowska, E. (2003). Information distance approach to Hammond postulate. *Chem. Phys. Lett.* 376: 33–39.

39 Nalewajski, R.F. and Parr, R.G. (2000). Information theory, atoms-in-molecules and molecular similarity. *PNAS* 97: 8879–8882. Nalewajski, R.F. and Parr, R.G. (2001). Information-theoretic thermodynamics of molecules and their Hirshfeld fragments. *J. Phys. Chem. A* 105: 7391–7400.

40 Nalewajski, R.F. (2002). Hirshfeld analysis of molecular densities: subsystem probabilities and charge sensitivities. *Phys. Chem. Chem. Phys.* 4: 1710–1721.

41 Nalewajski, R.F. and Broniatowska, E. (2003). Entropy displacement analysis of electron distributions in molecules and their Hirshfeld atoms. *J. Phys. Chem. A* 107: 6270–6280.

42 Parr, R.G., Ayers, P.W., and Nalewajski, R.F. (2005). What is an atom in a molecule? *J. Phys. Chem. A* 109: 3957–3959.

43 Nalewajski, R.F. and Broniatowska, E. (2007). Atoms-in-molecules from the stockholder partition of molecular two-electron distribution. *Theor. Chem. Acc.* 117: 7–27.

44 Heidar-Zadeh, F., Ayers, P.W., Verstraelen, T. et al. (2018). Information-theoretic approaches to atoms-in-molecules: Hirshfeld family of partitioning schemes. *J. Phys. Chem. A* 122: 4219–4245.

45 Hirshfeld, F.L. (1977). Bonded-atom fragments for describing molecular charge densities. *Theor. Chim. Acta* 44: 129–138.

46 Hammond, G.S. (1955). A correlation of reaction rates. *JACS* 77: 334–338.

47 Nalewajski, R.F. (2011). Through-space and through-bridge components of chemical bonds. *J. Math. Chem.* 49: 371–392.

48 Nalewajski, R.F. (2011). Chemical bonds from through-bridge orbital communications in prototype molecular systems. *J. Math. Chem.* 49: 546–561.

49 Nalewajski, R.F. (2011). On interference of orbital communications in molecular systems. *J. Math. Chem.* 49: 806–815.

50 Nalewajski, R.F. and Gurdek, P. (2011). On the implicit bond-dependency origins of bridge interactions. *J. Math. Chem.* 49: 1226–1237.
51 Nalewajski, R.F. (2012). Direct (through-space) and indirect (through-bridge) components of molecular bond multiplicities. *Int. J. Quantum Chem.* 112: 2355–2370.
52 Nalewajski, R.F. and Gurdek, P. (2012). Bond-order and entropic probes of the chemical bonds. *Struct. Chem.* 23: 1383–1398.
53 Nalewajski, R.F., Köster, A.M., and Escalante, S. (2005). Electron localization function as information measure. *J. Phys. Chem. A* 109: 10038–10043.
54 Becke, A.D. and Edgecombe, K.E. (1990). A simple measure of electron localization in atomic and molecular systems. *J. Chem. Phys.* 92: 5397–5403.
55 Silvi, B. and Savin, A. (1994). Classification of chemical bonds based on topological analysis of electron localization functions. *Nature* 371: 683–686.
56 Savin, A., Nesper, R., Wengert, S., and Fässler, T.F. (1997). ELF: the electron localization function. *Angew. Chem. Int. Ed. Engl.* 36: 1808–1832.
57 Nalewajski, R.F., de Silva, P., and Mrozek, J. (2010). Use of nonadditive Fisher information in probing the chemical bonds. *THEOCHEM* 954: 57–74.
58 Nalewajski, R.F. (2013). Exploring molecular equilibria using quantum information measures. *Ann. Phys.* 525: 256–268.
59 Nalewajski, R.F. (2014). On phase/current components of entropy/information descriptors of molecular states. *Mol. Phys.* 112: 2587–2601.
60 Nalewajski, R.F. (2014). Quantum information approach to electronic equilibria: molecular fragments and elements of non-equilibrium thermodynamic description. *J. Math. Chem.* 52: 1921–1948.
61 Nalewajski, R.F. (2014). On phase equilibria in molecules. *J. Math. Chem.* 52: 588–612.
62 Nalewajski, R.F. (2015). Phase/current information descriptors and equilibrium states in molecules. *Int. J. Quantum Chem.* 115: 1274–1288.
63 Nalewajski, R.F. (2015). Quantum information measures and molecular phase equilibria. In: *Advances in Mathematics Research*, vol. 19 (ed. A.R. Baswell), 53–86. New York: Nova Science Publishers.
64 Nalewajski, R.F. (2016). Complex entropy and resultant information measures. *J. Math. Chem.* 54: 1777–1782.
65 Nalewajski, R.F. (2017). Quantum information measures and their use in chemistry. *Curr. Phys. Chem.* 7: 94–117.
66 Nalewajski, R.F. (2018). Information equilibria, subsystem entanglement and dynamics of the overall entropic descriptors of molecular electronic structure. *J. Mol. Model.* (Chattaraj, P. issue) 24: 212–227.
67 Nalewajski, R.F. (2019). Equidensity orbitals in resultant-information description of electronic states. *Theor. Chem. Acc.* 138: 108–123.
68 Nalewajski, R.F. (2019). Resultant information description of electronic states and chemical processes. *J. Phys. Chem. A* (Geerlings, P. issue) 123: 45–60.
69 Nalewajski, R.F. (2019). Density matrix in determining electron communications and resultant information content in molecular states. In: *Understanding Density Matrices* (ed. N.V. Danielsen), 1–35. New York: Nova Science Publishers.
70 Nalewajski, R.F. (2020). Information-theoretic descriptors of molecular states and electronic communications between reactants. *Entropy* 22: 749–769.
71 Nalewajski, R.F. (2016). On entangled states of molecular fragments. *Trends Phys. Chem.* 16: 71–85.

72 Nalewajski, R.F. (2019). Entropy continuity, electron diffusion and fragment entanglement in equilibrium states. In: *Advances in Mathematics Research*, vol. 22 (ed. A.R. Baswell), 1–42. New York: Nova Science Publishers.

73 Nalewajski, R.F. (1980). Virial theorem implications for the minimum energy reaction paths. *Chem. Phys.* 50: 127–136.

74 Nalewajski, R.F. (2022). Kinetic-energy/information probes of chemical processes. In: *Advances in Chemistry Research*, vol. 73 (ed. J.C. Taylor), 1–60. New York: Nova Science Publishers.

75 Nalewajski, R.F. (2018). Phase description of reactive systems. In: *Conceptual Density Functional Theory and Its Applications in the Chemical Domain* (ed. N. Islam and S. Kaya), 217–249. Palm Bay: Apple Academic Press/CRC Press.

76 Dirac, P.A.M. (1967). *The Principles of Quantum Mechanics*. Oxford: Oxford University Press.

77 Nalewajski, R.F. (1994). Sensitivity analysis of charge transfer systems: in situ quantities, intersecting-state model and its implications. *Int. J. Quantum Chem.* 49: 675–703.

78 Nalewajski, R.F., Korchowiec, J., and Michalak, A. (1996). Reactivity criteria in charge sensitivity analysis. In: *Density Functional Theory IV: Theory of Chemical Reactivity*, Topics in Current Chemistry, vol. 183 (ed. R.F. Nalewajski), 25–141.

79 Nalewajski, R.F. and Korchowiec, J. (1997). *Charge Sensitivity Approach to Electronic Structure and Chemical Reactivity*. Singapore: World Scientific.

80 Nalewajski, R.F. (2014). Quantum information description of reactive systems. In: *Chemical Reactivity Theory*. Indian J. Chem. (Ghosh, S. K. issue), vol. 53A (ed. P.K. Chattaraj), 1010–1018.

81 Nalewajski, R.F. (2017). Chemical reactivity description in density-functional and information theories. In: *Chemical Concepts from Density Functional Theory*. Acta Phys.-Chim. Sin., vol. 33 (ed. S. Liu), 2491–2509.

82 Nalewajski, R.F. (2010). Charge response criteria of chemical reactivity: Fukui function indices and populational reference frames reflecting the inter-reactant charge coupling. *Int. J. Quantum Chem.* 61: 181–196.

83 Nalewajski, R.F. (2019). Understanding electronic structure and chemical reactivity: quantum-information perspective. In: *The Application of Quantum Mechanics to the Reactivity of Molecules. Appl. Sci.*, vol. 9 (ed. S. Sousa), 1262–1292.

84 Nalewajski, R.F. (2020). Phase equalization, charge transfer, information flows and electron communications in donor-acceptor systems. In: *The Application of Quantum Mechanics to the Reactivity of Molecules.*Appl. Sci., vol. 10 (ed. S. Sousa), 3615–3646.

85 Nalewajski, R.F. (1995). Charge sensitivity analysis as diagnostic tool for predicting trends in chemical reactivity. In: *Proceedings of the NATO ASI on Density Functional Theory (Il Ciocco, 1993)* (ed. R.M. Dreizler and E.K.U. Gross), 339–389. New York: Plenum.

86 Hohenberg, P. and Kohn, W. (1964). Inhomogeneous electron gas. *Phys. Rev.* 136B: 864–971.

87 Kohn, W. and Sham, L.J. (1965). Self-consistent equations including exchange and correlation effects. *Phys. Rev.* 140A: 1133–1138.

88 Levy, M. (1979). Universal variational functionals of electron densities, first-order density matrices, and natural spin-orbitals and solution of the *v*-representability problem. *PNAS* 76: 6062–6065.

89 Parr, R.G. and Yang, W. (1989). *Density-Functional Theory of Atoms and Molecules*. New York: Oxford University Press.

90 Dreizler, R.M. and Gross, E.K.U. (1990). *Density Functional Theory: An Approach to the Quantum Many-Body Problem*. Berlin: Springer.

91 Nalewajski, R.F. (ed.) (1996). *Density Functional Theory I–IV, Topics in Current Chemistry*, vol. 180–182. Heidelberg: Springer-Verlag.

92 Davydov, A.S. (1965). *Quantum Mechanics*. Oxford: Pergamon Press.

93 Cohen-Tannoudji, C., Diu, B., and Laloë, F. (1977). *Quantum Mechanics*. New York: Wiley.

94 Nalewajski, R.F. (2020). Interacting subsystems and their molecular ensembles. *Acad. J. Chem.* 5: 25–30.

95 Nalewajski, R.F. (2000). Manifestations of the maximum complementarity principle for matching atomic softnesses in model chemisorption systems. *Top. Catal.* 11: 469–485.

96 Chandra, A.K., Michalak, A., Nguyen, M.T., and Nalewajski, R.F. (1998). On regional matching of atomic softnesses in chemical reactions: two-reactant charge sensitivity study. *J. Phys. Chem. A* 102: 10182–10188.

97 Pearson, R.G. (1973). *Hard and Soft Acids and Bases*. Stroudsburg: Dowden, Hutchinson and Ross.

98 Parr, R.G. and Pearson, R.G. (1983). Absolute hardness: companion parameter to absolute electronegativity. *JACS* 105: 7512–7516.

99 Nalewajski, R.F. (1984). Electrostatic effects in interactions between hard (soft) acids and bases. *JACS* 106: 944–945.

100 Chattaraj, P.K., Lee, H., and Parr, R.G. (1991). HSAB principle. *JACS* 113: 1855–1856.

101 Gazquez, J.L. and Mendez, F. (1994). The hard and soft acids and bases principle: an atoms-in-molecules viewpoint. *J. Phys. Chem.* 98: 4591–4593.

102 Ayers, P.W. (2007). The physical basis of the hard/soft acid/base principle. *Faraday Discuss.* 135: 161–190.

103 Parr, R.G. and Yang, W. (1984). Density functional approach to the frontier-electron theory of chemical reactivity. *JACS* 106: 4049–4050.

104 Sanderson, R.T. (1951). An interpretation of bond lengths and a classification of bonds. *Science* 114: 670–672.

105 Sanderson, R.T. (1976). *Chemical Bonds and Bond Energy*, 2e. New York: Academic Press.

106 Nalewajski, R.F. (2023). Extreme physical information principle and property networking. In: *Advances in Mathematics Research*, vol. 33, 183–223. New York: Nova Science Publishers.

107 Nalewajski, R.F. (2023). Virial energy components, extreme physical information and property flows in communication networks. In: *Advances in Mathematics Research*, vol. 34, 39–70. New York: Nova Science Publishers.

10

Impacts of External Electric Fields on Aromaticity and Acidity for Benzoic Acid and Derivatives: Directionality, Additivity, and More

Meng Li[1], *Xinjie Wan*[1], *Xin He*[2], *Chunying Rong*[1], *Dongbo Zhao*[3], *and Shubin Liu*[4,5]

[1] *Key Laboratory of Chemical Biology and Traditional Chinese Medicine Research (Ministry of Education of China), Hunan Normal University, Changsha, Hunan, P.R. China*
[2] *Qingdao Institute for Theoretical and Computational Sciences, Institute of Frontier and Interdisciplinary Science, Shandong University, Qingdao, Shandong, P.R. China*
[3] *Institute of Biomedical Research, Yunnan University, Kunming, P.R. China*
[4] *Research Computing Center, University of North Carolina, Chapel Hill, NC, USA*
[5] *Department of Chemistry, University of North Carolina, Chapel Hill, NC, USA*

10.1 Introduction

The impact of external fields has attracted considerable research interest in the recent literature [1–17]. It is commonly believed that they can influence both structures [18–21], reactivities [22–25], and other electronic properties of a molecular system [26–29]. In this chapter, we highlight our recent studies about the impact of external electric fields on acidity and aromaticity for benzoic acid and its derivatives. This was done through altering the direction and strength of an external electric field applied to the molecule and then calculating its acidity and aromaticity values. Our results showed that the impact from the external field is unambiguously directional. The direction along with the carboxyl group is the most sensitive, but the one perpendicular to the benzene ring, though least sensitive, is most vulnerable and destructive. We also found that these impacts can be adequately accounted for by reactivity descriptors associated with frontier orbitals and partial charges [30]. Additionally, we found that multiply substituted systems can be satisfactorily assessed by singly substituted ones with the sum rule. We have employed numerous analytical tools to make sense of our computational results, including density functional theory (DFT), conceptual density functional theory (CDFT), and the information-theoretic approach (ITA).

10.2 Methodology

The two physiochemical properties are employed in this work to investigate the impact of external electric fields. The first one is aromaticity, which can be quantified by many descriptors [31–33], but for which the most widely used is nuclear-independent chemical shift (NICS) [34] derived from the electromagnetic effect due to the aromatic ring current. A more negative value of NICS is an indication of stronger aromaticity. It can be expressed as the sum of partial chemical shifts arising

Electron Density: Concepts, Computation and DFT Applications, First Edition.
Edited by Pratim Kumar Chattaraj and Debdutta Chakraborty.

from occupied molecular orbitals Ψ_{k0} at the chosen points located at or on top of the aromatic ring,

$$\text{NICS} = \frac{1}{2c^2}\sum_k^{\text{occ}} \left\langle \Psi_{k0} \left| \frac{\boldsymbol{r}\boldsymbol{r}_\text{N}\mathbf{I} - \boldsymbol{r}_\text{N} \otimes \boldsymbol{r}}{|\boldsymbol{r} - \boldsymbol{R}_\text{N}|^3} \right| \Psi_{k0} \right\rangle - \frac{2}{c}\sum_k^{\text{occ}} \left\langle \Psi_{k0} \left| \frac{(\boldsymbol{L}_\text{N})}{|\boldsymbol{r} - \boldsymbol{R}_\text{N}|^3} \right| \Psi_{k1} \right\rangle \tag{10.1}$$

where, $\boldsymbol{R}_\text{N}$ is the chosen vector point for calculation, $\boldsymbol{r}$ is the electron coordinate, $\boldsymbol{r}_\text{N} = \boldsymbol{r} - \boldsymbol{R}_\text{N}$ is the relative coordinate, and $\boldsymbol{L}_\text{N} = \boldsymbol{r}_\text{N} \times \nabla$ is the angular momentum, $\mathbf{I}$ is the unit matrix, and the first and second terms on the right side of Eq. (10.1) are diamagnetic and paramagnetic contributions to NICS, respectively. There are many different flavors of NICS in the literature, but in most cases, their results and trends are identical. In this work, we only consider the NICS value at the center of the ring, NICS(0). Besides NICS, there are other aromaticity descriptors as supporting evidence, such as the harmonic oscillator model of aromaticity (HOMA) [35], parametric delocalization index (PDI), [36] and the aromatic fluctuation index (FLU) [37]. In addition, we will also plot the gauge-including magnetically induced current (GIMIC) [38] diagrams to visualize the ring current difference from different external fields.

The second property, molecular acidity, can be quantified by $\text{p}K_\text{a}$ values as a measure of the strength of a Brønsted–Lowry acid. In principle, the $\text{p}K_\text{a}$ value of an acid, HA, is defined as the negative logarithm, $\text{p}K_\text{a} = -\log_{10} K_\text{a}$, of the equilibrium constant K_a of the acid dissociation reaction, $\text{HA} \rightleftharpoons \text{A}^- + \text{H}^+$, with $K_\text{a} = [\text{A}^-][\text{H}^+]/[\text{HA}]$, the equilibrium concentration of the three species in the dissociation reaction. In thermodynamics, the equilibrium constant K_a is related to the standard Gibbs free energy change $\Delta G°$ of the above reaction through the following equation,

$$\Delta G° = 2.303\, RT\, \text{p}K_a = -2.303\, RT \log_{10} K_\text{a} \tag{10.2}$$

where R is the gas constant, and T is the temperature in Kelvin. In practice, traditionally, to computationally obtain $\text{p}K_\text{a}$ values, one employs *ab initio* or DFT method through a thermodynamic cycle [39, 40], which is not only computationally demanding but also numerically inaccurate because of the errors introduced in the implicit solvent model and proton hydration calculations. Instead, under the CDFT framework, we proposed a new approach using molecular electrostatic potential (MEP) and natural atomic orbital (NAO) descriptors [41–44].

$$\text{p}K_\text{a} \propto \text{MEP}_\text{O} \equiv \sum_{\text{i}\neq\text{O}} \frac{Z_\text{i}}{|\mathbf{R}_\text{i} - \mathbf{R}_\text{O}|} - \int \frac{\rho(\mathbf{r})}{|\mathbf{r} - \mathbf{R}_\text{O}|} d\tau \tag{10.3}$$

where $\mathbf{R}_\text{O}$ is the acidic oxygen coordinate, $\mathbf{R}_\text{i}$ is the coordinate of other atoms, Z is the nuclear charge, and $\rho(\mathbf{r})$ is the electron density. Using this new method, we have predicted the acidity for a number of molecules with different acidity scales, such as substituted benzoic acid [45], benzenesulfinic acids [46], benzeneseleninic acids [47], phenol derivatives [47], alkyl carboxylic acids [47], amines, anilines [47], metal-water complexes [42], aqueous solutions trivalent rare earth metal cations [48], and so on.

To analyze our numerical results, we employ three theoretical tools. The first one is the energy partition analysis in DFT [49]. In DFT, the total energy difference ΔE between two states or conformations of a molecular system can be conventionally partitioned in the following manner,

$$\Delta E = \Delta T_\text{S} + \Delta V_\text{ne} + \Delta J + \Delta E_\text{xc} + \Delta V_\text{nn} = \Delta T_\text{s} + \Delta E_\text{e} + \Delta E_\text{xc} \tag{10.4}$$

where T_s, V_ne, J, V_nn, and E_xc stand for noninteracting kinetic energy, nuclear–electron attraction, classical electron–electron Coulombic repulsion, nuclear–nuclear repulsion, and exchange–correlation energy density functionals, respectively. V_ne, V_nn, and J are of the electrostatic nature, so they can be put together, yielding $\Delta E_e = \Delta V_{ne} + \Delta J + \Delta V_{nn}$. Earlier, we also proposed another total energy partition by using three independent physical-chemical effects [50]:

$$\Delta E = \Delta E_s + \Delta E_e + \Delta E_q \tag{10.5}$$

where the total electronic energy E of a molecular system is assumed to come from the contribution of three independent physiochemical effects, steric E_s, electrostatic E_e, and fermionic quantum due to the exchange–correlation effect E_q. These schemes have been successfully applied to numerous chemical processes and transformations for a better understanding of the nature and origin of their stability and reactivity [51–62].

The second theoretical framework that we employ in the present work is CDFT [63–68], where chemical potential μ defined by

$$\mu = \left(\frac{\delta E}{\delta \rho(r)}\right)_v = \left(\frac{\delta E}{\delta N}\right)_v = -\chi \tag{10.6}$$

where E is the total energy of a system with N the total number of electrons and $v(\mathbf{r})$ as the external potential, and χ is electronegativity, which can be approximated by the energy of frontier molecular orbitals, highest occupied molecular orbitals (HOMO) and lowest occupied molecular orbitals (LUMO), $\chi = -\mu \approx -\frac{1}{2}(\varepsilon_{\text{HOMO}} + \varepsilon_{\text{LUMO}})$, chemical hardness η defined by

$$\eta = \left(\frac{\partial^2 E}{\partial N^2}\right)_v = \left(\frac{\partial \mu}{\partial N}\right)_v \approx \varepsilon_{\text{LUMO}} - \varepsilon_{\text{HOMO}} \tag{10.7}$$

and the electrophilicity index ω defined by [69]

$$\omega = \frac{\mu^2}{2\eta} \tag{10.8}$$

We will make use of these global descriptors to rationalize our numerical results for the propensity change of the aromaticity property.

The third theoretical framework employed in this work for the analytical purpose of our results is ITA [67, 70], which borrowed the ideas from information theory by employing simple density functionals for the analytical purpose of electronic properties. These ITA quantities include Shannon entropy S_S, which measures the spatial delocalization of the electronic density distribution $\rho(\boldsymbol{r})$,

$$S_S = -\int \rho(\boldsymbol{r}) \ln \rho(\boldsymbol{r}) d\boldsymbol{r} \tag{10.9}$$

Fisher Information I_F,

$$I_F = \int \frac{|\nabla \rho(\boldsymbol{r})|^2}{\rho(\boldsymbol{r})} d\boldsymbol{r} \tag{10.10}$$

and the alternative definition of Fisher information I'_F,

$$I'_F = -\int \nabla^2 \rho(\boldsymbol{r}) \ln \rho(\boldsymbol{r}) d\boldsymbol{r} \tag{10.11}$$

where $\nabla \rho(\boldsymbol{r})$ is the density gradient and $\nabla^2 \rho(\boldsymbol{r})$ is the Laplacian. Fisher information gauges the sharpness or concentration of the electron density distribution. Another important quantity in ITA is the relative Shannon entropy I_G, also called Kullback–Leibler divergence, relative entropy, or information divergence, whose definition in the atoms-in-molecules representation is as follows,

$$I_G = \sum_A \int \rho_A(\boldsymbol{r}) \ln \frac{\rho_A(\boldsymbol{r})}{\rho_A{}^0(\boldsymbol{r})} d\boldsymbol{r} \tag{10.12}$$

where $\rho_A(\boldsymbol{r})$ is the electron density of Atom A and $\rho_A{}^0(\boldsymbol{r})$ is the reference electron density for the same atom. It is an asymmetric measure of the entropy difference between two probability distribution functions and the reference density $\rho_A{}^0(\boldsymbol{r})$ must satisfy the same normalization condition as $\rho_A(\boldsymbol{r})$,

$$\sum_A \int \rho_A(\boldsymbol{r}) d\boldsymbol{r} = \sum_A \int \rho_A{}^0(\boldsymbol{r}) d\boldsymbol{r} = N \tag{10.13}$$

There are three new relative functions that we recently introduced [71],

$$G_1 = \sum_A \int \nabla^2 \rho_A(\boldsymbol{r}) \ln \frac{\rho_A(\boldsymbol{r})}{\rho_A{}^0(\boldsymbol{r})} d\boldsymbol{r} \tag{10.14}$$

$$G_2 = \sum_A \int \rho_A(\boldsymbol{r}) \left[\frac{\nabla^2 \rho_A(\boldsymbol{r})}{\rho_A(\boldsymbol{r})} - \frac{\nabla^2 \rho_A{}^0(\boldsymbol{r})}{\rho_A{}^0(\boldsymbol{r})} \right] d\boldsymbol{r} \tag{10.15}$$

and

$$G_3 = \sum_A \int \rho_A(\boldsymbol{r}) \left[\nabla \ln \frac{\rho_A(\boldsymbol{r})}{\rho_A{}^0(\boldsymbol{r})} \right]^2 d\boldsymbol{r} \tag{10.16}$$

Furthermore, we have considered Ghosh–Berkowitz–Parr entropy S_{GBP}, Rényi entropy R_n, and relative Rényi entropy ${}^rR_n$ of order n, whose definitions and meanings are available elsewhere [71–76].

10.3 Computational Details

The origin of the uniform external electric field is set to be at the center of a benzene ring. The coordinate system points the y-axis along the carboxyl group and the z-axis perpendicular to the benzene ring. The strength of the external electric field applied along the x and y axes is gradually increased from −0.025 to +0.025 a.u. with a step size of 0.0025 a.u., whereas along the z-axis, it is from −0.01 to +0.01 a.u. with a step size of 5×10^{-4} a.u. (1 a.u. = 5.1422×10^{11} V/m). Our studies found that the field applied perpendicular to the benzene ring is much more vulnerable than other directions to the stability of the molecule, which can significantly distort the molecular structure. This is the reason why we reduce the strength of the electric field along the z-axis direction to a smaller range.

To study the substituting effect of different groups with the different nature of electron donation and acceptance, we choose fluorine (–F) as the representative of electron-donating groups and cyano(–CN) as the representative of electron-accepting groups [76], and then systematically replace them at the two meta- and one para-positions of benzoic acid to generate singly, doubly, and triply substituted benzoic acid derivatives (Scheme 10.1).

All structures were fully optimized at the DFT M062X/aug-cc-pVTZ level of theory using the Gaussian 16 package, version C01 [77] with tight self-consistent-field (SCF) convergence and ultrafine integration grids. All afterward calculations were performed at the same theory level as well. Full geometry optimization with no symmetry constraint imposed was conducted for all models followed by a single-point frequency calculation to make sure that there is no imaginary frequency for each optimized structure. The energy components were calculated by adding the keyword of *iop(5/33 = 1)*. A single-point calculation was ensued to calculate the MEP value on nuclei using the keyword of *pop = chelpg* followed by an natural bond orbital (NBO) analysis to obtain natural population analysis (NPA) with the keyword of *pop = nbo* [78]. NICS(0) values were computed to

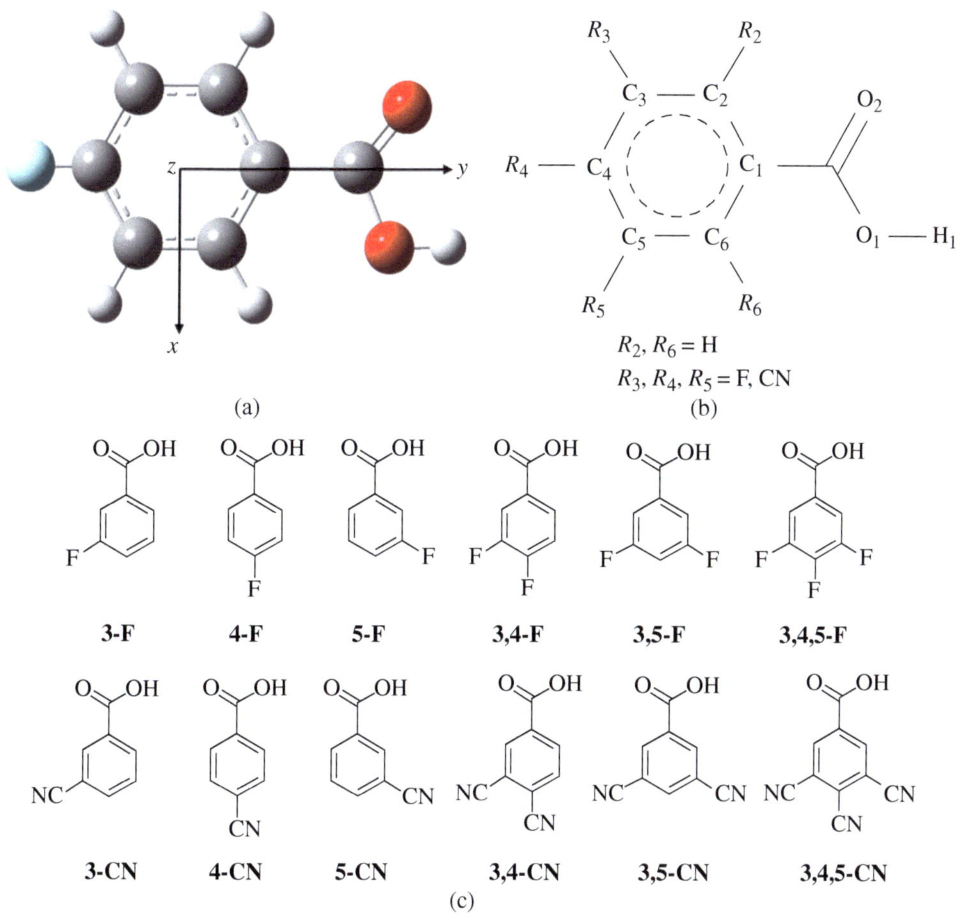

Scheme 10.1 The substituted benzoic acid derivatives and the coordination system for this study. (a) The origin is set to be at the center of the benzene ring, with the *y*-axis pointing along the carboxyl group and the *x*-axis parallel to the ring plane; (b) substituted benzoic acid derivative structures; and (c) the singly, doubly, and triply substituted benzoic acids studied in this work.

evaluate the global aromaticity by placing a dummy atom at the benzene ring center, whose coordination was determined by averaging the coordinates of all six carbon atoms on the benzene ring. To confirm and compare our NICS results, we used the GIMIC program [38] to visualize the ring current for the benzoic acid molecule with and without an external electric field applied. All ITA quantities, plus HOMA, PDI, and FLU values, were calculated by Multiwfn package, version 3.8 [79], with the checkpoint file generated by Gaussian calculations as the input file. The Hirshfeld partition scheme [80] was employed to calculate atomic contributions of all ITA quantities.

10.4 Results and Discussion

Figure 10.1 displays NICS(0) profiles as a function of the applied external electric field strength along three directions. All NICS(0) values are negative, indicating that the benzoic acid molecule is aromatic throughout. However, as the field strength increases along the x and y axes, so does

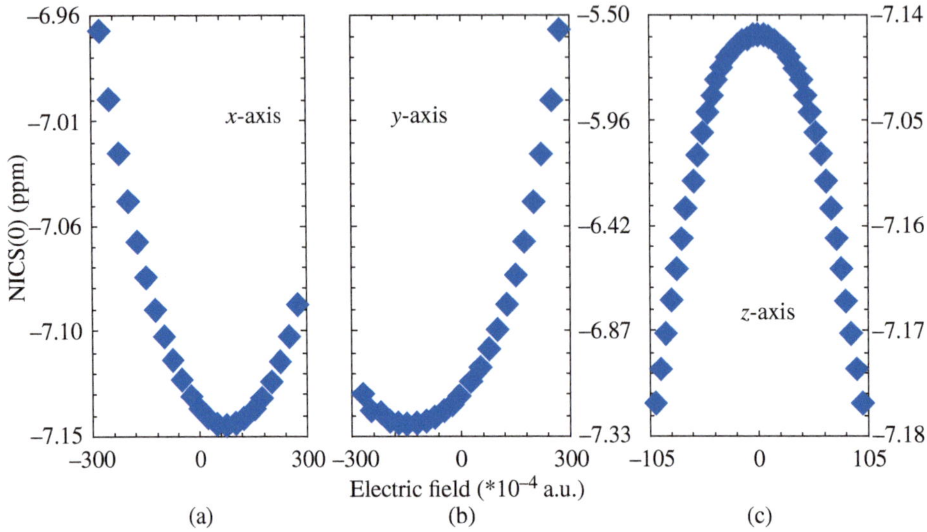

Figure 10.1 Profiles of aromaticity index NICS(0) values as a function of the applied external electric field along (a) *x*-axis; (b) *y*-axis; and (c) *z*-axis directions.

the NICS(0) value, suggesting that benzoic acid becomes less aromatic. On the *z*-axis, however, the trend is the opposite, even though the change of magnitude is much smaller than in other directions. In a nutshell, different directions of the field are seen to have different impacts on aromaticity. The *y*-axis produces the largest aromaticity changes, whereas the *z*-axis produces the smallest. This trend can readily be understood because the carboxyl group is attached in the *y*-axis direction, and the *z*-axis direction is least impactive due to the conjugation effect of the benzene ring. The asymmetrical behavior along the *x* and *y* directions is, again, due to molecular symmetry. As supporting evidence, we examined HOMA, FLU, and PDI aromaticity descriptors for the same system under same conditions. The same changing patterns as NICS(0) were confirmed. With these, we can clearly see that (i) in general, when an external field is applied along *x* and *y* axes, it reduces the aromaticity propensity of the benzoic acid molecule, but the trend in *z*-axis is opposite; (ii) the *y*-axis direction (i.e. along with the carboxyl acid group) is most sensitive to the external field, resulting in the largest change in aromaticity index values; (iii) the *z*-axis direction is least sensitive due to the aforementioned reasons; and (iv) molecular symmetry can be employed to explain the asymmetrical trend in *x* and *y* directions. To further verify these points, we plot GIMIC diagrams to visualize the ring current for the benzoic acid molecule with and without an external electric field applied, as shown in Figure 10.2. Notice that the current is running over the whole molecule, as shown from the Figure 10.2. The anticlockwise direction of the blue arrows indicates that they are aromatic, but the density of the ring current is discernibly less dense in Figure 10.2b when an external electric field is applied, confirming once again that it becomes less aromatic indeed.

To examine whether or not NICS(0) variation trends can ever be correlated by ITA quantities, shown in Table 10.1 are correlation coefficients (R) of NICS(0) data in three directions of the applied external field with eight ITA quantities, including Shannon entropy S_S, Fisher Information I_F, information gain I_G, our newly introduced G_2 and G_3 functions, second-order Rényi entropy R_2, and second- and third-order relative Rényi entropies ${}^rR_2$ and ${}^rR_3$. As can be seen from the

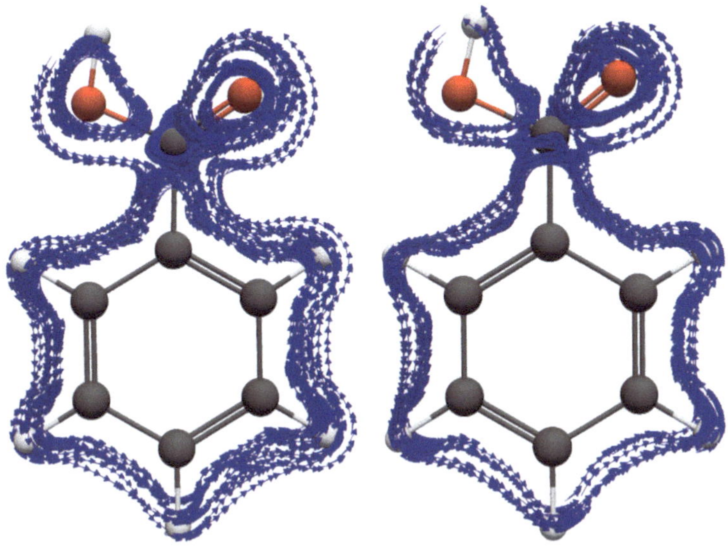

(a) (b)

Figure 10.2 The top view of the gauge-including magnetically induced current (GIMIC) represented by arrows to visualize the global electron delocalization for benzoic acid with (a) no electric field and (b) 0.025 a.u. electric field along +*y* direction. The same settings were applied to both diagrams when they were plotted (0.2 initial step length, 0.01 minimum step length and 0.5 maximum step length).

Table 10.1 The correlation coefficient (*R*) values of NICS(0) with eight ITA quantities.

ITA quantity	*x*-axis	*y*-axis	*z*-axis
S_S	0.436	0.330	−0.999
I_G	0.969	0.978	−0.998
I_F	−0.590	0.989	0.982
G_2	−0.710	−0.913	0.929
G_3	0.450	−0.984	−0.766
R_2	−0.965	−0.834	0.975
${}^rR_2$	0.938	0.969	−0.997
${}^rR_3$	0.922	0.983	−0.996

Table 10.1, for some ITA quantities such as S_S and I_F, strong correlations ($R > 0.9$) along one or two directions are observed, but not in all three directions. For I_G, ${}^rR_2$, and ${}^rR_3$, however, their correlations are strong in all three directions, suggesting that these ITA quantities are robust descriptors of NICS(0) changes in external electric fields for this system.

To find out the root cause of above changes in aromaticity in external fields, Figure 10.3 illustrates the change profile of chemical hardness and electrophilicity index from CDFT for benzoic acid as a function of the external electric field strength. As we know, hardness is a measure of

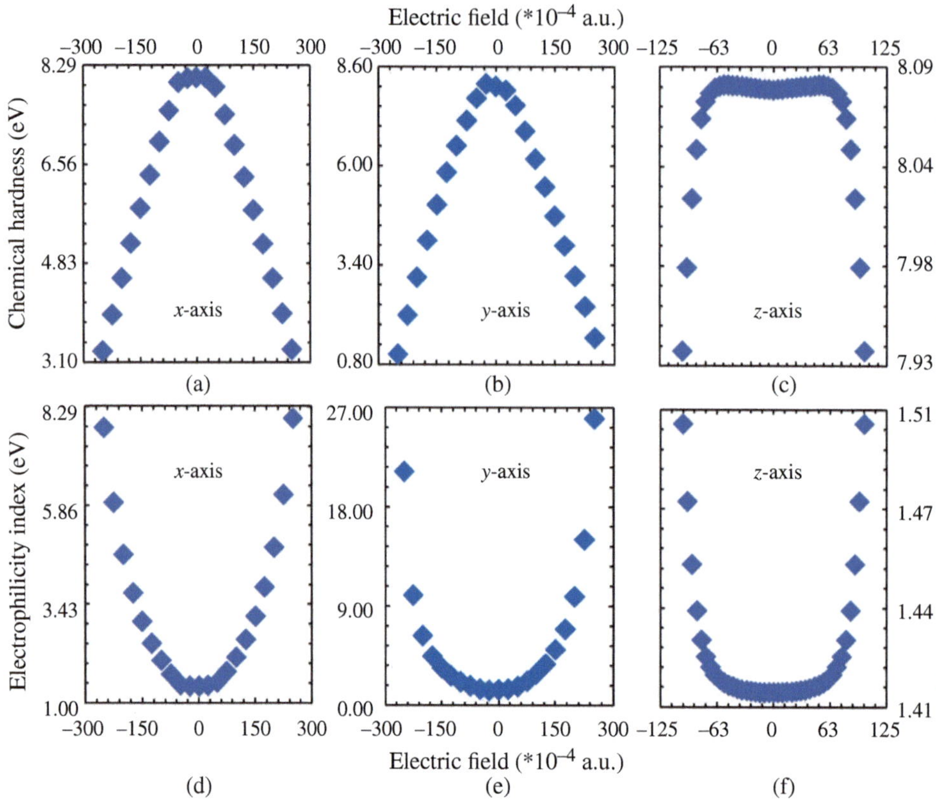

Figure 10.3 Changes in chemical hardness and electrophilicity index values of benzoic acid as a function of the electric field strength. Shown here are changes in all three directions. Chemical hardness: (a) *x*-axis; (b) *y*-axis; (c) *z*-axis; and electrophilicity index: (d) *x*-axis; (e) *y*-axis; (f) *z*-axis.

molecular stability. The larger the hardness, the more stable the molecule. On the other hand, electrophilicity index is a measure of reactivity. The larger the index, the more reactive the molecule. From the Figure 10.3, we can see that as the strength of the applied field increases, hardness becomes smaller, but the electrophilicity index increases. This unambiguous trend elucidates that the molecule becomes less stable and more reactive when an electric field is applied, so its aromaticity propensity becomes weaker, consistent with what we unveiled along the *x* and *y* axes in Figure 10.3.

Figure 10.4 shows the pK_a values calculated with MEP and NAO descriptors as a function of the applied external electric field. As can be seen from the Figure 10.4, an external electric field greatly affects the acidity of the benzoic acid molecule in both *x*-axis and *y*-axis directions, with the predicted pK_a value of benzoic acid varying from 1.30 to 5.70 in the *x*-axis direction and from 1.00 to 6.40 in the *y*-axis direction, where the carboxyl acid group is attached. In both directions, a strong linear relationship between the pK_a value and field strength is obtained. In the *z*-axis direction, which is perpendicular to the benzene ring, the pK_a value is predicted to be only slightly changed, from 3.40 to 3.59. It also shows a quadratic dependence on the electric field, as shown in Figure 10.3c, which is likely a result of the peculiarity of the density distribution in the *z*-axis direction. This result is consistent with what we obtained for aromaticity, indicating that the molecule is least sensitive yet most vulnerable to the external field in the *z*-axis direction.

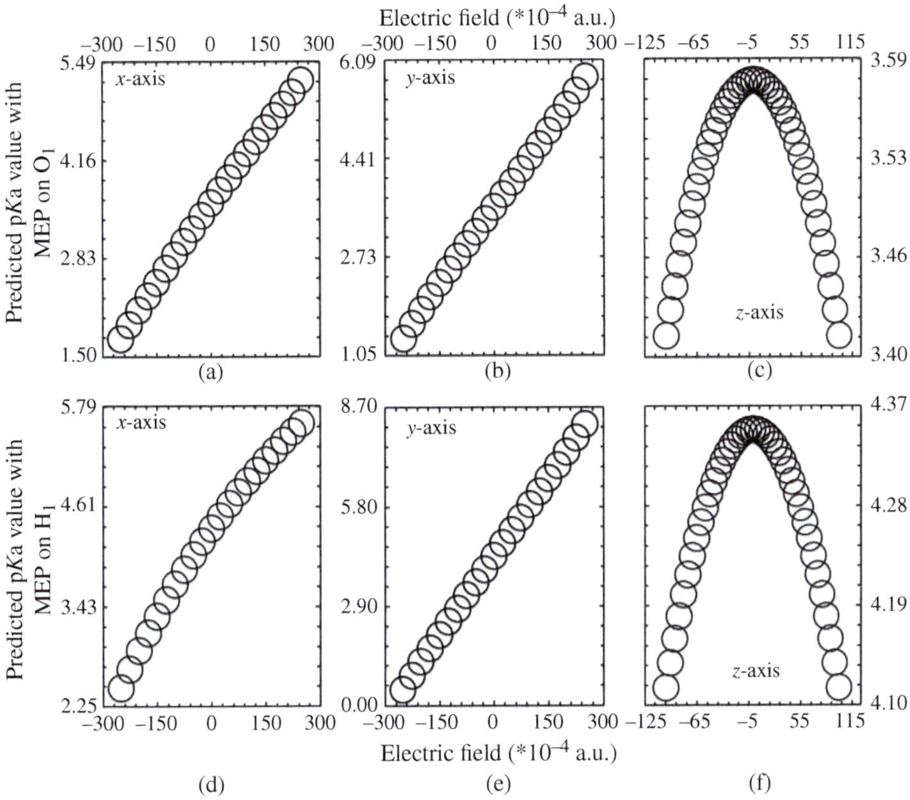

Figure 10.4 Predicted pK_a values as a function of the external electric field strength applied in three different directions. (a–c) are predicted pK_a values using molecular electrostatic potential (MEP) results from O_1 in Scheme 10.1, and (d–f) are valued using MEP data from H_1.

To justify the above acidity results, plotted in Figure 10.5 are strong linear correlations between MEP values on O_1 and H_1 (Scheme 10.1) and their partial changes from the NPA analysis as a function of the external electric field strength. As the field strength increases, so does the partial charges on O and H atoms, leading to the larger polarity of the O—H bond and thus, the increased molecular acidity. The only exception is the *z*-axis, where a quadric relationship is observed. In this case, MEP values and partial charges are still strongly correlated, as shown in Figure 10.5c,f. We further correlated our predicted pK_a values with ITA quantities, as shown in Table 10.2. Among the eight ITA quantities, I_F, I_G, G_2, ${}^rR_2$, and ${}^rR_3$ are found to be strongly correlated with pK_a values in all three directions. Compared with the correlation coefficients in Table 10.1, it becomes apparent that information gain I_G, the second-order relative Rényi entropy ${}^rR_2$ and the third-order of relative Rényi entropy ${}^rR_3$ are effective descriptors of both aromaticity and acidity propensities under external electric fields for the benzoic acid molecule.

To study the effect of group substitution on the aromatic ring of benzoic acid, we considered single, double, and triple substitutions with both fluorine and cyano groups. Figures 10.6 and 10.7 are profiles of the calculated pK_a values through the MEP descriptor on the O_1 atom, MEP_O, in Scheme 10.1 for singly, doubly, and triply fluorine and cyano substituted benzoic acids, respectively, as a function of the strength of the applied external electric field along (a) *x-axis* and (b) *y-axis*. From the Figure 10.6, the following three points are in order. First, it can be clearly seen that with the

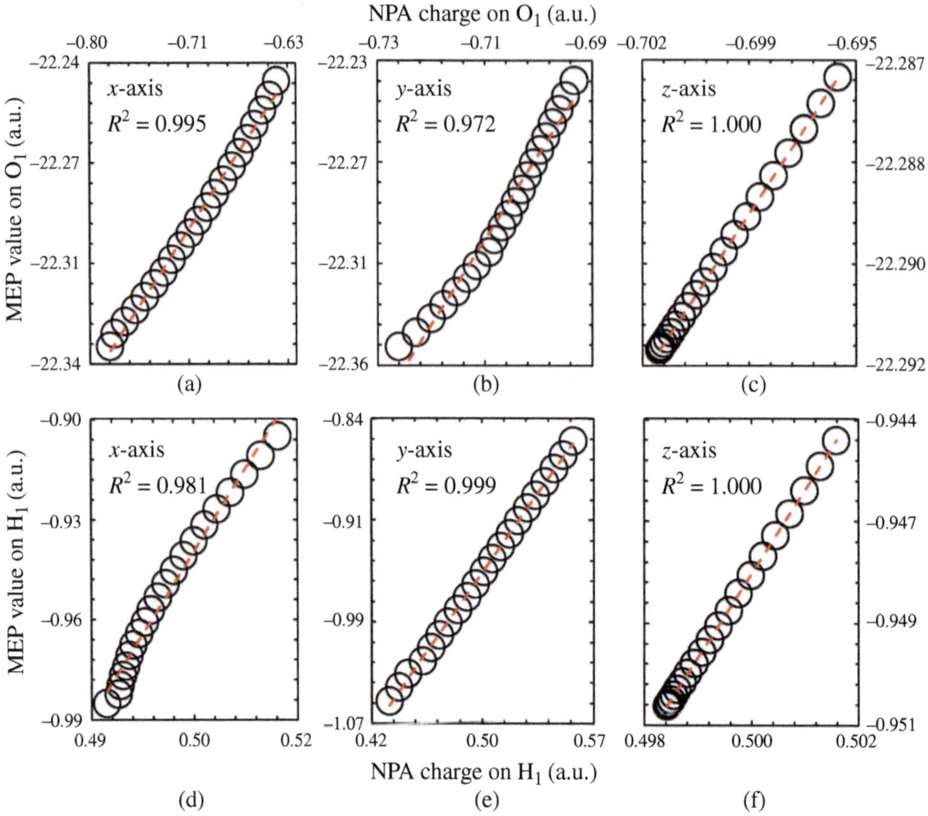

Figure 10.5 Strong linear correlations of MEP values with NPA partial charges on both O_1 and H_1 when the external electric field is applied to any of the three directions. (a–c) are for O_1 in Scheme 10.1, and (d–f) are for H_1 in Scheme 10.1.

Table 10.2 The correlation coefficient (R) of predicted pK_a values using MEP results from H_1 in Scheme 10.1 with eight ITA quantities.

ITA quantity	*x*-axis	*y*-axis	*z*-axis
I_F	0.950	0.930	0.988
I_G	−0.932	0.956	−0.999
G_1	−0.951	0.631	−0.959
G_2	0.980	−0.974	0.934
G_3	−0.885	−0.788	−0.765
R_2	0.937	−0.472	0.978
${}^rR_2$	−0.964	0.967	−0.998
${}^rR_3$	−0.975	0.934	−0.996

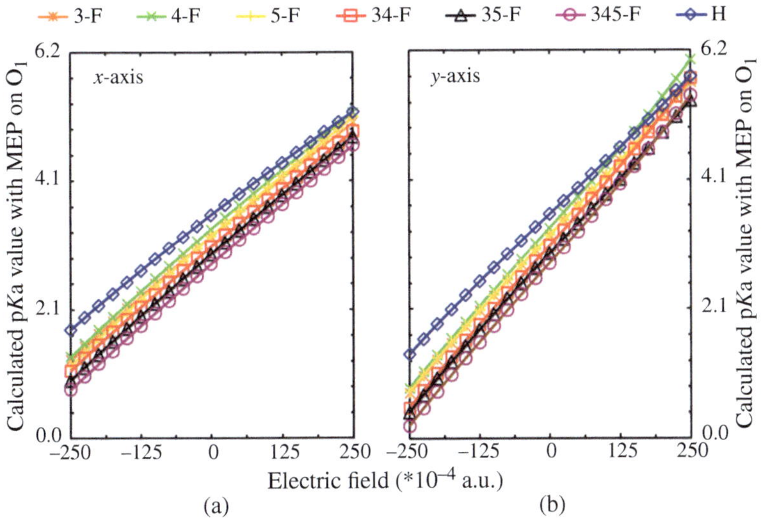

Figure 10.6 Profiles of calculated pK_a values using MEP from O_1 in Scheme 10.1 for singly, doubly, and triply fluorine-substituted benzoic acid derivatives as a function of the strength of the applied external electric field along (a) *x*-axis and (b) *y*-axis.

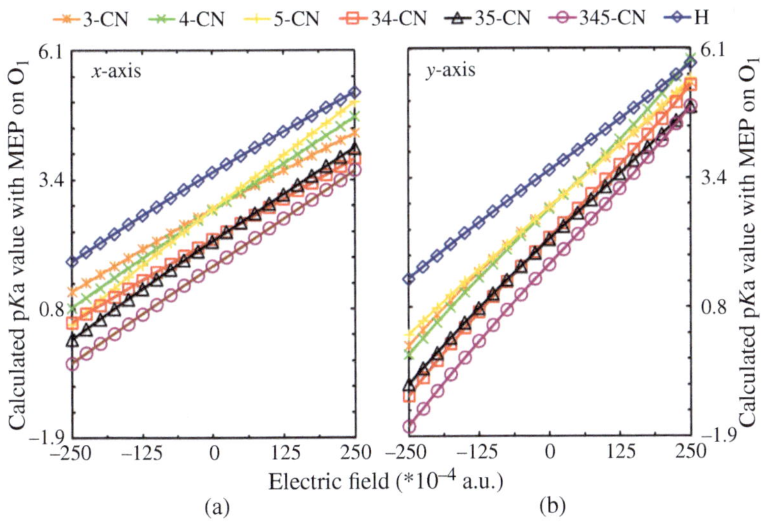

Figure 10.7 Profiles of calculated pK_a values using MEP results from O_1 in Scheme 10.1 for singly, doubly, and triply cyano-substituted benzoic acid derivatives as a function of the strength of the applied external electric field along (a) *x*-axis and (b) *y*-axis.

increase of the electric field strength from 250 to −250 (in the unit of 0.0001 a.u.), the acidity of these species monotonically increases along both *x*- and *y*-directions. In both cases, a strong linear relationship between the pK_a value and field strength is obtained. This linear and monotonical trend is the same as that of the benzoic acid molecule without substitution. Increasing the strength in the positive direction of the applied field leads to the weakened acidity of the fluorine-substituted benzoic acid, but when applied in the opposite direction, the acidity is enhanced. Second, the impact of the external field along the *y*-axis, which is parallel to the carboxyl group, is stronger than that along

Table 10.3 Statistical analysis results of the fitted pK_a values using the additivity effect for doubly and triply cyano-substituted benzoic acid derivatives under external electric filed along x- and y-axes, including correlation coefficient R^2, mean squared error (MSE), root mean square error (RMSE), and mean absolute error (MAE) between the calculated and fitted pK_a values.

	External field along the x-axis			External field along the y-axis		
	Fitted 3,4-CN	Fitted 3,5-CN	Fitted 3,4,5-CN	Fitted 3,4-CN	Fitted 3,5-CN	Fitted 3,4,5-CN
R^2	1.000	1.000	1.000	1.000	1.000	1.000
MSE	0.368	0.421	1.322	0.416	0.476	1.465
RMSE	0.606	0.649	1.150	0.645	0.690	1.211
MAE	0.602	0.649	1.149	0.618	0.681	1.175

the x-axis, yielding larger changes in pK_a values. The range of changes in the y-axis is 6.2 pK_a unit but in the x-axis, it is only 4.7 for the fluorine case. This directionality effect was also observed in our previous study for benzoic acid without substitution. Again, this difference originates in the different changes of the partial charge on the acidic atom. Third, as the number of fluorine substituents on the benzene ring increases, so does the acidity of the substituted benzoic acid derivative, regardless of the direction of the applied external field. As can be seen from the Figures 10.6 and 10.7, for both directions, doubly substituted derivatives are more acidic than the singly substituted ones, and the triply substituted species is the most acidic. This is due to the additivity effect. While the features of linearity and monotonicity are unique to acidity, directionality, and additivity are common to both acidity and aromaticity.

Using the cyano group as an illustrative example, Table 10.3 shows the result from the additivity effect of pK_a values. We display the statistical results in the table for x- and y-directions. As can be found from the statistical analysis results between calculated and fitted results in the table, using the sum rule from individual substituting groups, we can obtain reliable predictions of calculated pK_a results in both directions for multiply substituted species. As quality control measures, we examined this additivity effect for other acidity groups using the MEP descriptor from the dissociating proton and the same conclusions were obtained. Put together, and we are certain that this sum rule is applicable for any substituting groups, whether it is electron-donating or electron-withdrawing in nature and no matter which direction an external electric field is applied.

The profiles of NICS(0) values for cyano-substituted benzoic acid and fluorine, respectively, as a function of the external electric field's strength along the x- and y-axes (a and b) are displayed in Figures 10.8 and 10.9. We find that the more benzoic acid is substituted with these groups, the more aromatic they become. This is true for both directions in Figures 10.8 and 10.9. The impact of an electric field can be positive or negative, and it can also be linear or quadratic. The feature of linearity and monotonicity observed for acidity is no longer observed for aromaticity. On the other hand, we do find substantial differences between different axes, suggesting that when the field is applied to different directions, its impact is also different and thus confirms that the directionality characteristic from the impact of external electric fields is still in place. We also examined the additivity effect for aromaticity. It is also in place. For the sake of saving space, no details of these results are given here.

We also examined other aromaticity indexes such as HOMA, FLU, and PDI, and the same additivity effect is confirmed. A comparison of the fitted and computed indexes is presented in

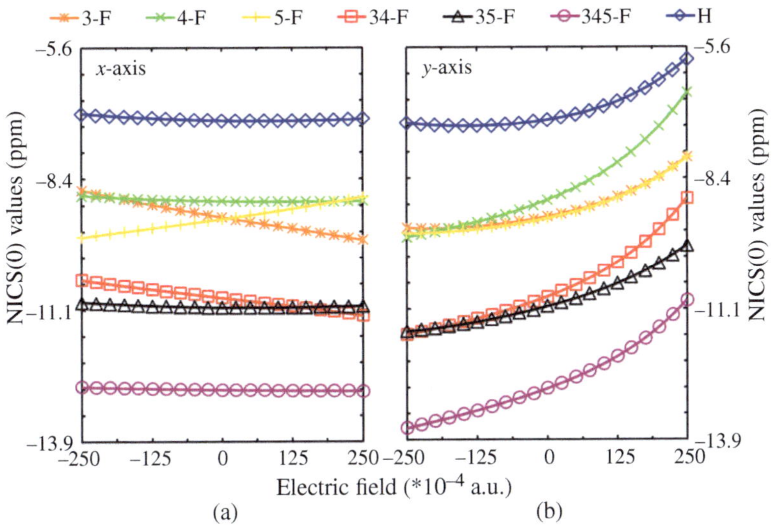

Figure 10.8 Profiles of the calculated aromaticity index, NICS(0), for singly, doubly, and triply fluorine-substituted benzoic acid derivatives as a function of the strength of the applied external electric field along the (a) *x*-axis and (b) *y*-axis.

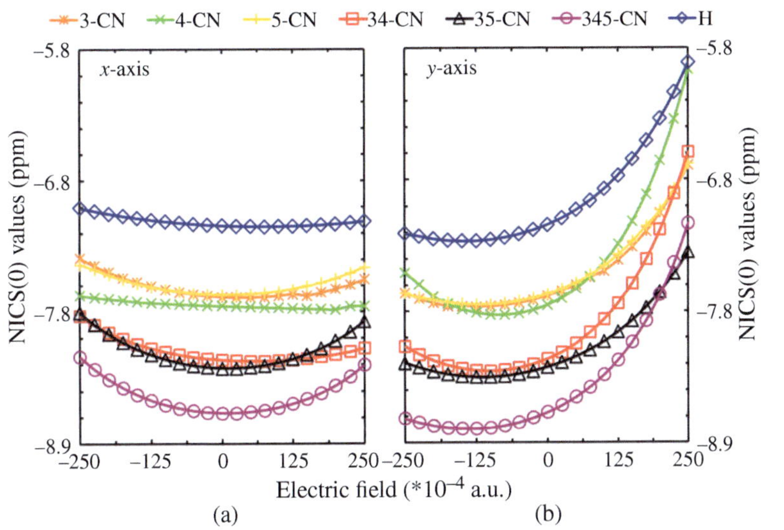

Figure 10.9 Profiles of the calculated aromaticity index, NICS(0), for singly, doubly, and triply cyano-substituted benzoic acid derivatives as a function of the strength of the applied external electric field along (a) *x*-axis; and (b) *y*-axis.

Figure 10.10. We regularly found robust linear connections between the computed and fitted values. Specifically, we found that the sum rule can reliably reproduce the calculated results for FLU and PDI indexes with correlation coefficients R^2 equal to 1.000. In the HOMA case, the correlation is a little worse, with $R^2 = 0.911$. These results suggest that the sum rule from individual groups is also applicable to other aromaticity indexes, thus confirming that the additivity effect of aromaticity to account for the impact from substituting groups should be a general one, independent of the indexes employed to quantify the property.

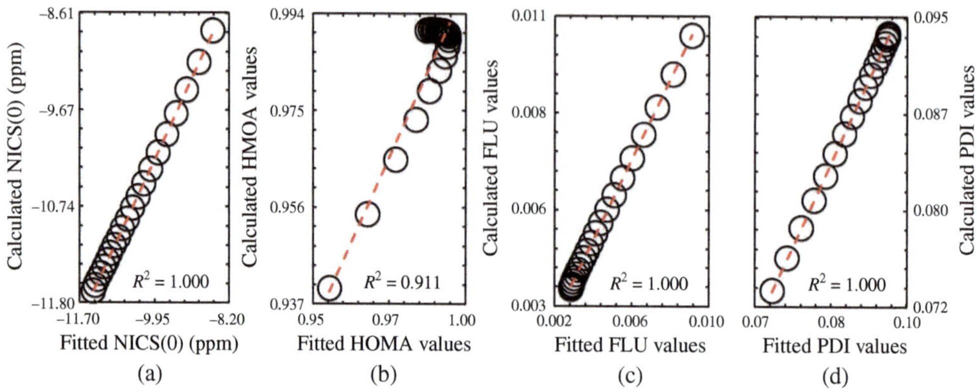

Figure 10.10 Strong linear correlations of the calculated value with the fitted value using the sum rule of four aromaticity indexes, including (a) NICS(0); (b) harmonic oscillator model of aromaticity (HOMA); (c) fluctuation index (FLU); and (d) parametric delocalization index (PDI) for the doubly substituted 3,4-F system in Scheme 10.1 with the external electric filed applied along the *y*-axis direction.

Table 10.4 The correlation coefficient (*R*) of NICS(0) and calculated pK_a values of F-substituted benzoic acids using MEP descriptor from O_1 in Scheme 10.1 with four information-theoretic quantities.

pK_{aO}	3-F	4-F	5-F	3,4-F	3,5-F	3,4,5-F
I_G	0.916	0.631	0.923	−0.258	0.556	−0.821
${}^rR_2$	0.899	0.692	0.907	0.337	0.718	−0.338
${}^rR_3$	0.853	−0.822	0.855	−0.778	−0.182	−0.944
I_F	0.876	0.804	0.884	0.818	0.816	0.841
R_3	0.906	0.972	0.941	0.985	0.962	0.992
G_1	−0.957	−0.975	−0.945	−0.990	−0.986	−0.994
G_2	0.941	0.993	0.940	0.998	0.992	0.998
NICS(0)	3-F	4-F	5-F	3,4-F	3,5-F	3,4,5-F
I_F	0.991	0.928	0.987	0.930	0.930	0.931
R_3	0.979	0.994	0.986	0.995	0.995	0.996
G_1	−0.993	−0.996	−0.996	−0.992	−0.996	−0.993
G_2	0.896	0.975	0.928	0.972	0.977	0.983

The applied external electric field is along the direction of the carboxyl group. The grey shared are the TWO descriptors (pK_a and NICS) in relation to MEP values.

Finally, we correlated acidity and aromaticity change propensities with information-theoretic quantities. Table 10.4 shows the correlation coefficient *R* of a few ITA quantities with pK_a and NICS(0) values for singly, doubly, and triply substituted benzoic acids with fluorine as the substituting group. Even though the above three ITA quantities still yield statistically significant correlations for some systems, as shown by the pK_a results in the Table 10.4 (also true for NICS(0), not shown), they are not always behaving the same way. We found that it is Fisher information I_F, third-order Rényi entropy R_3, and two new relative functions associated with Fisher information, G_1 and G_2, that provide strong correlations across all derivatives. This behavior difference from the substituting groups indicates that the impact of substituting groups under an external electric field will substantially alter the electron density distribution and, thus, the numerical values of ITA quantities. On the other hand, strong correlations of acidity and aromaticity properties with I_F, R_3, G_1, and G_2 for all derivatives suggest that these ITA quantities can be employed as reliable and robust descriptors.

10.5 Conclusions

In summary, in this work, using benzoic acid as an illustrative example, we systematically investigated the impact of the external electric field on aromaticity and acidity propensities through altering the direction and strength of an applied external electric field. We analyzed our results using DFT, CDFT, and ITA. Our results show that no matter which direction an external field is applied, it impacts both the aromaticity and acidity of the molecule. The direction with the carboxyl group attached is most sensitive to the external field, whereas the direction perpendicular to the benzene ring is the least sensitive yet most vulnerable. These change patterns can be satisfactorily rationalized by changes in reactivity descriptors associated with frontier orbitals and partial charges.

We also examined the influence of the substituting groups of benzoic acids. We systematically examined a total of 12 substituted benzoic acids with fluorine and cyano as the substituting group and gradually increased the strength of the applied external electric field from −250 to +250 with a step size of 25 (in the unit of 0.0001 a.u.) along the direction either parallel or perpendicular to the carboxyl group. Our results show that as the number of substituents increases, both acidity and aromaticity of the substituted benzoic acid are enhanced under the impact of an electric field, regardless of the direction of the applied electric field and the type of the substituent. We also found that, while there are unique features applicable to only one of these properties, directionality and additivity are common characteristics independent of the direction of the applied electric field and the type of the substituent as well.

The novel results from these studies should enrich the body of our knowledge about the impact of external fields for different kinds of electronic properties and henceforth provide the starting point for our future studies of this phenomenon in other molecular systems.

Acknowledgments

C.Y.R. acknowledges support from the Hunan Provincial Natural Science Foundation of China (Grant No. 2022JJ30373) and the Scientific Research Fund of the Hunan Provincial Education Department (Grant No. 22B0063). D.B.Z. is supported by the startup funding of Yunnan University and the Yunnan Fundamental Research Projects (Grant No. 202101AU070012) and the National Natural Science Foundation of China (Grant No. 22203071).

References

1 Geerlings, P. and De Proft, F. (2023). *J. Comput. Chem.* 44: 442–455.
2 Dutta, B.J. and Bhattacharyya, P.K. (2015). *Int. J. Quantum Chem.* 115: 1459–1466.
3 Mata, I., Molins, E., Alkorta, I., and Espinosa, E. (2009). *J. Chem. Phys.* 130: 044104.
4 Timerghazin, Q.K. and Talipov, M.R. (2013). *J. Phys. Chem. Lett.* 4: 1034–1038.

5 English, N.J. and Waldron, C.J. (2015). *Phys. Chem. Chem. Phys.* 17: 12407–12440.
6 Dittner, M. and Hartke, B. (2020). *J. Chem. Phys.* 152: 114106.
7 Gao, Q.F., Han, Y., Liang, P.Y., and Meng, J. (2020). *Phys. Chem. Chem. Phys.* 22: 6291–6299.
8 Zhang, R.F., Xia, W.X., Xu, X.S. et al. (2022). *J. Mol. Model.* 28: 185.
9 Ge, X.Y., Wu, X.D., Shi, Q.Y., and Wang, W.Z. (2022). *Nanomaterials* 12: 2067.
10 Linggawati, A., Arifin, K., and Yunus, R.M. (2022). *Int. J. Integr. Eng.* 14: 193–197.
11 Yuan, G., Cheng, Z.W., Cheng, Y.H. et al. (2022). *ACS Appl. Electron. Mater.* 4: 2897–2905.
12 Zhou, E., Wei, D.H., Wu, J. et al. (2022). *Phys. Chem. Chem. Phys.* 24: 17479–17484.
13 Li, Y.S., Liu, S.C., Liu, Q.D. et al. (2022). *Acta Phys. Sin.* 71: 026801.
14 Huo, J.R., Wei, H.C., Fu, L. et al. (2022). *Mater. Today Commun.* 31: 103544.
15 Sharma, N.K., Sahoo, S., Sahu, M.C. et al. (2022). *Surf. Interfaces* 29: 101817.
16 Dominikowska, J. and Palusiak, M. (2018). *ChemPhysChem* 19: 590–595.
17 Shaik, S., Mandal, D., and Ramanan, R. (2016). *Nat. Chem.* 8: 1091–1098.
18 Chen, L.J., Huan, Q.S., Tang, S.Y. et al. (2022). *J. Phys. Org. Chem.* 35: e4351.
19 Wu, X.W., Xu, J.H., Li, Y.Q. et al. (2022). *Phys. Chem. Chem. Phys.* 24: 12488–12500.
20 Wu, X.P. and Ren, D.S. (2022). *Molecules* 27: 1714.
21 An, H., Xiang, M., Abulimiti, B., and Zhen, J.Y. (2022). *Eur. Phys. J. D* 76: 86.
22 Dutta, B.J. and Bhattacharyya, P.K. (2014). *J. Phys. Chem. B* 118: 9573–9582.
23 Yu, S., Vermeeren, P., Hamlin, T.A., and Bickelhaupt, F.M. (2021). *Chemistry* 27: 5683–5693.
24 Shaik, S., Ramanan, R., Danovich, D., and Mandal, D. (2018). *Chem. Soc. Rev.* 47: 5125–5145.
25 Blyth, M.T., Noble, B.B., Russell, I.C., and Coote, M.L. (2020). *JACS* 142: 606–613.
26 Lu, T. and Chen, Q.X. (2021). *ChemPhysChem* 22: 386–395.
27 Stuyver, T., Danovich, D., Joy, J., and Shaik, S. (2020). *WIREs Comput. Mol. Sci.* 10: e1438.
28 Mandal, S. and Datta, A. (2022). *J. Phys. Org. Chem.* 7: e4327.
29 Silva, A.L.P. and Varela, J.D.G. (2022). *J. Nanopart. Res.* 24: 1.
30 Li, M., Wan, X.J., He, X. et al. (2023). *Phys. Chem. Chem. Phys.* 25: 2595–2605.
31 Yu, D.H., Rong, C.Y., Lu, T. et al. (2020). *Phys. Chem. Chem. Phys.* 22: 4715–4730.
32 Yu, D.H., Stuyver, T., Rong, C.Y. et al. (2019). *Phys. Chem. Chem. Phys.* 21: 18195–18210.
33 Yu, D.H., Rong, C.Y., Lu, T. et al. (2017). *Phys. Chem. Chem. Phys.* 19: 18635–18645.
34 Schleyer, P.V.R., Maerker, C., Dransfeld, A. et al. (1996). *JACS* 118: 6317–6318.
35 Kruszewski, J. and Krygowski, T. (1972). *Tetrahedron Lett.* 13: 3839–3842.
36 Güell, M., Matito, E., Luis, J.M. et al. (2006). *J. Phys. Chem. A* 110: 11569–11574.
37 Matito, E., Duran, M., and Sola, M. (2005). *J. Chem. Phys.* 122: 014109.
38 Sundholm, D., Fliegl, H., and Berger, R.J.F. (2016). *WIREs Comput. Mol. Sci.* 6: 639–678.
39 Jorgensen, W.L., Briggs, J.M., and Gao, J.L. (1987). *JACS* 109: 6857–6858.
40 Pliego, J.R. (2003). *Chem. Phys. Lett.* 367: 145–149.
41 Liu, S.B. and Pedersen, L.G. (2009). *J. Phys. Chem. A* 113: 3648–3655.
42 Liu, S.B., Schauer, C.K., and Pedersen, L.G. (2009). *J. Chem. Phys.* 131: 164107.
43 Burger, S.K., Liu, S.B., and Ayers, P.W. (2011). *J. Phys. Chem. A* 115: 1293–1304.
44 Huang, Y., Zhong, A.G., and Liu, S.B. (2011). *J. Nat. Sci. Hunan Normal Univ.* 34: 52.
45 Huang, Y., Liu, L.H., Liu, W.H. et al. (2011). *J. Phys. Chem. A* 115: 14697–14707.
46 Cao, X.F., Rong, C.Y., Zhong, A.G. et al. (2018). *J. Comput. Chem.* 39: 117–129.
47 Xiao, X.Z., Cao, X.F., Zhao, D.B. et al. (2020). *Acta Phys. Chim. Sin.* 36: 1906034.
48 Yu, D.H., Du, R.B., Xiao, J.C. et al. (2018). *J. Phys. Chem. A* 122: 700–707.
49 Parr, R.G. and Yang, W.T. (1989). *Density-Functional Theory of Atoms and Molecules*. New York: Oxford University Press.
50 Liu, S.B. (2007). *J. Chem. Phys.* 126: 204113.

51 Li, M., He, X., Wang, B. et al. (2020). *Front. Chem.* 8: 566.
52 Wang, B., Yu, D.H., Zhao, D.B. et al. (2019). *Chem. Phys. Lett.* 730: 451–459.
53 Li, M., He, X., Chen, J. et al. (2021). *J. Phys. Chem. A* 125: 1269–1278.
54 Wan, X.J., He, X., Li, M. et al. (2022). *J. Phys. Org. Chem.* 10: e4352.
55 He, X., Li, M., Yu, D.H. et al. (2021). *Theor. Chem. Acc.* 140: 123.
56 Zhao, D.B., He, X., Li, M. et al. (2021). *Phys. Chem. Chem. Phys.* 23: 24118–24124.
57 Zhao, D.B., Liu, S.Y., Rong, C.Y. et al. (2018). *ACS Omega* 3: 17986–17990.
58 Cao, X.F., Liu, S.Q., Rong, C.Y. et al. (2017). *Chem. Phys. Lett.* 687: 131–137.
59 Wu, Z.M., Rong, C.Y., Lu, T. et al. (2015). *Phys. Chem. Chem. Phys.* 17: 27052–27061.
60 Rong, C.Y., Zhao, D.B., Zhou, T.J. et al. (2019). *J. Phys. Chem. Lett.* 10: 1716–1721.
61 Zhao, D.B., Rong, C.Y., Jenkins, S. et al. (2013). *Acta Phys. Chim. Sin.* 29: 43–54.
62 Huang, Y., Zhong, A.G., Yang, Q.S., and Liu, S.B. (2011). *J. Chem. Phys.* 134: 084103.
63 Geerlings, P. and De Proft, F. (2008). *Phys. Chem. Chem. Phys.* 10: 3028–3042.
64 Liu, S.B. (2009). *Acta Phys. Chim. Sin.* 25: 590–600.
65 Geerlings, P., De Proft, F., and Langenaeker, W. (2003). *Chem. Rev.* 103: 1793–1873.
66 Geerlings, P., Chamorro, E., Chattaraj, P.K. et al. (2020). *Theor. Chem. Acc.* 139: 36.
67 Liu, S.B. (2022). *Conceptual Density Functional Theory: Towards a New Chemical Reactivity Theory*. Germany: Wiley-VCH.
68 Liu, S.B. (2016). *Acta Phys. Chim. Sin.* 32: 98–118.
69 Parr, R.G., Szentpály, L.V., and Liu, S.B. (1999). *JACS* 121: 1922–1924.
70 Rong, C.Y., Wang, B., Zhao, D.B., and Liu, S.B. (2020). *WIREs Comput. Mol. Sci.* 10: e1461.
71 Wang, B., Zhao, D.B., Lu, T. et al. (2021). *J. Phys. Chem. A* 125: 3802–3811.
72 Liu, S.B. (2019). *J. Chem. Phys.* 151: 141103.
73 Ghosh, S.K., Berkowitz, M., and Parr, R.G. (1984). *PNAS* 81: 8028–8031.
74 Liu, S.B., Rong, C.Y., Wu, Z.M., and Lu, T. (2015). *Acta Phys. Chim. Sin.* 31: 2057–2063.
75 Rong, C.Y., Lu, T., Chattaraj, P.K., and Liu, S.B. (2014). *Indian J. Chem. Sect. A* 53: 970–977.
76 Liu, S.B. (2014). *J. Chem. Phys.* 141: 194109.
77 Frisch, M.J., Trucks, G.W., Schlegel, H.B., et al. (2016). *Gaussian16, Revision C.01*. Gaussian, Inc.: Wallingford, CT.
78 Badenhoop, J.K. and Weinhold, F. (1997). *J. Chem. Phys.* 107: 5406–5421.
79 Lu, T. and Chen, F.W. (2012). *J. Comput. Chem.* 33: 580–592.
80 Hirshfeld, F.L. (1977). *Theor. Chem. Acc.* 44: 129–138.

11

A Divergence and Rotational Component in Chemical Potential During Reactions

Jean-Louis Vigneresse

GéoRessources, UMR CNRS 7539, Vandœuvre-lès-Nancy, 54706, France

11.1 Introduction

Hard-soft acid–base (HSAB) concepts [1] describe reactions between chemical elements, separating them into two categories, with acids and bases defined by their hardness or softness, its inverse. Quantification occurs through conceptual density functional theory (CDFT), providing numerical values of chemical descriptors for elements [2]. Those are based on energy considerations through the derivatives of the chemical potential with respect to their charge (N). Such energy concepts identified the forces acting on bonding elements, with the possibility of having covalent or electrostatic bonding; polarizability linked forces or weak noncovalent forces (dipole–dipole, π-interactions, van der Waals forces…). They gave place to reaction types, namely two of them, respectively frontier molecular orbital (FO) or charge-controlled (CC) reactions. The first category mainly considers soft-soft affinities, whereas hard-hard elements bonding would be CC [3]. Those concepts helped in understanding chemical reactions and the possible formation of new molecules. They are presently the basis of modern chemistry.

The HSAB and CDFT concepts have proven their efficiency in understanding chemical reactions by identifying the reactants and the probable reaction products. Such performance relies on the global properties linked with the chemical descriptors and the associated chemical principles that rule reactions. Local descriptors also allow detailed reactivity of chemical sites within chemical molecules and regioselectivity [2]. They are widely used to build new molecules with specific chemical properties. Such techniques represent about 90% of the actual chemical papers. However, understanding how reactions develop and their kinetics are presently less addressed. The different types of chemical elements (acids or bases) have been identified, as well as their relative strength (soft or hard). However, the energy involved during reactions remains partly unanswered, although recent advances, using the activation strain model or distortion/interaction model [4] have brought new insights into reaction mechanisms. This is partly due to the lack of examining simultaneously the chemical descriptors and the principles, joined with the difficulty in addressing complex molecules.

Transferring HSAB and CDFT concepts to geosciences problems [5] was a key point for the development of new techniques and ideas. Hence, methods derived from potential fields (electrostatic, gravity, and magnetics) make the data treatment of chemical potential more familiar. The frequent use of mapping a parameter as a function of coordinates, which are not necessarily spatial

Electron Density: Concepts, Computation and DFT Applications, First Edition.
Edited by Pratim Kumar Chattaraj and Debdutta Chakraborty.

coordinates, also provided new insights for examining chemical reactions. In particular, the charge and energy exchange during chemical reactions, pertain to similar concepts as examining the same changes during potential variations [5].

The present chapter is organized as follows. After a short introduction to the used chemical descriptors and the principles that rule chemical reactions, the charge and energy exchanges are quantified along theoretical reaction paths. In particular, the variations of the descriptors according to the charge changes are estimated to determine what exactly controls the reaction types.

11.2 Chemical Descriptors

Chemical descriptors are numerical values, often derived from the energy of the electronic cloud, the external potential of which is $v(r)$, considered as a system of N electrons. The first derivative of the energy when the other variables are fixed is the electronegativity (χ) [1], or the chemical potential (μ), changed of sign [6].

$$\chi = -\mu = -\left(\frac{\partial E}{\partial N}\right)_{v(r)} \tag{11.1}$$

The second derivative is the hardness (η), in the same conditions [1].

$$\eta = \frac{1}{S} = \frac{1}{2}\left(\frac{\partial^2 E}{\partial N^2}\right)_{v(r)} \tag{11.2}$$

Softness (S) is the inverse of hardness. The above descriptors constitute the basis of the HSAB concepts [7]. Electronegativity determines the power to attract an electron, whereas hardness indicates the resistance to a charge transfer.

A third descriptor, electrophilicity (ω), has been coined from the two preceding descriptors [8].

$$\omega = \frac{\mu^2}{2\eta} = \frac{\chi^2}{2\eta} \tag{11.3}$$

Electrophilicity measures the energy change when an element is ideally put in contact with a perfect electron donor (i.e. an electron reservoir with null hardness and null electronegativity). It can also be considered as a derivative of the chemical potential (see below).

At last, polarizability (α) links with the deformation of an element in the presence of an external electrostatic field [9]. It is an even-order derivative of the energy with respect to an applied electric field with components F_x, F_y, and F_z, common value α is the simple arithmetic averaging of the diagonal values $\alpha = (\alpha_{xx} + \alpha_{yy} + \alpha_{zz})/3$ in which

$$\sim \alpha_{a,b} = -\left(\frac{\partial^2 E}{\partial F_a \partial F_b}\right) \quad \text{with } a, b = x, y, z \tag{11.4}$$

The average value α never becomes zero. Its value, in atomic unit (a.u.), increases with decreasing hardness (in eV). This occurs when the element has a large charge and a small size for cations, and, conversely, a large charge and large size for anions. Hence, negative ions have excess electrons and a large ionic radius, making them highly polarizable. Polarizability links with the other descriptors [1, 10]. It also enhances the formation of covalent bonds since it represents the ability of a cation to distort an anion following Fajan's rules [10].

11.3 Charge and Energy Exchange

The amount of charge (dN) exchanged during chemical reactions can be estimated from the above descriptors [11], while the maximum charge exchange during reactions is limited by $N = \chi/\eta$. Considering the chemical descriptors (χ_A, η_A, ω_A) of a starting element A, yielding to element B with its own descriptors, the variations in descriptors write ($d\chi$, $d\eta$, $d\omega$). The effective charge exchange can be computed all along the reaction path. It provides the charge exchange (dN) and the energy (dE) exchange of the reaction [11].

$$dN = \frac{(\chi_A - \chi_B)}{(\eta_A + \eta_B)} = \frac{d\chi}{(2\eta_A + d\eta)} \tag{11.5}$$

and

$$dE = -\frac{(\chi_A - \chi_B)^2}{2(\eta_A + \eta_B)} = -\frac{d\chi^2}{2(2\eta_A + d\eta)} \tag{11.6}$$

Given a starting and a final element, with their own descriptors (χ, η, ω), it allows computing what would be the charge and energy exchange along the reaction. Conversely, given a starting reactant, the final product can be estimated after imposing some constraints on the energy of charge exchange. Because of the square value of electronegativity change, the energy exchange is always negative whether the electron flow develops from element A to B or from element B to A. The value of charge exchange, positive or negative, is required to infer the direction of electron transfer.

11.4 Fitness Landscape Diagrams

The above-defined descriptors are determined either experimentally [1] or through numerical computation using CDFT concepts [2]. Electronegativity and hardness can also be estimated from the finite half sum and half difference between the successive values of ionization potential (I) and electron affinity (A) of the considered elements. Their use for chemical reactions is nonpredictive and cannot inform on the reaction kinetics. Nevertheless, they indicate trends toward which reactions should evolve through maximum and minimum principles, all derived from energy considerations.

First, the chemical potential should be similar in all components at equilibrium, justifying the electronegativity equalization principle for compounds [12]. In particular, the principle justifies the use of geometrical averaging between descriptors for compounds or solvents [13].

Second, HSAB theory states that hard molecules arrange themselves so as to be as hard as possible [1], giving place to the maximum hardness principle (MHP).

Third, a minimum electrophilicity principle (MEP) has been coined [14], stating that reactions should evolve toward minimum electrophilicity. However, the principle has been questioned [14] and reaffirmed [15].

Finally, a minimum polarizability principle (MPP) states that molecules should form in the region of minimum polarizability, i.e. having the highest charge [8].

Considering the significance of the chemical descriptors, a similitude immediately develops between hardness (Eq. (11.2)) and electrophilicity (Eq. (11.3)). Hardness is the second derivative of energy, i.e. the first derivative of the chemical potential, corresponding to the tangential component of the acceleration. It indicates places of greater stability when it becomes large

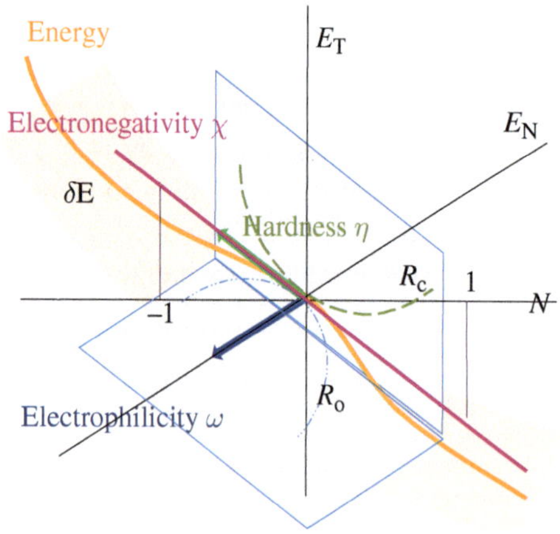

Figure 11.1 Geometric representation of the energy (E) surface as a function of the number (N) of electrons within the system. The chemical descriptors are the tangent (electronegativity, χ) as the first derivative and the components of the second derivative. Those are – the tangential component, analog to the radius of curvature (R_c) (hardness, η) and – the normal component (electrophilicity, ω), analog to the radius (R_o) of the osculating circle (normal to the plane of the drawing).

(i.e. lower potential value), fitting the MHP. The inverse of electrophilicity is the second derivative of the inverse of the energy.

$$\frac{1}{\omega} = \frac{2\eta}{\chi^2} = -2\left(\frac{\frac{\partial^2 E}{\partial N^2}}{\left(\frac{\partial E}{\partial N}\right)^2}\right) = -2\left(\frac{\partial\left(\frac{1}{\frac{\partial E}{\partial N}}\right)}{\partial N}\right) \tag{11.7}$$

It is analogous to the normal acceleration of a dynamic motion, also implying the MEP. While hardness is assimilated to the radius of curvature, electrophilicity can be assimilated to the osculating circle (Figure 11.1) [16]. It thus determines the lateral deviation of the energy curve (Figure 11.1).

The use of a diagram using the two descriptors as coordinates, or [ω–η] diagram, with similar units, could be interpreted as a mapping of the potential field, analog to a topographic map. The local maxima and minima would indicate zones of instability and stability. Both components of the acceleration reflect the steepest descent (or ascent) and possible slaloming toward potential wells or summits. By analogy with potential maps for evolution trends [17], those diagrams with coordinates [η–ω] are called fitness landscape diagrams. A [η–ω] diagram is preferred to a [χ–η] diagram [18] because it incorporates identical information on both coordinates, i.e. the components of the acceleration of energy.

11.5 Chemical Reactions

Chemical reactions between elements relate to the type of bonding, i.e. the forces in action controlling element associations. Three major processes control bonding, namely covalent, electrostatic and polarization energy [19]. Noncovalent bonds often imply lower energy and can be considered of lesser importance.

Electrostatic forces mainly rule frontier molecular orbital (FO) controlled reactions, linking with ionic bonds. Considering the Lewis theory of acids, the latter accepts pairs of electrons during reactions with no change in the oxidation numbers of any atoms. It links with the electronegativity concept and the polarizability of molecules, characterizing the partial covalent character of ionic bonding. Covalent bonding involves the sharing of electron pairs between atoms. For the soft–soft interactions, the covalent component of energy and the nature of frontier control are transparent.

Conversely, hard–hard interactions are CC, and the predominant component of energy is electrostatic [20]. Depending on the respective influence of each of those components, two reaction types have been identified [20]. In the case of full electrostatic control, or CC reactions, the reagents present a large charge, and the charge exchange is minimal ($dN \approx 0$). Conversely, during electron-transfer controlled reactions, or frontier-orbital controlled reactions (FO), the reactive sites have small charges, and the electron transfer between reagents is significant ($dN \approx -1$). A third case combines electrostatic and electron transfer control. A general-purpose reactivity indicator (κ) has been suggested, varying between 1 and −1, to quantify the balance between electrostatic and electron transfer [20].

FO-controlled reactions have the highest occupied molecular orbitals (HOMO) of the donor base atom and the lowest occupied molecular orbitals (LUMO) of the acceptor atom of the acid molecule. Such concepts of molecular orbitals link with electronegativity and hardness as the mean value and the difference between LUMO and HOMO energies. When the energy gap between HOMO and LUMO is large, no electron transfer occurs, and the CC interactions or ionic forces dominate in the resulting compounds. Conversely, when the energy gap is small, electron transfer occurs between the donor and the acceptor, and the resulting compound is covalent. In short, hard-hard interactions are CC, and soft–soft reactions are FO.

11.6 Examining the Charge Exchange

The problem states as the change from a starting element A with descriptors (χ_A, η_A, ω_A) to a final element B with (χ_B, η_B, ω_B). The charge exchange is defined along the path by the variations in χ and η, thus a path $p_{\chi\eta}(\zeta)$ between the two points within a [χ–η] diagram to follow the evolution of the charge exchange dN. This part has already been extensively explored in a preceding paper [21]. In a second round, a [ω–η] diagram involving the variations in ω and η along a path $p_{\omega\eta}(\zeta)$ would inform about the bulk direction of the path, through the acceleration components.

A note of caution is necessary since the differential descriptors are simply the change in the respective descriptors between elements A and B. In what follows, the coordinate along the reaction (ζ) is omitted for simplicity. All units for the descriptors are in eV.

11.6.1 Path $p_{\chi\eta}(\zeta)$ and Charge Exchange

The path of the reaction $p_{\chi\eta}(\zeta)$ between the two points in a [χ–η] diagram may adopt several conditions depending on the values of the differential descriptors ($d\chi$, $d\eta$, or $d\omega$). The problem is aborded from three different points of view. First, a trivial solution does not offer, as expected, a good insight. A second solution is presented through a mapping of the energy variation. It results from the computation using a given starting point and ending points on a grid. Finally, the equations (Eqs. (11.5) and (11.6)) are examined through an equation solver, providing variation curves when all parameters vary.

The simplest solution uses Eq. (11.5), stating that dN is a linear combination of $d\chi$. This trivial solution does not inform more on the path, nor does it provide insights into the variations in descriptors. Starting from the Eq. (11.5)

$$p_{\chi\eta}{}^2 = d\chi^2 + d\eta^2 \quad \text{with} \quad d\chi = \chi_A - \chi_B \text{ and } d\eta = \eta_A - \eta_B \tag{11.8}$$

$$p_{\chi\eta}{}^2 = A^2 + d\eta^2 \quad \text{with} \quad A = d\chi = (2\eta_A + d\eta)dN \tag{11.9}$$

and reorganizing, yielding a cubic equation with the two square terms positive.

$$p_{\chi\eta}{}^2 = d\eta^2(1 + dN^2) + 4dN^2\eta_A d\eta + 4dN^2\eta_A \quad (11.10)$$

It corresponds to the equation of a hyperbola in a $[\eta - p_{\chi\eta}]$ diagram since its determinant ($\delta = 4 + 4dN^2$) is always positive (Figure 11.2). The curves for various values of dN present two stationary points for $\eta_B = -2\eta_A$ with values $p_{\chi\eta} = \pm 2\eta_A$ (Figure 11.2), also corresponding to the value of $d\chi = 0$, i.e. $\omega_B = \pm\omega_A/2$. The curves show a progressive rotation of the asymptotes toward the $p_{\chi\eta}$ axis as the charge exchange dN increases. However, the various curves are not very informative on the way the path, and among them, the relative $d\chi$ or $d\eta$, vary as a function of dN.

Nevertheless, the length of the path varies with dN, first decreasing and then increasing after a minimum. The position of this minimum can be computed from the first derivative of Eq. (11.10) and writes

$$d\eta_{min} = -8\eta_A dN^2/(1 + 4dN^2) \quad (11.11)$$

It varies with dN^2, the charge exchange. It also depends on the starting point through the coordinate η_A. The minimum is obviously for $dN = 0$ and then increases asymptotically toward $-\eta_A$ (Figure 11.3).

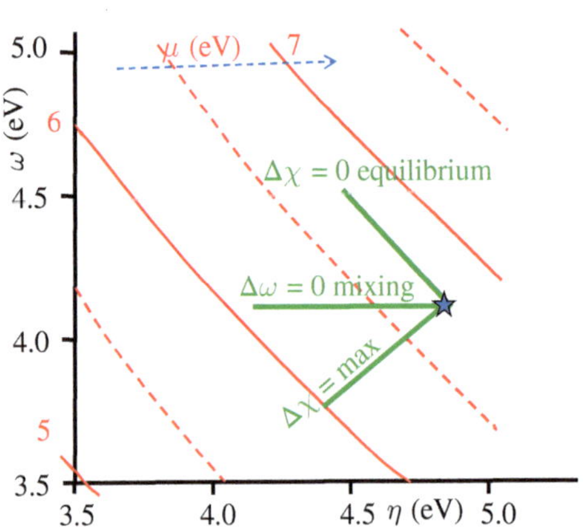

Figure 11.2 Cartoon showing the variations in $\Delta\chi$ (=0 or equilibrium), to $\Delta\chi$ max (steepest descent), and $\Delta\omega$ (mixing).

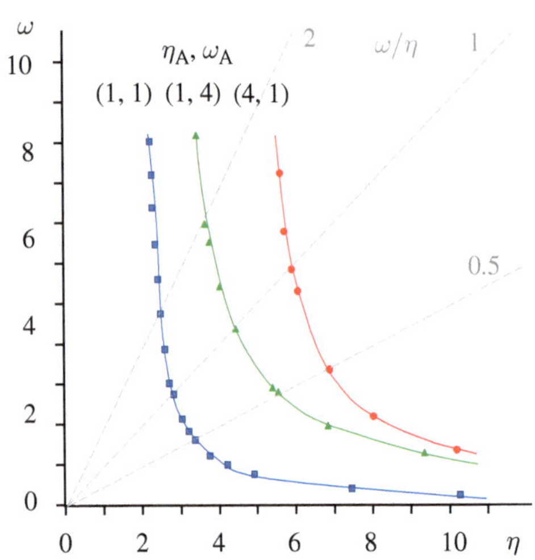

Figure 11.3 Position of the minimum values (η_B, ω_B) in a $[\eta-\omega]$ diagram. The traces of the minimum describe an increasing value with dN, the higher (η_B, ω_B) being associated with the highest dN values. Three cases are presented with different starting points (η_A, ω_A), respectively (1, 1) (squares), (1, 4) (triangles), and (4, 1) (circles). All have been relocated in the appropriate coordinates system.

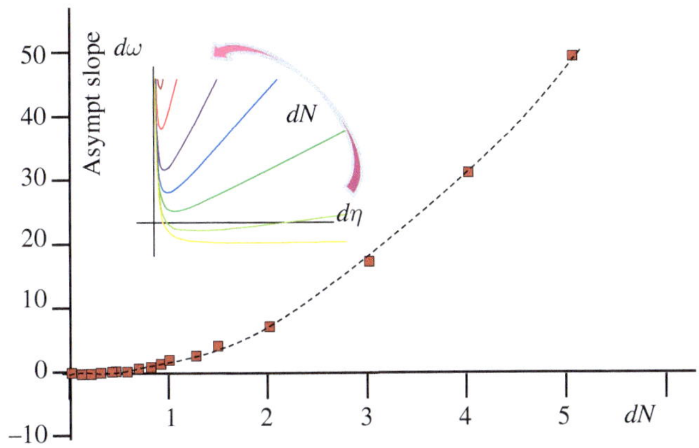

Figure 11.4 Rotation of the asymptotes toward the w axis when dN varies. The curve is computed using successive values of dN for an initial point (η_A, ω_A) located at (3, 3). The slope is estimated for values of $d\eta$ between 50 and 100 eV, not shown on the preceding curves, but representative of the asymptotic character of the curves. The dependence of the curve is $2dN^2$, as expected from their analytical equation (Eq. (11.16)).

Three specific paths (Figure 11.4) can be estimated and conform to usual potential theories [22]. However, they do not constitute real progress in the understanding of the charge exchange.

When $d\chi = 0$, it corresponds to a path along a potential line (equal chemical potential). It necessitates no charge exchange ($dN = 0$) and energy ($dE = 0$), and it corresponds to equilibrium curves (Figure 11.4).

Conversely, a path orthogonal to the equipotential lines presents the largest variation in charge and energy curves (Figure 11.4). It corresponds to the steepest descent along the potential surface. The charge naturally depends on the length of the reaction path, i.e. the change in chemical potential. The simplest path occurs when the initial point is on the line $\chi/\eta = 1$ (i.e. $\chi_A = \eta_A$), and the final point remains on that line ($d\chi = d\eta$). In this case, the charge exchange is given by

$$1/dN = 2 + 4\eta_A/d\eta \tag{11.12}$$

It results from those specific solutions that the curve of charge exchange depends on the starting point η_A, and of two linked values, the change in hardness ($d\eta$) and the total length of the reaction path.

11.6.2 Systematic Changes Depending on the Starting Points on $p_{\chi\eta}(\zeta)$

The analytical equation (Eq. (11.10)) relates the path $p_{\chi\eta}$ as a function of the charge exchange dN, but also on the starting point through η_A and η_B, using $d\eta = \eta_A - \eta_B$. Thus, the equation implies two variables and two other values, making it hard to be fully interpreted. The adopted solution has been through the use of a graphical equation solver [21]. Hence several sets of curves have been computed and plotted, varying one parameter for each run. The respective curves are all hyperbolas, as expected from the cubic equation (Eq. (11.10)). The minimum points were determined graphically, providing their respective coordinates, i.e. the coordinates in which the equation was computed. The respective values of the minimum points are then plotted, showing the trend of changes as a function of the initial input parameters [23]. The adopted solution is a "poor man solution" in that sense, it hardly can be quantified, nor does it inform on the path behavior (Figure 11.5).

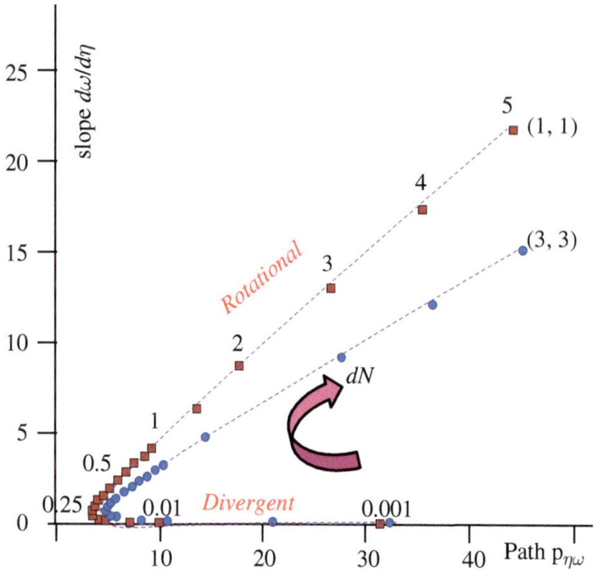

Figure 11.5 Respective evolution of the path length ($p_{\eta\omega}$) and its instantaneous slope ($s_{\eta\omega} = d\omega/d\eta$). Two curves are presented, computed for an original point with coordinates in eV (3, 3) (circles) and (1, 1) (squares). Both indicate two branches with different behaviors linked by an intermediate state. They all rely on *dN* values. For low *dN* (<0.25), the curve remains quasi-parallel to the $d\eta$ axis, and the path length decreases with *dN*. Conversely, for higher *dN* (>0.50), the path length increases, but its slope also increases linearly with *dN*. In between, a short segment connects the two branches, going through a minimum of the path length for *dN* about 0.23–0.26.

11.6.3 Specific Solutions Using a $p_{\eta\omega}$ Path

A global approach to the chemical reactions should consider total charge (*dN*) exchange as the global change in hardness and in electronegativity, resulting in a path length ($p_{\eta\omega}$), thus involving the descriptors of the initial point (χ_A, η_A, ω_A), and their variations ($d\omega, d\eta$) within a [ω–η] diagram. In the same way, the slope of the path $s_{\eta\omega}$ may be computed

$$p_{\eta\omega}{}^2 = d\eta^2 + d\omega^2 \tag{11.13}$$

$$s_{\eta\omega} = d\omega/d\eta \tag{11.14}$$

The equation of the path is simplified from the two aforementioned equations into

$$d\eta^2 p_{\eta\omega}{}^2 = s_{\eta\omega}{}^2 + 1 \tag{11.15}$$

or the equation of a hyperbola in a [$s_{\eta\omega}$–$p_{\eta\omega}$] diagram, with respective asymptotic curves, the axis of $p_{\eta\omega}$ and the constant line of slope $d\eta^{-2}$.

The general solution could be used for determining the change from low charge to high charge exchange. Hence, two contrasting behaviors are inferred from Eq. (11.15), depending on the shape of the path. A path under a nearly constant small slope ($s_{\eta\omega} \sim 0$) occurs for a $d\omega$ also close to zero, which is a variation mainly due to a change in η. It corresponds to a small charge exchange lower than 1. Conversely, a change with few variations in $d\eta$ develops under a large change in $d\omega$, under a progressively extending path, i.e. larger charge exchange *dN*.

11.7 Significance and Applications

The chemical potential is a central concept in thermodynamics because all of the thermodynamic properties of a material at a given temperature and pressure can be obtained from its value.

It is commonly defined as the derivative of the Gibbs free energy of the system, along with temperature (T) and pressure (P) in a gaseous phase, or temperature and stress (σ) in viscous bodies. The corresponding responses to such forces are, respectively, entropy (S), strain (ε) and a number of particles (N). The intensive quantity determines the magnitude of an energy change related to a change of the corresponding extensive quantity. Such a 3D representation of the energy variations should be preferred [22] to the common PT diagram used in a gaseous phase.

$$U = TS - \sigma\varepsilon + \mu N \tag{11.16}$$

The chemical potential helps in determining the energy variations when an element is submitted to a surrounding electronic cloud. This is similar to the effect of a flux of current when dealing with electrostatic material. The resulting effects are commonly described by electric and magnetic responses, i.e. each having a divergent or a rotational component. They are described by Maxwell equations [23]. Similarly, the response to the deformation of a viscous material can be decomposed into a pure shear, or compression, and a simple shear, i.e. with a rotational component [24]. The decomposition of the chemical potential into a divergence and a rotational component corresponds to the usual definition of the derivative of a gradient field. Such decomposition is observed in physics, electromagnetism, or deformation (Figure 11.6).

Nevertheless, the concept of chemical potential remains poorly understood, justifying its acceptance as being "elusive" [25], although it has been addressed by many authors [26]. As a measure of the tendency for electronic clouds to diffuse, the chemical potential measures the "rate of change" (not in time) of a thermo-dynamical function. Its temperature coefficient should be equal to the negative molar entropy, whereas its pressure coefficient should be equal to the molar volume [22].

The existence of two different ways for accommodating chemical reactions between elements bears consequences on their structural bonding. The discrepancy between the two types of reactions has been evidenced in a [$d\omega$–$d\eta$] diagram, coined from a fitness landscape diagram [ω–η]

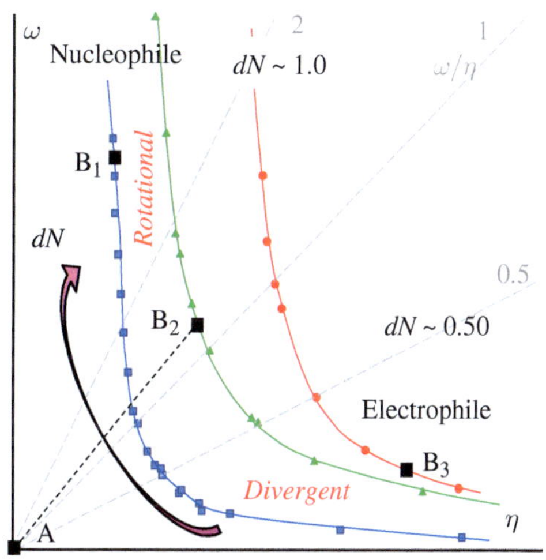

Figure 11.6 Cartoon illustrating the reaction path between element A (set at the origin) and element B for three different ending elements (B_1 to B_3). The charge exchange increases for AB_1 to AB_3, as shown by the three curves scaled in increasing dN. The field in which reactions develop is indicated by a gray shade. It is constrained by curves of equal chemical potential values. In each case, the variation in chemical potential ($d\mu$ or grad $\nabla\mu$) is a combination of electrophile (divergent) and nucleophile (rotational) components. They also indicated the maximum values of dN corresponding to $\omega/\eta = 0.5$, 1.0 and 2.0.

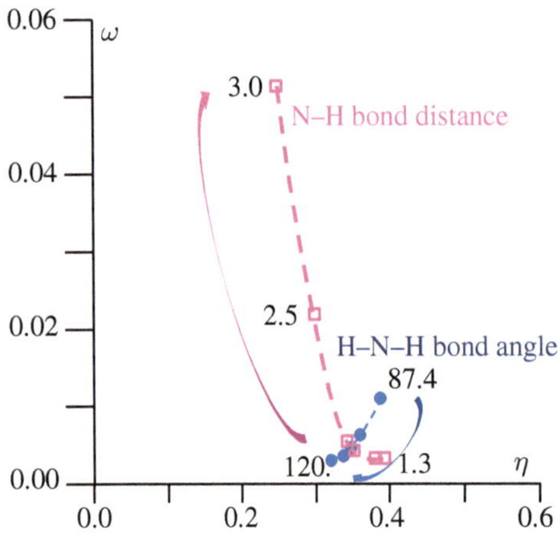

Figure 11.7 The position of the bonding distance for N–H and H–N–H bond angle derives from Pearson and Palke (1992) data set [29]. Unit values are given, as in the original paper in Hartree, equivalent to 27.2114 eV, whereas distortion in bohrs of the N–H bond length and angles N–H–N in degrees.

mapping two descriptors together [27]. The hyperbolic curves discriminate between reactions mainly varying along the two coordinates, namely electrophilicity and hardness. Both are akin to a second derivative of the energy with a link to either the curvature or the osculating circle of the energy curve, i.e. with the longitudinal or the normal component of the acceleration. A change in electrophilicity induces a change in the trajectory of the energy, traduced by a rotational component. Conversely, a change in hardness corresponds to an acceleration, or deceleration, in the energy path.

Turning back to chemistry, electronegativity and hardness both link to the first vertical ionization energy and electron affinity to the HOMO and LUMO energies, as their mean and their difference [28]. A change in hardness ($d\eta$) during a chemical reaction corresponds to a change in the HOMO–LUMO gap. Since electronegativity is their mean value, it is not affected. It is effectively a simple divergence or convergence of the orbitals. Conversely, a change in electrophilicity ($d\omega$) corresponds to a change in both hardness and electronegativity, modifying the gap and its mean value. It implies a modification of the orbitals, thus implying a rotational component, modifying the spatial position of the orbitals.

Surprisingly, this had been noted and experimentally measured in a seminal paper from Pearson's lab [29], in which they provide descriptors for N—H and H—N—H bonding in ammonia. In the first case, the bonding distance varies, inducing a neat reduction in hardness and consequent increase in electrophilicity (Figure 11.7). The increased rotation of the bond in the case of H-N-H corresponds to lower both descriptors.

The existence of a simple bonding associated with frontier orbital (FO) reactions corresponds to a simple bonding between elements. It thus has no reason for rotting. Conversely, when the charge exchange is large, giving place to a CC reaction, the electronic cloud needs to reorientate and deform to accommodate the new cloud formed by the reactants. It thus seems evident that such a reorientation should be accommodated by a rotational component. By the way, it seems more developed in large molecules as found in organic chemistry. The consequences of such rotation are the development of tautomers and rotamers during reactions, both presenting different chemical descriptors, and electrophilicity being more developed in nucleophile reactions (Table 11.1).

Table 11.1 Symmetric distortion of ammonia.

	-2μ	2η	$\omega = \mu^2/2\eta$
Pearson Palke	**https://doi.org/10.1021/j100187a020**		
Bond distance			**NH**
1.30	0.0930	0.7912	0.002,732,87
1.50	0.0895	0.7701	0.002,600,39
1.89	0.1142	0.7186	0.004,537,17
1.91	0.1169	0.7151	0.004,777,52
1.93	0.1197	0.7115	0.005,034,47
2.00	0.1300	0.6988	0.006,046,08
2.50	0.2279	0.5901	0.022,004,07
3.00	0.3169	0.4871	0.051,542,6
Bond angle			**HNH**
87.4	0.1907	0.7681	0.011,836,51
106.7	0.1169	0.7151	0.004,777,52
113.9	0.0945	0.6869	0.003,250,2
117.0	0.0859	0.6731	0.002,740,61
120.0	0.0782	0.6584	0.002,322,01

Values in Hartree, equivalent to 27.2114 eV, distorsion in Bohrs of the N—H bonds length, angles N-H-N in degrees.

11.8 Conclusions

The value of the charge exchanged during a chemical reaction depends on the variation in electronegativity and hardness between elements. From this simple relation, the reaction path is examined to find a transition between FO and CC reactions. Former indications gave a charge value dN close to 0.5 for such transitions. In fact, the present modeling suggests a nonlinear behavior of the path according to charge value. Two branches are observed. One of the lowest values of dN (<0.25) develops at low ω values, close to the η axis. It shows a decreasing length associated with no change in orientation. It corresponds to divergent behavior. A second branch presents a linear increase of its slope (ω/η) while the dN increases (>0.50) and a path close to the ω axis. It corresponds to rotational behavior. Both divergent and rotational components are commonly linked with a change in the chemical potential gradient.

Acknowledgments

The paper came out after a stay in 2009 at the Department of Chemistry, IIT Kharagpur, India, with a grant by the Center for Theoretical Studies (CTS). It allowed a fruitful introduction to DFT concepts and collaboration with Pratim Kumar Chattaraj and his students. Conversations with Laurent Truche (GéoRessources, Nancy; currently ISTERRE Grenoble) and Christophe Morell (Université de Lyon 1) inspired me to draft the ensuing papers. Warm thanks are also extended for helpful reviews that include remarks and linkages to the activation strain model.

References

1 (a) Pearson, R.G. (1968). Hard and soft acids and bases, HSAB, part 1: Fundamental principles. *J. Chem. Educ.* 45 (9): 581. https://doi.org/10.1021/ed076p267; (b) Pearson, R.G. (1968). Hard and soft acids and bases, HSAB, part II: Underlying theories. *J. Chem. Educ.* 45 (10): 643. https://doi.org/10.1021/ed045p581; (c) Chattaraj, P.K., Lee, H., and Parr, R.G. (1991). HSAB principle. *JACS* 113 (5): 1855–1856. https://doi.org/10.1021/ja00005a073; (d) Gazquez, J.L. and Mendez, F. (1994). The hard and soft acids and bases principle: an atoms in molecules viewpoint. *J. Phys. Chem.* 98 (17): 4591–4593. https://doi.org/10.1021/j100068a018; (e) Ayers, P.W. and Parr, R.G. (2000). Variational principles for describing chemical reactions: the Fukui function and chemical hardness revisited. *JACS* 122 (9): 2010–2018. https://doi.org/10.1021/ja9924039; (f) Miranda-Quintana, R.A., Kim, T.D., Cárdenas, C., and Ayers, P. (2017, 2017). The HSAB principle from a finite-temperature grand-canonical perspective. *Theor. Chem. Acc.* 136: 135. https://doi.org/10.1007/s00214-017-2167-y.

2 (a) Parr, R.G. and Yang, W. (1984). Density functional approach to the frontier-electron theory of chemical reactivity. *JACS* 106 (14): 4049–4050. https://doi.org/10.1021/ja00326a036; (b) Parr, R.G. and Yang, W. (1995). Density-functional theory of the electronic structure of molecules. *Annu. Rev. Phys. Chem.* 46 (1): 701–728. https://doi.org/10.1146/annurev.pc.46.100195.003413; (c) Geerlings, P., De Proft, F., and Langenaeker, W. (2003). Conceptual density functional theory. *Chem. Rev.* 103 (5): 1793–1874. https://doi.org/10.1021/cr990029p; (d) Pearson, R.G. (2009, 2009). Reactivity theory: a density hardness of closed systems. In: *Chemical Functional View* (ed. P.K. Chattaraj), 155–162. Boca Raton: CRC Press.

3 Chattaraj, P.K. (2001). Chemical reactivity and selectivity: local HSAB principle versus frontier orbital theory. *J. Phys. Chem. A* 105 (2): 511–513. https://doi.org/10.1021/jp003786w.

4 Bickelhaupt, F.M. and Houk, K.N. (2017). Analyzing reaction rates with the distortion/interaction-activation strain model. *Angew. Chem. Int. Ed.* 56 (34): 10070–10086. https://doi.org/10.1002/anie.201701486.

5 (a) Vigneresse, J.L. (2012). Chemical reactivity parameters (HSAB) applied to magma evolution and ore formation. *Lithos* 153: 154–164. https://doi.org/10.1016/j.lithos.2012.03.014; (b) Vigneresse, J.L., Truche, L., and Chattaraj, P.K. (2014). Metal (copper) segregation in magmas. *Lithos* 208: 462–470. https://doi.org/10.1016/j.lithos.2014.09.025.

6 Pauling, L. (1932). The nature of the chemical bond. IV. The energy of single bonds and the relative electronegativity of atoms. *JACS* 54 (9): 3570–3582. https://doi.org/10.1021/ja01348a011.

7 (a) Parr, R.G. and Chattaraj, P.K. (1991). Principle of maximum hardness. *JACS* 113 (5): 1854–1855. https://doi.org/10.1021/ja00005a072; (b) Pearson, R.G. (1993). The principle of maximum hardness. *Acc. Chem. Res.* 26 (5): 250–255. https://doi.org/10.1021/ar00029a004; (c) Chermette, H. (1999). Chemical reactivity indexes in density functional theory. *J. Comput. Chem.* 20 (1): 129–154. https://doi.org/10.1002/(SICI)1096-987X(19990115)20:1<129::AID-JCC13>3.0.CO;2-A.

8 (a) Parr, R.G., Szentpály, L.V., and Liu, S. (1999). Electrophilicity index. *JACS* 121 (9): 1922–1924. https://doi.org/10.1021/ja983494x; (b) Chattaraj, P.K., Giri, S., and Duley, S. (2010).

Electrophilicity equalization principle. *J. Phys. Chem. Lett.* 1 (7): 1064–1067. https://doi.org/10.1021/jz1001117; (c) Chattaraj, P.K., Fuentealba, P., Jaque, P., and Toro-Labbé, A. (1999). Validity of the minimum polarizability principle in molecular vibrations and internal rotations: an ab initio SCF study. *J. Phys. Chem. A* 103 (46): 9307–9312. https://doi.org/10.1021/jp9918656.

9 Ghanty, T.K. and Ghosh, S.K. (1996). A density functional approach to hardness, polarizability, and valency of molecules in chemical reactions. *J. Phys. Chem.* 100 (30): 12295–12298. https://doi.org/10.1021/jp960276m.

10 (a) Nagle, J.K. (1990). Atomic polarizability and electronegativity. *JACS* 112 (12): 4741–4747. https://doi.org/10.1021/ja00168a019; (b) Ghanty, T.K. and Ghosh, S.K. (1993). Correlation between hardness, polarizability, and size of atoms, molecules, and clusters. *J. Phys. Chem.* 97: 4951–4953. https://doi.org/10.1021/j100121a015; (c) Fajans, K. (1923). Struktur und Deformation der Elektronenhüllen in ihrer Bedeutung für die chemischen und optischen Eigenschaften anorganischer Verbindungen. *Naturwissenschaften* 11: 165–172. https://doi.org/10.1007/BF01552365.

11 Morell, C., Grand, A., and Toro-Labbé, A. (2006). Theoretical support for using the $\Delta f(r)$ descriptor. *Chem. Phys. Lett.* 425: 342–346. https://doi.org/10.1016/j.cplett.2006.05.003.

12 Sanderson, R.T. (1976). *Chemical Bonds and Bond Energy*, vol. 1976, 232. New York: Academic Press.

13 (a) Mortier, W.J., Ghosh, S.K., and Shankar, S. (1986). Electronegativity-equalization method for the calculation of atomic charges in molecules. *JACS* 108 (15): 4315–4320. https://doi.org/10.1021/ja00275a013; (b) Parr, R.G. and Bartolotti, L.J. (1982). On the geometric mean principle for electronegativity equalization. *JACS* 104: 3801–3803. https://doi.org/10.1021/ja00378a004.

14 Noorizadeh, S. and Shakerzadeh, E. (2010). Shannon entropy as a new measure of aromaticity, Shannon aromaticity. *Phys. Chem. Chem. Phys.* 12: 4742–4749. https://doi.org/10.1039/B916509F.

15 Morell, C., Labet, V., Grand, A., and Chermette, H. (2009). Minimum electrophilicity principle: an analysis based upon the variation of both chemical potential and absolute hardness. *Phys. Chem. Chem. Phys.* 11 (18): 3417–3423. https://doi.org/10.1039/B818534D.

16 Gray, A. (2006). *Modern Differential Geometry of Curves and Surfaces with Mathematica* (ed. A. Gray, E. Abbena, and S. Salamon), 111. Boca Raton, FL: CRC Press.

17 (a) Wright, W. (1932). *Proceedings of the Sixth International Congress of Genetics*, vol. 1, 356; (b) Kauffman, S.A. (1995). *At Home in the Universe: The Search for Laws of Self-Organization and Complexity*, 321. New York: Oxford University Press.

18 Das, R., Vigneresse, J.L., and Chattaraj, P.K. (2013). Redox and Lewis acid–base activities through an electronegativity-hardness landscape diagram. *J. Mol. Model.* 19: 4857–4864. https://doi.org/10.1002/qua.24706.

19 Anderson, J.S., Melin, J., and Ayers, P.W. (2007). Conceptual density-functional theory for general chemical reactions, including those that are neither charge-nor frontier-orbital-controlled. 1. Theory and derivation of a general-purpose reactivity indicator. *J. Chem. Theory Comput.* 3 (2): 358–374. https://doi.org/10.1021/ct600164j.

20 Vigneresse, J.L. (2018). Variations in chemical descriptors during reactions. *Chem. Phys. Lett.* 706: 577–585. https://doi.org/10.1016/j.cplett.2018.05.032.

21 Liu, A.J. and Nagel, S.R. (1998). Jamming is not just cool any more. *Nature* 396 (6706). https://doi.org/10.1038/23819.

22 Hobbs, B.E. and Ord, A. (2010). The mechanics of granitoid systems and maximum entropy production rates. *Philos. Trans. R. Soc. A* 368: 53–93. https://doi.org/10.1098/rsta.2009.0202.

23 (a) Feynman, R.P. (2011). *The Feynman Lectures on Physics*. Basic Books 1552 pp; (b) Hurray, P.G. (2009). *Maxwell's Equations*. Wiley 312 pp.

24 (a) Ramsay, J.G. and Huber, M.I. (1983). *The Techniques of Modern Structural Geology. Volume 1: Strain Analysis*. London: Academic Press 307 pp; (b) Gay, N.C. (1968). Pure shear and simple shear deformation of inhomogeneous viscous fluids. 1. Theory. *Tectonophysics* 5: 211–234. https://doi.org/10.1016/0040-1951(68)90065-6.

25 Baierlein, R. (2001). The elusive chemical potential. *Am. J. Phys.* 69: 423–434. https://doi.org/10.1119/1.1336839.

26 (a) Job, G. and Herrmann, F. (2006, 2006). Chemical potential—a quantity in search of recognition. *Eur. J. Phys.* 27: 353–371. https://doi.org/10.1088/0143-0807/27/2/018; (b) Chen, L.Q. (2019). Chemical potential and Gibbs free energy. *MRS Bull.* 44: 520–523. https://doi.org/10.1557/mrs.2019.162; (c) Cook, G. and Dickerson, R.H. (1995). Understanding the chemical potential. *Am. J. Phys.* 63 (8): 737–742. https://doi.org/10.1119/1.17844.

27 Duley, S., Vigneresse, J.L., and Chattaraj, P.K. (2012). Fitness landscapes in natural rocks system evolution: a conceptual DFT treatment. *J. Chem. Sci.* 124: 29–34. https://doi.org/10.1007/s12039-011-0187-z.

28 Pearson, R.G. (1986). Absolute electronegativity and hardness correlated with molecular orbital theory. *Proc. Natl. Acad. Sci.* 83: 8440–8441. https://doi.org/10.1021/j100187a020.

29 Pearson, R.G. and Palke, W.E. (1992). Support for a principle of maximum hardness. *J. Phys. Chem.* 96: 3283–3285. https://doi.org/10.1021/j100187a020.

12

Deep Learning of Electron Density for Predicting Energies: The Case of Boron Clusters

Pinaki Saha[1] *and Minh Tho Nguyen*[2,3]

[1] *Biocomputation group, School of Physics, Engineering and Computer Science, University of Hertfordshire, UK*
[2] *Laboratory for Chemical Computation and Modeling, Institute for Computational Science and Artificial Intelligence, Van Lang University, Ho Chi Minh City, Vietnam*
[3] *Faculty of Applied Technology, School of Technology, Van Lang University, Ho Chi Minh City, Vietnam*

12.1 Introduction

Nanoclusters or atomic aggregates constitute an interesting class of compounds in the field of modern materials. They have been extensively studied due to their utilization in data storage, optoelectronics and catalysis, among others. In view of the diversity of structural motifs of atomic clusters, knowledge on their geometric structures plays an important role in the determination and/or understanding of their physico-chemical properties [1–10]. The structures of nanoclusters can be derived computationally via a series of quantum chemical (QC) calculations. These calculations tend to be scaled poorly with reference to the size of the nanocluster.

The most used substitute for QC-based computations is molecular mechanics (MM)-based computations [11]. MM treats a chemical bond in a molecule akin to a spring and the atoms as point charges. The molecule is thus treated in a Newtonian framework. MM trades accuracy for processing speed. The solution of the classical mechanics equation does not require very high processing power as opposed to performing QM-based calculations. MM has been traditionally utilized for organic and bioorganic molecules, especially in molecular docking and molecular dynamics simulation.

In terms of accuracy and processing power, the machine learning (ML) paradigm can be considered between QC and MM methods. Given a dataset with input variables and output response (training dataset), ML algorithms approximate the relationship between the input and the response. Using such an approximate relationship, the ML algorithm tries to predict the response of unknown datasets containing input variables (test dataset). Several ML algorithms have been reported in the literature for molecular systems, with different levels of accuracy, such as quantitative structure–activity relationship (QSAR), SVM and neural networks, to name a few [12, 13]. Recently, neural network-based potentials have been reported in the literature [14–16]. Neural network potentials, also known as ML potentials with (ML Potentials), are computational models used to approximate the potential energy surfaces of molecular systems. They leverage the power of neural networks to learn the complex relationship between atomic positions and

Electron Density: Concepts, Computation and DFT Applications, First Edition.
Edited by Pratim Kumar Chattaraj and Debdutta Chakraborty.

their corresponding energies/forces. Some of the popularly used neural networks for molecular systems are:

1. *The Behler–Parrinello neural network (BPNN)*: BPNN [17, 18] employs a symmetry function-based approach to capture the local atomic environment and has been successful in describing the structural and thermodynamic properties of various materials. In a BPNN scheme, symmetry is preserved by mapping the coordinates onto a large set of two- and three-body symmetry functions. Fixing these symmetry functions is a painstaking endeavour for molecules containing many elements. BPNN is sensitive to the choice of input features used to describe the molecular system. Selecting the most relevant features can be a time-consuming and challenging task.
2. *The atom-centred symmetry functions (ACSF)*: ACSF [19, 20] is a neural network potential representing atomic environments using symmetry functions. These functions encode information about the distances and angles between atoms.

While neural network potentials have been proven powerful tools in computational chemistry, they have certain drawbacks including:

i. *Transferability*: Transferability refers to the ability of a neural network's potential to accurately describe new systems or environments that are not part of the training data.
ii. *Representation of long-range interactions*: In both ACSF and BPNN, these potentials focus on capturing short-range atomic environments and interactions, which may limit their ability to accurately describe long-range interactions. The latter may not be important for organic molecules, but for nanoclusters, like boron nanoclusters that we are primarily interested in, long-range interactions via extensive electron delocalization are primal interactions.

Boron clusters have been studied extensively due to their nonclassical structures, electronic features and chemical bonding phenomena. In fact, boron clusters can take a myriad of shapes with intriguing structural and bonding features, including the planar and quasi-planar forms, bowl, ribbon, tube, bilayer, cage, core–shell and fullerene structures [21–48]. The complex architecture of boron skeletons arises from the fact that the boron atom with an intrinsic electron deficiency can take part in both localised and delocalized bonding in different geometric shapes, leading to rich structural and bonding landscapes.

Small pure neutral B_n clusters with $n < 20$ tend to prefer a planar or quasi-planar structure, except for B_{14}, which is a fullerene type structure [21–24]. A transition from planarity to 3D shape is observed at B_{20}. A double-ring tube connecting two B_n strings in an antiprism manner was found to be the ground state of certain B_{2n} systems, including B_{20}, B_{22}, B_{24} and B_{26} clusters in the neutral state [21, 22]. More particularly, both B_{27}^+ and B_{42} exhibit a triple-ring tubular shape in which their geometry is constructed by superposing three B_9 and B_{14} strings in an anti-prism fashion, respectively [21, 44]. These results show that the tube is also an inherent structural motif of boron clusters besides the planar or quasi-planar one. Furthermore, both B_{28} and B_{29} sizes seem to have a cage-like structure rather than a planar or tubular shape, emphasizing the unpredictability of boron cluster structures [49].

Along with the planar/quasi-planar and tubular structures, larger cluster sizes favouring bowl-shaped geometries were also discovered. The B_{30} global minimum was characterized as a bowl, and its chemical bonding was rationalized by a disk aromaticity model [43]. Numerous previous studies demonstrated that the B_{32}, B_{33}^-, B_{34} and B_{36} clusters are classified into a bowl-like class, which contains either one pentagonal or one hexagonal hole [45, 46]. Similar to B_{30}, B_{36} was theoretically predicted to have a bowl shape stabilized by a disk aromaticity [39, 41, 44]. The following sizes B_{37}^- and B_{38}^- also exhibit bowl-shaped structures but with two hexagonal holes [50]. The appearance of several bowl-shaped structures led to an assumption that these geometries could act as precursors to boron sheets. The neutral B_{36} is in fact the smallest boron cluster exhibiting a sixfold symmetry and a hexagonal vacancy, and it can be viewed as a potential precursor to extended two-dimensional (2D) boron sheets.

The B_{40} fullerene was initially predicted via computational approaches and subsequently experimentally verified using photoelectron spectroscopy (PES) [44, 51]. While the B_{40} cage with a D_{2d} symmetry consists of four heptagonal rings and two hexagonal rings, the B_{40}^- anion is a quasi-planar structure with two hexagonal holes [44].

The B_{46} was predicted to be a cage with an octagonal hole, but subsequent calculations showed that a core–shell structure with four internal B atoms could be more stable. A double-layer structural pattern was suggested for B_n with n from 48 to 72 atoms [52, 53]. Such a structure has been corroborated by some experimental results in the case of B_{48}^- [54].

Both B_{68} and B_{74} were reported to exist as core–shell structures, whereas the B_{72} interestingly exhibits a tubular shape [55]. The B_{80} size has been known as a boron buckyball having the same number of 240 valence electrons as the C_{60} fullerene, and its shape and electronic structure are quite similar to those of C_{60}. The most interesting aspect of the B_{80} buckyball is that it contains 12 pentagonal and 20 hexagonal holes [55]. However, some core–shell isomers could be more stable, and this issue needs to be further examined. A quasi-planar structure was reported to be the global minimum for B_{84}, which, however, has a competing nearly degenerate tubular structure [56]. Various other structures, such as cage, amorphous and core–shell derivatives, were also reported for B_{84} [57]. Recently, we showed that the quasi-planar forms of the sizes B_{50}, B_{56} and B_{70} can be generated from topological leapfrog paths from smaller sizes, and the quasi-planar structures of B_{50}, B_{56} and B_{70}^{2-} correspond to their lowest-lying isomers [58–60]. Generally, boron clusters show extreme diversity in the structures they adopt; therefore, it is quite challenging to utilize MM methods due to a dearth of force fields that can accurately capture the complexity of bonding in boron. Neural network-based potentials fare no better as they cannot capture the long-range interactions.

12.2 Deep Learning of Electron Density

In this context we set out to look at other approaches in devising a new methodology involving neural networks. The neural network consisting of a single hidden layer can approximate any continuous function, and hence, they are also termed as universal function approximators. This makes neural networks particularly suitable for regression problems. It is also well known that deep convolutional networks can approximate any continuous function. Featureless learning is the unique ability of the convolutional neural network (CNN) to utilize training data, especially images, directly for modelling without feature extraction. Featureless learning is also expected to be useful for the creation of a regression model that predicts not only energy but also multiple properties of molecular systems. In this study, we study application of the neural network to predict energies of nanocluster. We opted for featureless learning using convolutional neural network trained on promolecule densities. It is widely known that the Hohenberg–Kohn theorem [61] states that the electron density of a molecule $\rho(\mathbf{r})$ uniquely determines the ground state wave function ψ, and thereby its ground state electronic energy. A unique energy functional $E[\rho(\mathbf{r})]$ is required for mapping the electron density to the ground state electronic energy (Eq. (12.1)):

$$E = E[\rho(\mathbf{r})] \tag{12.1}$$

With regard to density functional theory (DFT) calculations for molecules, a trial electron density is assigned to molecules and using suitable functional at a certain level of theory, the energy of the molecule is calculated. Instead of the utilization of using the functional, we can map the electron density to the ground state electronic energy using CNN (Eq. (12.2)):

$$\rho(r) \rightarrow E \tag{12.2}$$

To obtain the electron density, we are required to perform QC-based calculations. ML models are expected to predict on the fly for test datasets, and performing QC calculations is not viable for large datasets or for datasets consisting of large molecules or clusters. Recently, Zhao *et al.* reported a study regarding 3D-CNN, which uses the electron localization function (ELF) for the prediction of various properties for various materials [62]. The ELF does capture both the localization and delocalization of the electron density. Generation of an ELF map requires QC calculations. A speedy ML algorithm would require rapid calculations of electron density without explicitly using traditional QC calculation. Our solution to this problem is the construction of Hirshfeld partitioning-based promolecule densities [63] for the clusters/molecules examined.

For a better implementation of the promolecule density, we need to use the Hirshfeld partitioning scheme, which is a scheme for partitioning of the molecular electron density into its constituent atomic density. In this partitioning scheme, a promolecule density is assigned to a molecule by superposition of precalculated electron densities of the atomic constituents of the molecule considered. Then this electron density is partitioned by giving weightage to the atomic constituents proportionated to their contribution to the electron density. The promolecule density ($\rho^{\circ}(r)$) can be written as a weighted summation of atomic densities (12.3):

$$\rho^{\circ}(r) = \sum_{A} w_{A}(r)\, \rho_{A}{}^{\circ}(r) \tag{12.3}$$

where the weight is given by the following expression (12.4):

$$w_{A}(r) = \rho_{A}{}^{\circ}(r)/\rho^{\circ}(r) \tag{12.4}$$

The promolecule density is akin to the trial density utilized in DFT calculations. This density can be mapped to the ground state energy of a system by using an appropriate neural network model. The promolecule density is a 3D quantity and thus cannot be utilized in a CNN. CNNs are typically used for 2D images; in that regard, we must convert the 3D promolecule density to a 2D quantity. We can achieve this by projecting the 3D density to three 2D planes, namely *xy*, *yz* and *zx* planes. Via this 2D projection, we obtain three projected data points from a single 3D data point. This 2D projection thus leads to a data augmentation (artificial increase of the dataset size). The dataset required for our model is mainly sourced from the current literature (including reviews on nanoclusters), and QC calculations are carried out to ascertain whether the reported structures are truly global minimum. Because the CNN approaches tend to require a large dataset for accurate prediction, we are use a data augmentation technique to increase the data size. Data augmentation is a way of artificially creating new training data from existing training data. In our case, we can use molecular dynamics (MD) simulations to achieve data augmentation, in which the numerous conformers/poses generated by MD simulations are used for augmenting our dataset. In this study, we use *ab initio* molecular dynamics simulation, specifically atom-centred density matrix propagation (ADMP) molecular dynamics simulations, which is a robust and computationally efficient method of data augmentation as compared to the random generation of particles and computing the energies of the generated particles [64]. As far as we are aware, the prediction of total energies of molecular systems via ML algorithm using the promolecule density has not been reported yet in the literature.

As for a concrete application, we perform tests of our neural network models on boron nanoclusters. As stated above, boron clusters show a wide complexity in their structures that they can adopt a various geometrical motifs, and their bonding phenomenon is not straightforward as it contains extensive delocalized bonding in both 2D and 3D structures. Thus, boron clusters form a challenging dataset for ML problems. Following a successful implementation of a deep learning model on boron clusters, we plan to expand the model on other homonuclear clusters that are abundantly reported in the literature.

In this present study, we utilize pure boron clusters B_n ranging from the size B_4 to B_{40} for our ML-based study, due to the fact the structures for the B_n clusters in the size range of $n = 4-40$ have been well established [65]. Some larger clusters (B_n, with $n > 40$) still do not have very well-established global minimum structures. Large clusters can take up many configurations; therefore, it is quite difficult to ascertain the identity of their global minimum structures. Our present training data set for the B_n boron clusters is thus limited up to the size of B_{40}.

The ADMP calculations on boron clusters are performed using the PBE0 functional and the Gaussian 09 program [66] in conjunction with the 6-311+G basis set. Promolecular densities of the clusters are generated using the Crystal Explorer software [67]. Automation of the Crystal Explorer processes is done via Microsoft's Power Automate tool. The CNN models are created using the MATLAB R2021b suite. In-house bash/python scripts are also utilized for small data processing tasks.

12.3 Neural Networks for Neutral Boron Clusters

We utilize a CNN-based regression model that takes a regular CNN, which is succeeded by a regression layer at the end of the network. The feature extraction layers contain convolution, RELU and pooling layers.

CNNs are well known in the field of image recognition. The underlying assumption is that the complex geometrical features of an image are an ensemble of smaller and simpler patterns. CNNs use small filters that extract local features and progressively reduce the size of the input variable with each layer. The number of layers in CNNs increases with an increase in the size of the input image. Let us consider the input image X of size $N \times M$, as shown in Figure 12.1. Each convolution + RELU and pooling layer produces the feature maps of reduced dimension as compared to its input. Additionally, the pooling layers produce an output map that is invariant to small changes within the pooling window. The model is trained using a regression loss function, which is given by Eq. (12.5):

$$L = \frac{1}{N}\sum_{i}^{N}(y_i - \bar{y}_i)^2 \tag{12.5}$$

The model utilizes the Adam optimizer for the purpose of back-propagation. The model is trained on an AMD Ryzen-7 PRO 4750U CPU, 16GB RAM with RADEON graphics card using MATLAB R2021B.

The dataset used for training the model is generated from boron clusters, ranging from the size of B_4 to B_{40}. *Ab initio*-based molecular dynamics simulations (ADMP) are performed on all these clusters considered. Each ADMP trajectory provides a set of molecular clusters along with their corresponding energies. So, there are 37 clusters in the dataset, and each cluster generates 1020 structures along with their corresponding energies; thus, we obtain 37,740 structures. For each of these 37.740 structures, we generate the corresponding promolecule density for each structure

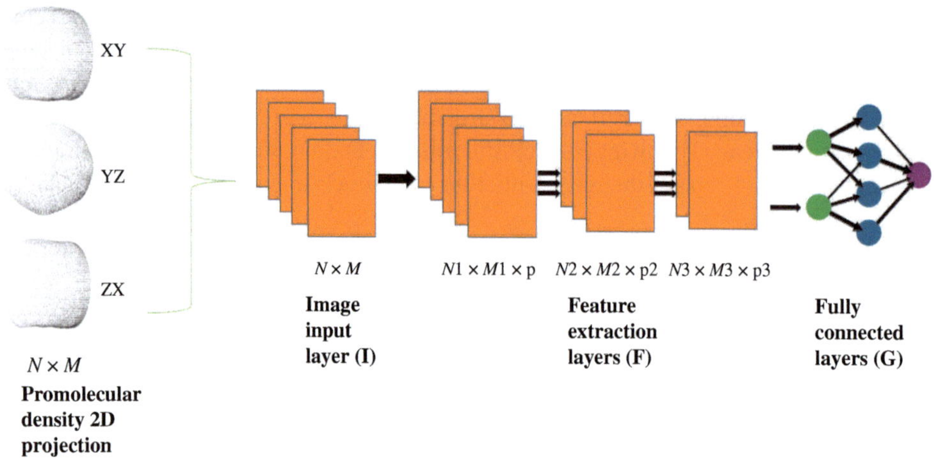

Figure 12.1 The promolecule density obtained from boron clusters is converted to three images, representing projection along the three axes. These images of $N \times M$ size are fed to the image input layer, which normalizes the image and feeds the normalized image to the feature extraction layer (F). The feature extraction layer consists of convolutional, RELU, and pooling layers to produce low dimensional feature representation of the original image. The features obtained are then normalized using batch normalization layers and finally fed to the classical neural network, i.e. the fully connected layer.

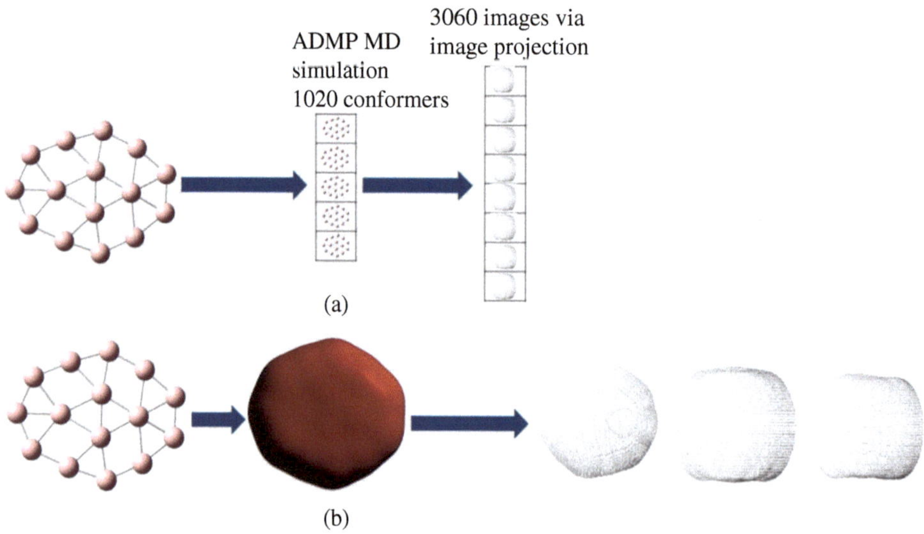

Figure 12.2 In (a), we can see the generation of conformers via MD simulation from boron clusters. From each of these conformers, promolecule density is generated (b) and projected to two-dimensional images in the XY, YZ and XZ planes, respectively.

using the Crystal Explorer (Figure 12.2a). The promolecule density is a 3D quantity and cannot be utilized by 2D CNN; of course, there are 3D CNNs that can directly work with the promolecule density, but we do not use 3D CNNs as they possess an inherent disadvantage; that is, they have high computational complexity and excessive memory usage. Therefore, we limit our study to 2D CNN to utilize the promolecule density for our model and we project this 3D quantity to three axes, **X**, **Y** and **Z**, to get 2D images in three orthogonal planes, respectively (Figure 12.2b).

Projection of the surface on the three orthogonal planes ensures that we are capturing maximum data from the 3D surface. Projection of the promolecule density to images is done via a pointcloud2image MATLAB script [68]. Accordingly, from 33.740 structures generated, we obtain 113,220 projected images; hence, we obtain a dataset of a size of 113.220.

The dataset obtained consists of projected images from planar, quasi-planar, bowl, tubular, cage and fullerene structures (cf. Table 12.1). A quite complicated growth pattern is observed as the sizes move from B_4 to B_{40}. Planar, quasi-planar and bowl structures are predominant (67.6%), followed by tubular (18.9%), cage (8.1%) and fullerene (5.4%) structures. The dataset is divided into the training, validation and test set in the following proportions of 60, 20 and 20%, respectively. We construct the test set to include the planar, quasi-planar, bowl, tubular fullerene and cage structures. Table 12.2 shows the distribution of training, validation and test datasets.

Accordingly, the training dataset consists of 22 boron clusters, which generate 67.320 data points. The validation dataset consists of eight boron clusters, which corresponds to 24.480 data points. The test dataset consists of seven boron clusters, which corresponds to 21.420 data points. The distribution of the datasets is presented in Table 12.3.

The model is trained using MATLAB, and the data are trained for 50 epochs. The initial learning rate is kept at 0.002, and the learning rate is lowered after 20 epochs, with a learning rate drop factor of 0.1. Early stopping is utilized to avoid overfitting. In early stopping, the variation between the evolving trend of the loss on training and validation is seen, then the training is stopped when the variation between training and validation increases by a large margin. The time required to train the model amounts to 1100 minutes (~18.3 hours).

Table 12.1 Distribution of structures of various boron clusters B_n ranging from size $n = 4 - 40$.

Planar, quasiplanar, bowl	Tubular	Fullerene	Cage
B_4, B_5, B_6, B_7, B_8, B_{10}, B_{11}, B_{12}, B_{13}, B_{15}, B_{16}, B_{17}, B_{18}, B_{19}, B_{21}, B_{23}, B_{25}, B_{28}, B_{29}, B_{30}, B_{33}, B_{35}, B_{36}, B_{37}, B_{38}	B_{20}, B_{22}, B_{24}, B_{26}, B_{27}, B_{32}, B_{39}	B_{14}, B_{40}	B_9, B_{31}, B_{34}

Table 12.2 Distribution of boron nanoclusters in the training, validation and test dataset.

Training dataset clusters	Validation dataset clusters	Test dataset clusters
B_7, B_8, B_{10}, B_{12}, B_{13}, B_{16}, B_{18}, B_{19}, B_{23}, B_{26}, B_{27}, B_{28}, B_{30}, B_{31}, B_{32}, B_{33}, B_{34}, B_{35}, B_{36}, B_{38}, B_{39}, B_{40}	B_4, B_6, B_9, B_{15}, B_{20}, B_{21}, B_{22}, B_{29}	B_5, B_{11}, B_{14}, B_{17}, B_{24}, B_{25}, B_{37}

Table 12.3 Distribution of the classes of structures for training, validation and test dataset.

	Planar, quasiplanar, bowl	Tubular	Fullerene	Cage
Training	68.2 %	18.2%	4.5%	9.1%
Validation	62.5%	25.0%		12.5%
Test	71.4%	14.3%	14.3%	

To further assess our model, we now utilize the following quantities, including the root-mean-square error (RMSE), the regression coefficient R^2, and the correlation coefficient R. The RMSE is ubiquitously used to assess the regression model, as it checks the difference between the value given by the model and the actual response variable. Therefore, a lower RMSE close to zero signifies a good model, but the RMSE is also dependent on the scale of the response variable. The response variable in our case is the ground state energy of the nanocluster. The ground state energy of nanocluster is obtained by averaging over the energy of all the conformers of a nanocluster.

The energy of a conformer A (E_{confA}) with respect to the ground state system's energy (E_{gs}) can be written as (12.6):

$$E_{\text{confA}} = E_{\text{gs}} + \Delta E_{\text{A}} \tag{12.6}$$

The average of all the conformers for a particular nanocluster can thus be written as:

$$E_{\text{avg\_conf}} = \left(\sum^{n} E_{\text{gs}} + \sum^{n} \Delta E_{n}\right) / n \tag{12.7}$$

$$E_{\text{avg\_conf}} = E_{\text{gs}} + \Delta E_{\text{avg}} \tag{12.8}$$

Therefore,

$$E_{\text{avg\_conf}} \approx E_{\text{gs}} \tag{12.9}$$

Overall, by taking the average of the energies predicted by the CNN model, we can evaluate the ground state total energy of a particular boron cluster in the test dataset (Figure 12.3).

The R^2 and R values are respectable, and the model can approximate the energy trends well enough, but the RMSE of 54.7 Hartree seems lacking. Consequently, this model may not be very suitable for the purpose of regression, but since it is excellent in terms of capturing the trend, we believe this model is better suited for classification-based tasks. The regression model can be converted to a classification model by using thresholds. For the classification task, we use a test dataset containing boron clusters ranging from B_{41} to B_{50}. The global minimum structures and a few of their higher energy constitutional isomeric structures are used in this task. We again perform the MD simulation, generating 50 conformers each for the global minimum and their higher energy isomeric counterparts (Table 12.4).

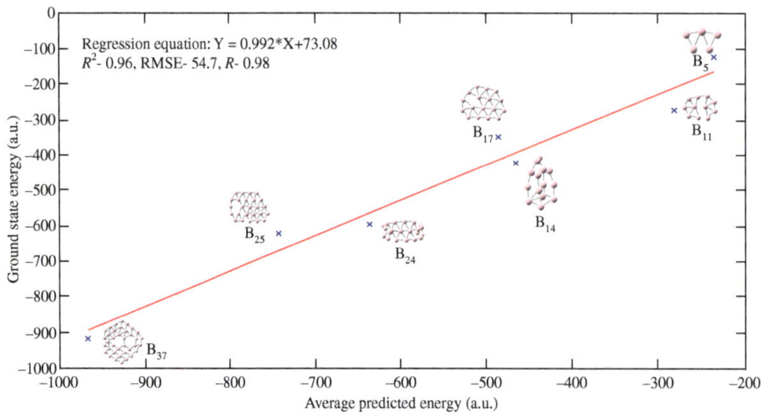

Figure 12.3 The performance on the test dataset is illustrated by the graph; we obtain a good R^2 and R value of 0.96 and 0.98, respectively, and a respectable RMSE of 54.7.

Table 12.4 Boron clusters used for the purpose of classification.

Boron cluster	Global minimum	Higher energy configurational isomer 1	Higher energy configurational isomer 2
B_{41}	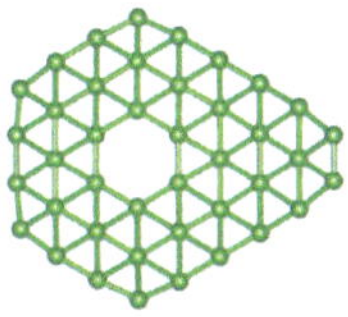	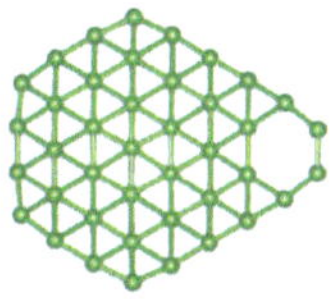	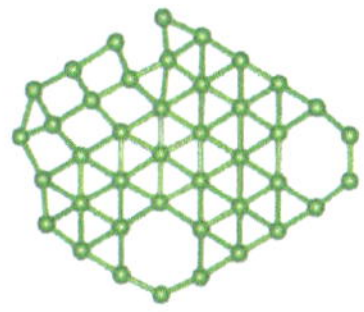
B_{42}	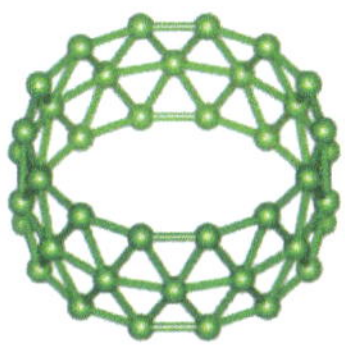	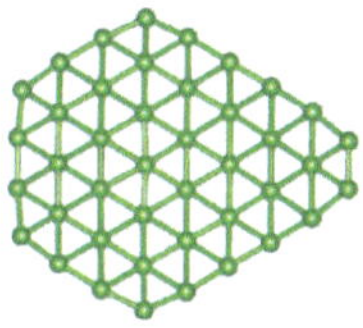	
B_{43}	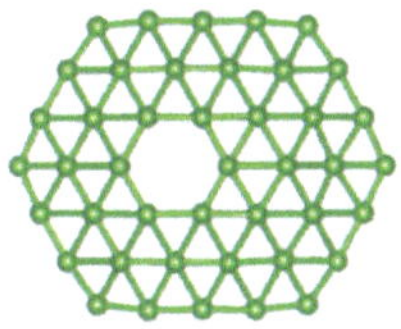	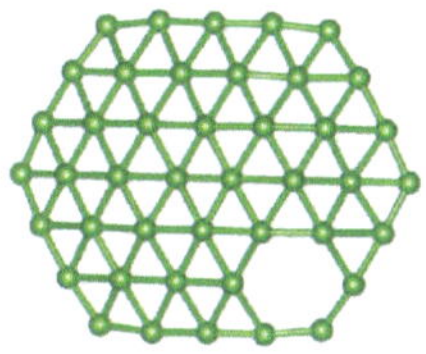	
B_{44}	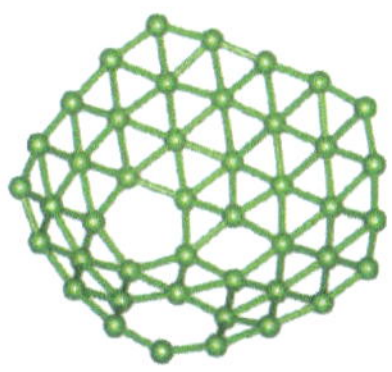	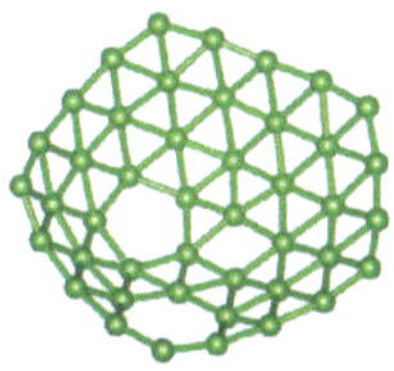	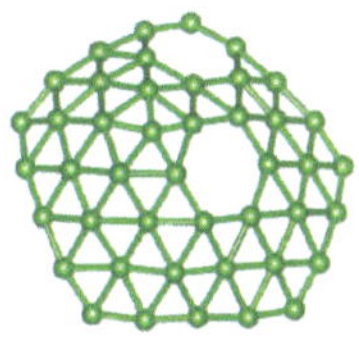
B_{45}	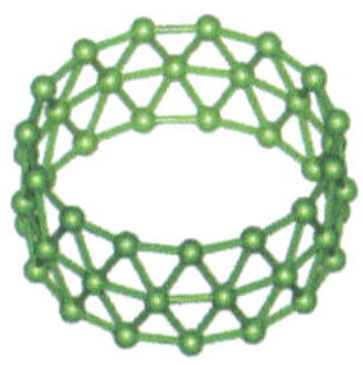	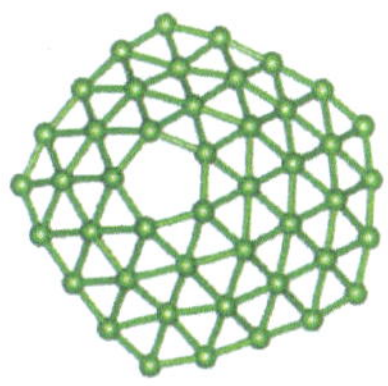	
B_{46}	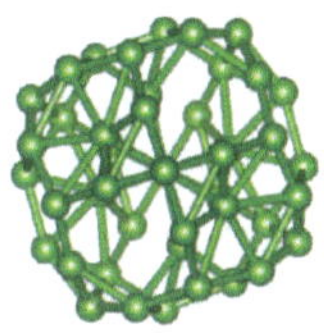	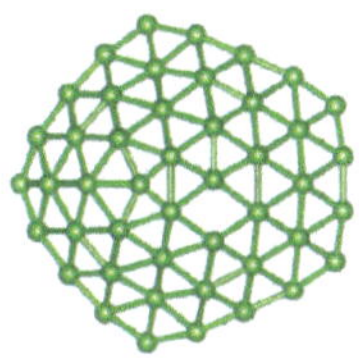	

(Continued)

Table 12.4 (Continued)

Boron cluster	Global minimum	Higher energy configurational isomer 1	Higher energy configurational isomer 2
B_{47}	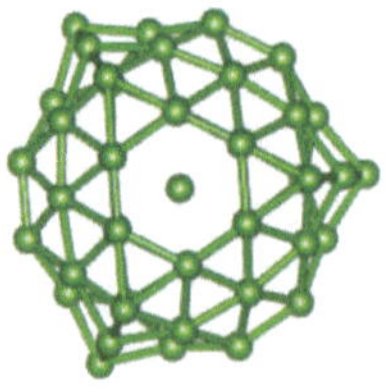	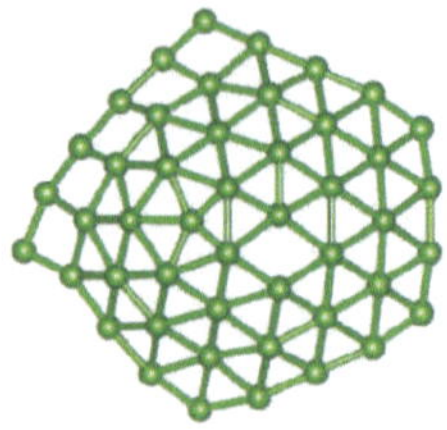	
B_{48}	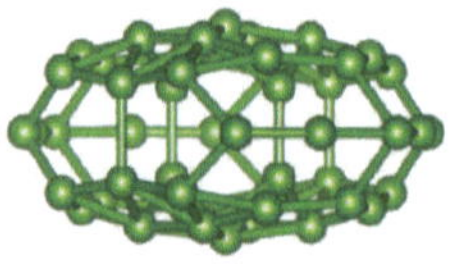	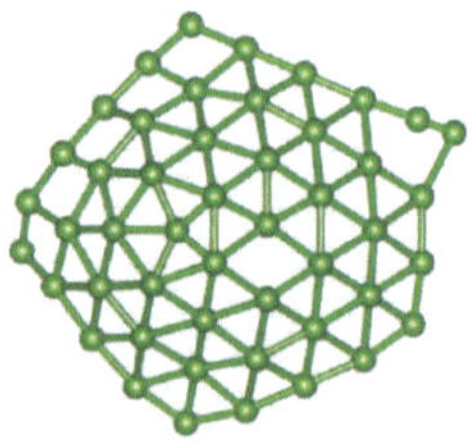	
B_{49}	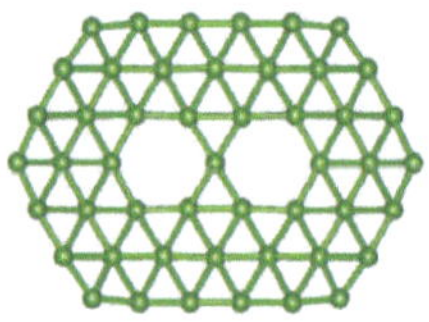	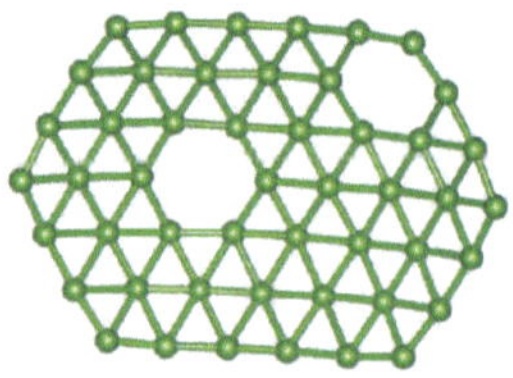	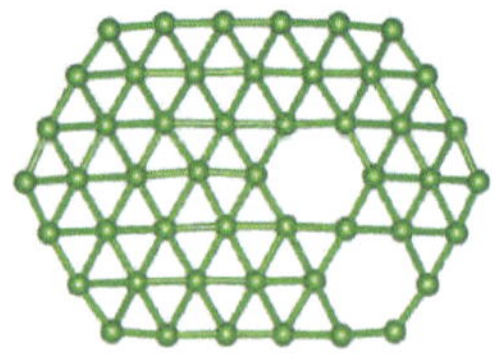
B_{50}	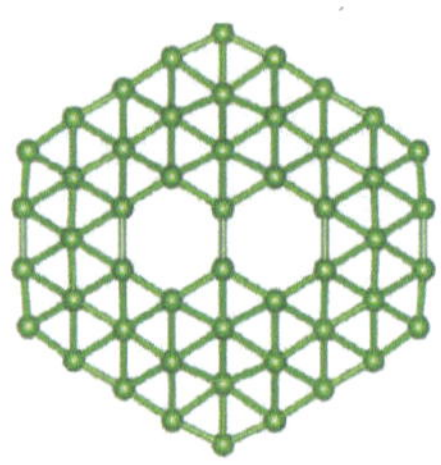	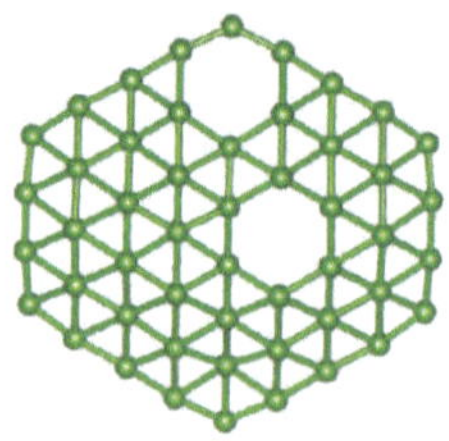	

The promolecule densities are subsequently calculated, and their 2D projection is utilized by CNN to obtain predicted energies. Consequently, we take the average of the energy over the 50 conformers for each constitutional isomer to reach the predicted ground state energy. Hence, we obtain ground state energy for each of the constitutional isomers of each nanocluster. The ground state energies of the isomers are then normalized using max–min normalization. The purpose of a max–min normalization is to ensure that the constitutional isomer with the lowest energy gets the value zero, while the isomer with the highest energy gets the value 1. The constitutional isomer having the value zero is then classified as the global minimum (Table 12.5). These are the normalized scores obtained for the classification task (Table 12.6).

Table 12.5 The process of classification using our CNN approach.

Cluster	MD simulation (50 conformers)	Promolecular density projections	CNN predicted energy	CNN predicted ground state energy	Min–max normalization $\frac{E - \min(E_{avg}, E_{avg'})}{E_{avg'} - E_{avg}}$
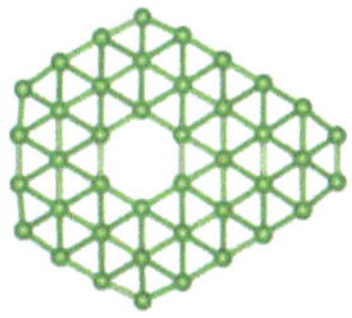	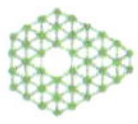		E_{conf1}	E_{avg}	$\text{Minmax}(E_{avg})$
			E_{conf2}		
B_{41} ground state			E_{conf3}		
			…		
			E_{conf4}		
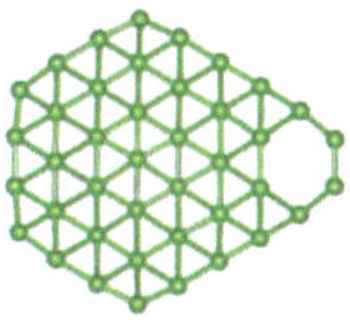	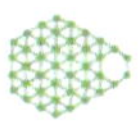		$E_{conf1'}$	$E_{avg'}$	$\text{Minmax}(E_{avg'})$
	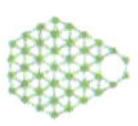		$E_{conf2'}$		
B_{41} constitutional isomer (higher energy)	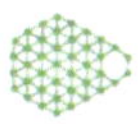		$E_{conf3'}$		
	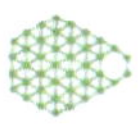		…		
		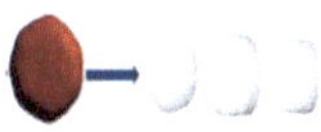	$E_{conf4'}$		

Consequently, we go on to perform binary classification with this data by taking the threshold t as greater than zero as inactive for the normalized score. Thus, for $t > 0$ the class inactive is assigned as is for the rest class active. Using this threshold, we obtain the following confusion matrix for our classification problem (Table 12.7):

The Mathew's correlation coefficient (MCC) is written in expression (12.10):

$$\text{MCC} = \{(\text{TP} \cdot \text{TN} - \text{FP} \cdot \text{FN})/\text{sqrt}[(\text{TP} + \text{FP}) \cdot (\text{TP} + \text{FN}) \cdot (\text{TN} + \text{FP}) \cdot (\text{TN} + \text{FN})]\} \times 100 \tag{12.10}$$

The MCC, in this case results in 82.3%, which is a respectable score. In this sense, the CNN classification paradigm is successful in separating the global minimum out of the different configurational isomers except for one misclassification. This CNN classification paradigm can then be further used to augment structure prediction algorithms.

Table 12.6 The normalized scores obtained for each given nanocluster.

	Predicted Minmax normalized score for global minimum	Predicted Minmax normalized score for constitutional isomer 1	Predicted Minmax normalized score for constitutional isomer 2
B_{41}	0	0.46	1
B_{42}	0	1	—
B_{43}	0	1	—
B_{44}	0	1	0.1
B_{45}	1	0	—
B_{46}	0	1	—
B_{47}	0	1	—
B_{48}	0	1	—
B_{49}	0	0.81	1
B_{50}	0	1	—

Table 12.7 This is the confusion matrix for the classification problem.

	Predicted actives	Predicted inactives
True actives	9 (TP)	1 (FP)
True inactives	1 (FN)	12 (TN)

TP, TN, FP and FN are true positives, true negatives, false positives, and false negatives, respectively.

12.4 Concluding Remarks

Traditionally, the mapping of the electron density of a molecule to its ground state energy has been carried out by using QC programs utilizing DFT functionals. In the present study, we attempt a different approach for mapping the electron density to the ground state total energy: *we utilize neural networks for mapping the electron density of a molecule to its ground state energy.*

Our approach consists of the utilization of a 2D CNN with projected images. The electron density is obtained via promolecule densities that are then projected to images along X, Y and Z axes. These images are utilized by the CNN model to predict energy. In the case of a 2D CNN, we obtain a featureless learning: the images are directly utilized by the CNN model to predict the energies of the boron clusters.

Boron nanoclusters ranging from B_4 to B_{40} are used for the study. The dataset of projected images is divided into training, test and validation sets (with a split percentage: 60 : 20 : 20%). The 2D CNN model obtained shows a good R^2 value of 0.96 and a good RMSE of 54.7 on the test set. Therefore, the model meets our goal for the utilization in exploration of PES. The model cannot predict the ground state energy with high accuracy as we are mapping the promolecule density to energy instead of the actual electron density. Hence, we transform this model into a classification model that can accurately predict which structures are the global minimum structure. This will be important for PES exploration of boron nanoclusters. We can quickly find the initial local minimum structures using our neural network. Overall, our approach shows that the mapping of the electron density of a molecule to its ground state energy is possible with the use of neural networks. Such neural network models can be augmented with the available structure generation programs which utilised Genetic algorithm and Particle Swarm optimisation algorithm for much faster prediction of the structures, using shorter computing times. The following step of this study, which could be more interesting but also more challenging, is a prediction of the geometric shape of the stable clusters.

Acknowledgements

This work was funded by VinGroup (Vietnam) and supported by VinGroup Innovation Foundation (VinIF) under the project code VinIF.2020.DA21. MTN is also grateful to Van Lang University for their support.

References

1 Goedecker, S., Hellmann, W., and Lenosky, T. (2005). Global minimum determination of the Born–Oppenheimer surface within density functional theory. *Phys. Rev. Lett.* 95: 055501.
2 Nouemo, S., Tchoffo, F., Ndjaka, J.M.B., and Domngang, S. (2016). Global minima of iron clusters described by Gupta potential. *J. Taibah Univ. Sci.* 10: 430–436.
3 Wang, Y., Lv, J., Zhu, L., and Ma, Y. (2012). CALYPSO: a method for crystal structure prediction. *Comput. Phys. Commun.* 183: 2063–2070.
4 Glass, C.W., Oganov, A.R., and Hansen, N. (2006). USPEX—evolutionary crystal structure prediction. *Comput. Phys. Commun.* 175: 713–720.
5 Entel, P., Gruner, M.E., Rollmann, G. et al. (2008). First-principles investigations of multimetallic transition metal clusters. *Philos. Mag.* 88: 2725–2738.
6 Dai, H., Rinzler, A.G., Nikolaev, P. et al. (1996). Single-wall nanotubes produced by metal-catalyzed disproportionation of carbon monoxide. *Chem. Phys. Lett.* 260: 471–475.
7 Elliott, J.A., Hamm, M., and Shibuta, Y. (2009). A multiscale approach for modeling the early stage growth of single and multiwall carbon nanotubes produced by a metal-catalyzed synthesis process. *J. Chem. Phys.* 130: 034704.
8 Berry, R.S. and Wales, D.J. (1989). Melting, spinodals, and clusters. *Phys. Rev. Lett.* 63: 1156.
9 Shubuta, Y. and Suzuki, T. (2007). Melting and nucleation of iron nanoparticles: a molecular dynamics study. *Chem. Phys. Lett.* 445: 265–270.
10 Billas, I.M., Châtelain, A., and de Heer, W.A. (1997). Magnetism of Fe, Co and Ni clusters in molecular beams. *J. Magn. Magn. Mater.* 168: 64–84.
11 Vanommeslaeghe, K. and Guvench, O. (2014). Molecular mechanics. *Curr. Pharm. Des.* 20: 3281–3292.
12 Sukumar, N., Krein, M., Luo, Q., and Breneman, C. (2012). MQSPR modeling in materials informatics: a way to shorten design cycles? *J. Mater. Sci.* 47: 7703–7715.
13 Gasteiger, J. and Zupan, J. (2017). Atomistic calculations and materials informatics: a review. *Curr. Opin. Sol. State Mat. Sci.* 21: 167–176.
14 Gasteiger, J. and Zupan, J. (1993). Neural networks in chemistry. *Angew. Chem. Int. Ed.* 32: 503–527.

15 Baskin, I.I., Winkler, D., and Tetko, I.V. (2016). A renaissance of neural networks in drug discovery. *Expert Opin. Drug Discovery* 11: 785–795.

16 Ekins, S. (2016). The next era: deep learning in pharmaceutical research. *Pharm. Res.* 33: 2594–2603.

17 Behler, J. and Parrinello, M. (2007). Generalized neural-network representation of high-dimensional potential-energy surfaces. *Phys. Rev. Lett.* 98: 146401.

18 Chmiela, S., Tkatchenko, A., Sauceda, H.E. et al. (2017). Machine learning of accurate energy-conserving molecular force fields. *Sci. Adv.* 3: e1603015.

19 Behler, J. (2011). Atom-centered symmetry functions for constructing high-dimensional neural network potentials. *J. Chem. Phys.* 134: 074106.

20 Eckhoff, M. and Behler, J. (2021). High-dimensional neural network potentials for magnetic systems using spin-dependent atom-centered symmetry functions. *npj Comput. Mater.* 7: 170.

21 Pham, H.T., Duong, L.V., Pham, B.Q., and Nguyen, M.T. (2013). The 2D-to-3D geometry hopping in small boron clusters: the charge effect. *Chem. Phys. Lett.* 577: 32–40.

22 Kiran, B., Bulusu, S., Zhai, H.J. et al. (2005). Planar-to-tubular structural transition in boron clusters: B_{20} as the embryo of single-walled boron nanotubes. *PNAS* 102: 961–971.

23 Boustani, I. (1994). Systematic LSD investigation on cationic boron clusters: B_n (n: 2–14). *Int. J. Quantum Chem.* 52: 1081–1091.

24 Boustani, I. (1995). A comparative study of ab initio SCF-CI and DFT. Example of small boron clusters. *Chem. Phys. Lett.* 233: 273–279.

25 Li, W.L., Romanescu, C., Galeev, T.R. et al. (2012). Transition-metal-centered nine-membered boron rings: M@B_9 and M@B_9 (M = Rh, Ir). *JACS* 134: 165.

26 Zhai, H.J., Alexandrova, A.N., Birch, K.A. et al. (2003). Hepta- and octacoordinate boron in molecular wheels of eight- and nine-atom boron clusters: observation and confirmation. *Angew. Chem. Int. Ed.* 42: 6004–6008.

27 Zhai, H.J., Kiran, B., Li, J., and Wang, L.S. (2003). Hydrocarbon analogues of boron clusters—planarity, aromaticity and antiaromaticity. *Nat. Mater.* 2: 827–837.

28 Ma, J., Li, Z., Fan, K., and Zhou, M. (2003). Density functional theory study of the B_6, B_6^+, B_6^-, and B_6^{2-} clusters. *Chem. Phys. Lett.* 372: 708–714.

29 Alexandrova, A.N., Zhai, H.J., Wang, L.S., and Boldyrev, A.I. (2004). Molecular wheel B_8^{2-} as a new inorganic ligand. Photoelectron spectroscopy and ab initio characterization of LiB_8. *Inorg. Chem.* 43: 3552–3560.

30 Li, Q.S. and Gong, L.F. (2004). Novel pyramidal MB_7 (M = Li, Na, K, Rb, or Cs) species: structure and aromaticity. *J. Phys. Chem. A* 108: 4322–4330.

31 Aihara, J.I., Kanno, H., and Ishida, T. (2005). Aromaticity of planar boron clusters confirmed. *JACS* 127: 13324–13331.

32 Chen, Q., Chen, T.T., Li, H.R. et al. (2019). B_{31}^- and B_{32}^-: chiral quasi-planar boron clusters. *Nanoscale* 11: 9698–6706.

33 Boustani, I. (1997). New quasi-planar surfaces of bare boron. *Surf. Sci.* 370: 355–362.

34 Li, W.L., Zhao, Y.F., Hu, H.S. et al. (2014). B_{30}^-: a quasiplanar chiral boron cluster. *Angew. Chem. Int. Ed.* 53: 5540–5544.

35 Li, R., You, X.R., and Zhai, H.J. (2020). Are all planar and quasi-planar boron clusters aromatic? Counter examples of island or global π antiaromaticity from chemical bonding analysis. *Phys. Chem. Chem. Phys.* 22: 25084–25092.

36 Krishnan, S. and Senthilkumar, K. (2022). Opto-electronic properties of quasi-planar boron clusters – a DFT investigation. *Chem. Phys. Lett.* 139914.

37 Popov, I.A., Piazza, Z.A., Li, W.L. et al. (2013). Theoretical prediction of a new class of aromatic aluminum clusters $Al_3(-/0/+)$. *J. Chem. Phys.* 139: 144307.

38 Pan, S., Barroso, J.F., Jalife, S. et al. (2019). A combined photoelectron spectroscopy and ab initio study of the quasi-planar B_{24}^- cluster. *Acc. Chem. Res.* 522: 2732–2740.

39 Piazza, Z.A., Hu, H.S., Li, W.L. et al. (2014). Planar hexagonal B_{36} as a potential basis for extended single-atom layer boron sheets. *Nat. Commun.* 5: 1–6.

40 Sergeeva, A.P., Popov, I.A., Piazza, Z.A. et al. (2014). Understanding boron through size-selected clusters: structure, chemical bonding, and fluxionality. *Acc. Chem. Res.* 47: 1349–1359.

41 Chen, Q., Gao, Y., Zhang, X. et al. (2014). Quasi-planar aromatic B_{36} and B_{36}^- clusters: all-boron analogues of coronene. *Phys. Chem. Chem. Phys.* 16: 18282–18291.

42 Becker, R. and Chkhartishvili, L. (2015). Dipole moment of quasi-planar boron clusters. *Nano Stud.* 11: 29–36.

43 Tai, T.B., Duong, L.V., Pham, H.T. et al. (2014). A disk-aromatic bowl cluster B 30: toward formation of boron buckyballs. *Chem. Commun.* 50: 1558–1562.

44 Pham, H.T., Duong, L.V., Tam, N.M. et al. (2014). The boron conundrum: bonding in the bowl B_{30} and B_{36}, fullerene B_{40} and triple ring B_{42} clusters. *Chem. Phys. Lett.* 608: 295–304.

45 Tai, T.B. and Nguyen, M.T. (2015). The B_{32} cluster has the most stable bowl structure with a remarkable heptagonal hole. *Chem. Commun.* 51: 7677–7681.

46 Li, R., You, X.R., Wang, K., and Zhai, H.J. (2018). Nature of bonding in bowl-like B_{36} cluster revisited: concentric $(6\pi + 18\pi)$ double aromaticity and reason for the preference of a hexagonal hole in a central location. *Chem. Asian J.* 13: 1148–1153.

47 Li, S.D., Miao, C.Q., Guo, J.C., and Ren, G.M. (2006). Transition metal–boron complexes BnM: from bowls (n = 8–14) to tires (n = 14). *J. Comput. Chem.* 27: 1858–1868.

48 Bai, H., Chen, Q., Miao, C.Q. et al. (2013). Ribbon aromaticity in double-chain planar $B_nH_{22}^-$ and $Li_2B_nH_2$ nanoribbon clusters up to n = 22: lithiated boron dihydride analogues of polyenes. *Phys. Chem. Chem. Phys.* 15: 18872–18880.

49 Tai, T.B. and Nguyen, M.T. (2015). Electronic structure and photoelectron spectra of B_n with n = 26–29: an overview of structural characteristics and growth mechanism of boron clusters. *Phys. Chem. Chem. Phys.* 17: 13672–13679.

50 Chen, Q., Tian, W.J., Feng, L.Y. et al. (2017). B_{31}^- and B_{32}^-: chiral quasi-planar boron clusters. *Nanoscale* 9: 4550–4556.

51 Zhai, H.J., Zhao, Y.F., Li, W.L. et al. (2014). Observation of an all-boron fullerene. *Nat. Chem.* 6: 727–731.

52 Wu, X., Sai, L., Zhou, S. et al. (2020). Competition between tubular, planar and cage geometries: a complete picture of structural evolution of B_n (n = 31–50) clusters. *Phys. Chem. Chem. Phys.* 22: 12959–12967.

53 Yan, Q.Q., Pei, L., and Li, S.D. (2021). Predicting bilayer B_{50}, B_{52}, B_{56}, and B_{58}: structural evolution in bilayer B_{48}–B_{72} clusters. *J. Mol. Model.* 27: 1–11.

54 Chen, W.J., Ma, Y.Y., Chen, T.T. et al. (2021). B_{48}^-: a bilayer boron cluster. *Nanoscale* 13: 3868–3875.

55 Zhao, J., Wang, L., Li, F., and Chen, Z. (2010). B_{80} and other medium-sized boron clusters: core–shell structures, not hollow cages. *J. Phys. Chem. A* 114: 9969–9977.

56 Rahane, A.B. and Kumar, V. (2015). B_{84}: a quasi-planar boron cluster stabilized with hexagonal holes. *Nanoscale* 7: 21822–21828.

57 Prasad, D.L. and Jemmis, E.D. (2008). Stuffing improves the stability of fullerene like boron clusters. *Phys. Rev. Lett.* 100: 165504.

58 Pham, H.T., Muya, J.T., Buendia, F. et al. (2019). Formation of the quasi-planar B_{50} boron cluster: topological path from B_{10} and disk aromaticity. *Phys. Chem. Chem. Phys.* 21: 7039–7049.

59 Buendia, F., Pham, H.T., Barquera-Lozada, J.E. et al. (2022). Formation of the quasi-planar B_{56} boron cluster: topological path from B_{12} and disk aromaticity. *Phys. Chem. Chem. Phys.* 24: 21822–21834.

60 Saha, P., Buendia, F., Duong, L.V., and Nguyen, M.T. (2023). A topological path to the formation of the quasi-planar B_{70} boron cluster and its dianion. *Phys. Chem. Chem. Phys.* in press (2024).

61 Hohenberg, P. and Kohn, W. (1964). Inhomogeneous electron gas. *Phys. Rev. E* 136: B864.

62 Zhao, Y., Yuan, K., Liu, Y. et al. (2020). Predicting elastic properties of materials from electronic charge density using 3D deep convolutional neural networks. *J. Phys. Chem. C* 124: 17262–172670.

63 Hirshfeld, F.L. (1977). Bonded-atom fragments for describing molecular charge densities. *Theor. Chim. Acta* 44: 129–139.

64 Schlegel, H.B., Millam, J.M., Iyengar, S.S. et al. (2001). Ab initio molecular dynamics: propagating the density matrix with Gaussian orbitals. *J. Chem. Phys.* 114: 9758–9768.

65 Castleman, A.W. and Jena, P. (2006). Clusters: a bridge across the disciplines of environment, materials science, and biology. *Proc. Nat. Acad. Sci.* 103: 10554–10562.

66 Frisch, M.J., Trucks, G.W., Schlegel, H.B., et al. (2009). *Gaussian 09, rev. A.1*. Gaussian Inc: Wallingford, CT.

67 Spackman, P.R., Turner, M.J., McKinnon, J.J. et al. (2021). CrystalExplorer: a program for Hirshfeld surface analysis, visualization and quantitative analysis of molecular crystals. *J. Appl. Crystallogr.* 54: 1006–1011.

68 Vahid, B. (2023). pointcloud2image (x,y,z,numr,numc). MATLAB Central File Exchange. https://www.mathworks.com/matlabcentral/fileexchange/55031-pointcloud2image-x-y-z-numr-numc (accessed 26 April 2022).

13

Density-Based Description of Molecular Polarizability for Complex Systems

Dongbo Zhao[1], *Xin He*[2], *Paul W. Ayers*[3] *and Shubin Liu*[4,5]

[1] *Institute of Biomedical Research, Yunnan University, Kunming, P.R. China*
[2] *Qingdao Institute for Theoretical and Computational Sciences, Institute of Frontier and Interdisciplinary Science, Shandong University, Qingdao, Shandong, P.R. China*
[3] *Department of Chemistry and Chemical Biology, McMaster University, Hamilton, Canada*
[4] *Research Computing Center, University of North Carolina, Chapel Hill, NC, USA*
[5] *Department of Chemistry, University of North Carolina, Chapel Hill, NC, USA*

13.1 Introduction

Polarizability (α_{iso}) is a fundamental quantity in physics, chemistry, biology, and especially materials science [1, 2]. It has been found to be correlated with many fundamental electronic structure characteristics, of which are electronegativity [3], nuclear quadrupole resonance [4], chemical hardness/softness [5–10], the van der Waals radius [11], etc. Many attempts have been devoted to the computation of α_{iso} [12–21]. However, for macromolecular and excited-stated systems, only sporadic studies have been reported in the literature because solving the coupled-perturbed Hartree–Fock (CPHF) [22, 23] or Kohn–Sham (CPKS) [24] equations is time-consuming to obtain the α_{iso} results, normally computationally intractable for macromolecular systems, such as proteins. For large systems in the ground state, both density- [25, 26] and fragment-based [27] linear-scaling algorithms are available for predicting the α_{iso} results. Among these methods, the generalized energy-based fragmentation (GEBF) [28–31] method is a prominent approach. Additionally, recent progress includes the machine-learning (ML)-based [32–34] acceleration techniques and linear regression-based [35] predictions. For excited-state systems of small molecules, the prediction of α_{iso} values is still challenging. This is mainly because an accurate and efficient computational methods for predicting excited-state geometries [36–38] have only been recently developed.

In this chapter, we aim to introduce an alternative way of predicting the α_{iso} values of macromolecular systems [39, 40] and small molecules [41] in their first excited states, as shown in Scheme 13.1. To this end, we borrow some key density-based quantities, such as Shannon entropy [42] and Fisher information [43], from the information-theoretic approach (ITA) [44]. We first construct a set of molecules as the training set, based on which we establish linear relationships between the α_{iso} values and ITA quantities. Then, one can predict the α_{iso} values of a molecule in the test set once the total molecular wavefunction (or electron density) is obtained. For proteins whose molecular wavefunction is a computational bottleneck for ITA quantities, we combine the GEBF method and ITA quantities to predict the α_{iso} data in the framework of density functional theory (DFT) [45, 46]. Our results showcase that (i) ITA quantities can be used for predicting both the ground-state and the first excited-state α_{iso} results; (ii) the GEBF-ITA protocol is a promising

Electron Density: Concepts, Computation and DFT Applications, First Edition.
Edited by Pratim Kumar Chattaraj and Debdutta Chakraborty.

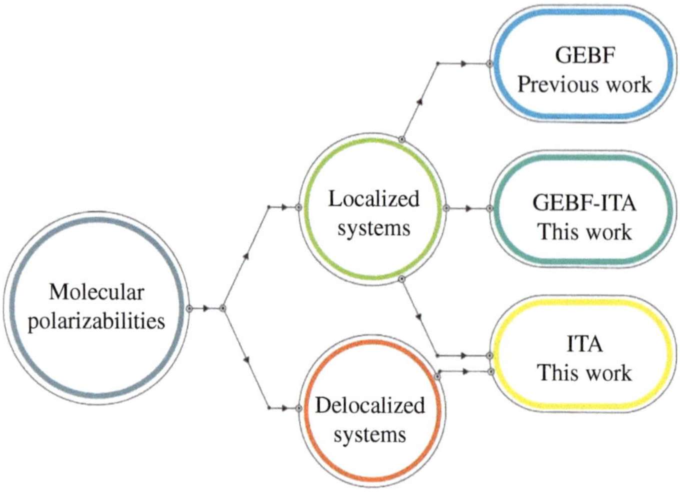

Scheme 13.1 Several protocols were used in this work to predict molecular polarizabilities of both delocalized and localized systems in the ground state and excited state. For both delocalized and localized systems, one can predict the polarizabilities via linear regression equations of ITAs when the molecular wave function is available. For localized systems, one can directly predict the polarizability of a macromolecule with the previous GEBF method. Alternatively, one can predict the polarizabilities of all subsystems generated from GEBF and obtain the molecular polarizability via a linear combination of subsystem polarizabilities. Source: Reprinted with permission from Zhao et al. [40]. Copyright 2022 Royal Chemical Society.

theoretical model in accurately and efficiently predicting the α_{iso} data of macromolecular systems; (iii) good transferability is also observed for ITA quantities.

The remainder of this chapter is organized as follows. In Section 13.2, we will outline the main theoretical foundations. In Section 13.3, we will present the main results. Finally, we will summarize our findings and provide future perspectives in Section 13.4.

13.2 Methodology and Computations

13.2.1 Information-Theoretic Approach (ITA) Quantities

A list of ITA quantities is introduced in this chapter, including Shannon entropy (S_S) [42], Fisher information (I_F) [43], information gain (I_G, also called relative Shannon entropy or Kullback–Leibler divergence) [47], relative Fisher information (G_1, G_2, and G_3) [48, 49], Ghosh–Berkowitz–Parr entropy (S_{GBP}) [50], Onicescu information energy of second and third order (E_2 and E_3) [51], the relative Rényi entropy of order two and three (${}^{r}R_2$ and ${}^{r}R_3$) [51]. To obtain atomic contributions, the Hirshfeld partition scheme [44, 52–54] is often adopted in the literature. The detailed definitions of all these quantities and the partition scheme are shown below:

$$S_S = -\int \rho(\mathbf{r}) \ln \rho(\mathbf{r}) d\mathbf{r} \tag{13.1}$$

$$I_F = \int \frac{|\nabla \rho(\mathbf{r})|^2}{\rho(\mathbf{r})} d\mathbf{r} \tag{13.2}$$

$$I_G = \int \rho(\mathbf{r}) \ln \frac{\rho(\mathbf{r})}{\rho_0(\mathbf{r})} d\mathbf{r} \tag{13.3}$$

$$G_1 = \sum_{\mathrm{A}} \int \nabla^2 \rho_{\mathrm{A}}(\mathbf{r}) \ln \frac{\rho_{\mathrm{A}}(\mathbf{r})}{\rho_{\mathrm{A}}^0(\mathbf{r})} d\mathbf{r} \tag{13.4}$$

$$G_2 = \sum_{\mathrm{A}} \int \rho_{\mathrm{A}}(\mathbf{r}) \left[\frac{\nabla^2 \rho_{\mathrm{A}}(\mathbf{r})}{\rho_{\mathrm{A}}(\mathbf{r})} - \frac{\nabla^2 \rho_{\mathrm{A}}^0(\mathbf{r})}{\rho_{\mathrm{A}}^0(\mathbf{r})} \right] d\mathbf{r} \tag{13.5}$$

$$G_3 = \sum_{\mathrm{A}} \int \rho_{\mathrm{A}}(\mathbf{r}) \left[\nabla \ln \frac{\rho_{\mathrm{A}}(\mathbf{r})}{\rho_{\mathrm{A}}^0(\mathbf{r})} \right]^2 d\mathbf{r} \tag{13.6}$$

$$S_{\mathrm{GBP}} = -\int \frac{3}{2} k \rho(\mathbf{r}) \left[c + \ln \frac{t(\mathbf{r}; \rho)}{t_{\mathrm{TF}}(\mathbf{r}; \rho)} \right] d\mathbf{r} \tag{13.7}$$

$$E_2 = \int \rho^2(\mathbf{r}) d\mathbf{r} \tag{13.8}$$

$$E_3 = \frac{1}{2} \int \rho^3(\mathbf{r}) d\mathbf{r} \tag{13.9}$$

$$R_r^2 = \ln \int \frac{\rho^2(\mathbf{r})}{\rho_0(\mathbf{r})} d\mathbf{r} \tag{13.10}$$

$$R_r^3 = \frac{1}{2} \ln \int \frac{\rho^3(\mathbf{r})}{\rho_0^2(\mathbf{r})} d\mathbf{r} \tag{13.11}$$

$$\rho_{\mathrm{A}}(\mathbf{r}) = \omega_{\mathrm{A}}(\mathbf{r}) \rho(\mathbf{r}) = \frac{\rho_{\mathrm{A}}^0(\mathbf{r})(\mathbf{r} - \mathbf{R}_{\mathrm{A}})}{\sum_{\mathrm{B}} \rho_{\mathrm{B}}^0(\mathbf{r})(\mathbf{r} - \mathbf{R}_{\mathrm{A}})} \rho(\mathbf{r}) \tag{13.12}$$

where $\rho(\mathbf{r})$, $\nabla\rho(\mathbf{r})$, and $\nabla^2\rho(\mathbf{r})$ are the electron density, its gradient, and its Hessian, respectively, $\rho_0(\mathbf{r})$ is the reference state density, subjected to the same conditions as those of $\rho(\mathbf{r})$. An alternative expression of Eq. (13.2) has been proposed [55] by one of the present authors. In Eqs. (13.4)–(13.6), atomic contributions from a given atom A are considered. In Eq. (13.7), k is the Boltzmann constant, $c = (5/3) + \ln(4\pi c_{\mathrm{K}}/3)$, and $c_{\mathrm{K}} = (3/10)(3\pi^2)^{2/3}$. $t(\mathbf{r}; \rho)$ and $t_{\mathrm{TF}}(\mathbf{r}; \rho)$ are the noninteracting kinetic energy density and the Thomas–Fermi (TF) kinetic energy density, respectively, while the former is a simple function of electron density, $t_{\mathrm{TF}}(\mathbf{r}; \rho) = c_{\mathrm{K}}\rho^{5/3}(\mathbf{r})$, the latter is expressed as a function of orbital densities, $t(\mathbf{r}; \rho) = \sum_i \frac{1}{8} \frac{\nabla\rho_i \cdot \nabla\rho_i}{\rho_i} - \frac{1}{8}\nabla^2\rho$. In Eq. (13.12), $\rho_{\mathrm{A}}(\mathbf{r})$ denotes the Hirshfeld density for atom A in a molecule, $\omega_{\mathrm{A}}(\mathbf{r})$ is a sharing function for atom A, $\rho_{\mathrm{B}}^0(\mathbf{r})(\mathbf{r} - \mathbf{R}_{\mathrm{A}})$ is the density of atom B centered at $\mathbf{R}_{\mathrm{A}}$, and $\rho(\mathbf{r})$ is the actual electron density in a molecule. The stockholder approach normally uses the so-called promolecular density as input, which is typically obtained as a summation of spherically averaged atomic densities. Since partitioning schemes cannot be uniquely defined, Becke's fuzzy atom approach [56] and Bader's zero-flux atoms-in-molecules method [57] are two alternative schemes in the literature. In addition, recent progress of ITA quantities can be found in our two reviews [58, 59]. One particularly interesting topic is homochirality [60] and the principle of chirality hierarchy [61].

13.2.2 The GEBF Method

Originally designed for approximating the total energy of a macromolecule with good accuracy and efficiency, the GEBF method has witnessed tremendous success in predicting the energetics, structures, and molecular properties in the past decade or so. For a large molecule or molecular aggregate, the total energy is obtained as a linear combination of a few small embedded subsystems, as shown in Eq. (13) [28–31],

$$E_{\mathrm{mol}} = \sum_m C_m E_m - \left[\left(\sum_m C_m \right) - 1 \right] \sum_{\mathrm{A}} \sum_{\mathrm{B>A}} \frac{Q_{\mathrm{A}} Q_{\mathrm{B}}}{R_{\mathrm{AB}}} \tag{13.13}$$

$$\Omega_{\mathrm{mol}} = \sum_m C_m \Omega_m \tag{13.14}$$

where E_m and C_m are the total energy and the corresponding coefficient of the mth subsystem, Q_A, is the atomic charge on atom A, and R_{AB} denotes the interatomic distance. In Eq. (13.14), one can obtain the total molecular property (Ω_{mol}) via a linear combination of the corresponding molecular properties (Ω_m) in all subsystems. This equation has been successfully applied to obtain the dipole moment [29], chemical shielding tensor [62–64], etc. The success of Eq. (13.14) is due to the spatial translation invariance of α_{iso}. With this symmetry property, all the α_{iso} data can be translated from origins on each subsystem calculation to the same molecular origin.

Next, it is important to discuss the construction of subsystems in GEBF. For a protein molecule, each amino acid residue is routinely considered as a fragment. We cut a single C—C bond between two neighboring residues. Hydrogen atoms are added to have closed-shell electronic structures. Centered at each fragment with a distance parameter (4.0 Å by default), we first build a series of *primitive* subsystems whose coefficient (C_m) is set to be +1. Then, we need to build some smaller *derivative* subsystems to eliminate the overlap of primitive subsystems. The coefficient of each derivative subsystem is a unique integer, as automatically determined by the principle of inclusion and exclusion. Note that the maximum number of fragments in a subsystem is normally set to eight to further reduce the computational cost without compromising accuracy. For atomic charges, standard single-point calculations on primitive subsystems are carried out to obtain natural population analysis (NPA) [65] charges. All quantum chemical calculations of subsystems are performed with the GEBF method in the LSQC [66] program. Both Gaussian 16 [67] and Multiwfn 3.8 [68] are used to obtain the polarizabilities (α_{iso}), wavefunctions, and ITA quantities.

13.3 Results and Discussion

We have constructed five classes of inorganic, organic, and biological systems (see Ref. [40] and references therein), and they were fully optimized at the M06-2X/def2-SVP [69, 70] level. Molecular polarizabilities, volumes, quadrupole moments, and wavefunctions were obtained at the M06-2X/def2-TZVP [69, 70] level. The correlation coefficients between all ITA variables, molecular volumes (Vol), quadrupole moments (Θ_{iso}), and molecular polarizabilities (α_{iso}) are included in Tables 13.1 and 13.2. It is clear that the overall trend is satisfactory, with only a few exceptions, indicating that ITA quantities can be widely applied to various inorganic, organic, and biological systems. Particularly, we have observed that ITA quantities such as S_S, S_{GBP}, ${}^2R_r$, and G_3 should be good indicators of α_{iso}. For example, we have established strong correlations between the α_{iso} results and (i) molecular volume, (ii) S_{GBP}, (iii) ${}^2R_r$, and (iv) G_3 for a total of 20 amino acids, 400 dipeptides, and 8000 tripeptides as shown in Figure 13.1. Can these linear regressions be used to predict the α_{iso} results of even larger peptides or even proteins (see Scheme 13.2)? The answer is apparently positive!

Figure 13.2 compares the calculated and predicted α_{iso} data of delocalized fullerenes (C_{100} and C_{180}) and localized systems (15 tetrapeptides and 10 proteins) at the M06-2X/def2-TZVP level. The total molecular wavefunction is used as an input for ITA quantities. The overall trend is satisfactory both qualitatively and quantitively, with molecular volume and G_3 showing good performance in reproducing conventional α_{iso} results of larger peptides and even proteins. Furthermore, we aim to check if the subsystem wavefunction can be used to predict the subsystem α_{iso} results, leading to the total α_{iso} results of large proteins via a linear combination. Shown in Table 13.3 is a collection of α_{iso} of 11 proteins obtained either from GEBF or GEBF-ITA. With the GEBF data as a reference, one can easily find that the α_{iso} data predicted from GEBF-ITA are in good agreement. It is interesting to mention that the predicted α_{iso} from GEBF-G3 only deviates 5% from GEBF. However, the other

Table 13.1 Correlation coefficients (R^2) between molecular polarizabilities (α_{iso}), and molecular volumes (Vol), quadrupole moments (Θ_{iso}) and all ITA quantities for inorganic systems at the M06-2X/def2-TZVP level.

Systems	S_S	I_F	S_{GBP}	E_2	E_3	${}^rR_2$	${}^rR_3$	G_1	G_2	G_3	I_G	Vol	Θ_{iso}
Be[a)]	0.997	0.997	0.997	0.997	0.997	0.997	0.996	0.990	0.993	0.996	0.832	0.970	0.981
Mg[b)]	0.970	0.984	0.984	0.984	0.984	0.983	0.983	0.939	0.937	0.957	0.877	0.973	0.986
Na[c)]	0.994	0.995	0.995	0.995	0.995	0.995	0.995	0.998	0.997	0.996	0.976	0.978	0.994
S[d)]	0.994	0.995	0.995	0.995	0.995	0.995	0.995	0.990	0.980	0.992	0.978	0.984	0.995
Au[e)]	0.996	0.996	0.996	0.996	0.996	0.996	0.996	0.989	0.986	0.994	0.895	0.991	0.992
HF[f)]	0.974	0.994	0.994	0.995	0.995	0.994	0.993	0.826	0.987	0.989	0.693	0.959	0.935
H[g)]	1.000	1.000	1.000	1.000	1.000	1.000	1.000	0.690	0.998	1.000	0.999	0.994	0.998
$H^+ \cdot (H_2O)$[h)]	1.000	1.000	1.000	1.000	1.000	1.000	1.000	0.607	1.000	1.000	0.999	0.994	0.992
CO_2[i)]	1.000	1.000	1.000	1.000	1.000	1.000	1.000	1.000	0.998	1.000	1.000	0.995	1.000
$Cl^- \cdot (NaCl)_n$[j)]	1.000	1.000	1.000	1.000	1.000	1.000	1.000	0.987	0.903	0.999	0.995	0.981	0.985
Fullerene[k)]	0.985	0.985	0.985	0.985	0.985	0.985	0.985	0.985	0.985	0.985	0.983	0.981	0.986
BN[l)]	1.000	1.000	1.000	1.000	1.000	1.000	1.000	1.000	0.999	1.000	1.000	0.995	0.999
ZnS[m)]	0.998	0.998	0.998	0.998	0.998	0.998	0.998	0.996	0.249	0.994	0.998	0.999	0.998

a) Beryllium clusters with sizes varying from 3 to 25, $n = 23$.
b) Magnesium clusters with sizes ranging from 3 to 28, $n = 19$.
c) Sodium clusters with sizes varying from 13 to 80, $n = 81$.
d) Sulfur clusters with sizes ranging from 2 to 18, $n = 17$.
e) Gold clusters with sizes varying from 2 to 50, $n = 21$.
f) HF clusters (both linear and cyclic) with sizes varying from 3 to 8, $n = 12$.
g) Water clusters with sizes varying from 2 to 20, $n = 19$.
h) Protonated water clusters with sizes varying from 2 to 21, $n = 20$.
i) Carbon dioxide (CO_2) clusters with sizes varying from 4 to 40, $n = 37$.
j) $Cl^- \cdot (NaCl)_n$ with sizes varying from 1 to 35, $n = 35$.
k) Fullerene cages with sizes varying from 20 to 76, $n = 21$.
l) Boron nitride (BN) cages with sizes varying from 12 to 48, $n = 5$.
m) Zinc sulfide (ZnS) cages with sizes varying from 12 to 108, $n = 8$.

Table 13.2 Correlation coefficients (R^2) between molecular polarizabilities (α_{iso}), and molecular volumes (Vol), quadrupole moments (Θ_{iso}) and all ITA quantities for organic and biological systems at the M06-2X/def2-TZVP level.

Systems	S_S	I_F	S_{GBP}	E_2	E_3	${}^rR_2$	${}^rR_3$	G_1	G_2	G_3	I_G	Vol	Θ_{iso}
Polyene[a)]	0.980	0.980	0.980	0.980	0.980	0.980	0.980	0.980	0.980	0.980	0.980	0.980	0.980
Polyyne[a)]	0.982	0.982	0.982	0.982	0.982	0.982	0.982	0.982	0.979	0.982	0.982	0.986	0.975
Acene[b)]	0.991	0.991	0.991	0.991	0.991	0.991	0.991	0.991	0.991	0.991	0.991	0.991	0.991
All-trans[c)]	0.992	0.992	0.992	0.992	0.992	0.992	0.992	0.992	0.994	0.992	0.992	0.986	0.991
Push–pull[d)]	0.998	0.998	0.998	0.998	0.998	0.998	0.998	0.998	0.998	0.998	0.998	0.988	0.999
Octane[e)]	0.952	0.864	0.903	0.048	0.004	0.311	0.317	0.383	0.070	0.770	0.298	0.192	0.007
Base pairs[f)]	0.867	0.788	0.884	0.613	0.076	0.885	0.885	0.027	0.214	0.962	0.582	0.628	0.602
$(C_6H_6)_n$[g)]	1.000	1.000	1.000	1.000	1.000	1.000	1.000	0.964	0.915	1.000	0.999	0.997	1.000
Amino acids[h)]	0.822	0.612	0.941	0.087	0.000	0.945	0.949	0.869	0.788	0.930	0.861	0.821	0.962
Dipeptides[i)]	0.799	0.628	0.944	0.090	0.000	0.946	0.949	0.806	0.727	0.899	0.196	0.870	0.069
Tripeptides[j)]	0.798	0.629	0.945	0.090	0.000	0.947	0.757	0.799	0.721	0.900	0.193	0.848	0.021
α-Helix[k)]	0.999	0.999	0.999	0.999	0.999	0.999	0.999	0.861	0.945	1.000	1.000	0.988	0.998
3_{10}-Helix[l)]	0.999	0.999	0.999	0.999	0.999	1.000	1.000	0.947	0.862	1.000	0.999	0.989	0.998
β-Sheet[m)]	1.000	1.000	1.000	1.000	1.000	1.000	1.000	1.000	1.000	1.000	1.000	0.996	1.000

a) Linear alkenes with varying numbers of monomers, $n = 11$.
b) Linear polycyclic aromatic hydrocarbon molecules with varying number of monomers from 2 to 10, $n = 9$.
c) All-trans-polymethineimine with a varying number of monomers from 1 to 30, $n = 12$.
d) Push-pull molecules with varying numbers of monomers from 1 to 10, $n = 10$.
e) All possible isomers of octane, $n = 24$.
f) Base pairs taken from Ref. [74], $n = 30$.
g) Benzene clusters with varying numbers of monomers from 4 to 30, $n = 27$.
h) Capped amino acids, $n = 20$.
i) Capped dipeptides, $n = 400$.
j) Capped tripeptides, $n = 8000$.
k) α-Helix with varying numbers of alanine units, $n = 6$.
l) 3_{10}-Helix with varying numbers of alanine units, $n = 6$.
m) β-Sheet with varying numbers of alanine units, $n = 6$.

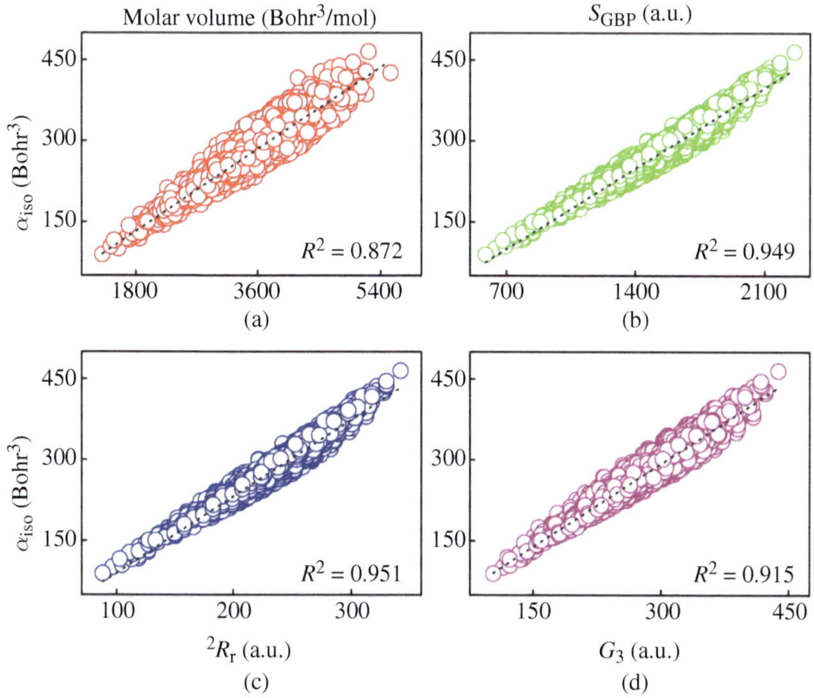

Figure 13.1 Strong correlations between molecular polarizability and (a) molecular volume, (b) S_{GBP}, (c) ^2R_r, and (d) G_3 for a total of 20 amino acids, 400 dipeptides, and 8000 tripeptides. Computations were carried out at the M06-2X/def2-TZVP//M06-2X/def2-SVP level. Source: Reprinted with permission from Ref. [40]. Copyright 2022 Royal Chemical Society.

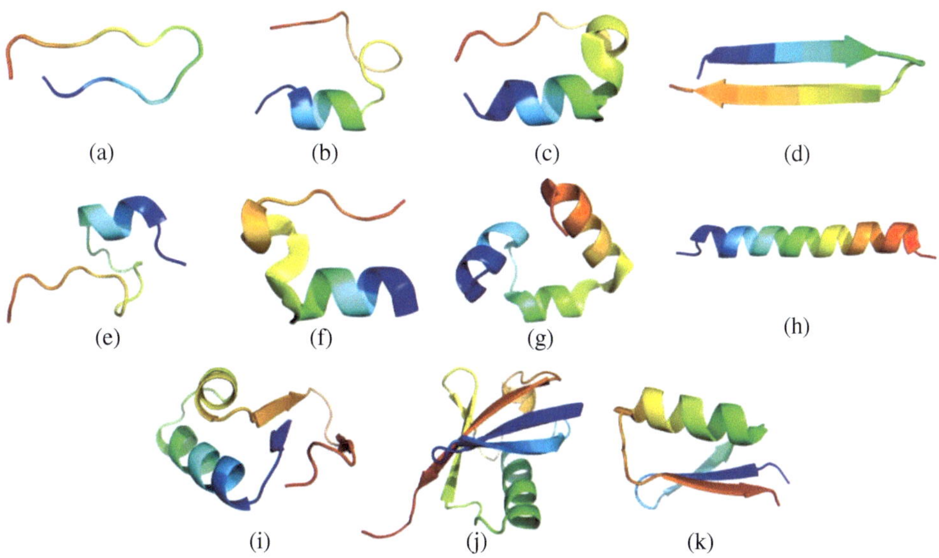

Scheme 13.2 All proteins studied in this work. PDB IDs are (a) 4TTK, sunflower trypsin inhibitor-1, (b) 2LDJ, Trp-cage miniprotein with D-amino acid, (c) 2LL5, cyclo-TC1 Trp-cage, (d) 2LYE, the cyclic cystine ladder motif of θ-defensins, (e) 1SP7, the cystine-rich C-terminal domain of Hydra minicollagen, (f) 1L2Y, Trp-cage miniprotein, (g) 2F4K, chicken villin subdomain, (h) 6PHM, glucagon analog fully composed of D-amino acids, (i) 1CBN, a hydrophobic protein, (j) 1UBQ, human erythrocytic ubiquitin, and (k) 5JI4, solution structure of the *de novo* mini protein, respectively. Source: Reprinted with permission from Ref. [40]. Copyright 2022 Royal Chemical Society.

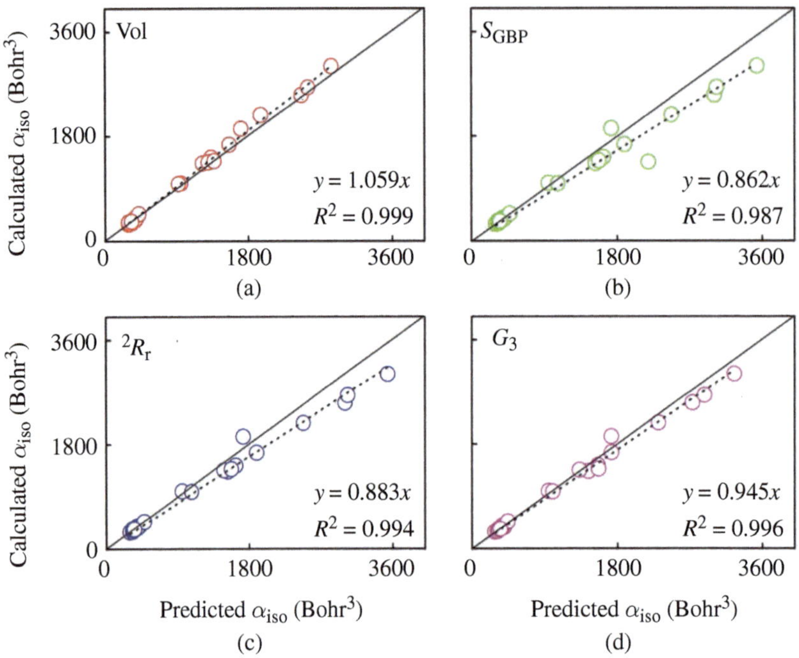

Figure 13.2 Comparison of conventional and predicted molecular polarizabilities of delocalized fullerenes (C_{100} and C_{180}) and localized systems (15 tetrapeptides and 10 proteins) at the M06-2X/def2-TZVP level. The *y*-axis of (a) and (c) spans to (b) and (d). Source: Reprinted with permission from Zhao et al. [40]. Copyright 2022 Royal Chemical Society.

Table 13.3 Prediction of a_{iso} (in Bohr^3) based on the subsystem wavefunctions of 11 protein systems.

	GEBF-ITA				
Systems	**Vol**	S_{GBP}	${}^rR_2$	G_3	**GEBF**
4TTK	865.5	1067.5	1069.2	1000.3	982.7
2LDJ	1351.6	1628.5	1533.8	1559.6	1438.8
2LL5	1301.5	1531.4	1631.9	1439.7	1332.2
2LYE	1184.9	1487.3	1483.8	1332.0	1345.4
5JI4	2177.5	3105.8	3109.9	2800.7	2513.6
1SP7	1412.4	1790.3	1791.8	1662.6	1661.3
1L2Y	1224.0	1472.9	1475.2	1450.4	1379.7
2F4K	2616.0	2932.1	2941.3	2822.5	2663.7
6PHM	1740.2	2474.1	2478.5	2311.1	2143.7
1CBN	3226.5	3433.3	3235.6	3057.4	3007.2
1UBQ	5589.6	5640.1	5659.4	5471.4	5574.8
MPE[a)]	−7.7	11.5	11.1	4.3	
MAPE[b)]	9.1	11.5	11.1	4.8	

a) **MPE**: mean percentage error.
b) **MAPE**: mean absolute percentage error.
Source: Reprinted with permission from Zhao et al. [40]. Copyright 2022 Royal Chemical Society.

three quantities, molecular volume, S_{GBP}, and ${}^2R_r$, cause a slightly larger deviation (~10%) of GEBF. Taken in concert, we have clearly shown that ITA or GEBF-ITA can be used to predict the α_{iso} results of large molecules in the ground state.

Before moving forward, we have to mention that (i) we have only numerically discovered these linear correlations, and theoretical proof is still lacking currently; (ii) we are not the first to use the electron density to predict the α_{iso} results. Jayatilaka et al. [71] have obtained the α_{iso} results with reasonably good accuracy compared with CPHF results. Up to now, we have only touched upon one side of a coin, the ground-state α_{iso}, and we will take a look at the other side, the excited-state α_{iso}

A series of 43 molecular systems (see Ref. [41] and references therein) with excited-state intramolecular proton transfer (ESIPT) properties were constructed, as shown in Schemes 13.3–13.5. The ground- (S_0) and excited-state (S_1) structural relaxation was carried out at the CAM-B3LYP/6-311+G(d) [72, 73] and TD-CAM-B3LYP/6-311+G(d) [36–38, 72, 73] level, respectively. Collected in Tables 13.4 and 13.5 are the correlation coefficients between the S_0/S_1 α_{iso} data and ITA quantities, molecular volumes, and quadrupole moments. The main conclusions still hold for S_0 and S_1, except for some unexpected outliers resulting from Br-containing species. Additionally, the transition density matrix can also be used as an input for ITA quantities to correlate with the S_0 and S_1 α_{iso} data. Due to the limited space, these results are not shown. Based on these linear correlations resulting from the training set, one can predict the α_{iso} data of molecules in the test set. The anticipated α_{iso} results for every molecule in Scheme 13.5 are displayed in

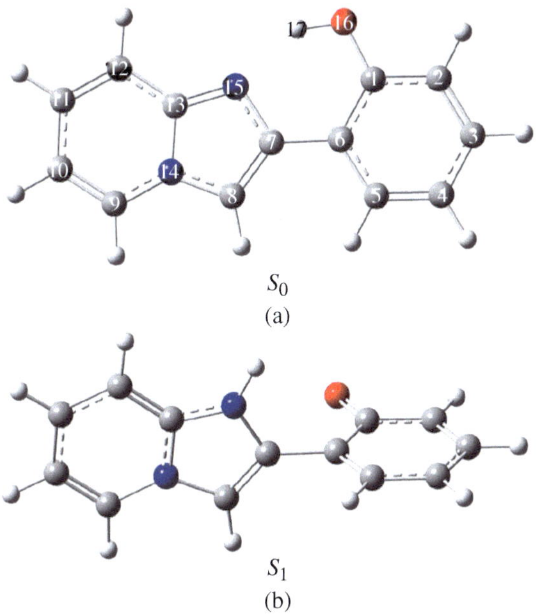

Scheme 13.3 (a) The ground-state and (b) excited-state 2-(2′-hydroxyphenyl)benzimidazole (HBI) structure and the corresponding atomic numbering. A total of 27 substituted HBI structures are generated, including **1**: HBI, **2**: 3-Br-HBI, **3**: 3-Et_2N-HBI, **4**: 3-HO-HBI, **5**: 3-MeO-HBI, **6**: 4-F-HBI, **7**: 4-Cl-HBI, **8**: 4-Br-HBI, **9**: 4-CN-HBI, **10**: 4-Me-HBI, **11**: 4-MeO-HBI, **12**: 10-Cl-HBI, **13**: 10-Br-HBI, **14**: 10-CN-HBI, **15**: 10-Me-HBI, **16**: 10-CF_3-HBI, **17**: 10-Ph-HBI, **18**: 12-Ph-HBI, **19**: 10-(*p*-MeO-Ph)-HBI, **20**: 10-(*p*-MeCO_2-Ph)-HBI, **21**: 12-(*p*-MeO-Ph)-HBI, **22**: 12-(*p*-MeCO_2-Ph)-HBI, **23**: 4-Me-10-Cl-HBI, **24**: 4-Me-10-CF_3-HBI, **25**: 10-Ph-12-Ph-HBI, **26**: 2-($CH_2CH_2CH{=}CH_2$)-8-(CH_2CH_2Ph)-HBI, **27**: 2-F-3-F-4-F-5-F-HBI. Color code: hydrogen in white, carbon in gray, nitrogen in blue, and oxygen in red. Source: Reprinted with permission from Ref. [41]. Copyright 2023 MDPI.

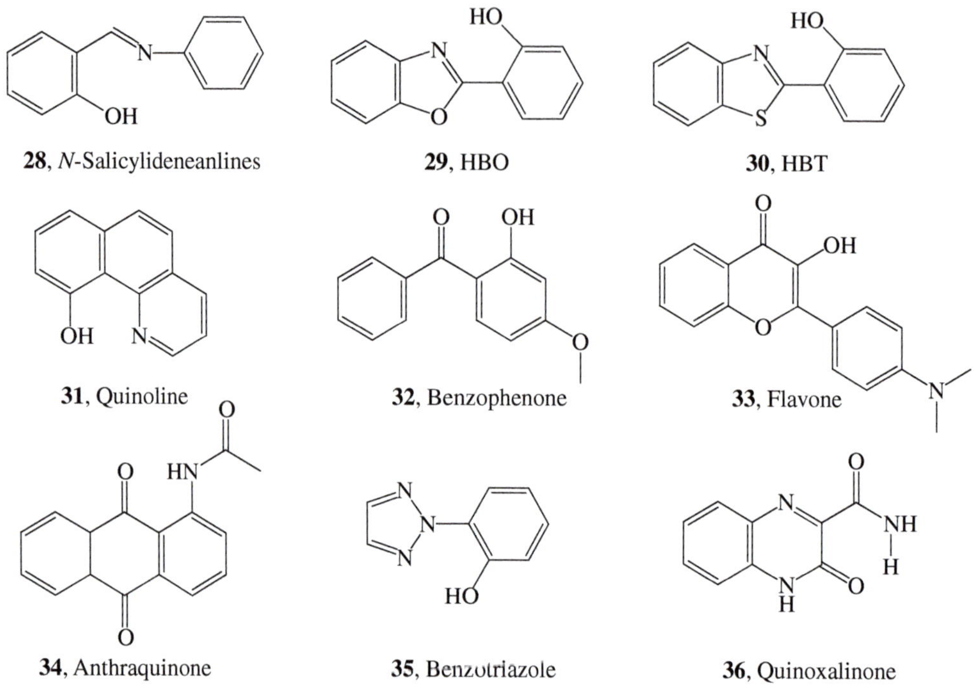

Scheme 13.4 Some representative examples of commonly used excited-state intramolecular proton transfer (ESIPT) fluorophores. Source: Reprinted with permission from Ref. [41]. Copyright 2023 MDPI.

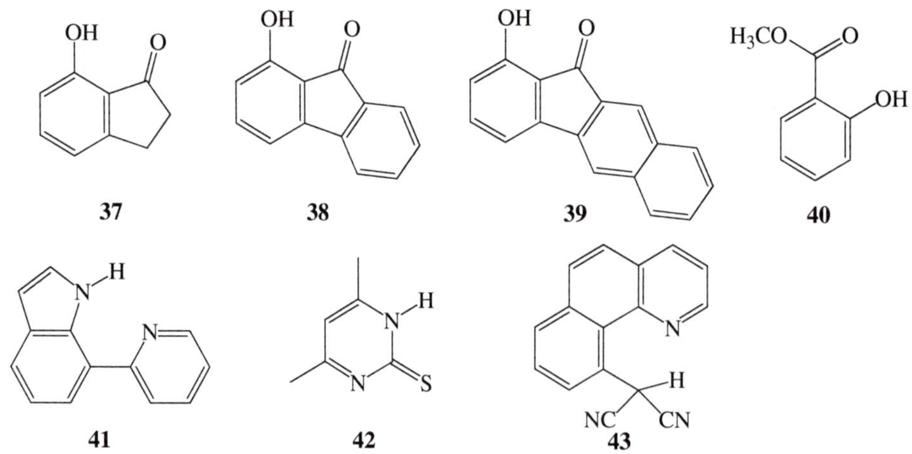

Scheme 13.5 Seven molecules were used to validate the transferability of the linear regression models. **37**: 7-hydroxy-2,3-dihydro-1*H*-inden-1-one, **38**: 1-hydroxy-9*H*-fluoren-9-one, **39**: 1-hydroxy-11*H*-benzo[*b*]flu oren-11-one, **40**: methyl salicylate, **41**: 7-(2-pyridyl)indole, **42**: 4,6-dimethyl-2-mercaptopyrimidine, **43**: 10-malononitrile-benzo[*h*]quinoline. Source: Reprinted with permission from Zhao et al. [41]. Copyright 2023 MDPI.

Table 13.4 Correlation coefficient (R^2) between the isotropic molecular polarizability (α_{iso}, in Bohr^3) and ITA quantities (in a.u.), molecular volume (Vol, in Bohr^3/mol), and the isotropic quadrupole moment (Θ_{iso}, in a.u.) at S_0 for all the systems in Schemes 13.3 and 13.4.

Index	α_{iso}	S_S	I_F	S_{GBP}	${}^rR_2$	${}^rR_3$	I_G	G_1	G_2	G_3	Vol	Θ_{iso}
1	175.68	96.35	4343.89	746.75	112.64	117.78	1.38	−35.34	24.07	139.78	1715.75	−87.39
2	200.87	20.66	13967.27	973.91	146.61	151.68	1.36	−34.93	23.70	141.74	2073.70	−103.29
3	237.32	142.90	5692.80	1016.86	154.03	161.91	2.10	−50.98	35.00	196.05	2441.40	−115.43
4	182.97	97.74	4793.46	801.56	120.76	126.07	1.45	−34.44	22.49	147.78	1855.73	−91.79
5	197.76	107.94	5045.63	855.45	129.06	135.01	1.60	−39.10	25.75	159.82	1886.74	−95.32
6	175.11	92.94	4917.98	801.81	120.62	125.71	1.37	−34.55	22.69	144.27	1765.67	−93.87
7	188.87	81.70	6524.29	855.22	128.61	133.70	1.37	−34.93	22.13	142.27	1959.26	−100.99
8	197.51	20.60	13967.06	973.87	146.60	151.67	1.36	−34.95	23.67	141.84	1850.09	−101.48
9	194.07	102.37	4928.42	829.28	124.78	130.18	1.46	−36.68	23.12	151.22	1897.41	−106.85
10	189.36	106.46	4595.48	800.72	120.93	126.64	1.53	−39.13	25.97	150.98	1931.31	−93.18
11	194.81	107.96	5045.77	855.48	129.06	135.02	1.60	−39.13	23.71	159.81	1994.38	−95.26
12	191.68	81.72	6524.35	855.23	128.60	133.66	1.36	−34.98	22.67	142.29	1957.16	−105.11
13	200.29	20.66	13967.37	973.93	146.60	151.67	1.36	−35.01	22.86	141.85	1980.91	−111.96
14	198.88	102.42	4928.71	829.32	124.77	130.16	1.45	−36.79	23.62	151.19	1696.49	−112.11
15	190.18	106.46	4595.43	800.71	120.93	126.64	1.53	−39.12	24.61	151.02	1842.17	−94.41
16	191.22	96.42	6318.70	965.73	144.80	150.24	1.46	−37.81	23.40	165.13	2064.71	−117.92
17	255.95	135.18	5826.22	1017.73	153.69	160.90	1.92	−52.71	36.65	192.16	2401.28	−125.79
18	246.58	135.11	5826.00	1017.67	153.68	160.90	1.92	−52.74	36.93	192.52	2494.60	−119.81
19	277.67	146.79	6528.00	1126.45	170.10	178.13	2.14	−56.48	38.53	212.18	2819.11	−137.35
20	295.24	153.19	7225.20	1222.09	184.32	192.76	2.25	−58.90	39.85	226.76	2761.18	−154.13
21	268.01	146.71	6527.76	1126.38	170.10	178.12	2.14	−56.47	39.11	212.53	2467.93	−128.04
22	283.05	153.12	7225.00	1222.03	184.32	192.76	2.25	−58.93	39.71	227.00	2913.80	−139.45
23	205.46	91.83	6775.95	909.20	136.89	142.52	1.51	−38.70	26.44	153.26	1963.41	−110.94
24	205.03	106.53	6570.28	1019.70	153.09	159.11	1.61	−41.55	26.57	176.20	2201.13	−123.83
25	327.20	173.93	7308.31	1288.65	194.73	204.01	2.46	−70.08	50.17	244.74	3066.91	−156.33
26	320.10	190.84	7330.19	1328.47	201.30	211.70	2.76	−74.61	52.80	255.30	3332.26	−160.67
27	175.19	82.72	6640.05	966.86	144.50	149.36	1.31	−32.94	20.03	158.50	*3218.41*	−108.98
28	173.51	96.75	4012.72	706.25	106.66	111.88	1.39	−37.76	26.11	130.76	1856.23	−84.26
29	169.68	91.91	4453.45	747.21	112.55	117.54	1.33	−36.02	24.66	137.43	1800.53	−87.68
30	186.83	80.69	5927.71	800.75	120.52	125.45	1.31	−36.59	25.86	133.70	2002.44	−95.13
31	163.90	90.16	4002.93	692.24	104.49	109.37	1.29	−36.26	25.07	129.50	1582.16	−82.37
32	176.33	103.65	4822.91	815.19	123.03	128.93	1.58	−41.53	28.16	148.35	1952.13	−95.39
33	243.77	128.34	5902.42	1004.58	151.72	158.98	1.94	−49.38	32.40	187.62	2418.28	−113.49
34	204.56	113.06	5648.71	937.42	141.25	147.59	1.70	−44.88	30.55	170.08	2071.58	−114.39
35	116.08	68.78	3451.98	571.00	85.89	89.56	0.99	−21.64	13.14	104.71	1404.36	−66.08
36	135.67	74.85	4147.84	666.78	100.16	104.28	1.15	−22.46	12.48	118.31	1405.54	−78.67
R^2		0.560	0.068	0.890	0.897	0.908	0.922	0.936	0.912	0.938	0.705	0.893

Table 13.5 Correlation coefficient (R^2) between the isotropic molecular polarizability (α_{iso}, in Bohr^3) and ITA quantities (in a.u.), molecular volume (Vol, in Bohr^3/mol), and the isotropic quadrupole moment (Θ_{iso}, in a.u.) at S_1 for all the systems in Schemes 13.3 and 13.4.

Index	α_{iso}	S_S	I_F	$^rR_2$	$^rR_3$	G_2	G_3	Vol	Θ_{iso}
1	166.08	96.94	4345.68	112.73	118.02	23.88	138.16	1797.33	−89.93
2	189.81	21.28	13969.19	146.67	151.87	23.61	140.08	2016.04	−114.26
3	224.23	143.51	5694.61	154.12	162.16	34.15	194.24	2493.81	−126.71
4	174.81	98.36	4795.23	120.83	126.27	22.33	146.07	1709.27	−96.38
5	187.84	108.57	5047.40	129.14	135.22	25.53	158.13	1977.08	−101.96
6	166.88	93.55	4919.65	120.68	125.87	22.56	142.61	1732.69	−97.69
7	182.88	82.29	6525.95	128.66	133.83	23.33	140.53	1814.22	−105.58
8	188.20	21.19	13968.88	146.65	151.80	22.97	140.28	1849.59	−110.07
9	182.03	103.04	4930.44	124.83	130.34	24.17	149.51	1973.71	−110.22
10	178.88	107.05	4597.05	121.02	126.88	26.58	149.31	1997.79	−96.99
11	182.81	108.54	5047.21	129.14	135.22	25.57	158.15	1950.36	−99.73
12	178.54	82.29	6526.00	128.69	133.90	23.14	140.43	1956.40	−103.35
13	186.41	21.24	13968.99	146.69	151.92	23.40	139.90	1829.44	−105.81
14	185.73	102.96	4930.10	124.85	130.37	24.33	149.38	1967.61	−111.23
15	179.03	107.06	4597.29	121.01	126.87	26.78	149.25	1905.76	−94.93
16	178.64	96.95	6320.18	144.88	150.45	23.97	163.21	1970.62	−111.79
17	240.42	135.77	5827.95	153.76	161.09	36.81	190.49	2259.46	−117.96
18	253.40	135.66	5827.52	153.75	161.08	36.75	190.81	2360.15	−120.97
19	260.31	147.38	6529.77	170.17	178.31	39.03	210.40	2749.56	−124.40
20	280.48	153.76	7226.84	184.39	192.94	40.08	225.02	2771.27	−136.33
21	267.99	147.29	6529.43	170.17	178.33	38.50	210.79	2513.32	−128.47
22	327.99	153.50	7225.08	184.32	192.78	39.42	226.35	2727.21	−138.46
23	190.90	92.40	6777.37	136.99	142.77	25.46	151.82	1979.58	−110.61
24	191.30	107.05	6571.54	153.18	159.32	26.69	174.50	2133.33	−119.08
25	331.10	174.16	7307.99	194.72	204.02	48.71	244.76	3175.10	−168.67
26	309.58	191.50	7332.09	201.39	211.92	53.05	253.29	3230.21	−164.60
27	190.21	83.28	6640.35	144.47	149.29	19.82	157.57	1816.70	−115.25
28	218.02	97.17	4012.63	106.64	111.81	26.47	129.86	1682.95	−83.82
29	212.42	92.41	4453.25	112.53	117.45	23.89	136.29	1811.12	−88.93
30	173.32	81.32	5929.80	120.61	125.71	25.00	132.03	1756.26	−92.75
31	170.34	90.61	4002.95	104.46	109.29	25.03	128.51	1684.98	−83.54
32	178.67	104.22	4824.82	123.06	129.03	27.87	146.61	1942.93	−94.35
33	301.59	128.75	5902.73	151.65	158.79	32.41	186.62	2379.78	−111.36
34	202.06	113.48	5649.74	141.23	147.54	29.60	169.10	2107.86	−113.38
35	148.12	69.41	3452.63	85.88	89.58	13.17	103.81	1270.25	−67.10
36	129.10	74.99	4148.18	100.12	104.16	12.54	117.98	1486.28	−83.28
R^2		0.552	0.025	0.729	0.740	0.821	0.824	0.818	0.672

Source: Reprinted with permission from Zhao et al. [41]. Copyright 2023 MDPI.

Table 13.6 Comparison of predicted α_{iso} (in Bohr^3/mol) at S_0 and conventional calculated data for all the systems in Scheme 13.5.

Index	α_{iso}	S_{GBP}	$^rR_2$	$^rR_3$	I_G	G_1	G_2	G_3	Vol	Θ_{iso}
37	105.36	109.32	109.77	110.45	143.92	146.72	154.26	120.04	137.25	115.53
38	155.68	151.14	151.26	151.64	169.53	185.83	187.63	158.87	163.68	147.97
39	216.91	196.27	196.48	197.01	207.58	229.76	236.35	203.48	205.75	184.62
40	101.70	113.02	113.08	113.47	140.59	136.98	143.60	120.15	136.21	100.86
41	166.55	150.90	151.41	152.11	174.87	185.60	194.70	166.02	168.11	144.96
42	111.10	102.29	102.60	102.89	130.30	116.57	125.36	97.52	125.61	106.82
43	196.65	193.10	192.90	193.28	200.66	216.50	224.05	196.26	199.29	253.81
MPE[a)]		−2.38	−2.22	−1.88	14.82	17.96	22.97	2.17	11.36	0.18
MAPE[b)]		6.64	6.61	6.56	16.05	17.96	22.97	7.58	12.83	10.89

a) **MPE**: mean percentage error.
b) **MAPE**: mean absolute percentage error.

Table 13.7 Comparison of predicted α_{iso} (in Bohr^3/mol) at S_1 and conventional calculated data for all the systems in Scheme 13.5.

Index	α_{iso}	$^rR_2$	$^rR_3$	G_2	G_3	Vol	Θ_{iso}
37	118.81	113.52	113.85	152.35	120.93	116.81	114.30
38	192.21	153.15	153.33	189.82	159.90	158.02	132.22
39	290.76	196.25	196.61	235.54	204.05	179.41	170.00
40	117.99	116.65	116.65	139.41	121.02	119.35	100.33
41	148.92	153.43	154.11	194.56	165.60	167.43	159.88
42	111.22	106.39	105.78	122.68	97.10	128.12	107.88
43	205.36	192.73	192.65	215.87	196.92	197.54	227.59
MPE[a)]		−9.41	−9.36	10.32	−6.84	−4.69	−10.90
MAPE[b)]		10.27	10.35	16.10	11.28	12.91	16.10

a) **MPE**: mean percentage error.
b) **MAPE**: mean absolute percentage error.

Tables 13.6 and 13.7, using traditional estimated α_{iso} data as a point of reference. One can see that the predicted results are in reasonably good agreement with those from conventional calculations. This relatively larger deviation should be understandable as the molecules in Scheme 13.5 bear little resemblance to those in Schemes 13.3 and 13.4. Finally, we use a multivariable strategy to predict the excitation energies via S_0 or S_1 ITA quantities, as shown in Table 13.8. One can see that the overall accuracy is comparable to the approximate density functional. To summarize, we have observed that ITA quantities can be used to predict both the ground- and excited-state α_{iso} and excitation energies with good accuracy.

Table 13.8 Comparison of the excitation energies (in eV), as predicted with the multivariant linear regression equations for all the systems in Scheme 13.5, with conventional data as a reference.

	Excitation energies		
Index	**Conv.**	**Mod_1[a)]**	**Mod_2[b)]**
37	4.28	2.24	2.25
38	3.52	2.67	2.55
39	3.53	2.64	2.51
40	4.55	3.00	3.45
41	4.12	3.13	3.33
42	3.71	3.30	NA[e)]
43	4.28	5.14	5.65
MUE[c)]		−0.84	−0.76
MSE[d)]		1.08	1.21

a) Based on the S_0 ITA quantities with the S_0 molecular wavefunction as an input for all the systems in Schemes 13.3 and 13.4. The multivariant linear regression equation is:
$\lambda_{\text{fit}} = 0.306 \times S_S + 0.003{,}04 \times I_F + 0.661 \times S_{GBP} - 4.210 \times {}^{r}R_2 - 0.363 \times {}^{r}R_3 + 0.102 \times G_1 - 0.0499 \times G_2 - 0.114 \times G_3 + 3.490 \times I_G$.

b) Base on the S_1 ITA quantities with the transition density matrix as an input for all the systems in Schemes 13.3 and 13.4. The multivariant linear regression equation is:
$\lambda_{\text{fit}} = 0.290 \times S_S + 0.003{,}17 \times I_F + 1.037 \times S_{GBP} - 14.075 \times {}^{r}R_2 + 6.914 \times {}^{r}R_3 + 0.104 \times G_1 - 0.0494 \times G_2 - 0.109 \times G_3 - 16.163 \times I_G$.

c) **MUE**: mean unsigned error.

d) **MSE**: mean signed error.

e) NA means not available. S_{GBP}, G_1, and I_G are not numerically obtained.

Source: Reprinted with permission from Zhao et al. [41]. Copyright 2022 Royal Chemical Society.

13.4 Conclusions and Perspectives

In this chapter, we have demonstrated that ITA quantities can be used to correlate with the α_{iso} data of various types of inorganic, organic, and biological systems. Based on these linear relationships, one can predict the α_{iso} result of even larger systems once the molecular wavefunction (or electron density) is obtained. This protocol can be applied to localized and delocalized systems either in the ground or excited state. For macromolecules like proteins, we employ the GEBF method to partition the whole system into small subsystems and obtain the subsystems α_{iso} results via ITA quantities. This GEBF-ITA protocol is verified to be both accurate and efficient. Since only electron density is used as an input, it is anticipated that more advanced electronic structure calculations should be possible with simple protocols. We mention in passing that for condensed-phase systems, such as molecular solids and liquid water, similar strategies have been adopted and one can use even Hartree–Fock densities to predict MP2 or CCSD polarizabilities. Work along this line is underway, and the results will be presented elsewhere.

Acknowledgment

This work is supported by the National Natural Science Foundation of China (Grant No. 22203071), the Yunnan Fundamental Research Projects (Grant No. 202101AU070012), the start-up funding of Yunnan University, the High Level Talents Special Support Plan, and NSERC, Canada Research Chairs, and the Digital Research Alliance of Canada.

References

1 Tian, T., Scullion, D., Hughes, D. et al. (2020). Electronic polarizability as the fundamental variable in the dielectric properties of two-dimensional materials. *Nano Lett.* 20 (2): 841–851.

2 Rupasinghe, T.P., Hutchins, K.M., Bandaranayake, B.S. et al. (2015). Mechanical properties of a series of macro- and nanodimensional organic cocrystals correlate with atomic polarizability. *J. Am. Chem. Soc.* 137 (40): 12768–12771.

3 Nagle, J.K. (1990). Atomic polarizability and electronegativity. *J. Am. Chem. Soc.* 112 (12): 4741–4747.

4 Machmer, P. (1968). Empirical correlation between the molecular electric field gradients and the average molecular polarizabilities of some halogen compounds. *Nature* 217 (5124): 165–166.

5 Hati, S. and Datta, D. (1994). Hardness and electric dipole polarizability. Atoms and clusters. *J. Phys. Chem.* 98 (41): 10451–10454.

6 Vela, A. and Gázquez, J.L. (1990). A relationship between the static dipole polarizability, the global softness, and the Fukui function. *J. Am. Chem. Soc.* 112 (4): 1490–1492.

7 Krishtal, A., Senet, P., and Alsenoy, C.V. (2009). Local softness, softness dipole, and polarizabilities of functional groups: application to the side chains of the 20 amino acids. *J. Chem. Phys.* 131 (4): 044312.

8 Ghanty, T.K. and Ghosh, S.K. (1993). Correlation between hardness, polarizability, and size of atoms, molecules, and clusters. *J. Phys. Chem.* 97: 4951–4953.

9 Simon-Manso, Y. and Fuentealba, P. (1998). On the density functional relationship between static dipole polarizability and global softness. *J. Phys. Chem. A* 102: 2029–2032.

10 Ayers, P. W. The Physical Basis of the Hard/Soft Acid/Base Principle. *Faraday Discuss.* 2007, 135, 161–190. https://doi.org/10.1039/b606877d

11 Tkatchenko, A., Fedorov, D.V., and Gori, M. (2021). Fine-structure constant connects electronic polarizability and geometric van-der-Waals radius of atoms. *J. Phys. Chem. Lett.* 12 (39): 9488–9492.

12 Miller, K.J. (1990). Additivity methods in molecular polarizability. *J. Am. Chem. Soc.* 112 (23): 8533–8542.

13 Miller, K.J. (1990). Calculation of the molecular polarizability tensor. *J. Am. Chem. Soc.* 112 (23): 8543–8551.

14 Tkatchenko, A. and Scheffler, M. (2009). Accurate molecular Van Der Waals interactions from ground-state electron density and free-atom reference data. *Phys. Rev. Lett.* 102 (7): 073005.

15 Szabó, P., Góger, S., Charry, J. et al. (2022). Four-dimensional scaling of dipole polarizability in quantum systems. *Phys. Rev. Lett.* 128 (7): 070602.

16 Huang, K., Cai, X., Li, H. et al. (2022). Variational quantum computation of molecular linear response properties on a superconducting quantum processor. *J. Phys. Chem. Lett.* 13 (39): 9114–9121.

17 Nakagawa, Y.O., Chen, J., Sudo, S. et al. (2023). Analytical formulation of the second-order derivative of energy for the orbital-optimized variational quantum eigensolver: application to polarizability. *J. Chem. Theory Comput.* 19 (7): 1998–2009.

18 Wouters, S., Limacher, P.A., Van Neck, D., and Ayers, P.W. (2012). Longitudinal static optical properties of hydrogen chains: finite field extrapolations of matrix product state calculations. *J. Chem. Phys.* 136 (13): 134110.

19 Mohammed, A.A.K., Limacher, P.A., and Ayers, P.W. (2017). Predicting optimal finite field strengths for calculating the first and second hyperpolarizabilities using simple molecular descriptors. *Chem. Phys. Lett.* 682: 160–167.

20 Patel, A.H.G., Mohammed, A.A.K., Limacher, P.A., and Ayers, P.W. (2017). Finite field method for nonlinear optical property prediction using rational function approximants. *J. Phys. Chem. A* 121 (28): 5313–5323.

21 Mohammed, A.A.K., Limacher, P.A., and Champagne, B. (2013). Finding optimal finite field strengths allowing for a maximum of precision in the calculation of polarizabilities and hyperpolarizabilities. *J. Comp. Chem.* 34 (17): 1497–1507.

22 McWeeny, R. (1960). Some recent advances in density matrix theory. *Rev. Mod. Phys.* 32 (2): 335–369.

23 Langhoff, P.W., Karplus, M., and Hurst, R.P. (1966). Approximations to Hartree–Fock perturbation theory. *J. Chem. Phys.* 44 (2): 505–514.

24 Colwell, S.M., Murray, C.W., Handy, N.C., and Amos, R.D. (1993). The determination of hyperpolarisabilities using density functional theory. *Chem. Phys. Lett.* 210 (1–3): 261–268.

25 Niklasson, A.M. and Challacombe, M. (2004). Density matrix perturbation theory. *Phys. Rev. Lett.* 92 (19): 193001.

26 Weber, V., Niklasson, A.M.N., and Challacombe, M. (2004). *Ab initio* linear scaling response theory: electric polarizability by perturbed projection. *Phys. Rev. Lett.* 92 (19): 193002.

27 Collins, M.A. and Bettens, R.P. (2015). Energy-based molecular fragmentation methods. *Chem. Rev.* 115 (12): 5607–5642.

28 Li, S., Li, W., and Fang, T. (2005). An efficient fragment-based approach for predicting the ground-state energies and structures of large molecules. *J. Am. Chem. Soc.* 127 (19): 7215–7226.

29 Li, W., Li, S., and Jiang, Y. (2007). Generalized energy-based fragmentation approach for computing the ground-state energies and properties of large molecules. *J. Phys. Chem. A* 111 (11): 2193–2199.

30 Li, S., Li, W., and Ma, J. (2014). Generalized energy-based fragmentation approach and its applications to macromolecules and molecular aggregates. *Acc. Chem. Res.* 47 (9): 2712–2720.

31 Li, W., Dong, H., Ma, J., and Li, S. (2021). Structures and spectroscopic properties of large molecules and condensed-phase systems predicted by generalized energy-based fragmentation approach. *Acc. Chem. Res.* 54 (1): 169–181.

32 Grisafi, A., Wilkins, D.M., Csányi, G., and Ceriotti, M. (2018). Symmetry-adapted machine learning for tensorial properties of atomistic systems. *Phys. Rev. Lett.* 120 (3): 036002.

33 Wilkins, D.M., Grisafi, A., Yang, Y. et al. (2019). Accurate molecular polarizabilities with coupled cluster theory and machine learning. *Proc. Natl. Acad. Sci. U.S.A.* 116 (9): 3401–3406.

34 Nguyen, V.H.A. and Lunghi, A. (2022). Predicting tensorial molecular properties with equivariant machine learning models. *Phys. Rev. B* 105 (16): 165131.

35 Amin, M., Samy, H., and Küpper, J. (2019). Robust and accurate computational estimation of the polarizability tensors of macromolecules. *J. Phys. Chem. Lett.* 10 (11): 2938–2943.

36 Bauernschmitt, R. and Ahlrichs, R. (1996). Treatment of electronic excitations within the adiabatic approximation of time dependent density functional theory. *Chem. Phys. Lett.* 256 (4–5): 454–464.

37 Furche, F. and Ahlrichs, R. (2002). Adiabatic time-dependent density functional methods for excited state properties. *J. Chem. Phys.* 117 (16): 7433–7447.

38 Liu, J. and Liang, W. (2011). Analytical Hessian of electronic excited states in time-dependent density functional theory with Tamm-Dancoff approximation. *J. Chem. Phys.* 135 (1): 014113.

39 Zhao, D., Liu, S., and Chen, D. (2022). A density functional theory and information-theoretic approach study of interaction energy and polarizability for base pairs and peptides. *Pharmaceuticals* 15 (8): 938.

40 Zhao, D., Zhao, Y., He, X. et al. (2023). Efficient and accurate density-based prediction of macromolecular polarizabilities. *Phys. Chem. Chem. Phys.* 25 (3): 2131–2141.

41 Zhao, D., He, X., Ayers, P.W., and Liu, S. (2023). Excited-state polarizabilities: a combined density functional theory and information-theoretic approach study. *Molecules* 28 (6): 2576.

42 Shannon, C.E. (1948). A mathematical theory of communication. *Bell Syst. Technol. J.* 27 (3): 379–423.

43 Fisher, R.A. (1925). Theory of statistical estimation. *Math. Proc. Cambridge Philos. Soc.* 22 (5): 700–725.

44 Nalewajski, R.F. and Parr, R.G. (2000). Information theory, atoms in molecules, and molecular similarity. *Proc. Natl. Acad. Sci. U.S.A.* 97 (16): 8879–8882.

45 Parr, R.G. and Yang, W. (1989). *Density Functional Theory for Atoms and Molecules*. London: Oxford University.

46 Teale, A.M., Helgaker, T., Savin, A. et al. (2022). DFT exchange: sharing perspectives on the workhorse of quantum chemistry and material science. *Phys. Chem. Chem. Phys.* 24 (47): 28700–28781.

47 Kullback, S. (1997). *Information Theory and Statistics*. Mineola, NY: Dover Publications.

48 Liu, S. (2019). Identity for Kullback-Leibler divergence in density functional reactivity theory. *J. Chem. Phys.* 151 (14): 141103.

49 Wang, B., Zhao, D., Lu, T. et al. (2021). Quantifications and applications of relative Fisher information in density functional theory. *J. Phys. Chem. A* 125 (17): 3802–3811.

50 Ghosh, S.K., Berkowitz, M., and Parr, R.G. (1984). Transcription of ground-state density-functional theory into a local thermodynamics. *Proc. Natl. Acad. Sci. U.S.A.* 81 (24): 8028–8031.

51 Liu, S., Rong, C., Wu, Z., and Lu, T. (2015). Rényi entropy, Tsallis entropy and Onicescu information energy in density functional reactivity theory. *Acta Phys. Chim. Sin.* 31 (11): 2057–2063.

52 Hirshfeld, F.L. (1977). Bonded-atom fragments for describing molecular charge densities. *Theoret. Chim. Acta* 44 (2): 129–138.

53 Heidar-Zadeh, F., Ayers, P.W., Verstraelen, T. et al. (2018). Information-theoretic approaches to atoms-in-molecules: Hirshfeld family of partitioning schemes. *J. Phys. Chem. A* 122 (17): 4219–4245.

54 Parr, R.G., Ayers, P.W., and Nalewajski, R.F. (2005). What is an atom in a molecule? *J. Phys. Chem. A* 109: 3957–3959.

55 Liu, S. (2007). On the relationship between densities of Shannon entropy and Fisher information for atoms and molecules. *J. Chem. Phys.* 126 (19): 191107.

56 Becke, A.D. (1988). A multicenter numerical integration scheme for polyatomic molecules. *J. Chem. Phys.* 88 (4): 2547–2553.

57 Bader, R.F.W. (1990). *Atoms in Molecules: A Quantum Theory*. Oxford, England: Oxford University Press.

58 Rong, C., Wang, B., Zhao, D., and Liu, S. (2020). Information-theoretic approach in density functional theory and its recent applications to chemical problems. *Wiley Interdiscip. Rev.: Comput. Mol. Sci.* 10 (4): e1461.

59 Rong, C., Zhao, D., He, X., and Liu, S. (2022). Development and applications of the density-based theory of chemical reactivity. *J. Phys. Chem. Lett.* 13 (48): 11191–11200.

60 Liu, S. (2020). Homochirality originates from handedness of helices. *J. Phys. Chem. Lett.* 11 (20): 8690–8696.

61 Liu, S. (2021). Principle of chirality hierarchy in three-blade propeller systems. *J. Phys. Chem. Lett.* 12 (36): 8720–8725.

62 Zhao, D., Song, R., Li, W. et al. (2017). Accurate prediction of NMR chemical shifts in macromolecular and condensed-phase systems with the generalized energy-based fragmentation method. *J. Chem. Theory Comput.* 13 (11): 5231–5239.

63 Zhao, D., Shen, X., Cheng, Z. et al. (2020). Accurate and efficient prediction of NMR parameters of condensed-phase systems with the generalized energy-based fragmentation method. *J. Chem. Theory Comput.* 16 (5): 2995–3005.

64 Zhao, D., Liao, K., Hong, B. et al. (2023). Accurate and efficient prediction of vibrational circular dichroism spectra of condensed-phase systems with the generalized energy-based fragmentation method. *Electron. Struct.* 5 (1): 014001.

65 Reed, A.E., Weinstock, R.B., and Weinhold, F. (1985). Natural population analysis. *J. Chem. Phys.* 83 (2): 735–746.

66 Li, W., Chen, C., Zhao, D., and Li, S. (2015). LSQC: low scaling quantum chemistry program. *Int. J. Quantum Chem.* 115 (10): 641–646.

67 Frisch, M.J., Trucks, G.W., Schlegel, H.B., et al. (2016). *Gaussian 16, Revision C.01*. Gaussian Inc.: Wallingford, CT.

68 Lu, T. and Chen, F. (2012). Multiwfn: a multifunctional wavefunction analyzer. *J. Comput. Chem.* 33 (5): 580–592.

69 Zhao, Y. and Truhlar, D.G. (2008). The M06 suite of density functionals for main group thermochemistry, thermochemical kinetics, noncovalent interactions, excited states, and transition elements: two new functionals and systematic testing of four M06-class functionals and 12 other functionals. *Theor. Chem. Acc.* 120 (1): 215–241.

70 Weigend, F. and Ahlrichs, R. (2005). Balanced basis sets of split valence, triple zeta valence and quadruple zeta valence quality for H to Rn: design and assessment of accuracy. *Phys. Chem. Chem. Phys.* 7 (18): 3297–3305.

71 Jayatilaka, D., Jha, K.K., and Munshi, P. (2018). Is it reasonable to obtain information on the polarizability and hyperpolarizability only from the electron density? *Aust. J. Chem.* 71 (4): 295–306.

72 Yanai, T., Tew, D.P., and Handy, N.C. (2004). A new hybrid exchange-correlation functional using the Coulomb-attenuating method (CAM-B3LYP). *Chem. Phys. Lett.* 393 (1–3): 51–57.

73 McLean, A.D. and Chandler, G.S. (1980). Contracted Gaussian-basis sets for molecular calculations. 1. 2nd row atoms, $Z = 11$–18. *J. Chem. Phys.* 72 (10): 5639–5648.

74 Zhao, D., Liu, S., and Chen, D.A. (2022). Density functional theory and information-theoretic approach study of interaction energy and polarizability for base pairs and peptides. *Pharmaceuticals* 15: 938. https://doi.org/10.3390/ph15080938.

14

Conceptual Density Functional Theory-Based Study of Pure and TMs-Doped CdX (X = S, Se, Te; TMs = Cu, Ag, and Au) Nano Cluster for Water Splitting and Spintronic Applications

Prabhat Ranjan[1], Preeti Nanda[2], Ramon Carbó-Dorca[3,4], and Tanmoy Chakraborty[5,6]

[1] *Department of Mechatronics Engineering, Manipal University Jaipur, Jaipur, India*
[2] *Department of Electronics and Communication Engineering, IIIT Bhagalpur, Bhagalpur, India*
[3] *Institute of Computational Chemistry and Catalysis, University of Girona, Spain*
[4] *Ronin Institute, Montclair, USA*
[5] *Department of Chemistry and Biochemistry, Sharda School of Basic Sciences and Research, Sharda University, Greater Noida, India*
[6] *SVKM'S NMIMS (Deemed to be University), Mumbai, India*

14.1 Introduction

Semiconductors of group II–VI, with their excellent electronic and optical properties, have generated a lot of interest among the scientific community. These materials have shown potential applications in optoelectronics as well as spintronic devices, especially in spin transistors, spin field-effect transistors, spin light emitting diodes, magneto-resistive random access memory, and resonant tunneling diodes [1, 2]. The physicochemical properties of semiconductor nanomaterials can be enhanced by doping with transition metals (TMs) [2]. As TMs are significantly easier to dissolve in II–VI semiconductors as compared to III–V materials, the TMs doped in II–VI semiconductor compounds exhibit zinc-blende arrangement, giving a strong possibility to be utilized in magneto-optical and spintronic devices [3–5]. Cadmium sulfide (CdS) is a binary semiconducting material and a member of II–VI family of chalcogenide compounds [6]. CdS has potential technological applications for instance in optoelectronics, piezo-electronics, solar cells, and biological staining and diagnostics due to its large energy gap [7–13]. The optoelectronic properties of CdS are altered by doping it with TMs such as Fe [14], Ni [15], Mn [16–21], Cu [22], and Ga [23]. Al-Zahrani et al. [14] reported CdS incorporated with Fe by using an experimental technique and found that the optical energy gap upsurges with an increase in irradiation levels. The authors stated that CdS incorporated with iron shows potential applications in several industries. Thool et al. [15] investigated nickel-doped CdS material by using an experimental method and found that Ni atoms significantly contributed to the development of CdS designs. Saikia et al. [16] examined Mn-doped CdS nanomaterials and found that doping of Mn atoms helps to improve the optoelectronic properties of overall system. Gao et al. [18] investigated CdS nanomaterials incorporated with Mn atoms by using a one-pot technique. An enhancement in luminous property is marked after addition of Mn atom in CdS nanocrystal, which has applications in internal as well as external illumination and can also be used in display units. Shen et al. [19] examined CdS doped with Mn atoms for solar cell applications by using an experimental technique. Authors found that Mn-doped with CdS enhances the light absorbance as well as the absorption range. System

Electron Density: Concepts, Computation and DFT Applications, First Edition.
Edited by Pratim Kumar Chattaraj and Debdutta Chakraborty.

Mn incorporated with CdS $Cd_{1-x}Mn_xS$ is studied with the help of first principle approach [20]. The author reported significant luminous and magneto-optical Mn:CdS compound results from Mn-doped CdS, which provide p-type semiconductors d-states close to the top of the valence band boundary. Nabi et al. [21] examined Mn:CdS, Cr:CdS and co-doping of Mn/Cr with CdS by using the density functional theory (DFT) technique. The authors found that Mn/Cr doped with CdS system is more stable as compared to Mn:CdS and Cr:CdS and suitable for spintronic application. Gupta et al. [24] examined Ni, Pd, Pt, and Au-doped CdS clusters. The authors demonstrated that energy gap of these metallic clusters reduced when it was placed on the CdS surface, however, an exception occurred for Pt clusters. Muthusamya et al. [22] examined copper-doped CdS with the help of a chemical bath deposition technique. They found that incorporation of copper in CdS helps in the improvement of a structural parameter as well as an energy gap. Sukkabot [1] investigated optoelectronic and magnetic properties of TMs doping in CdS with the help of the DFT method. It is found that doping of TMs in CdS changes its semiconductor characteristics to either metallic or half-metallic characteristics. Aimouch et al. [2] studied CdS with Mn and CdS with (Mn, Cr) by using a theoretical approach and found that the optical energy gap rises after doping of Mn as well as (Mn, Cr) in CdS. Doping of TMs in CdS provides a suitable result to minimize the excess reactivity potential for oxygen evolution reaction (OER) and hydrogen evolution reaction (HER) in comparison with the pure and bulk CdS surface [25]. Deka and Kalita [26] reported doping of TMs – Fe, Co, Ni, and Mn in CdS nanocrystals and demonstrated that doping of TMs in CdS upturns the photocatalytic responses. Gandouzi et al. [27] reported enhancement in the electronic and optical behavior of CdS after doping with Ag. Heiba et al. [28] studied TMs – Mn, Co, Cu, Zn doped CdS nanomaterials and reported that doping of these TMs results in the enhancement of the photocatalytic response. Additionally, they indicated that copper-doped CdS nanomaterial provides maximum photoconductivity in the IR spectrum. Zarhri et al. [29] studied magnetic characteristics of TMs doped CdSe: CdSe:V, CdSe:Cr, CdSe:Mn, CdSe:Fe, CdSe:Co, CdSe:Ni, and CdSe:Cu by using DFT technique. Shinde et al. [30] studied Fe and Mn-doped with CdSe nanostructure and demonstrated that optoelectronic properties of CdSe, Fe-CdSe, and Mn-CdSe are suitable for photovoltaic devices. Xie and Liu [31] investigated magnetic behavior of TMs doped CdTe materials. They demonstrated that Cr-CdTe, V-CdTe, and Mn-CdTe have half-metallic ferromagnetic characteristics. Chavan et al. [32] reported TMs doped CdTe materials and found that doping of TMs in CdTe increases the gap between HOMO and LUMO, which results in the improvement of the stability of overall system. Recently, Xia et al. [33] studied doping of TMs in CdX and ZnX (X = S, Se, Te) by using DFT technique and found that the doped system provides effective results to be used as catalysts. They also stated that CdS:Ni, CdSe:Cu, CdTe:Cu, CdTe:Cu, ZnS:Cu, and ZnSe:Cu materials are potential candidates for water-splitting applications.

From the above discussion, it is clear that TMs doped CdS, CdSe, and CdTe materials are of immense importance from technological point of view, especially in the domain of photocatalytic, solar cell, optoelectronic and spintronic. In this report, we have studied CdX as well as TMs – Cu, Ag, and Au-doped CdX (X = S, Se, Te) nanocluster by using the conceptual density functional theory (CDFT) approach. CDFT-based descriptors of pure and doped CdX clusters are calculated.

14.2 Methodology

DFT is a crucial computational method for comprehending the electronic structure of molecular systems. Several domains of engineering and sciences capitalize on the DFT technique [34]. In this work, pure CdX, as well as TM atoms – Cu, Ag, and Au-doped CdX (X = S, Se, Te) nanocluster are investigated on the DFT framework. Geometry optimization of these systems is performed with Gaussian 16 and Gauss View 6.0 [35]. For geometry optimization, functional local spin density approximation (LSDA)/LANL2DZ is used. Exchange correlation – LSDA is effectively applied by the researchers for the geometry optimization of cadmium and zinc chalcogenide materials [36, 37]. All the optimized cluster has real vibrational frequencies.

Ionization potential (I) and electron affinity (A) of CdX, as well as Cu, Ag, and Au-doped CdX nanoclusters (X = S, Se, Te) are calculated by using Koopmans's approach, which is as [38]:

$$I = -\varepsilon_{\text{HOMO}} \tag{14.1}$$

$$A = -\varepsilon_{\text{LUMO}} \tag{14.2}$$

CDFT-based global descriptors – molecular hardness, softness, electronegativity, as well as electrophilicity index (ω) are considered by using I and A, which are as under:

Molecular Hardness,

$$\eta = \frac{I - A}{2} \tag{14.3}$$

Softness,

$$S = \frac{1}{2\eta} \tag{14.4}$$

Electronegativity,

$$\chi = -\mu = \frac{I + A}{2} \tag{14.5}$$

Here, μ denotes the chemical potential

Electrophilicity Index,

$$\omega = \frac{\mu^2}{2\eta} \tag{14.6}$$

14.3 Results and Discussion

14.3.1 Electronic Properties and CDFT-Based Descriptors

In this section, we have discussed the electronic properties as well as CDFT-based descriptors of pure CdX and TMs (Cu, Ag, and Au)-doped CdX (X = S, Se, Te) materials. The computed CDFT-based descriptors and HOMO–LUMO energy gap of CdX and doped-CdX are displayed in Table 14.1.

The evolution of a donor–acceptor system is implied to depend critically on the molecular orbitals of the donor and acceptor in their exchange of charge and the emergence of bonds [39–41]. Frontier molecular orbitals – HOMO and LUMO is a crucial metrics to examine the electronic behavior of nanoclusters. It shows how little energy is desirable for an electron to transition from an occupied to an unoccupied orbital. The chemical stability of the molecular system is significantly impacted with the energy gap. Data exhibit that the HOMO–LUMO energy gap of CdX is found among 0.053–0.295 eV. For doped-CdX clusters, the HOMO–LUMO energy gap is found between 0.640 and 1.837 eV. Our computed data agree with the demonstrated result of TMs doped CdS material [1, 2, 33]. Among the studied compounds, CuCdS, CuCdSe, AgCdSe, and CuCdTe have HOMO–LUMO gap as 1.319, 1.294, 1.837, and 1.255 eV correspondingly, which indicate that these materials may be suitable as HER catalysts for water splitting reaction and

Table 14.1 HOMO–LUMO energy gap and CDFT-based descriptors of CdX and doped-CdX (X = S, Se, Te) nanocluster.

Species	HOMO–LUMO gap (eV)	Hardness (eV)	Softness (eV)	Electronegativity (eV)	Electrophilicity Index (eV)	Dipole moment (D)
CdS	0.295	0.147	3.393	5.586	105.870	6.213
CuCdS	1.319	0.659	0.758	4.566	15.812	3.718
AgCdS	1.144	0.572	0.874	4.669	19.057	3.904
AuCdS	0.714	0.357	1.401	5.174	37.510	3.219
CdSe	0.055	0.027	18.194	2.187	87.052	3.229
CuCdSe	1.294	0.647	0.773	4.527	15.836	3.487
AgCdSe	1.837	0.919	0.544	3.392	6.263	2.172
AuCdSe	0.669	0.335	1.494	5.093	38.749	2.860
CdTe	0.053	0.027	18.847	2.284	98.306	4.511
CuCdTe	1.258	0.629	0.795	4.484	15.975	3.122
AgCdTe	1.092	0.546	0.915	4.550	18.954	3.131
AuCdTe	0.640	0.320	1.562	4.979	38.716	2.386

spintronic applications. According to Li et al. [42] TMs doped CdS with an energy gap of more than 1.23 eV are proper for photocatalytic H_2S and H_2O rupturing. The smallest quantity of energy needed to separate water into hydrogen and oxygen is 1.23 eV, hence, the photocatalysts theoretical energy gap needs to be greater than this value [33]. Walter et al. [43] stated that there will be energy lost as a result of the electron transport mechanism, and the kinetic excess potential will also be able to surpass the barriers for the HER and the OER, the photocatalysts energy gap needs to be greater than 1.8 eV.

Data shows that the energy gap increases after incorporation of TMs (Cu, Ag, and Au) in CdS, CdSe, and CdTe clusters. The HOMO–LUMO gap of doped CdX clusters decreases from Cu to Ag to Au, except for AgCdSe. Data exhibits that HOMO–LUMO energy gap for CdX and doped-CdX nano cluster decreases from S to Se to Te, except for AgCdSe. The computed results follow a similar trend as reported previously [44–49]. The cluster with the largest energy gap will exhibit responses peripheral deviations the least. It illustrates that a cluster with a sizable HOMO–LUMO gap requires a substantial potential to drive electrons from occupied to unoccupied molecular orbital. The density of states (DOS) of CdX and doped-CdX nanoclusters are presented in Figure 14.1.

Chemical hardness is an essential component in understanding chemical systems kinetics, binding and molecular composition [50–52]. Molecular hardness is a measure of a systems electronic stability and firmness. A molecules hardness value is decreased and vice versa while it transports from a steady to an unsteady state [41]. The molecular hardness of CdX systems varies from 0.027 to 0.147 eV, whereas for doped-CdX systems, it swings between 0.320 and 0.919 eV. AgCdSe and AuCdTe show maximum and least hardness for doped clusters, respectively. Molecular hardness reduces from S to Se to Te, except AgCdSe. When compared to molecular hardness, softness has a reciprocal relation. For CdX and doped-CdX nanoclusters, softness varies between 3.393–18.847 and 0.544–1.562 eV, respectively. Softness value upsurges as of S to Se to Te, except AgCdSe. Conversely, for doped clusters softness value increases from Cu to Ag to Au, whereas hardness follows a reverse trend, except for AgCdSe.

Species	Density of states (DOS)
CdS	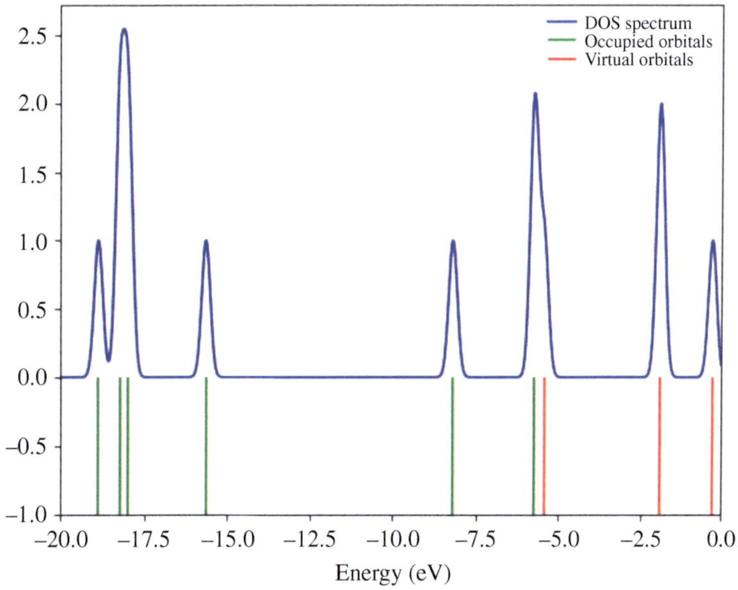
CuCdS	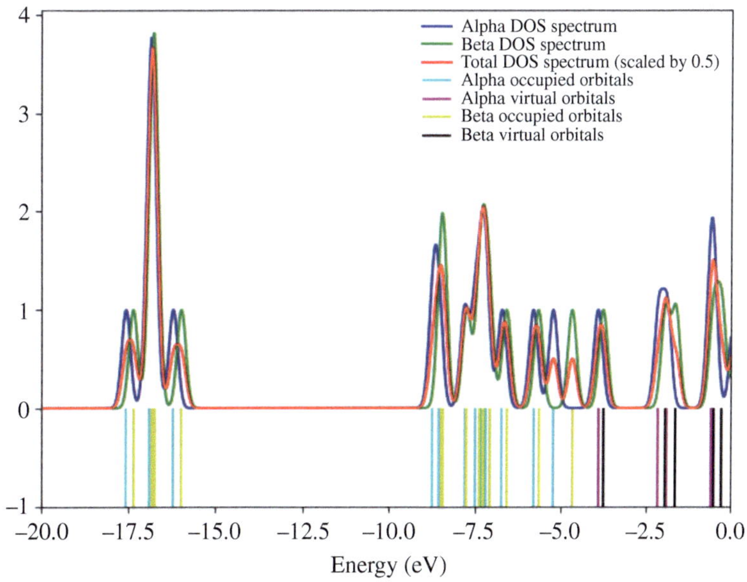

Figure 14.1 DOS of CdX and doped-CdX nanoclusters.

Species	Density of states (DOS)
AgCdS	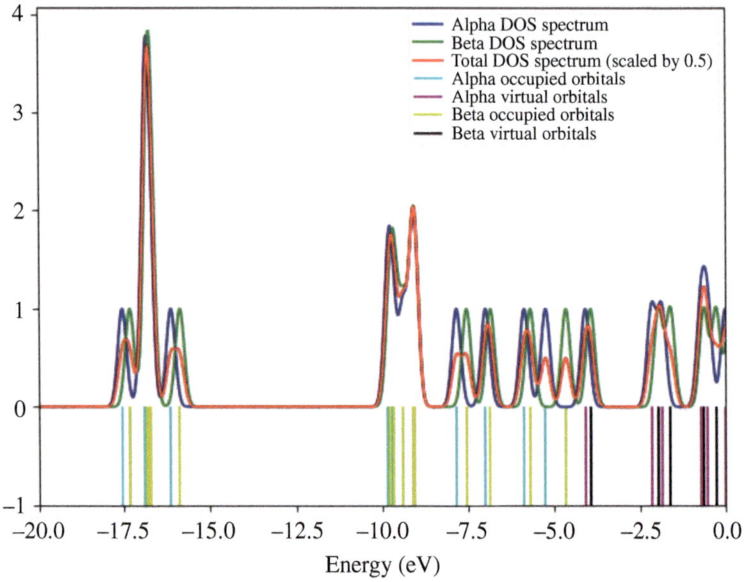
AuCdS	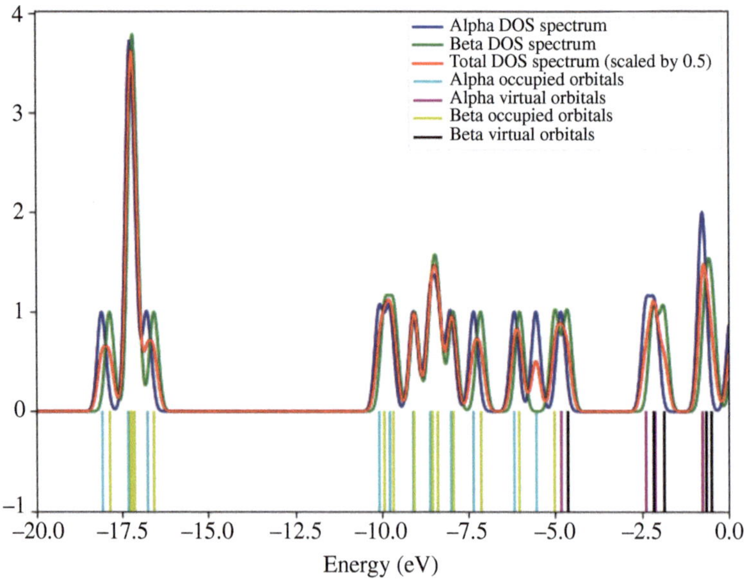

Figure 14.1 (*Continued*)

<table>
<tr><th>Species</th><th>Density of states (DOS)</th></tr>
<tr><td>CdSe</td><td>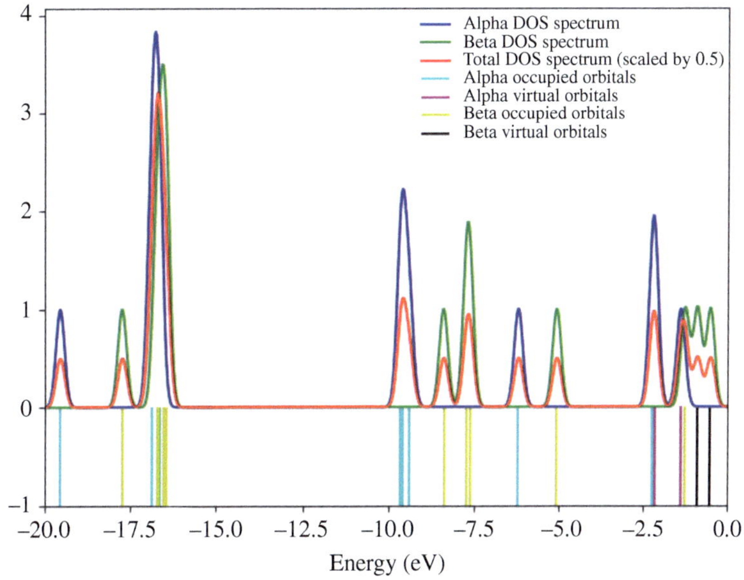
</td></tr>
<tr><td>CuCdSe</td><td>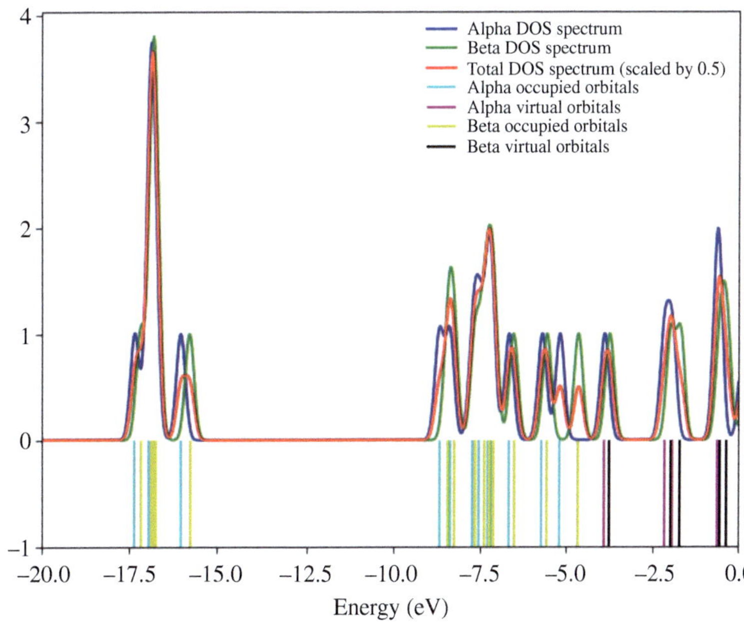
</td></tr>
</table>

Figure 14.1 (*Continued*)

Species	Density of states (DOS)
AgCdSe	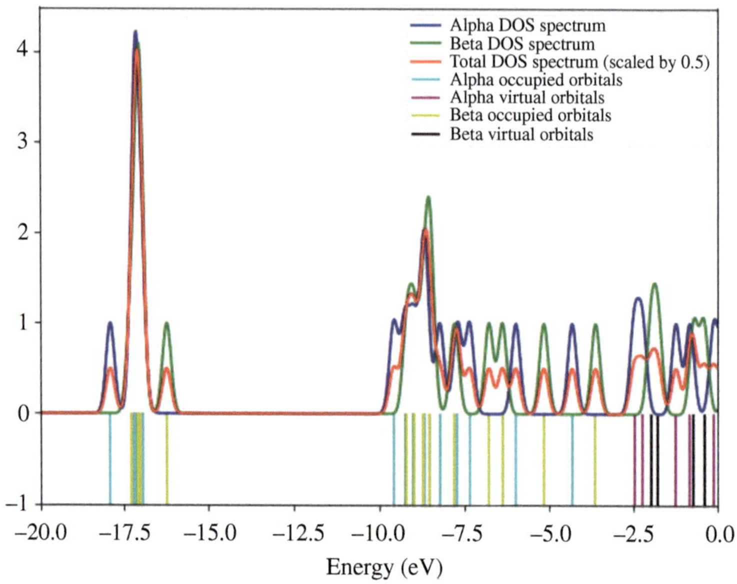
AuCdSe	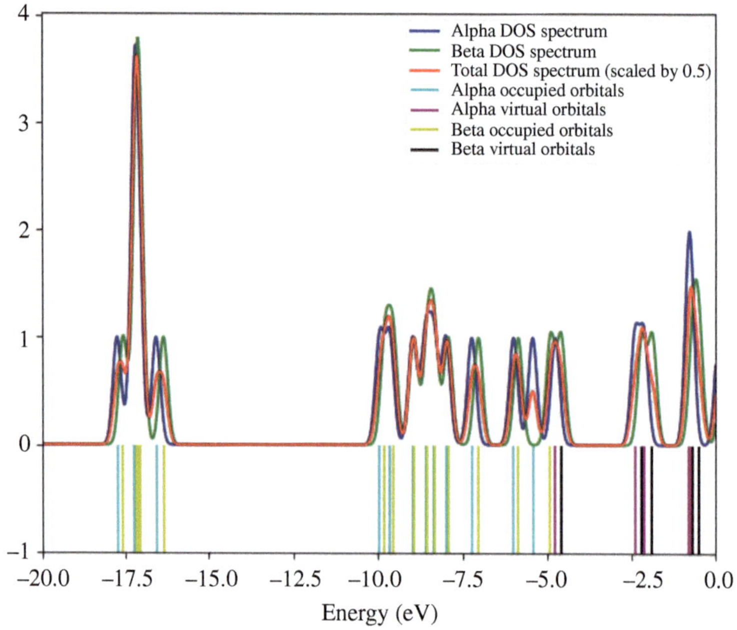

Figure 14.1 (*Continued*)

<table>
<tr><th>Species</th><th>Density of states (DOS)</th></tr>
<tr><td>CdTe</td><td>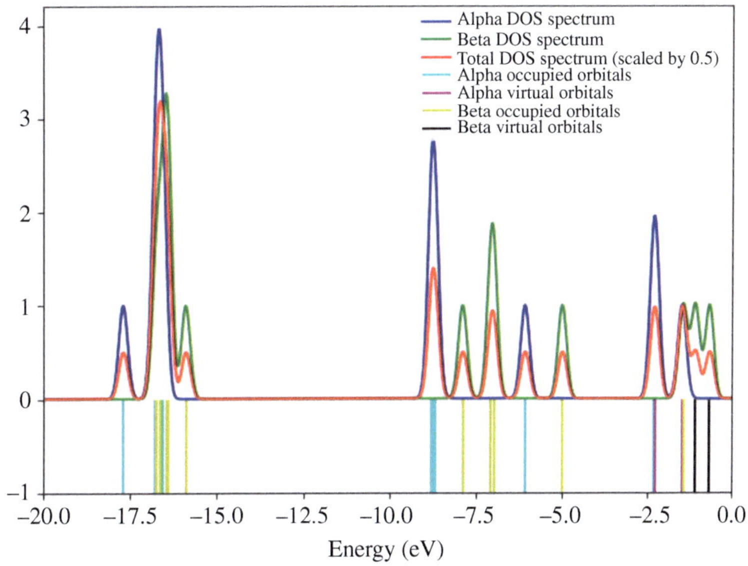
</td></tr>
<tr><td>CuCdTe</td><td>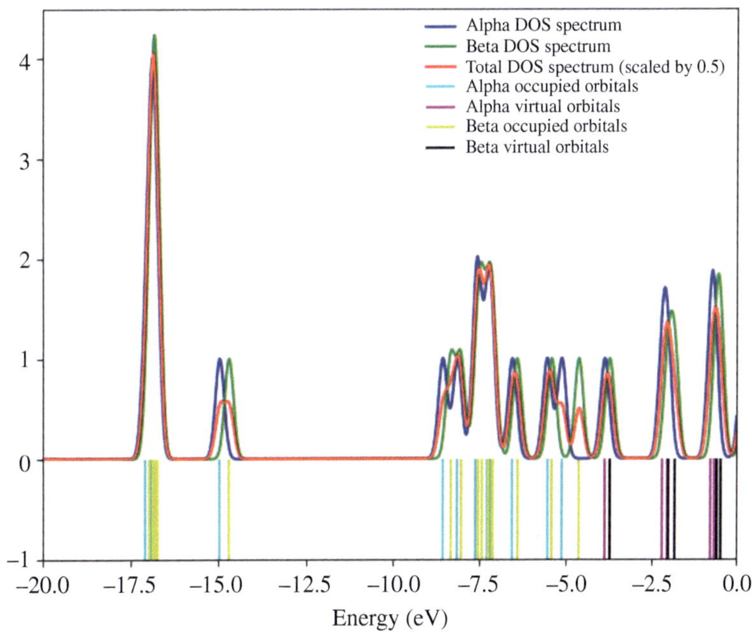
</td></tr>
</table>

Figure 14.1 *(Continued)*

Species	Density of states (DOS)
AgCdTe	

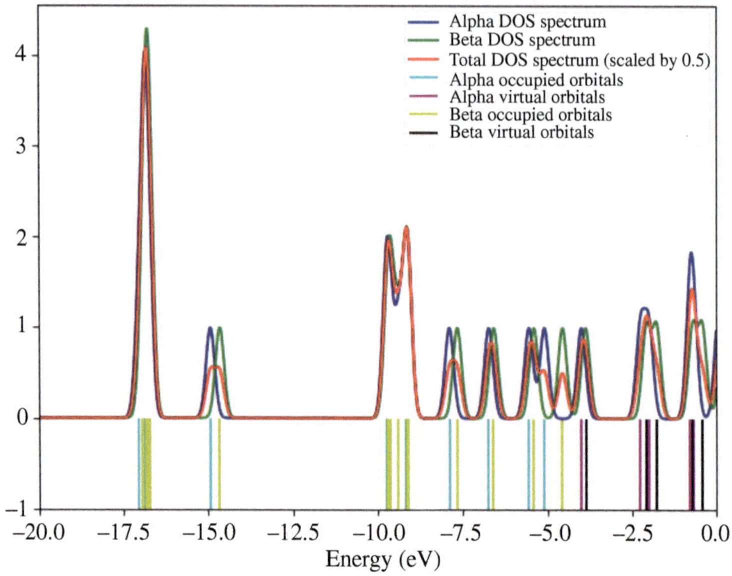

AuCdTe

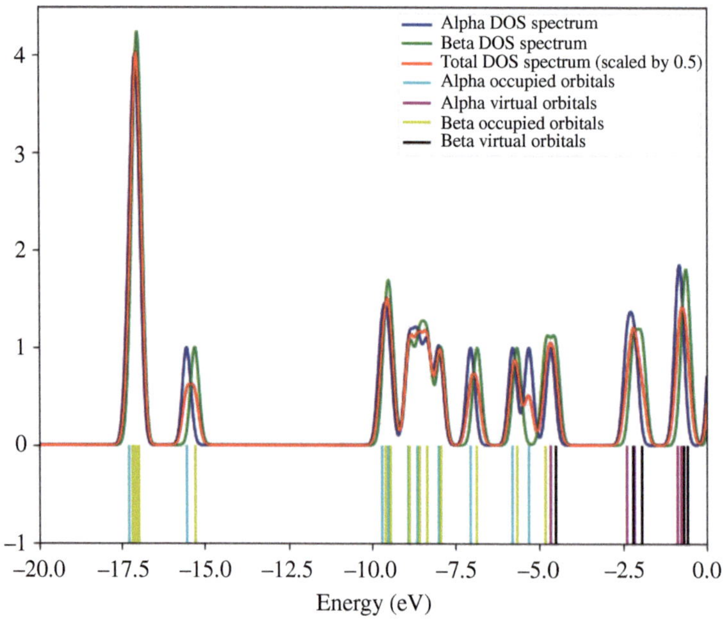

Figure 14.1 (*Continued*)

To comprehend the process of charge transfer across the donor and acceptor, it is crucial to understand the materials theory of electronegativity [53–58]. For the studied molecules, the electronegativity of CdX and doped CdX nanoclusters are observed among 2.187–5.586 and 3.392–5.174 eV correspondingly. In the case of doped clusters, the highest and lowest electronegativity values are found for AuCdS and AgCdSe clusters, respectively. The electronegativity of CdX and doped-CdX clusters decreases from S to Se to Te, except AgCdSe. The electronegativity value for doped clusters increases from Cu to Ag to Au, except AgCdSe.

The ionization energy and electron affinity have an influence on the electrophilicity index, which is a metric used to calculate the quantity of energy dissipated because of a substantial movement of electrons amid donor–acceptor interface [59, 60]. The electrophilicity index of CdX, as well as doped-CdX nanoclusters, fluctuates between 87.052–105.870 and 6.263–38.749 eV correspondingly. Among the doped CdX system, the highest and lowest electrophilicity index is witnessed for AuCdSe and AgCdSe respectively. The result exhibits that electrophilicity index of doped clusters follows an upward trend from Cu to Ag to Au, except AgCdSe.

The dipole moment (in Debye) of these clusters is presented in Table 14.1. In the case of pure CdX clusters, the dipole moment is observed for CdS, CdSe, and CdTe as 6.213, 3.229 and 4.511 D, respectively. In the case of doped clusters, AgCdSe and AgCdS provide minimum and maximum values of dipole moment, respectively.

14.4 Conclusion

In this report, pure CdX and TMs Cu, Ag, and Au-doped CdX (X = S, Se, Te) nanoclusters are studied with the help of CDFT methodology. CDFT-based global descriptors of these clusters are computed. According to the results, the HOMO–LUMO energy gap of CdX is between 0.053 and 0.295 eV, while they are between 0.640 and 1.837 eV for TMs-doped CdX clusters. It exhibits enhancement in the energy gap after the addition of TMs in CdX clusters. For the doped cluster, energy gap of CuCdS, CuCdSe, AgCdSe, and CuCdTe are witnessed as 1.319, 1.294, 1.837, and 1.255 eV correspondingly, which are within the range necessary for HER catalysts for water splitting processes and spintronic technologies. Clusters AgCdSe and AuCdTe show maximum and least hardness for doped clusters, respectively. Electronegativity of CdX and doped CdX nanoclusters are found between 2.187–5.586 and 3.392–5.174 eV, respectively. In the case of doped clusters, the highest and lowest electronegativity values are found for AuCdS and AgCdSe clusters correspondingly, whereas largest and least electrophilicity index is found for AuCdSe and AgCdSe respectively. For TMs-doped clusters, the minimum and maximum dipole moment are detected for AgCdSe and AgCdS, respectively. Data exhibits that HOMO–LUMO energy gap, molecular hardness and electronegativity for CdX and doped-CdX nano cluster decreases from S to Se to Te, except AgCdSe. The HOMO–LUMO energy gap and hardness of TMs-doped CdX clusters decreases from Cu to Ag to Au. In contrast, the electronegativity and electrophilicity index follow the opposite pattern, except for AgCdSe. The calculated energy gaps of TMs-CdX are consistent with the reported data.

Acknowledgments

Dr. Prabhat Ranjan is thankful to Manipal University Jaipur for providing computational resources as well as research facilities.

Dr. Tanmoy Chakraborty is thankful to Sharda University for providing research facilities and computational resources.

Funding

Dr. Prabhat Ranjan and Dr. Tanmoy Chakraborty would like to thank the Science and Engineering Research Board (SERB), Department of Science and Technology, Government of India, under Grant No. (CRG/2022/002539) and (CRG/2020/002951) respectively.

References

1 Sukkabot, W. (2020). *IEEE Trans. Magn.* 56: 1.
2 Aimouch, D.E., Meskine, S., Birsan, A. et al. (2018). *Mater. Chem. Phys.* 213: 249.
3 Amin, B. and Ahmad, I. (2009). *J. Appl. Phys.* 106: 093710.
4 Ahmad, I. (2013). *Comput. Mater. Sci* 68: 55.
5 Arif, S., Ahmad, I., and Amin, B. (2012). *Int. J. Quantum Chem.* 112: 882.
6 Georgobiani, A.N. (1974). *Usp. Fiz. Nauk* 113: 129.
7 Al-Douri, Y., Waheb, J.H., Ameri, M. et al. (2013). *Int. J. Electrochem. Sci.* 8: 10688.
8 Al-Douri, Y., Khasawneh, Q., Kiwan, S. et al. (2014). *Energy Convers. Manage.* 82: 238.
9 Ameri, M., Mesbah, S., Al-Douri, Y. et al. (2014). *Acta Phys. Pol. A* 125: 1110.
10 Al-Douri, Y. and Reshak, A.H. (2015). *Optik* 126: 5109.
11 Ibraheam, A.S., Al-Douri, Y., and Azman, A.H. (2016). *Optik* 127: 8907.
12 Al-Douri, Y., Ameri, M., Bouhemadou, A., and Khenata, R. (2016). *Microsyst. Technol.* 22: 2529.
13 Bruchez, M. Jr., Moronne, M., Gin, P. et al. (1998). *Science* 281: 2013.
14 Al-Zahrani, J.H., El-Hagary, M., and El-Taher, A. (2015). *Mater. Sci. Semicond. Process.* 39: 74.
15 Thool, G.S., Sraveen, K., Singh, A.K. et al. (2015). *J. Alloys Compd.* 649: 553.
16 Saikia, B.J., Nath, B.C., Borah, C., and Dolui, S.K. (2015). *J. Lumin.* 168: 178.
17 Patel, N.H., Deshpande, M.P., Bhatt, S.V., and Chaki, S. (2014). *Adv. Mater. Lett.* 5: 671.
18 Gao, X., Zhuo, N., Liao, C. et al. (2015). *Opt. Mater. Express* 5: 2164.
19 Shen, T., Tian, J., Lv, L. et al. (2016). *Electrochim. Acta* 191: 62.
20 Nabi, A. (2016). *Comput. Mater. Sci* 112: 210.
21 Nabi, A., Akhtar, Z., Iqbal, T. et al. (2017). *J. Semicond.* 38: 073001.
22 Muthusamya, M. and Muthukumaran, S. (2015). *Optik* 126: 5200.
23 Yanga, J., Liua, R., Huanga, S. et al. (2014). *Catal. Today* 224: 104.
24 Gupta, S.S. and van Huis, M.A. (2019). *J. Phys. Chem. C* 123: 9298.
25 Garg, P., Nair, A.S., Rawat, K.S., and Pathak, B. (2019). *J. Phys. Chem. C* 123: 13419.
26 Deka, K. and Kalita, M.P.C. (2018). *J. Alloys Compd.* 757: 209.
27 Gandouzi, M., Alshammari, A.S., Khan, Z.R., and Bouzidi, M. (2021). *Mater. Sci. Semicond. Process.* 129: 105794.
28 Heiba, Z.K., Mohamed, M.B., Farag, N.M. et al. (2021). *J. Mater. Sci.: Mater. Electron.* 32: 1850.
29 Zarhri, Z., Ez-Zahraouy, H., El Amraoui, Y. et al. (2015). *J. Supercond. Novel Magn.* 28: 2155.
30 Shinde, S.K., Dubal, D.P., Ghodake, G.S., and Fulari, V.J. (2014). *RSC Adv.* 4: 33184.
31 Xie, W.H. and Liu, B.G. (2004). *J. Appl. Phys.* 96: 3559.
32 Chavan, K.T., Chandra, S., and Kshirsagar, A. (2022). *Mater. Today Commun.* 30: 103104.
33 Xia, F., Shu, L., Wen, Y. et al. (2022). *Mater. Adv.* 3: 5772.

34 Hafner, J., Wolverton, C., and Ceder, G. (2006). *MRS Bull.* 31: 659.
35 Frisch, M.J., Trucks, G.W., Schlegel, H.B., et al. (2016). *Gaussian 16, Revision C.01*. Gaussian, Inc.: Wallingford, CT.
36 Gurel, H.H., Akinci, O., and Unlu, H. (2015). *Superlattices Microstruct.* 51: 725.
37 Akinci, O., Gurel, H.H., and Unlu, H. (2009). *Thin Solid Films* 517: 2431.
38 Parr, R.G. and Yang, W. (1989). *Density-Functional Theory of Atoms and Molecules*. Oxford: Oxford University Press.
39 Fujimoto, H., Kato, S., Yamabe, S., and Fukui, S. (1974). *J. Chem. Phys.* 60: 572.
40 Fujimoto, H., Kato, S., Yamabe, S., and Fukui, S. (1974). *JACS* 96: 2024.
41 Ghosh, D.C. and Bhattacharyya, S. (2004). *Int. J. Mol. Sci.* 5: 239.
42 Li, Y., Bahamon, D., Sinnokot, M., and Vega, L.F. (2022). *npj Comput. Mater.* 8: 229.
43 Walter, M.G., Warren, E.L., McKone, J.R. et al. (2010). *Chem. Rev.* 110: 6446.
44 Saloni, S., Ranjan, P., and Chakraborty, T. (2023). *Mater. Sci. Semicond. Process.* 164: 107608.
45 Ranjan, P., Kumar, P., Chakraborty, T. et al. (2020). *Mater. Chem. Phys.* 241: 122346.
46 Ranjan, P., Kumar, P., Surolia, P.K., and Chakraborty, T. (2021). *Thin Solid Films* 717: 138469.
47 Ranjan, P. and Chakraborty, T. (2021). *Mater. Sci. Semicond. Process.* 127: 105745.
48 Ranjan, P., Surolia, P.K., and Chakraborty, T. (2021). *Pure Appl. Chem.* 93: 591.
49 Saloni, S., Ranjan, P., and Chakraborty, T. (2023). *J. Mol. Graphics Modell.* 124: 108534.
50 Chattaraj, P.K. and Sengupta, S. (1999). *J. Phys. Chem.* 103: 6122.
51 Parr, R.G. and Chattaraj, P.K. (1991). *JACS* 113: 1854.
52 Ghosh, D.C. (1984). *Indian J. Pure Appl. Phys.* 22: 346.
53 Sanderson, R.T. (1951). *Science* 114: 670.
54 Sanderson, R.T. (1952). *Science* 116: 41.
55 Sanderson, R.T. (1955). *Science* 121: 207.
56 Sanderson, R.T. (1952). *J. Chem. Ed.* 29: 539.
57 Sanderson, R.T. (1954). *Science* 31: 238.
58 Sanderson, R.T. (1960). *Chemical Periodicity*. New York: Reinhold Publishing Corporation.
59 Parr, R.G., Szentpaly, L.V., and Liu, S. (1999). *JACS* 121: 1922.
60 Chattaraj, P.K. and Duley, S. (2010). *J. Chem. Eng. Data* 55: 1882.

15

"Phylogenetic" Screening of External Potential Related Response Functions

Paweł Szarek

Navi-Chem®, Warsaw, Poland

15.1 Introduction

The conceptual density functional theory (CDFT) offers an understanding of chemical processes by means of quantum mechanical observables – energy, density, and their derivatives. This field identifies and formulates the response functions, which give precise definitions to general concepts in chemistry, such as electronegativity, hardness, and softness, and at the same time provide their practical approximations [1, 2]. The response functions emerge from differentiability of energy functional, $E[N, v(r)]$ (E for short), with respect to electron number, N, and external potential, $v(r)$. In general, the "N" derivatives yield global quantities, characterizing the whole system, while "$v(r)$" derivatives produce "local" quantities, which change from point to point. The electron number is related to electron counting formalism assigning a number of valence electrons to individual atoms (or a molecule). The ∂N defines "electron-deficient" systems with fewer electrons or "hypervalent" with additional electrons, as compared to the hypothetical assignment of charge to free existing species. Since these two states are likely to be more prone to transformations, the variation in N is an important tool for identifying the reactivity. The second variable, the external potential includes electron-ion potential and any potential due to an applied electric field [2]. Nevertheless, specific types of responses related to external potential ingredients are commonly distinguished. For example, the polarizabilities, such as the first-order response of the dipole moment to external electric fields, $\vec{F}$, constitute an explicit class of functions. Therefore, usually, the $v(r)$ implicates potential from solely the nuclear charges, in the absence of other electric fields:

$$v(r) = -\sum_i \frac{Z_i}{|r - R_i|} \tag{15.1}$$

Direct perturbations with respect to the charge of ith nucleus, Z_i, yield a third class of external potential linked sensitivities, which will be introduced in a moment. The Z_i, $v(r)$ and $\vec{F}$ constitute a group of related external variables of Coulomb nature, i.e. the "Z," "Z/r," and "Z/r^2" like series, (note that $\vec{F}$ tends to be a uniform external electric field, although it might be generalized to an electric field of any kind [3]). The charge, potential and electric fields are in general related through

Electron Density: Concepts, Computation and DFT Applications, First Edition.
Edited by Pratim Kumar Chattaraj and Debdutta Chakraborty.

Maxwell's equations, Gauss's law and Poisson's equation,[1] where:

$$\vec{F} = -\nabla\Phi \tag{15.2}$$

and

$$\nabla \cdot \vec{F} = -\nabla^2\Phi = \frac{\rho}{\varepsilon} \tag{15.3}$$

Notably, emerging from the Hohenberg–Kohn theorems [10] and Kohn–Sham equations [11], the first energy derivative over the external potential defines electron density, $\rho(r)$ [12].

$$\rho(r) = \left(\frac{\partial E}{\partial v(r)}\right)_N \tag{15.4}$$

Higher order response functions wrt N or $v(r)$ might be characterized as chemical potential, μ [13], responses (i.e. chemical hardness, η [14]), and the density-density response functions (for instance, the linear response function, $\chi(r, r')$ – correlation function of the many-body theory [15]), or the Fukui function, $f(r)$ [16–19].

$$\mu = \left(\frac{\partial E}{\partial N}\right)_v \tag{15.5}$$

$$\eta = \left(\frac{\partial^2 E}{\partial N^2}\right)_v = \left(\frac{\partial \mu}{\partial N}\right)_v \tag{15.6}$$

$$\chi(r, r') = \left(\frac{\partial^2 E}{\partial v(r)\partial v(r')}\right)_N \tag{15.7}$$

$$f(r) = \frac{\partial^2 E}{\partial v(r)\partial N} = \left(\frac{\partial \mu}{\partial v(r)}\right)_N = \left(\frac{\partial \rho(r)}{\partial N}\right)_v \tag{15.8}$$

Corresponding response functions are presented in Figure 15.1. The response functions on the left-hand side of the scheme division line, $\partial/\partial N$ sensitivities, are well-established and thoroughly analyzed [1, 20, 21]. This chapter focuses on v-dependence and the class of response functions related to particular changes in external potential.

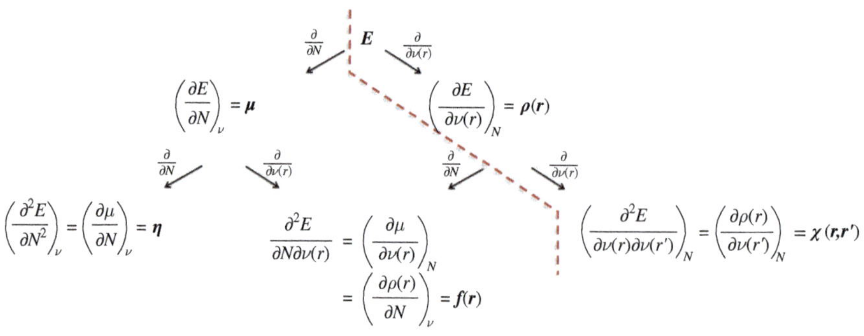

Figure 15.1 Energy sensitivities in the canonical ensemble.

---

1 In the context of Eqs. (15.2) and (15.3), there is another external potential related ingredient that comes from Faraday's low of induction and should be considered with respect to magnetic sources in atomic and molecular systems, such as electron and nuclear magnetic moments, or the response to external magnetic fields. The effects associated with the external magnetic fields [4] and electron spin, via variation in spin-state energetics (i.e. energy functional representation in terms of the spin number N_s, constituting a generalized framework of spin-polarized DFT) [5–9], to some extent have already been taken into account. However any magnetic and spin polarized perturbations are out of scope of this chapter.

15.2 Alchemical Approach

Over a decade ago, new concepts for exploring chemical space appeared at the forefront of CDFT [22–26]. The so-called alchemical derivatives of the system energy with respect to changes in one or more nuclear charges are becoming of increasing importance in chemistry and physics [27]. The alchemical, "Z"-dependent derivatives might be seen as analogue of the "N" derivatives, as well as a subclass of external potential sensitivities. Accordingly, the first-order derivative, $(\partial E/\partial Z_i)_N$, called alchemical potential, represents the electronic part of the molecular electrostatic potential at the position of ith nucleus, $V_e(R_i)$ [25, 28, 29].

$$\left(\frac{\partial E}{\partial Z_i}\right)_N = -\int \frac{\rho(r)}{|r-R_i|}dr = V_e(R_i) \tag{15.9}$$

In the mixed second-order derivative $(\partial^2 E/\partial Z_i \partial Z_j)_N$ the linear response function is concealed [30],

$$\left(\frac{\partial^2 E}{\partial Z_i \partial Z_j}\right)_N = \iint \left(\frac{\partial E}{\partial v(r)\partial v(r')}\right)_N \frac{\partial v(r)}{\partial Z_i}\frac{\partial v(r')}{\partial Z_j}drdr' = \iint \frac{1}{|r-R_i|}\frac{1}{|r'-R_j|}\chi(r,r')drdr' \tag{15.10}$$

while the single argument second-order sensitivity, $\left(\partial^2 E/\partial Z_i^2\right)_N$, has been identified with the electronic part of the alchemical hardness, η_{al} [22, 30]

$$\eta_{al} = \left(\frac{\partial^2 E}{\partial Z_i^2}\right)_N = -\int \frac{1}{|r-R_i|}\left(\frac{\partial \rho(r)}{\partial Z_i}\right)_N dr = \left(\frac{\partial V_e(R_i)}{\partial Z_i}\right)_N \tag{15.11}$$

In the above, after employing the chain rule, it has been used that the partial derivative of external potential, Eq. (15.1), with respect to nuclear charge of ith atom, reduces to

$$\frac{\partial v(r)}{\partial Z_i} \Rightarrow -\frac{1}{|r-R_i|} \tag{15.12}$$

15.3 The "Family Tree"

Before going deeper into the subject, the genealogy of relevant response functions associated with the three electric variables (related via Poisson's equations) will be presented. Figure 15.2 summarizes the responses of an energy functional to perturbations related to the external potential "components." The right-hand part of the canonical ensemble scheme in Figure 15.1 diverges into a family of energy functional derivatives over nuclear charge, $\partial/\partial Z$, external potential, $\partial/\partial v(r)$, and electric field, $\partial/\partial\vec{F}$, which corresponds to "Z," "Z/r" and "Z/r^2" like a series of perturbations. In upper half of the diagram (Figure 15.2), the global response functions are presented, while the lower half shows the local response functions.

Among (I) the global first-order responses (i) the electric dipole moment, D (i.e. sensitivity wrt electric field), and (ii) the electronic part of the overall electrostatic potential at the position of ith nucleus, $V_e(R_i)$ (i.e. sensitivity wrt nuclear charge), appear. (II) The second-order global responses include: (i) the electric dipole polarizability, α, (ii) the electronic part of alchemical hardness, η_{al}, and (iii) the alchemical characteristic radius, R_{al}. The latter is a direct analogue with $R = \sqrt{\alpha\eta}$ – the characteristic radius [31–33], defined as a product of electric dipole polarizability and chemical hardness. Additionally, while the η_{al} is a second-order derivative with respect to charge variation

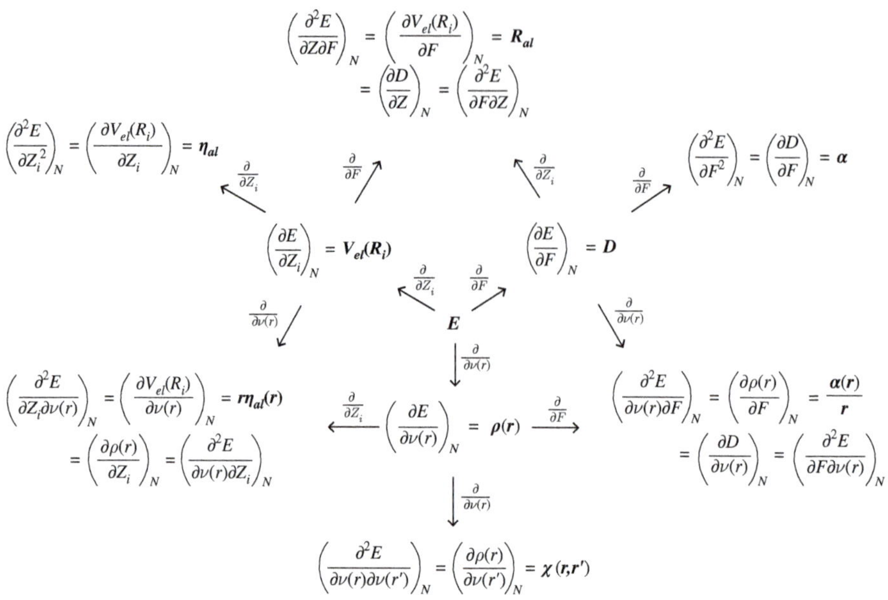

Figure 15.2 Summary and interpretation of external potential related energy derivatives over nuclear charge, $\frac{\partial}{\partial Z}$, external potential, $\frac{\partial}{\partial v(r)}$, and electric field, $\frac{\partial}{\partial F}$, thus sensitivities over “Z”, “Z/r” and “Z/r^2” like series.

on a single nucleus (diagonal elements of alchemical hardness matrix), the mixed second-order derivative over the nuclei charges Z_i and Z_j are related to linear response function, see Eq. (15.10), and constitutes off-diagonal elements.

Subsequently, (III) the local responses comprise at first order: (i) the electron density, $\rho(r)$ – an external potential derivative, and (IV) at the second order: (i) the linear response function, $\chi(r, r')$, (ii) the first moment of the local alchemical hardness, $r \cdot \eta_{\text{al}}(r)$, and (iii) the “polarizability potential” function, $\alpha(r)/r$ (which might also be regarded as the first moment of local softness).

15.4 First-order Sensitivities

Among the responses to three variables presented in Figure 15.2, the electron density (Eq. (15.4)) is fundamental sensitivity, $\left(\frac{\partial E}{\partial v(r)}\right)_N$, by means of which all others might be expressed. The $\rho(r)$ adjusts itself to the external potential field, $v(r)$, to be in a minimum of energy E.

Perturbation of the system by i.e. uniform external electric field leads to the next first-order energy derivative identified as electric dipole moment, D, which comes from Hellmann–Feynman theorem.

$$\left(\frac{\partial E}{\partial \vec{F}}\right)_N = -D = -\int r\delta\rho(r)dr \tag{15.13}$$

Here $\delta\rho(r)$ is a difference between electron density affected by an electric field and electron density of unperturbed system, thus, Eq. (15.13) gives the electric field-induced dipole moment. From perturbation theory, the dipole moment is defined via the summation of electron and nuclear charges [34–37]. The relation in Eq. (15.13) has been obtained by applying the chain rule, using

Eq. (15.2) and gradient theorem, followed by integration over ∂r and differentiation wrt $\partial \vec{F}$, which yields

$$\frac{dv(r)}{d\vec{F}} = -\int 1dr = -r \tag{15.14}$$

In general, the sources of external electric fields are arbitrary. Therefore uniform electric field might be substituted by one arising from, e.g. a point charge, a dipole or any higher multipole basis. In particular, conceptually, a point charge source of an electric field might even originate at the ith nucleus, $Z_i \pm \delta Z$. In such a case, an effective (isotropic) equivalent of a dipole moment might be defined. Its magnitude would be proportional to the average charge displacement around the source. The concept might be compared to a standing wave in a two-dimensional system, or in three-dimensional case like isolated atoms or ions, its interpretation is very similar to electronic stress tensor [38, 39] or "electronic pressure" around the nucleus.

$$\left(\frac{\partial E}{\partial \vec{F}(Z_i)}\right)_N = -\frac{1}{3}\int |r - R_i|\delta\rho(r)dr \tag{15.15}$$

Despite the convenience of using a uniform external electric field, its employment is an idealization of physical situations. In reality, the electric field might result from any electric potential generated by the environment, such as other reactants, catalysts, surfaces, solvents, and more remote electromagnetic sources. Especially enzyme active sites are known to provide sophisticated electric surroundings (resulting in, i.e. transition state stabilization [40–42]). Therefore, exploration of v-dependence using CDFT is challenging in its predictive capabilities, as indicated in recent prospective [21].

Concluding the first-order sensitivities from Figure 15.2, the alchemical potential is energy response to direct variation in nuclear charge, as parallel with chemical potential defined in Eq. (15.5). It is equal to the electronic potential at the position of the respective nucleus. The relation with electron density is established by employing the chain rule and Eq. (15.12),

$$\left(\frac{\partial E}{\partial Z_i}\right)_N = V_e(R_i) = \int \frac{1}{|r - R_i|}\left(\frac{\partial E}{\partial v(r)}\right)_N dr = \frac{1}{4\pi\varepsilon_0}\int \frac{\rho(r)}{|r - R_i|}dr \tag{15.16}$$

The alchemical potential is basically the electron–nuclear attraction energy, $V_{ne} = Z\mu_{al}$, normalized by atomic number [43].

15.5 Second-Order Sensitivities

The next level of energy sensitivities consists of six functions, which can be divided into three global and three local responses. Within this group, single variable and mixed partial derivatives appear. The transformation of variables reveals direct and indirect relations between these derivatives, which improves physical interpretation and conceptual understanding of energy functional evolution.

15.5.1 Electric Dipole Polarizability

The global electric dipole polarizability, α, characterizes system susceptibility for electric field-induced charge separation. Hence, the magnitude of electric dipole moment produced per electric field of unit strength is expressed through:

$$\left(\frac{\partial^2 E}{\partial \vec{F}^2}\right)_N = \left(\frac{\partial D}{\partial \vec{F}}\right)_N = \int r\left(\frac{\partial \rho(r)}{\partial \vec{F}}\right)_N dr = \int r\left(\frac{\partial^2 E}{\partial v(r)\partial \vec{F}}\right)_N dr = \int r\left(\frac{\partial D}{\partial v(r)}\right)_N dr = \alpha \tag{15.17}$$

Note that the global response function $\left(\frac{\partial^2 E}{\partial \vec{F}^2}\right)_N$ is related to integral of the first moment of $\left(\frac{\partial^2 E}{\partial v(r)\partial \vec{F}}\right)_N$, which is the neighboring local sensitivity – the polarizability potential. Similarly, the former can be expressed through the second moment of another derivative, a kernel $\left(\frac{\partial^2 E}{\partial v(r)\partial v(r')}\right)_N$. Respectively, the polarizability is associated with linear response function [44–48]:

$$\alpha_{i,j} = -\iint r_i r_j{}' \chi(r, r')drdr' \tag{15.18}$$

The i, j are Cartesian coordinates. For this reason, the name "polarizability kernel" has also been used in the context of $\chi(r, r')$ [45].

15.5.2 "Polarizability Potential" – Local Polarization

The first moment of $\left(\frac{\partial^2 E}{\partial v(r)\partial \vec{F}}\right)_N$ (the integrant in Eq. (15.17)) defines the local polarization density, while the quantity behind the derivative itself might be interpreted as polarizability potential[2]:

$$\left(\frac{\partial^2 E}{\partial v(r)\partial \vec{F}}\right)_N = \left(\frac{\partial \rho(r)}{\partial \vec{F}}\right)_N = \frac{\alpha(r)}{r} \tag{15.19}$$

or by employing the local density approximation [49], it might be recognized that the polarizability potential corresponds to the first moment of local softness:

$$\frac{\alpha(r)}{r} = rs(r) \tag{15.20}$$

Indeed, local softness [50], $s(r) = \eta^{-1} f(r)$, might be derived from a polarizability kernel at constant chemical potential [44]

$$s(r) = -\int \left(\frac{\partial \rho(r)}{\partial v(r')}\right)_\mu dr' \tag{15.21}$$

The integrant is called the softness kernel (the inverse of the hardness kernel), which is defined via the Berkowitz–Parr relation [44]

$$\left(\frac{\partial \rho(r)}{\partial v(r')}\right)_\mu = s(r, r') \tag{15.22}$$

and

$$s(r, r') = s(r)f(r') - \left(\frac{\partial \rho(r)}{\partial v(r')}\right)_N \tag{15.23}$$

Local density approximation gives a relation between local softness and polarizability potential $s(r) = \alpha(r)/r^2$. Since $s(r)$ might be obtained by integrating $s(r, r')$, thus

$$s(r, r') = \frac{ds(r)}{dr'} \cong \frac{d}{dr'}\left[\frac{\alpha(r)}{r^2}\right] \tag{15.24}$$

While $\chi(r, r')$ might also be approximated [45] by

$$\left(\frac{\partial \rho(r)}{\partial v(r')}\right)_N \cong -2s(r)f(r') \tag{15.25}$$

---

2 The expression is simplified, corresponding to the case of spherically symmetric systems like atoms or ions.

Combining relations in Eqs. (15.23)–(15.25), the linear response kernel is expressed with polarizability as a reversed relation to one in Eq. (15.18):

$$\chi(r, r') \approx \frac{2}{3}s(r, r') = \frac{2}{3}\frac{d}{dr'}\left[\frac{\alpha(r)}{r^2}\right] = \frac{2}{3}\frac{d}{dr'}\left[\frac{1}{r}\left(\frac{\partial\rho(r)}{\partial\vec{F}}\right)_N\right] \tag{15.26}$$

Keep in mind the simplified notation of this equation, while polarizability is represented by a tensor, or at least a principal axes vector. The local density approximation has been validated [46] for the softness and linear response kernels. Although neglecting electron correlation, it provides a fairly good estimate [45]. It is because the largest contribution to the linear response function comes from frontier orbitals. Equation (15.26) might be rewritten using the Leibniz rule, as:

$$\chi(r, r') \approx \frac{2}{3}\left(\frac{1}{r}\frac{d}{dr}\left[\frac{d\rho}{d\vec{F}}\right] - \frac{1}{r^2}\left(\frac{d\rho}{d\vec{F}}\right)\right) \tag{15.27}$$

Remarkably the sum of the components on right-hand side of above equation leads to the approximation of the first moment of alchemical hardness,

$$\left(\frac{d\rho}{dZ}\right) \approx \frac{1}{3}\left(\frac{1}{r}\frac{d}{dr}\left[\frac{d\rho}{d\vec{F}}\right] + \frac{1}{r^2}\left(\frac{d\rho}{d\vec{F}}\right)\right) \tag{15.28}$$

15.6 Alchemical Hardness

Second-order sensitivity of energy in relation to nuclear charge variation, $\left(\frac{\partial^2 E}{\partial Z_{i,j}^2}\right)$, defines alchemical hardness, η_{al}, Eq. (15.11), as a counterpart of the second-order energy response to electron number, $\eta = \partial^2 E/\partial N^2$, Eq. (15.6). The perturbation might involve a single nucleus or two different centers. Therefore, in polyatomic species, the alchemical hardness becomes a matrix with diagonal and off-diagonal elements. The off-diagonal elements, $\partial^2 E/\partial Z_i \partial Z_j$, include both a linear and a second-order nuclear–nuclear repulsion contribution [51]

$$\Delta V_{\mathrm{nn}} = \Delta Z_i \sum_{j\neq i} \frac{Z_j}{|R_i - R_j|} + \frac{\Delta Z_i \Delta Z_j}{|R_i - R_j|} \tag{15.29}$$

as compared with diagonal, $\partial^2 E/\partial Z_i^2$, terms:

$$\Delta V_{\mathrm{nn}} = \Delta Z_i \sum_{j\neq i} \frac{Z_j}{|R_i - R_j|} \tag{15.30}$$

Nonlinear terms are responsible for correlation and dispersion contributions to alchemical energy [51]. The generalized nuclear–nuclear repulsion energy for molecules experiencing remote electric fields from the environment (i.e. other molecules) is given by [3, 52]

$$V_{\mathrm{nn}} = \frac{1}{2}\sum_{i=1,i\neq j}^{\mathrm{atoms}}\sum_{j=1}^{\mathrm{atoms}} \frac{Z_i Z_j}{|R_i - R_j|} = \frac{1}{32\pi^2}\iint_{r\neq r'} \frac{\nabla^2 v(r)\nabla^2 v(r')}{|r - r'|} dr dr' \tag{15.31}$$

It is based on the assumption that electrostatic forces are exerted upon electrons by an effective "external" charge distribution, which, combined with Poisson's equation, leads to an integral on the right side of Eq. (15.31) [3, 52].

Moreover, it has been recognized that the off-diagonal elements of alchemical hardness represent effective "screened" potential [53] and hence might reflect average permittivity between two nuclei.

At this junction, it is worth mentioning that among numerous classical electrostatic approaches to "ordinary" chemical hardness, the simple screening model yields hardness expression [54].

$$\eta = \frac{1}{4}\int \frac{\rho(r)}{r}dr \tag{15.32}$$

which is much alike the electric potential definition for a continuous charge distribution,

$$\phi(r) = \frac{1}{4\pi\varepsilon_0}\int \frac{\rho(r')}{|r-r'|}dr' \tag{15.33}$$

In general, the core nature of hardness can be attributed to Coulomb interactions, particularly to electric potential generated by charge density associated with frontier orbitals [55]. Thus, the hardness kernel defined as a second-order derivative of Hohenberg–Kohn functional [56, 57], might be expressed as the major classical Coulomb part [58–60], and corrected by a term taking into account the kinetic exchange and correlation contributions. The Coulomb integral approximation allows recognition of classical Coulombic interaction energy in the hardness.

$$\eta = \frac{I-A}{2} \approx \frac{J}{2} = \frac{1}{2}\iint \frac{\phi_1^*(r_1)\phi_2^*(r_2)\phi_1(r_1)\phi_2(r_2)}{|r_1-r_2|}dr_1dr_2 \tag{15.34}$$

Furthermore, the Fukui function minimizes hardness functional [55], which leads to chemical hardness representation by means of the frontier density electron–electron repulsion, similar with Coulomb integral approximation.

$$\eta = \iint \frac{f(r)f(r')}{|r-r'|}drdr' \tag{15.35}$$

Analogously, the alchemical hardness accommodates nuclear–nuclear repulsion contribution. If one pays attention to Eqs. (15.23), (15.25), and (15.35), it might be observed that

$$s(r)f(r') = \frac{f(r)f(r')}{2\eta} \tag{15.36}$$

Therefore, one might notice direct relations between hardness and $\chi(r, r')$, and implied interrelations with polarizability (permittivity). These Coulomb-like formulations indicate that introducing the Taylor expansion of $|r-r'|^{-1}$ it is possible to express the hardness (or local hardness) with multiple moments of charge distribution similarly to the electric potential formulation [61].

The off-diagonal elements $\partial^2E/\partial Z_i\partial Z_j$ of alchemical hardness response function are analogues of

$$\left(\frac{\partial^2 E}{\partial Z_i \partial Z_j}\right)_\mu = -\iint \frac{1}{|r-R_i|}\frac{1}{|r'-R_j|}s(r,r')drdr' \tag{15.37}$$

which is connected with $\chi(r, r')$ via Fukui response kernel (accounting for polarization effects in Frontier density), i.e.:

$$\left(\frac{\partial^2 \mu}{\partial Z_i \partial Z_j}\right)_N = -\iint \frac{1}{|r-R_i|}\frac{1}{|r'-R_j|}f(r,r')drdr' \tag{15.38}$$

where [62]

$$\left(\frac{\partial^2}{\partial v(r)\partial v(r')}\left[\frac{\partial E}{\partial N}\right]_v\right)_N = \left(\frac{\partial^2 \mu}{\partial v(r)\partial v(r')}\right)_N = \left(\frac{\partial f(r)}{\partial v(r')}\right)_N = f(r,r') \tag{15.39}$$

and [63]

$$f(r,r') = -\left(\frac{\partial \chi(r,r')}{\partial N}\right)_v = \left(\frac{\partial s(r,r')}{\partial N}\right)_v - \frac{\partial}{\partial N}[s(r)f(r')] \tag{15.40}$$

15.6.1 Local Alchemical Hardness

The integrant in global alchemical hardness expression, Eq. (15.11), i.e. $|r - R_i|^{-1}(\partial\rho(r)/\partial Z_i)_N$, might be considered a nuclear charge counterpart of Fukui potential (also known as hardness potential) [64, 65], which is a potential related to frontier electron density – represented via Fukui function [16, 66], $f(r)$:

$$\frac{1}{r}\left(\frac{\partial\rho(r)}{\partial N}\right)_v = \frac{f(r)}{r} \tag{15.41}$$

The above equation shows similarity to expressions in Eqs. (15.32) and (15.33) and quantifies the electron–nuclear attraction of frontier electron density to the effective external potential of unit nuclear charge. The negative of Fukui potential at ith nucleus has been identified as the electron-transfer term [67, 68] in Taylor expansion of the ground state energy change due to variation in $v(r)$ and N [51]. The integral of Fukui potential over the whole space might be estimated via $(\partial V_{\mathrm{el}}(R_i)/\partial N)_v$ [67, 69, 70] and is defined by second-order energy derivative wrt N and Z_i.

$$\frac{\partial^2 E}{\partial N \partial Z_i} = -\int \frac{f(r)}{|r - R_i|} dr = \left(\frac{\partial\mu}{\partial Z}\right)_N = \left(\frac{\partial V_{\mathrm{el}}(R_i)}{\partial N}\right)_v = \frac{V_{\mathrm{e}}^{N+1}(R_i) - V_{\mathrm{e}}^{N-1}(R_i)}{2} \tag{15.42}$$

The Fukui potential is one of the classical electrostatic approaches to chemical hardness, $\eta = \int r^{-1} f(r) dr$.[3] Substituting $\partial\rho/\partial F$ with the first moment of local softness, $rs(r)$, in Eq. (15.28), the $\partial\rho/\partial Z$ sensitivity might be estimated through Fukui potential:

$$\frac{\partial\rho(r)}{\partial Z_i} = r\eta_{\mathrm{al}}(r) \approx \frac{1}{3}\frac{1}{\eta}\left(\frac{f(r)}{r} + \frac{1}{2}\frac{\partial f(r)}{\partial r}\right) \tag{15.43}$$

Similarly, following Eqs. (15.27) and (15.28), the linear response kernel could also be represented, using the difference of the components in the expression above. Obviously, such approximations will display unique character originating from the nature of partial derivatives involved and the lack of correlation. Nevertheless, frontier density provides the largest contribution to the linear response function, therefore such approximation is adequate.

There is a lot of ambiguity around precise expression for local chemical hardness [71, 72]. Additionally, at first sight, the definition of local chemical hardness, $\eta(r) = \left(\frac{\partial\mu}{\partial\rho(r)}\right)_v$, seems to contradict (via inverse-like formulation) with the integrant of Eq. (15.11), i.e. $\frac{1}{r}\left(\frac{\partial\rho(r)}{\partial Z}\right)_N$. However, if employing hardness kernel definition, where the kernel is approximated by $\eta(r, r') = |r - r'|^{-1}$, the local chemical hardness [73], which, in fact, is equal to global hardness,

$$\eta \equiv \eta(r) = \int \eta(r, r') f(r') dr' = \int \frac{1}{|r - r'|}\left(\frac{\partial\rho(r')}{\partial N}\right)_v dr' \tag{15.44}$$

well corresponds to alchemical hardness expression in Eq. (15.11). Note the Fukui potential nature of integrant in Eq. (15.44). However, despite analogies, the Fukui function, $f(r) = d\rho(r)/dN$, and the corresponding alchemical derivative over Z (i.e. $r\eta_{\mathrm{al}}(r) = d\rho(r)/dZ$), differ substantially in their

3 However, although the Fukui potential is a potential-like function, which is a derivative of the chemical potential with respect to a charge distribution, is not the global inverse to the local softness and it does not integrate to the chemical hardness. It might be suggested that by analogy the local hardness approximation (the major local hardness term) based on Fukui potential should also be treated as polarization related, i.e. like the alchemical hardness is described.

natures. Foremost, while the former integrates to 1, the latter balances to 0.

$$\int \left(\frac{\partial \rho(r)}{\partial N}\right)_{v} dr = 1 \tag{15.45}$$

$$\int \left(\frac{\partial \rho(r)}{\partial Z_i}\right)_{N} dr = 0 \tag{15.46}$$

It is also interesting to compare the Z derivative of electron density to another v-related derivative, i.e. over the electric field, $\vec{F}$, which also integrates to 0 (such as the $v(r)$ sensitivity itself [74], $\int \partial\rho(r)/\partial v(r')dr' = \int \chi(r, r')dr' = 0$).

$$\int \left(\frac{\partial \rho(r)}{\partial F}\right)_{N} dr = 0 \tag{15.47}$$

Further, not by coincidence, the alchemical hardness is sometimes referred to as polarization term [53]. The fact that electron density v-related sensitivities produce 0 expected value has some dipper implications, for example, an approximate relation between local alchemical hardness moment and polarizability for radial distribution function in spherical, atomic-like systems reads:

$$4\pi r^2 \left(\frac{\partial \rho(r)}{\partial Z_i}\right)_{N} \cong 4\pi r^2 \left[\frac{1}{r^2}\frac{\partial \alpha(r)}{\partial r}\right] \tag{15.48}$$

Furthermore, the first-order energy derivative with respect to an electric field, $(\partial E/\partial F)_N$, is identified as the electric dipole moment, D, and the first moment of electron density over the electric field derivative (the r product with integrant in Eq. (15.47)) constitutes the electric dipole polarizability:

$$\int r\left(\frac{\partial \rho(r)}{\partial F}\right)_{N} dr = \alpha \tag{15.49}$$

The first moment of the complementary alchemical derivative is related to capacitance [75],[4] hence proportional to $\sim\eta^{-1}$:

$$-\int |r - R_i|\left(\frac{\partial \rho(r)}{\partial Z_i}\right)_{N} dr = C \tag{15.50}$$

Although it is not 1 to 1 correspondence, the second moment of local capacitance yields polarizability. The non-alchemical local capacitance, $c(r) = \partial\rho(r)/\partial v_{\text{eff}}(r)$, in atomic and molecular systems is defined as:

$$c(r) = -2e^2 \frac{\rho^2(r)}{\nabla^2\rho(r) - \frac{\nabla\rho(r)\nabla\rho(r)}{\rho(r)}} \tag{15.51}$$

$$C = \int c(r)dr \tag{15.52}$$

The radial distribution function of local capacitance is related to local alchemical hardness through integral on $[0, r']$ intervals:

$$\int_0^r \left(\frac{\partial \rho(r')}{\partial Z_i}\right)_{N} dr' = 4\pi r^2 c(r) \tag{15.53}$$

---

4 Although the softness and capacitance are similar functions, the fundamental difference between them is that capacitance grows proportionally to system size, and only for isolated atoms or ions capacitance approximately equals softness, both locally and globally. Moreover locally the capacitance is positive-definite function, while local softness might assume both positive as well as negative values depending on definition.

or equivalently, the alchemical derivative might be estimated with capacitance and polarizability derivatives:

$$\left(\frac{\partial\rho(r')}{\partial Z_i}\right)_N \cong \frac{1}{r^2}\frac{\partial}{\partial r}[r^2 c(r)]_{Z_i,N} \cong \frac{1}{r^2}\left(\frac{\partial\alpha(r)}{\partial r}\right)_{Z_i,N} \tag{15.54}$$

15.7 Alchemical Characteristic Radius

The last of the global second-order alchemical sensitivity, with respect to Z_i and $\vec{F}$ variables, is alchemical radius:

$$R_{\text{al}} = \left(\frac{\partial^2 E}{\partial Z_i \partial \vec{F}}\right)_N = \left(\frac{\partial V_{\text{el}}(R_i)}{\partial \vec{F}}\right)_N = \left(\frac{\partial D}{\partial Z_i}\right)_N = \left(\frac{\partial^2 E}{\partial \vec{F} \partial Z_i}\right)_N \tag{15.55}$$

In isolated atomic-like systems, this derivative calculated using global electric dipole moment and global alchemical potential sensitivities are equal to 0, i.e. $\left(\frac{\partial D}{\partial Z_i}\right)_N = 0$ and $\left(\frac{\partial V_{\text{el}}(R_i)}{\partial \vec{F}}\right)_N = 0$. In polyatomic species, alchemical radius is a 3×3 tensor showing the average strength and direction of charge density extent around the ith nucleus. It is mainly related to the change in sharpness of the electron density cusp, as compared to the deformation of "tail" density variation (the core versus frontier density). For this reason, the alchemical sensitivity above resembles the definition of characteristic radius that is based on fundamental response functions, i.e. product of chemical hardness and electric dipole polarizability, $R = \sqrt{\alpha\eta}$, which in terms of energy derivatives is [31–33]

$$R^2 = \alpha\eta = \left(\frac{\partial^2 E}{\partial \vec{F}^2}\right)_N \left(\frac{\partial^2 E}{\partial N^2}\right)_v = \left(\frac{\partial D}{d\vec{F}}\right)_N \left(\frac{\partial \mu}{\partial N}\right)_v \tag{15.56}$$

The R might be directly related to the linear response function via polarizability expression, Eq. (15.18). On the other hand, based on the local density approximation, it is expressed as

$$R^2 = \int r^2 f(r)dr - \left(\int rf(r)dr\right)^2 \tag{15.57}$$

and noticing that

$$f(r) = \left(\frac{\partial\rho(r)}{\partial N}\right)_v = \left(\frac{\partial^2 E}{\partial N \partial v(r)}\right) \tag{15.58}$$

which now looks more alike the $\left(\frac{\partial^2 E}{\partial Z_i \partial \vec{F}}\right)_N$ sensitivity, particularly in the case of spherically symmetric systems (i.e. single atoms or ions), where R^2 reduces to $\int r^2 f(r)dr$. Impartially, the Fukui function integrals in Eq. (15.57) might be replaced with the quadrupole and dipole moments derivatives with respect to electron number, leading to [32]

$$R^2 = \frac{1}{3}\sum_{i=x,y,z}\left|\frac{\partial Q_{ii}}{\partial N} - \left(\frac{\partial D_i}{\partial N}\right)^2\right| \tag{15.59}$$

The $\partial D/\partial N$ appearing in the Eqs. (15.57) and (15.59) is analogue to $\partial D/\partial Z$, which defines the alchemical radius. Although conceptually equivalent, yet with only basic exchange–correlation contributions, Eqs. (15.57) and (15.59) are just approximations of Eq. (15.56) and produce proportional but distinct results.

The non-alchemical radius defined through Eqs. (15.57) and (15.59) might be interpreted as the standard deviation of the frontier electron density spread out around its center of mass [31].

Notably, definition in Eq. (15.56) represents a mathematical limit based on energy evolution rather than a physical boundary (in a similar manner to how it is decided for electron diameter) [32]. In the above context, it is postulated to interpret the alchemical derivatives, R_{al}, defined in Eq. (15.55) in terms of characteristic diameter as well, since it corresponds to the change in sharpness of the electron density cusp, which must be accompanied by respective tail adjustments.

Taking in to account the formulations of non-alchemical radius, the alchemical version might be redefined in corresponding fashion, denoted as $\tilde{R}_{al}$, thus by substituting $\eta \rightarrow \eta_{al}$, or $d\rho(r)/dN \rightarrow d\rho(r)/dZ$ in local density approximation based formula, Eq. (15.58).

$$\tilde{R}_{al}^2 = \alpha\eta_{al} = \left(\frac{\partial^2 E}{\partial \vec{F}^2}\right)_N \left(\frac{\partial^2 E}{\partial Z_i^2}\right)_N \tag{15.60}$$

In polynuclear systems, due to the matrix character of η_{al} and available combinations of Z_i and Z_j, the $\tilde{R}_{al}$ will also be defined as a square matrix, with diagonal and off-diagonal elements. Although the physical interpretation of $R = \sqrt{\alpha\eta}$ is the size of the chemical system, i.e. atomic or ionic radius, the proposed reformulation complicates understanding of alchemical counterpart or radius. Additionally, the η_{al} assume negative values, thus diagonal elements of alchemical characteristic radius by Eq. (15.60) would be imaginary numbers.

The alchemical sensitivities allow for a local variant of previous equation. Using respective local sensitivities in the product, i.e. alchemical hardness first moment, and polarizability potential, one obtains a relation

$$\eta_{al}(\mathbf{r}')\alpha(\mathbf{r}') = \left(\frac{\partial^2 E}{\partial Z_i \partial v(r')}\right)_N \left(\frac{\partial^2 E}{\partial v(r')\partial F}\right)_N = \int \frac{1}{r^2}\chi(r, r')dr \tag{15.61}$$

which integrates to $\sim R_{al}^{-2}$. This local representation directs attention to integrands of $V_{el}(R_i) = \int r^{-1}\rho(r)dr$ and $D = \int r\rho(r)dr$, where (recall Eq. (15.55)):

$$\left(\frac{\partial V_{el}(R_i)}{\partial F}\right)_N \longrightarrow -\frac{1}{3}\int \frac{1}{|r - R_i|}\frac{\partial \rho(r)}{\partial F}dr \neq 0 \tag{15.62}$$

and

$$\left(\frac{\partial D}{\partial Z}\right)_N \longrightarrow -\int |r - R_i|\frac{\partial \rho(r)}{\partial Z_i}dr \neq 0 \tag{15.63}$$

The integrant in Eq. (15.62), following from local density approximation, is local capacitance $r^{-2}\alpha(r) = c(r)$ (or approximately the local softness in atomic-like cases), and the value of integral in Eq. (15.63) produces the negative of the global capacitance. The fundamental difference between softness and capacitance is that global capacitance is proportional to system size.

Transmutation of local density approximation-based formula, Eq. (15.58), results in:

$$\tilde{R}_{al}^2 = \int r^2\left(\frac{\partial \rho(r)}{\partial Z_i}\right)_N dr - \left(\int r\left(\frac{\partial \rho(r)}{\partial Z_i}\right)_N dr\right)^2 = R_{al}^2 \tag{15.64}$$

Equation (15.64) matches well the results of R_{al} defined via Eq. (15.55).

Since the alchemical radius definitions presented in this section are based on various assumptions and approximations, the discussion on this sensitivity remains open, especially on the matters of relations with alternative definitions, as well as physical interpretation and its applications in predicting reactivity.

15.8 Linear Response Function

The $\chi(r,r') = \left(\frac{\partial^2 E}{\partial v(r)\partial v(r')}\right)_N$ has appeared multiple times throughout the formulations of other response functions. It contains information not only on the other sensitivities of the energy functional but also it is tightly interconnected with remaining kernels. The fundamental role of this second-order derivative is emphasized in the Berkowitz–Parr relation, Eq. (15.23), which indicates that all information about a system, including all second-order response functions, can be obtained from $\chi(r, r')$. The closest relation is established with the softness kernel, which is equivalent of the $\chi(r, r')$ in the grand canonical ensemble. The relations between linear response function and other kernels might be represented via [76]:

$$s(r,r') = f(r,r') - \chi(r,r') \tag{15.65}$$

for softness kernel, and with Fukui response kernel defined as

$$f(r,r') = \frac{d\chi(r,r')}{dN} = \frac{s(r)s(r')}{S} \tag{15.66}$$

and hardness kernel

$$\eta(r,r') = -\chi^{-1}(r,r') + \frac{\partial\mu}{\partial\rho(r)} \tag{15.67}$$

where $\chi^{-1}(r, r')$ is the inverse of the linear response function

$$\chi^{-1}(r,r') = \frac{\partial v(r)}{\partial\rho(r')} \tag{15.68}$$

Note that hardness kernel is the inverse of softness kernel, i.e. $s(r, r') = \eta^1(r, r')$. In contrast to softness and Fukui response kernels, which integrate into local softness

$$\int s(r,r')dr = \int f(r,r')dr = s(r) \tag{15.69}$$

the linear response function integrates to 0 ($\chi(r, r')$ is negative semidefinite), which means that the response kernel is not invertible in any real sense (the generalized inverse approach is proposed in Ref. [74]).

$$\int \chi(r,r')dr = 0 \tag{15.70}$$

Numerous reports (to mention just a few) summarize properties [15, 27, 47, 51, 77–79] and applications [74, 80–84] of $\chi(r, r')$ among other kernels. The linear response kernel has abilities to characterize electron delocalization, correlation, and polarization effects [80–82]. However, while the softness kernel is the key quantity associated with Kohn's nearsightedness of electronic matter [85], the linear response function is not nearsighted [86].

Moreover the $\chi(r, r')$ is not easy to compute. The most useful practical approximation is defined within the second-order perturbation theory [80]. Its simplified version, applicable to closed-shell systems, is given by expression [45, 47, 87]:

$$\frac{d\rho(r)}{dv(r')} = -4\sum_{i,a}\frac{\phi_i(r)\phi_a(r)\phi_a(r')\phi_i(r')}{\varepsilon_a - \varepsilon_i} \tag{15.71}$$

The capacitance [75] mentioned on several occasions in this chapter exhibits analogies with the linear response function (or softness kernel) and links to hardness and polarizability. The semi-direct associations of capacitance with linear response function, $\chi(r, r')$, also might be anticipated. The $\chi(r, r') = \left(\frac{\partial n(r)}{\partial v(r')}\right)_N$ on a general level, represents the charge over the potential derivative, which bears a resemblance to differential definition of nonlinear capacitance. On the other hand, this is also similar to softness kernel definition, where $\rho(r)$ derivative is taken with respect to intrinsic potential, $u(r') = v(r') - \mu$. However, chemical capacitance is defined locally, in a small volume element, which stores the chemical energy, thus relates to $v_{\text{eff}}(r)$. Furthermore, capacitance density is proportional to local susceptibility, and very accurate polarizabilities can be obtained directly from the local capacitance, similarly as from χ (r, r'). Fundamental difference between softness and capacitance is that global capacitance increase proportionally with system size.

15.9 Conclusions

The response functions related to particular changes in external potential compose a set of related functions that have been recognized in known properties of a system. Supplementing the list of basic reactivity parameters, most of the "alchemical" functions present the analogues of their "chemical" counterparts. Only the linear response function appears consistently in both families of derivatives since it contains all the information about a system necessary to derive other sensitivities. It has been shown that practical calculation of the responses discussed in this chapter employs expressions involving the electron density, see Figure 15.3, as a fundamental property, which is used to estimate aforementioned descriptors for quantification of reactivity and chemical principles.

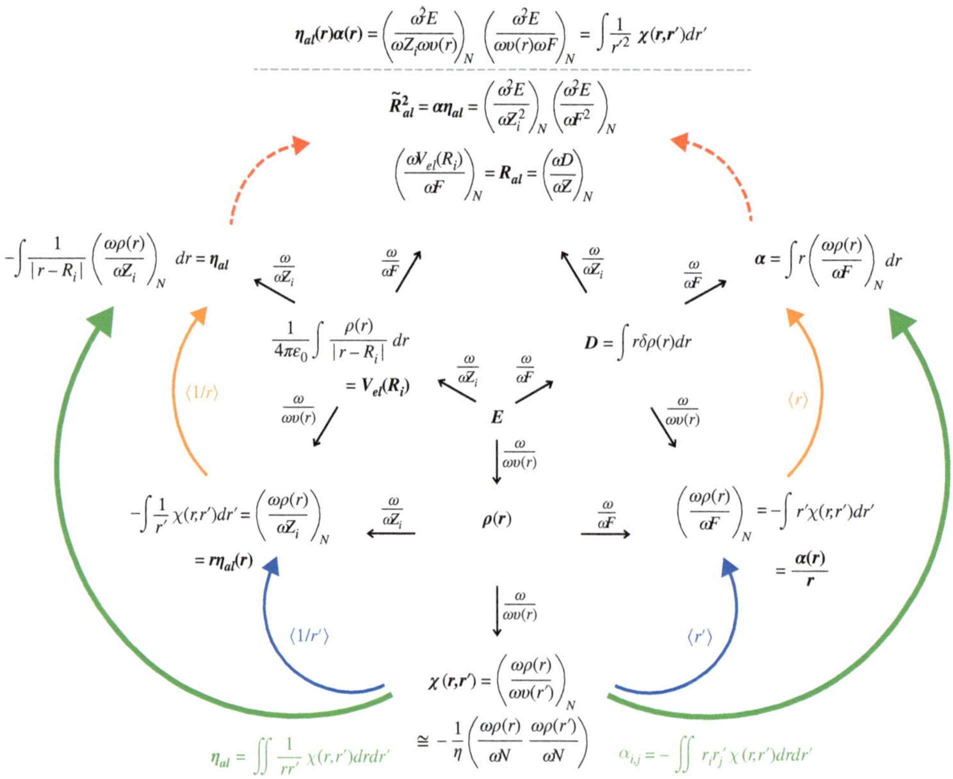

Figure 15.3 Electron density in expressions of v-related family of response functions.

References

1 Geerlings, P., De Proft, F., and Langenaeker, W. (2003). Conceptual density functional theory. *Chem. Rev.* 103: 1793–1873.

2 Parr, R.G. and Yang, W. (1989). *Density-Functional Theory of Atoms and Molecules*, International Series of Monographs on Chemistry. London, New York: Oxford University Press.

3 Ayers, P.W. and Parr, R.G. (2001). Variational principles for describing chemical reactions. Reactivity indices based on the external potential. *JACS* 123: 2007–2017.

4 Francotte, R., Irons, T.J.P., Teale, A.M. et al. (2022). Extending conceptual DFT to include external variables: the influence of magnetic fields. *Chem. Sci.* 13: 5311–5324.

5 Galvan, M., Vela, A., and Gazquez, J.L. (1988). Chemical reactivity in spin-polarized density functional theory. *J. Phys. Chem.* 92: 6470–6474.

6 Vargas, R., Cedillo, A., Garza, J., and Galván, M. (2002). *Reviews of Modern Quantum Chemistry*, vol. 2 (ed. R.G. Parr and K.D. Sen), 936. Singapore: World Scientific.

7 Chamorro, E., De Proft, F., and Geerlings, P. (2005). Hardness and softness reactivity kernels within the spin-polarized density-functional theory. *J. Chem. Phys.* 123: 154104.

8 Chamorro, E. and Pérez, P. (2005). Condensed-to-atoms electronic Fukui functions within the framework of spin-polarized density-functional theory. *J. Chem. Phys.* 123: 114107.

9 Pérez, P., Chamorro, E., and Ayers, P.W. (2008). Universal mathematical identities in density functional theory: results from three different spin-resolved representations. *J. Chem. Phys.* 128: 204108.

10 Hohenberg, P. and Kohn, W. (1964). Inhomogeneous electron gas. *Phys. Rev.* 136 (3B): B864–B871.

11 Kohn, W. and Sham, L.J. (1965). Self-consistent equations including exchange and correlation effects. *Phys. Rev.* 140: A1133–A1138.

12 Nalewajski, R.F. and Parr, R.G. (1982). Legendre transforms and Maxwell relations in density functional theory. *J. Chem. Phys.* 77: 399–407.

13 Parr, R.G., Donnelly, R.A., Levy, M., and Palke, W.E. (1978). Experimental and theoretical investigations on copper corrosion inhibition by cefixime drug in 1 M HNO_3 solution. *J. Chem. Phys.* 68: 3801–3807.

14 Parr, R.G. and Pearson, R.G. (1983). Absolute hardness companion parameter to absolute electronegativity. *JACS* 105: 7512–7516.

15 Senet, P. (1996). Nonlinear electronic responses, Fukui functions and hardnesses as functionals of the ground-state electronic density. *J. Chem. Phys.* 105: 6471–6489.

16 Parr, R.G. and Yang, W. (1984). Density functional approach to the frontier-electron theory of chemical reactivity. *JACS* 106: 4049–4050.

17 Fukui, K., Yonezawa, Y., and Shingu, H. (1952). Molecular orbital theory of reactivity in aromatic hydrocarbons. *J. Chem. Phys.* 20: 722–725.

18 Parr, R.G. and Parr, J.B. (1999). Kenichi Fukui: recollections of a friendship. *Theor. Chem. Acc.* 102: 4–6.

19 Kato, S. (2000). Perspective on "A molecular orbital theory of reactivity in aromatic hydrocarbons". *Theor. Chem. Acc.* 103: 219–220.

20 Pucci, R. and Angilella, G.G.N. (2022). Density functional theory, chemical reactivity, and the Fukui functions. *Found. Chem.* 24: 59–71.

21 Geerlings, P., Chamorro, E., Chattaraj, P.K. et al. (2020). Conceptual density functional theory: status, prospects, issues. *Theor. Chem. Acc.* 139: 36.

22 von Lilienfeld, O.A. and Tuckerman, M.E. (2006). Molecular grand-canonical ensemble density functional theory and exploration of chemical space. *J. Chem. Phys.* 125: 154104.

23 von Lilienfeld, O.A. and Tuckerman, M.E. (2007). Alchemical variations of intermolecular energies according to molecular grand-canonical ensemble density functional theory. *J. Chem. Theory Comput.* 3: 1083–1090.

24 von Lilienfeld, O.A. (2009). Accurate ab initio energy gradients in chemical compound space. *J. Chem. Phys.* 131: 164102.

25 von Lilienfeld, O.A. (2013). First principles view on chemical compound space: gaining rigorous atomistic control of molecular properties. *Int. J. Quantum Chem.* 113: 1676–1689.

26 Chang, K.Y.S., Fias, S., Ramakrishnan, R., and von Lilienfeld, O.A. (2016). Covalent bonding from alchemical linear response density functional theory. *J. Chem. Phys.* 144: 174110.

27 Geerlings, P., Fias, S., Boisdenghien, Z., and De Proft, F. (2014). Conceptual DFT: chemistry from the linear response function. *Chem. Soc. Rev.* 43: 4989–5008.

28 Bonaccorsi, R., Scrocco, E., and Tomasi, J. (1970). Molecular SCF calculations for the ground state of some three-membered ring molecules: $(CH_2)_3$, $(CH_2)_2NH$, $(CH_2)_2NH_2{}^+$, $(CH_2)_2O$, $(CH_2)_2S$, $(CH)_2CH_2$, and N_2CH_2. *J. Chem. Phys.* 52: 5270–5284.

29 von Lilienfeld, O.A., Lins, R.D., and Rothlisberger, U. (2005). Variational particle number approach for rational compound design. *Phys. Rev. Lett.* 95: 153002.

30 Lesiuk, M., Balawender, R., and Zachara, J. (2012). Higher order alchemical derivatives from coupled perturbed self-consistent field theory. *J. Chem. Phys.* 136: 034104.

31 Szarek, P. and Grochala, W. (2014). Most probable distance between the nucleus and HOMO electron: the latent meaning of atomic radius from the product of chemical hardness and polarizability. *J. Phys. Chem. A* 118: 10281–10287.

32 Szarek, P., Chlebicki, A., and Grochala, W. (2019). Atomic/ionic radius as mathematical limit of system energy evolution. *J. Phys. Chem. A* 123: 682–692.

33 Szarek, P., Witkowski, M., and Woźniak, A.P. (2019). Unconventional look at the diameters of quantum systems: could the characteristic atomic radius be interpreted as a reactivity measure? *J. Phys. Chem. C* 123: 11572–11580.

34 Taketa, H., Huzinaga, S., and O-ohata, K. (1966). Gaussian-expansion methods for molecular integrals. *J. Phys. Soc. Jpn.* 21: 2313–2324.

35 McMurchie, L.E. and Davidson, E.R. (1978). One- and two-electron integrals over Cartesian Gaussian functions. *J. Comput. Phys.* 26: 218–231.

36 Obara, S. and Saika, A. (1986). Efficient recursive computation of molecular integrals over cartesian gaussian functions. *J. Chem. Phys.* 84: 3963–3974.

37 Gill, P.M.W. (1994). Molecular integrals over Gaussian basis functions. *Adv. Quantum Chem.* 25: 141–205.

38 Tachibana, A. (2010). Energy density concept: a stress tensor approach. *J. Mol. Struct. THEOCHEM* 943: 138–151.

39 Szarek, P. and Tachibana, A. (2019). Chapter 4: Contemporary analysis of the influence of adsorbents on the structure, stability, and reactivity of main group nanoparticles using regional density functional theory. In: *Harnessing Nanoscale Surface Interactions* (ed. D.J. Henry), 97–127. Elsevier.

40 Ranaghan, K.E., Ridder, L., Szefczyk, B. et al. (2004). Transition state stabilization and substrate strain in enzyme catalysis: ab initio QM/MM modelling of the chorismate mutase reaction. *Org. Biomol. Chem.* 2: 968–980.

41 Szarek, P., Dyguda-Kazimierowicz, E., Tachibana, A., and Sokalski, W.A. (2008). Physical nature of intermolecular interactions within cAMP-dependent protein kinase active site: differential transition state stabilization in phosphoryl transfer reaction. *J. Phys. Chem. B* 112 (37): 11819–11826.
42 Chudyk, E. (2013). Differential transition state stabilization – nonempirical analysis. In: *Encyclopedia of Biophysics* (ed. G.C.K. Roberts), 459–460. Berlin, Heidelberg: Springer.
43 Balawender, R., Holas, A., De Proft, F. et al. (2018). Chapter 20: Alchemical derivatives of atoms: a walk through the periodic table. In: *Many-Body Approaches at Different Scales* (ed. G.G.N. Angilella and C. Amovilli), 227–251. Springer Cham.
44 Berkowitz, M. and Parr, R.G. (1988). Molecular hardness and softness, local hardness and softness, hardness and softness kernels, and relations among these quantities. *J. Chem. Phys.* 88: 2554–2557.
45 Ayers, P.W. (2007). The physical basis of the hard/soft acid/base principle. *Faraday Discuss.* 135: 161–190.
46 Szarek, P. and Komorowski, L. (2011). Modeling the electron density kernels. *J. Comput. Chem.* 32: 1721–1724.
47 Ayers, P.W. (2001). Strategies for computing chemical reactivity indices. *Theor. Chem. Acc.* 106: 271–279.
48 Mahan, G.D. and Subbaswamy, K.R. (1990). *Local Density Theory of Polarisability*. New York: Springer.
49 Vela, A. and Gázquez, J.L. (1990). A relationship between the static dipole polarizability, the global softness, and the Fukui function. *JACS* 112: 1490–1492.
50 Yang, W. and Parr, R.G. (1985). Hardness, softness, and the Fukui function in the electronic theory of metals and catalysis. *PNAS* 82: 6723–6726.
51 Muñoz, M. and Cárdenas, C. (2017). How predictive could alchemical derivatives be? *Phys. Chem. Chem. Phys.* 19: 16003–16012.
52 Johnson, P.A., Bartolotti, L.J., Ayers, P.W. et al. (2012). Chapter 21: Charge density and chemical reactions: a unified view from conceptual DFT. In: *Modern Charge-Density Analysis* (ed. C. Gatti and P. Macchi), 715–764. Dordrecht: Springer.
53 Ayers, P.W., Liu, S.B., and Li, T.L. (2009). Chargephilicity and chargephobicity: two new reactivity indicators for external potential changes from density functional reactivity theory. *Chem. Phys. Lett.* 480: 318–321.
54 Larson, E.G. and Smith, D.W. (1971). A density matrix analysis of slater's hyper-hartree-fock method. *Int. J. Quantum Chem.* S4: 87–94.
55 Ayers, P.W. and Parr, R.G. (2000). Variational principles for describing chemical reactions: the Fukui function and chemical hardness revisited. *JACS* 122: 2010–2018.
56 Liu, S., De Proft, F., and Parr, R.G. (1997). Simplified models for hardness kernel and calculations of global hardness. *J. Phys. Chem. A* 101: 6991–6997.
57 Von Weiszäcker, C.F. (1935). Zur Theorie der Kernmassen. *Z. Angew. Phys.* 96: 431–458.
58 Komorowski, L., Lipinski, J., and Pyka, M.J. (1993). Electronegativity and hardness of chemical groups. *J. Phys. Chem.* 97: 3166–3170.
59 Balawender, R., Komorowski, L., and Roszak, S. (1997). Acidic and basic molecular hardness in LCAO approximation. *Int. J. Quantum Chem.* 61: 499–505.
60 Pariser, R. (1953). An improvement in the π-electron approximation in LCAO MO theory. *J. Chem. Phys.* 21: 568–569.

61 Grochala, W. and Szarek, P. (2023). Chapter 10: Lessons from the maximum hardness principle. In: *Chemical Reactivity*, Vol. 2: Approaches and Applications (ed. S. Kaya, L. von Szentpály, G. Serdarŏglu, and L. Guo), 277–312. Elsevier.

62 Fuentealba, P. and Parr, R.G. (1991). Higher-order derivatives in density-functional theory, especially the hardness derivative $\partial\eta/\partial N$. *J. Chem. Phys.* 94: 5559–5564.

63 Contreras, R., Domingo, L.R., Andrés, J. et al. (1999). Nonlocal (pair site) reactivity from second-order static density response function: gas- and solution-phase reactivity of the acetaldehyde enolate as a test case. *J. Phys. Chem. A* 103: 1367–1375.

64 Harbola, M.K., Parr, R.G., and Lee, C. (1991). Hardnesses from electrostatic potentials. *J. Chem. Phys.* 94: 6055–6056.

65 Mineva, T., Parvanov, V., Petrov, I. et al. (2001). Fukui indices from perturbed Kohn-Sham orbitals and regional softness from mayer atomic valences. *J. Phys. Chem. A* 105: 1959–1967.

66 Yang, W., Parr, R.G., and Pucci, R. (1984). Electron density, Kohn-Sham frontier orbitals, and Fukui functions. *J. Chem. Phys.* 81: 2862–2863.

67 Cárdenas, C., Tiznado, W., Ayers, P., and Fuentealba, P. (2011). The Fukui potential and the capacity of charge and the global hardness of atoms. *J. Phys. Chem. A* 115: 2325–2331.

68 Cárdenas, C. (2011). The Fukui potential is a measure of the chemical hardness. *Chem. Phys. Lett.* 513: 127–129.

69 Berkowitz, M. (1987). Density functional approach to frontier controlled reactions. *JACS* 109: 4823–4825.

70 Ayers, P.W. and Levy, M. (2000). Perspective on density functional approach to the frontier-electron theory of chemical reactivity. *Theor. Chem. Acc.* 103: 353–360.

71 Guégan, F., Lamine, W., Chermette, H., and Morell, C. (2018). Comment on "Revisiting the definition of local hardness and hardness kernel" by C.A. Polanco-Ramirez, M. Franco-Pérez, J. Carmona-Espíndola, J. L. Gázquez and P. W. Ayers, *Phys. Chem. Chem. Phys.* 2017, 19, 12355. *Phys. Chem. Chem. Phys.* 20: 9006–9010.

72 Polanco-Ramírez, C.A., Franco-Pérez, M., Carmona-Espíndola, J. et al. (2017). Revisiting the definition of local hardness and hardness kernel. *Phys. Chem. Chem. Phys.* 19: 12355–12364.

73 Berkowitz, M., Ghosh, S.K., and Parr, R.G. (1985). On the concept of local hardness in chemistry. *JACS* 107: 6811–6814.

74 Liu, S., Li, T., and Ayers, P.W. (2009). Potentialphilicity and potentialphobicity: reactivity indicators for external potential changes from density functional reactivity theory. *J. Chem. Phys.* 131: 114106.

75 Szarek, P. (2016). Capacitance, the next of kin to chemical softness and density of states, an unexpected perk of being the "middle child". *J. Phys. Chem. C* 120: 17175–17183.

76 Fuentealba, P. (1998). Reactivity indices and response functions in density functional theory. *J. Mol. Struct. THEOCHEM* 433: 113–118.

77 Fias, S., Ayers, P.W., De Proft, F., and Geerlings, P. (2022). Properties of the density functional response kernels and its implications on chemistry. *J. Chem. Phys.* 157: 114102.

78 Sablon, N., De Proft, F., Ayers, P.W., and Geerlings, P. (2010). Computing second-order functional derivatives with respect to the external potential. *J. Chem. Theory Comput.* 6: 3671–3680.

79 Geerlings, P. and De Proft, F. (2008). Conceptual DFT: the chemical relevance of higher response functions. *Phys. Chem. Chem. Phys.* 2008 (10): 3028–3042.

80 Sablon, N., De Proft, F., and Geerlings, P. (2010). The linear response kernel: inductive and resonance effects quantified. *J. Phys. Chem. Lett.* 1: 1228–1234.

81 Sablon, N., De Proft, F., and Geerlings, P. (2010). The linear response kernel of conceptual DFT as a measure of electron delocalisation. *Chem. Phys. Lett.* 498: 192–197.

82 Angyan, J.G. (2011). Linear response and measures of electron delocalization in molecules. *Curr. Org. Chem.* 15: 3609–3618.

83 Balawender, R., Lesiuk, M., De Proft, F. et al. (2019). Exploring chemical space with alchemical derivatives: alchemical transformations of H through Ar and its ions as a proof of concept. *Phys. Chem. Chem. Phys.* 21: 23865–23879.

84 Balawender, R., Lesiuk, M., De Proft, F., and Geerlings, P. (2018). Exploring chemical space with alchemical derivatives: BN-simultaneous substitution patterns in C60. *J. Chem. Theor. Comput.* 14: 1154–1168.

85 Kohn, W. (1996). Density functional and density matrix method scaling linearly with the number of atoms. *Phys. Rev. Lett.* 76: 3168–3171.

86 Fias, S., Heidar-Zadeh, F., Geerlings, P., and Ayers, P.W. (2017). Chemical transferability of functional groups follows from the nearsightedness of electronic matter. *PNAS* 114: 11633–11638.

87 Senet, P. (2009). Reactivity and polarisability responses. In: *Chemical Reactivity Theory* (ed. P.K. Chattaraj), 331–362. Boca Raton, FL: CRC Press.

16

On the Nature of Catastrophe Unfoldings Along the Diels–Alder Cycloaddition Pathway

Leandro Ayarde-Henríquez[1], Cristian Guerra[2], Mario Duque-Noreña[2], Patricia Pérez[2], Elizabeth Rincón[3] and Eduardo Chamorro[2]

[1] *School of Physics, Trinity College Dublin, Dublin 2, Ireland*
[2] *Departmento de Química, Centro de Química Teórica & Computacional (CQT&C) and Department of Chemistry. Facultad de Ciencias Exactas, Universidad Andrés Bello, Avenida República 275, Santiago, Chile*
[3] *Facultad de Ciencias, Instituto de Ciencias Químicas, Universidad Austral de Chile, Valdivia, Chile*

16.1 Introduction

The so-called bonding evolution theory (BET) [1] has gained popularity as a powerful methodology for providing a deeper understanding of how and where bonding situations driving chemical reactions occur along a given coordinate [2–11].

Within the framework of BET [1], Thom's elementary catastrophe theory (CT) essentials [7, 11–17] are applied to characterize the changes that the gradient field of the electron localization function (ELF, $\eta(\boldsymbol{r}; c_\alpha)$), experiences under variations of "control parameters" $\{c_\alpha\}$ chosen, for instance, to be a subset of the atomic framework [18–21]. Hence, within the BET approach, one assumes that the dynamics of the $\eta(\boldsymbol{r}; c_\alpha)$ scalar function is gradient-like. The CT characterizes how the equilibria of such a gradient system change upon changing control parameters. A picture of bonding evolution emerges thus from solutions of the equilibria for the gradient dynamical system (GDS) of the smooth potential $\eta(\boldsymbol{r}; c_\alpha)$ [15, 22]. ELF carries all the information concerning the degree of electron localization as a consequence of Pauli repulsion [23]. Therefore, and under suitable conditions, the changes within the *phase-space portrait* of $\eta(\boldsymbol{r}; c_\alpha)$, i.e., in the number or type of the collection of all nondegenerate or Morse critical points (CPs) and the associated set of connecting gradient trajectories, can be formally described through parametric unfolding polynomials [7, 11–17, 21]. The CPs or equilibrium points are those at which $\nabla\eta(\boldsymbol{r}; c_\alpha) = 0$. An unambiguous description of the relevant rupture/formation of bonds (as described by ELF) arises associated with the structure of the CPs describing electron pair reorganization along the studied pathway [7, 11, 16, 24, 25]. The characterization of the evolution of the form or structure of the extension of electron pair (de)localization becomes a useful tool for justifying chemical transformations, given that chemical bonding is strongly tied to the electron pair notion (e.g., Lewis-like models).

Recently, we have emphasized the significance of considering the essential ideas underpinning the mathematical basis of the BET methodological framework in studying reacting systems in both the ground as well as excited electronic states. The proper assignment of the elementary catastrophe featuring a given chemically significant event will strongly depend on fully considering the formal foundations of both GDS and CT frameworks. CT provides a rigorous mathematical framework

Electron Density: Concepts, Computation and DFT Applications, First Edition.
Edited by Pratim Kumar Chattaraj and Debdutta Chakraborty.

enabling a rationale for discontinuous transitions between the possible alternative structurally stable states of a system [12–15]. Hence, essentially, two simple facts have to be considered using the BET framework: (i) degenerate, non-hyperbolic, or non-Morse type critical points (DCPs) lead to catastrophes if specific conditions are satisfied [12–15]; and (ii) at DCPs, the determinant of the stability matrix or Hessian matrix vanishes, i.e. one or more of the eigenvalues is zero. Indeed, the determinant of the stability matrix associated with each of the two approaching nondegenerate CPs will always decrease essentially because space flattens under such a circumstance. Moreover, we have stressed the importance of considering both the behavior of the determinant of the Hessian matrix and the associated relative distance of complete sets of CPs of ELF along the reaction path [86, 88]. The proper application of CT will demand the capacity to identify the control parameters as the associated underlying relationships. We have evidence that the analysis of sequences of molecular graphs based on single changes in the number and type of CPs, as exemplified in most insofar applications [7, 16, 24, 29–32], constitutes indeed an imprecise procedure that might yield an improper characterization of the bifurcation catastrophes.

Moreover, we should emphasize that the assignment of the type of catastrophe performed on the exclusive consideration of the appearance/disappearance of ELF maxima CPs is simply incorrect. Such a process should be complementized by examining additional properties of the Morse CPs in the neighborhood of the catastrophe. This means that considering all types of CPs (maxima, saddles, and minima) is necessary for a comprehensive understanding and adequate characterization of the nature of catastrophes. [1].

Over the last 25 years, applications of BET have been extensively documented across a wide range of chemical applications in the literature [16, 29, 30, 33–40]. Pericyclic reactions have been a favorite target of analysis for the BET methodology and very recent examples include for instance the examination of bonding in the cyclocondensation reaction between ethyl acetate 2-oxo-2-(4-oxo-4*H*-pyrido[1.2-*a*]pyrimidin-3-yl) polyazaheterocycle and ethylenediamine [41], the Schonberg rearrangement of *O*,*O*-diphenyl carbonothioate to *O*,*S*-diphenyl carbonothioate [42], the thione-to-thiol rearrangement via [1,3]-sigmatropic process [43], the seven-membered internal nitronates formation via cycloaddition reactions involving Z-C-aryl-*N*-methylnitrones and E-2-aryl-1-cyano-1-nitroethenes [44], the 1,3-Cyclohexadiene ring opening [45], the photochemically-induced 1,3-*cis*-butadiene [46], the double proton transfer (DPT) reaction between pyrazole and guanidine [47], the Newman–Kwart rearrangement [48], the complex reaction between methylidyne radical, $CH(X^2\Pi)$ and cyclopentadiene, C_5H_6 [2], the thermal Curtius rearrangement of syn-syn and syn–anti conformers of oxalyl diazide [49], the [3+2] cycloaddition (32CA) reactions of *N*-benzylcyclohexylnitrone with methyl-3-butenoate [50] and that of C-cyclopropyl-*N*-methylnitrone with styren [51], the nature of the chlorine atom transferred between ammonia and hypochlorous acid [38], the dehydrochlorination of the 3,3-diphenyl-4-(trichloromethyl)-5-nitropyrazoline [52], the Diels–Alder (DA) reaction between cyclopentadiene and carbon dioxide [53], the degenerate Cope rearrangement of hypostrophene [54], the mechanism of the DA reactions between perfluorobicyclo[2.2.0]hex-1(4)-ene and bicyclo[2.2.0]hex-1(4)-ene with benzene and naphthalene [55], the [4+1] cycloaddition (41CA) and [2+1] cycloaddition (21CA) reactions of the α-halogeno-nitroalkenes with carbene [56], the formation of the four-membered cyclic nitronates and the retro 32CA reaction of the 4*H*-[1,2]oxazete 2-oxide [57], the photochemically activated Paternò–Büchi reaction mechanism [27], the photochemical addition of two ethylenes [28], the 32CA reaction of the 2-nitroprop-1-ene with diphenyldiazomethane and the nitrogen elimination from 3,3-diphenyl-5-methyl-5-nitro-δ(1)-pyrazoline [58], the prototypical DA reaction between 1,3-butadiene and ethylene [26], the reaction mechanisms of 32CA between *N*-methylphenylnitrone and styrene catalyzed by the chromium tricarbonyl complex [59], the reactions between carbonyl oxides and ammonia/water [60], the strain promoted 32CA reaction of benzyl azide with cyclooctyne [61], the pyrolysis of cubane via its cage opening to reach bicyclooctatriene and then thermal rearrangement of bicyclooctatriene to produce [8]annulene [62], the polymerization reaction between nitroethene with (*Z*)-C,N-diphenylnitrone and with methyl vinyl ether [63], the molecular mechanism of the reaction of (*E*)-1,1,1-trifluoro-3-nitrobut-2-ene with 3,3-dimethyl-2-morpholinobutene [64], the 32CA reaction of push-pull nitronates with nitroethylene [65], the 32CA reaction of acetonitrile oxide with 7-bromo-oxanorborn-5-en-2-one [66], the Aza DA reaction of dinitro tetrazine with TME [67], the Binding of ThioTEPA to Purine Bases [68], the oxaphosphetane formation along the Wittig reaction [21], the 1,3-dipolar rearrangement between acetonitrile oxide and (1*S*,2*R*,4*S*)-2-cyano-7-oxabicyclo[2.2.1]hept-5-en-2-yl acetate derivatives [3], the 32CA reactions of 1-pyrroline-1-oxide with *N*-vinyl nucleobases leading to bicyclic N,O nucleoside analogs [69], the prediction of activation enthalpies in organic and organometallic reaction systems [85], the analysis of the radicaloid behavior of benzene in electronically excited states [87], gaining mechanistic insights into arynes' production [89], the study of imines photoinduced formation [90]. The accumulated evidence seems to point out that the assignation of unfoldings characterizing meaningful chemical events (i.e. bond-forming/bond-breaking processes) is far from being merely a straightforward technical issue within the BET methodological framework. In essence, the characterization of the local character of the ELF function provides a way for the proper identification of the associated elementary catastrophes. An appealing quest emerges while using BET in relation to correctly characterizing the control space, which constitutes a fundamental goal when applying the elementary CT [12–15]. We focus here on exploring further relationships between the symmetry of the molecules and the nature of the unfoldings. We will review three simple applications recently studied in our group, namely the DA reaction, the ring opening of 1,3-cyclohexadiene, and the C—C bond rupture of the ethane molecule. Such reaction systems offer different types of molecular symmetry as target examples.

16.2 Molecular Symmetry and Elementary Catastrophe Unfoldings

By following recent findings [26–28, 40, 45, 46, 48], and on the basis that, by construction, ELF properly incorporates the symmetry of the wave function, it can be expected that the evolution of its CPs along a given reaction pathway, must explicitly be dependent on the local electronic characteristics of specific characteristics of molecular space. Such regional features can certainly be related to substituent effects, the approaching reaction partner, or even selective solvation factors. We here present results concerning how the catastrophe signatures change in the function of the electronic nature of the reactants by selecting simple reaction cases as illustrative examples.

16.2.1 The Case of Normal- and Inverse-Electron-Demand Diels–Alder Reactions

As evidenced by the brief summary of recent BET applications mentioned above, the understanding of the pericyclic processes [70, 71] impact widely on several areas of chemistry [72–77]. The nature of "concerted" mechanisms [78–81] will remain the subject of intense debate regarding the local rearrangement of electron density along the observed single kinetic step [21, 29–31, 70, 82–84]. The case of the DA mechanisms has been, in fact focus of renewed attention within the aim of identifying the nature of catastrophe bifurcations [12–15], as exemplified by the analysis of the prototypical cycloaddition involving the 1,3-butadiene and ethylene [26, 29, 30]. This process was first explored within the BET formalism by Andrés and coworkers two decades ago [29, 30]. The electron rearrangements lead to the ring closure in terms of *cusp-type* functions. However, after reexamining such a reaction system by *exploiting both Hessian at potentially degenerate CPs and*

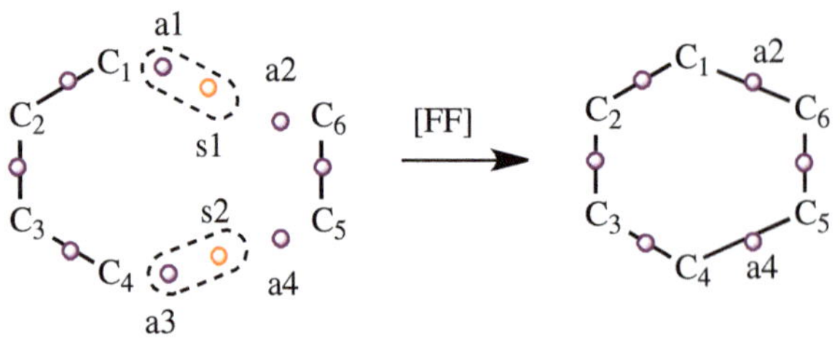

Scheme 16.1 Concurrent fold catastrophes [FF] are associated with the change in the ELF phase portrait at the onset of the ring closure in the cycloaddition process between 1,3-butadiene and ethylene. Fold catastrophes involve attractors arising at the terminal carbons belonging to the most electron-rich species. See Ref. [26] for the entire sequence of topographical changes along the intrinsic reaction coordinate and further details.

their relative distance, new and intriguing results were obtained since *no flag for a cusp unfolding was detected* [26]. Our methodological approach to exploring the emerging bifurcation catastrophes in the ELF field clearly shows that for each of the new CC bonds regions, only two CPs (an attractor *ai* and a saddle of index one *si*) indeed collide originating a wandering point (*wp*) featuring a *fold* signature [12–15]. The third potentially non-hyperbolic point (the other maxima) remains in the phase space, constituting the attractor for the basins V(C,C) associated with the new sigma bonds as depicted in Scheme 16.1. Thus, ELF bifurcation at the origin of the formation of two new disynaptic basins $V(C_1,C_6)$ and $V(C_2,C_5)$ originates finally from the annihilation of colliding maximum and saddle points occurring along the C_1C_6 and C_4C_5 bonding regions, i.e. $a1 + s1 \rightarrow wp$, and $a3 + s2 \rightarrow wp$, respectively. The attractors a2 and a4 do not degenerate, remaining in the topographic map.

It should be stressed that this fact has been found to be *independent of several different levels of theory* tested by calculations. *The fold catastrophe involves the attractors that originate from the most electron-rich fragment, namely the 1,3-butadiene species, in the case of the prototypical [2+4] DA reaction* [26].

It becomes interesting to test this finding by examining the effect of changing substituents to modulate the electron-demand type for the DA process [40]. As an example of an enhanced direct electron demand example, the reaction between 1-methoxy 1,3-butadiene and 1,2 dicyano ethylene has also been examined. Scheme 16.2 depicts selected frames of the sequence of topographic changes leading to the 3-methoxy-4-cyclohexene-1,2-dicarbonitrile cycloadduct. It is observed that

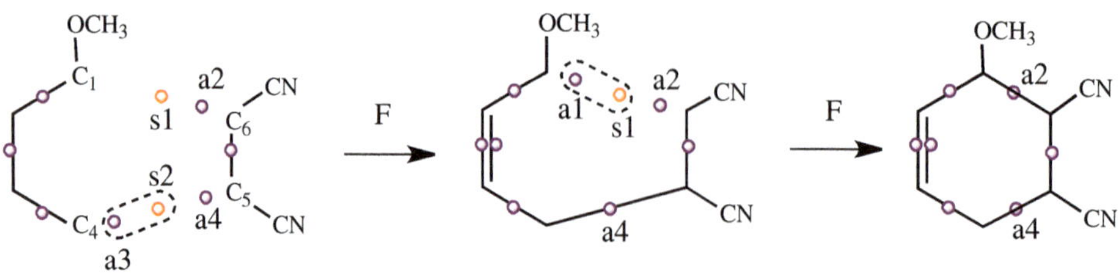

Scheme 16.2 Consecutive fold catastrophes are associated with the change in the ELF phase portrait at the onset of the ring closure in the cycloaddition reaction between 1-methoxy 1,3-butadiene and 1,2 dicyano ethylene, an example of an enhanced direct electron-demand Diels Alder (DA) process. Fold catastrophes involve attractors arising at the terminal carbons belonging to the most electron-rich species. See Ref. [40] for the entire sequence of topographical changes along the intrinsic reaction coordinate and further details.

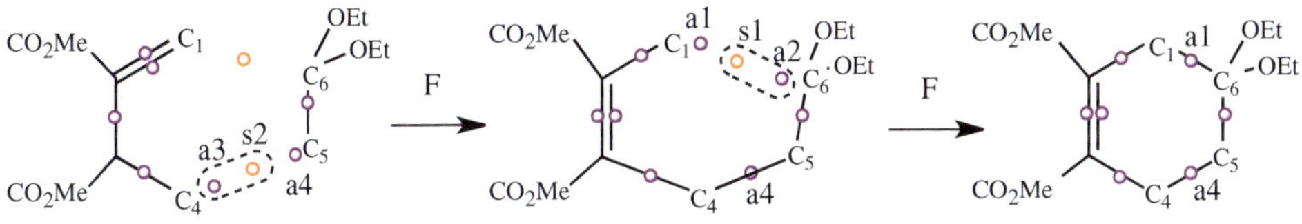

Scheme 16.3 Consecutive fold catastrophes are associated with the change in the ELF phase portrait at the onset of the ring closure in the cycloaddition reaction between 1,4-dimethyl 2,3-bis(methylene)butanedioate and 1,1-diethoxyethene, an example of an inverse electron-demand DA process. Fold catastrophes involve attractors arising at the terminal carbons belonging to the most electron-rich species. See Ref. [40] for the entire sequence of topographical changes along the intrinsic reaction coordinate and further details.

each maximum near the C_1 and C_4 centers in the butadiene fragment (i.e. a1 and a3) collides with a saddle point (i.e., s1 and s2), and both annihilate into a noncritical point (wp) [12–15]; the other ELF maxima in the neighborhood of the C_6 and C_5 centers (i.e. a2 and a4) remains as the attractors associated to the new disynaptic basins corresponding to the new single C_1C_2 and C_4C_5 bonds, respectively. These facts are consistent with the same parametric function (i.e., a fold) [15], which describes the topological changes associated with both CC bond formation yielding to the ring closure. In contrast to the case of the reaction between the 1,3-butadiene and ethylene [26], chemical substitution removes the previously observed simultaneity of the ring closure process. However, as observed for the nonactivated prototypical reaction, *the fold catastrophes involve the attractors that originate at the terminal carbons in the most electron-rich fragment* [40].

Let us now consider an example for the inverse electron-demand process by using the cycloaddition reaction between the 1,4-Dimethyl 2,3-bis(methylene)butanedioate and the 1,1-diethoxyethene species (Scheme 16.3). The substitution pattern, including both ethoxy groups, breaks the synchronicity of the new CC sigma bond formations. Nonetheless, no flag of a cusp polynomial is again observed. The ring closure takes place via two consecutive fold-type catastrophes. In such a case of a strongly deactivated diene and a strongly activated dienophile, the folds describing the electron reorganization leading to the ring closure occur near each rich-electron moiety. The first fold event, associated with a new sigma bond, originates in the locality of the C_4 in the diene fragment, whereas the second one occurs near the C_6 center in the dienophile fragment [40].

These results regarding both inverse- and normal-electron demand examples, as well as the prototypical DA cycloaddition reaction, evidence that *fold catastrophes will involve attractors arising at the terminal carbons belonging to the most electron-rich species*. As it is well known, the different substituents affect the synchronicity of the process, yielding the new CC bonds. Despite that fact, it seems apparent that for combinations of unsymmetrically substituted reagents, the *fold polynomial reasonably characterizes the formation of CC sigma bonds between the reagents involved in the Diels–Alder reaction*. Further systematic research is due to test such emerging working hypotheses [26, 40]. Note that the key common characteristic of the discussed reacting systems is the characterization in terms of a *heterolytic nature* of the bonding, which, in turn, is conditioned by a significant difference in the electron density induced by the substituents at the reaction centers (Scheme 16.4).

One might expect to detect some flags of a cusp unfolding (or even higher dimensional ones) in studying chemical events associated with forming/breaking bonds only when highly symmetrical fragments are implied. The next two examples explore such a possibility.

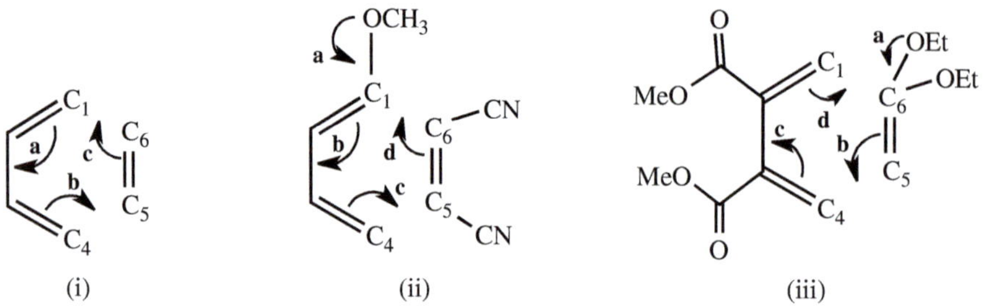

Scheme 16.4 Schematic representation of the flow of electrons in examples of the DA reaction: (i) the prototypical slow direct electronic demand DA reaction between 1,3 butadiene and ethylene; (ii) an enhanced direct electronic demand DA reaction between an activated diene and a deactivated dienophile; (iii) an inverse electronic demand DA reaction involving a deactivated diene and an activated dienophile [26, 40].

16.2.2 The C—C Bond Breaking in a High Symmetry Environment

Let us revisit the example presented by Krokidis, Noury, and Silvi in their seminal work introducing the BET methodology [1]. The dissociation of ethane was characterized in terms of the cusp unfolding. The attractor a1 defining the disynaptic basin $V(C_1,C_2)$ degenerates and splits into three new Morse-type CPs (i.e. two maxima a2 and a3 and the saddle point of index one, *s*) after the C—C bond is stretched enough, as depicted in Scheme 16.5. The high symmetry of the electron-pair density imposes identical values of the Hessian at both Morse-type maxima, and the difference in the relative distance between pairs of points (attractor, saddle) is negligible, as displayed in Figure 16.1.

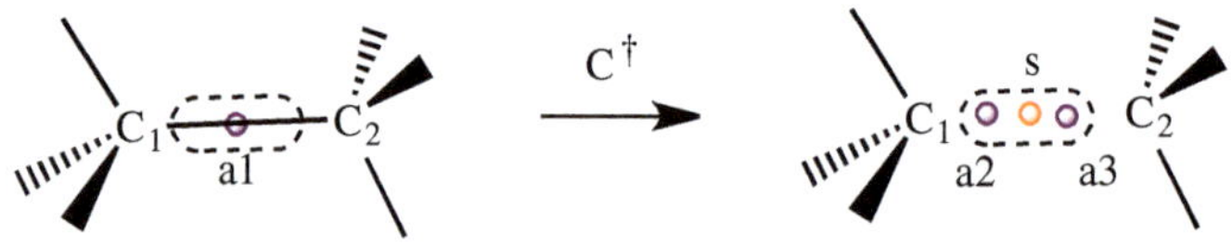

Scheme 16.5 A cusp bifurcation characterizes the CC bond breaking of ethane into two radical methyl fragments. From a topologic viewpoint, the sigma bond between carbons breaks once the attractor a1 splits, through a cusp catastrophe, into three hyperbolic CPs, namely, maxima a2 and a3 and the saddle point of index one *s* [1].

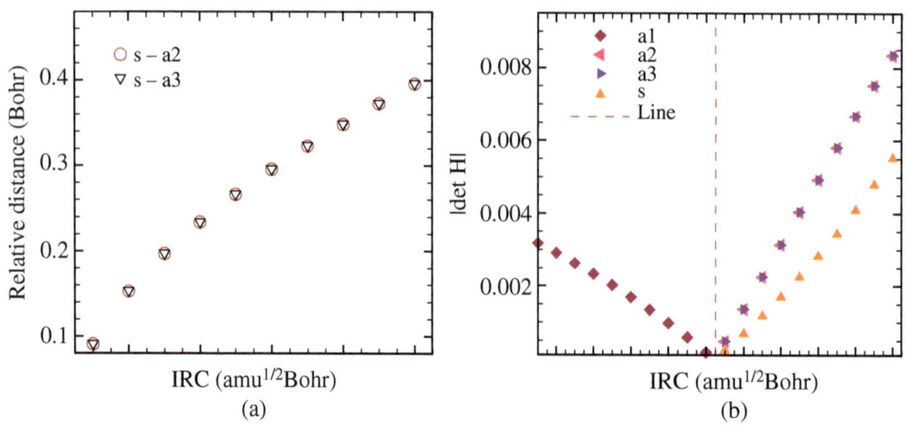

Figure 16.1 (a) Relative distance s–a2 and s–a3 between pairs of CPs (s, a2) and (s, a3), respectively. (b) The absolute value of the determinant of the Hessian matrix was evaluated at a1, a2, a3, and s1 critical points. The dashed red line is an arbitrary way to indicate the location for the splitting of a1.

16.2.3 The Photochemical Ring Opening of 1,3-Cyclohexadiene

Within the context of revisiting reactive processes occurring in electronically excited states, we have recently reported on the electro-opening of 1,3-cyclohexadiene (CHD) into 1,3,5-cyclohexatriene (HT) [45]. The topological characterization of key electron reorganizations along the reaction pathway was performed by following the determinant of the Hessian matrix calculated at potentially degenerate CPs. Once the reacting system leaves the conical intersection, it moves along the 2^1A surface on which the C_5C_6 bond breaks (Scheme 16.6). Note the perfectly symmetrical environment featuring the environment of the C_5C_6 bonding region in the CHD system. The electron population of the basin (linked to the degenerate attractor representing the breaking sigma bond) depletes as the electron-pair density flows. It accumulates within the basin associated with each new hyperbolic attractor that appears near both the C_5 and the C_6 centers. The remarkable *equality exhibited by the determinant of the Hessian and electron population values at the corresponding basins for the new maxima (a2 and a3) constitutes solid flags of a cusp catastrophic process*, i.e. $C^{\dagger}: a1 \rightarrow a2 + s1 + a3$. Computational evidence points out that the C_5C_6 breaks along the 2^1A, leading to symmetrically redistribution of pairing density over each C_5 and C_6 atom. The electron pair rearrangement implies crossing the $2^1A/1^1A$ conical intersection toward the ground state where the system reaches the stable HT product [45].

This textbook photochemical reaction reveals the relevance of the electron density symmetry throughout the sigma bond cleavage. The difference in the charge density near both reaction centers is not high enough to induce a measurable heterolytic character in the process. In consequence, a *homolytic event is observed*, which is clearly indicated from both statistical (electron population associated with the basins of the new attractors) and topological (Hessian value at the new maxima) metrics. It is worth noting that an observer in a frame fixed to the saddle point of index one (s1) located between the new attractors sees both maxima moving away with the *same velocity* because the *relative distance* between pairs of such CPs varies at the same ratio. Therefore, on a reverse-time consideration, the same observer would argue that both maxima collide at the same time with the saddle point. This physical description of the bonding situation based on the ELF topography constitutes a solid argument on the *impossibility* of characterizing the chemical event using a cubic polynomial, i.e., the *fold* function, since it involves *two CPs*. The *fundamental theorem of algebra* thus indicates a fourth-degree function as the correct one. Since the *Hessian* at the degenerate attractor shows that only *one eigenvalue changes significantly along the reaction path* and considering that the *cusp* polynomial has (at most) *three CPs*, there is no doubt that a cusp signature should be the appropriate unfolding [45].

We should note on this point that from a topologic perspective, the ring opening of 1,3-cyclohexadiene [45] and ethane dissociation [1] are different from the DA reactions [26].

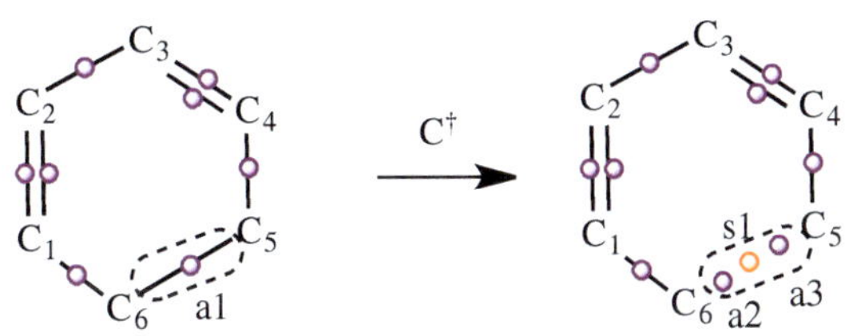

Scheme 16.6 A cusp bifurcation is a topological signature for the C_5C_6 bond breaking in the ring opening of the 1,3-cyclohexadiene occurring over the 2^1A surface since the attractor associated with the $V(C_5,C_6)$ single bond degenerates into three new Morse-type CPs, namely two attractors (a2 and a3) and a saddle point of index one s1, representing the onset of the sigma C5C6 sigma bond. See Ref. [45] for a detailed discussion concerning the sequence of topographical changes along the photochemical process.

In the examined DA cases, we have the Hessian at potentially degenerate CPs and the relative distance between them shows that only *two CPs are indeed non-hyperbolic* points; therefore, these two CPs merge and disappear. The other attractor becomes not involved in the topographic change. In contrast, in the high symmetric CC bond rupture, the change in the ELF topography encompasses *three CPs*, which suggests that *our working hypothesis is, at least, reasonable* [12–15]. Moreover, the capability of unfoldings in distinguishing and describing these subtle differences is clear, which is deeply related to the electron density symmetry around reacting centers. The *fold* function *accurately* describes the rupture/formation of sigma bonds in *low symmetry reaction systems*, whereas the *cusp is the suitable* unfolding to be *potentially* used in systems exhibiting a *high degree of symmetry* (e.g. dissociation of ethane). However, the new single bond between two approaching identical fragments can still be described via a fold under certain conditions, for instance, the face-to-face cycloaddition of two ethylenes in the excited state leading to a rhomboidal transition state geometry [26–28, 40, 45, 46, 48]. Thus, *the existence of a high symmetry is only a necessary condition* to detect any cusp flag. The sufficient condition is then the *persistence of the electron density symmetry with respect to a plane perpendicular to the forming/breaking sigma bond throughout this chemical event*. This statement guarantees an equally distributed electron density at both sides of the plane, which resembles a specular reflection. It provides a straightforward tool for assigning *a priori* (i.e. without any calculation) the adequate unfolding describing a key chemical event.

16.3 Concluding Remarks

Based on the above observations, one should expect that the fold polynomial characterizes most chemical processes since the detection of any of its flags demands no special condition. Accumulative results evidence the *significance* of strictly *following root ideas* of CT and (gradient) dynamical systems for describing key chemical events (e.g. rupture/formation of bonds) along a reaction pathway. That is, the consideration that assigning the unfolding describing a meaningful electron reorganization, e.g. the rupture/formation of a single bond, carefully evaluating the *Hessian matrix* at potentially degenerate CP and the *relative distance* between them. The *fold* polynomial *characterizes* the forming/breaking processes of sigma bonds in systems exhibiting an asymmetric pattern of substitution in the associated reaction center. In contrast, the *cusp* function seems to accurately *describe* such relevant chemical changes if the symmetry is *high enough*. However, the existence of this degree of symmetry is only a *necessary condition*; the *sufficient* one happens to be the *persistence of the pairing electron density symmetry with respect to a plane perpendicular to the forming/breaking sigma bond throughout the chemical event*. This statement guarantees that each density, on both sides of the plane, behaves as the *specular-like image* of the other. These findings provide a simple tool for *predicting* the correct unfolding in any case prior to the reaction occurring just by *visually* inspecting the system geometry. We hope these results can further stimulate the discussion of research [26–28, 40, 45, 46, 48] concerning the detailed identification, characterization, and analysis of chemical processes and reactivity within the methodological framework that offers BET approach [1] based on the combined use of topological analysis of ELF and the CT [26–28, 40, 45, 46, 48].

Acknowledgments

The authors thank UNAB and the ANID/CONICYT PhD scholarships awarded to C. Guerra and L. Ayarde-Henríquez, respectively. Also, we are indebted to the Fondo Nacional de Ciencia y Tecnología (FONDECYT–ANID, Chile) for the continuous financial and academic support provided through Projects Nos. 1231018 (EC), 1181582 (EC), 1180348 (PP), and 1221383 (PP).

References

1 Krokidis, X., Noury, S., and Silvi, B. (1997). Characterization of elementary chemical processes by catastrophe theory. *J. Phys. Chem. A* 101: 7277–7282.

2 Andres, J., Safont, V.S., Oliva, M. et al. (2022). A bonding evolution theory study of the reaction between methylidyne radical, CH($X^2\Pi$), and cyclopentadiene, C_5H_6. *Int. J. Quantum Chem.* 122: 21.

3 Adjieufack, A.I., Nana, C.N., Ketcha-Mbadcam, J. et al. (2020). Deciphering the curly arrow representation and electron flow for the 1,3-dipolar rearrangement between acetonitrile oxide and (1*S*,2*R*,4*S*)-2-cyano-7-oxabicyclo 2.2.1 hept-5-en-2-yl acetate derivatives. *ACS Omega* 5: 22215–22225.

4 Adjieufack, A.I., Bake, M.M., Mbadcam, J.K. et al. (2019). How effectively bonding evolution theory retrieves and visualizes curly arrows: the cycloaddition reaction of cyclic nitrones. *Int. J. Quantum Chem.* 119: 14.

5 Gonzalez-Navarrete, P., Andres, J., and Safont, V.S. (2018). A bonding evolution analysis for the thermal Claisen rearrangement: an experimental and theoretical exercise for testing the electron density flow. *Phys. Chem. Chem. Phys.* 20: 535–541.

6 Oliva, M., Safont, V.S., Gonzalez-Navarrete, P., and Andres, J. (2017). Electronic structure and rearrangements of anionic ClMg(η^2-O_2C)- and ClMg(η^2-CO_2)– complexes: a quantum chemical topology study. *Theor. Chem. Acc.* 136: 11.

7 Andres, J., Gonzalez-Navarrete, P., Safont, V.S., and Silvi, B. (2017). Curly arrows, electron flow, and reaction mechanisms from the perspective of the bonding evolution theory. *Phys. Chem. Chem. Phys.* 19: 29031–29046.

8 Adjieufack, A.I., Ndassa, I.M., Patouossa, I. et al. (2017). On the outside looking in: rethinking the molecular mechanism of 1,3-dipolar cycloadditions from the perspective of bonding evolution theory. The reaction between cyclic nitrones and ethyl acrylate. *Phys. Chem. Chem. Phys.* 19: 18288–18302.

9 Andres, J., Berski, S., and Silvi, B. (2016). Curly arrows meet electron density transfers in chemical reaction mechanisms: from electron localization function (ELF) analysis to valence-shell electron-pair repulsion (VSEPR) inspired interpretation. *Chem. Commun.* 52: 8183–8195.

10 Viciano, I., Gonzalez-Navarrete, P., Andres, J., and Marti, S. (2015). Joint use of bonding evolution theory and QM/MM hybrid method for understanding the hydrogen abstraction mechanism via cytochrome P450 aromatase. *J. Chem. Theory Comput.* 11: 1470–1480.

11 Andres, J., Gracia, L., Gonzalez-Navarrete, P., and Safont, V.S. (2015). Chemical structure and reactivity by means of quantum chemical topology analysis. *Comput. Theor. Chem.* 1053: 17–30.

12 Thom, R. (1972). *Structural Stability and Morphogenesis: An Outline of a General Theory of Models*. Westview Press.

13 Poston, T. and Stewart, I. (1978). *Catastrophe Theory and Its Applications*. London/San Francisco/Melbourne: Pitman Publishing Limited.

14 Castrigiano, D.P.L. and Hayes, S.A. (2018). *Catastrophe Theory*, 2e, 33487–32742. Boca Raton, FL: CRC Press/Taylor & Francis Group.

15 Gilmore, R. (1993). *Catastrophe Theory for Scientists and Engineers*. Dover Publications.

16 Andres, J., Gonzalez-Navarrete, P., and Safont, V.S. (2014). Unraveling reaction mechanisms by means of quantum chemical topology analysis. *Int. J. Quantum Chem.* 114: 1239–1252.

17 Domingo, L.R., Rios-Gutierrez, M., Pérez, P., and Chamorro, E. (2016). Understanding the $2n+2n$ reaction mechanism between a carbenoid intermediate and CO_2. *Mol. Phys.* 114: 1374–1391.

18 Savin, A., Nesper, R., Wengert, S., and Fassler, T.F. (1997). ELF: the electron localization function. *Angew. Chem. Int. Ed.* 36: 1809–1832.

19 Silvi, B. and Savin, A. (1994). Classification of chemical bonds based on topological analysis of electron localization functions. *Nature* 371: 683–686.

20 Becke, A.D. and Edgecombe, K.E. (1990). A simple measure of electron localization in atomic and molecular-systems. *J. Chem. Phys.* 92: 5397–5403.

21 Chamorro, E., Duque-Norena, M., Gutierrez-Sanchez, N. et al. (2020). A close look to the oxaphosphetane formation along the Wittig reaction: a 2 + 2 cycloaddition? *J. Org. Chem.* 85: 6675–6686.

22 Arnold, V.I., Afrajmovich, V.S., Ol'yashenko, Y.S., and Shil'nikov, L.P. (1999). *Bifurcation Theory and Catastrophe Theory*. Berlin-Heildelberg, Germany: Springer-Verlag.

23 Bader, R.F.W. (1994). *Atoms in Molecules. A Quantum Theory*. New York, USA: Oxford University Press Inc.

24 Andres, J., Berski, S., Domingo, L.R. et al. (2011). Describing the molecular mechanism of organic reactions by using topological analysis of electronic localization function. *Curr. Org. Chem.* 15: 3566–3575.

25 Fuentealba, P., Chamorro, E., and Santos, J.C. (2007). *Theoretical and Computational Chemistry*, vol. 19 (ed. A. Toro-Labbé), 57–85. Elsevier.

26 Ayarde-Henriquez, L., Guerra, C., Duque-Norena, M. et al. (2021). Are there only fold catastrophes in the Diels–Alder reaction between ethylene and 1,3-butadiene? *J. Phys. Chem. A* 125: 5152–5165.

27 Guerra, C., Ayarde-Henriquez, L., Duque-Norena, M., and Chamorro, E. (2021). Unraveling the bonding nature along the photochemically activated Paterno-Büchi reaction mechanism. *ChemPhysChem* 22: 9.

28 Guerra, C., Ayarde-Henriquez, L., Duque-Norena, M. et al. (2021). On the nature of bonding in the photochemical addition of two ethylenes: C–C bond formation in the excited state? *Phys. Chem. Chem. Phys.* 23: 20598–20606.

29 Berski, S., Andres, J., Silvi, B., and Domingo, L.R. (2003). The joint use of catastrophe theory and electron localization function to characterize molecular mechanisms. A density functional study of the Diels–Alder reaction between ethylene and 1,3-butadiene. *J. Phys. Chem. A* 107: 6014–6024.

30 Berski, S., Andres, J., Silvi, B., and Domingo, L.R. (2006). New findings on the Diels–Alder reactions. An analysis based on the bonding evolution theory. *J. Phys. Chem. A* 110: 13939–13947.

31 Domingo, L.R. and Saez, J.A. (2009). Understanding the mechanism of polar Diels–Alder reactions. *Org. Biomol. Chem.* 7: 3576–3583.

32 Gonzalez-Navarrete, P., Domingo, L.R., Andres, J. et al. (2012). Electronic fluxes during Diels–Alder reactions involving 1,2-benzoquinones: mechanistic insights from the analysis of electron localization function and catastrophe theory. *J. Comput. Chem.* 33: 2400–2411.

33 Domingo, L.R., Mazarei, E., Seif, A. et al. (2021). Closer investigation of the kinetics and mechanism of spirovinylcyclopropyl oxindole reaction with $^3\Sigma^-$ g-O_2 by topological approaches and unraveling the role of the I_2 catalyst. *J. Phys. Chem. A* 125: 6913–6926.

34 Merino, P., Chiacchio, M.A., Legnani, L., and Tejero, T. (2017). BET & ELF quantum topological analysis of neutral 2-aza-Cope rearrangement of γ-alkenyl nitrones. *Molecules* 22: 14.

35 Berski, S. and Latajka, Z. (2011). A mechanism of the 1,3-dipolar cycloaddition between the hydrogen nitryl HNO_2 and acetylene HCCH: the electron localization function study on evolution of the chemical bonds. *Int. J. Quantum Chem.* 111: 2378–2389.

36 Krokidis, X., Goncalves, V., Savin, A., and Silvi, B. (1998). How malonaldehyde bonds change during proton transfer. *J. Phys. Chem. A* 102: 5065–5073.

37 Michalski, M., Gordon, A.J., and Berski, S. (2021). Theoretical insights and quantitative prediction of the nature of boron-chalcogen (O, S, Se, Te) interactions using the electron density and the electron localisation function (ELF). *Polyhedron* 210: 8.

38 Michalski, M. and Berski, S. (2021). Understanding the molecular mechanism of the chlorine atom transfer between ammonia and hypochlorous acid with electron localisation function (ELF). *Mol. Phys.* 119: 12.

39 Berski, S. and Latajka, Z. (2006). Quantum chemical topology description of the hydrogen transfer between the ethynyl radical and ammonia (C_2H center dot + NH_3) – the electron localization function study. *Chem. Phys. Lett.* 426: 273–279.

40 Ayarde-Henriquez, L., Guerra, C., Duque-Norena, M. et al. (2022). On the notation of catastrophes in the framework of bonding evolution theory: the case of normal and inverse electron demand Diels–Alder reactions. *ChemPhysChem* https://doi.org/10.1002/cphc.202200343.

41 Maraf, M.B., Idrice, A.A., Pelagie, M.A.M. et al. (2022). Decoding the reaction mechanism of the cyclocondensation of ethyl acetate2-oxo-2-(4-oxo-4*H*-pyrido 1.2-*a* pyrimidin-3-yl) polyazaheterocycle and ethylenediamine using bond evolution theory. *J. Comput. Chem.* 43: 972–985.

42 Karimi, F.S., Seif, A., and Zahedi, E. (2022). Kinetics and molecular mechanism of the Schonberg rearrangement. *Comput. Theor. Chem.* 1208: 8.

43 Karimi, F.S., Seif, A., and Zahedi, E. (2022). Understanding the kinetics and topological events within the thione-to-thiol rearrangement of xanthates. *ChemistrySelect* 7: 12.

44 Kacka-Zych, A. and Jasinski, R. (2022). Mechanistic aspects of the synthesis of seven-membered internal nitronates via stepwise 4+3 cycloaddition involving conjugated nitroalkenes: molecular electron density theory computational study. *J. Comput. Chem.* 43: 1221–1228.

45 Guerra, C., Ayarde-Henriquez, L., Duque-Norena, M., and Chamorro, E. (2022). On electron pair rearrangements in photochemical reactions: 1,3-cyclohexadiene ring opening. *J. Phys. Chem. A* 126: 395–405.

46 Guerra, C., Ayarde-Henriquez, L., Duque-Norena, M., and Chamorro, E. (2022). Photochemically induced 1,3-butadiene ring-closure from the topological analysis of the electron localization function viewpoint. *ChemPhysChem* https://doi.org/10.1002/cphc.202200217,10.

47 Geoffroy-Neveux, A., Labet, V., and Alikhan, M.E. (2022). Influence of an oriented external electric field on the mechanism of double proton transfer between pyrazole and guanidine: from an asynchronous plateau transition state to a synchronous or stepwise mechanism. *J. Phys. Chem. A* 126: 3057–3071.

48 Ayarde-Henriquez, L., Guerra, C., Duque-Norena, M., and Chamorro, E. (2022). Unraveling the role of the electron-pair density symmetry in reaction mechanism patterns: the Newman-Kwart rearrangement. *New J. Chem.* 46: 12002–12009.

49 Taherian, R., Chahkandi, B., and Zahedi, E. (2021). A comprehensive theoretical analysis of Curtius rearrangement of syn-syn and syn-anti conformers of oxalyl diazide. *J. Mol. Graphics* 109: 13.

50 Mohammad-Salim, H.A., Basheer, H.A., Abdallah, H.H. et al. (2021). A molecular electron density theory study for 3+2 cycloaddition reactions of *N*-benzylcyclohexylnitrone with methyl-3-butenoate. *New J. Chem.* 45: 262–267.

51 Mohammad-Salim, H.A. (2021). Understanding the reactivity of C-cyclopropyl-*N*-methylnitrone participating in 3+2 cycloaddition reactions towards styrene with a molecular electron density theory perspective. *J. Mex. Chem. Soc.* 65: 129–140.

52 Kula, K., Kacka-Zych, A., Lapczuk-Krygier, A. et al. (2021). Experimental and theoretical mechanistic study on the thermal decomposition of 3,3-diphenyl-4-(trichloromethyl)-5-nitropyrazoline. *Molecules* 26: 12.

53 Kula, K., Kacka-Zych, A., Lapczuk-Krygier, A., and Jasinski, R. (2021). Analysis of the possibility and molecular mechanism of carbon dioxide consumption in the Diels–Alder processes. *Pure Appl. Chem.* 93: 427–446.

54 Khojandi, M., Zahedi, E., Seif, A. et al. (2021). A theoretical study on the degenerate Cope rearrangement of hypostrophene using the RRKM theory and topological approaches. *ChemistrySelect* 6: 1607–1615.

55 Kacka-Zych, A. and Perez, P. (2021). Perfluorobicyclo[2.2.0]hex-1(4)-ene as uniquc partner for Diels–Alder reactions with benzene: a density functional theory study. *Theor. Chem. Acc.* 140: 16.

56 Kacka-Zych, A. (2021). Understanding the uniqueness of the stepwise 4+1 cycloaddition reaction between conjugated nitroalkenes and electrophilic carbene systems with a molecular electron density theory perspective. *Int. J. Quantum Chem.* 121: 14.

57 Kacka-Zych, A. (2021). The molecular mechanism of the formation of four-membered cyclic nitronates and their retro (3+2) cycloaddition: a DFT mechanistic study. *Molecules* 26: 12.

58 Fryzlewicz, A., Kacka-Zych, A., Demchuk, O.M. et al. (2021). Green synthesis of nitrocyclopropane-type precursors of inhibitors for the maturation of fruits and vegetables via domino reactions of diazoalkanes with 2-nitroprop-1-ene. *J. Cleaner Prod.* 292: 16.

59 Adjieufack, A.I., Ongagna, J.M., Tchidjo, J.F.K., and Ndassa, I.M. (2021). Topological unraveling of the 3+2 cycloaddition (32CA) reaction between *N*-methylphenylnitrone and styrene catalyzed by the chromium tricarbonyl complex using electron localization function and catastrophe theory. *New J. Chem.* 45: 20342–20351.

60 Adjieufack, A.I., Bake, M.M., Nguimkeu, C.N. et al. (2021). Exploring the sequence of electron density along the chemical reactions between carbonyl oxides and ammonia/water using bond evolution theory. *ChemPhysChem* 22: 1792–1801.

61 Acharjee, N. and Mondal, S. (2021). A molecular electron density theory study to understand the strain promoted 3+2 cycloaddition reaction of benzyl azide and cyclooctyne. *Indian J. Chem.* 60: 62–71.

62 Seif, A., Domingo, L.R., Zahedi, E. et al. (2020). Unraveling the kinetics and molecular mechanism of gas phase pyrolysis of cubane to 8 annulene. *RSC Adv.* 10: 32730–32739.

63 Kacka-Zych, A. and Jasinski, R. (2020). A DFT study on the molecular mechanism of the conjugated nitroalkenes polymerization process initiated by selected unsaturated nucleophiles. *Theor. Chem. Acc.* 139: 18.

64 Kacka-Zych, A. and Jasinski, R. (2020). Molecular mechanism of hetero Diels–Alder reactions between (*E*)-1,1,1-trifluoro-3-nitrobut-2-enes and enamine systems in the light of molecular electron density theory. *J. Mol. Graphics* 101: 12.

65 Kacka-Zych, A. (2020). Push-pull nitronates in the 3+2 cycloaddition with nitroethylene: molecular electron density theory study. *J. Mol. Graphics* 97: 14.

66 Idrice, A.A., Karelle, D.L., Blaise, L.N.R. et al. (2020). Unraveling the sequence of electron flows along the reaction mechanism by quantum topological tools: the 32CA reaction of acetonitrile oxide with 7-bromo-oxanorborn-5-en-2-one. *J. Mol. Graphics* 96: 9.

67 Domingo, L.R., Rios-Gutierrez, M., and Pérez, P. (2020). A molecular electron density theory study of the participation of tetrazines in aza-Diels–Alder reactions. *RSC Adv.* 10: 15394–15405.

68 Cherni, E., Adjieufack, A.I., Champagne, B. et al. (2020). Density functional theory investigation of the binding of thioTEPA to purine bases: thermodynamics and bond evolution theory analysis. *J. Phys. Chem. A* 124: 4068–4080.

69 Acharjee, N. (2020). Unravelling the regio- and stereoselective synthesis of bicyclic N,O-nucleoside analogues within the molecular electron density theory perspective. *Struct. Chem.* 31: 2147–2160.

70 Houk, K.N., Gonzalez, J., and Li, Y. (1995). Pericyclic reaction transition-states – passions and punctilios, 1935–1995. *Acc. Chem. Res.* 28: 81–90.

71 Wiest, O., Montiel, D.C., and Houk, K.N. (1997). Quantum mechanical methods and the interpretation and prediction of pericyclic reaction mechanisms. *J. Phys. Chem. A* 101: 8378–8388.

72 Ischay, M.A., Lu, Z., and Yoon, T.P. (2010). 2+2 Cycloadditions by oxidative visible light photocatalysis. *JACS* 132: 8572–8574.

73 Mahatthananchai, J., Dumas, A.M., and Bode, J.W. (2012). Catalytic selective synthesis. *Angew. Chem. Int. Ed.* 51: 10954–10990.

74 Bickelhaupt, F.M. and Houk, K.N. (2017). Analyzing reaction rates with the distortion/interaction-activation strain model. *Angew. Chem. Int. Ed.* 56: 10070–10086.

75 Shaik, S., Mandal, D., and Ramanan, R. (2016). Oriented electric fields as future smart reagents in chemistry. *Nat. Chem.* 8: 1091–1098.

76 Mavroskoufis, A., Rajes, K., Golz, P. et al. (2020). N-Heterocyclic carbene catalyzed photoenolization/Diels–Alder reaction of acid fluorides. *Angew. Chem. Int. Ed.* 59: 3190–3194.

77 Cullen, W., Misuraca, M.C., Hunter, C.A. et al. (2016). Highly efficient catalysis of the Kemp elimination in the cavity of a cubic coordination cage. *Nat. Chem.* 8: 231–236.

78 Domingo, L.R., Aurell, M.J., Perez, P., and Contreras, R. (2002). Quantitative characterization of the global electrophilicity power of common diene/dienophile pairs in Diels–Alder reactions. *Tetrahedron* 58: 4417–4423.

79 Jiao, H.J. and Schleyer, P.V. (1998). Aromaticity of pericyclic reaction transition structures: magnetic evidence. *J. Phys. Org. Chem.* 11: 655–662.

80 Dewar, M.J.S. and Jie, C.X. (1992). Mechanisms of pericyclic reactions – the role of quantitative theory in the study of reaction mechanisms. *Acc. Chem. Res.* 25: 537–543.

81 Beno, B.R., Houk, K.N., and Singleton, D.A. (1996). Synchronous or asynchronous? An "experimental" transition state from a direct comparison of experimental and theoretical kinetic isotope effects for a Diels–Alder reaction. *JACS* 118: 9984–9985.

82 Domingo, L.R., Rios-Gutierrez, M., Silvi, B., and Perez, P. (2018). The mysticism of pericyclic reactions: a contemporary rationalisation of organic reactivity based on electron density analysis. *Eur. J. Org. Chem.* 2018: 1107–1120.

83 Domingo, L.R., Rios-Gutierrez, M., Chamorro, E., and Perez, P. (2016). Aromaticity in pericyclic transition state structures? A critical rationalisation based on the topological analysis of electron density. *ChemistrySelect* 1: 6026–6039.

84 Domingo, L.R., Chamorro, E., and Perez, P. (2010). Understanding the mechanism of non-polar Diels–Alder reactions. A comparative ELF analysis of concerted and stepwise diradical mechanisms. *Org. Biomol. Chem.* 8: 5495–5504.

85 Ayarde-Henríquez, L., Guerra, C., Duque-Noreña, M., and Chamorro, E. (2023). A simple topology-based model for predicting the activation barriers of reactive processes at 0 K. *Physical Chemistry Chemical Physics*, 25(20), 14274–14284. https://doi.org/10.1039/d3cp01008b.

86 Chamorro, E., Guerra, C., Ayarde-Henríquez, L. et al. (2023). New insights from a bonding evolution theory based on the topological analysis of the electron localization function. In *Chemical Reactivity* (p. 465–481). Elsevier. https://doi.org/10.1016/b978-0-32-390257-1.00023-1.

87 Guerra, C., Ayarde-Henríquez, L., Chamorro, E., & Ensuncho, A. (2023). Uncovering Triradicaloid Structures in S1 Benzene Photochemistry**. *ChemPhotoChem*, 7(5). Portico. https://doi.org/10.1002/cptc.202200263.

88 Ayarde-Henríquez, L., Guerra, C., Duque-Noreña, M., and Chamorro, E. (2023). Revisiting the bonding evolution theory: a fresh perspective on the ammonia pyramidal inversion and bond dissociations in ethane and borazane. *Physical Chemistry Chemical Physics*, 25(40), 27394–27408. https://doi.org/10.1039/d3cp03572g.

89 Guerra, C., Ayarde-Henríquez, L., Rodriguez-Nuñez, Y. A., et al. (2023). Mechanistic insights into benzyne formation via 1,2-di-iodobenzene photolysis. *New Journal of Chemistry*, 47(46), 21270–21275. https://doi.org/10.1039/d3nj04022d.

90 Guerra, C., Ayarde-Henríquez, L., Rodríguez-Núñez, Y. A. et al. (2023). Elucidating the N–N and C–N Bond-breaking Mechanism in the Photoinduced Formation of Nitrile Imine. *ChemPhysChem*, 24(11). Portico. https://doi.org/10.1002/cphc.202200867.

17

Designing Principles for Ultrashort H···H Nonbonded Contacts and Ultralong C—C Bonds

Nilangshu Mandal and Ayan Datta

School of Chemical Sciences, Indian Association for the Cultivation of Science, Kolkata, India

17.1 Introduction

Understanding the concept of chemical bonding is of utmost importance toward the advancement of chemical sciences [1, 2]. In this aspect, the origin and consequences of chemical interactions play crucial roles in various domains such as inorganic, biochemistry, structural chemistry, solid matter, structural biology, and physical chemistry [3–6]. Over the last few years, researchers have witnessed a renewed interest in using such qualitative ideas to explore the limits of a few interactions less common than classical hydrogen bond or ion pairing [7–9]. Further, the directionality and strength of the chemical bonds offer validation for the reactivity rules of molecules and solids [10, 11]. Apart from its significance as a conceptual model to connect molecular structure with reactivity, the chemical bond acts as a tool to rationalize observations from magnetism and photochemical reactions [12, 13]. Well-known models for conceptualizing the over-the-barrier processes like the Evans–Polyani (EP) principle (linear-energy relationship) or more advanced activation-strain analyses are convenient tools to analyze activation barriers to determine the origin of reactivity and selectivities, depending on the relative strength of the chemical bonds being formed or ruptured along the reaction coordinate [14, 15]. Further, the strength of a chemical bond relies on both the nature of the atom pairs and the extent of delocalization across the topology, as well as the perturbations due to the remote functionalization in the molecule.

The physical origin of the chemical bond was, for a long time, a mystery not only for chemists but in natural science in general. It was clear that among the four elementary forces, only electrostatic forces could account for the strong interatomic interactions that create a chemical bond. However, the attraction between neutral species could hardly be reconciled with the classical Coulomb law. According to the quantum-theoretical approach by Heitler and London, the electrostatic interactions between atoms have to be termed in the form of mathematical functions relating to quantum objects rather than classical particles. Chemical bonding comes from the interference of spatially extended wave functions, which may either have a positive or negative sign. Covalent bonding originates from the interference of the wave functions of the interacting species. Chemical bonding occurs even when only one electron occupies the new functions, such as in H_2^+ [16]. The second electron usually strengthens the chemical bond further, such as in H_2, but there are molecules where the second electron even weakens the bond. For example, Li_2 has a lower bond dissociation energy (BDE) than Li_2^+ [17].

Electron Density: Concepts, Computation and DFT Applications, First Edition.
Edited by Pratim Kumar Chattaraj and Debdutta Chakraborty.

In this regard, a more qualitative Badger's empirical rule provides insightful guidelines to assess the strength of the chemical bond from its bond length [18]. Several efforts aiming at a polyatomic generalization of the Badger rule have been reported, even though only with limited success, they have reinforced the belief of chemists in the use of bond lengths as convenient bond strength indicators [19]. In the last three decades, several observations have satisfied the bond length–bond strength (BLBS) relationship that is in line with the Badger empirical rule, and such findings are denoted as *bond anomalies* [20].

17.1.1 The Art of the Chemical Bond

Chemical bonding enables scientists to utilize the 100-plus elements of the periodic table as a template and combine them in numerous ways to form chemical compounds and materials. The empirical observatory approach in experimental chemistry is still a predominant method to fabricate new molecules and materials [21, 22]. Nevertheless, it might be highly inefficient if we know only the desired properties but have little knowledge of how to tune them by atomic manipulations. With the advent of modern computational architectures, quantum chemical methods have become major tools for designing materials *a priori* [23]. Understanding a chemical structure and associated properties usually starts with chemical bonding. Equipped with chemical knowledge, one can design molecules with specific types of bonds aiming at specific properties. Vice versa, we decode chemical bonding to understand how specific bonds contribute to the properties of interest.

17.1.2 Designing and Decoding Chemical Bond

In recent years, researchers have continued to find new ways to get atoms to interact and stick together. They also continue to evolve methods for breaking and reforming bonds during chemical reactions to improve the synthesis, properties, and performance of existing chemicals and materials. It is necessary to probe new ways of manipulating bonds that help to improve chemical synthesis, material design, and property analysis.

When we design chemical bonds, it is essential to demonstrate that species possessing such proposed bonding do exist even if they are fleetingly stable. It is textbook knowledge that if two atoms approach each other close enough, they feel repulsion and become energetically destabilized. A qualitative rule of thumb for atoms that are assumed to be "non-bonded" is that van der Waals (vdW) radii are used to distinguish whether the two atoms are close enough (penetrating each other's vdW sphere) to have destabilizing interactions that are sometimes also called as steric repulsions. However, even for the simplest atom, namely hydrogen, recent studies explain that two hydrogen atoms from two different molecules may well penetrate each other's vdW sphere. Yet, the two molecules form a stable aggregate [24].

The study of weak intermolecular contacts, embracing several types of noncovalent interactions traditionally termed vdW or dispersion forces, has fascinated chemists in recent years due to their relevance in many aspects of supramolecular and biological chemistry as well as in crystal engineering [25–28]. Regardless of the abundance and variety of these interactions, some of them are not well understood yet, especially regarding their nature and the factors that affect their strength. Short dihydrogen contacts are the type of noncovalent interactions that have attracted much interest in the last decades [29, 30]. Those contacts can occur between two H atoms with the same or different atomic charges, resulting in homopolar and heteropolar dihydrogen interactions, respectively.

More recently, the concept and understanding of hydrogen bonding have been extended to include proton–hydride interactions as X–H· · ·H–Y interaction [31]. In the late 1960s, this type of interaction was first recognized by Brown et al., who reported Lewis acid-base adducts containing an L·BH_3 (L = Me_3N, Et_3N, etc.) moiety in weak H· · ·H interactions in the presence of a proton donor, like MeOH [32, 33]. A subsequent survey of the Cambridge structural database (CSD) for short H· · ·H contacts in N–B complexes (1.7–2.2 Å) revealed a distinct preference for a bent geometry for these interactions, in contrast to the more linear disposition of their X–H· · ·Y counterparts [34]. The NH· · ·H–B angles generally fall between 95° and 120°, and N–H· · ·HB angles are closer to 150–170°. Such a geometry is consistent with the hydridic nature of the B–H moiety, as the bent configuration allows the approach of the N–H group to maximize the electrostatic component of the interaction. Furthermore, a series of calculations exploring the interaction of metal hydrides with HF revealed that the strength of these dihydrogen bonds is highly correlated with their H· · ·H distances [35, 36].

On the other hand, covalent bond lengths, such as 1.54 Å for C(sp^3)–C(sp^3), are basic parameters in chemistry [37]. The carbon–carbon single bond distance remains consistent, with very few deviations observed due to unconjugated aromatic or p-conjugating units. Compounds that show remarkable deviation from standard values have attracted attention [38–42]. For example, a longer C—C bond was predicted for hexaphenylethane (HPE) as a result of large steric repulsion among its six phenyl groups [43]. Based on the "expandability" of the extremely expanded C—C bonds found in other types of HPEs, much longer C—C bonds could be realized by proper molecular design [44]. From a theoretical viewpoint, the limit for the C—C bond length can be estimated by supposing a linear correlation between bond length and BDE. However, it has been known for years that through-bond coupling can weaken the bond without changing the bond length [45]. Furthermore, the Lennard-Jones-type potential may be more appropriate to describe the BDE of an expanded C—C bond than a simple linear correlation. Thus, the change in BDE with an increase in the length of pre-strained weak C—C bonds (>1.7 Å) should be smaller than the energy required to lengthen a normal C—C bond (1.54 Å) to the same degree.

Therefore, finding new and unusual examples of molecules where H· · ·H nonbonded distances and elongated d(C—C) bond lengths are new challenges in terms of synthesis. In addition to the need to elucidate the origin of bond elongation or compression, we must also explore novel reactivities that involve such weakened or strengthened bonding contacts to shed light on the boundary between long bonds and short nonbonded contacts.

17.2 Governing Factors for Ultrashort H· · ·H Nonbonded Contacts

In recent years, one of the most fascinating aspects that excited chemists is the limit to which two nonbonded hydrogen atoms can be enforced into proximity. In 2016, Hoffmann and Ashcroft found r(H-atom) = 1.54 Å based on the electron density of atoms with a cut-off of 0.001 electron/$Bohr^3$ [46]. In particular, recent reports dealing with ultrashort H· · ·H nonbonded distances d(H· · ·H) < 3.08 Å revealed their significance from a structural and theoretical perspective. Further, d(H· · ·H) distance below the limit of 3.08 Å, provides a strong indication of an attractive interaction between two nonbonded H-atoms. In 1964, Bondi reported the crystallographic vdW radii of a hydrogen atom is 1.20 Å [47]. However, recent studies demonstrate that two hydrogen atoms from two different molecules may well penetrate each other's vdW sphere and facilitate the formation of a stable aggregate [48]. One might envision that the interaction of two nonbonded hydrogen atoms is destabilizing below a certain inter-proton distance. But, such "local" destabilization does not certainly prohibit the formation of stable molecular structures with short or ultrashort H· · ·H distances. Accordingly, from an experimental perspective, many stable molecules have been synthesized during the last few decades with ultrashort nonbonded H· · ·H contacts below 2.0 Å [49–53]. In the last decade, the "world record," H· · ·H contact ~1.56 Å, was reported using both experimental and computational investigations on *in,in*-bis(hydrosilane) [50]. This is much smaller than the sum of the vdW radii of two hydrogen atoms ~2 × 1.20 Å = 2.40 Å. Therefore, the quest for achieving compressed H· · ·H is of fundamental interest and has a large impact on the stabilization of molecular conformations and aggregates.

In this chapter, we aim to demonstrate some design strategies that are useful to open the door for constructing a proper set of molecules. Future studies on the nature of the so-called nonbonded hydrogen–hydrogen interactions are now an active and controversial area of research.

17.2.1 London Dispersion Interaction

The implication of stabilizing dispersion effects within sterically hindered hydrocarbons is well recognized. In general, noncovalent interactions between neutral, sterically hindered organic molecules comprise a strong stabilizing contribution from dispersion forces that, in many systems, turn the "steric repulsion" into a "steric attraction." Figure 17.1 reemphasizes the importance of accounting for dispersion effects even when modeling molecular architectures exhibiting intramolecular and intermolecular H· · ·H nonbonded contacts.

Using a unique molecular design approach, Pascal and coworkers reported examples of intramolecular short H· · ·H nonbonded contacts in macrocyclic systems, like *in,in*-bis(hydrosilane) [50]. Further computational investigations suggest that the in,in-form is stabilized by weak

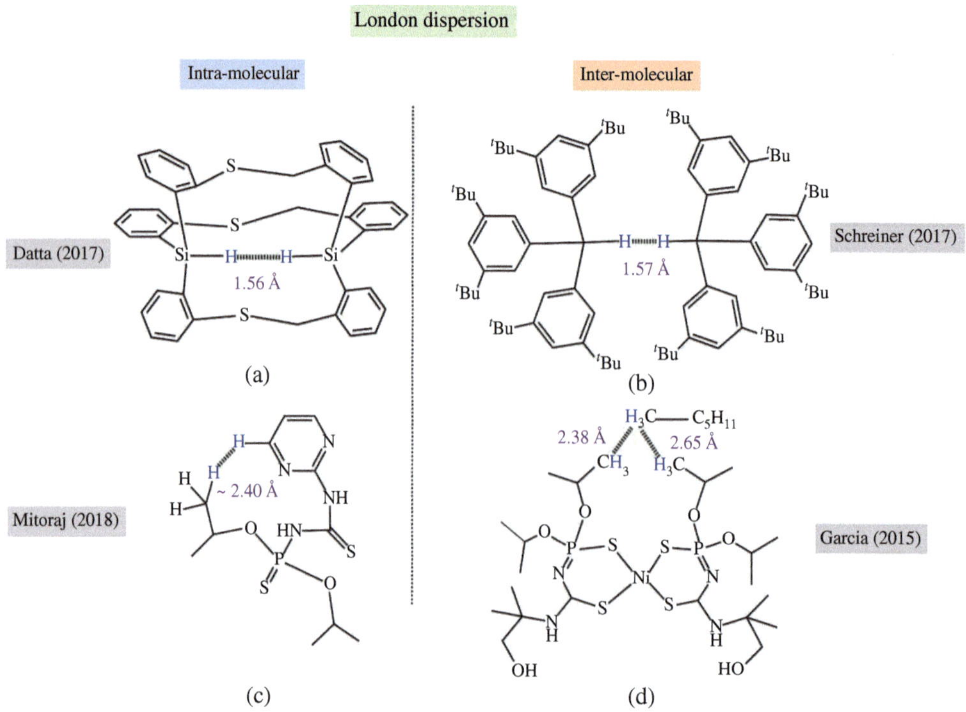

Figure 17.1 Figures (a, b) and (c, d) display H· · ·H nonbonded contacts, stabilized by London dispersion.

vdW attraction, that is, London dispersion (see Figure 17.1a) [54]. Using experimental and computational studies, Mitoraj et al. reported that monomers of five thiourea derivatives [RNHC(S)NHP(S)(O*i*Pr)$_2$] [55]. In Figure 17.1b, the thiourea monomer revealed homopolar intramolecular C–H· · ·H–C non-bonded contact and stabilized by London dispersion interaction. Schreiner reported intermolecular London dispersion interactions between contacting *tert*-butyl substituents surrounding the central contact to deliver the decisive energetic contributions to enable this remarkable bonding situation in the H· · ·H non-bonded regime (Figure 17.1c) [56]. In addition, Garcia and coworkers established intermolecular close H· · ·H nonbonded distances between *n*-Hexane and nickel-thiourea complex (see Figure 17.1d) [57]. They have used the extended transition state (ETS) combined with natural orbitals for chemical valence (ETS-NOCV) analyses to interpret the plausible stabilization arising from close C–H· · ·H–C contacts formed between *n*-hexane and neighboring Ni-containing units [58, 59]. Therefore, London dispersion interaction plays a crucial role in designing molecular frameworks that can exhibit short H· · ·H nonbonded contacts.

17.2.2 Polarity and Charge Separation

Recently, the design of ultrashort hydrogen–hydrogen nonbonded distances with electronically different H-atoms has attracted much attention to the scientific community. Moreover, the concept of hydrogen bonding has been extended to include proton–hydride interactions. In this case, a hydridic moiety, rather than a nonbonding electron pair, serves as the acceptor in an E–H· · ·H–X interaction, where E = transition-metal ion or main-group element and X = N or O with high electronegativity, the two H-atoms are oppositely polarized which brings them closer through electrostatic interactions.

For example, Crabtree reported a series of metal hydride complexes containing X–H· · ·H–Ir (X = O or N) hydrogen bonds. An iridium hydride amide complex tautomerized to the iminol form on account of a stabilizing Ir–H· · ·H–O interaction, and the iridium hydride complex with aminopyridines involves Ir–H· · ·H–N contact (see Figure 17.2a,b) [60]. A much weaker Ir–H· · ·H–O interaction (d(H· · ·H) = 2.4 Å, based on neutron diffraction) was reported in 1986 by Milstein et al. [61], and an Ir–H· · ·H–O example (d(H· · ·H) = 1.8 Å) independently discovered by Ramachandran and Morris [62].

In contrast, it is quite counter-intuitive to deliberate an attractive scenario for close homopolar dihydrogen contacts. In Figure 17.2c, Echeverría and coworkers have shown that a short P–H· · ·H–P distance of 2.05 Å is possible in a P_3H_5 dimer complex with a basis set superposition error (BSSE) corrected the binding energy of −4.5 kcal/mol [63]. Nevertheless, Bader and Trujillo reported that nonpolar C–H moieties can engage in mutually stabilizing C–H· · ·H–C interactions (see Figure 17.2d) [64, 65]. The partial charges associated with the hydrogen atoms of the nonpolar moieties are often small but not necessarily of opposite signs, suggesting that homopolar dihydrogen bonding is not dominated by electrostatics, but rather is underpinned by a substantial contribution from vdW attraction or London dispersion, as we described earlier.

17.2.3 Conformations and Orientations

Molecular architecture modeling with suitable molecular conformations is crucial for designing chemical motifs exhibiting short H· · ·H nonbonded contacts. This phenomenon has lots of applications in crystal engineering and supramolecular chemistry. In the case of triphenylmethane and all-*meta tert*-butyl triphenylmethane dimers, $(TPM)_2$ and $(T^tBuPM)_2$, Schreiner and

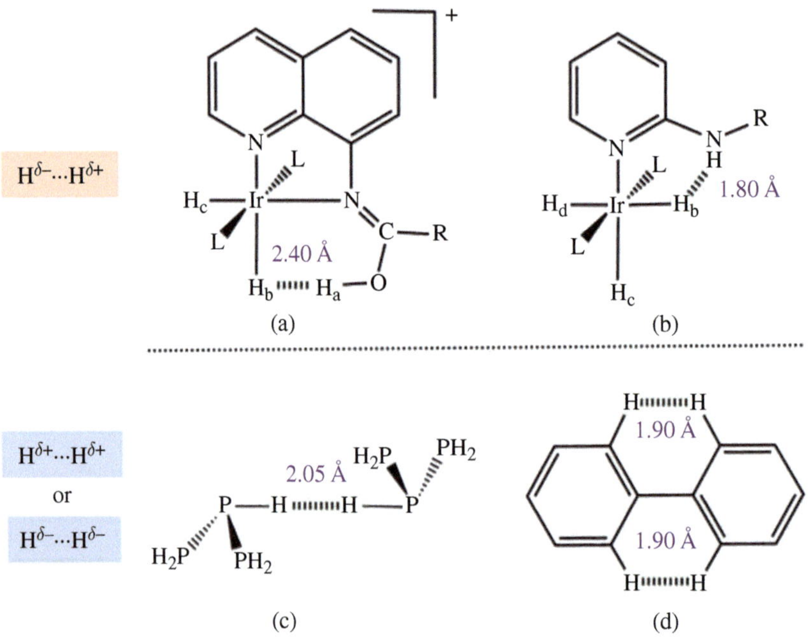

Figure 17.2 Figures (a, b) and (c, d) exemplify heteropolar and homopolar H···H nonbonded contacts.

coworkers reported three different orientations like *head-to-head*, *head-to-tail*, and *tail-to-tail* for these molecules. However, *head-to-head* orientation is mostly dominant and provides promising stabilization through C–H···H–C interactions in both gas and solid phases [66].

Impressive examples of very short *intramolecular* C–H···H–C contacts are bowl-like structures such as *exo,exo*-tetracyclo-dodecane derivatives with d(H···H) = 1.71 Å, and the "half-cage" pentachloro-pentacyclododecane with a short H···H contact of 1.62 Å (see Figure 17.3a,b) [51, 67]. In addition, Pascal and coworkers reported a series of cyclophanes. Interestingly, the in,in-forms were only experimentally detected and also possess exceptionally short H···H contact distances that can be analyzed by "through-space" NMR spin–spin coupling measurements [49]. In Figure 17.3c,d, such sterically congested molecules are presented. The helicene-like compound contains two hydrogen atoms that are quite close in space yet separated by seven bonds. However, the hydrogen atoms are not as close as in the compound, as shown in Figure 17.3c. This helicene-like molecule provides a second example of a C–H···H–C close contact, d(H···H) = 2.13 Å (from X-ray structure) whereas, *in,in*-cyclophane reflects d(H···H) = 1.70 Å, shown in Figure 17.3d. Therefore, molecular conformation plays an important role in stabilizing molecules with close H···H contacts.

17.2.4 Iron Maiden Effect

The terminology of molecular "iron maiden" was first introduced by Pascal [68]. However, this effect is well recognized when less reactive or normally unreactive functional groups are enforced to proximity towards an aromatic molecule or another functional group. Later, Baldridge and Seigel reported a series of bis-tryptycene cyclophanes, like 1-(CH><HC), 1-(CH><HSi), and 1-(SiH><HSi) [69]. These cyclophanes were investigated concerning the molecular compression

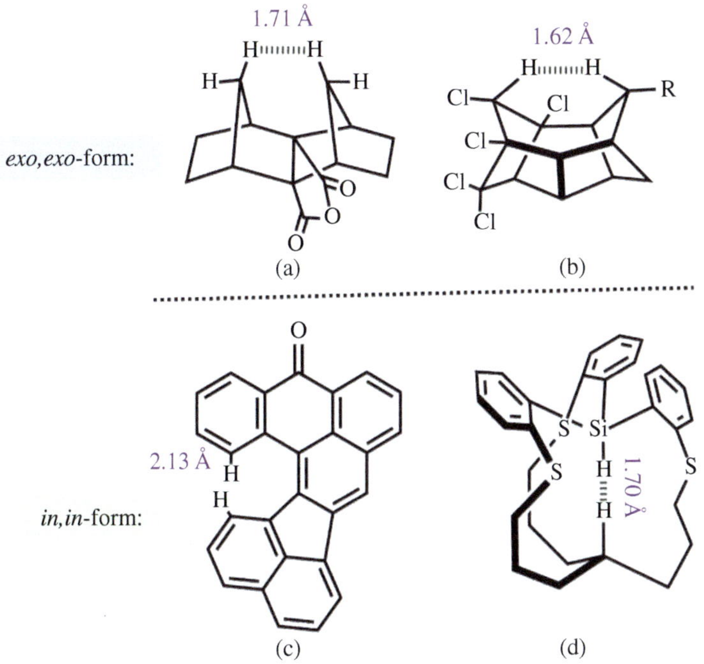

Figure 17.3 Conformation-dependent short H· · ·H contacts in (a, b) *exo,exo*- and (c, d) *in,in*-conformations.

exerted upon "*endo*" hydrogen to achieve close hydrogen–hydrogen nonbonded contacts (see Figure 17.4a–c). Computational investigations of these structures and properties highlighted several molecular engineering principles relevant to molecular machines or molecular presses. Interestingly, such methodology is indeed useful for further reduction of H· · ·H nonbonded contact. For example, 1-(CH><HSi) with a stable positive definite stationary point displays an H· · ·H contact of 1.41 Å.

Another interesting cyclophane, shown in Figure 17.4d, reveals an intramolecular H· · ·H nonbonded distance of 1.38 Å [54]. This type of molecular iron maidens can be further tuned by the stereoelectronic effect of the functional groups, like -CN. However, the parent analog, *in,in*-bis(hydrosilane), contains only bulky phenyl groups. Therefore, steric inclusion is not the sole deciding factor, and rather, neighbor substituents also play a crucial role in enforcing the hydrogens at proximity.

17.3 Elongation Strategies for C—C Bonds

Organic molecules with ultralong C—C bonds have been known for almost 40 years. Strained molecules such as sterically congested polycyclic aromatic hydrocarbons as well as cyclic π-conjugated compounds often exhibit unusual parameters and thus have attracted much attention due to their potential applications [70–72]. In 1986, Mislow reported the crystal structure of the HPE derivative, reflecting a carbon–carbon bond length of 1.67 Å whereas the C—C bond in ethane is 1.54 Å [73]. Several efforts have been made to design molecules with unusual C—C bond lengths since then [74–77]. Such compounds exhibit a range of C—C bond lengths

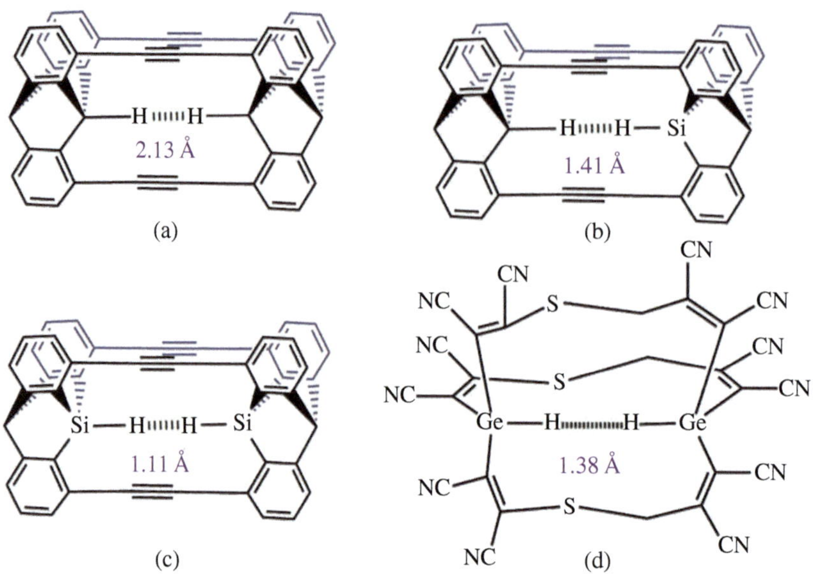

Figure 17.4 (a–d) Common molecular "iron maidens."

between 1.60–1.99 Å and beyond 2.00 Å [78]. These sets of molecules have also attracted the attention of computational chemists. Furthermore, density functional theory (DFT) has been successfully applied to investigate the chemical nature and bond length properties, stabilities of 1,1,1,2,2,2-hexakis(3,5-di-*tert*-butylphenyl)ethane and related compounds [79].

In this context, based on the "expandability" of the extremely elongated C—C bonds observed in molecular frameworks, a much longer carbon–carbon bond could be realized by proper "molecular design." To create novel molecular architectures with an unprecedented C—C bond beyond the typical limit of 2.00 Å, some important key points are demonstrated below for molecular design so that the compounds could maintain their chemical stability despite the presence of a very weak C—C bond.

17.3.1 Steric Crowding Effect

Steric effects are the consequence of space required to accommodate the atoms and groups within a molecule and originate from the overlapped electron densities. To understand chemical reactivity and bonding, one needs to understand attractive interactions, that is vdW interactions (also known as London dispersion). There is a long-time debate on the higher stability of branched alkanes compared to linear alkanes. This is due to the "steric attraction," and this analogy can be applied to elongate the carbon–carbon bond. For example, this steric crowding effect works well for the design of diamondoid molecules, see Figure 17.5. These sets of molecules possess elongated C—C bonds of up to 1.72 Å. In Figure 17.5a bond length of 1.67 Å was experimentally determined. Interestingly, this molecule is sterically much more crowded – hexakis(3,5-di-*t*-butylphenyl)ethane (where, R = *t*-butyl) [73].

Further, Schreiner and coworkers demonstrated C—C bond elongation strategy to overcome the bond weakening repulsive interactions is to design a structure with pronounced London dispersion. Such attractive interaction can be imposed by controlling the number of lengths of hydrogen-hydrogen nonbonded contacts in diamondoid adducts [80].

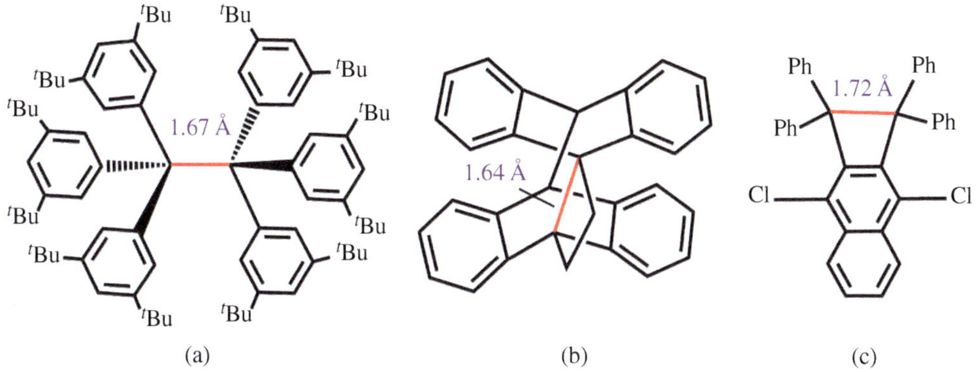

Figure 17.5 Thermally unstable hydrocarbons with elongated C–C bonds in (a) hexakis(3,5-di-*t*-butylphenyl)ethane, (b, c) related compounds.

17.3.2 Core–Shell Strategy and Scissor Effect

To create hydrocarbons with remarkably long C—C bonds beyond normal values, the intramolecular core–shell strategy is quite useful [77]. The intramolecular core–shell strategy applied to the HPE analogs also favors bond expansion as the "shell" structure is directly connected with the "core." Further, the "scissor effect" is a reliable and important structural modification for further weakening the elongated bond [81].

For example, the greater bond length for the C—C bonds of HPE analogs can be rationalized by considering the angle strain due to the ethano bridge at the opposite peri position, see Figure 17.6a,b. Therefore, it is easier to apply the scissor effect by incorporating the fused ring at the terminal part of the shell. On the other hand, due to the eclipsed position of these rigid structures, they gain greater "eclipseness," which is the driving force for realizing the ultralong carbon–carbon bonds. In general, a more eclipsed molecule in a crystal environment exhibits more elongation of the C—C bond than an analog with a less eclipsed geometry. Another beautifully designed molecule is dispiro(acridan)-substituted pyracene (DSAP), which remains almost eclipsed conformation and reveals an ultralong C—C bond length of 1.80 Å (see Figure 17.6c).

Further, a novel hydrocarbon, acenaphthene, has a high C—C bond length in the order of 1.73 Å. Such elongation is due to the shorter bridge at the opposite peri position exerting a greater angle stain on the naphthalene nucleus, see Figure 17.6d. Therefore, the core–shell strategy and the scissor effect are important tools for designing molecules with close hydrogen-hydrogen non-bonded contacts.

17.3.3 Negative Hyperconjugation Effect

A different engineering strategy to stretch a C—C bond is to populate electrons in the antibonding, σ*(C—C), which effectively reduces the bond order and produces a weaker and longer C—C bond. This is one of the most successful applications of negative hyperconjugation to model ultralong C—C bond and has been realized for 1,2-disubstituted-*ortho*-carboranes (see Figure 17.7a). However, 1,2-diamino-carboranes can exist in three different conformational isomers: in,in-, out,out- and in,out-forms, respectively [82]. All the isomers reveal elongated carbon–carbon bonds in the range of 1.83–1.87 Å. In Figure 17.7b, different substitutions are used at both C-sites to observe their effects on the C—C bond length.

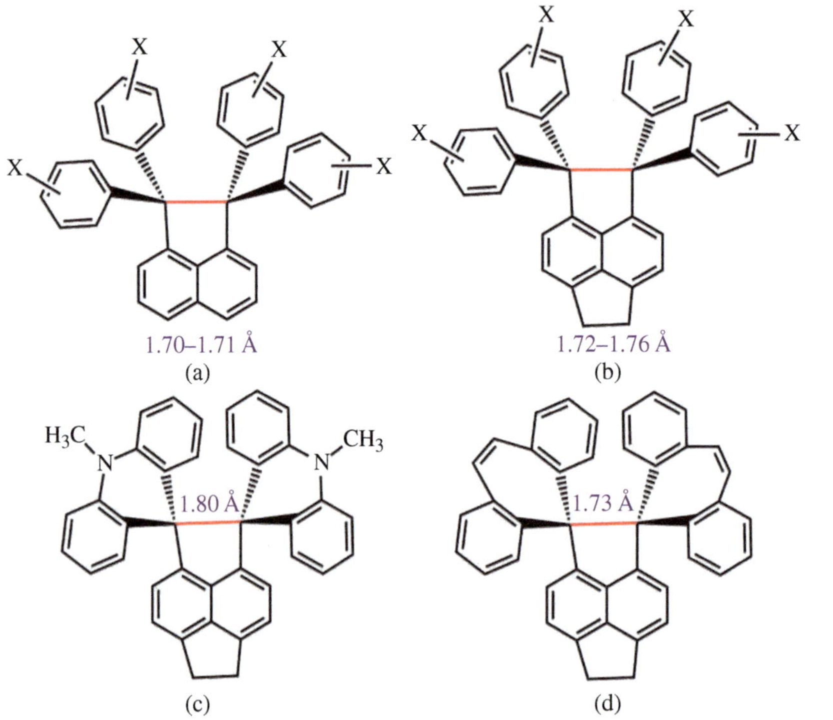

Figure 17.6 (a, b) HPE analogs, (c) dispiro(acridan)-substituted pyracene (DSAP), and (d) acenaphthene molecules with ultralong C–C bonds.

The excellent donor ability of the dimethylamino ($-N(CH_3)_2$) and oxide ($-O^-$) groups increases the C—C bond lengths beyond 2.0 Å [83]. On the other hand, a poor donor group like $-NF_2$ shrinks C–C to 1.68 Å, which is correlated with much smaller and inequivalent $E^2(N\rightarrow\sigma^*_{(C-C)}) = 13.7$ and 1.3 kcal/mol, respectively.

The two $-NH_2$ groups in the in,out-form of diamino-carborane lead to two different pyramidalization angles of N-atoms. This is further evident from the magnitude of their $E^2(N\rightarrow\sigma^*_{(C-C)})$ and has important implications for these structures. In addition, Müller and coworkers reported some

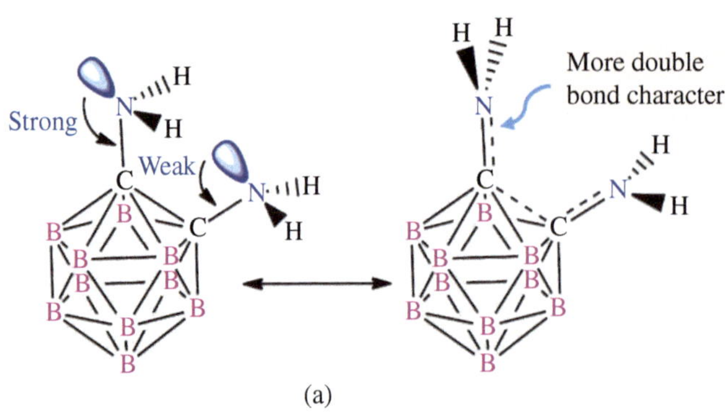

Figure 17.7 (a) Negative hyperconjugation and (b) expandability of C–C bond lengths in carboranes.

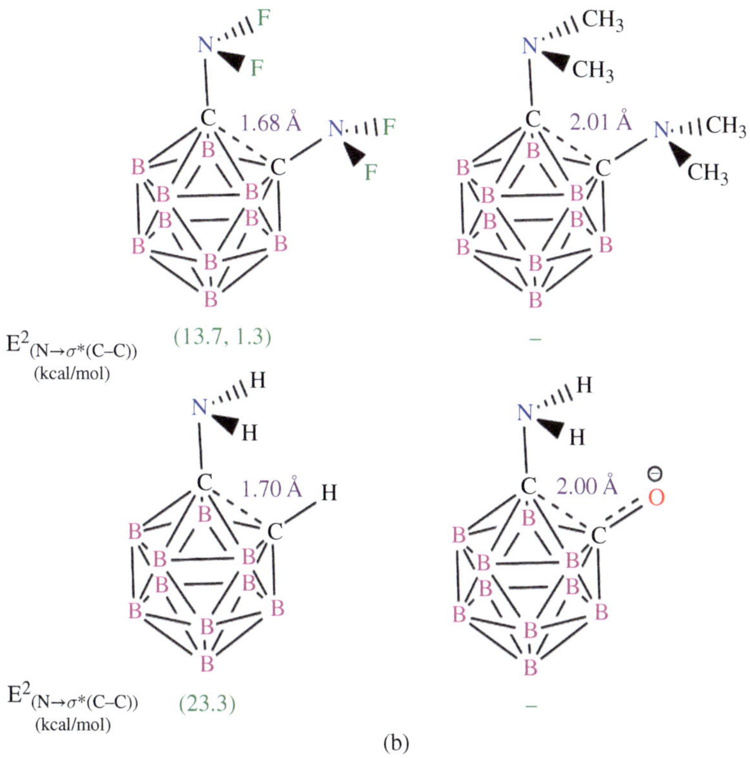

Figure 17.7 (*Continued*)

secondary amine structures of 1,2-diamino-carborane. They have experimentally reported a carborane derivative with an elongated C—C bond of 1.93 Å [75]. Therefore, through-bond interaction like negative hyperconjugation plays a crucial role in the design of molecular architecture with a sufficiently large C—C bond.

17.4 Concluding Remarks

This chapter aims to provide a detailed account of recent advances in our understanding of unusually short and long chemical bonds. The quest for designing molecules with ultralong C–C and ultrashort H···H nonbonded contacts is limitless. It has emerged as one of the important topics to advance various fields of chemical science. The use of weak molecular interactions to rationally design frameworks exhibiting unusual bonds and chemical properties has been a subject of intense interest in many research groups. Further, design principles for obtaining ultrashort H···H nonbonded contacts like London dispersion, hetero-polarity, conformations, and iron maiden effects are useful for designing complex supermolecular frameworks with vast applications. Developing compounds with ultrashort H···H contacts for dihydrogen insertion into the C=C/C=O/C=N/N=N bonds can have important applications for directed synthesis. Therefore, the rational design rules would guide chemists to achieve even shorter nonbonded distances.

On the other hand, the fascination took hold when the scientists stumbled upon a surprisingly long but stable carbon–carbon single bond. The search for innovative effects like steric attraction, negative hyperconjugation, and core–shell strategy also harnesses the conformational rigidity of the framework, which kinetically arrests these structures with elongated C–C for their isolation. In this regard, Roald Hoffmann stated that "the strategies for elongating carbon–carbon bonds are more important for chemistry than records." (https://cen.acs.org/physical-chemistry/chemical-bonding/fight-over-longest-carbon-carbon/97/i10) To understand the nature of covalent bonds, chemists have to understand extreme cases, and one of those extremes is length. To summarize, the design approaches for ultrashort H· · ·H and ultralong C—C bonds have attracted significant interest recently, and we hope to see rapid growth in this exciting research area with the development of more advanced methodologies and their applications.

Acknowledgments

NM thanks CSIR for SRF. AD thanks the TRC-DST and SERB grant nos. DIA/2018/000013 and CRG/2020/000301 for partial funding.

References

1 Frenking, G. and Shaik, S. (2014). *The Chemical Bond: Fundamental Aspects of Chemical Bonding*. Weinheim: Wiley-VCH.
2 Gray, B.H. (1994). *Chemical Bonds: An Introduction to Atomic and Molecular Structure*. Mill Valley, CA: University Science Books.
3 Lund-Johansen, F., de la Rosa Carrillo, D., Mehta, A. et al. (2016). MetaMass, a tool for meta-analysis of subcellular proteomics data. *Nat. Methods* 13: 837–840.
4 Chien, P. and Gierasch, L.M. (2014). Challenges and dreams: physics of weak interactions essential to life. *Mol. Biol. Cell* 25: 3474.
5 Elemans, J.A.A.W., Rowan, A.I.E., and Nolte, R.J.M. (2003). Mastering molecular matter. Supramolecular architectures by hierarchical self-assembly. *J. Mater. Chem.* 13: 2661.
6 Bader, R.F.W. (1975). Molecular fragments or chemical bonds. *Acc. Chem. Res.* 8: 34.
7 Warshel, A., Sharma, P.K., Kato, M. et al. (2006). Electrostatic basis for enzyme catalysis. *Chem. Rev.* 106: 3210.
8 Herschlag, D. and Pinney, M.M. (2018). Hydrogen bonds: simple after all? *Biochemistry* 57: 3338.
9 Batsanov, S. (2018). Weak interactions in crystals: old concepts, new developments. *Acta Crystallogr.* E74: 570.
10 Albright, T.A., Burdett, J.K., and Whangbo, M.-H. (2013). *Orbital Interactions in Chemistry*, 2e. Hoboken, USA: Wiley.
11 Hoffmann, R. (1989). *Solids and Surfaces: A Chemist's View of Bonding in Extended Structures.* Wiley-VCH.
12 Goodenough, J.B. (1963). *Magnetism and the Chemical Bond*. New York: Interscience (Wiley).
13 Garavelli, M. (2006). Computational organic photochemistry: strategy, achievements and perspectives. *Theor. Chem. Acc.* 116: 87–105.
14 Liu, F., Yang, Z., Yu, Y. et al. (2017). Bimodal Evans–Polanyi relationships in dioxirane oxidations of sp^3 C–H: nonperfect synchronization in generation of delocalized radical intermediates. *JACS* 139: 16650–16656.

15 Bickelhaupt, F.M. and Houk, K.N. (2017). Analyzing reaction rates with distortion/interaction-activation strain model. *Angew. Chem. Int. Ed.* 56: 10070–10086.

16 Kutzelnigg, W. (1973). The physical mechanism, of the chemical bond. *Angew. Chem. Int. Ed. Engl.* 12: 546–562.

17 Huber, K.P. and Herzberg, G. (1979). *Molecular Spectra and Molecular Structure IV. Constants of Diatomic molecules*. New York: Springer US.

18 Boyer, M.A., Marsalek, O., Heindel, J.P. et al. (2019). Beyond Badger's rule: the origins and generality of the structure–spectra relationship of aqueous hydrogen bonds. *J. Phys. Chem. Lett.* 10: 918–924.

19 Kurtia, E., Matsuura, H., and Ohno, K. (2004). Relationship between force constants and bond lengths for CX (X = C, Si, Ge, N, P, As, O, S, Se, F, Cl and Br) single and multiple bonds: formulation of Badger's rule for universal use. *Spectrochim. Acta Part A* 60: 3013.

20 Kraka, E., Setiawan, D., and Cremer, D. (2016). Re-evaluation of the bond length-bond strength rule: the stronger bond is not always the shorter bond. *J. Comput. Chem.* 37: 130–142.

21 Sun, C. and Xue, D. (2016). Chemical bonding theory of single crystal growth and its application to crystal growth and design. *CrystEngComm* 18: 1262–1272.

22 Sergeeva, A.P., Popav, I.A., Piazza, Z.A. et al. (2014). Understanding boron through size-selected clusters: structure, chemical bonding, and fluxionality. *Acc. Chem. Res.* 47: 1349–1358.

23 Karelson, M. and Lobanov, V.S. (1996). Quantum chemical descriptors in QSAR/QSPR studies. *Chem. Rev.* 96: 1027–1043.

24 Danovich, D., Shaik, S., Neese, F. et al. (2013). Understanding the nature of the CH···HC interactions in alkanes. *J. Chem. Theory Comput.* 9: 1977.

25 Zhang, S. (2003). Fabrication of novel biomaterials through molecular self-assembly. *Nat. Biotechnol.* 21: 1171.

26 Desiraju, G.R. (2002). Hydrogen bridges in crystal engineering: interactions without borders. *Acc. Chem. Res.* 35: 565.

27 Desiraju, G.R., Vittal, J.J., and Ramanan, A. (2011). *Crystal Engineering*. World Scientific Singapore.

28 Schneider, H.-J. and Yatsmirskii, A. (1999). *Principles and Methods in Supramolecular Chemistry*. Chichester: Wiley.

29 Alkorta, I., Rozas, I., and Elguero, J. (1998). Non-conventional hydrogen bond. *Chem. Soc. Rev.* 27: 163–170.

30 Alkorta, I., Elguero, J., and Grabowski, S.J. (2008). How to determine whether intramolecular H···H interactions can be classified as dihydrogen bonds. *J. Phys. Chem. A* 112: 2721.

31 Custelcean, R. and Jackson, J.E. (2001). Dihydrogen bonding: structures, energetics, and dynamics. *Chem. Rev.* 101: 1963–1980.

32 Brown, M.P. and Heseltine, R.W. (1968). Co-ordinated BH_3 as a proton acceptor group in hydrogen bonding. *Chem. Commun.* 1551–1552.

33 Brown, M.P., Heseltine, R.W., Smith, P.A., and Walker, P.J. (1970). An infrared study of co-ordinated BH_3 and BH_2 groups as proton acceptors in hydrogen bonding. *J. Chem. Soc. A* 410–414.

34 Richardson, T.B., de Gala, S., Crabtree, R.H., and Siegbhan, P.E.M. (1995). Unconventional hydrogen bonds: intermolecular B–H···H–N Interactions. *JACS* 117: 12875–12876.

35 Grabowski, S.J. (1999). Study of correlations for dihydrogen bonds by quantum-chemical calculations. *Chem. Phys. Lett.* 312: 542–547.

36 Grabowski, S.J. (2000). High-level ab initio calculations of dihydrogen-bonded complexes. *J. Phys. Chem. A* 104: 5551–5557.

37 Allen, F.H., Kennard, O., Watson, D.G. et al. (1987). Tables of bond lengths determined by X-ray and neutron diffraction. Part 1. Bond lengths in organic compounds. *J. Chem. Soc., Perkin Trans. 2* 12: S1–S19.

38 Hopf, H. (2000). *Classics in Hydrocarbon Chemistry*. Weinheim: Wiley-VCH.

39 Halton, B. (ed.) (2000). *Advances in Strained and Interesting Organic Molecules*. Stamford, CT: JAI Press.

40 Komarov, I.V. (2001). Organic molecules with abnormal geometric parameters. *Russ. Chem. Rev.* 70: 991–1016.

41 Tsuji, T. (2005). *Modern Cyclophane Chemistry* (ed. R. Gleiter and H. Hopf), 81–104. Weinheim: Wiley-VCH.

42 Tani, K. and Stoltz, B.M. (2006). Synthesis and structural analysis of 2-quinuclidonium tetrafluoroborate. *Nature* 441: 731–734.

43 Hounshell, W.D., Dougherty, D.A., Hummel, J.P., and Mislow, K. (1977). Structure of hexaphenylethane and congeners as determined by empirical force field calculations. *JACS* 99: 1916–1924.

44 Takeda, T., Kawai, H., Herges, R. et al. (2009). Negligible diradical character for the ultralong C—C bond in 1,1,2,2-tetraarylpyracene derivatives at room temperature. *Tetrahedron Lett.* 50: 3693–3697.

45 Battersby, R., VernonClark, R., and Siegel, J.S. (1997). Does π–σ–π through-bond coupling significantly increase C—C bond lengths? *JACS* 119: 7048–7054.

46 Rahm, M., Hoffmann, R., and Ashcroft, N.W. (2016). Atomic and ionic radii of elements 1–96. *Chem. Eur. J.* 22: 14625–14632.

47 Batsanov, S.S. (2001). Van der Waals radii of elemets. *Inorg. Mater.* 37: 871–885.

48 Echeverría, J., Aullón, G., Danovich, D. et al. (2011). Dihydrogen contacts in alkanes are subtle but not faint. *Nat. Chem.* 3: 323.

49 Pascal, R.A. Jr., Zong, J., Dougherty, K.J. et al. (2014). Experimental and computational studies of molecules with close, non-bonded hydrogen–hydrogen contacts: common computational methods grossly overestimate some 'through-space' NMR scalar coupling constants. *Tetrahedron* 70: 7518–7526.

50 Zong, J., Mague, J.T., and Pascal, R.A. (2013). Exceptional steric congestion in an *in,in*-bis(hydrosilane). *JACS* 135: 13235.

51 Ermer, O., Mason, S.A., Anet, F.A.L., and Miura, S.S. (1985). Ultrashort nonbonded hydrogen...hydrogen distance in a half-cage pentacyclododecane. *JACS* 107: 2330.

52 Ermer, O. (1977). Structure of two highly strained tetracyclododecanes. *Angew. Chem. Int. Ed.* 16: 798.

53 Jørgenson, F.S. and Paddon-Row, M.N. (1983). On the presence of extremely short H…H non-bonded distances, and orbital interactions through eight σ bonds or through two methylene groups in tetramethanonaphthacenes. *Tetrahedron Lett.* 24: 5415.

54 Mandal, N., Pratik, S.M., and Datta, A. (2017). Exploring ultrashort hydrogen-hydrogen nonbonded contacts in constrained molecular cavities. *J. Phys. Chem. B* 121: 825–834.

55 Mitoraj, M.P., Babashkina, M.G., Isaev, A.Y. et al. (2018). London dispersion forces in crystal packing of thiourea derivatives. *Cryst. Growth Des.* 18 (9): 5385–5397.

56 Rosel, S., Quanz, H., Logemann, C. et al. (2017). London dispersion enables the shortest intermolecular hydrocarbon H· · ·H contact. *JACS* 139: 7428–7431.

57 Safin, D.A., Babashkina, M.G., Robeyns, K. et al. (2015). Influence of the homopolar dihydrogen bonding C–H· · ·H–C on coordination geometry: experimental and theoretical studies. *Chem. Eur. J.* 21: 16679–16687.

58 Mitoraj, M.P. and Michalak, A. (2007). Natural orbitals for chemical valence as descriptors of chemical bonding in transition metal complexes. *J. Mol. Model.* 13: 347–355.

59 Mitoraj, M.P., Michalak, A., and Ziegler, T. (2009). A combined charge and energy decomposition scheme for bond analysis. *J. Chem. Theory Comput.* 5: 962–975.

60 Lee, J.C. Jr., Peris, E., Rheingold, A.L., and Crabtree, R.H. (1994). An unusual type of H···H interaction: Ir–H···H–O and Ir–H···H–N hydrogen bonding and its involvement in s-bond metathesis. *JACS* 116: 11014–11019.

61 Milstein, D., Calabrese, C., and Williams, I.D. (1986). Formation, structures, and reactivity of cis-hydroxy-, cis-methoxy-, and cis-mercaptoiridium hydrides. Oxidative addition of water to Ir(I). *JACS* 108: 6387.

62 Lough, A.J., Park, S., Ramachandran, R., and Morris, R.H. (1994). Switching on and off a new intramolecular hydrogen-hydrogen interaction and the heterolytic splitting of dihydrogen. Crystal and molecular structure of $[Ir\{H(\eta^1\text{-}SC_5H_4NH)\}_2(PCy_3)_2]BF_4{\cdot}2.7CH_2Cl_2$. *JACS* 116: 8356–8357.

63 Yourdkhani, S., Jablonski, M., and Echeverría, J. (2017). Attractive PH···HP interactions revealed by state-of-the-art *ab initio* calculations. *Phys. Chem. Chem. Phys.* 19: 28044–28055.

64 Matta, C.F., Hernández-Trujillo, J., Tang, T.-H., and Bader, R.F.W. (2003). Hydrogen–hydrogen bonding: a stabilizing interaction in molecules and crystals. *Chem. Eur. J.* 9: 1940–1951.

65 Landeros-Rivera, B., Jancik, V., Moreno-Esparza, R. et al. (2021). Non-covalent iteractions in the biphenyl crystal: is the planar conformer a transition state? *Chem. Eur. J.* 27: 11912–11918.

66 Maué, D., Strebert, P.H., Bernhard, D. et al. (2021). Dispersion-bound isolated dimers in the gas phase: observation of the shortest intermolecular CH···H–C distance via stimulated Raman spectroscopy. *Angew. Chem. Int. Ed.* 60: 11305–11309.

67 Alvarez, S. (2013). A cartography of the van der Waals territories. *Dalton Trans.* 42: 8617.

68 Pascal, R.A. Jr. (2004). Molecular "iron maidens": ultrashort nonbonded contacts in cyclophanes and other crowded molecules. *Eur. J. Org. Chem.* 2004: 3763–3771.

69 Baldridge, K.K. and Seigel, J.S. (2015). Molecular compression chambers: exceeding the Euler modulus. *J. Phys. Org. Chem.* 28: 223–225.

70 Povie, G., Segawa, Y., Nishihara, T. et al. (2017). Synthesis of a carbon nanobelt. *Science* 356: 172–175.

71 Cheung, K.Y., Gui, S., Deng, C. et al. (2019). Synthesis of armchair and chiral carbon nanobelts. *Chem* 5: 838–847.

72 Xiao, Y., Mague, J.T., Schmehl, R.H. et al. (2019). Dodecaphenyltetracene. *Angew. Chem. Int. Ed.* 58: 2831–2833.

73 Kahr, B., Van Engen, D., and Mislow, K. (1986). Length of the ethane bond in hexaphenylethane and its derivatives. *JACS* 108: 8305–8307.

74 Fokin, A.A., Chernish, L.V., Gunchenko, P.A. et al. (2012). Stable alkanes containing very long carbon-carbon bonds. *JACS* 134: 13641–13650.

75 Li, J., Pang, R., Li, Z. et al. (2019). Exceptionally long C–C single bonds in diamino-*o*-carborane as induced by negative hyperconjugation. *Angew. Chem. Int. Ed.* 58: 1397–1401.

76 Ishigaki, Y., Shimajiri, T., Takeda, T. et al. (2018). Longest C–C single bond among neutral hydrocarbons with a bond length beyond 1.8 Å. *Chem* 4: 795–806.

77 Kuroda, Y., Kobayashi, M., and Taketsugu, T. (2019). Dispersion interaction and crystal packing realize an ultralong C–C single bond: a theoretical study on dispirobis(10-methylacridan) derivatives. *Chem. Lett.* 48: 137–139.

78 Kubo, T., Suga, Y., Hashizume, D. et al. (2021). Long carbon–carbon bonding beyond 2 Å in tris(9-fluorenylidene)methane. *JACS* 143: 14360–14366.

79 Grimme, S. and Schreiner, P.R. (2011). Steric crowding can stabilize a labile molecule: solving the hexaphenylethane riddle. *Angew. Chem. Int. Ed.* 50: 12639–12642.
80 Schreiner, P.R., Chernish, L.V., Gunchenko, P.A. et al. (2011). Overcoming lability of extremely long alkane carbon–carbon bonds through dispersion forces. *Nature* 447: 308–310.
81 Balasubramaniyan, V. (1966). Peri interaction in naphthalene derivatives. *Chem. Rev.* 66: 567–641.
82 Mandal, N. and Datta, A. (2020). Molecular designs for expanding the limits of ultralong C–C bonds and ultrashort H· · ·H non-bonded contacts. *Chem. Commun.* 56: 15377–15386.
83 Mandal, N., Pal, A.K., Gain, P. et al. (2020). Transition-state-like planar structures for amine inversion with ultralong C–C bonds in diamino-*o*-carborane and diamino-*o*-dodecahedron. *JACS* 142 (11): 5331–5337.

18

Accurate Determination of Materials Properties: Role of Electron Density

Anup Pramanik[1], *Sourav Ghoshal*[2], *Santu Biswas*[2], *Biplab Rajbanshi*[2] *and Pranab Sarkar*[2]

[1] *Department of Chemistry, Sidho-Kanho-Birsha University, Purulia, India*
[2] *Department of Chemistry, Visva-Bharati University, Santiniketan, India*

18.1 Introduction

Determination of materials properties is one of the central goals in both physics and chemistry communities. Of course, it involves the determination of both geometric structure and electronic properties as they are strongly interrelated. Sophisticated experimental techniques such as X-ray diffractometer and different electron microscope measurements can determine the accurate geometry and electronic properties. Nevertheless, it requires electronic structure calculations to have concrete overview on the energy and associated properties for better understanding the materials properties and their reactivities [1]. Electronic structure calculations provide a wealth of knowledge about materials properties relating to the electron density and distribution of electronic states along a particular electronic degrees of freedom. Although accurate determination of electronic structure is really a difficult task, especially for materials consisting of a large number of atoms, recent advanced level theories have succeeded to do that. Density functional theory (DFT) is one of such theories, which can provide electronic structure properties of materials, even in the experimental regime, without much loosing the accuracy within reasonable computational cost. The central theme of DFT is to introduce electron density (ρ) as the basis for electronic structure calculation. Although we need to solve the eigenvalue equation for having energy and wavefunction, the use of the electronic density as variable is advantages in many aspects. First, it depends only on three spatial coordinates. Second, distribution of electronic density in polyatomic systems is somehow intuitive; third and most importantly, it can be experimentally measured in X-ray diffraction experiments. That is why DFT has become an essential tool for describing the materials properties.

Knowing electron density is of fundamental importance for explaining chemical reactions too. Studying chemical reactions have always been the central task of chemical community; homogeneous and heterogeneous catalysis play significant role there. Although both experimental and theoretical chemists contribute to this interdisciplinary area, computational chemistry, in particular, has a significant role for designing materials for catalysis as well as providing insights into the mechanistic pathways, which in turn facilitate for developing newer and advanced catalytic systems. Here also DFT plays a crucial role by providing accurate materials properties, which are the key factors for controlling a chemical reaction. Beyond molecules, nanomaterials are always in the front-lines for catalytic reactions because of their tunable optoelectronic properties associated with their sizes, shapes, compositions, surface functionalizations, etc. Computing such properties of nanomaterials, which generally comprise quite a large number of atoms, are now

Electron Density: Concepts, Computation and DFT Applications, First Edition.
Edited by Pratim Kumar Chattaraj and Debdutta Chakraborty.

being possible by the advent of sophisticated DFT techniques like tight-binding-based density functional theory (DFTB). Such a theory can provide reasonable structure–property relationship for larger nanoclusters, even in the experimental regime.

Now, the difficulty arises while obtaining the minimum energy structure of an unknown nanocluster. Searching for the global minimum structure for nanocluster is in fact very difficult job because of the involvement of a large number of degrees of freedom in nanocluster geometry, which in fact can provide thousands of local minima. However, people have developed a number of ways to achieve those structures by using basin hopping and genetic algorithm (GA) techniques [2, 3]. Our group has developed GA-based density-functional tight-binding method for searching the lowest energy structures of nanoclusters of different shapes and sizes [4]. Besides, there are numerous approaches based on DFT, which can also predict the global minimum structures [5–7]. In general, DFT is a good compromise between accuracy and computational cost while studying the geometry and electronic structure of nanoclusters for catalytic activities. In this chapter, we are going to provide a revenue on the advancement of DFT for determining accurate materials properties and their applicability and impact on the thrust areas of modern science like sustainable materials, energy storage, conversion, and catalysis.

18.2 Materials Properties: Structure and Electronic Properties

18.2.1 Classification of Materials

The term 'materials property' signifies a collection of intensive properties of a material starting from physical nature to mechanical, thermal, magnetic, electrical, optical properties, and so on. However, depending upon the physical nature of the material i.e. whether it is amorphous or crystalline, its properties vary extensively. Again for crystalline bulk material, the properties abruptly vary with the dimensionality of the material. The main reason of such variation is the electronic confinement. As we will be mainly confined on the electronic properties and associated phenomena, our focus would be on the materials properties arising from that dimensionality. Depending upon the dimensions of the materials, they can be broadly categorized into – three dimensional (3D) or bulk nanomaterials and materials of lower dimensions while the later one may be sub-categorized into 2D materials or quantum wells, 1D materials or quantum wires and 0D materials or quantum dots (QDs). For lower dimensional materials, electronic motion is confined in at least one of the dimensions. In particular, the length of such dimension typically ranges in between 1 and 100 nm and thus these materials are called nanomaterials. Because of the electronic confinement the electrical, chemical, optical, and other physical properties of such materials can be vastly different from properties of same material at extremely low (atomic) or higher (bulk) dimensions.

0D nanomaterials have electronic confinement in all the three directions. Metallic nanoparticles, including gold and silver QDs and semiconductor QDs are the perfect examples of this kind of nanomaterials. Most of these materials are spherical in size and the diameter of these particles are in the range of 1–50 nm. Carbon-based nanomaterials, fullerenes also belong to this group. Other than sphere, cubes and polygons shapes are also found for this kind of nanomaterials. In 1D materials, electronic movements are restricted in any two dimensions and thus only one dimension is unrestricted. These include nanowires, nanorods, and nanotubes. These materials are long (several micrometer in length), but with diameter of only a few nanometer. Nanowire and nanotubes of metals, oxides and other materials, carbon nanotubes (CNTs), etc., are few examples of this kind of materials. In 2D materials, two dimensions are unrestricted for electronic movements. These include different kinds of nanofilms such as coatings and thin-film-multilayers, nano sheets or nano-walls. The area of the nano films can be large (several square micrometer) but the thickness is always in nano scale range. Graphene is the most famous material in this category; besides, 2D transition-metal dichalcogenides, carbides, etc., are also currently known. None of the three dimensions for 3D materials are electronically confined. These include bulk materials composed of metallic block to the most famous recent materials in this category, metal–organic or covalent–organic frameworks.

Apart from these, there are several materials composed of more than one confined materials, called nanocomposites. These are materials that incorporate nanosized particles into a matrix of standard material or hybrid materials consisting different nanoscale objects. The later category, in general, includes inorganic (different QDs, nanowires/nanorods, sheets, etc.), organic (small or polymeric organic molecules, fullerenes and their derivatives, CNTs, etc.), organic–organic or inorganic–inorganic (especially QD assemblies) composites, at least one of which is in the nano scale diameter. The resultant materials provide a drastic improvement in properties that includes mechanical strength, toughness, and electrical or thermal conductivity. In mechanical terms, nanocomposites differ from conventional composite materials due to the exceptionally high surface to volume ratio of the reinforcing phase and/or its exceptionally high aspect ratio. From the electronic energy term, these materials show extraordinary behaviour due to incorporation of new energy band in between the precise energy gap of the constituting materials. Thus, nanocomposites are promising materials for solar energy harvesting processes [8–11].

It is quire obvious that the properties of nanomaterials are governed by the quantum mechanical rules. The transition from bulk to nanometer-sized dimension leads to discretization in energy bands of these materials, known as quantum confinement effect. Naturally, this effect is the most prominent for the materials where all the dimensions are confined, i.e. 0D QDs or nanoclusters. Nanoclusters show dramatically different electrical, optical, and chemical properties from those of the respective bulk materials. Different elements in a nanocluster form different bonds, leading to the formation of different structures. These bonds and structures contribute to their unique physical and chemical properties. The ability of a nanocluster to react with any species should depend on cluster size. The cluster size also controls the stability and magnetic behaviours. In some instances entirely new behaviours are observed for the clusters, which are not seen in the corresponding bulk materials. However, depending upon the morphology and composition, nanoclusters are classified into different groups each of which has different characteristics.

Metal nanoclusters include s-block metals, e.g. alkali and alkaline earth metals. Here, the bonding is metallic, involving mainly the valence s-orbitals; sp-metals, e.g. aluminum nanocluster for which the bonding involves some covalent character and transition metal nanoclusters, where greater degree of covalency and directionality are observed in the bonding. Semiconductor nanoclusters are composed of elements like C, Si, and Ge, which are also known as bulk semiconductors. However, these elements form semiconductor nanoclusters comprising single element or heteroatomic elements, e.g. Ga_xAs_y. The bonding in between the elements here is mainly polar covalent bond. Besides, semiconductor nanoclusters include different metal or non-metal chalcogenides like II–VI and IV–VI semiconductors. Semiconductor metal oxides are of special interest in this series, the most common examples being TiO_2, Al_2O_3, SnO_2, ZnO, etc. Heteroatomic nanoclusters composed of elements with large difference in electronegativity form ionic nanoclusters. Here, bonding is predominately ionic (electrostatic); e.g. alkali metal halides.

Indeed, in many ways, these nanoclusters represent a unique state-of-matter of their own. The structures of small metal clusters remain, however, largely a mystery and very little is known of the complex relationship between geometrical structure and reactivity. In any way, owing to their reactivity and unusual properties, nanoclusters are of tremendous technological interest in numerous areas of applied science like optoelectronics, energy conversion, and catalysis.

18.2.2 Electronic Properties of Materials

Now, the question is how one can obtain such materials properties theoretically. The answer is based on the electronic structure calculations of the materials. The simplest model of such calculations can be thought by using free electron gas theory within restricted boundary conditions. One should calculate the ground state properties of n-electron system confined to a volume V, which is the volume of unit cell. According to Sommerfeld theory a single electron can be described by a wave function $\psi(r)$, which follow the time independent Schrödinger equation

$$\hat{H}\psi(r) = -\frac{\hbar^2}{2m}\left(\frac{\partial^2}{\partial x^2} + \frac{\partial^2}{\partial y^2} + \frac{\partial^2}{\partial z^2}\right)\psi(r) = -\frac{\hbar^2}{2m}\nabla^2\psi(r) = \varepsilon\psi(r) \tag{18.1}$$

where electron is considered to be confined to the volume V. For cubical system, $V = L^3$, L being the length of the box and applying boundary condition $\psi(x, y, z + L) = \psi(x, y, z)$, we get the general solution

$$\psi_k(r) = \frac{1}{\sqrt{V}}e^{ik\cdot r} \tag{18.2}$$

where k is any position independent vector. The associated energy is expressed as

$$\varepsilon(k) = \frac{\hbar^2 k^2}{2m} \tag{18.3}$$

However, for interacting electrons the situation becomes much more complicated. Furthermore, as in a perfect crystal the atoms are arranged in a regular periodic array, the electrons must feel periodic potential with a periodicity of underlying Bravais lattice, i.e. $U(r + R) = U(R)$ for all Bravais lattice vectors R. Note that the many electron model for crystalline solid can be simplified to one electron model by considering $U(r)$ as effective one electron potential. Thus, the Schrödinger equation for independent electron (Bloch electron) could be written as

$$\hat{H}\psi(r) = \left(-\frac{\hbar^2}{2m}\nabla^2 + U(r)\right)\psi(r) = \varepsilon\psi(r) \tag{18.4}$$

Now, Bloch's theorem states that the eigenstates $\psi(r)$ for the one electron Hamiltonian can be written as

$$\psi_{nk}(r) = e^{ik\cdot r}u_{nk}(r) \tag{18.5}$$

where $u_{nk}(r + R) = u_{nk}(r)$ for all R in the Bravais lattice. Note that, here k is the wave vector and n is index of the solutions to the Schrödinger equation for a given value of k. The above equation can be simplified to

$$\psi_{nk}(r + R) = e^{ik\cdot R}\psi_{nk}(r) \tag{18.6}$$

Extending this idea, one can get a full range of solutions for eigenstates and eigenvalues as a periodic function of k, which leads to a description of the energy levels of an electron in a periodic potential as a continuous function, called band structure of the solid. For larger system, handling such equations with so many variables is extremely difficult. However, DFT is successfully implemented there for obtaining accurate band structure and eigenstates. In the subsequent section, we will show how the concept of DFT makes easier for solving those equations starting from many electron atom to molecule to crystalline solid materials.

18.3 Molecules to Materials, Essential Role of Electron Density

Atoms, molecules, crystals, and all other forms of the matter in material science comprise of interacting nuclei and electrons. For calculating electronic properties thus, one needs to handle with interacting particle model for electrons. The basis idea starts from solving time-independent Schrödinger equation for many body system, which gives energies and wave functions, called stationary states. The equation can be represented as

$$\hat{H}\Psi(\{r_i\},\{R_I\}) = E\Psi(\{r_i\},\{R_I\}) \tag{18.7}$$

where $\Psi(\{r_i\},\{R_I\})$ is the wave function of the system, which is function of nuclear coordinates, R_I and electronic coordinates, r_i. $\hat{H}$ and E are the Hamiltonian and energy of the system, respectively. The Hamiltonian, $\hat{H}$ of a system having N_e number of electrons and N_I number of nuclei consist of five operators contributing five different energies and represented as (in a.u.)

$$\hat{H} = -\frac{1}{2}\sum_{i=1}^{N_e}\nabla_i^2 - \frac{1}{2}\sum_{I=1}^{N_I}\frac{\nabla_I^2}{M_I} - \sum_{i=1}^{N_e}\sum_{I=1}^{N_I}\frac{Z_I}{r_{iI}} + \sum_{i=1}^{N_e}\sum_{j>i}^{N_e}\frac{1}{r_{ij}} + \sum_{I=1}^{N_I}\sum_{J>I}^{N_J}\frac{Z_I Z_J}{R_{IJ}} \tag{18.8}$$

where Z_I denotes the atomic number, M_I is the mass of the nucleus I and r_i and R_I are the positions of the electron i and nucleus I, respectively. The first and second terms of Eq. (18.8) are the kinetic energy operators for electrons and nuclei, respectively. The third term denotes the attractive potential between nuclei and electrons. The fourth and fifth terms describe the repulsive interactions between electrons and nuclei, respectively. Once the Hamiltonian, $\hat{H}$ is defined for a system, everything should be known, in principle, by solving the second order differential equation. However, in practice, due to the presence of cross terms like r_{iI}, r_{ij} and R_{IJ} in the $\hat{H}$ makes it impossible to solve the Schrödinger equation. So, approximations are needed in order to solve the equation; the very first one is the Börn–Oppenheimer (BO) approximation, which states that because the nuclei are much heavier than electrons, the electrons move very fast as compared to nuclei. Thus, the nuclear coordinates remain unchanged in the time frame of electronic movement and hence the kinetic energy terms for the nuclei can be omitted from Eq. (18.8). So, the Hamiltonian can now be written as

$$\hat{H} = -\frac{1}{2}\sum_{i=1}^{N_e}\nabla_i^2 - \sum_{i=1}^{N_e}\sum_{I=1}^{N_I}\frac{Z_I}{r_{iI}} + \sum_{i=1}^{N_e}\sum_{j>i}^{N_e}\frac{1}{r_{ij}} + \sum_{I=1}^{N_I}\sum_{J>I}^{N_J}\frac{Z_I Z_J}{R_{IJ}} \tag{18.9}$$

The first three terms of Eq. (18.9) are associated with the electronic motions and their interactions and thus, as a whole they are referred to electronic Hamiltonian ($\hat{H}_{el}$):

$$\hat{H}_{el} = -\frac{1}{2}\sum_{i=1}^{N_e}\nabla_i^2 - \sum_{i=1}^{N_e}\sum_{I=1}^{N_I}\frac{Z_I}{r_{iI}} + \sum_{i=1}^{N_e}\sum_{j>i}^{N_e}\frac{1}{r_{ij}} \tag{18.10}$$

while the last term of Eq. (18.9) is for nuclei–nuclei repulsion and is fixed for a particular nuclear configuration, which can be written as

$$V_{NN} = \sum_{I=1}^{N_I}\sum_{J>I}^{N_J}\frac{Z_I Z_J}{R_{IJ}} \tag{18.11}$$

Noteworthy, that constant term just changes the energy eigenvalue additively. Thus, the electronic motion can be effectively decoupled from nuclear motion and the electronic Schrödinger equation can be written as

$$\hat{H}_{el}\psi_{el}(\{r_i\},R_I) = E_{el}\psi_{el}(\{r_i\},R_I) \tag{18.12}$$

The total energy of the system now can be expressed as

$$E_{\text{tot}} = E_{\text{el}} + V_{\text{NN}} \tag{18.13}$$

There are an infinite number of nuclear configurations and for each of these we have to solve the electronic Schrödinger equation to get electronic wavefunctions and eigenvalues. Thus, within BO approximation, the electronic wavefunctions and energies depend parametrically on nuclear coordinate.

In spite of this great simplification offered by BO approximation, solving electronic Schrödinger equation is still very difficult task. It is due to the intrinsic spin properties of the electrons. In fact, electronic degrees of freedom consists of four coordinates, three spatial coordinates and one spin coordinate. According to Pauli exclusion principle, no two electrons in a same system can have identical four coordinates. The motion of one electron is always affected by other electrons of the system, resulting 'correlation energy'. Across the years, several methods have been developed based on Hartree–Fock (HF) Theory [12, 13], where all-electron wavefunction is approximated by determinantal wave function (Slater determinant) that contains the single-electronic wavefunction, the energy of which is then minimized variationally. However, the Hartree–Fock method does not take into account the correlation between the electrons. A number of post-HF treatments have also been developed for accounting proper correlation energy such as configuration-interaction (CI) [14] and coupled-cluster (CC) [15]. Although such methods are accurate enough, they need to calculate a large number of integrals and hence demand very high computational cost. Practically, they become inoperative for the systems larger than few atoms. As a consequence, from the very beginning, there have been attempts for studying the quantum many-body problem in terms of a simpler and physically meaningful variable such as the electron density (ρ). This attempt had been successful with the development of DFT by Walter Kohn and his group.

18.3.1 The Density Functional Theory (DFT)

The main feature of the DFT is to introduce electron density (ρ) as a central variable. The use of electronic density, instead of electronic wavefunction, as variational parameter has three main advantages: (i) it depends only on three spatial coordinates, (ii) distribution of electronic density in polyatomic systems is somehow intuitive and (iii) it can be experimentally measured in X-ray diffraction experiments. The basic ideas of DFT are pillared on the works of Hohenberg, Kohn, and Sham [16, 17] and are referred as the Hohenberg–Kohn theorems. In next section, we provide the essential details behind DFT.

18.3.2 The Hohenberg–Kohn Theorems

Hohenberg and Kohn proved that for a system having a nondegenerate ground state, the ground-state energy, wavefunction and other properties can be uniquely determined by the ground state electron probability density, ρ_0, which is a function of only three variables x, y and z [16]. As density is function of coordinates (x, y, z) and energy is a function of density, the ground-state electronic energy is represented in a functional of ρ_0 as $E_0 = E_0[\rho_0]$.

The electronic Hamiltonian ($\hat{H}_{\text{el}}$) is the sum of kinetic energy operator ($\hat{T}$), electron–electron repulsive potential ($\hat{V}_{\text{ee}}$) and electron–nuclear attractive potential ($\hat{V}_{\text{Ne}}$) operators. Thus, average electronic energy can be represented as the sum of averages of three contributing energy terms i.e. $E = \overline{T} + \overline{V}_{\text{ee}} + \overline{V}_{\text{Ne}}$. Each of these average values is calculated in terms of ground-state electronic density, ρ_0. Therefore, for ground state

$$E_0 = \overline{T}[\rho_0] + \overline{V}_{ee}[\rho_0] + \overline{V}_{Ne}[\rho_0] \tag{18.14}$$

In Eq. (18.10) the electron–nuclei attractive potential $\overline{V}_{Ne} = \sum_{i=1}^{N_e} \upsilon(\mathbf{r}_i)$, where $\upsilon(\mathbf{r}_i) = -\sum_{I=1}^{N_I} \frac{Z_I}{\mathbf{r}_{iI}}$. The electron–nuclei attraction energy can now be represented as

$$\overline{V}_{Ne} = \int \rho_0(\mathbf{r})\upsilon(\mathbf{r})d\mathbf{r} \tag{18.15}$$

where $\upsilon(\mathbf{r})$ is nuclear attraction potential energy for an electron located at point r and it is called external potential. Thus, we have

$$E_0 = E_\upsilon[\rho] = \int \rho_0(\mathbf{r})\upsilon(\mathbf{r})dr + \overline{T}[\rho_0] + \overline{V}_{ee}[\rho_0] = \int \rho_0(\mathbf{r})\upsilon(\mathbf{r})d\mathbf{r} + F[\rho_0] \tag{18.16}$$

where $F[\rho_0] \equiv \overline{T}[\rho_0] + \overline{V}_{ee}[\rho_0]$.

18.3.3 The Hohenberg–Kohn Variational Theorems

Hohenberg and Kohn proved that for any trial density $\rho_{tr}(\mathbf{r})$, $\int \rho_{tr}(\mathbf{r})d\mathbf{r} = n$ (n is the number of total electrons) and $\rho_{tr}(\mathbf{r}) \geq 0$ for all $\mathbf{r}$ and the following condition is always satisfied-

$$E_0 \leq E[\rho_{tr}] \tag{18.17}$$

When the trial density is equal to the true ground state density, then the equality holds

$$E_0 = E[\rho_0] \tag{18.18}$$

Thus, the true ground-state electron density (ρ_0) minimizes the energy functional $E[\rho_{tr}]$. According to the Hohenberg–Kohn theorem, it is possible to calculate all the ground-state properties from ground-state electron density without knowing the wavefunction, but it does not tell how to find density without knowing wavefunction, nor does it tell how to calculate energy from density as the functional $F[\rho_0]$ is not known. Kohn and Sham proposed a practical method for finding density and then finding energy from density [17].

18.3.4 The Kohn–Sham (KS) Method

Kohn and Sham considered a fictitious reference system consisting of n noninteracting electrons and each electron experiences same external potential $\upsilon_s(\mathbf{r}_i)$, which gives rise to electron probability density, $\rho_s(\mathbf{r})$. Thus, for non-interacting electronic model, the Hamiltonian of the reference system becomes

$$\hat{H}_s = \sum_{i=1}^{n} \left[-\frac{1}{2}\nabla_i^2 + \upsilon_s(\mathbf{r}_i)\right] \equiv \sum_{i=1}^{n} \hat{h}_i^{KS}, \qquad \hat{h}_i^{KS} \equiv -\frac{1}{2}\nabla_i^2 + \upsilon_s(\mathbf{r}_i) \tag{18.19}$$

where $\hat{h}_i^{KS}$ is the one electron Kohn–Sham Hamiltonian.

However, in reality there is interaction between electrons, which affect their kinetic energy as well as potential energy. Thus, Kohn–Sham defined the difference between average ground-state electronic kinetic energy of the real (interacting) system and the reference system of noninteracting electrons ($\Delta\overline{T}[\rho]$), which is represented as

$$\Delta\overline{T}[\rho] \equiv \overline{T}[\rho] - \overline{T}_s[\rho] \tag{18.20}$$

Similarly, for potential energy

$$\Delta\overline{V}_{ee}[\rho] \equiv \overline{V}_{ee}[\rho] - \frac{1}{2}\iint \frac{\rho(\mathbf{r}_1)\rho(\mathbf{r}_2)}{r_{12}} d\mathbf{r}_1 d\mathbf{r}_2 \tag{18.21}$$

where r_{12} is the distance between points (x_1, y_1, z_1) and (x_2, y_2, z_2). The term $\frac{1}{2}\iint \frac{\rho(\mathbf{r}_1)\rho(\mathbf{r}_2)}{r_{12}} d\mathbf{r}_1 d\mathbf{r}_2$ is the classical expression for the electrostatic interelectronic repulsion energy between the electrons, which are smeared out into a continuous distribution of charge with electron density ρ.

Now, using Eqs. (18.20) and (18.21), Eq. (18.16) becomes

$$E_v[\rho] = \int \rho(\mathbf{r})v(\mathbf{r})d\mathbf{r} + \overline{T}_s[\rho] + \frac{1}{2}\iint \frac{\rho(\mathbf{r}_1)\rho(\mathbf{r}_2)}{r_{12}} d\mathbf{r}_1 d\mathbf{r}_2 + \Delta\overline{T}[\rho] + \Delta\overline{V}_{ee}[\rho] \tag{18.22}$$

The sum of the unknown functional $\Delta\overline{T}[\rho]$ and $\Delta\overline{V}_{ee}[\rho]$ is defined as an another functional, exchange–correlation energy functional ($E_{xc}[\rho]$), which is represented as

$$E_{xc}[\rho] \equiv \Delta\overline{T}[\rho] + \Delta\overline{V}_{ee}[\rho] \tag{18.23}$$

Finally, the electronic energy of a system can be written as

$$E_0 = E_v[\rho] = \int \rho(\mathbf{r})v(\mathbf{r})d\mathbf{r} + \overline{T}_s[\rho] + \frac{1}{2}\iint \frac{\rho(\mathbf{r}_1)\rho(\mathbf{r}_2)}{r_{12}} d\mathbf{r}_1 d\mathbf{r}_2 + E_{xc}[\rho] \tag{18.24}$$

The first three terms in the energy expression are easy to calculate from density and include major contribution to the ground-state energy. The remaining term has small contribution to the electronic energy, but it is difficult to evaluate accurately. Thus, the accuracy of Kohn–Sham DFT method is based on good approximation of exchange–correlation functional.

In order to calculate the energy, we have to find out the ground-state electron density first. The ground-state wavefunction $\psi_{s,0}$ of the reference system is the Slater determinant of the lowest energy Kohn–Sham orbital u_i^{KS} of the reference system, where the spatial part $\phi_i^{KS}(r_i)$ of each orbital is an eigenfunction of the one-electron operator $\hat{h}_i^{KS}$:

$$\psi_{s,0} = |u_1^{KS} u_2^{KS} \ldots u_n^{KS}|, \quad u_i^{KS} = \phi_i^{KS}(\mathbf{r}_i)\sigma_i \tag{18.25}$$

$$\hat{h}_i^{KS}\phi_i^{KS} = \varepsilon_i^{KS}\phi_i^{KS} \tag{18.26}$$

where σ_i is a spin function and the ε_i^{KS} are Kohn–Sham orbital energies. It can be readily proved that the electron probability density of n-particle system is given by

$$\rho = \rho_s = \sum_{i=1}^{n} |\phi_i^{KS}|^2 \tag{18.27}$$

Thus, it is possible to calculate energy from density if we can find Kohn–Sham orbitals. Now, we can show that the orthonormal Kohn–Sham orbitals, which minimize the electronic energy for the ground-state energy, satisfy the following relation:

$$\left[-\frac{1}{2}\nabla_1^2 - \sum_I \frac{Z_I}{r_{1I}} + \int \frac{\rho(\mathbf{r}_2)}{r_{12}} d\mathbf{r}_2 + v_{xc}(1)\right] \phi_i^{KS}(1) = \varepsilon_i^{KS}\phi_i^{KS}(1) \tag{18.28}$$

where v_{xc} is the exchange–correlation potential, which can be obtained as

$$v_{xc}(\mathbf{r}) \equiv \frac{\delta E_{xc}[\rho(\mathbf{r})]}{\delta\rho(\mathbf{r})} \tag{18.29}$$

If $E_{xc}[\rho]$ is known, v_{xc} can be found from its functional derivative, which in turn provides ϕ_i^{KS} and thus, the ground-state density is obtained. However, the crucial step is to find ρ and E_0 because no one knows what is the exact form of exchange–correlation functional $E_{xc}[\rho]$. The accuracy of DFT calculation depends on the choice of exchange–correlation functional. In the subsequent section, we provide a short discussion on different types of exchange–correlation functionals which are commonly used in DFT.

18.3.5 Local Density Approximation

The simplest and most commonly used approximation to treat the exchange–correlation functional within the DFT is the local-density approximation (LDA). In this approximation, the exchange–correlation energy at every point of space is considered as that for homogeneous electron gas. The exchange–correlation energy in the LDA is expressed by

$$E_{xc}[\rho] \approx E_{xc}^{\text{LDA}}[\rho] = \int \rho(\mathbf{r})\varepsilon_{xc}^{\text{hom}}(\rho(\mathbf{r}))d\mathbf{r} \tag{18.30}$$

where $\varepsilon_{xc}^{\text{hom}}(\rho(\mathbf{r}))$ is the exchange–correlation energy per particle in a homogenous electron gas of density $\rho(\mathbf{r})$. The exchange–correlation potential is given by

$$v_{xc}^{\text{LDA}}(\mathbf{r}) = \frac{\delta E_{xc}^{\text{LDA}}[\rho]}{\delta\rho(\mathbf{r})} = \varepsilon_{xc}^{\text{hom}}(\rho(\mathbf{r})) + \rho(\mathbf{r})\frac{\partial\varepsilon_{xc}^{\text{hom}}}{\partial\rho(\mathbf{r})} \tag{18.31}$$

Thus, the Kohn–Sham equation within the LDA appears-

$$\left[-\frac{1}{2}\nabla^2 + v(\mathbf{r}) + \int \frac{\rho(\mathbf{r}')}{|r - r'|}d\mathbf{r}' + v_{xc}^{\text{LDA}}(\mathbf{r})\right]\phi_i^{\text{KS}} = \varepsilon_i^{\text{KS}}\phi_i^{\text{KS}} \tag{18.32}$$

The function $\epsilon_{xc}^{\text{hom}}(\rho(\mathbf{r}))$ can be divided into exchange and correlation parts:

$$\varepsilon_{xc}^{\text{hom}}(\rho(\mathbf{r})) = \varepsilon_{x}^{\text{hom}}(\rho(\mathbf{r})) + \varepsilon_{c}^{\text{hom}}(\rho(\mathbf{r})) \tag{18.33}$$

The exchange part is given by the Thomas–Fermi–Dirac [18] exchange energy functional:

$$\varepsilon_{x}^{\text{hom}}(\rho(\mathbf{r})) = -\frac{3}{4}\left(\frac{3}{\pi}\right)^{1/3}\rho(\mathbf{r})^{1/3} \tag{18.34}$$

On the other hand, the correlation part can be derived from Quantum Monte Carlo (QMC) simulation as performed by Ceperley and Alder [19]. The obtained values have been interpolated to provide an analytic form for $\varepsilon_{c}^{\text{hom}}(\rho(\mathbf{r}))$. Extensive works have been done by Perdew and Zunger [20]; according to them, for non-spin systems, the parameterization can be given as

$$\begin{aligned}\varepsilon_{c}^{\text{hom}} &= -0.1423(1 + 1.0529\sqrt{r_s} + 0.3334r_s)^{-1} \quad \text{if} \quad r_s \geq 1\\ &= -0.0480 + 0.0311\ln r_s - 0.0116r_s + 0.002r_s\ln r_s \quad \text{if} \quad 0 \leq r_s \ll 1\end{aligned} \tag{18.35}$$

where $r_s = (4\pi\rho(r)/3)^{-1/3}$ is the so-called Wigner–Seitz radius.

Although LDA works good for bulk solids, it becomes poor where the homogeneity of electron gas is not maintained properly. As for example, for semiconductors, LDA often underestimates the band gap.

18.3.6 Generalized Gradient Approximation

To have a better description for E_{xc}, we must go beyond homogenous electron gas model. Apart from doped or defected bulk metals any molecular system is generally very different from homogeneity of electron density. In fact, any real system is spatially inhomogeneous. To account the inhomogeneity of the electron density, the xc energy is expressed in terms of the density and gradient of the density, and hence the name, generalized gradient approximations (GGA). This leads to a large improvement over LDA, resulting with high degree of accuracy. In this approach, the xc functional is represented by electron density and its first derivative (Laplacian) as

$$E_{\mathrm{xc}}^{\mathrm{GGA}}[\rho(\mathbf{r})] = \int F_{\mathrm{xc}}[\rho(\mathbf{r}), \nabla\rho(\mathbf{r})]d\mathbf{r} \tag{18.36}$$

where F_{xc} is some function of the electron density and its gradient. Over the years, several forms of F_{xc} have been proposed within GGA and the first-rate examples are the Becke exchange [21] and Perdew correlation [22] (BP), Becke exchange and Lee–Yang–Parr correlation (BLYP) [23], Perdew and Wang (PW91) [24] and the Perdew–Burke–Ernzerhof (PBE) [24] functionals, etc. Incorporation of GGA instead of LDA has several advantages such as it improves atomization and surface energies, improves energies and geometries of H-bonded systems, etc. However, it has still some limitations; what was correct in LDA is worsened in GGA, the exchange potential still decays exponentially into vacuum regions, and it still gives incorrect dissociation limit.

18.3.7 Meta-GGA and Hybrid Functionals

Potentially more accurate than the GGA functionals are the meta-GGA functionals, a natural development after the GGA. Meta-GGA DFT functional in its original form includes the second derivative of the electron density whereas GGA includes only the density and its first derivative in the xc potential.

$$E_{\mathrm{xc}}^{\text{meta-GGA}}[\rho] = \int \rho(\mathbf{r})\varepsilon_{\mathrm{xc}}[\rho(\mathbf{r}), \nabla\rho(\mathbf{r}), \tau]d\mathbf{r} \tag{18.37}$$

where

$$\tau(\mathbf{r}) = \frac{1}{2}\sum_{i}^{\text{occupied}} |\nabla\varphi_i(\mathbf{r})|^2 \tag{18.38}$$

Some examples include the exchange functionals B98 [25], TPSS [26, 27], VSXC [28] and the correlation functional KCIS [29–31]. For improving the xc functionals, another approximation was developed by mixing a certain amount of exact HF exchange into the functionals. Based on this strategy, the developed xc functionals are commonly known as hybrid functionals (HF-DFT).

$$E_{\mathrm{xc}}^{\mathrm{hyb}} = E_{\mathrm{x}}^{\mathrm{HF}} + E_{\mathrm{xc}}^{\mathrm{DFT}} \tag{18.39}$$

Herein, we mention a very popular Becke's three parameter Lee–Yang–Parr (LYP) functional (B3LYP)

$$E_{\mathrm{xc}}^{\mathrm{B3LYP}} = E_{\mathrm{x}}^{\mathrm{LDA}} + a_0(E_{\mathrm{x}}^{\mathrm{HF}} - E_{\mathrm{x}}^{\mathrm{LDA}}) + a_{\mathrm{x}}(E_{\mathrm{x}}^{\mathrm{GGA}} - E_{\mathrm{x}}^{\mathrm{LDA}}) + E_{\mathrm{c}}^{\mathrm{LDA}} + a_{\mathrm{c}}(E_{\mathrm{c}}^{\mathrm{GGA}} - E_{\mathrm{c}}^{\mathrm{LDA}}) \tag{18.40}$$

In a similar manner, different other functionals can also be developed by simply varying the functional components and their weightage. Since its foundation, B3LYP functional has experienced an unprecedented success for predicting ground-state geometry and energy for various types of systems. However, the poor consideration of dispersion interactions makes B3LYP relatively unreliable for composite systems. It also underestimates the energy barriers for multicomponent reactions [32]. To overcome these problems, new functionals of the type 'MXX' have recently been developed by Truhlar and coworkers [33, 34], which give satisfactory results in many cases. The first series of these functionals are the M05 [33] and the M05-2X [34]. The other series of functionals are M06, M06-2X [35], M08 [36], and so on. The recent developments in the M06 series are M06-L [37] (without exact exchange) and M06-HF [38] which utilizes 100% of exact exchange energy.

18.4 Further Approximations in DFT

18.4.1 The Density Functional Tight-Binding Theory

The DFTB method is an approximation to Kohn–Sham DFT, which reduces the Kohn–Sham equations to a form of tight binding related to the Harris functional [39]. Although the original approximation limits interactions to a non-self-consistent two centre Hamiltonian between confined atomic states, latter on a second-order expansion of the Kohn–Sham energy has enabled a charge self-consistent treatment where residual charges (Mulliken) over the atoms are solved self-consistently. Further developments have been continued to the third order in charge fluctuations as well as with respect to spin fluctuations. Although it is an approximate method but it avoids any empirical parameterization by calculating the Hamiltonian and overlap matrices out of DFT-derived local orbitals. The method includes *ab initio* concepts in relating the Kohn–Sham orbitals but the Hamiltonian matrix elements are restricted to a two-centre representation.

The total energy of a N_i electron system in the field of N_I nuclei may be written as a functional of a charge density $\rho(\mathbf{r})$ within DFT,

$$E = \sum_{i}^{N_{\text{occ}}} \langle \phi_i \mid -\frac{\nabla^2}{2} + V_{\text{ext}} + \frac{1}{2} \int \frac{\rho(\mathbf{r}')}{\mid r - r' \mid} d\mathbf{r}' \mid \phi_i \rangle + E_{\text{xc}}[\rho(\mathbf{r})] + \frac{1}{2} \sum_{I,J}^{N} \frac{Z_I Z_J}{\mid r_I - r_J \mid} \tag{18.41}$$

where the first sum is over occupied Kohn–Sham orbitals $\{\phi_i\}$, second term is the exchange–correlation energy and the last term is the ion–ion interaction, E_{IJ}. Following the ideas of Foulkes and Haydock [40], one can rewrite the total energy in order to transform the leading matrix elements. First, we substitute the charge density in Eq. (18.41) by a superposition of a reference or input density $\rho_0' = \rho_0(\mathbf{r}')$ and a small fluctuation $\delta\rho' = \delta\rho(\mathbf{r}')$, $\int d\mathbf{r}'$ is expressed by $\int'$:

$$\begin{aligned} E = & \sum_{i}^{N_{\text{occ}}} \langle \phi_i \mid -\frac{\nabla^2}{2} + V_{\text{ext}} + \int' \frac{\rho_0'}{\mid r - r' \mid} + v_{\text{xc}}[\rho_0] \mid \phi_i \rangle \\ & - \frac{1}{2} \int\int' \frac{\rho_0'(\rho_0 + \delta\rho)}{\mid r - r' \mid} - \int v_{\text{xc}}[\rho_0](\rho_0 + \delta\rho) \\ & + \frac{1}{2} \int\int' \frac{\delta\rho'(\rho_0 + \delta\rho)}{\mid r - r' \mid} + E_{\text{xc}}[\rho_0 + \delta\rho] + E_{IJ} \end{aligned} \tag{18.42}$$

The second term in this equation is the double counting of the new Hartree, the third term is the new exchange correlational contribution and the fourth term comes from dividing the full Hartree energy in Eq. (18.41) into a part related to ρ_0 and $\delta\rho$.

The exchange–correlation energy (E_{xc}) can now be expanded up to second order with respect density fluctuations by the advent of Lagrangian transformation

$$E_{\text{xc}}[\rho] = E_{\text{xc}}[\rho_0] + \delta\rho \left(\frac{dE_{\text{xc}}}{d\rho} \right)_{\rho_0} + \frac{1}{2} (\delta\rho)^2 \left(\frac{d^2 E_{\text{xc}}}{d\rho^2} \right)_{\rho_0} \tag{18.43}$$

where,

$$\int v_{\text{xc}}[\rho]\rho = E_{\text{xc}}[\rho]$$

So, one can write-

$$E_{\text{xc}}[\rho_0] + \delta\rho \left(\frac{dE_{\text{xc}}}{d\rho} \right)_{\rho_0} = \int v_{\text{xc}}[\rho_0](\rho_0 + \delta\rho_0) \tag{18.44}$$

By using the Eqs. (18.43) and (18.44), Eq. (18.42) can be written as,

$$
\begin{aligned}
E = & \sum_{i}^{N_{occ}} \langle \phi_i \mid \hat{H}_0 \mid \phi_i \rangle - \frac{1}{2} \iint' \frac{\rho_0' \rho_0}{\mid r - r' \mid} + E_{xc}[\rho_0] \\
& - \int v_{xc}[\rho_0]\rho_0 + E_{IJ} \\
& + \frac{1}{2} \iint' \left(\frac{1}{\mid r - r' \mid} + \left. \frac{\delta^2 E_{xc}}{\delta\rho\delta\rho'} \right|_{\rho_0} \right) \delta\rho\delta\rho'
\end{aligned}
\tag{18.45}
$$

18.4.2 Self-Consistent-Charge Density-Functional Tight-Binding (SCC-DFTB) Method

Apart from some minor approximations such as neglecting the crystal field corrections and the truncation error for the expansion of $E_{xc}[\rho(\mathbf{r})]$, Eq. (18.45) is just another form of the Kohn–Sham total energy expression. Hence, Eq. (18.45) can be resolved with the same accuracy and at similar computational resources like other Kohn–Sham DFT implementations. Further approximations in Eq. (18.45) lead to a more computationally efficient scheme, the self-consistent charge density-functional tight-binding (SCC-DFTB) method, the full details of which can be found elsewhere [41–44].

18.5 Solar Cell Materials, Interfacial Charge Transfer Phenomena

So far we have discussed about basic principles of DFT and the way out of its practical implementation. However, the real essence could actually be taken by its application to real or at least some model systems providing some data, which can be directly compared with some experimental values. In the forthcoming section, we will show a more concrete way how such theories could be employed to the materials systems. As a prototypical example, we take here the materials, which can convert solar energy to electricity and apply DFT methodologies to compute some parameters relating to the efficacy of such materials for solar energy harvesting process.

Rapid depletion of fossil fuels has now pushed mankind to think of economically viable renewable energy sources for further sustaining the society. It has been proposed that the reserves of fossil fuels throughout the world could only last for another 50–60 years [45]. Solar, wind, hydro, biomass and geothermal are the main renewable energy sources among which solar energy is the most promising one [46]. The direct conversion of solar energy into electrical energy is done by photovoltaic cell or solar cell. The first generation solar cells were made of monocrystalline silicon [47, 48] but their high fabrication cost makes them commercially inaccessible. Furthermore, the performance of crystalline solar cells is affected by temperature and consequently it affects their efficiency. To overcome those problems, people have thought about thin film based solar cells which are referred to second generation of the solar cell. These are mainly made of amorphous silicon materials, however, cadmium telluride, copper indium gallium selenide are also the various categories of thin film solar cells. The main obstacle of the thin film solar cell is again its low efficiency. The solar cell constituted of dyes, commonly known as dye sensitized solar cell (DSSC) is considered as the third generation of the solar cell whose efficiency is in between thin films and crystalline solar cells [49]. DSSC was first fabricated by O'Regan and Grätzel in 1991 [50] by mimicking the natural photosynthesis that occurs in plant. Although the efficiency of DSSCs is not up to the mark, many advantageous effects like low cost, easy preparation process, etc.,

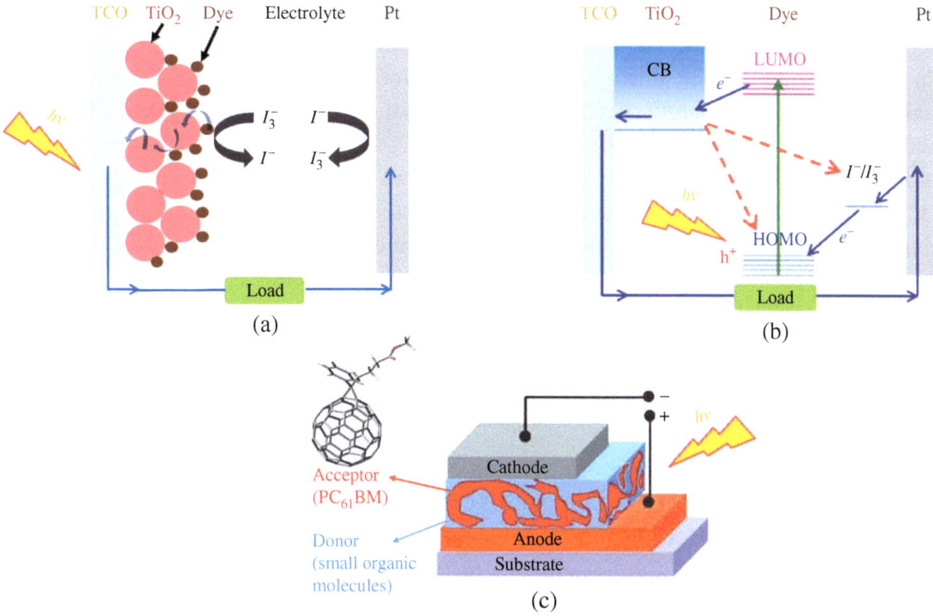

Figure 18.1 Schematic representations of a typical DSSC (a), mechanism of energy conversion (b), and model of a donor–acceptor solar cell (c).

make them quite attractive. Furthermore, tunable optoelectronic properties of various dyes open a new revenue in solar cell industry with various possibilities. Over the last two decades, DFT has made remarkable contribution for developing the basic understanding and the bottlenecks for improving the efficacy and efficiency of DSSC.

In DSSCs, the sensitizer is one of the key component for harvesting solar radiation and converting it into electrical current. Besides this, a DSSC consists of a semiconductor coated photo anode, a counter electrode (cathode), and an electrolyte as could be found in Figure 18.1a. The mechanism of a typical DSSC is also shown in Figure 18.1b, while Figure 18.1c represents a typical DSSC in bulk heterojunction model. When photon from the sunlight strikes on the DSSC surface, the dye molecule gets excited and produces a electron–hole pair, known as exciton which subsequently dissociates to produce a hot electron at the lowest unoccupied molecular orbital (LUMO) leaving a hole at the highest occupied molecular orbital (HOMO) of the dye. The minimum energy required to dissociate the exciton is called exciton binding energy. However, the excited electron is then injected into the conduction band of the semiconductor. Thereafter, the injected electron travels through the semiconductor material toward the transparent conductive electrode, and reaches to an external load in the form of electrical energy. Finally, the cycle is completed via the regeneration of the electron in the dye through the counter electrode and electrolyte [51].

The most crucial steps of a photovoltaic cell are the injection and regeneration processes as described above. These two processes are facilitated by suitable energy level alignment of the frontier molecular orbitals of the dye, conduction band of the semiconductor and electrochemical potential of the electrolyte. During this whole electron dynamic process, the electron injection efficiency to the conduction band of the semiconductor surface and the electron collection efficiency at the transparent conductive electrode directly determine the short circuit current density (J_{SC}) of the DSSC. The potential difference between the Fermi level of the semiconductor surface and the redox potential of the electrolyte gives the open circuit photovoltage (V_{OC}). These two parameters are important factors for determining the energy conversion efficiency of a DSSC. In between, some retarding processes are also occurred, recombination of electron and hole being the main obstacle. Additionally, during the charge–transfer processes, there must be structural relaxations (reorganization energy) that are to be compensated in terms of energy loss which also affects the overall power conversion efficiency.

For understanding all those processes one need to have a concrete knowledge on the ground and excited state properties of the individual as well as the composite materials. Here, we provide the basis things that are necessary for demonstrating the facts and findings. As excited state properties are well understood on the basis of time-dependent density functional theory (TDDFT), we present here the basic ideas of TDDFT, first.

18.5.1 The Time-Dependent Density Functional Theory

We have already introduced that the ground-state properties of a system can readily be determined with the help of DFT, which relies on the unique 1:1 correspondence between ground-state density and potential as postulated in Hohenberg–Kohn theorem. However, for obtaining the excited states of a system, one must go beyond Kohn–Sham DFT formalism thereby introducing TDDFT.

We begin with the time-dependent Schrödinger equation as

$$i\frac{\delta}{\delta t}\Psi_j(r,t) = \hat{H}\Psi_j(r,t) \tag{18.46}$$

where the time-dependent Hamiltonian ($\hat{H}$) is now represented by the sum of the kinetic-energy operator ($\hat{T}$) and electron–electron repulsion part $\hat{V}_{\text{ee}}$, similar to that in the static Hamiltonian (Eq. (18.8)) and an additional explicitly time-dependent external potential ($\hat{V}_{\text{Ne}}(t) = \sum_{j=1}^{N} v(r_j, t)$) such that it looks

$$\hat{H} = \hat{T} + \hat{V}_{\text{ee}} + \hat{V}_{\text{Ne}}(t) \tag{18.47}$$

The time-dependent Schrödinger equation formally represents an initial value problem. However, we define an initial time, t_0 and start with a given many-body wavefunction, $\Psi(t_0) = \Psi_0$ which then propagates with time.

According to Runge–Gross theorem [52], for a given initial state, the time-dependent density is a unique functional of the external potential. Now, if the external potential is expanded in a Taylor series with respect to t_0, we have

$$v(r,t) = \sum_{k=0}^{\infty} \frac{v_k(r)}{k!}(t - t_0)^k \tag{18.48}$$

For such potential, the existence of the unique 1:1 correspondence can be proven, which ensures that the density is a functional of the external potential and the initial wavefunction. Thus, we can write

$$\rho(r,t) = \rho[v, \Psi_0](r,t) \leftrightarrow v(r,t) = v[\rho, \Psi_0](r,t) \tag{18.49}$$

The formal dependence on the initial state, Ψ_0 goes away if the system starts from the ground-state, i.e. $\Psi_0 = \Psi_{\text{GS}}$, since the ground-state wavefunction is a functional of the density. So, $v(r,t)$ can be expressed as a functional of density only and consequently, the time-dependent Hamiltonian also becomes a density functional.

For a non-interacting system with N single-particle orbitals, the time-dependent density, $\rho(r,t)$ can be written as

$$\rho(r,t) = \sum_{j=1}^{N} |\phi_j(r,t)|^2 \tag{18.50}$$

where, the orbitals $\phi_j(r,t)$ follow the time-dependent Kohn–Sham equation,

$$i\frac{\delta}{\delta t}\phi_j(r,t) = \left[-\frac{1}{2}\nabla^2 + v_s(r,t)\right]\phi_j(r,t)$$

Here, the time-dependent effective potential, $v_s(r,t)$ is the sum of time-dependent external potential ($v(r,t)$), Hartree potential ($v_H(r,t)$) and time-dependent xc potential ($v_{xc}(r,t)$). Now, to apply TDDFT in practical problem, it requires a suitable approximation for $v_{xc}(r,t)$, the time-dependent Kohn–Sham equations need to be solved numerically and thus, the physical observable of interest could be obtained from the time-dependent density ($\rho(r,t)$).

18.5.2 TDDFT and Linear Response

The situation becomes little bit easier when a system of interest does not deviate strongly from its initial state and it is then considered as a weakly perturbed system. In that case, solution of the full time-dependent Schrödinger or Kohn–Sham equation is not necessary, instead, perturbation theory is sufficient for treating the phenomena. The most prominent example is the linear response theory [53, 54]. The goal of the linear response theory is to directly evaluate the change of a certain variable (density in our case) by means of first order perturbative correction without having the knowledge of change in wavefunction.

Now, if $v_1(r,t)$ is treated as a small perturbation, which causes some time-dependent change in the system, then $\rho(r,t)$ can be expanded into

$$\rho(r,t) = \rho_0(r) + \rho_1(r,t) + \rho_2(r,t) + \cdots \tag{18.51}$$

where ρ_0 is the ground-state density and the other terms (ρ_1, ρ_2, etc.) are contributed from linear and higher order density response, respectively. As the perturbation is weak, the equation can be truncated after linear density response which can be expressed as,

$$\rho_1(r,t) = \int_{-\infty}^{+\infty} dt' \int d^3r'\, \chi(r,t,r',t')v_1(r',t') \tag{18.52}$$

where χ is the density–density response function, which is defined as

$$\chi(r,t,r',t') = -i\theta(t-t')\langle\Psi_{GS}|[\hat{\rho}(r,t-t'),\hat{\rho}(r')]|\Psi_{GS}\rangle$$

Now, one can obtain the frequency-dependent response

$$\rho_1(r,\omega) = \int d^3r'\, \chi(r,r',\omega)v_1(r',\omega) \tag{18.53}$$

by performing Fourier transformation of the response function as below

$$\chi(r,r',\omega) = \lim_{\eta\to 0^+}\sum_{n=1}^{\infty}\left\{\frac{\langle\Psi_{GS}|\hat{\rho}(r)|\Psi_n\rangle\langle\Psi_n|\hat{\rho}(r')|\Psi_{GS}\rangle}{\omega-\Omega_n+i\eta} - \frac{\langle\Psi_{GS}|\hat{\rho}(r')|\Psi_n\rangle\langle\Psi_n|\hat{\rho}(r)|\Psi_{GS}\rangle}{\omega+\Omega_n+i\eta}\right\} \tag{18.54}$$

Here, $\Omega_n = E_n - E_0$ represents the nth excitation energy of the system. The above equation indicates that the response function has the poles at the exact excitation energies of the system. So, if any perturbation, $v_1(r, \omega)$ is applied to the system with the same frequency as that of the excitation energy, the response of the system would be very prominent.

In principle, the linear density response can be calculated exactly by using TDDFT as the response of the non-interacting Kohn–Sham system to an effective perturbation in terms of χ_s, the density–density response function of the Kohn–Sham system.

$$\rho_1(r,t) = \int dt' \int d^3r' \chi_s(r,t,r',t') v_{1s}(r',t') \tag{18.55}$$

Then, the effective perturbation is given as the sum of the real external perturbation and the linearized Hartree and xc potential.

$$v_{1s}(r,t) = v_1(r,t) + \int d^3r' \frac{\rho_1(r',t)}{|r-r'|} + \int dt' \int d^3r' f_{xc}(r,t,r',t')\rho_1(r',t') \tag{18.56}$$

Here, f_{xc} is the xc kernel, which is defined as the functional derivative of the time-dependent xc potential with respect to the time-dependent density, which is evaluated at the ground-state density,

$$f_{xc}(r,t,r',t') = \frac{\delta v_{xc}[\rho](r,t)}{\delta \rho(r',t')} \mid_{\rho_0(r)}$$

Therefore, the frequency dependent response is given by,

$$\rho_1(r,\omega) = \int d^3r' \chi_s(r,r',\omega) v_{1s}(r',\omega) \tag{18.57}$$

and

$$v_{1s}(r,\omega) = v_1(r,\omega) + \int d^3r' \left\{ \frac{1}{|r-r'|} + f_{xc}(r,r',\omega) \right\} \rho_1(r',\omega) \tag{18.58}$$

So, the Kohn–Sham response function can be written in terms of the occupation number (f) of the orbital in the Kohn–Sham ground-state.

$$\chi_s(r,r',\omega) = \lim_{\eta \to 0^+} \sum_{jk}^{\infty} (f_k - f_j) \frac{\phi_j(r)\phi_j^*(r')\phi_k(r)\phi_k^*(r')}{\omega - (\varepsilon_j - \varepsilon_k) + i\eta} \tag{18.59}$$

18.5.3 Excitation Energy and Excited State Properties

In TDDFT, electronic excitation is represented as a dynamical process where the system experiences a transition in between two eigenstates. Here, the excitation energy describes the rearrangements of probability density during the electronic transition. Now, in absence of any external perturbation, the density is represented by

$$\rho_1(r,\omega) = \int d^3r' \chi_s(r,r',\omega) \int d^3r'' f_{Hxc}(r',r'',\omega)\rho_1(r'',\omega) \tag{18.60}$$

where f_{Hxc} is the combined Hartree-xc kernel, which is defined as,

$$f_{Hxc}(r,r',\omega) = \frac{1}{|r-r'|} + f_{xc}(r,r',\omega)$$

Equation (18.60) has a trivial solution $\rho_1 = 0$ for all frequencies (ω), however, at some special frequencies (Ω), there also exist nontrivial solutions where the density response is finite and self-sustained. Those particular frequencies correspond to the excitation energies of the system associated with the eigenmodes, $\rho_1(r, \Omega)$.

Now, in case of real Kohn–Sham orbitals and frequency-independent xc kernel, one can calculate excitation energies and the eigenmodes ($Z_{i'l'}$) by using Casida equation [53, 54]

$$\sum_{i'l'} \left[\delta_{ii'}\delta_{ll'}(\omega_{il}^2 - \Omega^2) + 2\sqrt{\omega_{il}\omega_{i'l'}}K_{il,i'l'}\right] Z_{i'l'} = 0 \tag{18.61}$$

where

$$K_{il,i'l'}(\omega) = \int d^3r \int d^3r' \phi_i^*(r) \times f_{\text{Hxc}}(r, r', \omega)\phi_{i'}(r')\phi_{l'}^*(r')$$

and i, i' and l, l' correspond to occupied and unoccupied Kohn–Sham orbitals.

TDDFT within linear response formalism has been implemented successfully, where the key parameter for performance of the technique lies on the suitability of the xc kernel. The adiabatic hybrid kernels, especially the family of B3LYP or CAM-B3LYP are extensively used for predicting the excitation energy and related phenomena.

One can compute the excited state properties of nanocomposites of donor–acceptor (D–A) architecture as a model of photovoltaic cells by using TDDFT methodologies. Apart from the excitation energy, it requires the consideration of electronic coupling between the donor and acceptor moieties in the ground as well as excited states to evaluate the effective charge transfer interactions. The electronic coupling matrix at the donor–acceptor heterojunction is generally evaluated by using the generalized Mulliken–Hush (GMH) model [55] and the finite field method on the excitation energies [56]. Allowing singlet excitations in the TDDFT calculations provides the excited states ($|S_n\rangle$) in terms of configuration state function, Φ_i^l as

$$|S_n\rangle = \sum_{i\to l} c_i^l \Phi_i^l \tag{18.62}$$

where the summation runs over all occupied (i) and unoccupied (l) molecular orbitals (ϕ). Now, Lu and Chen [57, 58] showed that the density distribution of electrons and holes could be represented as

$$\rho^{\text{elec}}(r) = \sum_{i\to l}(c_i^l)^2\phi_i(r)\phi_i(r) + \sum_{i\to l}\sum_{j\neq i\to l} c_i^l c_j^l \phi_i(r)\phi_j(r) \tag{18.63}$$

and

$$\rho^{\text{hole}}(r) = \sum_{i\to l}(c_i^l)^2\phi_l(r)\phi_l(r) + \sum_{i\to l}\sum_{i\to m\neq l} c_i^l c_j^l \phi_l(r)\phi_m(r) \tag{18.64}$$

where the first term of the right-hand side of the above equations signifies the local term and the second one denotes the cross term to the corresponding electron and hole densities. So, the charge density difference (CDD) between the excited state and the ground state could be represented by

$$\Delta\rho(r) = \rho^{\text{elec}}(r) - \rho^{\text{hole}}(r) \tag{18.65}$$

There appear two more important parameters: extent of overlap between electron and hole (S) and the charge–transfer distance (l^{CT}). S is simply defined by

$$S = \int \min(\rho^{\text{hole}}(r), \rho^{\text{elec}}(r))dr \tag{18.66}$$

while the l^{CT} is defined by the spatial distance between the barycenters of the electron density depletion and enhancement [59, 60].

For TDDFT excited state wavefunction, applying Slater–Condon rule, one may write transition density matrix as below

$$T(r; r') = \sum_i \sum_l c_i^l \phi_i(r)\phi_l(r') \tag{18.67}$$

Now, taking the diagonal terms only, the transition density becomes

$$T(r) = \sum_i \sum_l c_i^l \phi_i(r)\phi_l(r) \tag{18.68}$$

This $T(r)$ can be treated as a normal real space function, which can be visualized in terms of isosurface. The transition density contains information about the spatial location of the excitation [61] and is directly related to the transition dipole moment. The transition dipole moment in a particular direction (say $\mu_{\mathrm{tr_x}}$) is related to the component of transition density ($T_{\mathrm{x}}(x) = -xT(r)$) by

$$\mu_{\mathrm{tr_x}} = \int T_{\mathrm{x}}(r)dr \tag{18.69}$$

18.5.3.1 Exciton Binding Energy

The exciton binding energy is one of the key parameters that determine the interfacial charge–transfer capability and hence photovoltaic performance of a composite material. It is defined as the energy requirement for the dissociation of exciton to free polarons through an intermediate state of bound polaron [62]. So, it is basically a two-step process. However, the excitonic dissociation is directly related to the charge separation in organic solar cells. It can be viewed as the Coulomb interaction energy between the cation of donor and the anion of the acceptor and is calculated as

$$E_{\mathrm{Col}} = \sum_{\mathrm{d}\in \mathrm{D,a}\in \mathrm{A}} \epsilon \frac{q_{\mathrm{d}} q_{\mathrm{a}}}{r_{\mathrm{da}}} \tag{18.70}$$

where q_{d} and q_{a} are the partial charges accumulated on donor cation and acceptor anion, respectively, r_{da} is the distance between these charge centres, and ϵ is the dielectric constant. Theoretically, $\mathrm{E_{Coul}}$ can be estimated as the difference between the electronic and optical band-gap energy [63]. It should be pointed out that, the electronic band-gap can easily be obtained from the energy offset between HOMO and LUMO, while the optical gap is taken as the first singlet excitation energy that can be computed from TDDFT calculation.

18.5.3.2 Reorganization Energy

Another important parameter in this regard is the reorganization energy (λ), which basically consists of two terms, inner reorganization energy (λ_{i}) and outer reorganization energy (λ_{o}) [64]. The first one arises from the change in equilibrium geometry of the donor and acceptor sites upon charge re-distribution in the excited state, which can be estimated theoretically as follows [65, 66]. We consider that the initial ground state equilibrium geometry of the D–A composite is transformed into a product geometry (Q_{P}) of charge–transfer state ($\mathrm{D^+A^-}$) through an initially excited reactant geometry (Q_{R}) where the donor is in the photochemically excited state ($\mathrm{D^*A}$) as shown in Figure 18.2. Due to difference in reactant and product geometry, they will follow different potential energy surfaces and thus two kinds of inner reorganizations could be considered. λ_{i1} corresponds to the difference in energy of the reactant at product geometry (Q_{P}) and its equilibrium reactant geometry (Q_{R}), whereas λ_{i2} corresponds to the difference in energy of the product at reactant geometry and its equilibrium product geometry. Therefore, the inner reorganization energy (λ_{i}) is the average of those λ_{i1} and λ_{i2}.

$$\lambda_{\mathrm{i}} = (\lambda_{\mathrm{i1}} + \lambda_{\mathrm{i2}})/2 \tag{18.71}$$

$$\lambda_{\mathrm{i1}} = [E^{\mathrm{D}^*}(Q_{\mathrm{P}}) + E^{\mathrm{A}}(Q_{\mathrm{P}})] - [E^{\mathrm{D}*}(Q_{\mathrm{R}}) + E^{\mathrm{A}}(Q_{\mathrm{R}})] \tag{18.72}$$

$$\lambda_{\mathrm{i2}} = [E^{\mathrm{D}^+}(Q_{\mathrm{R}}) + E^{\mathrm{A}^-}(Q_{\mathrm{R}})] - [E^{\mathrm{D}^+}(Q_{\mathrm{P}}) + E^{\mathrm{A}^-}(Q_{\mathrm{P}})] \tag{18.73}$$

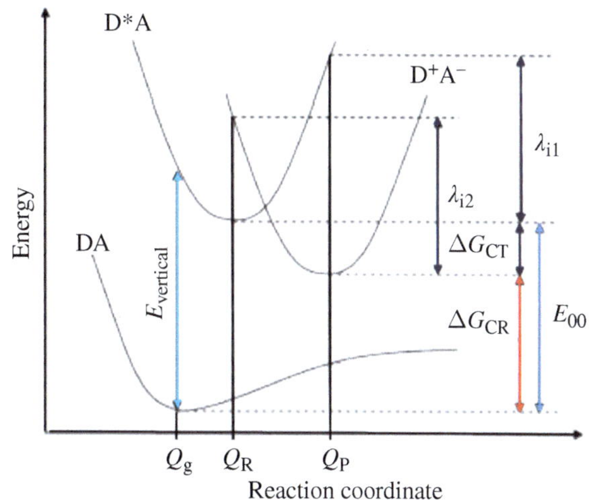

Figure 18.2 Schematic representation of energy profile diagram showing charge transfer from the ground to charge–transfer state through photochemically excited state. The different reorganization energies and free energy changes are also indicated.

Alternatively, λ_i could be considered as the sum of electron and hole reorganization energy (λ_e, λ_h, respectively)

$$\lambda = \lambda_e + \lambda_h \tag{18.74}$$

where the two terms, λ_e, and λ_h could be simplified as below

$$\lambda_e = E(A^-) - E(A) \tag{18.75}$$

$$\lambda_h = E(D) - E(D^+) \tag{18.76}$$

where $E(A^-)$ and $E(A)$ are the energies of the neutral acceptor A at the anionic geometry and optimal ground-state geometry, respectively, and $E(D)$ and $E(D^+)$ are the energies of the radical cation at the neutral geometry and optimal cation geometry, respectively.

On the other hand, outer reorganization energy, λ_o mainly arises due to electronic and nuclear polarization/relaxation of the surrounding medium. So, it is very much dependent on the polarity of the solvent medium and can be expressed as

$$\lambda_o = \Delta q^2 \left[\frac{1}{2d_D} + \frac{1}{2d_A} - \frac{1}{d_{DA}}\right]\left[\frac{1}{\epsilon_{op}} - \frac{1}{\epsilon_0}\right] \tag{18.77}$$

where d_D and d_A are the radius of the donor and acceptor, respectively and it has been reported that d_{DA} is the simple sum of d_D and d_A and ϵ_{op} is the optical dielectric constant and ϵ_0 is the zero-frequency dielectric constant.

18.5.3.3 The Rates of Charge Transfer and Recombination Processes

The dynamics of charge transfer is of crucial importance for determining photovoltaic performance. The rates of charge transfer and charge recombination can be evaluated on the basis of Marcus theory [67] as follows:

$$k = \sqrt{\frac{4\pi^3}{h^2 \lambda k_B T}} |V_{DA}|^2 \exp(-\frac{(\Delta G + \lambda)^2}{4\lambda k_B T}) \tag{18.78}$$

where λ is the reorganization energy, V_{DA} is the charge–transfer integral between donor and acceptor states, and ΔG is the free energy change for the corresponding electron transfer processes at a particular temperature T. We define ΔG_{CT} for charge transfer and ΔG_{CR} for charge recombination.

Now, the charge–transfer integral between the states S_0 and S_n can be expressed in terms of GMH model [55]

$$V_{\mathrm{DA}} = \frac{\mu_{\mathrm{tr}} \Delta E}{\sqrt{(\Delta\mu)^2 + 4(\mu_{\mathrm{tr}})^2}} \tag{18.79}$$

where μ_{tr} is the calculated transition dipole moment along the major axis, $\Delta\mu$ is the difference between dipole moments of the two states, S_0 and S_n, and ΔE is the vertical excitation energy. Now, the value of $\Delta\mu$ can be calculated by the use of finite field method on the excitation energy [68, 69], where the variation of excitation energy, E_{ex} with static electric field (F) in a specified direction (having maximum value of transition dipole moment, here along x axis) is represented by

$$E_{\mathrm{ex}}(F) = E_{\mathrm{ex}}(0) - \Delta\mu F - \frac{1}{2}\Delta\alpha F^2 \tag{18.80}$$

Here, $\Delta\alpha$ is the change in the polarizability. Now, for very weak electric field the equation can be simplified to a linear one whose negative slope gives the value of $\Delta\mu$.

As we have already introduced, the free energy changes for charge transfer and charge recombination are ΔG_{CT} and ΔG_{CR}, respectively. Now, the ΔG_{CR} can be estimated as the difference in ionization potential of the donor (E_{IP}(D)) and electron affinity of the acceptor (E_{EA}(A)) as

$$\Delta G_{\mathrm{CR}} = E_{\mathrm{IP}}(\mathrm{D}) - E_{\mathrm{EA}}(\mathrm{A}) \tag{18.81}$$

As a general approximation, the E_{IP}(D) and E_{EA}(A) are the HOMO and LUMO energies of the donor and acceptor counter parts, respectively [70], thus the value of ΔG_{CR} can easily be estimated as shown in Figure 18.2. Now, with the help of ΔG_{CR}, one can calculate the value of ΔG_{CT} using Rehm–Weller equation [71]

$$\Delta G_{\mathrm{CT}} = -\Delta G_{\mathrm{CR}} - \Delta E_{0-0} - E_{\mathrm{b}} \tag{18.82}$$

where ΔE_{0-0} and E_{b} are the energy of the lowest excited state of free-base donor and the exciton binding energy, respectively. Concrete results relating to such kinds of D–A systems could be found in the recent articles [72–75].

18.6 Concluding Remarks

Summarizing, the present chapter deals with aspects and utilities of DFT (and TDDFT) for computing the materials properties in the realm of solar energy conversion by D–A composite materials. We emphasized on the underlying theoretical details, computational strategies, and detailed analysis for predicting such properties. It has been demonstrated that accurate electronic structure calculations are the fingerprint for predicting materials properties and that can be done with the help of DFT even for the materials of experimental regime. Electron density is the fundamental property that establishes the whole concept of DFT, being a single variable for determining the structure property relationship. It can be utilized for explaining the ground-state properties, charge transfer and transport phenomena, chemical reactivity as well as excited state properties like absorption/emission phenomena, photochemical reactions and so on. The mechanism of electron transfer phenomena and their dynamical behaviour in solar cell have been explained on the basis of electronic energy level alignment, the corresponding energy gaps in the D–A architecture and the corresponding excited state properties. Nevertheless, DFT has also some limitations, especially in incorporating exchange–correlation energies. Further developments are needed for handling this particular issue. It is believed that proper development of density functionals, universally applicable for all systems, could resolve this issue. Incorporation of environmental effects such as salvation, variable temperature, pressure, etc. are still far from reality in the view of computational materials properties using DFT. The system size is also another constraint for comparing materials properties obtained from theoretical and experimental results. However, latest component of quantum computing may dramatically reduce the computational time, which will certainly strengthen DFT for predicting more accurate materials properties.

Acknowledgements

The authors gratefully acknowledge the financial support from BRNS, DAE, Govt. of India through sponsored research grant (sanction no. 37(2)/14/07/2018-BRNS). P.S. acknowledges UGC, Govt. of India for providing him with UGC Mid-Career Award, 2021. A.P. sincerely acknowledges UGC, Govt. of India, for the UGC-BSR Research Start-Up-Grant (No.F.30-557/2021(BSR)).

References

1 Sarkar, P. and Bhattacharyya, S.P. (2022). *Understanding Properties of Atoms, Molecules and Materials*. CRC Press.
2 Chen, X., Zhao, Y.-F., Wang, L.-S., and Li, J. (2017). Recent progresses of global minimum searches of nanoclusters with a constrained Basin-Hopping algorithm in the TGMin program. *Comput. Theor. Chem.* 1107: 57–65.
3 Roberts, C., Johnston, R.L., and Wilson, N.T. (2000). A genetic algorithm for the structural optimization of Morse clusters. *Theor. Chem. Acc.* 104: 123–130.
4 Pal, S., Sharma, R., Goswami, B. et al. (2009). A search for lowest energy structures of ZnS quantum dots: genetic algorithm tight-binding study. *J. Chem. Phys.* 130: 214703.
5 Lin, Z., Slee, T., and Mingos, D. (1990). A structural jellium model of cluster electronic structure. *Chem. Phys.* 142: 321–334.
6 Koch, E. and Gunnarsson, O. (1996). Density dependence of the electronic supershells in the homogeneous jellium model. *Phys. Rev. B* 54: 5168.
7 Galvão, B.R. and Viegas, L.P. (2019). What electronic structure method can be used in the global optimization of nanoclusters? *J. Phys. Chem. A* 123: 10454–10462.
8 Saha, S., Sarkar, S., Pal, S., and Sarkar, P. (2013). Tuning the energy levels of ZnO/ZnS core/shell nanowires to design an efficient nanowire-based dye-sensitized solar cell. *J. Phys. Chem. C* 117: 15890–15900.
9 Sarkar, S., Rajbanshi, B., and Sarkar, P. (2014). Understanding the electronic structure of CdSe quantum dot-fullerene (C60) hybrid nanostructure for photovoltaic applications. *J. Appl. Phys.* 116: 114303.
10 Sarkar, R., Sarkar, S., Pramanik, A. et al. (2016). Isoelectronically doped CdSe/Te nanoalloys as alternative solar cell materials: insight from computational analysis. *RSC Adv.* 6: 86494–86501.
11 Mandal, B., Sarkar, S., and Sarkar, P. (2015). Theoretical studies on understanding the feasibility of porphyrin-sensitized graphene quantum dot solar cell. *J. Phys. Chem. C* 119: 3400–3407.
12 Lieb, E.H. and Simon, B. (1977). The Hartree-Fock theory for coulomb systems. *Commun. Math. Phys.* 53: 185–194.
13 Szabo, A. and Ostlund, N.S. (2012). *Modern Quantum Chemistry: Introduction to Advanced Electronic Structure Theory*. Courier Corporation.

14 Shavitt, I. (1977). The method of configuration interaction. In: *Methods of Electronic Structure Theory*, Modern Theoretical Chemistry, vol. 3 (ed. H.F. Schaefer), 189–275. Boston, MA: Springer, Springer.

15 Bartlett, R.J. and Musiał, M. (2007). Coupled-cluster theory in quantum chemistry. *Rev. Mod. Phys.* 79: 291.

16 Hohenberg, P. and Kohn, W. (1964). Inhomogeneous electron gas. *Phys. Rev.* 136: B864.

17 Kohn, W. and Sham, L.J. (1965). Self-consistent equations including exchange and correlation effects. *Phys. Rev.* 140: A1133.

18 Martin, R.M. (2004). *Electronic Structure: Basic Theory and Practical Methods*. Cambridge University Press.

19 Ceperley, D.M. and Alder, B. (1980). Ground state of the electron gas by a stochastic method. *Phys. Rev. Lett.* 45: 566.

20 Perdew, J.P. and Zunger, A. (1981). Self-interaction correction to density-functional approximations for many-electron systems. *Phys. Rev. B* 23: 5048.

21 Becke, A.D. (1988). Density-functional exchange-energy approximation with correct asymptotic behavior. *Phys. Rev. A* 38: 3098.

22 Perdew, J.P. (1986). Density-functional approximation for the correlation energy of the inhomogeneous electron gas. *Phys. Rev. B* 33: 8822.

23 Becke, A.D. (1988). Correlation energy of an inhomogeneous electron gas: a coordinate-space model. *J. Chem. Phys.* 88: 1053–1062.

24 Perdew, J.P., Burke, K., and Ernzerhof, M. (1996). Generalized gradient approximation made simple. *Phys. Rev. Lett.* 77: 3865.

25 Schmider, H.L. and Becke, A.D. (1998). Optimized density functionals from the extended G2 test set. *J. Chem. Phys.* 108: 9624–9631.

26 Tao, J., Perdew, J.P., Staroverov, V.N., and Scuseria, G.E. (2003). Climbing the density functional ladder: nonempirical meta-generalized gradient approximation designed for molecules and solids. *Phys. Rev. Lett.* 91: 146401.

27 Kanai, Y., Wang, X., Selloni, A., and Car, R. (2006). Testing the TPSS meta-generalized-gradient-approximation exchange-correlation functional in calculations of transition states and reaction barriers. *J. Chem. Phys.* 125: 234104.

28 Van Voorhis, T. and Scuseria, G.E. (1998). A novel form for the exchange-correlation energy functional. *J. Chem. Phys.* 109: 400–410.

29 Rey, J. and Savin, A. (1998). Virtual space level shifting and correlation energies. *Int. J. Quantum Chem.* 69: 581–590.

30 Toulouse, J., Savin, A., and Adamo, C. (2002). Validation and assessment of an accurate approach to the correlation problem in density functional theory: the Kriger–Chen–Iafrate–Savin model. *J. Chem. Phys.* 117: 10465–10473.

31 Krieger, J.B., Chen, J., Iafrate, G.J., and Savin, A. (1999). Construction of an accurate self-interaction-corrected correlation energy functional based on an electron gas with a gap. In: *Electron Correlations and Materials Properties* (ed. A. Gonis, N. Kioussis, and M. Ciftan), 463–477. Boston, MA: Springer.

32 Gonzales, J.M., Cox, R.S., Brown, S.T. et al. (2001). Assessment of density functional theory for model SN2 reactions: $CH_3X + F^-$ ($X = F$, Cl, CN, OH, SH, NH_2, PH_2). *J. Phys. Chem. A* 105: 11327–11346.

33 Zhao, Y., Schultz, N.E., and Truhlar, D. (2005). Exchange-correlation functional with broad accuracy for metallic and nonmetallic compounds, kinetics, and noncovalent interactions. *J. Chem. Phys.* 123: 161103.

34 Zhao, Y., Schultz, N.E., and Truhlar, D.G. (2006). Design of density functionals by combining the method of constraint satisfaction with parametrization for thermochemistry, thermochemical kinetics, and noncovalent interactions. *J. Chem. Theor. Comput.* 2: 364–382.

35 Zhao, Y. and Truhlar, D.G. (2008). The M06 suite of density functionals for main group thermochemistry, thermochemical kinetics, noncovalent interactions, excited states, and transition elements: two new functionals and systematic testing of four M06-class functionals and 12 other functionals. *Theor. Chem. Acc.* 120: 215–241.

36 Zhao, Y. and Truhlar, D.G. (2008). Exploring the limit of accuracy of the global hybrid meta density functional for main-group thermochemistry, kinetics, and noncovalent interactions. *J. Chem. Theor. Comput.* 4: 1849–1868.

37 Zhao, Y. and Truhlar, D.G. (2006). A new local density functional for main-group thermochemistry, transition metal bonding, thermochemical kinetics, and noncovalent interactions. *J. Chem. Phys.* 125: 194101.

38 Zhao, Y. and Truhlar, D.G. (2006). Density functional for spectroscopy: no long-range self-interaction error, good performance for Rydberg and charge-transfer states, and better performance on average than B3LYP for ground states. *J. Phys. Chem. A* 110: 13126–13130.

39 Harris, J. (1985). Simplified method for calculating the energy of weakly interacting fragments. *Phys. Rev. B* 31: 1770.

40 Foulkes, W.M.C. and Haydock, R. (1989). Tight-binding models and density-functional theory. *Phys. Rev. B* 39: 12520.

41 Porezag, D., Frauenheim, T., Köhler, T. et al. (1995). Construction of tight-binding-like potentials on the basis of density-functional theory: application to carbon. *Phys. Rev. B* 51: 12947.

42 Elstner, M., Porezag, D., Jungnickel, G. et al. (1998). Self-consistent-charge density-functional tight-binding method for simulations of complex materials properties. *Phys. Rev. B* 58: 7260.

43 Seifert, G. (2007). Tight-binding density functional theory: an approximate Kohn–Sham DFT scheme. *J. Phys. Chem. A* 111: 5609–5613.

44 Aradi, B., Hourahine, B., and Frauenheim, T. (2007). DFTB+, a sparse matrix-based implementation of the DFTB method. *J. Phys. Chem. A* 111: 5678–5684.

45 Li, B., Wang, L., Kang, B. et al. (2006). Review of recent progress in solid-state dye-sensitized solar cells. *Sol. Energy Mater. Sol. Cells* 90: 549–573.

46 Fukurozaki, S., Zilles, R., and Sauer, I. (2013). Energy payback time and CO2 emissions of 1.2 kWp photovoltaic roof-top system in Brazil. *Int. J. Smart Grid Clean Energy* 2: 1–6.

47 Goodrich, A., Hacke, P., Wang, Q. et al. (2013). A wafer-based monocrystalline silicon photovoltaics road map: utilizing known technology improvement opportunities for further reductions in manufacturing costs. *Sol. Energy Mater. Sol. Cells* 114: 110–135.

48 Richhariya, G., Kumar, A., Tekasakul, P., and Gupta, B. (2017). Natural dyes for dye sensitized solar cell: a review. *Renewable Sustainable Energy Rev.* 69: 705–718.

49 Ludin, N.A., Mahmoud, A.A.-A., Mohamad, A.B. et al. (2014). Review on the development of natural dye photosensitizer for dye-sensitized solar cells. *Renewable Sustainable Energy Rev.* 31: 386–396.

50 O'Regan, B. and Grätzel, M. (1991). A low-cost, high-efficiency solar cell based on dye-sensitized colloidal TiO2 films. *Nature* 353: 737–740.

51 Stathatos, E. (2012). Dye sensitized solar cells: a new prospective to the solar to electrical energy conversion. Issues to be solved for efficient energy harvesting. *J. Eng. Sci. Technol. Rev.* 5: 9–13.

52 Runge, E. and Gross, E.K. (1984). Density-functional theory for time-dependent systems. *Phys. Rev. Lett.* 52: 997.

53 Casida, M.E. (2009). Time-dependent density-functional theory for molecules and molecular solids. *J. Mol. Struct. THEOCHEM* 914: 3–18.

54 Casida, M.E. and Wesołowski, T.A. (2004). Generalization of the Kohn–Sham equations with constrained electron density formalism and its time-dependent response theory formulation. *Int. J. Quantum Chem.* 96: 577–588.

55 Voityuk, A.A. (2006). Estimation of electronic coupling in π-stacked donor-bridge-acceptor systems: correction of the two-state model. *J. Chem. Phys.* 124: 064505.

56 Kjellberg, P., He, Z., and Pullerits, T. (2003). Bacteriochlorophyll in electric field. *J. Phys. Chem. B* 107: 13737–13742.

57 Lu, T. and Chen, F. (2012). Quantitative analysis of molecular surface based on improved marching tetrahedra algorithm. *J. Mol. Graph. Model.* 38: 314–323.

58 Lu, T. and Chen, F. (2012). Multiwfn: a multifunctional wavefunction analyzer. *J. Comput. Chem.* 33: 580–592.

59 Le Bahers, T., Adamo, C., and Ciofini, I. (2011). A qualitative index of spatial extent in charge-transfer excitations. *J. Chem. Theor. Comput.* 7: 2498–2506.

60 Ciofini, I., Le Bahers, T., Adamo, C. et al. (2012). Through-space charge transfer in rod-like molecules: lessons from theory. *J. Phys. Chem. C* 116: 11946–11955.

61 Jespersen, K.G., Beenken, W.J., Zaushitsyn, Y. et al. (2004). The electronic states of polyfluorene copolymers with alternating donor-acceptor units. *J. Chem. Phys.* 121: 12613–12617.

62 Emelianova, E., Van der Auweraer, M., and Bässler, H. (2008). Hopping approach towards exciton dissociation in conjugated polymers. *J. Chem. Phys.* 128: 224709.

63 Scholes, G.D. and Rumbles, G. (2006). Excitons in nanoscale systems. *Nat. Mater.* 5: 683–696.

64 Brédas, J.-L., Beljonne, D., Coropceanu, V., and Cornil, J. (2004). Charge-transfer and energy-transfer processes in π-conjugated oligomers and polymers: a molecular picture. *Chem. Rev.* 104: 4971–5004.

65 Lemaur, V., Steel, M., Beljonne, D. et al. (2005). Photoinduced charge generation and recombination dynamics in model donor/acceptor pairs for organic solar cell applications: a full quantum-chemical treatment. *J. Am. Chem. Soc.* 127: 6077–6086.

66 Song, P., Li, Y., Ma, F. et al. (2013). External electric field-dependent photoinduced charge transfer in a donor–acceptor system for an organic solar cell. *J. Phys. Chem. C* 117: 15879–15889.

67 Marcus, R.A. (1993). Electron transfer reactions in chemistry: theory and experiment (Nobel lecture). *Angew. Chem. Int. Ed. Engl.* 32: 1111–1121.

68 Tong, M., Sheng, C.-X., and Vardeny, Z.V. (2007). Nonlinear optical spectroscopy of excited states in polyfluorene. *Phys. Rev. B* 75: 125207.

69 Mukamel, S., Tretiak, S., Wagersreiter, T., and Chernyak, V. (1997). Electronic coherence and collective optical excitations of conjugated molecules. *Science* 277: 781–787.

70 Zhang, X., Chi, L., Ji, S. et al. (2009). Rational design of d-PeT phenylethynylated-carbazole monoboronic acid fluorescent sensors for the selective detection of α-hydroxyl carboxylic acids and monosaccharides. *J. Am. Chem. Soc.* 131: 17452–17463.

71 Kavarnos, G.J. and Turro, N.J. (1986). Photosensitization by reversible electron transfer: theories, experimental evidence, and examples. *Chem. Rev.* 86: 401–449.

72 Biswas, S., Pramanik, A., Ahmed, T. et al. (2016). Superiority of D–A–D over D–A type of organic dyes for the application in dye-sensitized solar cell. *Chem. Phys. Lett.* 649: 23–28.

73 Biswas, S., Pramanik, A., and Sarkar, P. (2017). Effect of additional donor group on the charge transfer/recombination dynamics of a photoactive organic dye: a quantum mechanical investigation. *Comput. Theor. Chem.* 1103: 38–47.

74 Biswas, S., Pramanik, A., Pal, S., and Sarkar, P. (2017). A theoretical perspective on the photovoltaic performance of S,N-heteroacenes: an even–odd effect on the charge separation dynamics. *J. Phys. Chem. C* 121: 2574–2587.

75 Biswas, S., Pramanik, A., and Sarkar, P. (2018). Origin of different photovoltaic activities in regioisomeric small organic molecule solar cells: the intrinsic role of charge transfer processes. *J. Phys. Chem. C* 122: 14296–14303.

19

A Conceptual DFT Analysis of Mechanochemical Processes

Ruchi Jha[1], Shanti Gopal Patra[2,3], Debdutta Chakraborty[4], and Pratim Kumar Chattaraj[4]

[1] *Advanced Technology Development Centre, Indian Institute of Technology Kharagpur, Kharagpur, India*
[2] *Department of Chemistry, Indian Institute of Technology Kharagpur, Kharagpur, India*
[3] *Department of Chemistry, National Institute of Technology Silchar, Silchar, India*
[4] *Department of Chemistry, Birla Institute of Technology, Mesra, Ranchi, Jharkhand, India*

19.1 Introduction

Conventionally, a chemical reaction is performed through the application of heat, light, and electricity to activate the molecules. The field of chemistry where mechanical force is employed to perform chemical reactions is called mechanochemistry [1–3]. The most simple mechanochemical reaction is the use of mortar and pestle to react mercury sulfide with metallic copper to form metallic mercury and copper sulfide [1]. In the year 1891, Ostwald coined the term “mechanochemistry” while discussing methods of excitation of a chemical process in the “Textbook of General Chemistry.” In the IUPAC “Gold Book,” mechanochemistry is defined as a “chemical reaction induced by the direct absorption of mechanical energy” [4]. Various reactions have been studied using mechanochemistry and have emerged as a promising field for cheap and green synthesis [5]. Common mechanochemical procedures involve using ball mills [5], ultrasound bath [6], and atomic force microscope (AFM) [1, 7]. The theoretical understanding of mechanochemistry only started two decades ago [3]. In quantum chemistry, recent developments in geometry optimizations through implicit [8, 9] and explicit [10–13] application of force leads to the understanding of the effect of mechanochemical force on structure, energy, and spectroscopic properties.

As discussed, several methods can be employed to transfer external force to the molecular level. In this regard, the sophisticated AFM method has been employed to break the chemical bond or protein unfolding. A vital muscle component, namely titin, has a seven-stranded β-barrel that requires a force of 150 and 300 pN to unfold the β-structure [14, 15]. AMF was utilized to study the disulfide bond cleavage in the presence of an enzyme by external mechanical force [16, 17]. The force versus extension curves in biological molecules were studied using a combined statistical mechanics model and quantum mechanical calculations [18]. Researchers have also used single molecular force spectroscopy (SMFS) in this regard [5]. In chemistry, ultrasonic baths are generally used to clean reaction vessels and to make homogeneous mixtures [6]. However, an ultrasonic probe can produce strong vibration and pressure due to gas bubbles and heat [19]. Eventually, the mechanical force generated in an ultrasonic bath can be transferred to a molecule to perform sonochemistry [20]. Chemical reactions such as aliphatic nucleophilic substitution, aromatic

Electron Density: Concepts, Computation and DFT Applications, First Edition.
Edited by Pratim Kumar Chattaraj and Debdutta Chakraborty.

electrophilic substitution, disulfide metathesis, and diradical transition state trapping have been performed using sonochemistry [13, 21–23].

The ball milling process provides environmentally benign, solvent-free, energy-efficient, cheap, and green chemical synthesis [24, 25]. In a ball milling process, the collisional impact among the balls produces a mechanical force that is transferred to the molecules, and selective bonds are ruptured to perform a chemical reaction. The solvent-free synthesis of tetra-substituted porphyrins is a typical example of a ball-milling chemical response [26]. Another system for the ball milling study is the breaking of dichalcogenide bonds. In this regard, Ranu and coworkers have performed the synthesis of organic chalcogenides employing the ball milling process [27–29].

Typically, mechanochemistry deals with the application of external force. However, a suitable molecule can develop force to its environment in molecular machines. Often, these molecules are triggered by a light that undergoes a spatial change in a photoswitch [30]. In a chemical environment, a mechanical force is developed due to photoisomerization. Hugel and coworkers have developed azobenzene-based single-molecule optomechanical devices as molecular actuators [31, 32]. Related devices having rotational motion are also developed [33]. Although conclusions obtained from experiments are limited to the studied molecules, quantum mechanochemistry can be utilized on various molecules to develop general concepts in mechanochemistry. Thus, in the last two decades, systematic development of mechanochemical processes have been developed in terms of their understanding through quantum mechanochemistry.

In this book chapter, we discuss the general development of the theoretical methods under an external force field. The most used electronic structure principles in computing geometry, energy, and change in certain properties with external forces are discussed. Finally, a summary is done based on the application of quantum mechanochemistry. Very recently, force-dependent conceptual density functional theory (CDFT) descriptors have been developed. The mechanochemical force-dependent CDFT descriptors on small molecules are discussed. The confinement effect on the mechanical force is also emphasized. The dynamical aspect of the external mechanical force is discussed as an outlook.

19.2 Theoretical Background

19.2.1 The Constrained Geometries Simulate External Force (COGEF)

In Layman's terms, mechanical energy is the sum of kinetic and potential energies. The application of mechanical force can alter the state of any physical system. The mechanical force can be applied in many ways, for example, by stretching.

The application of mechanical force on chemical systems gives rise to a new category of reactions called mechanochemical reactions. In these reactions, the energy barrier is crossed by the application of mechanical force in contrast to other types of reactions where thermal energy or electrical energy is utilized by the reactants to form products.

In order to study the mechanochemical processes computationally, various methods have been suggested. One of the widely used methods is the COGEF method. COGEF stands for Constrained Geometries Simulate External Force. In this method, the external pulling force is simulated by varying fixed geometrical parameters. This method has become popular due to its simplicity.

External force application on molecular systems alters the nuclear configuration, and the molecule is deformed. The gradient of nuclear energy at the deformed geometry provides valuable information about the mechanical force required to bring that change. In this method, the potential energy surface (PES) scan is done by fixing the bond length as a parameter on which the external force is to be simulated. The bond length is then varied in steps, and the whole molecule is allowed to relax.

For a diatomic molecule, the potential energy of the system as a function of the distance between the nuclei of the molecule can be given by the Morse potential, which is given by the following equation:

$$V_{\mathrm{M}}(r) = a\left(1 - e^{-b(r-r_{\mathrm{e}})}\right)^2 \tag{19.1}$$

Here, $(r - r_{\mathrm{e}})$ is the bond distance difference between the distorted and equilibrium geometry. a is the equilibrium bond dissociation energy, and b is a parameter that is associated with the curvature of the potential.

As the potential is a minimum at the equilibrium bond distance, a slight deviation from the equilibrium causes an internal restoring force to act on the system that tends to bring back the orientation to the initial minimum geometry. This restoring force can be calculated by taking the derivative of the Morse potential. Hence,

$$F_{\mathrm{i}} = \frac{\partial V_{\mathrm{M}}}{\partial r} = 2abe^{-b(r-r_{\mathrm{e}})}\left(1 - e^{-b(r-r_{\mathrm{e}})}\right) \tag{19.2}$$

At the distorted geometry, a new equilibrium structure is generated. Now, the net force on this orientation is also zero. Therefore, the external force must be equal in magnitude to the internal force and opposite in direction. So,

$$F_{\mathrm{e}} = -F_{\mathrm{i}} = -2abe^{-b(r-r_{\mathrm{e}})}\left(1 - e^{-b(r-r_{\mathrm{e}})}\right) \tag{19.3}$$

Now, the total potential becomes the sum of the Morse potential V_{M} and the external potential V_{e}, which is given by,

$$V_{\mathrm{e}} = -\int F_{\mathrm{e}} dr \tag{19.4}$$

And the total potential V_{T} is given as

$$V_{\mathrm{T}} = V_{\mathrm{M}} + V_{\mathrm{e}} = a\left(1 - e^{-b(r-r_{\mathrm{e}})}\right)^2 - \int F_{\mathrm{e}} dr \tag{19.5}$$

From Eq. (19.3), we can calculate the maximum force. This maximum force gives quantitative insight into the strength of the bond considered. The larger the maximum force, the stronger the bond, and vice versa. From Eq. (19.3) it can be easily calculated,

$$F_{\mathrm{max}} = \frac{ab}{2} \tag{19.6}$$

In order to understand the theory of mechanochemistry involved in the ball milling experiment, Bettens et al. [34] have described a few CDFT-based reactivity descriptors. It is well known that in CDFT, the energy of a system is expressed as a unique function of a total number of electrons N and the external potential, $v(\mathbf{r})$, which is the attractive potential due to electrons and nuclei. Applying external force to the system perturbs it, and the change in the energy can be expressed as a Taylor series expansion given by:

$$\Delta E = \left(\frac{\partial E}{\partial N}\right)_{v(\mathbf{r})} \Delta N + \int \left(\frac{\delta E}{\delta v(\mathbf{r})}\right)_N \Delta v(\mathbf{r}) d\mathbf{r} + \frac{1}{2}\left(\frac{\partial^2 E}{\partial^2 N}\right)_{v(\mathbf{r})} \Delta N^2 + \int \left(\frac{\partial \delta E}{\partial N \delta v(\mathbf{r})}\right) \Delta N \Delta v(\mathbf{r}) d\mathbf{r}$$
$$+ \frac{1}{2}\iint \left(\frac{\delta^2 E}{\delta v(\mathbf{r})\delta v(\mathbf{r}')}\right)_N \Delta v(\mathbf{r})\Delta v(\mathbf{r}') d\mathbf{r} d\mathbf{r}' + \ldots \tag{19.7}$$

A few reactivity indices have been defined, which gives us insights into the reactivity of the chemical system under consideration. These indices are as follows [35–40]:

$$\mu = -\chi = \left(\frac{\partial E}{\partial N}\right)_{v(\boldsymbol{r})} = -\left(\frac{I+A}{2}\right) \tag{19.8}$$

$$\eta = \frac{1}{S} = \left(\frac{\partial^2 E}{\partial N^2}\right)_{v(\boldsymbol{r})} = I - A \tag{19.9}$$

$$\omega = \frac{\mu^2}{2\eta} \tag{19.10}$$

Here, μ is the chemical potential, χ is the electronegativity, η is the hardness, S is the softness, ω is the electrophilicity, A is vertical electron affinity, and I is vertical ionization energy. When we consider external force, the Taylor functional expansion (Eq. (19.7)) gets modified, and the first-order derivatives of CDFT indices with respect to external force are chosen as the mechanochemistry-based CDFT reactivity descriptors. These descriptors are:

$$\left(\frac{\partial I}{\partial F_{\rm e}}\right)_{F_{\rm e}=0}, \left(\frac{\partial A}{\partial F_{\rm e}}\right)_{F_{\rm e}=0}, \left(\frac{\partial \chi}{\partial F_{\rm e}}\right)_{F_{\rm e}=0}, \left(\frac{\partial \eta}{\partial F_{\rm e}}\right)_{F_{\rm e}=0}, \left(\frac{\partial S}{\partial F_{\rm e}}\right)_{F_{\rm e}=0}, \text{ and } \left(\frac{\partial \omega}{\partial F_{\rm e}}\right)_{F_{\rm e}=0}.$$

19.2.2 External Force is Explicitly Included (EFEI)

The chemistry of many-particle systems changes when external mechanical forces are applied, which is explained by the fact that mechanical forces affect the system's PES. We obtain the corresponding change in PES as a function of the external force by using the external force as the control parameter rather than some structural parameter used in the COGEF technique. This method, where external force is explicitly included, is abbreviated as EFEI [11].

In the EFEI approach, the potential is demonstrated by,

$$V_{\rm EFEI}(\vec{x}, \vec{F_0}) = V_{\rm BO}(\vec{x}) - \vec{F_0}\vec{q}(\vec{x}) \tag{19.11}$$

Here, $V_{\rm BO}(\vec{x})$ is the Born–Oppenheimer potential as a function of $\vec{x}$, $\vec{q}(\vec{x})$ is the generalized coordinate in terms of $\vec{x}$, $\vec{x}$ is the distance between the atoms and $\vec{F_0}$ is the constant external force. As the net force on the system is zero at equilibrium, the derivative of Eq. (19.11) is zero, and the net external force is equal to the internal force, which is given by the derivative of $V_{\rm BO}(\vec{x})$.

So,

$$\nabla_{\vec{x}} V_{\rm EFEI}(\vec{x}, \vec{F_0}) = 0 \tag{19.12}$$

and solving this gives us $\vec{x_0}$. Now, as an internal force $\vec{F}$ cancels out the external force $\vec{F_0}$ exactly at x_0, we have,

$$\vec{F} = -\nabla_q V_{\rm BO}(x_0) \text{ and we get } \vec{q}(\vec{x_0}) = \vec{q_0} \tag{19.13}$$

Including the external force explicitly gives us the distortion $\vec{x_0}(\vec{F_0})$ in terms of the constant external force $\vec{F_0}$ and this method is exact, self-consistent, and fully nonlinear.

19.3 Results and Discussions

19.3.1 General Consideration

Kauzmann and Erying [41] in 1940 studied the effect of external pulling force with a Morse-type potential. When a force acts on a molecule, the chemical reactivity is typically increased or decreased. The bond dissociation energy of a chemical bond decreases when an external force stretches the bond. The deformed PES, $V_F(r)$ of a single bonded molecule in the presence of an external stretching force, F_0 can be expressed as:

$$V_F(r) = V_{\text{ab initio}}(r) - F_0\Delta r \tag{19.14}$$

where $F_0\Delta r$ is the energy due to F_0 and displacement Δr that is subtracted from the total energy, $V_{\text{ab initio}}(r)$. Equation (19.14) represents the one-dimensional (1-D) mechanochemistry and is the foundation of quantum mechanochemical studies for the calculation of deformed geometries under external force. The lifetime of bond rupture decreases with the application of a force field [42]. The probability of bond rupture is small in the absence of external as evident from the steepness of potential well force; however, it decreases rapidly in the presence of external force [43]. The activation energy of bond rupture changes upon the application of force. Considering Morse potential, it was shown that temperature plays a significant role in the bond rupture. Room temperature can provide only ~0.6 kcal/mol. A stretching vibration in a bond produces periodic elongation and compression, and at the highest amplitude, the bond breaks apart. In an AFM setup, Lo et al. [44] have studied the effect of temperature (13 and 37 °C) on the force required to rupture the biotin–avidin bond (Figure 19.1).

The mechanically relevant degrees of freedom are identified through force analysis. The nuclear gradient vector is obtained automatically using quantum chemical analysis in geometry optimization using force analysis tools. The same is applicable to molecular dynamics (MD) simulations. The cartesian coordinate system is used in the partitioning scheme. However, using Wilson's B-matrix, one can easily transform it into an internal coordinate system [45, 46]. An element **B** is defined as:

$$\mathbf{B}_{ij} = \frac{\partial q_i}{\partial x_i} \tag{19.15}$$

The internal coordinate $\vec{g_q}$ can be obtained from the cartesian coordinates $\vec{g_x}$ with the generalized transpose B-matrix $(\mathbf{B}^t)^+$:

$$\vec{g_q} = (\mathbf{B}^t)^+\vec{g_x} \tag{19.16}$$

Now, it is essential to discuss the definition and meaning of the force applied to a bond. Let us consider two bonded atoms stretched apart by a force of 2 nN, which acts outwardly along the

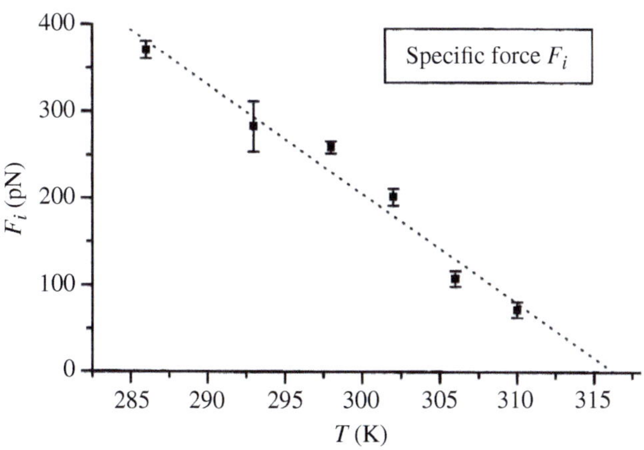

Figure 19.1 An AFM experiment measured biotin–avidin unbinding forces as a temperature function. Source: Reprinted from Lo et al. [44]. Copyright 2002 American Chemical Society.

connecting bond. Here, a force of 2 nN is operating on each atom; hence, a total force of 4 nN is stretching the bond. This definition of force operates on an isolated molecule in the gas or solution phase. Another definition of mechanochemical force derived from the AFM experiment is more meaningful and sounding [47–49]. When a molecule is attached to a glass surface, and the cantilever of an AFM instrument exerts a force of 2 nN, then an opposite force is also operated. Here, the molecule does not move, and it is defined as the pulling force. From the classical definition, force is the derivative of energy (E) with respect to distance (r):

$$E = \int -\frac{dE}{dr}dr = \int F\,dr \tag{19.17}$$

In mechanochemistry, it is advantageous to express the results in terms of mechanical work as it is a scalar quantity while the force is a vector quantity.

19.3.2 Constrained Geometries Simulate External Force (COGEF)

In a molecule, the nuclear positions are changed due to the application of mechanochemical force. Thus, the nuclear configuration and mechanical force are correlated; hence, in a deformed geometry from the nuclear configuration, one can calculate the force. From this realization, Beyer [9] developed the COGEF methodology. Typically, in a COGEF method, a relaxed geometry optimization is performed by fixing the bond between the atom on which the force is being applied. Varying the "r" produces the COGEF potential curve. The pulling scenario in the $SiH_3CH_2CH_3$ molecule with the variation of terminal H–H distance ($r_{H\text{-}H}$) is shown in Figure 19.2. The bond-breaking point is determined from the inflection point in the PES. The study agrees well with the relevant studies on the n-butane system. In the $SiH_3CH_2CH_3$ molecule among the four bonds (H—Si, Si—C,

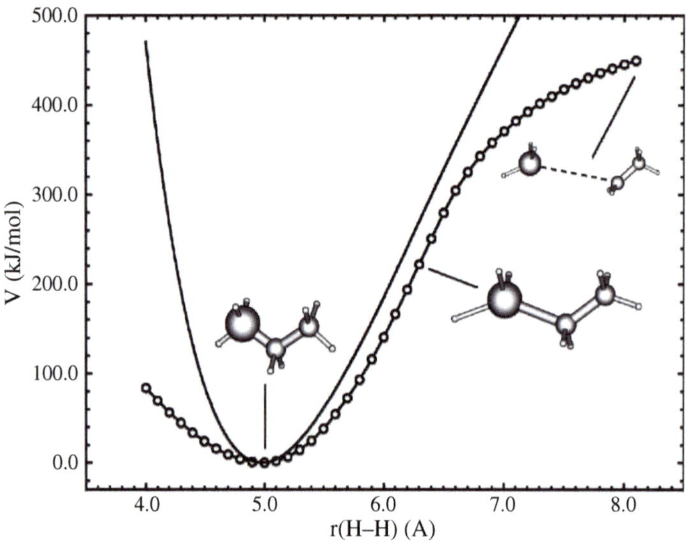

Figure 19.2 COGEF potential of the molecule $SiH_3CH_2CH_3$, in which the distance r between the terminal hydrogen atoms was varied between 4 and 8 Å, calculated at the B3LYP/D95(d,p) level of theory. A Morse potential with the same maximum slope as the COGEF potential is included (solid line). From left to right, the structures indicate the equilibrium geometry, the geometry at maximum force, and the geometry at a force that exceeds the bond rupture force. Source: Reprinted with permission from Beyer [9]. Copyright 2000 American Institute of Physics.

Table 19.1 Bond length r_e (Å), bond dissociation energy a (kJ/mol), length corresponding to maximum force r_{max} (Å), and maximum force F_{max} (nN), calculated in selected diatomic molecules at the B3LYP level of theory.

Bond	r_e (Å)	a (kJ/mol)	r_{max} (Å)	F_{max} (nN)
H—H	0.743	471.6	1.19	8.31
	0.743	460.8	1.17	7.87
H—F	0.928	572.2	1.29	10.79
	0.922	585.6	1.29	10.83
H—Cl	1.288	434.3	1.71	7.35
	1.281	442.1	1.72	7.46
F—F	1.400	163.5	1.75	7.89
	1.395	151.7	1.76	7.67
Cl—F	1.666	240.4	2.09	6.50
	1.642	253.8	2.07	6.67
Cl—Cl	2.045	212.6	2.49	4.60
	2.010	230.8	2.47	5.06
O=O	1.221	508.6	1.57	14.89
	1.203	518.9	1.56	14.98
N≡N	1.114	897.5	1.43	26.27
	1.091	960.7	1.41	27.14

To assess basis set effects, both the standard D95(d,p) polarized double-zeta basis set and the large 6-311++G(3df,3pd) basis set, including two diffuse and four polarization functions, were used. The first row shows the D95(d,p), and the second row shows the 6-311++G(3df,3pd) results.
Source: Reprinted with permission from Beyer [9]. Copyright 2000 American Institute of Physics.

C—C, and C—H), the weakest Si—C bond is cleaved, and the relevant energy is 220 kJ/mol. All the bonds were proposed to be activated by 55 kJ/mol. The values of force required to break the bond in some diatomic molecules are summarized in Table 19.1. A large force is needed for the triple (N≡N) and double (O=O) bonds as these bonds are short and strong. The force required for the Si—Si bond (3.28 nN) is less than that of the C—C bond (6.92 nN), owing to their strength.

Further, for the same type of bond, the force is dependent on the chemical environment. For example, the force required to rupture the O—H bond in H_2O and HOCl are 9.77 and 9.18 nN, respectively. In a COGEF method, it should be noted that the force is dependent on the choice of coordinate since the r changes with a change in coordinate. For example, in water, instead of stretching the O—H bond, if H—H is stretched, then the force changes from 9.77 to 10.27 nN, which is not trivial.

In a mechanochemical process, applying the COGEF method is beneficial in finding the bond rupture force. The COGEF method has been used extensively since its inception in the last two decades. It has played a significant role in developing the field of quantum mechanochemistry. One demerit of the method is that it does not consider the thermal effect in the calculation. As a result, the bond rupture force is different from the experimental one (in general higher). Further, it also depends on the method and basis sets for the geometry optimization. The primary strength of the method is its simplicity.

19.3.2.1 Mechanochemical CDFT Reactivity Descriptors and Their Application to Diatomic Molecules

Applying the well-known COGEF method discussed in Section 19.3.2 Bettens et al. [34] have quantified the mechanical strength of small diatomic molecules. The dissociation energy decreases as an external force is applied to a bond. Further increase in external force causes to decrease in the dissociation energy to a point where dissociation energy becomes zero. At that point, the external force applied is the maximum (F_{max}). This F_{max} gives us information about the mechanical strength of any bond. Higher the value of F_{max}, stronger is the bond and vice versa. As a result of applied external force, the CDFT-based descriptors also change. An analysis based on the mechanochemical response functions of the CDFT-based descriptors has been done. It was observed that these 21 diatomic molecules tend to follow a set of rules. In Table 19.2, the Morse parameters a and b, along with F_{max}, have been provided.

Table 19.2 a, b, and F_{max} for small diatomic molecules.

Molecules	a (kJ/mol)	b (A^{-1})	F_{max} (nN)
H_2	452.6	1.950	7.327
HF	664.1	2.048	11.292
HCl	527.1	1.720	7.527
HBr	487.9	1.612	6.530
F_2	303.2	2.446	6.156
Cl_2	375.7	1.614	5.036
Br_2	344.3	1.498	4.283
FCl	404.9	1.898	6.379
FBr	413.8	1.802	6.192
ClBr	350.6	1.597	4.647
N_2	1339.2	2.343	26.051
P_2	790.6	1.500	9.845
PN	825.3	1.997	13.684
BN	761.2	2.022	12.781
O_2	845.5	2.174	15.264
S_2	706.9	1.443	8.466
SO	771.3	1.851	11.856
CO	1272.5	2.150	22.717
LiF	336.1	1.447	4.037
NaCl	225.7	1.143	2.142
MgO	476.2	1.334	5.274

Source: Reprinted with permission from Beyer [9]. Copyright 2000 American Institute of Physics.

As can be seen from Table 19.2 above, the force required to break the N—N bond of N_2 is the highest, which is obvious due to the presence of the triple bond, and F_{max} for NaCl is the lowest as it is highly ionic and can dissociate into Na^+ and Cl^- ions easily.

Six global reactivity descriptors based on CDFT that are vertical ionization energy (I), vertical electron affinity (A), electronegativity (χ), hardness (η), and electrophilicity (ω) have been calculated, and their first-order response w.r.t. the external force has been calculated as mechanochemical reactivity indices. These values have been provided in Table 19.3.

For these calculations, the forces are taken from $F_{max}/2$ to $-F_{max}/2$. The bond equilibrium distances of the diatoms in their neutral state and their cationic and anionic states have been provided in Table 19.4. Six rules have been devised based on the differences in equilibrium bond length of the neutral and the ionic species and the first-order response of the reactivity descriptors w.r.t. small external force (treated as perturbation).

These rules are summarized below:

$$\text{Rule 1: } \left(\frac{\partial I}{\partial F_{ext}}\right)_{F_{ext=0}} > 0 \text{ if } (r_{neutral} - r_{cation}) > 0$$

Table 19.3 Mechanochemical reactivity descriptors for the diatomic molecules.

Systems	$\left(\frac{\partial I}{\partial F_{ext}}\right)_{F_{ext}=0}$	$\left(\frac{\partial A}{\partial F_{ext}}\right)_{F_{ext}=0}$	$\left(\frac{\partial \chi}{\partial F_{ext}}\right)_{F_{ext}=0}$	$\left(\frac{\partial \eta}{\partial F_{ext}}\right)_{F_{ext}=0}$	$\left(\frac{\partial S}{\partial F_{ext}}\right)_{F_{ext}=0}$	$\left(\frac{\partial \omega}{\partial F_{ext}}\right)_{F_{ext}=0}$
H_2	−0.4037	0.0276	−0.1881	−0.4313	1.0196	−0.0410
HF	−0.1351	0.0327	−0.0509	−0.1672	0.4436	−0.0058
HCl	−0.0706	0.0516	−0.0095	−0.1222	0.5214	0.0084
HBr	−0.0589	0.0723	0.0067	−0.1312	0.6586	0.0169
F_2	0.2014	0.8709	0.5362	−0.6695	1.9945	0.3621
Cl_2	0.2099	0.7473	0.4786	−0.5374	3.4234	0.3655
Br_2	0.1467	0.5460	0.3464	−0.3992	3.3898	0.3069
FCl	0.1796	0.5921	0.3858	−0.4124	2.0071	0.2647
FBr	0.1574	0.5171	0.3373	−0.3597	2.1888	0.2547
ClBr	0.2064	0.7045	0.4555	−0.4981	3.7181	0.3741
N_2	−0.0372	0.1332	0.0480	−0.1457	0.3343	0.0252
P_2	−0.1473	0.1464	−0.0005	−0.2937	2.2166	0.0437
PN	0.0147	0.0940	0.0545	−0.0792	0.4265	0.0373
BN	−0.1282	0.0025	−0.0629	−0.1307	1.4116	−0.0030
O_2	0.1801	0.2090	0.1945	−0.0289	0.1319	0.0975
S_2	0.1538	0.1997	0.1767	−0.0459	0.5217	0.1306
SO	0.1278	0.1446	0.1362	−0.0168	0.1346	0.0834
CO	0.0309	0.0568	0.0439	−0.0259	0.0797	0.0196
LiF	−0.4419	0.0893	−0.1763	−0.5312	3.3022	−0.0186
NaCl	−0.4534	0.1941	−0.1297	−0.6475	7.0596	0.0344
MgO	−0.1065	0.0838	−0.0114	−0.1902	4.3190	0.0512

Source: Reprinted with permission from Bettens et al. [34]. Copyright 2019 Royal Society of Chemistry.

Table 19.4 The bond equilibrium distances of the neutral, cationic and anionic species, along with the differences in the bond lengths of their ionic and neutral species.

Systems	$r_{neutral}$	r_{cation}	r_{anion}	$r_{(neutral\text{-}cation)}$	$r_{(neutral\text{-}anion)}$
H_2	0.743	1.058	0.752	−0.315	−0.009
HF	0.918	1.001	0.935	−0.083	−0.017
HCl	1.277	1.316	1.307	−0.039	−0.03
HBr	1.411	1.443	1.461	−0.032	−0.05
F_2	1.396	1.290	1.920	0.106	−0.524
Cl_2	2.010	1.907	2.603	0.103	−0.593
Br_2	2.294	2.200	2.844	0.094	−0.55
FCl	1.634	1.540	2.168	0.094	−0.534
FBr	1.753	1.674	2.245	0.079	−0.492
ClBr	2.151	2.055	2.718	0.096	−0.567
N_2	1.097	1.114	1.177	−0.017	−0.08
P_2	1.900	1.904	1.985	−0.004	−0.085
PN	1.493	1.487	1.551	0.006	−0.058
BN	1.279	1.354	1.281	−0.075	−0.002
O_2	1.202	1.109	1.339	0.093	−0.137
S_2	1.903	1.832	2.019	0.071	−0.116
SO	1.491	1.430	1.584	0.061	−0.093
CO	1.129	1.113	1.171	0.016	−0.042
LiF	1.586	2.066	1.637	−0.48	−0.051
NaCl	2.409	2.869	2.537	−0.46	−0.128
MgO	1.781	1.828	1.821	−0.047	−0.04

Source: Reprinted with permission from Bettens et al. [34]. Copyright 2019 Royal Society of Chemistry.

This rule means that if the cationic bond distance is greater than the neutral bond distance, the application of external stretching force favors the lowering of vertical ionization energy (I) and *vice versa*.

$$\text{Rule 2: } \left(\frac{\partial A}{\partial F_{\text{ext}}}\right)_{F_{\text{ext}=0}} > 0 \text{ if } (r_{\text{neutral}} - r_{\text{anion}}) < 0$$

This rule is understandable in similar terms to rule 1. If the anionic bond distance is greater than the neutral bond distance, the application of external stretching force will tend to increase the vertical electron affinity (A).

$$\text{Rule 3: } \left(\frac{\partial I}{\partial F_{\text{ext}}}\right)_{F_{\text{ext}=0}} < 0 \text{ if the electron is removed from a bonding orbital}$$

This can be understood on the basis of bond order. If an electron is removed from a bonding orbital, that will decrease bond order, increasing bond length. Thus, the application of external pulling force, which tends to increase the bond length, decreases the vertical ionization energy (I).

$$\text{Rule 4: } \left(\frac{\partial A}{\partial F_{\text{ext}}}\right)_{F_{\text{ext}=0}} > 0 \text{ if the electron is added to an antibonding orbital}$$

This rule is similar to the previous rule. If an electron is added to an antibonding orbital, the bond order decreases, which leads to an elongation of the bond. Thus, the application of external forces tends to increase the electron affinity.

$$\text{Rule 5: } \left(\frac{\partial \eta}{\partial F_{\text{ext}}}\right)_{F_{\text{ext}=0}} < 0$$

As chemical hardness can be approximated as the difference in the energies of the lowest unoccupied molecular orbital (LUMO) and the highest occupied molecular orbital (HOMO), the application of external force tends to lower the LUMO and uplift the HOMO and thus decrease the energy gap. Hence, for all these diatomic systems, the hardness decreases with the application of the external stretching force.

It would be interesting to explore the effect of confinement on the mechanical strength of a few of these molecules. As confinement leads to interesting results, a few diatomic systems inside the C_{60} fullerene cage have been encapsulated, and F_{max} for the systems with and without confinement have been calculated. It was found that the mechanical strength of covalent bonds of a few systems increased many folds depending on the nature of the systems considered, while in other systems, the strength decreased.

As the rules devised for mechanochemical reactivity descriptors hold for the representative diatomic systems, these rules could be tried for real mechanochemical reactions. One of the ways in which these reactions can be conducted experimentally is the ball milling experiment. Many of the reactions have already been performed by Ranu and coworkers [27], and a theoretical investigation of these reactions is in progress. Since the systems considered in those reactions were not diatomic, specific bonds involved in those reactions were considered. The mechanical forces required to break those bonds were calculated, and the rules discussed earlier have been testified. It was found that as these complex systems were not diatomic, all the plans did not follow all the said rules.

19.3.3 Understanding Ball Milling Mechanochemical Processes with DFT Calculations and Microkinetic Modeling

As mechanochemistry has many potential applications primarily due to its solvent-free approach toward product formation, it has gained popularity among environmental scientists as it leads to sustainable and green chemistry. To gain insight into the mechanistic foundations of such reactions, Pladerall et al. [50] have studied a few experimental ball-milling reactions computationally by employing DFT calculations. Reaction times were reproduced in a series of Diels–Alder reactions and the synthesis of sulfonylguanidines.

As it is well known that the Diels–Alder reaction is the pericyclic reaction that progresses through a single cyclic transition state, a series of these reactions were examined by Anderson and Mack [51] under ball-milling conditions with varying frequencies. The study was based on all combinations between two dienophiles and three dienes (Figure 19.3). From Figure 19.3 we can say that the combination of two dienophiles and three dienes can produce six different systems, and all these systems were submitted to ball-milling with varying frequencies. The frequencies that were chosen for this were 15, 18, and 21 Hz. Each of these 18 had different yields, and the yields after three hours were reported. The reported yield and the computed values have been provided in Figure 19.4.

Dienes	R¹	R²	Dienophiles
H9:	H	H	**MA:** Maleic anhydride
MA9:	H	CH_3	**BQ:** Benzoquinone
DMA:	CH_3	CH_3	

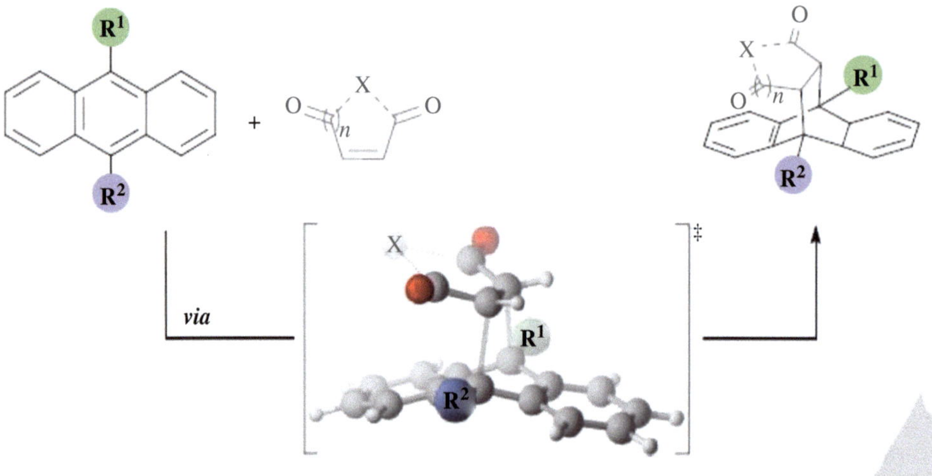

Figure 19.3 Set of Diels–Alder reactions in the study. Source: Reprinted with permission from Pladevall et al. [50]. Copyright 2021 Wiley VCH.

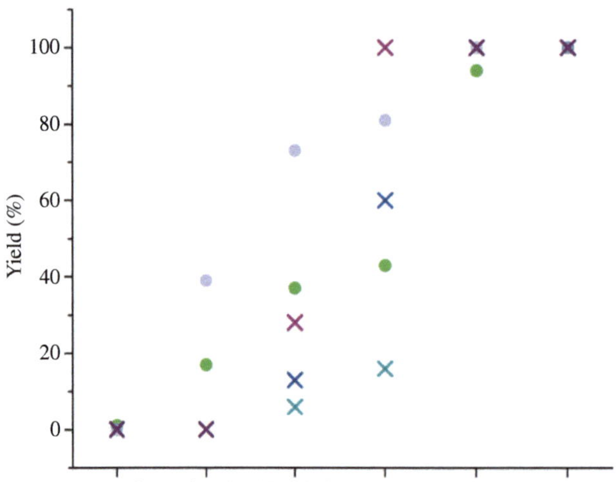

	H9-BQ	H9-MA	MA9-BQ	DMA-BQ	MA9-MA	DMA-MA
● Vacuum	0	0	6	16	100	100
● Continuum	0	0	13	60	100	100
× 15 Hz HS	0	0	28	100	100	100
× 18 Hz HS	0	17	37	43	94	100
× 21 Hz HS	0	39	73	81	100	100

Figure 19.4 Reported experimental yields (after three hours of reaction time) and the computed values for the Diels–Alder reactions. Source: Reprinted with permission from Pladevall et al. [50]. Copyright 2021 Wiley VCH.

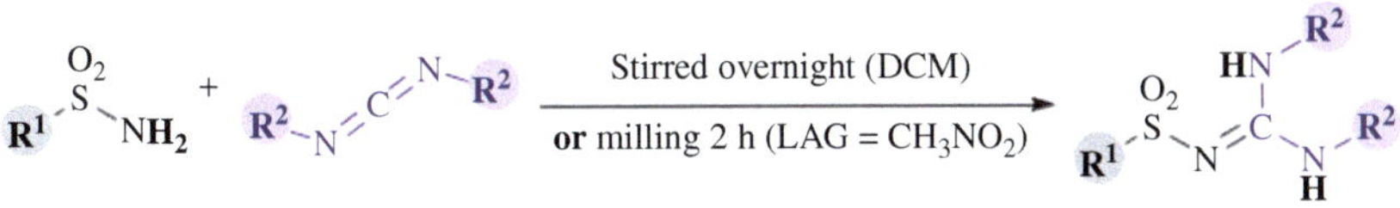

	R^1	R^2	
N1:	CF_3	*t*Bu	*Yield 92%*
N2:	*p*-Tolyl	*i*Pr	
N3:	*p*-Tolyl	*t*Bu	*No reaction*
N4:	CF_3	*p*-Tolyl	

Figure 19.5 *N*-Sulfonylguanidines synthesis reactions under study. Source: Reprinted with permission from Pladevall et al. [50]. Copyright 2021 Wiley VCH.

The free energy barriers were computed, including entropic corrections, so that these could be converted into rate constants for the microkinetic model to get a quantitative reproduction of experimental reaction times. For the computations, benchmarking was done, and B3LYP-D3 functional had been chosen in conjunction with The 6-311++G(d,p) basis set. As can be seen from Figure 19.5, some of the reactions are sensitive to the frequency of the ball mill, so it can be stated that in mechanochemistry, the reaction time can be modulated with frequency, and this has the advantage over other traditional methods. In order to incorporate the dielectric environment for these reactions, two options were considered. First, a process in a vacuum, and second, a medium with a dielectric as the average of that of the two reactants. For both these approaches, the results have been summarized in Figure 19.4.

As Diels–Alder reactions are insensitive towards dielectric environments due to negligible change in charge distribution during the process, both hypotheses produce similar results. The computed yields can be affected by the initial concentration that has been estimated by density value extrapolation from the literature, which implies a high concentration of the reactants. Ideally, this means a complete homogenization has been achieved, and thus, a "maximum" frequency has been used.

A good agreement has been observed between the experimental and the computed yield values for the fastest systems (with 100% yield) and the slowest systems (with 0% yield). The intermediate yield values could not be reproduced due to the exponential dependence of milling frequency on yield in the Eyring equation and because the maximum frequency had been taken. Second, the role of temperature has not been taken into consideration.

For the computational study of the synthesis of *N*-sulfonylguanidine, the experimental results summarized in Figure 19.5 have been taken. *N*-Sulfonylguanidine compounds have gained popularity due to their various applications in the field of pharmaceuticals and herbicides. Usually, synthesizing such compounds requires much time to complete, and in most cases, catalysts are needed. Figure 19.5 shows four sets of reactants (N1, N2, N3 and N4) were tested. Without a catalyst, only N1 reacts. This reaction took place in solution as well as by ball milling, and a significant reduction in reaction time has been observed when using mechanochemical conditions. The computational method focuses mainly on this particular synthesis to account for its considerable time reduction from 12 hours in the solution phase to 2 hours in the ball mill. The lack of reactivity in other systems was also accounted for.

A schematic of the free energy profile for these reactions has been shown in Figure 19.6. Energy data have been excluded from the schematic as different values were obtained for different dielectric environments. It can be observed from the schematic that intermediate I2 gets stabilized more due to the presence of FMSA sulphonamide in set N1. This difference in stabilization energy is relatively

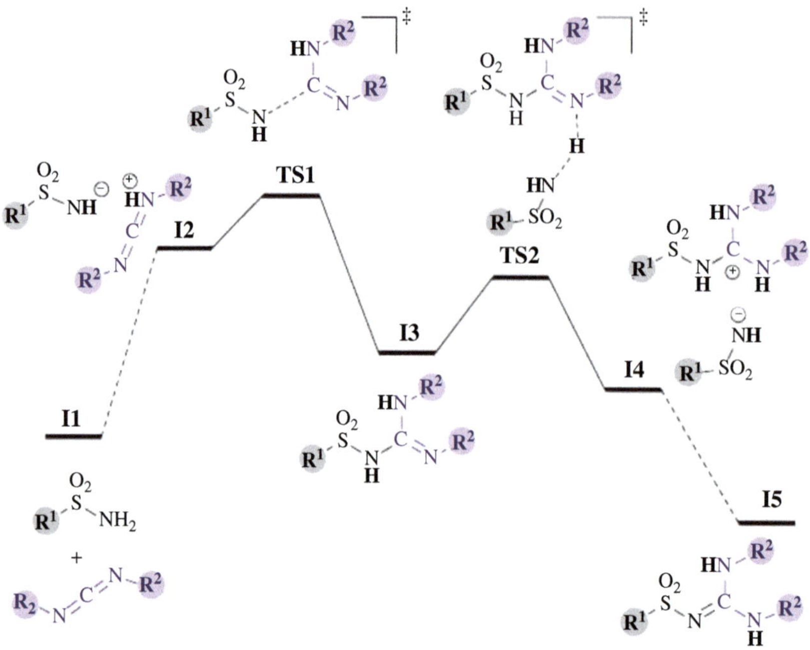

Figure 19.6 Schematic representation of the mechanism for the synthesis of *N*-sufonylguanidines. Source: Reprinted with permission from Pladevall et al. [50]. Copyright 2021 Wiley VCH.

high (9 kcal/mol) compared to *p*-tolylsulfonamide used in set N3. Due to this stabilizing effect, only the N1 set gets toward completion. The role of substituents is in deciding the barrier heights due to inductive effects. The rate-determining step is the C—N bond formation through TS1.

To have an idea about the kinetics of the reaction, the energy difference between TS1 and the reactants can be considered. There is a charge separation in intermediate I2, which can also be seen slightly in TS1. To estimate the reaction times, the microkinetic model can be considered. The results are tabulated in Table 19.5. The results of N2 and N4 were similar to that of N3 and hence not shown in Table 19.5. In this case, the dielectric environment is also simulated, just as in the case of Diels–Alder reaction.

Table 19.5 The time necessary for the quantitative conversion (99%) was obtained by the application of a kinetic model to the reaction of tBuCD with FMSA and TolSA.

	Media (ϵ)	Conc. (mol/l)	Predicted time (h)	Exp. time (h)
N1	DCM (8.9)	0.7	11 : 47	12
	Vacuum (1)	2.9	0 : 15	2
	Solid average (16.7)	2.9	1 : 08	2
N2	DCM (8.9)	0.7	9.95×10^8	>99
	Vacuum (1)	2.9	1.13×10^6	>99
	Solid average (16.7)	2.9	2.03×10^5	>99

Source: Reprinted with permission from Pladevall et al. [50]. Copyright 2021 Wiley VCH.

It can be seen that experimental data for the reaction time are not exact. From Table 19.5, it can be observed that there is an excellent agreement between the experimental and the computed data. For the N3 set, the reaction time, calculated to be very high, clearly shows an agreement with the experiment. For the N1 set, the computed reaction time comes out to be 11 : 47 hours which is in excellent agreement with the experimental reaction time of 12 hours. There is a slight discrepancy in the mechanochemical reaction when two different dielectric approaches are considered. For the vacuum case computation, the predicted time comes out to be 0.15 hours, which is relatively less than the actual experimental reaction time of 2 hours. For the solid average dielectric medium, the reaction time is computed to be 1 : 08 hours which is a better result as compared to the vacuum case.

So, it can be stated that mechanochemical conditions introduce a change in the dielectric environment of the reacting species. Thus, this reduction in reaction time can be associated with the difference in the concentration of reacting species due to the high effective reactant concentration accessible in the ball mill. So, it can be said that the standard DFT tool can be used to predict, design, and optimize mechanochemical reactions.

19.3.4 Explicit Force

Another way of applying force is explicitly and tensional stretching force. Examples of such explicit forces are force-modified potential energy surface (FMPES) [10], EFEI [11], and enforced geometry optimization (EGO) [12]. This method implies mechanical force directly and not indirectly by varying the distance. In these processes, the calculation of geometries and transition states and, hence activation energy barrier is performed by applying force in each direction. Ong et al. have generalized the 1D picture of Kauzmann and Eyring to multi-dimensions [10].

$$V_{\text{FMPES}}(r) = V_{\text{ab initio}}(r) + \sum_{i}^{N_{\text{attach}}} F_0 \left(\left\| r_i^{\text{fix}} - r_i \right\| - \left\| r_i^{\text{fix}} - r_i^0 \right\| \right) \tag{19.18}$$

The point in the molecule where an external force is applied is called an attachment point (AP). The Born–Oppenheimer PES $V_{\text{ab initio}}(r)$ is calculated by adding the potential at each point. F_0 is the external force, r_i^0, r_i^{fix} are the invariant and pulling point (PP). In a mechanochemical reaction, the pathways were rationalized, and Woodward–Hoffmann's rules were overcome using the FMPES method [10]. Under the thermal condition, the electrocyclic reaction is allowed in a conrotatory fashion ($4n\pi$ electrons), while the reaction proceeds in a disrotatory manner under a photochemical condition [52–54]. In the case of the mechanochemical ring opening of cyclobutene, these rules are circumvented. It was shown that under the mechanochemical condition, the forbidden disrotatory pathway proceeds through less activation barrier (Figure 19.7). Recently it was discussed whether it is meaningful to discuss the Woodward–Hoffmann rules in the context of mechanochemical reaction.

19.3.5 Dynamical Aspect of Mechanochemistry

In the last several decades, enormous development has been made in the field of MD simulation. Large proteins and polymers are studied using MD simulation using classical force fields. In the MD field, an essential example of force analysis is the force distribution analysis (FDA), which was used to study the mechanochemical properties of proteins [55]. Later, it was shown that protein molecules fail to undergo a conformational change, and hence, geometry alone is not sufficient for the quantification of mechanical stress. With ample conformational flexibility, the FDA approach does not converge. The β-sheets are highly resistant to mechanical force, and hence, their

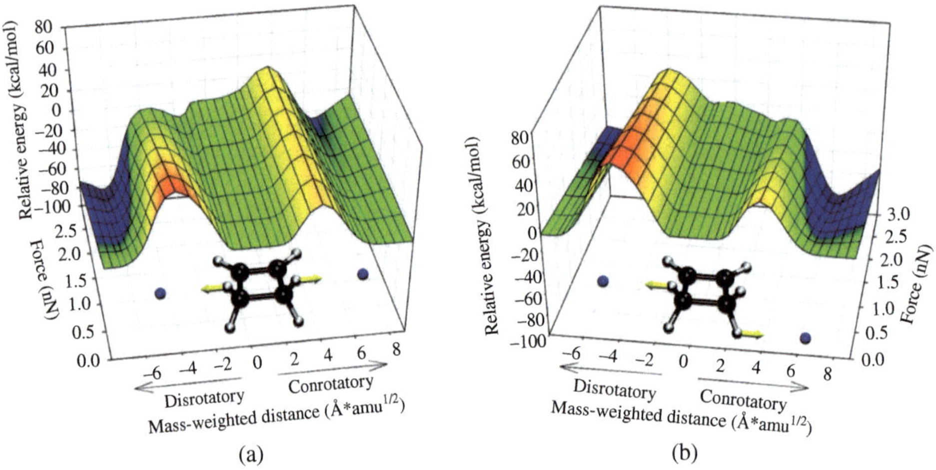

Figure 19.7 Different pulling scenarios facilitate the electrocyclic ring opening of cyclobutene. The fixed pulling points (PP) are symbolized by blue spheres. The yellow arrows indicate the molecular attachment points (AP). (a) In this pulling scenario, the thermally forbidden disrotatory pathway can be preferred if mechanical force is pulling two hydrogen atoms on the same side of the molecule apart. (b) The activation energy of the thermally allowed conrotatory pathway can be lowered even further if the force is stretching two hydrogen atoms on different sides of the molecule apart. Source: Reprinted from Ong et al. [10]. Copyright 2009 American Chemical Society.

lengths were determined using the FDA method [56–59]. The gain in rupture force was calculated using FDA, and the vertical deflection of strain was found through hydrogen bonding (Figure 19.8) [59]. Further studies were done on allosteric pathways [60, 61], protein unfolding [62], and other mechanochemical properties [63].

Mechanochemistry arises due to the application of force along the reaction coordinate, thereby affecting the topography of the concerned PES. This, in turn, affects the rate of the chemical reaction. Quapp and Bofill [64] utilized classical trajectories in order to analyze the salient features of mechanochemistry within reduced dimensional model potentials. They showed that when the force is such that the saddle point changes to a shoulder, then the mechanochemical action is fulfilled, and the activation barrier is decreased. This point is called the barrier breakdown point. On the original PES (i.e. without the application of external forces), different corridors are formed by different classical trajectories that describe the action of different forces. The valley ridge inflection points (VRI) represent the border regions of the corridors. The perturbed generic structures of triatomic (ABC) and tetratomic (AABB) are studied in the context of external perturbation due to force [65]. Hase and coworkers [66] computed the PES of stretched *n*-butane. They showed that the nature of the effective PES (under the influence of the external force) depends on the magnitude of stretching. If the force is less than that required to reach the VRI points on the C–C stretch potentials and the C–C torsions, then only one potential energy minimum is found. Herein, each bond gets elongated only. On the contrary, for stretching past these inflection points, the PES becomes bifurcated into three minima, thereby exhibiting interesting post-transition state effects.

However, the topography of the PES alone cannot decipher the actual reaction mechanism, and one needs to consider the corresponding phase-space structures in detail in order to shed some light on the reaction mechanism. Moreover, quantum dynamical features such as tunneling, interference, and resonances have not been accounted for in the aforementioned classical trajectory studies. Thus, one might pose the following question: how do quantum dynamical features

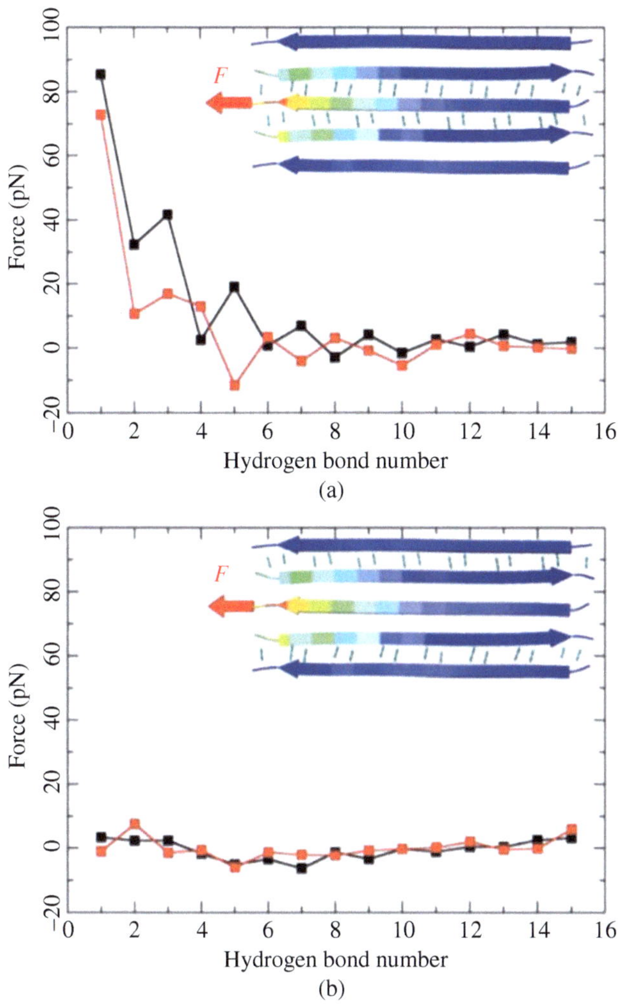

Figure 19.8 Force distribution patterns in the hydrogen bonds of a model β-sheet. (a) Forces in the hydrogen bonds between the central strand, on which the force is applied, and the two adjacent strands. Red and black indicate the upper and lower strand hydrogen bonds, respectively. In the β-sheet structure, high and low tension is symbolized by red and blue, respectively. (b) Analogous plots of the force distribution among the hydrogen bonds between the outermost strands of the β-sheet. Source: Reprinted with permission from Xiao et al. [59]. Copyright 2011 Royal Society of Chemistry.

influence the reaction dynamics when considered within the context of mechanochemistry? To this end, Bohm's quantum theory of motion (QTM) [67] provides an ideal theoretical framework for understanding the quantum phase-space structures on the model potentials. Within the formalism of QTM, the wave and particle characters of a physical system could be treated in a synchronous way, thus providing greater conceptual insights into the underlying quantum dynamics. As per QTM formalism, the wave function acts as a pilot-wave guiding ensemble element [67]. Thus, it becomes possible to bring in the hydrodynamic analogy within the QTM formalism, thereby attributing physical reality to the individual quantum trajectories [68–71]. The critical component in Bohm's QTM is the quantum potential term, which brings all the nonlocal aspects of quantum mechanics into consideration. The nonlinear nature of the quantum potential makes QTM an ideal theoretical framework for studying nonlinear dynamics [69–71] in the quantum domain. For the benefit of the reader, we outline the basic theoretical framework of QTM. The Bohmian hydrodynamic formulation [67–71] could be arrived at with the consideration of the polar form of the wave function $\psi(r,t) = R(r,t)\exp\left(\frac{iS(r,t)}{\hbar}\right)$. Substitution of this wave function into the time-dependent Schrödinger equation (TDSE) followed by separation of the real and imaginary parts provides the following continuity equation for density and a quantum Hamilton–Jacobi equation.

$$\frac{\partial \rho(r,t)}{\partial t} + \nabla \cdot \left(\rho \frac{1}{m} \nabla S\right) = 0 \tag{19.19}$$

$$-\frac{\partial S(r,t)}{\partial t} = \frac{1}{2m}(\nabla S)^2 + V(r,t) + Q(\rho; r,t) \tag{19.20}$$

$R(r, t)$ and $S(r, t)$ are real-valued functions.

The probability distribution associated with the concerned wave function has the form $\rho(r, t) = R(r, t)^2$. Upon taking the gradient of Eq. (19.20), one obtains a “Newton-like” equation of motion as follows:

$$m\frac{dv}{dt} = -\nabla(V + Q) = f_c + f_q \tag{19.21}$$

In the transformed Lagrangian framework [68], the equations of motion (Eqs. (19.19) and (19.20)) could be expressed as below:

$$\left(\frac{\partial}{\partial t} + \upsilon \cdot \nabla\right)\rho = \frac{d\rho}{dt} = -\rho\nabla \cdot \upsilon \tag{19.22}$$

$$\frac{dS}{dt} = \frac{1}{2}mv \cdot \upsilon - Q - V \tag{19.23}$$

The critical factor that introduces all the nonlocal attributes of the quantum domain into QTM ensembles is the quantum potential term [67], which could be expressed as follows:

$$Q(\rho; r,t) = -\frac{\hbar^2}{2m}\frac{1}{R}\nabla^2 R = -\frac{\hbar^2}{2m}\rho^{-\frac{1}{2}}\nabla^2\rho^{\frac{1}{2}} \tag{19.24}$$

QTM trajectories could be initiated according to the guidance [67] equation:

$$\frac{dr}{dt} = \frac{1}{m}\nabla S(r,t) \tag{19.25}$$

Therefore, it would be interesting to apply QTM to study the nonlinear dynamics of the model potentials mentioned above and decipher the influence of the quantum phase space structure in affecting the reaction dynamics.

On the other hand, classical trajectory studies fail to account for changes in chemical bonding due to the neglect of quantum mechanical effects. In this regard, one might consider the quantum fluid density functional theory (QFDFT) [72–79] for analyzing the changes in chemical bonding within a time-dependent framework. Details regarding QFDFT could be found elsewhere [72–79]. Within the framework of QFDFT, the current density and time-dependent density are evaluated. By making use of the time-dependent density, one could calculate various density-based reactivity descriptors [80], for example. It is also possible to compute information theory-based descriptors such as Shannon entropy. To incorporate the effects of external mechanical forces within the QFDFT framework, one could consider the potential described by Eq. (19.1), for example. One could introduce this potential term “by hand” within the external potential term within the QFDFT framework and solve the pertinent integrodifferential equation: generalized nonlinear Schrodinger equation (GNLSE). Thus, by following the aforementioned formulations, one could study the reaction dynamics in detail.

19.4 Concluding Remarks

Understanding mechanochemical reactions through quantum chemistry is a blooming field. In 2000, Beyer published a very straightforward and easy-to-implement method in mechanochemistry called constrained geometries simulate external force (COGEF). In the simplest case of a diatomic molecule, the bond length increases when an external force is applied to the two atoms. Thus, it can be said that the bond length represents the amount of external force. In this method, typically, geometry optimization is done by scanning the bond length between two atoms. The Morse potential represents the PES. When sufficient force is applied, the bond rupture takes place. Using the COGEF method, the change in CDFT-based descriptors, namely, ionization energy, electron affinity, hardness, electrophilicity, electronegativity etc., were studied in some diatomic molecules. The Diels–Alder reaction under ball milling conditions is discussed. The Diels–Alder reaction was studied in a series of reactions using DFT calculations. The time required for the completion of reactions and yields was reproduced. The free energy barrier was computed using a microkinetic model with entropic corrections. The ball milling synthesis of *N*-sulfonylguanidines was also discussed under ball milling conditions. The relevant free energy diagram was calculated. Using explicit forces, the electrocyclic ring opening reaction of cyclobutene was justified. It was shown that under the mechanochemical process, the Woodward–Hoffmann rule is relaxed. Finally, the dynamic aspect of mechanochemical reactions was discussed using the MD simulation and Newton trajectory (NT) methods.

Acknowledgments

We are delighted to dedicate this chapter to Prof. B.C. Ranu, “The Green Chemist of India.” PKC would like to thank DST, New Delhi, for the J.C. Bose National Fellowship. SGP thanks IIT Kharagpur for a postdoctoral fellowship. RJ thanks IIT Kharagpur for SRF. DC thanks SERB, New Delhi (File No. SRG/2022/001280) for the financial assistance.

References

1 Takacs, L. (2013). The historical development of mechanochemistry. *Chem. Soc. Rev.* 42 (18): 7649. https://doi.org/10.1039/c2cs35442j.

2 Stauch, T. and Dreuw, A. (2017). Quantum chemical strain analysis for mechanochemical processes. *Acc. Chem. Res.* 50 (4): 1041–1048. https://doi.org/10.1021/acs.accounts.7b00038.

3 Stauch, T. and Dreuw, A. (2016). Advances in quantum mechanochemistry: electronic structure methods and force analysis. *Chem. Rev.* 116 (22): 14137–14180. https://doi.org/10.1021/acs.chemrev.6b00458.

4 McNaught, A.D., Wilkinson, A., and IUPAC (1997). *Compendium of Chemical Terminology (the “Gold Book”)*, 2e. Oxford, UK: Blackwell Scientific Publications.

5 James, S.L., Adams, C.J., Bolm, C. et al. (2012). Mechanochemistry: opportunities for new and cleaner synthesis. *Chem. Soc. Rev.* 41 (1): 413–447. https://doi.org/10.1039/C1CS15171A.

6 Cravotto, G., Gaudino, E.C., and Cintas, P. (2013). On the mechanochemical activation by ultrasound. *Chem. Soc. Rev.* 42 (18): 7521. https://doi.org/10.1039/c2cs35456j.

7 Neuman, K.C. and Nagy, A. (2008). Single-molecule force spectroscopy: optical tweezers, magnetic tweezers and atomic force microscopy. *Nat. Methods* 5 (6): 491–505. https://doi.org/10.1038/nmeth.1218.

8 Nikitina, E.A., Khavryutchenko, V.D., Sheka, E.F. et al. (1999). Deformation of poly(dimethylsiloxane) oligomers under uniaxial tension: quantum chemical view. *J. Phys. Chem. A* 103 (51): 11355–11365. https://doi.org/10.1021/jp990221v.

9 Beyer, M.K. (2000). The mechanical strength of a covalent bond calculated by density functional theory. *J. Chem. Phys.* 112 (17): 7307–7312. https://doi.org/10.1063/1.481330.

10 Ong, M.T., Leiding, J., Tao, H. et al. (2009). First principles dynamics and minimum energy pathways for mechanochemical ring opening of cyclobutene. *JACS* 131 (18): 6377–6379. https://doi.org/10.1021/ja8095834.

11 Ribas-Arino, J., Shiga, M., and Marx, D. (2009). Understanding covalent mechanochemistry. *Angew. Chem. Int. Ed.* 48 (23): 4190–4193. https://doi.org/10.1002/anie.200900673.

12 Wolinski, K. and Baker, J. (2009). Theoretical predictions of enforced structural changes in molecules. *Mol. Phys.* 107 (22): 2403–2417. https://doi.org/10.1080/00268970903321348.

13 Boulatov, R. (ed.) (2016). *Polymer Mechanochemistry*. Springer Cham https://doi.org/10.5860/CHOICE.195523.

14 Beyer, M.K. and Clausen-Schaumann, H. (2005). Mechanochemistry: the mechanical activation of covalent bonds. *Chem. Rev.* 105 (8): 2921–2948. https://doi.org/10.1021/cr030697h.

15 Rief, M., Gautel, M., Oesterhelt, F. et al. (1997). Reversible unfolding of individual titin immunoglobulin domains by AFM. *Science* 276 (5315): 1109–1112. https://doi.org/10.1126/science.276.5315.1109.

16 Wiita, A.P., Perez-Jimenez, R., Walther, K.A. et al. (2007). Probing the chemistry of thioredoxin catalysis with force. *Nature* 450 (7166): 124–127. https://doi.org/10.1038/nature06231.

17 Alegre-Cebollada, J., Perez-Jimenez, R., Kosuri, P., and Fernandez, J.M. (2010). Single-molecule force spectroscopy approach to enzyme catalysis. *J. Biol. Chem.* 285 (25): 18961–18966. https://doi.org/10.1074/jbc.R109.011932.

18 Cui, S., Yu, Y., and Lin, Z. (2009). Modeling single chain elasticity of single-stranded DNA: a comparison of three models. *Polymer* 50 (3): 930–935. https://doi.org/10.1016/j.polymer.2008.12.012.

19 Ashokkumar, M., Lee, J., Kentish, S., and Grieser, F. (2007). Bubbles in an acoustic field: an overview. *Ultrason. Sonochem.* 14 (4): 470–475. https://doi.org/10.1016/j.ultsonch.2006.09.016.

20 Patra, S.G., Sathiyan, K., Meistelman, M., and Zidki, T. (2020). Green synthesis of M0 nanoparticles (M = Pd, Pt, and Ru) for electrocatalytic hydrogen evolution. *Isr. J. Chem.* 60 (5–6): 630–637. https://doi.org/10.1002/ijch.201900175.

21 Ando, T., Sumi, S., Kawate, T. et al. (1984). Sonochemical switching of reaction pathways in solid–liquid two-phase reactions. *J. Chem. Soc., Chem. Commun.* 7: 439–440. https://doi.org/10.1039/C39840000439.

22 Fritze, U.F. and Von Delius, M. (2016). Dynamic disulfide metathesis induced by ultrasound. *Chem. Commun.* 52 (38): 6363–6366. https://doi.org/10.1039/c6cc02034h.

23 Lenhardt, J.M., Ong, M.T., Choe, R. et al. (2010). Trapping a diradical transition state by mechanochemical polymer extension. *Science* 329 (5995): 1057–1060. https://doi.org/10.1126/science.1193412.

24 Boldyreva, E. (2013). Mechanochemistry of inorganic and organic systems: what is similar, what is different? *Chem. Soc. Rev.* 42 (18): 7719. https://doi.org/10.1039/c3cs60052a.

25 Friščić, T., James, S.L., Boldyreva, E.V. et al. (2015). Highlights from faraday discussion 170: challenges and opportunities of modern mechanochemistry, Montreal, Canada, 2014. *Chem. Commun.* 51 (29): 6248–6256. https://doi.org/10.1039/C5CC90113H.

26 Shy, H., Mackin, P., Orvieto, A.S. et al. (2014). The two-step mechanochemical synthesis of porphyrins. *Faraday Discuss.* 170: 59–69. https://doi.org/10.1039/C3FD00140G.

27 Mukherjee, N., Chatterjee, T., and Ranu, B.C. (2013). Reaction under ball-milling: solvent-, ligand-, and metal-free synthesis of unsymmetrical diaryl chalcogenides. *J. Org. Chem.* 78 (21): 11110–11114. https://doi.org/10.1021/jo402071b.

28 Mukherjee, N., Chatterjee, T., and Ranu, B.C. (2015). Transition metal- and solvent-free synthesis of unsymmetrical diaryl sulfides and selenides under ball-milling. *Arkivoc* 2016 (2): 53–61. https://doi.org/10.3998/ark.5550190.p009.238.

29 Ranu, B.C., Chatterjee, T., and Mukherjee, N. (2014). Chapter 1: Carbon–heteroatom bond forming reactions and heterocycle synthesis under ball milling. In: *Ball Milling Towards Green Synthesis: Applications, Projects, Challenges* (ed. A. Stolle and B. Ranu), 1–33. Royal Society of Chemistry https://doi.org/10.1039/9781782621980-00001.

30 Hugel, T., Holland, N.B., Cattani, A. et al. (2002). Single-molecule optomechanical cycle. *Science* 296 (5570): 1103–1106. https://doi.org/10.1126/science.1069856.

31 Barrett, C.J., Mamiya, J.I., Yager, K.G., and Ikeda, T. (2007). Photo-mechanical effects in azobenzene-containing soft materials. *Soft Matter* 3 (10): 1249–1261. https://doi.org/10.1039/b705619b.

32 Bandara, H.M.D. and Burdette, S.C. (2012). Photoisomerization in different classes of azobenzene. *Chem. Soc. Rev.* 41 (5): 1809–1825. https://doi.org/10.1039/c1cs15179g.

33 Koumura, N., Zijlstra, R.W.J., van Delden, R.A. et al. (1999). Light-driven monodirectional molecular rotor. *Nature* 401 (6749): 152–155. https://doi.org/10.1038/43646.

34 Bettens, T., Alonso, M., Geerlings, P., and De Proft, F. (2019). Implementing the mechanical force into the conceptual DFT framework: understanding and predicting molecular mechanochemical properties. *Phys. Chem. Chem. Phys.* 21 (14): 7378–7388. https://doi.org/10.1039/C8CP07349J.

35 Parr, R.G., Donnelly, R.A., Levy, M., and Palke, W.E. (1978). Electronegativity: the density functional viewpoint. *J. Chem. Phys.* 68 (8): 3801–3807. https://doi.org/10.1063/1.436185.

36 Parr, R.G. and Pearson, R.G. (1983). Absolute hardness: companion parameter to absolute electronegativity. *JACS* 105 (26): 7512–7516. https://doi.org/10.1021/ja00364a005.

37 Chattaraj, P.K., Sarkar, U., and Roy, D.R. (2006). Electrophilicity index. *Chem. Rev.* 106 (6): 2065–2091. https://doi.org/10.1021/cr040109f.

38 Chattaraj, P.K. and Roy, D.R. (2007). Update 1 of: electrophilicity index. *Chem. Rev.* 107 (9): PR46–PR74. https://doi.org/10.1021/cr078014b.

39 Chattaraj, P.K., Giri, S., and Duley, S. (2011). Update 2 of: electrophilicity index. *Chem. Rev.* 111 (2): PR43–PR75. https://doi.org/10.1021/cr100149p.

40 Sarkar, U. and Chattaraj, P.K. (2021). Reactivity dynamics. *J. Phys. Chem. A* 125 (10): 2051–2060. https://doi.org/10.1021/acs.jpca.0c10788.

41 Kauzmann, W. and Eyring, H. (1940). The viscous flow of large molecules. *JACS* 62 (11): 3113–3125. https://doi.org/10.1021/ja01868a059.

42 Evans, E. (2001). Probing the relation between force—lifetime—and chemistry in single molecular bonds. *Annu. Rev. Biophys. Biomol. Struct.* 30 (1): 105–128. https://doi.org/10.1146/annurev.biophys.30.1.105.

43 Freund, L.B. (2009). Characterizing the resistance generated by a molecular bond as it is forcibly separated. *Proc. Natl. Acad. Sci. U.S.A.* 106 (22): 8818–8823. https://doi.org/10.1073/pnas.0903003106.

44 Lo, Y.S., Simons, J., and Beebe, T.P. (2002). Temperature dependence of the biotin-avidin bond-rupture force studied by atomic force microscopy. *J. Phys. Chem. B* 106 (38): 9847–9852. https://doi.org/10.1021/jp020863.

45 Wilson, E.B., Decius, J.C., Cross, P.C., and Sundheim, B.R. (1955). Molecular vibrations: the theory of infrared and Raman vibrational spectra. *J. Electrochem. Soc.* 102 (9): 235C. https://doi.org/10.1149/1.2430134.

46 Bakken, V. and Helgaker, T. (2002). The efficient optimization of molecular geometries using redundant internal coordinates. *J. Chem. Phys.* 117 (20): 9160–9174. https://doi.org/10.1063/1.1515483.

47 Stauch, T. and Dreuw, A. (2016). Knots “choke off” polymers upon stretching. *Angew. Chem. Int. Ed.* 55 (2): 811–814. https://doi.org/10.1002/anie.201508706.

48 Stauch, T. and Dreuw, A. (2016). Stiff-stilbene photoswitch ruptures bonds not by pulling but by local heating. *Phys. Chem. Chem. Phys.* 18 (23): 15848–15853. https://doi.org/10.1039/C6CP02395A.

49 Stauch, T., Günther, B., and Dreuw, A. (2016). Can strained hydrocarbons be “forced” to be stable? *J. Phys. Chem. A* 120 (36): 7198–7204. https://doi.org/10.1021/acs.jpca.6b05461.

50 Pladevall, B.S., Aguirre, A., and Maseras, F. (2021). Understanding ball milling mechanochemical processes with DFT calculations and microkinetic modeling. *ChemSusChem* 14 (13): 2763–2768. https://doi.org/10.1002/cssc.202100497.

51 Andersen, J.M. and Mack, J. (2017). Decoupling the Arrhenius equation via mechanochemistry. *Chem. Sci.* 8 (8): 5447–5453. https://doi.org/10.1039/C7SC00538E.

52 Woodward, R.B. and Hoffmann, R. (1969). The conservation of orbital symmetry. *Angew. Chem. Int. Ed. Engl.* 8 (11): 781–853. https://doi.org/10.1002/anie.196907811.

53 Hoffmann, R. and Woodward, R.B. (1965). Selection rules for concerted cycloaddition reactions. *JACS* 87 (9): 2046–2048. https://doi.org/10.1021/ja01087a034.

54 Woodward, R.B. and Hoffmann, R. (1965). Stereochemistry of electrocyclic reactions. *JACS* 87 (2): 395–397. https://doi.org/10.1021/ja01080a054.

55 Stacklies, W., Vega, M.C., Wilmanns, M., and Gräter, F. (2009). Mechanical network in titin immunoglobulin from force distribution analysis. *PLoS Comput. Biol.* 5 (3): https://doi.org/10.1371/journal.pcbi.1000306.

56 Eom, K., Li, P., Makarov, D.E., and Rodin, G.J. (2003). Relationship between the mechanical properties and topology of cross-linked polymer molecules: parallel strands maximize the strength of model polymers and protein domains. *J. Phys. Chem. B* 107 (34): 8730–8733. https://doi.org/10.1021/jp035178x.

57 Hoffmann, T., Tych, K.M., Brockwell, D.J., and Dougan, L. (2013). Single-molecule force spectroscopy identifies a small cold shock protein as being mechanically robust. *J. Phys. Chem. B* 117 (6): 1819–1826. https://doi.org/10.1021/jp310442s.

58 He, C., Lamour, G., Xiao, A. et al. (2014). Mechanically tightening a protein slipknot into a trefoil knot. *JACS* 136 (34): 11946–11955. https://doi.org/10.1021/ja503997h.

59 Xiao, S., Stacklies, W., Debes, C., and Gräter, F. (2011). Force distribution determines optimal length of β-sheet crystals for mechanical robustness. *Soft Matter* 7 (4): 1308–1311. https://doi.org/10.1039/c0sm00510j.

60 Louet, M., Seifert, C., Hensen, U., and Gräter, F. (2015). Dynamic allostery of the catabolite activator protein revealed by interatomic forces. *PLoS Comput. Biol.* 11 (8): 1–17. https://doi.org/10.1371/journal.pcbi.1004358.

61 Zhou, J., Bronowska, A., Le Coq, J. et al. (2015). Allosteric regulation of focal adhesion kinase by PIP2 and ATP. *Biophys. J.* 108 (3): 698–705. https://doi.org/10.1016/j.bpj.2014.11.3454.

62 Baldauf, C., Schneppenheim, R., Stacklies, W. et al. (2009). Shear-induced unfolding activates von Willebrand factor A2 domain for proteolysis. *J. Thromb. Haemost.* 7 (12): 2096–2105. https://doi.org/10.1111/j.1538‐7836.2009.03640.x.

63 Wang, B., Xiao, S., Edwards, S.A., and Gräter, F. (2013). Isopeptide bonds mechanically stabilize Spy0128 in bacterial pili. *Biophys. J.* 104 (9): 2051–2057. https://doi.org/10.1016/j.bpj.2013.04.002.

64 Quapp, W. and Bofill, J.M. (2018). Mechanochemistry on the Müller-Brown surface by Newton trajectories. *Int. J. Quantum Chem.* 118 (8): e25522. https://doi.org/10.1002/qua.25522.

65 Quapp, W., Bofill, J.M., and Ribas-Ariño, J. (2018). Toward a theory of mechanochemistry: simple models from the very beginnings. *Int. J. Quantum Chem.* 118 (23): 1–19. https://doi.org/10.1002/qua.25775.

66 Lourderaj, U., McAfee, J.L., and Hase, W.L. (2008). Potential energy surface and unimolecular dynamics of stretched *n*-butane. *J. Chem. Phys.* 129: 094701. https://doi.org/10.1063/1.2969898.

67 Holland, P.R. (1993). *The Quantum Theory of Motion*. Cambridge: Cambridge University Press.

68 Lopreore, C.L. and Wyatt, R.E. (1999). Quantum wave packet dynamics with trajectories. *Phys. Rev. Lett.* 82: 5190. https://doi.org/10.1103/PhysRevLett.82.5190.

69 Sengupta, S. and Chattaraj, P.K. (1996). The quantum theory of motion and signatures of chaos in the quantum behaviour of a classically chaotic system. *Phys. Lett. A* 215: 119–127. https://doi.org/10.1016/0375-9601(96)00240-X.

70 Chakraborty, D., Kar, S., and Chattaraj, P.K. (2015). Orbital free DFT versus single density equation: a perspective through quantum domain behavior of a classically chaotic system. *Phys. Chem. Chem. Phys.* 17: 31516–31529. https://doi.org/10.1039/C5CP00995B.

71 Chakraborty, D. and Chattaraj, P.K. (2014). Quantum equivalence of a driven triple-well Van der Pol oscillator: a QTM study. *Chem. Phys.* 438: 7–15. https://doi.org/10.1016/j.chemphys.2014.04.005.

72 Deb, B.M. and Chattaraj, P.K. (1989). Density-functional and hydrodynamical approach to ion-atom collisions through a new generalized nonlinear Schrödinger equation. *Phys. Rev. A* 39: 1696. https://doi.org/10.1103/PhysRevA.39.1696.

73 Deb, B.M., Chattaraj, P.K., and Mishra, S. (1991). Time-dependent quantum-fluid density-functional study of high-energy proton-helium collisions. *Phys. Rev. A* 43: 1248–1257. https://doi.org/10.1103/PhysRevA.43.1248.

74 Chattaraj, P.K. and Maiti, B. (2004). Regioselectivity in the chemical reactions between molecules and protons: a quantum fluid density functional study. *J. Phys. Chem. A* 108: 658–664. https://doi.org/10.1021/jp035156a.

75 Chattaraj, P.K. and Sengupta, S. (1999). Chemical hardness as a possible diagnostic of the chaotic dynamics of Rydberg atoms in an external field. *J. Phys. Chem. A* 103: 6122–6126. https://doi.org/10.1021/jp990242p.

76 Chattaraj, P.K. and Sengupta, S. (1997). Dynamics of chemical reactivity indices for a many-electron system in its ground and excited states. *J. Phys. Chem. A* 101: 7893–7900. https://doi.org/10.1021/jp971408u.

77 Chattaraj, P.K. and Poddar, A. (1999). Chemical reactivity and excited-state density functional theory. *J. Phys. Chem. A* 103: 1274–1275. https://doi.org/10.1021/jp983821n.

78 Sarkar, U., Khatua, M., and Chattaraj, P.K. (2012). A tug-of-war between electronic excitation and confinement in a dynamical context. *Phys. Chem. Chem. Phys.* 14: 1716–1727. https://doi.org/10.1039/C1CP22862E.

79 Khatua, M. and Chattaraj, P.K. (2013). Molecular reactivity dynamics in a confined environment. *Phys. Chem. Chem. Phys.* 15: 5588–5614. https://doi.org/10.1039/C3CP43511C.

80 Chakraborty, D. and Chattaraj, P.K. (2021). Conceptual density functional theory based electronic structure principles. *Chem. Sci.* 12: 6264–6279. https://doi.org/10.1039/D0SC07017C.

20

Molecular Electron Density and Electrostatic Potential and Their Applications

Shyam V.K. Panneer[1,2], Masiyappan Karuppusamy[1,2,3], Kanagasabai Balamurugan[1,2,3], Sathish K. Mudedla[2], Mahesh K. Ravva[4] and Venkatesan Subramanian[1,5]

[1] *Centre for High Computing, CSIR-Central Leather Research Institute (CSIR-CLRI), Chennai, India*
[2] *Inorganic and Physical Chemistry Laboratory, CSIR-Central Leather Research Institute (CSIR-CLRI), Chennai, India*
[3] *Academy of Scientific and Innovative Research (AcSIR), Ghaziabad, India*
[4] *Department of Chemistry, SRM University – AP, Amaravati, India*
[5] *Department of Chemistry, Indian Institute of Technology–Madras, Chennai, India*

20.1 Introduction

Bader et al. have developed the concept of the theory of atoms in molecules (AIM) based on molecular electron density (MED) [1]. They have explored its applications in various fields [2]. Currently, in the literature, this topic is referred to as the quantum theory of atoms in molecules (QTAIM). It is based on the topographical analysis of electron density (ED) ($\rho(\mathbf{r})$) and its gradient ($\nabla\rho(\mathbf{r})$) and Laplacian ($\nabla^2\rho(\mathbf{r})$). The objectives of QTAIM are to describe the structure, covalent and noncovalent interactions, and associated chemical reactivity of molecules. The pioneering contributions of Bader and coworkers have paved the way for the analysis of other scalar fields, such as bare nuclear potential (BNP) [3] and molecular electrostatic potential (MESP) [4]. The topographical analysis of MED and MESP has received widespread attention from researchers to gain insight into covalent and noncovalent interactions and chemical reactivity [4a, 5]. The electron density is defined as

$$\rho(\boldsymbol{r};\boldsymbol{X}) = N\int d\tau'\psi^*(\boldsymbol{x};\boldsymbol{X})\psi(\boldsymbol{x};\boldsymbol{X}) \tag{20.1}$$

The QTAIM provides insights on chemical bonding and molecular structure through the bond path (BP) and critical points (CPs). A CP is a point at which the gradient of the electron density vector vanishes, $\nabla\rho(\mathbf{r}) = 0$. It has been shown that CPs and the associated value of the electron density at the CP yield profound information on the molecular structure, chemical bonding, and interaction. The Laplacian ($\nabla^2\rho(\mathbf{r})$) is a measure of local concentration of electron density. It can have positive or negative values depending on the accumulation or depletion of electron density at a particular point. It is possible to characterize the covalent and noncovalent interactions involving molecules and clusters with the help of the QTAIM [1]. Although we can obtain rich chemical information from the topological analysis of MED, it does not provide a clear picture of nonbonded electrons, i.e. lone pairs in a molecule [5b]. There is no local maximum in the lone pair region of electron density. However, it is found that the topological analysis of Laplacian of electron density yields an overall description of electron density distribution in molecules [2e]. Further, it is possible

Electron Density: Concepts, Computation and DFT Applications, First Edition.
Edited by Pratim Kumar Chattaraj and Debdutta Chakraborty.

to understand the profound physical origin of valence shell electron pair repulsion theory (VSEPR) with the help of the Laplacian of electron density [2e, 6].

With a view to unravel the connection between molecular structure and associated chemical reactivity, Bonaccorsi et al. have employed the MESP for locating protonation sites in DNA base pairs [4a]. In a similar direction, Politzer and coworkers have made numerous significant contributions to MESP [4d–g, 7]. They have described bonding, chemical reactivity and biological activity, inductive effect, resonance, etc., with the aid of MESP. Specifically, they have mapped the positive and negative regions of MESP of a molecule with its molecular surface. Gadre and his coworkers have made several outstanding contributions to this field in unraveling various fundamental concepts of chemical reactivity and interaction using topological analysis of MESP as a tool [5b–e, 8]. They have shown that the uniqueness of MESP can provide crucial information on the lone pair regions in a molecule [9]. Along with Gadre and coworkers, we have used MESP and electrostatic potential for intermolecular complexation (EPIC) to compare the hydration of hydrogen-bonded cytosine dimers with stacked cytosine dimers [5f]. The following sections discuss basic ideas of topography analysis of MED and MESP.

20.2 Topography Analysis of Scalar Fields

20.2.1 Molecular Electron Density

It is based on the determination and characterization of the scalar field with the help of isosurfaces, contours, gradient paths, and CPs. In the case of electron density, the CP is defined as the point at which $\nabla\rho = 0$. It is defined as

$$\nabla\rho(x, y, z) = \boldsymbol{i}\frac{\partial\rho}{\partial x} + \boldsymbol{j}\frac{\partial\rho}{\partial y} + \boldsymbol{k}\frac{\partial\rho}{\partial z} = 0 \tag{20.2}$$

The CPs are characterized by the calculation of eigenvalues of the following Hessian matrix.

$$\nabla^2\rho(\boldsymbol{r}) = \begin{bmatrix} \frac{\partial^2\rho(\boldsymbol{r})}{\partial x^2} & \frac{\partial^2\rho(\boldsymbol{r})}{\partial x\partial y} & \frac{\partial^2\rho(\boldsymbol{r})}{\partial x\partial z} \\ \frac{\partial^2\rho(\boldsymbol{r})}{\partial y\partial x} & \frac{\partial^2\rho(\boldsymbol{r})}{\partial y^2} & \frac{\partial^2\rho(\boldsymbol{r})}{\partial y\partial z} \\ \frac{\partial^2\rho(\boldsymbol{r})}{\partial z\partial x} & \frac{\partial^2\rho(\boldsymbol{r})}{\partial z\partial y} & \frac{\partial^2\rho(\boldsymbol{r})}{\partial z^2} \end{bmatrix} \tag{20.3}$$

$$\nabla^2\rho(\boldsymbol{r}_c) = \frac{\partial^2\rho(\boldsymbol{r}_c)}{dx'^2} + \frac{\partial^2\rho(\boldsymbol{r}_c)}{dy'^2} + \frac{\partial^2\rho(\boldsymbol{r}_c)}{dz'^2} = \lambda_1 + \lambda_2 + \lambda_3 \tag{20.4}$$

where λ_1, λ_2, and λ_3 are eigenvalues of the Hessian matrix. The possible signatures of eigenvalues are (+,+,+), (+,+,−) (+,−,−), and (−,−,−), which determine the nature of maxima/minima/saddle point of the electron density distribution.

In the topography analysis, a nondegenerate CP is characterized by (ω, σ) where ω is the number of nonzero eigenvalues, and the signature (σ) is the algebraic sum of the signs of eigenvalues of the Hessian matrix. There are four possible values for CPs of rank three. They are

(a) (3, −3): all the curvatures are negative, and ρ is a local maximum at $\boldsymbol{r}_c$.
(b) (3, −1): two curvatures are negative, and ρ is a maximum at $\boldsymbol{r}_c$ in the plane defined by their corresponding axes, and ρ is a minimum at $\boldsymbol{r}_c$ along the third axis perpendicular to the above-mentioned plane.

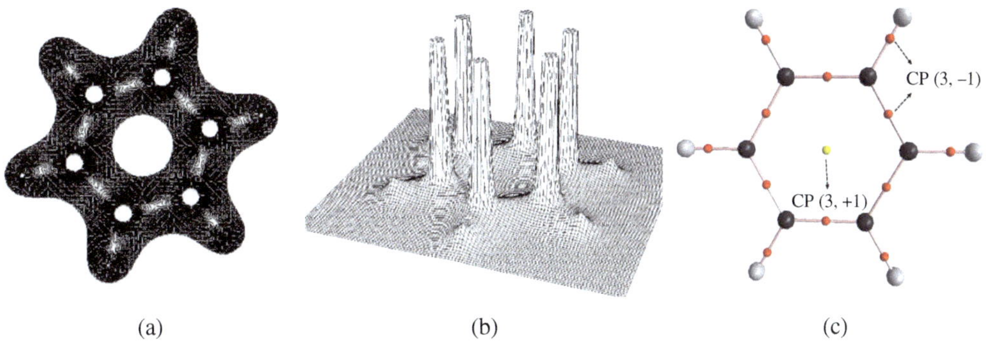

Figure 20.1 (a) Electron density contour map, (b) Relief map, and (c) Molecular graph of Benzene. (Reproduced from Ref. [5a] with permission).

(c) (3, +1): two curvatures are positive and ρ is a minimum at CP in the plane defined by their corresponding axes, and ρ is a maximum at CP along the third axis perpendicular to the above-mentioned plane.
(d) (3, +3): all the curvatures are positive, and ρ is a local minimum at $\boldsymbol{r}_{\mathbf{c}}$.

The topological electron density distribution of benzene (**BZ**) is depicted in Figure 20.1, which shows that (3, −3) CP occurs at the nuclear positions. In QTAIM, the nucleus of an atom is considered to be an attractor of the gradient of the ED distribution of molecules. It consists of a region of three-dimensional space called a basin. Hence, an atom can be considered to be the unification of the attractor and its accompanying basin. For example, **BZ** (Figure 20.1) has a (3, −1) CP between each adjacent nucleus. A (3, −1) CP is referred to as a bond critical point (BCP). The gradient path pairs start from each (3, −1) CP and end at the neighboring attractor, describing a line that is defined as the atomic interaction line. It connects any two adjacent nuclei through the electron density distribution, and hence ρ is maximum along this line than any other neighboring lines. The atomic interaction line of an equilibrium geometry implies that the atoms in a molecule are bonded and called the bond path. The (3, −1) CP and its corresponding atomic interaction line give information about the electron density distribution between any two bonded nuclei. A molecular graph consists of all these topological features. Other CPs of rank three arise due to the unique geometry of the molecule, which can be deduced by the corresponding bond paths. For example, **BZ** has bond paths that form a ring, indicating its cyclic structure. A ring critical point (RCP) is represented by (3, +1), which lies on the inside of the ring. Similarly, a cage critical point (CCP) (3, +3) can be found in the bond paths enclosing the interior of a molecule with ring surfaces.

20.2.2 Topology of Molecular Electrostatic Potential

The MESP $V(\mathbf{r})$, at $\mathbf{r}$, due to a MED $\rho(\mathbf{r})$, is defined as

$$V(\mathbf{r}) = \sum_{A}^{N} \frac{Z_A}{|r - R_A|} - \int \frac{\rho(\mathbf{r}')d^3\mathbf{r}'}{|r - \mathbf{r}'|} \tag{20.5}$$

In Eq. (20.5), N represents the total number of nuclei in the molecule, and Z_A denotes the charge of the nucleus located at R_A. The first term in the equation is due to nuclear contributions. In the literature, it is referred to as BNP. The second term arises due to electronic contributions. Similar to topological features of electron density, the topological analysis of electrostatic potential $\nabla V(\boldsymbol{r})$

is the identification of CPs at which $E = -\nabla V(\mathbf{r}) = 0$. Since $\nabla V(\mathbf{r})$, topological analysis of MESP contains further information related to the electric field arising from nuclear and electronic charge distributions. It can be seen from the definition that both nuclear and electronic distributions contribute to the scalar field of MESP. Therefore, MESP can have both positive and negative values. Thus, MESP topography analysis clearly describes valence electron charge distribution and chemical bonding along with lone pairs. Similar to MED analysis, the rank of the CPs is determined by the Hessian matrix as given below.

$$\nabla^2 V(\mathbf{r}) = \begin{bmatrix} \frac{\partial^2 V(\mathbf{r})}{\partial x^2} & \frac{\partial^2 V(\mathbf{r})}{\partial x \partial y} & \frac{\partial^2 V(\mathbf{r})}{\partial x \partial z} \\ \frac{\partial^2 V(\mathbf{r})}{\partial y \partial x} & \frac{\partial^2 V(\mathbf{r})}{\partial y^2} & \frac{\partial^2 V(\mathbf{r})}{\partial y \partial z} \\ \frac{\partial^2 V(\mathbf{r})}{\partial z \partial x} & \frac{\partial^2 V(\mathbf{r})}{\partial z \partial y} & \frac{\partial^2 V(\mathbf{r})}{\partial z^2} \end{bmatrix} \tag{20.6}$$

As stated above, the CP of rank three is equal to one of the four possible combinations such as (3, −3), (3, −1), (3, +1), and (3, +3). The CP (3, −3) represents maxima. It cannot be observed other than in nuclear locations. The CP (3, +3) represents local minima in the electrostatic potential distribution. It is evident from previous studies that a negative valued (3, +3) minimum in MESP indicates the regions of electron concentration in free radicals, anions, and neutral molecules [5b, 8, 10]. In addition, Gadre and coworkers have shown that negative valued local minima, (3, +3) CPs represent π-bonds and lone pairs [5b]. This information is not discernible from the topography analysis of MED. Both (3, +1) and (3, −1) are saddle points. Similar to the topography of MED, (3, −1) CP is the BCP that occurs between two bonded nuclei. The occurrence of (3, +1) CP of MESP is observed in the ring systems, which is akin to that of the topography of electron density. However, a (3, +1) CP is observed between two (3, +3) CPs, which is additional information not found in the analysis of MED. Therefore, the presence of (3, +1) CP in the topography analysis of MESP does not always indicate the ring structure in contrast to the topography analysis of MED. In the last three decades, analysis of MESP has been widely used to understand the various concepts in chemistry. Some of the applications of topography analysis of MED and MESP are presented in Section 20.3.

20.3 Usefulness of MESP and MED Analysis for Understanding Weak Interactions

20.3.1 MESP and MED Topography Analysis of Oligomers of Conjugated Polymers and their Interaction with PCBM Acceptors

Studies on conjugated polymers and their interactions with partner molecules continue to be the current topic of interest due to their applications in organic electronics such as organic solar cells (OSCs), organic light-emitting diodes (OLEDs), and field effect transistors (FETs). The performance of OSCs strongly depends on conjugated polymers, which contain the donor (D), π-bridge, and acceptor (A) units. It is well known that fullerene derivatives have been used as an electron acceptor along with the donor molecules/oligomers/polymers in the development of solar cells. It is important to know how these donor units interact with fullerene derivatives to achieve the required local ordering and associated morphology. Both local ordering and morphology play a vital role in the charge conductivity and stability of OSCs during illumination. The interaction between two building blocks in the active layer is governed by noncovalent interactions, including electrostatic and dispersive interactions. These features predominantly derive from local ordering, charge carrier conduction, and, thus, related optical processes. Most conjugated molecules contain thiophene (**TH**) as a primary bridging unit. Benzene has also been employed wherever there is a need for charge separation. Both charge continuity and separation are essential for the development of new organic electronics-based devices. In this context, fundamental concepts such as conjugation, delocalization, and aromaticity are employed to understand the various electronic and optical properties of these molecular systems.

Previous studies highlighted the usefulness of MESP and MED in unraveling conjugated systems [10, 11]. In addition to the electron density isosurface, its topology shows regions of electron density concentration and depletion. It has been elicited in numerous studies that topographical features of MED quantify the strength of interaction between two molecules. Hence, by examining the MESP topography and isosurfaces, it is possible to fine-tune the overall electrostatics and conductivity by incorporating suitable functional groups. In the following paragraph, a detailed discussion on topographical analysis of electron density and MESP of isolated conjugated molecules and conjugated molecules-fullerene complexes has been presented.

We have selected two donor polymers, poly(2,5-bis(3-dodecyl-2-yl)thieno[3,2-b]thiophene) (**PBTTT**) [12] and poly[2,6-(4,4-bis(2-ethylhexyl)-4*H*-cyclopenta[2,1-*b*;3,4-*b''*]dithiophene)-*alt*-4,7-(2,1,3-benzothiadiazole)] (**PCPDTBT**) [13] along with ([6,6]-phenyl-C_{61}-butyric acid methyl ester) (**PCBM**) [14] acceptor for the study. **PBTTT** is made up of two donor (D) units, viz., TH and thienothiophene (TT), and hence can be considered as a D-polymer. **PCPDTBT** comprises both donor and acceptor (A) moieties with an alternating D-A architecture. It is referred to as DA-polymer. The model systems consist of monomer, dimer and trimer units of **PBTTT** and **PCPDTBT** interacting noncovalently with the **PCBM** acceptor (fullerene derivative). All the model systems were optimized using density functional theory (DFT) based on M06L/6-31G** [15] level of theory employing the Gaussian 09 and 16 suite of programs [16]. The methyl groups have been used instead of long alkyl side chains to reduce computational costs. The wave functions from the calculations were used for the generation of MESP and MED topographies using the DAMQT 3.0 software package [17].

The optimized geometries and the corresponding MESP topographies for all the isolated model systems and complexes are provided in Figure 20.2. It can be seen that the electrostatic potential for the isolated oligomers of D-polymer is observed along the oligomer backbone, indicating the presence of conjugation and associated conductivity across the backbone. However, in the case of DA-polymer models, the electrostatic potential predominantly appears on the acceptor moiety. It shows a "clamp" pattern due to alternating D and A moiety, which can facilitate its interaction with **PCBM**.

Comparison between MESP of D-polymer and DA-polymer reveals key information. (i) the electrostatic potential appears between D and A units, which is absent in D-polymer. The presence of S···N interactions between adjacent D and A in DA-polymer results in the appearance of electrostatic potential. These potentials reduce the structural disorder along the polymer backbone and increase the charge conductivity. Also, as D-polymer shows the electrostatic potential only on the above and below the surface of the polymer backbone, the fullerene prefers to bind only on the polymer backbone. However, the DA-polymer has electrostatic potential, a "clamp" pattern in addition

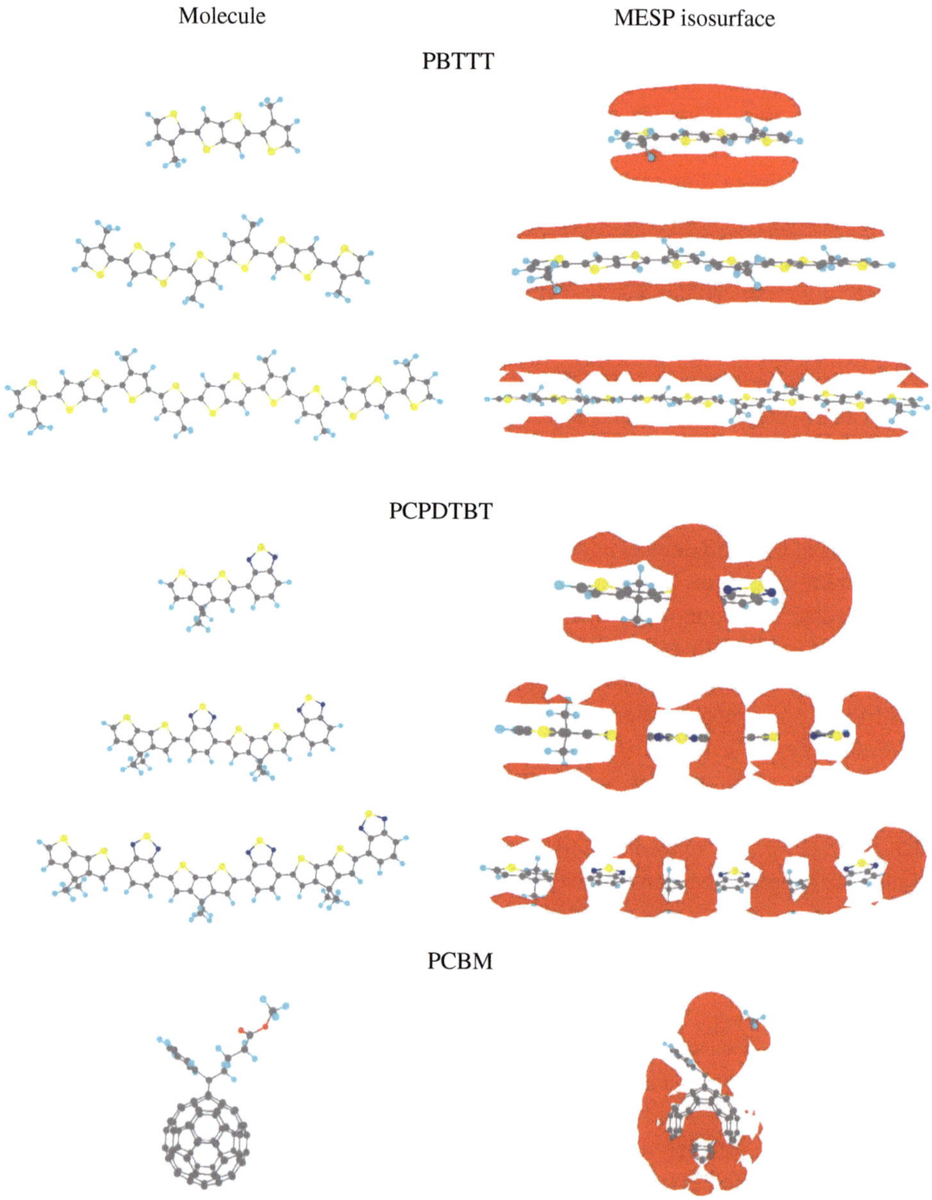

Figure 20.2 The optimized geometries of the isolated molecules considered for the study at M06L/6-31G** level of theory and the corresponding MESP isosurfaces calculated at M06L/6-311+G** level of theory. The isosurface values for polymer molecules and PCBM are −0.01 and −0.004 a.u., respectively.

to the potential above and below the surface of the polymer backbone. Thus, one can expect a face-on type interaction between the polymer backbone and fullerene and a side-on interaction between the polymer backbone and fullerene derivatives.

The MESP topography of isolated **PCBM** shows the possible involvement of the interaction of its ester and fullerene moieties with donor polymers. The presence of electrostatic potential on **PCBM** (π-cloud) and the various polymers reveals the role of π–π interactions between the two systems. The optimized geometries and the corresponding MESP topographies of the model trimer-PCBM

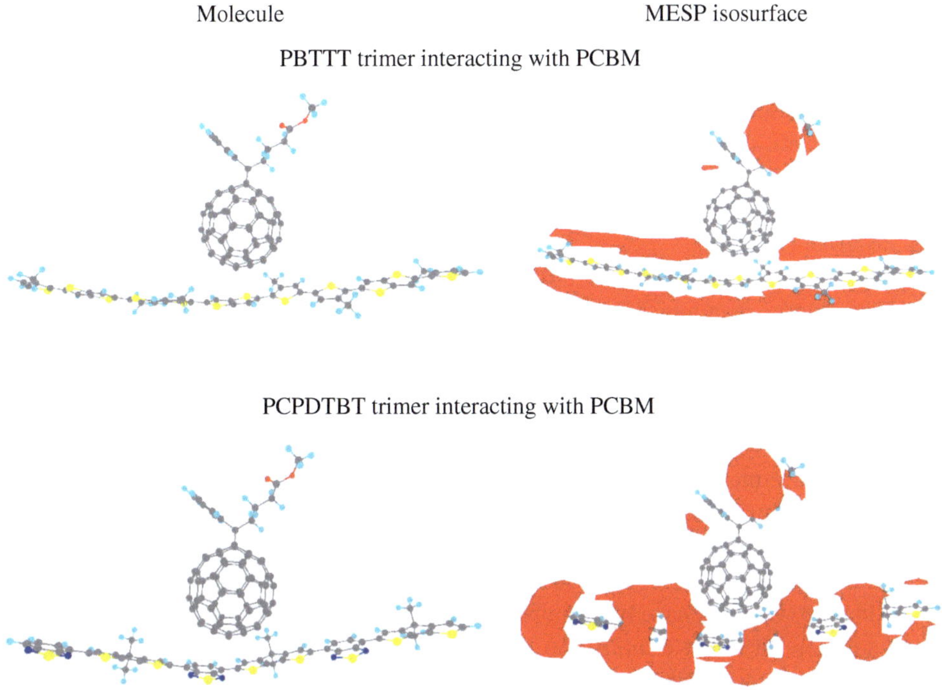

Figure 20.3 The optimized geometries of the model polymer donor-PCBM acceptor complexes considered for the study at the M06L/6-31G** level of theory and the corresponding MESP isosurfaces were calculated at the M06L/6-311+G** level of theory. The isosurface values for the molecules are −0.01 a.u.

complexes are presented in Figure 20.3. Although there may be a number of conformations plausible for the interaction of a donor polymer with **PCBM**, we have considered only one conformation due to the limitations of the scope of the present chapter. The stacking distances and the calculated interaction energies of the two model complexes considered for the study are presented in Table 20.1. The geometry of the **PBTTT-PCBM** complex shows that the interaction distance between the **PBTTT** donor and **PCBM** acceptor ranges from 3.295 to 3.548 Å. It can be noted that the bottom six-membered ring of **PCBM** interacts with the central five-membered ring of **TT** and the neighboring **TH** unit. Locally, this π–π (**BZ-TH** and **BZ-TT**) interaction is responsible for the stability of the complex. Due to this interaction, the dihedral angle of C–C–S–C of the **PBTTT** trimer twists to about ~16°, which induces a bending of the interaction region of the donor towards **PCBM** by 10°. In the case of the **PCPDTBT-PCBM** noncovalent complex, the donor molecule itself contains two moieties: cyclopentadithiophene (**CDT**), which is a donor building block and benzothiadiazole (**BTD**), an acceptor unit. Hence, two different interactions may be expected: (i) **PCBM-CDT**

Table 20.1 The stacking distances and interaction energies of the model complexes considered for the study were calculated at the M06L/6-311++G** level of theory.

Molecule	Stacking distances (Å)	Interaction energy (kcal/mol)
PBTTT-PCBM	3.29–3.55	−9.26
PCPDTBT-PCBM	3.15–3.60	−10.48

(acceptor-donor) and (ii) **PCBM-BTD** (acceptor-acceptor), which are depicted in Figure 20.3. The interaction distance (Table 20.1) for this complex varies from 3.152 to 3.569 Å. Close analysis reveals that the **CDT** segment of the trimer bends toward PCBM with an angle of ~9° about its molecular plane. On the other side of the oligomer, the **BTD** segment bends toward **PCBM** and subtends at an angle of ~14° with reference to the plane of the donor molecule. These intricate geometrical changes may affect the local ordering and related optoelectronic properties. However, as mentioned earlier, an ensemble of conformations contributes to the local ordering and morphological structure of the bulk material.

The molecular density topographies of the isolated and complex systems are depicted in Figures 20.4 and 20.5 as mentioned in several investigations, the electron density at (3, −1) CPs in the interaction regions is a descriptor of the strength of the interaction. In the case of the **PBTTT-PCBM** complex, the molecular graphs show the noncovalent interaction between the two units with five CPs of (3, −1) type. The electron density values in these regions are found to be 0.002,04–0.0109 a.u. Also, the field lines reiterate the interaction of a single six-membered ring of **PCBM** with two five-membered rings of the polymer, as found from the MESP analysis. From Table 20.1, it can be known that the calculated zero-point energy corrected interaction energies at the M06-L/6-311++G** level of theory for this complex is −9.26 kcal/mol. The five noncovalent interactions between the **PBTTT** and **PCBM** are responsible for the stabilization of the system.

Similarly, three (3, −1) CPs have been found for the trimer model of the **PCPDTBT-PCBM** complex. In addition, the close examination of the electron density field lines of this complex highlights the presence of several noncovalent interactions involving **PCBM** and the D/A moieties of the polymer. It is interesting to note that, unlike the D-type polymer-PCBM complex, for the DA-type polymer-PCBM system, two six-membered rings of **PCBM** were found to interact with D and A moieties of the polymer donor. The electron density values for these CPs are calculated to be 0.005,76–0.008,77 a.u. The counterpoise corrected interaction energy between the participating units at the same level of theory is found to be −10.48 kcal/mol. We have also calculated the interaction of monomer, dimer, and trimer models of polymer with **PCBM**. Overall results show that the interaction energies exhibit a linear relationship with the sum of electron densities at the BCPs in the noncovalent interaction regions between the polymer donor and **PCBM** acceptor. This study clearly elicits the usefulness of MED and MESP in understanding the strength of interaction and charge conductivity involving conjugated systems.

20.3.2 Interaction of Small Molecules with Models of Single-Walled Carbon Nanotube and Graphene

20.3.2.1 Interaction of Nucleobases with Carbon Nanomaterials

Carbon-based nanomaterials have fascinated many researchers due to their extraordinary properties and concomitant applications in different fields. From the standpoint of applications in biology and medicine, the interaction of biomolecules with single-walled carbon nanotube (SWCNT), graphene (Graph) and its derivatives has also received considerable interest [18]. We have investigated how DNA bases interact with graphene using electronic structure theory-based methods [19]. The interaction between these two systems has been characterized with the help of both topography analysis of MED and MESP. The optimization and interaction energy calculations were performed at the M06-2X/6-31+G** level of theory [20]. The theory of atoms in molecules (AIM) was employed using the AIM 2000 package to characterize the nature of the interaction between graphene models and NBs [21].

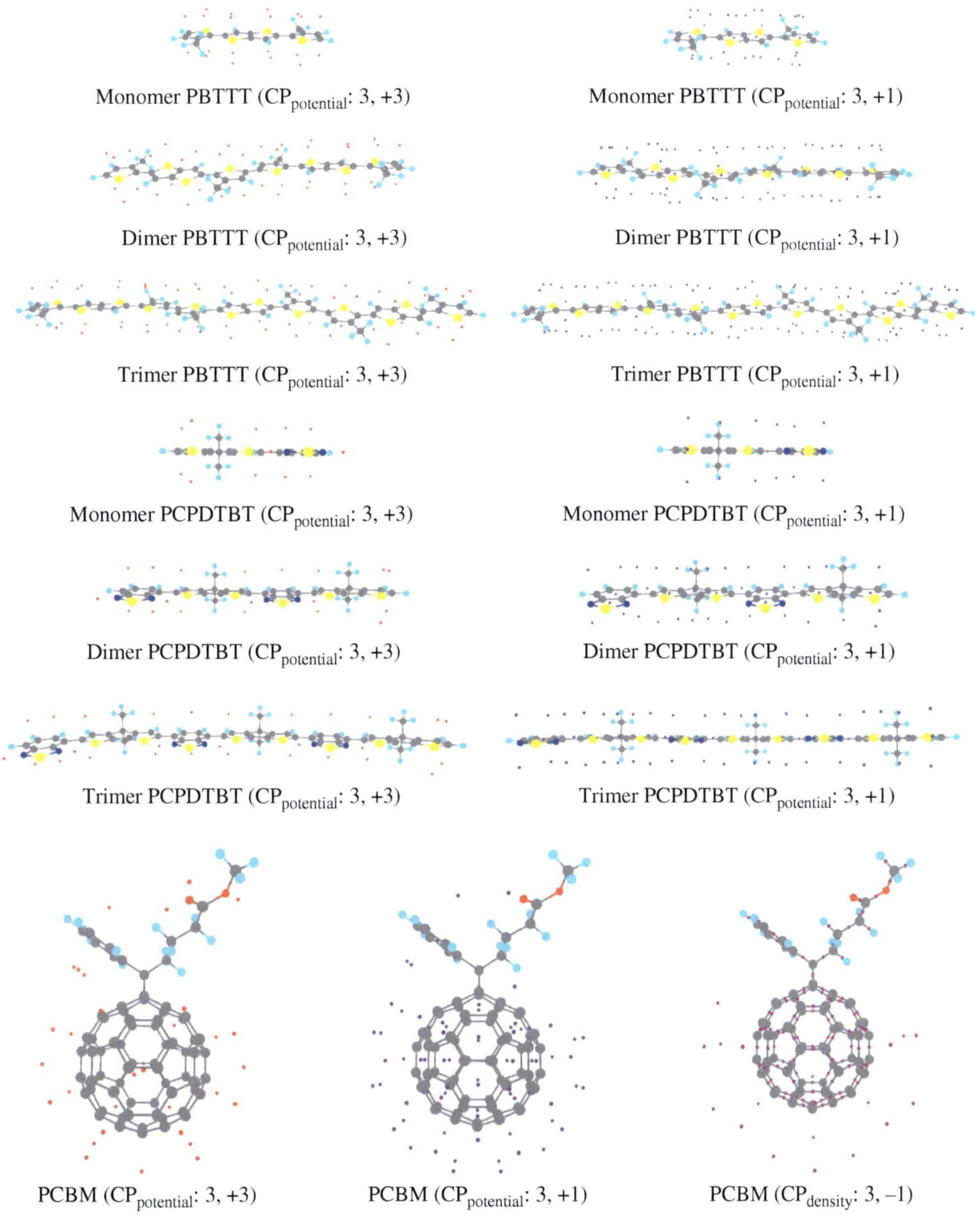

Figure 20.4 The calculated values of MESP minima corresponding to (3, +3) and (3, +1) critical points (CPs) and MED minima corresponding to (3, −1) CPs for the isolated molecules at the M06-L/6-311+G** level of theory. The small red, violet, and pink dots represent (3, +3), (3, +1), and (3, −1) CPs, respectively.

Nucleobases interact with carbon nanomaterials through noncovalent interactions such as π–π stacking, NH–π, CO–π, and CH–π. The trend in binding strength of nucleobases with carbon nanomaterials varies as G > A > T > C > U. The strength of the interaction is high for Graph on comparing to SWCNT. Single-molecule detection can be done with the help of variations in the conductivity of Graph upon adsorption of nucleobases. The weak dispersion interaction (π–π stacking) is responsible for the sequencing of DNA using Graph. The strength of interaction

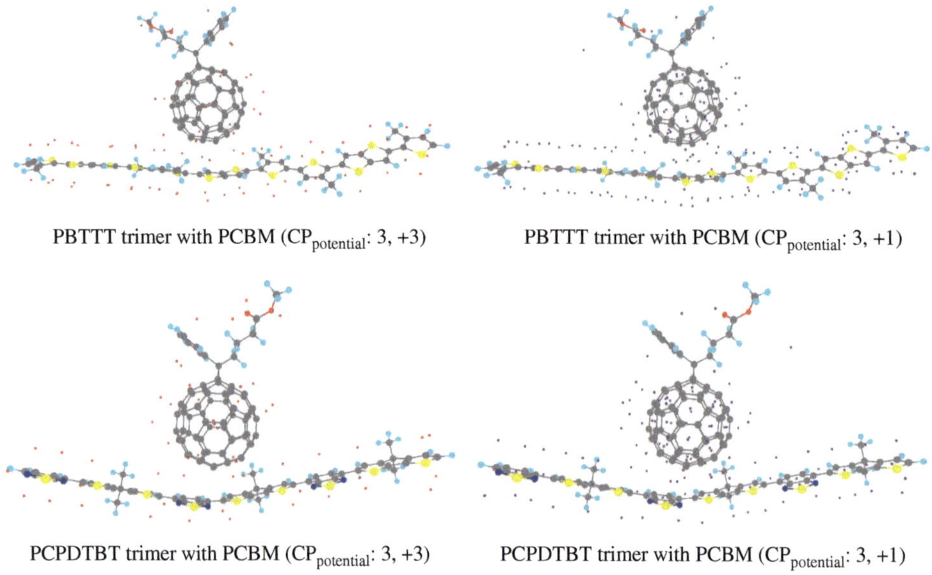

Figure 20.5 The calculated values of MESP minima corresponding to (3, +3) and (3, +1) CPs and MED minima corresponding to (3, −1) CPs for the PBTTT-PCBM and PCPDTBT-PCBM complexes at M06-L/6-311+G** level of theory. The small red, violet, and pink dots represent (3, +3), (3, +1), and (3, −1) CPs, respectively.

of nucleobases can be increased with the doping of graphene. We have explored the interaction of nucleobases with silicon-doped graphene (SiGr) using DFT calculations. Silicon atom is projected out of the plane of graphene due to its large size and it is electropositive in nature. The protruded silicon atom preserves its sp^3 hybridization. Nucleobases can interact through specific atoms (Oxygen and Nitrogen), which are in negative potential (Figure 20.6). The active sites (atoms that are in negative potential) of nucleobases have been made to interact with silicon in parallel and perpendicular directions to the surface of doped graphene. Nucleobases interact through specific atoms in addition to traditional π–π stacking. Figure 20.7 shows that the interactions, including Si–O, Si–N, and NH–π stabilize the nucleobases on the surface of silicon-doped graphene. The binding strength of nucleobases is higher than that of graphene. The order of interaction energies is as follows C > G > A > T > U, and it is different from graphene. The interaction energies are provided in Table 20.2. The introduction of silicon significantly changes the binding strength and pattern of nucleobases with graphene.

The reactivity of graphene is modulated by the presence of defects. Five and five-seven-membered ring defects were introduced in silicon-doped graphene. Nucleobases interacted with five-membered ring defective silicon doped graphene (5SiGr) and five-seven fused ring defective silicon doped graphene (57SiGr). The results show that the nucleobases are stabilized by both π–π stacking and Si· · ·O (N) interactions in all the models. The stability of nucleobases is higher for defective silicon-doped graphene than silicon-doped and pristine graphene. The order of interaction energy for these graphene models is found to be 5SiGr > 57SiGr > SiGr > Gr. The order of interaction energies of nucleobases is observed to be C > G > A > T > U, which is similar to silicon-doped graphene. Defects in silicon-doped graphene further enhance the binding strength of nucleobases on the surface of graphene.

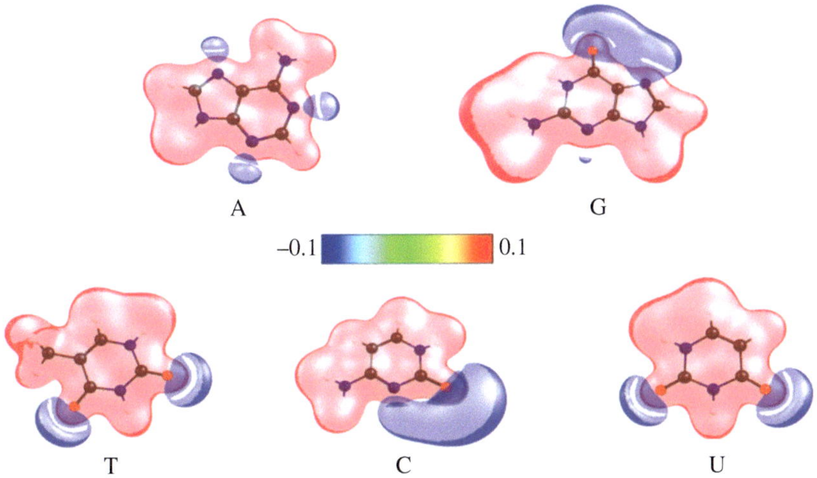

Figure 20.6 The calculated MESP isosurface of NBs (A, G, T, C and U) at isovalue 0.06 a.u.

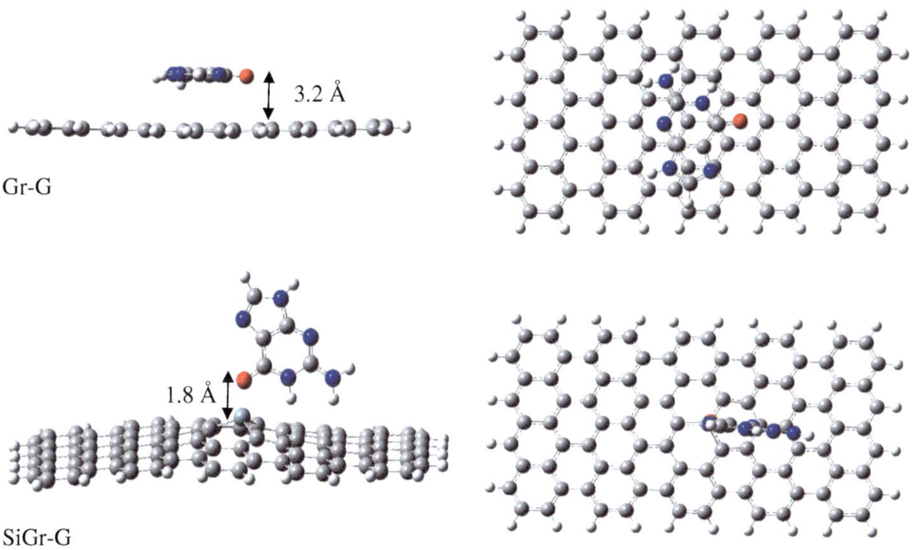

Figure 20.7 The optimized geometries of the complex of guanine with graphene and silicon-doped graphene.

The molecular graph for guanine interaction is presented in Figure 20.8. The sum of electron density at BCP is shown in Table 20.3. The sum of electron density is directly proportional to the interaction energies in both Gr and SiGr complexes. The SiGr-NBs have higher values than the Gr-NBs complexes at BCP. The $\nabla^2\rho(\boldsymbol{r}_c)$ values at BCPs in SiGr and Gr complexes with NBs are positive. These values illustrate the presence of noncovalent interaction between SiGr (Gr) and NBs. The electron density $\rho(\boldsymbol{r}_c)$ at the BCP is of the order of (0.1 a.u.) for covalent bonds, while in the case of noncovalent interactions, it is 1 order lower (0.01 a.u. or even less). The calculated $\rho(\boldsymbol{r}_c)$ values (0.07–0.08 a.u.) at BCPs involving Si and O (N) atoms in SiGr-NBs complexes lie in between the values corresponding to the noncovalent and covalent interactions. Further, to understand the

Table 20.2 The calculated IEs of NBs with graphene models (Gr and SiGr) at M06-2X/6-31+G** level of theory.

Model	NB	IE (kcal/mol)
	A	−16.2
Gr	G	−19.5
	T	−15.9
	C	−15.4
	U	−13.5
	A	−35.9
	G	−43.4
SiGr	T	−33.0
	C	−44.3
	U	−31.8

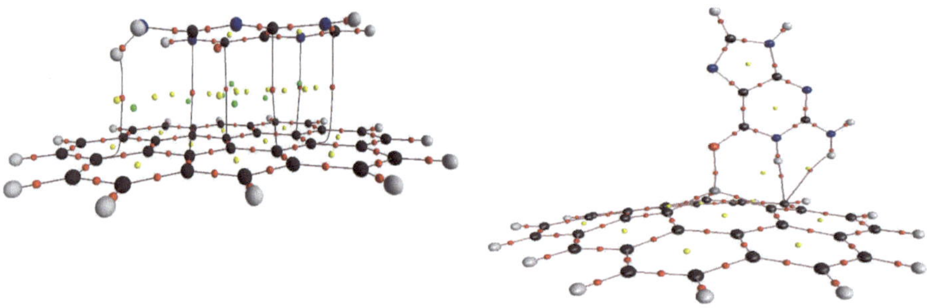

Gr-G SiGr-G

Figure 20.8 Molecular graphs of different complexes of Gr and SiGr with Guanine. BCPs are represented by red color dots.

nature of Si· · ·O (N) interaction, the total energy density (H_c) at BCP was calculated by using the following equation.

$$H_c = G_c + V_c \tag{20.7}$$

where G_c is the kinetic energy density, and V_c is the potential energy density at BCP. The negative H_c value indicates the accumulation of charge density at BCP, a case of a covalent bond. The more negative the H_c value, greater the covalent character of interaction. The positive H_c values represent the noncovalent interaction. The H_c values of Si· · ·O (N) are listed in Table 20.4. From Table 20.4, it can be seen that the H_c values are negative. The interaction of Si· · ·O (N) is partially electrostatic and partially covalent in nature. Therefore, NBs interact with SiGr through both noncovalent, partially covalent and electrostatic interactions.

To understand further the factors which predominantly responsible for the interaction, energy decomposition analysis (EDA) was performed for the SiGr-NBs and Gr-NBs. The EDA decomposes the interaction energy between two monomers into several components, such as electrostatic energy, orbital energy, Pauli repulsion, and dispersion energy. The results from the analysis of EDA

Table 20.3 The calculated sum of the electron density ($\rho(r_c)$) at BCP of complexes of Gr and SiGr with NBs (a.u.).

Model	NB	Electron density ($\rho(r_c)$)
	A	0.076
	G	0.085
Gr	T	0.075
	C	0.072
	U	0.011
	G	0.122
	A	0.115
SiGr	T	0.112
	C	0.127
	U	0.111

are presented in Table 20.5. Several reports have proved that the interaction of NBs on the surface of Gr is stabilized by dispersion. It can be observed from Table 20.5 that the dispersion interaction stabilizes the Gr-NBs, whereas the electrostatic interaction is responsible for the stabilization of SiGr-NBs. The electrostatic energies are directly proportional to the IE values. In addition to electrostatic energies, orbital interaction is also important for the stabilization of NBs on SiGr. The role of dispersion interaction is less in the stabilization of SiGr-NBs. Therefore, both electrostatic and partially covalent interactions are primary determinants in the formation of complexes of NBs with SiGr.

Carbon nanomaterials have been used as electrodes for the detection of modified nucleobases (xanthine [X], caffeine [CAF], uric acid [UA], and hypoxanthine [HX]), which acts as biological indicators for several diseases. The strength of adsorption of molecules on graphene is important for their electrochemical detection. We have performed a systematic investigation on the adsorption process of modified nucleobases on the surface-doped graphene in order to increase the strength of interaction between graphene and modified nucleobases [22]. Graphene models were doped with boron, nitrogen and silicon atoms. The modified bases also interact through π–π stacking with pristine graphene, boron, nitrogen, and silicon-doped graphene. In addition

Table 20.4 The calculated Laplacian of electron density ($\nabla^2\rho(r_c)$) and total energy density (H_c) at BCPs of Si···O(N) interaction (a.u.).

Complex	BCP between	$\nabla^2\rho(r_c)$	H_c
SiGr-A	Si···N	0.076	−0.028
SiGr-G	Si···O	0.136	−0.017
SiGr-T	Si···O	0.112	−0.016
SiGr-C	Si···O	0.139	−0.017
SiGr-U	Si···O	0.108	−0.016

Table 20.5 The energy decomposition analysis of the complexes formed by NBs with Gr and SiGr.

Model	NB	Ele. stat. energy (kcal/mol)	Pau. rep. energy (kcal/mol)	Orb. int. energy (kcal/mol)	Disp. energy (kcal/mol)	Inter. energy (kcal/mol)
Gr	A	−11.65	28.26	−5.51	−28.08	−16.97
	G	−14.39	35.86	−9.37	−32.34	−20.25
	T	−11.38	30.64	−7.08	−28.69	−16.51
	C	−11.73	29.37	−7.41	−26.22	−16.00
	U	−9.86	26.60	−6.38	−24.62	−14.27
SiGr	A	−142.61	225.42	−100.96	−16.11	−34.26
	G	−150.34	246.98	−123.49	−13.90	−40.75
	T	−134.11	219.85	−104.24	−12.44	−30.95
	C	−154.31	249.32	−123.59	−12.45	−41.04
	U	−131.30	216.23	−102.32	−12.37	−29.77

Ele. stat. energy: electrostatic energy; Orb. int. energy: orbital interaction energy; Pau. rep. energy: Pauli repulsion energy; Disp. energy: dispersion energy; Inter. energy: interaction energy.

to π–π stacking and NH–π, Si–O (N) partial electrostatic and covalent interactions are found in the case of silicon-doped graphene. For graphene, the order of interaction energies is found as CAF > UA > X > HX. The strength of interaction tremendously increases for modified nucleobases with silicon-doped graphene when compared to pristine graphene, and the energetics follow the order is X > CAF > UA > HX.

20.3.2.2 Interaction of Chlorobenzene with Carbon Nanomaterials

Carbon nanomaterials are applied in various fields including environmental remediation. Hence, we have investigated the adsorption of the homologues series of chlorobenzenes (CBs) onto the surface of the (5,5) armchair SWCNT and Graph sheet using DFT-based calculations [23]. The optimization of the SWCNT–CBs and Graph–CBs complexes was carried out using the two-layer approach employing the ONIOM (M06-2X/6-31+G**:PM6) method [24]. The CBs were treated at M06-2X/6-31+G** level. Similarly, the interacting surfaces of SWCNT and Graph with CBs were treated at the M06-2X/6-31+G** level. The remaining atoms of SWCNT and Graph were treated at a low level using a semiempirical PM6 Hamiltonian. The IEs were calculated using the supermolecule approach at the M06-2X/6-31+G** level. The AIM 2000 package was employed to characterize the nature of the interaction between carbon nanomaterials and CBs [21]. The interaction energy results are provided in Table 20.6, and the concomitant interaction between the CBs and SWCNT/Graph has also been characterized with the help of Bader's theory of AIM (Table 20.7). Results showed the enhancement of the interaction propensity of CBs with the SWCNT and graphene with the increase in the chlorine content. In addition, the adsorption of CBs on graphene is more favorable when compared with SWCNT. The correlation plot between the degree of chlorination of CB and CCPs and a sum of BCPs (substituents of CB) on the nanosurfaces are provided in Figure 20.9. Findings revealed that the planar surface of the graphene facilitates the direct through-space interaction of the chlorine atoms of CBs with the aromatic surface, as shown

Table 20.6 Interaction energies of CBs on adsorption to CNT and Graph and difference in interaction energy (ΔE) between the respective CNT-CBs and Graph-CBs complexes. (Reproduced from Ref. [23] with permission from the American Chemical Society).

S. No	Compound	Interaction energy (kcal/mol)		$\|\Delta E\|$ (kcal/mol) (Graph-CNT)
		CNT	Graph	
1	CB	−6.55	−9.68	3.13
2	1,2-C2B	−7.86	−11.22	3.36
3	1,3-C2B	−7.90	−11.04	3.14
4	1,4-C2B	−7.95	−11.23	3.27
5	1,2,3-C3B	−9.39	−12.00	2.61
6	1,2,4-C3B	−9.19	−12.49	3.30
7	1,3,5-C3B	−8.92	−12.07	3.15
8	1,2,3,4-C4B	−10.45	−13.61	3.16
9	1,2,3,5-C4B	−10.19	−13.13	2.94
10	1,2,4,5-C4B	−10.19	−13.62	3.42
11	1,2,3,4,5-C5B	−11.46	−14.63	3.17
12	1,2,3,4,5,6-C6B	−12.61	−14.91	2.30

Table 20.7 Calculated total electron density ($\rho(r_c)$) of CCPs and sum of BCPs for the complex involving CNT-CB's and Graph-CBs in a.u. (Reproduced from Ref. [23] with permission from the American Chemical Society).

S. No	Compound	CNT		Graph	
		CCP $\rho(r_c)$	Sum of WBCP of substituent's (r_c)	CCP $\rho(r_c)$	Sum of WBCP of substituent's $\rho(r_c)$
1	CB	0.0041	0.0066	0.0043	0.0066
2	1,2-C2B	0.0042	0.0129	0.0043	0.0134
3	1,2,3-C3B	0.0043	0.0129	0.0042	0.0200
4	1,2,3,4-C4B	0.0046	0.0128	0.0042	0.0326
5	1,2,3,4,5-C5B	0.0049	0.0192	0.0040	0.0341
6	1,2,3,4,5,6-C6B	0.0049	0.0148	0.0039	0.0420

in Figure 20.10. The nature of the interaction between the two systems has been characterized with the help of the theory of AIM.

In addition, the MESP isosurfaces of various systems have been obtained. It is found that the MESP isosurface of monocholorobenzene has a considerable amount of π electron cloud above and below the ring. However, with an increase in chlorination, there is a gradual change in the π electron density. As expected, the overall negative potential above and below the ring is gradually absent for hexachlorinated benzene. The results are illustrated in Figure 20.11. The role of electrostatic complementarity between the SWCNT and CBs is evident from the MESP analysis.

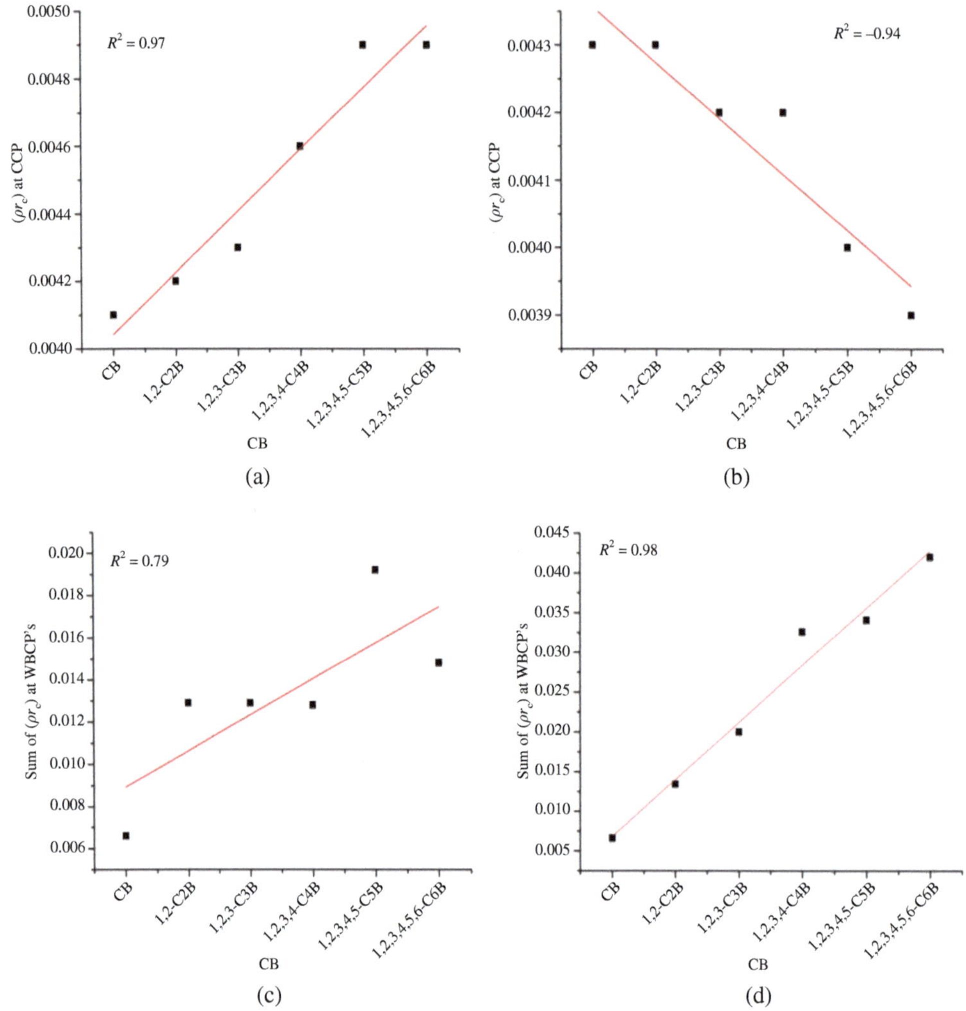

Figure 20.9 The linear relationship between the CCP of CNT-CB (a) and Graph-CB (b) complexes, a sum of BCPs (substituents of CB) of CNT-CB (c) and Graph-CB (d) complexes. (Reproduced from Ref. [23] with permission from the American Chemical Society).

20.3.2.3 Interaction of Carbohydrates with Carbon Nanomaterials

The interaction between the fragments of carbon nanotube (CNT) and graphene (COR) with different carbohydrates has been investigated using MP2 and M05-2X methods to assess and compare the role of curvature on the interaction [25]. Three carbohydrates, β-D-glucose, β-D-galactose, and β-D-xylose, with different degrees of hydrophobic nature, have been considered. Calculations have been performed on intermolecular complexes comprising COR-carbohydrates and CNT-carbohydrates. Findings from electronic structure calculations combined with Bader's analysis revealed that CH$\cdots\pi$ interaction is the predominant one in the stabilization of the carbohydrate-CNT and carbohydrate-COR complexes. Furthermore, the importance of OH$\cdots\pi$ and lone pair$\cdots\pi$ (lp$\cdots\pi$) interactions are also evident from the results. The calculated BEs for the various carbohydrate-CNT and carbohydrate-COR complexes at M05-2X with dual

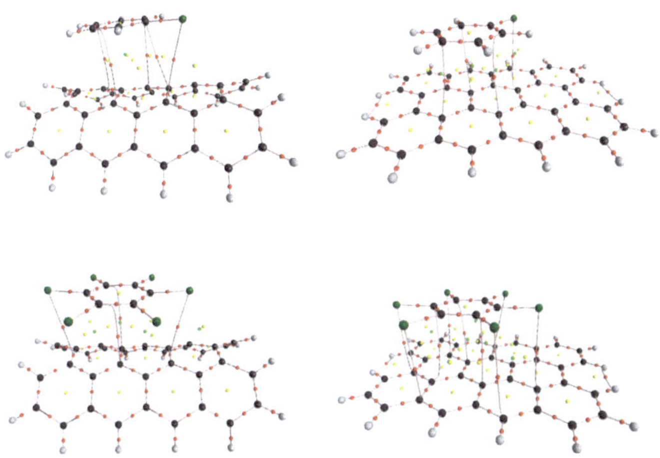

Figure 20.10 AIM topography of the optimized complexes of CB and 1,2,3,4,5,6C6B with CNT and Graph systems BCPs are represented by red color dots, and CCPs are represented by green color dots, respectively. (Reproduced from Ref. [23] with permission from the American Chemical Society).

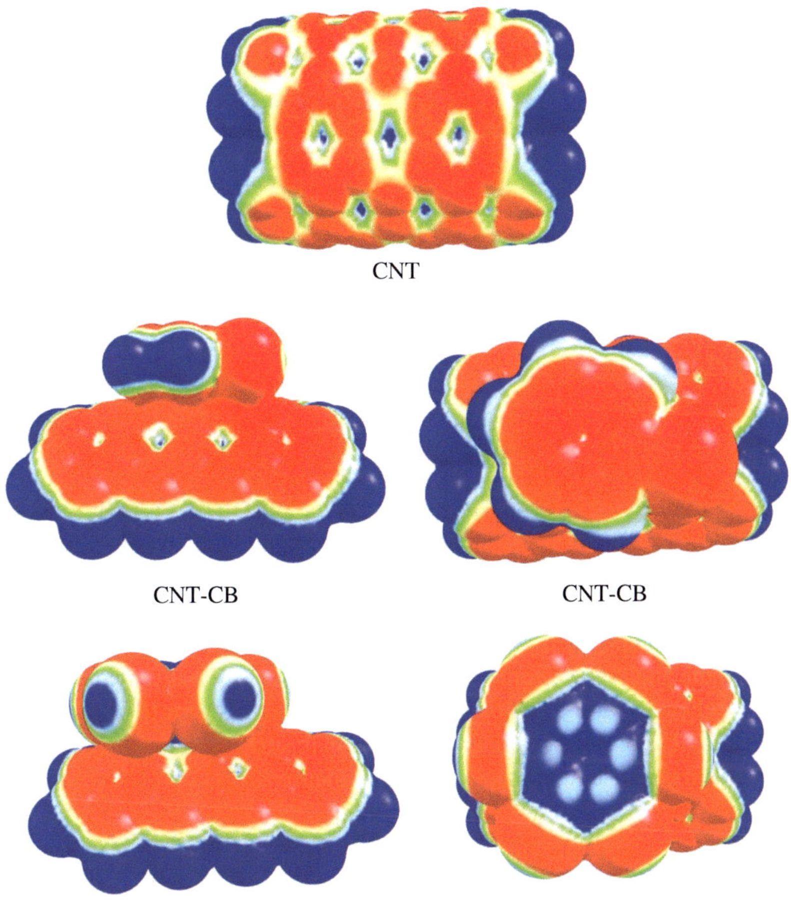

Figure 20.11 Molecular electrostatic potential (MESP) of (i) CNT, CNT-CB, and CNT-C6B Complex (top), and (ii) Graph, Graph-CB and Graph-C6B Complex (bottom). (Reproduced from Ref. [23] with permission from the American Chemical Society).

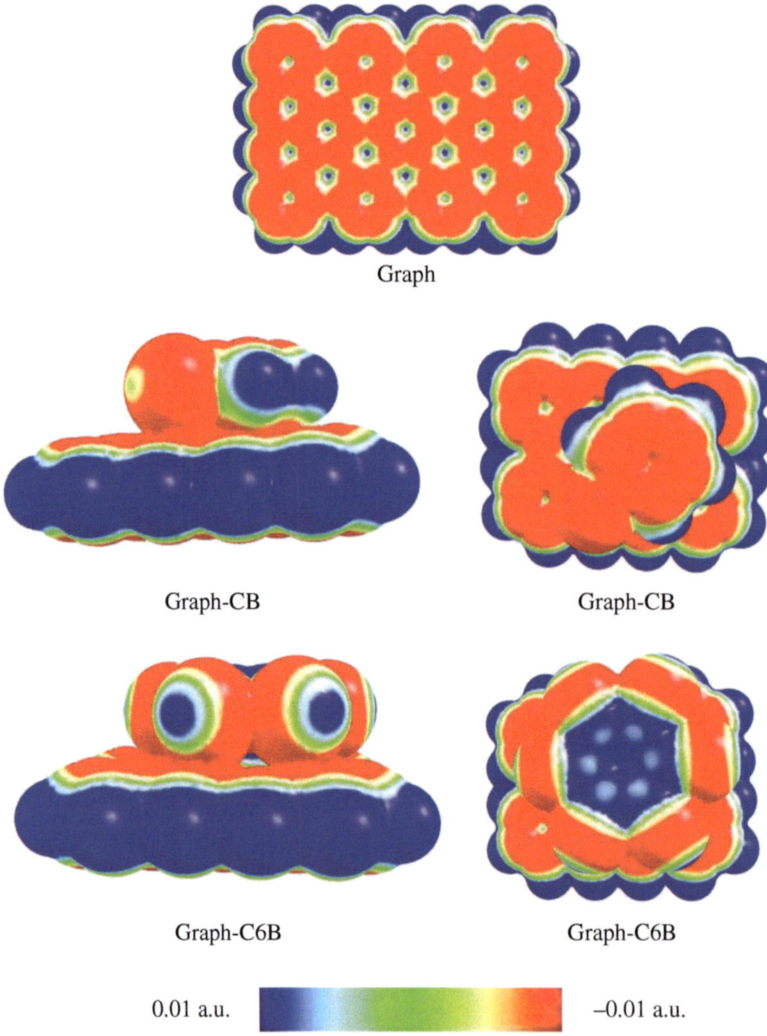

Figure 20.11 (*Continued*)

basis set (aug-cc-pVTZ for carbohydrate + cc-pVTZ for both CNT and COR) vary from −2.52 to −5.14 and from −4.14 to −8.04 kcal/mol, respectively. The corresponding BEs obtained from MP2/6-311++G**//M05-2X/6-31+G** level of calculation ranges from −4.92 to −9.93 and from −6.75 to −12.53 kcal/mol. Scrutiny of the energetics of all the complexes elucidates that the electron correlation energy (dispersion energy) significantly contributes to the stability of these complexes. It is found from the analysis of geometrical parameters and BEs that the interplay of the orientation of the X–H (where X= C and O) bond to the π-surface is crucial for the recognition and further stabilization. MESP isosurfaces of curved and planar surfaces have provided the difference between the π-electron distributions, as illustrated in Figure 20.12.

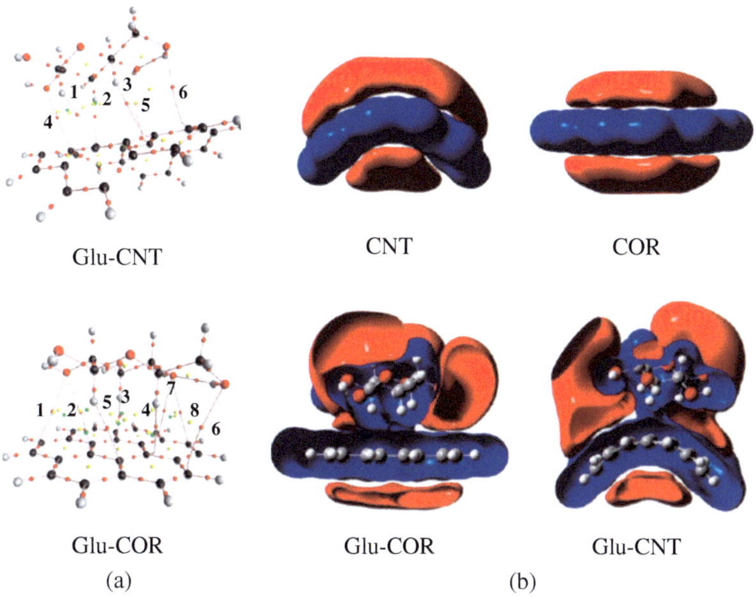

Figure 20.12 (a) Molecular graphs of carbohydrate (glucose) – CNT and carbohydrate (glucose) – COR complexes. Numerical numbers highlight the number of bond paths that exist between glucose molecules and the surface of carbon nanomaterial. (b) MESP (isosurface value = 0.0015 a.u.) carbohydrate (glucose) – CNT and carbohydrate (glucose) – COR complexes. Negative and positive potentials are represented by red and blue colors, respectively. (Reproduced from Ref. [25] with permission from the American Chemical Society).

20.4 Conclusion

Overall, in this chapter, we have summarized our studies on how MED and MESP analyses can be used to describe the weak interactions in molecular complexes. Examining the MESP topology of π-conjugated materials helps in gaining insights into the extent of π-delocalization and conductivity across the backbone of the conjugated polymer. Furthermore, the features of MESP and MED aid in estimating the strength of interaction between conjugated polymers and other molecular systems. Such details enable us to understand the local molecular packing of pure conjugated polymers and blends of conjugated polymers. It is worth noting that the intermolecular interactions between conjugated polymers and acceptor materials dictate the performance of the device and stability. Thus, the molecular level information on molecular packing is critical for improving the device performance and understanding the issues related to the stability of organic solar cells. Also, we have used MED and MESP analyses to describe the weak interactions in the complexes made up of nanomaterials and various types of small molecules, viz., nucleobases, polychlorinated aromatic molecules, and carbohydrates. Molecular fragments of pristine carbon-based and Si-doped carbon nanomaterials are considered as model systems for nanomaterials. We demonstrated the curvature of nanomaterials plays a crucial role in the strength of interactions between small molecules and nanomaterial complexes as the curvature brings uneven distribution of electron density above and below the surface. Also, the MED and MESP analyses yield all possible interactions between small molecules and nanomaterials. Using these analyses, we could identify $\pi\cdots\pi$, $C{-}H\cdots\pi$, $O{-}H\cdots\pi$,

Si···O, Si···N, and lone pair···π interactions. These results are highly relevant in developing energy storage materials, sensors, and nano-based drug delivery systems. In conclusion, MED and MESP analyses are powerful tools for predicting and assessing the weak interactions between small molecular clusters and bulk materials.

Acknowledgment

The authors (PS, MK, KB, MS, and VS) acknowledge CSIR-CLRI and CSIR-4PI for the supercomputing facility. MK gratefully acknowledges the Council of Scientific and Industrial Research (CSIR) for the CSIR-SRF fellowship (Grant No.: 31/006(0470)/2020-EMR-I, dated 12/10/2020). CSIR-CLRI Communication No.: 1829.

Conflict of Interest

The authors declare no conflict of interest regarding the publication of this chapter, financial, and/or otherwise.

References

1 (a) Bader, R.F.W. (1990). *Atoms in Molecules – A Quantum Theory*. Oxford, UK: Oxford University Press. (b) Popelier, P.L.A. (2000). *Atoms in Molecules – An Introduction*. Harlow, UK: Pearson Education.

2 (a) Bader, R.F.W. and MacDougall, P.J. (1985). Toward a theory of chemical reactivity based on the charge density. *JACS* 107 (24): 6788–6795. (b) Bader, R.F.W. (1991). A quantum theory of molecular structure and its applications. *Chem. Rev.* 91 (5): 893–928. (c) Bader, R.F.W. and Laidig, K.E. (1992). The definition of a chemical bond, of molecular structure and of its change as exemplified by the structure diagram for $C_4H_7^+$. *J. Mol. Struct. THEOCHEM* 261: 1–20. (d) Bader, R.F.W. (2009). Bond paths are not chemical bonds. *J. Phys. Chem. A* 113 (38): 10391–10396. (e) Popelier, P.L.A. (2000). On the full topology of the Laplacian of the electron density. *Coord. Chem. Rev.* 197 (1): 169–189. (f) Bader, R.F.W. (2010). Definition of molecular structure: by choice or by appeal to observation? *J. Phys. Chem. A* 114 (28): 7431–7444.

3 (a) Tal, Y., Bader, R.F.W., and Erkku, J. (1980). Structural homeomorphism between the electronic charge density and the nuclear potential of a molecular system. *Phys. Rev. A* 21 (1): 1–11. (b) Gadre, S.R. and Bendale, R.D. (1986). On the similarity between molecular electron densities, electrostatic potentials and bare nuclear potentials. *Chem. Phys. Lett.* 130 (6): 515–521.

4 (a) Bonaccorsi, R., Pullman, A., Scrocco, E., and Tomasi, J. (1972). The molecular electrostatic potentials for the nucleic acid bases: adenine, thymine, and cytosine. *Theor. Chim. Acta* 24 (1): 51–60. (b) Scrocco, E. and Tomasi, J. (1973). The electrostatic molecular potential as a tool for the interpretation of molecular properties. In: *New Concepts II*, 95–170. Berlin, Heidelberg: Springer. (c) Scrocco, E.A. and Tomasi, J. (1978). Electronic molecular structure, reactivity and intermolecular forces: an heuristic interpretation by means of electrostatic molecular potentials. In: *Advances in Quantum Chemistry*, vol. 11 (ed. P.O. Lowdin). New York: Academic Press. (d) Politzer, P., Donnelly, R.A., and Daiker, K.C. (1973). Molecular electrostatic potentials. Mechanistic aspects of electrophilic attack on furan. *J. Chem. Soc., Chem. Commun.* 17:

617–618. (e) Politzer, P., Abrahmsen, L., and Sjoberg, P. (1984). Effects of amino and nitro substituents upon the electrostatic potential of an aromatic ring. *JACS* 106 (4): 855–860. (f) Murray, J.S., Paulsen, K., and Politzer, P. (1994). Molecular surface electrostatic potentials in the analysis of non-hydrogen-bonding noncovalent interactions. *Proc. Indian Acad. Sci. – Chem. Sci.* 106 (2): 267–275. (g) Sen Murray, J.S.K. (1996). *Molecular Electrostatic Potentials: Concepts and Applications*, vol. 3. Amsterdam, The Netherlands: Elsevier. (h) Shirsat, R.N. and Shridhar, R. (2000). *Electrostatics of Atoms and Molecules*. Hyderabad, India: Universities Press. (i) Pathak, R.K. and Gadre, S.R. (1990). Maximal and minimal characteristics of molecular electrostatic potentials. *J. Chem. Phys.* 93 (3): 1770–1773.

5 (a) Kumar, P.S.V., Raghavendra, V., and Subramanian, V. (2016). Bader's theory of atoms in molecules (AIM) and its applications to chemical bonding. *J. Chem. Sci.* 128 (10): 1527–1536. (b) Gadre, S.R., Suresh, C.H., and Mohan, N. (2021). Electrostatic potential topology for probing molecular structure, bonding and reactivity. *Molecules* 26 (11): 3289. (c) Pingale, S.S., Gadre, S.R., and Bartolotti, L.J. (1998). Electrostatic insights into the molecular hydration process: a case study of crown ethers. *J. Phys. Chem. A* 102 (49): 9987–9992. (d) Gadre, S.R. and Kulkarni, A.D. (2000). Electrostatics for exploring hydration patterns of molecules: 2. Formamide. *Ind. J. Chem. A* 39: 50–59. (e) Gadre, S.R., Babu, K., and Rendell, A.P. (2000). Electrostatics for exploring hydration patterns of molecules. 3. Uracil. *J. Phys. Chem. A* 104 (39): 8976–8982. (f) Sivanesan, D., Babu, K., Gadre, S.R. et al. (2000). Does a stacked DNA base pair hydrate better than a hydrogen-bonded one?: an ab initio study. *J. Phys. Chem.y A* 104 (46): 10887–10894. (g) Prakash, M., Subramanian, V., and Gadre, S.R. (2009). Stepwise hydration of protonated carbonic acid: a theoretical study. *J. Phys. Chem. A* 113 (44): 12260–12275. (h) Gadre, S.R. and Suresh, C.H. (1997). Electronic perturbations of the aromatic nucleus: Hammett constants and electrostatic potential topography. *J. Org. Chem.* 62 (8): 2625–2627. (i) Suresh, C.H. and Gadre, S.R. (1998). A novel electrostatic approach to substituent constants: doubly substituted benzenes. *JACS* 120 (28): 7049–7055. (j) Suresh, C.H. and Gadre, S.R. (2007). Electrostatic potential minimum of the aromatic ring as a measure of substituent constant. *J. Phys. Chem. A* 111 (4): 710–714.

6 (a) Bader, R.F.W., Gillespie, R.J., and MacDougall, P.J. (1988). A physical basis for the VSEPR model of molecular geometry. *JACS* 110 (22): 7329–7336. (b) Bader, R.F.W. (1985). Atoms in molecules. *Acc. Chem. Res.* 18 (1): 9–15.

7 Politzer, P., Daiker, K.C., and Donnelly, R.A. (1976). Molecular electrostatic potentials: a new approach to the study of the metabolic and carcinogenic activities of hydrocarbons. *Cancer Letters* 2 (1): 17–23.

8 Yeole, S.D. and Gadre, S.R. (2011). Topography of scalar fields: molecular clusters and π-conjugated systems. *J. Phys. Chem. A* 115 (45): 12769–12779.

9 Kumar, A., Gadre, S.R., Mohan, N., and Suresh, C.H. (2014). Lone pairs: an electrostatic viewpoint. *J. Phys. Chem. A* 118 (2): 526–532.

10 Pingale, S.S. (2011). Molecular electrostatic potential for exploring π-conjugation: a density-functional investigation. *Phys. Chem. Chem. Phys.* 13 (33): 15158–15165.

11 (a) Anjalikrishna, P.K., Suresh, C.H., and Gadre, S.R. (2019). Electrostatic topographical viewpoint of π-conjugation and aromaticity of hydrocarbons. *J. Phys. Chem. A* 123 (46): 10139–10151. (b) Elango, M., Subramanian, V., Rahalkar, A.P. et al. (2008). Structure, energetics, and reactivity of boric acid nanotubes: a molecular tailoring approach. *J. Phys. Chem. A* 112 (33): 7699–7704. (c) Khedkar, J.K., Pinjari, R.V., and Gejji, S.P. (2011). Electronic structure, molecular electrostatic potentials, vibrational spectra in substituted calix[n]arenes (n = 4, 5) from density functional theory. *J. Phys. Chem. A* 115 (38): 10624–10637. (d) Suresh, C.H., Lincy, T.L., Mohan,

N., and Rakhi, R. (2015). Aromatization energy and strain energy of buckminsterfullerene from homodesmotic reactions. *J. Phys. Chem. A* 119 (25): 6683–6688. (e) Boschetto, G., Krompiec, M., and Skylaris, C.-K. (2018). Insights into the charge-transfer mechanism of organic photovoltaics: effect of domain size. *J. Phys. Chem. C* 122 (30): 17024–17034.

12 Zhang, Y., Steyrleuthner, R., and Bredas, J.-L. (2016). Charge delocalization in oligomers of poly(2,5-bis(3-alkylthiophene-2-yl)thieno[3,2-*b*]thiophene) (PBTTT). *J. Phys. Chem. C* 120 (18): 9671–9677.

13 Etzold, F., Howard, I.A., Forler, N. et al. (2015). Sub-ns triplet state formation by non-geminate recombination in PSBTBT:PC_{70}BM and PCPDTBT:PC_{60}BM organic solar cells. *Energy Environ. Sci.* 8 (5): 1511–1522.

14 Liu, T. and Troisi, A. (2011). Absolute rate of charge separation and recombination in a molecular model of the P3HT/PCBM interface. *J. Phys. Chem. C* 115 (5): 2406–2415.

15 Zhao, Y. and Truhlar, D.G. (2006). A new local density functional for main-group thermochemistry, transition metal bonding, thermochemical kinetics, and noncovalent interactions. *J. Chem. Phys.* 125 (19): 194101.

16 Frisch, M.J., Trucks, G.W., Schlegel, H.B. et al. (2016). *Gaussian 16 Rev. A.01*. Wallingford, CT.

17 Kumar, A., López, R., Martínez, F. et al. (2022). DAMQT 3: advanced suite for the analysis of molecular density and related properties in large systems. *Comput. Phys. Commun.* 279: 108460.

18 (a) Balamurugan, K., Singam, E.R.A., and Subramanian, V. (2011). Effect of curvature on the α-helix breaking tendency of carbon based nanomaterials. *J. Phys. Chem. C* 115 (18): 8886–8892. (b) Umadevi, D. and Narahari Sastry, G. (2015). Graphane versus graphene: a computational investigation of the interaction of nucleobases, aminoacids, heterocycles, small molecules (CO_2, H_2O, NH_3, CH_4, H_2), metal ions and onium ions. *Phys. Chem. Chem. Phys.* 17 (45): 30260–30269.

19 Mudedla, S.K., Balamurugan, K., Kamaraj, M., and Subramanian, V. (2016). Interaction of nucleobases with silicon doped and defective silicon doped graphene and optical properties. *Phys. Chem. Chem. Phys.* 18 (1): 295–309.

20 Zhao, Y. and Truhlar, D.G. (2008). The M06 suite of density functionals for main group thermochemistry, thermochemical kinetics, noncovalent interactions, excited states, and transition elements: two new functionals and systematic testing of four M06-class functionals and 12 other functionals. *Theor. Chem. Acc.* 120 (1): 215–241.

21 Biegler-Konig, F., Schonbohm, J., Derdau, R., Bayles, D., and Bader, R.F.W. (2000). *AIM 2000*. Version 1. Bielefeld, Germany.

22 Mudedla, S.K., Balamurugan, K., and Subramanian, V. (2014). Computational study on the interaction of modified nucleobases with graphene and doped graphenes. *J. Phys. Chem. C* 118 (29): 16165–16174.

23 Balamurugan, K. and Subramanian, V. (2013). Adsorption of chlorobenzene onto (5,5) armchair single-walled carbon nanotube and graphene sheet: toxicity versus adsorption strength. *J. Phys. Chem. C* 117 (41): 21217–21227.

24 Svensson, M., Humbel, S., Froese, R.D.J. et al. (1996). ONIOM: a multilayered integrated MO + MM method for geometry optimizations and single point energy predictions. A test for Diels–Alder reactions and $Pt(P(t\text{-}Bu)_3)_2 + H_2$ oxidative addition. *J. Phys. Chem.* 100 (50): 19357–19363.

25 Kumar, R.M., Elango, M., and Subramanian, V. (2010). Carbohydrate-aromatic interactions: the role of curvature on XH···π interactions. *J. Phys. Chem. A* 114 (12): 4313–4324.

21

Origin and Nature of Pancake Bonding Interactions: A Density Functional Theory and Information-Theoretic Approach Study

Dongbo Zhao[1], *Xin He*[2] *and Shubin Liu*[3,4]

[1] *Institute of Biomedical Research, Yunnan University, Kunming, Yunnan, P.R. China*
[2] *Qingdao Institute for Theoretical and Computational Sciences, Institute of Frontier and Interdisciplinary Science, Shandong University, Qingdao, Shandong, P.R. China*
[3] *Research Computing Center, University of North Carolina, Chapel Hill, NC, USA*
[4] *Department of Chemistry, University of North Carolina, Chapel Hill, NC, USA*

21.1 Introduction

Pancake bonding interactions, also known as radical–pair interactions, have a long history [1, 2] and aroused great interest in the past few decades because they are extensively involved in supramolecular chemistry [3], materials science [4], and energy science [5–7]. Many structures possess intriguing pancake bonding properties, including intermolecular imide-fused polycyclic aromatic hydrocarbons [8], TCNE [9–12], phenalenyl [12, 13], olympicenyl [14], thiadiazolyl/selenadiazolyl [15], thiatriazine [16], viologen [17, 18], and intra-molecular helical structures [19]. Then, what causes the aggregation? It is a crucial and controversial issue in the literature. For example, Kertesz and coworkers proposed that singly occupied molecular orbital (SOMO) overlap [15, 19, 20] governs the preferred configurations. Additionally, dispersion forces [10, 12, 15, 17, 21–23] are typically involved in rationalizing molecular stability. Moreover, a few reports have indicated that the charge–transfer effect makes some contributions [24, 25]. Aromatic properties [26], and formation mechanisms [27–30], have also been reported.

In this work, we aim to dissect the molecular stability of phenalenyl and olympicenyl dimers and their derivatives in different states, as shown in Figure 21.1. To make that happen, we employ two total energy partition schemes from density functional theory (DFT) [31–33] to decompose the interaction energy into its components and determine which one dictates the overall stability. We also use a few simple density-based quantities from the information-theoretic approach (ITA) to gain insights into the dominant forces. Here, we choose some commonly used ITA functions, such as the Shannon entropy [34], Fisher information [35], Ghosh–Berkowitz–Parr entropy [36], relative Shannon entropy [37], relative Fisher information [38, 39], Onicescu information energy [40], and relative Rényi entropy [40]. Some other theoretical tools, such as noncovalent interaction (NCI) [41] and strong covalent interactions (SCI) [42] are employed to explore the bonding and interaction patterns of the dimer. Our computational work has revealed that dispersion forces are absent between the two monomers, as evidenced by the NCI data. This finding is in stark contrast to a commonly held belief that dispersion forces are responsible for the aggregation. More importantly, we have unraveled that it is the exchange–correlation effect that is highly responsible for molecular

Electron Density: Concepts, Computation and DFT Applications, First Edition.
Edited by Pratim Kumar Chattaraj and Debdutta Chakraborty.

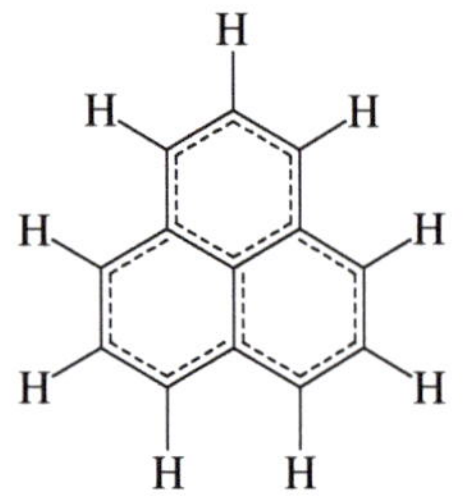

1, Phenalenpy (PLY)

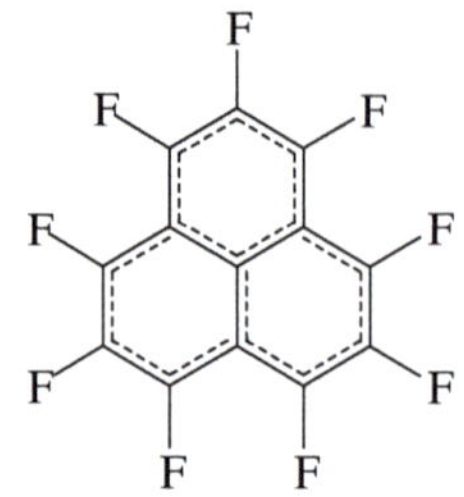

2, Perfluoro-phenalenyl (PF-PLY)

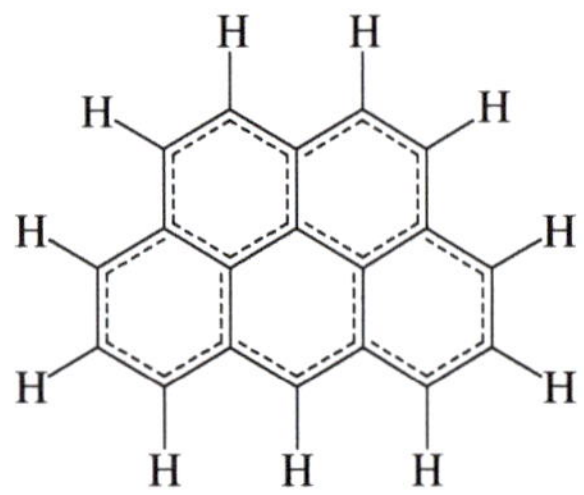

3, Olympicenyl (OPY)

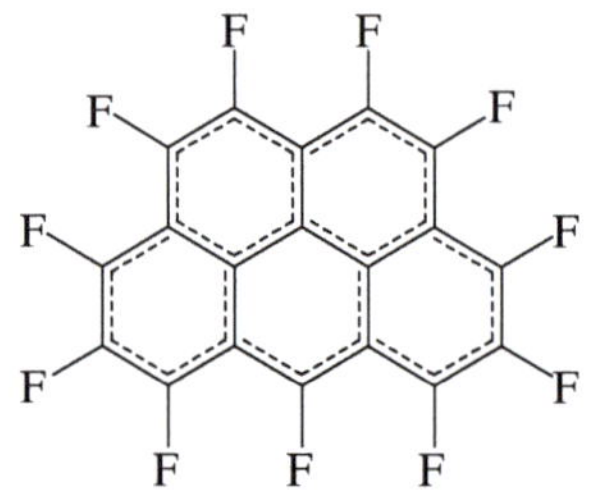

4, Perfluoro-olympicenyl (PF-OPY)

Figure 21.1 Schematic representation of four molecular systems that form the basis of this work.

stability. This observation is different from previous orbital-based understanding. The implication of this work is straightforward: deceptively simple systems typically have something unexpected.

21.2 Methodology

21.2.1 Interaction Energy and Its Components in DFT

In this work, we aim to dissect the molecular stability of a pancake dimer. To make that happen, we use the Kohn–Sham DFT, whose total energy can be expressed as [31, 32],

$$\Delta E[\rho] = \Delta T_S[\rho] + \Delta E_e[\rho] + \Delta E_{xc}[\rho] \tag{21.1}$$

where T_S, E_e, and E_{xc} are the noninteracting kinetic, electrostatic, and exchange–correlation energies, respectively. The electrostatic energy E_e includes three independent components: the nuclear–electron attraction, V_{ne}; the classical interelectron Coulombic repulsion, J; and the nuclear–nuclear repulsion, V_{nn}. The last term E_{xc} consists of exchange (E_x) and correlation (E_c) components.

In addition, we have proposed an alternative way to partition the total energy difference into its components as [33],

$$\Delta E[\rho] = \Delta E_s[\rho] + \Delta E_e[\rho] + \Delta E_q[\rho] \tag{21.2}$$

where E_e is the same as in Eq. (21.1), E_s stands for the steric effect and E_q signifies the contribution originating from the Fermionic quantum effect (due to the exchange–correlation interaction). In Eq. (21.2), the energy component associated with the steric effect E_s has been proved to be simply the Weizsäcker kinetic energy

$$E_s[\rho(\mathbf{r})] = T_W[\rho(\mathbf{r})] = \frac{1}{8}\int \frac{|\nabla\rho(\mathbf{r})|^2}{\rho(\mathbf{r})} d\mathbf{r} \tag{21.3}$$

where $\rho(\mathbf{r})$ and $\nabla\rho(\mathbf{r})$ are the electron density and its gradient, respectively. After combining Eqs. (21.1) and (21.2), one has the definition of Eq. (21.4).

$$\Delta E_q[\rho] = \Delta E_{xc}[\rho] + \Delta T_s[\rho] - \Delta E_s[\rho] \tag{21.4}$$

Of note, this new formulation has its distinct physical meaning and corresponding physical state. It has been applied to several molecular systems and phenomena whose results are consistent with our chemical intuition and conventional wisdom. Its further implementation for condensed-phase systems will be in our forthcoming work.

21.2.2 Information-Theoretic Approach Quantities

Shannon entropy S_S [34] characterizes the spatial delocalization of the electron density and Fisher information I_F [35] measures the sharpness or localization of the same. They are defined as Eqs. (21.5) and (21.6), respectively

$$S_S = -\int \rho(\mathbf{r}) \ln \rho(\mathbf{r}) d\mathbf{r} \tag{21.5}$$

and

$$I_F = \int \frac{|\nabla\rho(\mathbf{r})|^2}{\rho(\mathbf{r})} d\mathbf{r} \tag{21.6}$$

Ghosh–Berkowitz–Parr (GBP) entropy S_{GBP} [36] is associated with both electron density and kinetic energy density, as given in Eq. (21.7),

$$S_{GBP} = -\int \frac{3}{2} K\rho(\mathbf{r}) \left[c + \ln \frac{t(\mathbf{r};\rho)}{t_{TF}(\mathbf{r};\rho)} \right] d\mathbf{r} \tag{21.7}$$

where $t(\mathbf{r};\ \rho)$ and $t_{TF}(\mathbf{r};\ \rho)$ are the kinetic and Thomas–Fermi (TF) kinetic energy density, respectively. The former is related to the total kinetic energy T_S via $\int t(\mathbf{r};\ \rho) d\mathbf{r} = T_S$ and $t_{TF}(\mathbf{r};\ \rho)$ is defined as $t_{TF}(\mathbf{r};\ \rho) = c_K \rho^{5/3}(\mathbf{r})$. Of note, K is the Boltzmann constant, $c = (5/3) + \ln(4\pi c_K/3)$ and $c_K = (3/10)(3\pi^2)^{2/3}$ are also constants, and $t(\mathbf{r};\ \rho)$ is cast in the form of density and its derivative, $t(\mathbf{r};\ \rho) = \sum_i \frac{1}{8}\frac{\nabla\rho_i \cdot \nabla\rho_i}{\rho_i} - \frac{1}{8}\nabla^2\rho$. Of note, different forms of the kinetic-energy density are used in different contexts [43–50], but within the mathematical framework of Ghosh, Berkowitz, and Parr, this is the maximum-entropy choice [36].

Additional ITA quantities have been introduced as new reactivity descriptors in conceptual density functional theory (CDFT) [51–54]. One example is Onicescu information energy of order n [40]

$$E_n = \frac{1}{n-1} \int \rho^n(\mathbf{r}) d\mathbf{r} \tag{21.8}$$

relative Rényi entropy of order n [40]

$$R_n^r = \frac{1}{n-1} \ln \left[\int \frac{\rho^n(\mathbf{r})}{\rho_0^{n-1}(\mathbf{r})} d\mathbf{r} \right] \tag{21.9}$$

and information gain [37] (also called Kullback–Leibler divergence or relative Shannon entropy) I_G is given in Eq. (21.10)

$$I_G = \int \rho(\mathbf{r}) \ln \frac{\rho(\mathbf{r})}{\rho_0(\mathbf{r})} d\mathbf{r} \tag{21.10}$$

where $\rho_0(\mathbf{r})$ is the reference-state density satisfying the same normalization condition as $\rho(\mathbf{r})$.

Recently [38], one of the present authors proposed another three functions G_1, G_2, and G_3. Their analytical forms are available, and here, we mainly focus on G_3, as shown below:

$$G_3 = \sum_A \int \rho_A(\mathbf{r}) \left[\nabla \ln \frac{\rho_A(\mathbf{r})}{\rho_A^0(\mathbf{r})} \right]^2 d\mathbf{r} \tag{21.11}$$

The quantifications and applications of Eq. (21.11) can be found in Refs. [38, 39, 55, 56]. Note that during the past decade, we have attempted to seamlessly glue the DFT and information theory together, as electron density can be a linker between these two theories. The progress and applications can be found in our recent reviews [57, 58]. In addition, Hirshfeld's stockholder approach [59–63] is often adopted to partition atoms in a molecule, as defined in Eq. (21.12),

$$\rho_A(\mathbf{r}) = \omega_A(\mathbf{r})\rho(\mathbf{r}) = \frac{\rho_A^0(\mathbf{r})(\mathbf{r} - \mathbf{R}_A)}{\sum_B \rho_B^0(\mathbf{r} - \mathbf{R}_A)} \rho(\mathbf{r}) \tag{21.12}$$

Here $\rho_A(\mathbf{r})$ is the Hirshfeld density for atom A in a molecule, $\omega_A(\mathbf{r})$ is a sharing function for atom A, $\rho_B^0(\mathbf{r} - \mathbf{R}_A)$ is the density of atom B centered at the nuclear position vector $\mathbf{R}_A$, and $\rho(\mathbf{r})$ is the actual molecular density. The sum of all the nuclei-centered free atom densities, typically spherically averaged ground-state atomic densities, is termed the promolecular density. While the stockholder approach is natural in the context of ITA because it is also based on information-theoretic arguments, other molecular partitioning schemes like Becke's fuzzy atom approach [64] and Bader's zero-flux atoms-in-molecules (AIM) method [65] can also be used.

21.3 Computational Details

We present the first systematic study of energetically competitive conformations of four dimerized phenalenyl derivatives: phenalenyl (PLY, **1**), perfluoro-phenalenyl (PFPLY, **2**), triphenylphenalenyl (TPPLY, **3**), and tris(perfluorophenyl)-phenalenyl (TPFPPLY, **4**) (see Figure 21.1). All DFT calculations were performed with the Gaussian 16 [66] package with default tight self consistent field (SCF) convergence criteria and ultrafine integration grids. The structures were fully relaxed at the (U)M05-2X/6-311G(d) [67, 68] level, followed by vibrational analysis. On top of the optimized dimers, we checked the stability of wavefunctions at the M05-2X/6-31G level to substantially reduce the computational costs without qualitatively compromising the results. The molecular wavefunctions were then obtained at the M05-2X/Def2-TZVP [67, 69] level, as well as the energetic components as obtained with the Gaussian keyword IOp(5/33 = 1). Each molecular wavefunction file was then fed into the Multiwfn 3.8 [70] program to obtain all the ITA quantities, NCI and ESP cube files for further visualization.

21.4 Results and Discussion

We first look at the bonding properties of neutral $\mathbf{1}_2$ and $\mathbf{2}_2$ using the SCI function, as shown in Figure 21.2. No discernible difference is observed for the charged counterpart. The monomeric

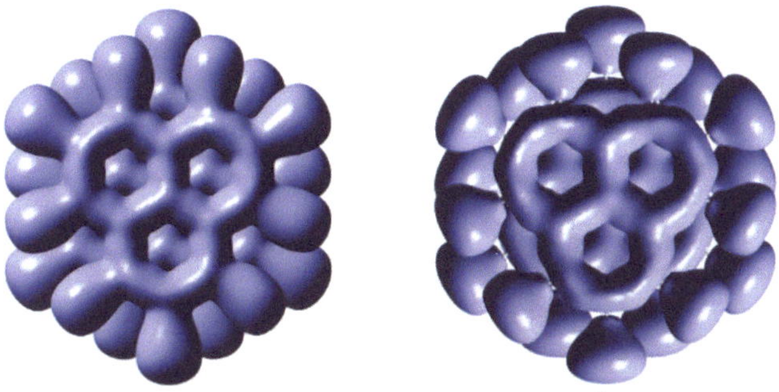

Figure 21.2 SCI isosurfaces for neutral dimers (**1_2** and **2_2**) at the M05-2X/6-311G(d) level. The isovalue of 1.0 a.u. was used.

motif of **1_2** has a delocalized electron distribution. When all the H atoms in **1_2** are replaced with F atoms in **2_2**, electrons are mainly found in two regions: one surrounds the F atoms because of its high electronegativity, and the other is the phenalenyl ring. No isosurfaces are observed between two monomers of each dimer, indicating no strong covalent interactions exist. Similar results are observed for **3_2** and **4_2**, and thus, they are not shown here.

Shown in Figure 21.3 are the reduced density gradient (RDG) results for the neutral, cationic, and anionic **1_2** and **2_2**. For both **1_2** and **2_2**, adding or removing an electron does not change the pattern of the corresponding intermolecular interactions, indicating they can potentially be used as building blocks of materials with great damage tolerance. For **1_2**, no such a "spike" region [41] exists near zero density, indicating strong interactions. However, such a spike region for **2_2** suggests weak interactions within a dimer. Then, where does this kind of weak interaction come from? To this end, we use the RDG isosurfaces to dissect its origin, as shown in Figure 21.4. The red areas in a six-membered ring center represent repulsive interactions, while green areas signify attractive interactions. It is clearly shown that an attractive surface lies between two monomers for both **1_2** and **2_2**, which should be the primary root source of molecular stability. Careful inspection of the RDG isosurface for **2_2** leads to a conclusion that attractive intermolecular interactions between two adjacent F atoms in a plane or from two planes should be responsible for the spike region in Figure 21.3. This observation is in line with our previous work [71]. Of note, similar results have been obtained for **3_2** and **4_2**, thus the results are not shown. Collectively, weak intermolecular interactions may be one of the primary sources of stabilizing the dimer.

Next, we asked where the strong interactions between two monomers come from. Since the dimers in Figure 21.1 have diradical characteristics, spin density distributions of neutral **1_2** and **2_2** obtained at the M05-2X/6-311G(d) level are plotted in Figure 21.5. It is clearly shown that extra electrons are mainly localized at C atoms, with a small portion of electrons on F atoms. Though the signs of wave functions are different (blue and green), its square is always nonnegative. In this sense, this kind of interatomic interaction should be repulsive. Similar observations are obtained for **3_2** and **4_2**, thus the results are not shown. Since persistent radical–pair interactions are local, it makes more sense to look into the global electrostatic potential (ESP) surface. Shown in Figure 21.6 are the ESP isosurfaces of neutral, cationic, and anionic **1_2** and **2_2**, mapping to the corresponding total electron density. It is clear that for **1_2**, the most electrophilic sites are at the H atoms, while for **2_2**, they are mainly situated at the C atoms due to the introduction of F atoms. In addition, the removal or addition of an electron influences the whole delocalized region. Similar results are also obtained for **3_2** and **4_2**.

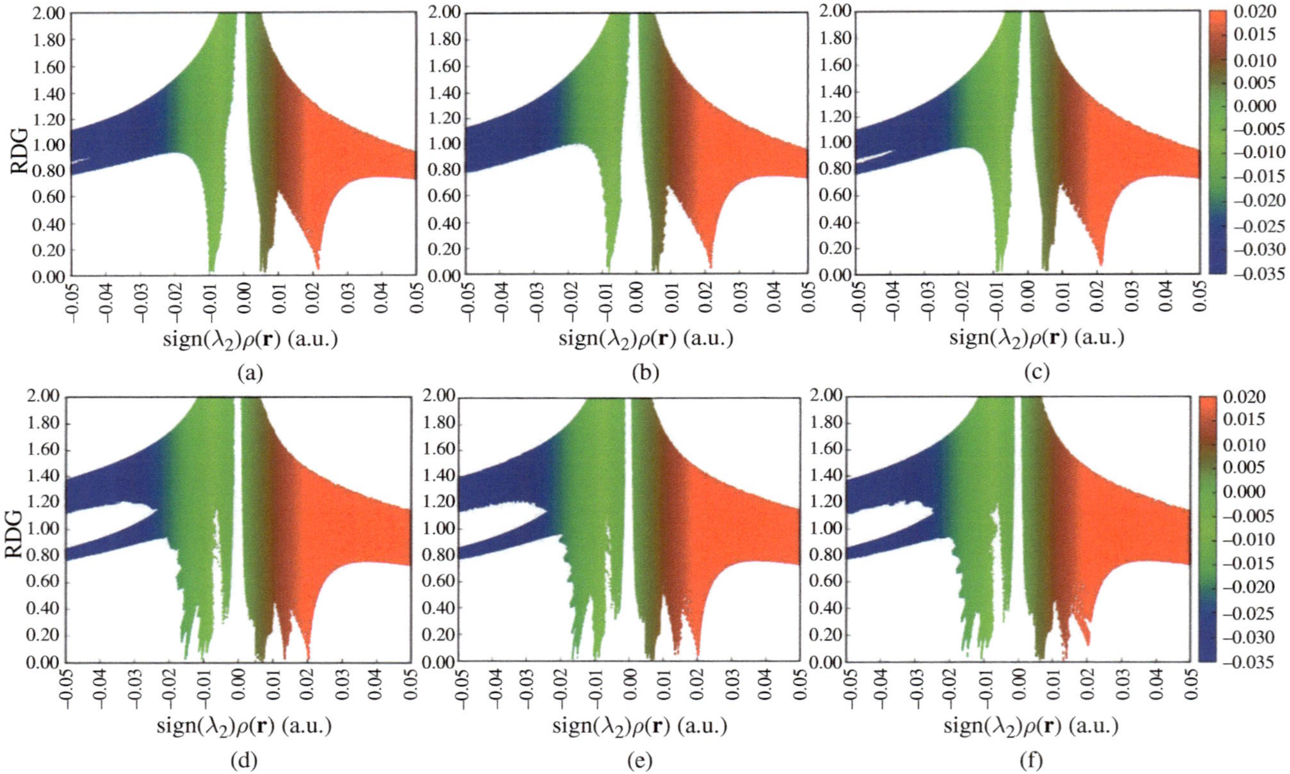

Figure 21.3 Diagrams of the reduced density gradient (RDG) versus sign(λ_2)ρ(**r**) for neutral (left), cationic (middle), and anionic (right) $\mathbf{1}_2$ (upper panel) and $\mathbf{2}_2$ (lower panel).

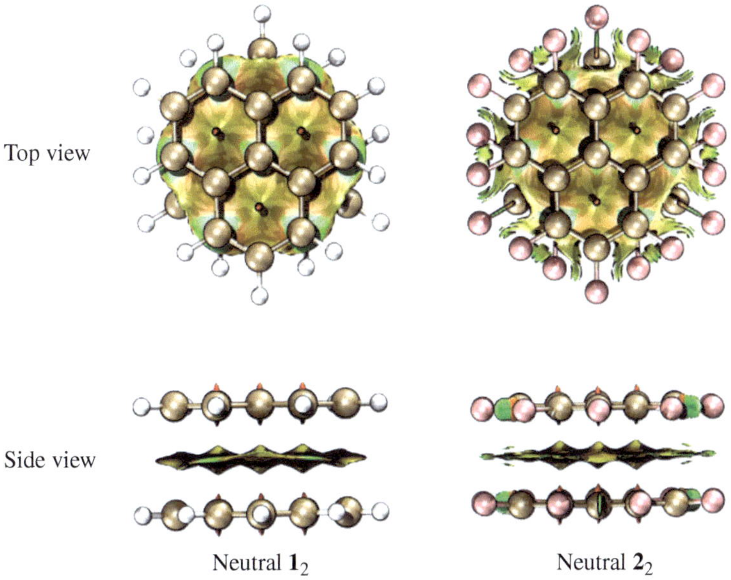

Figure 21.4 Reduced density gradient isosurfaces at the M05-2X/6-311G(d) level for $\mathbf{1}_2$ and $\mathbf{2}_2$ from both top and side views. Red areas are repulsive interactions, whereas blue regions represent attractive interactions. The isovalue of 0.5 a.u. was used.

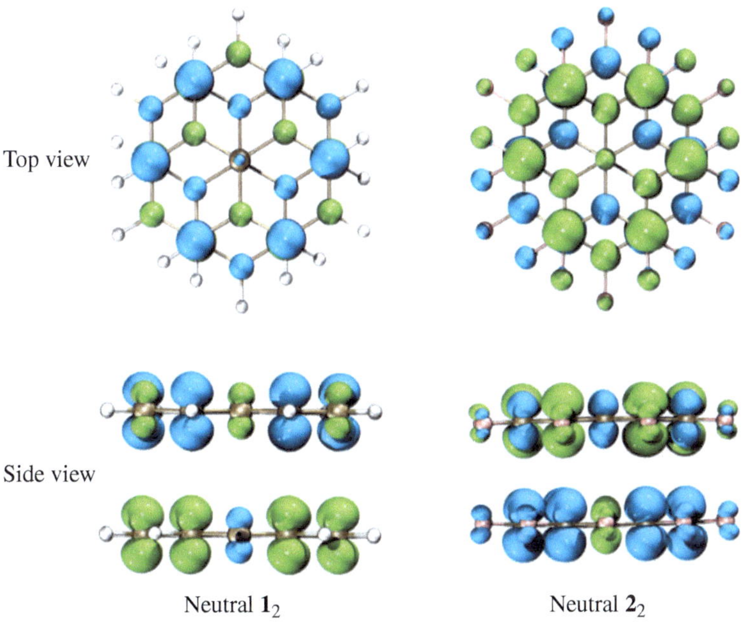

Figure 21.5 Spin density maps at the M05-2X/6-311G(d) level for $\mathbf{1}_2$ and $\mathbf{2}_2$ from both top and side views. The isovalue 0.0004 a.u. was used.

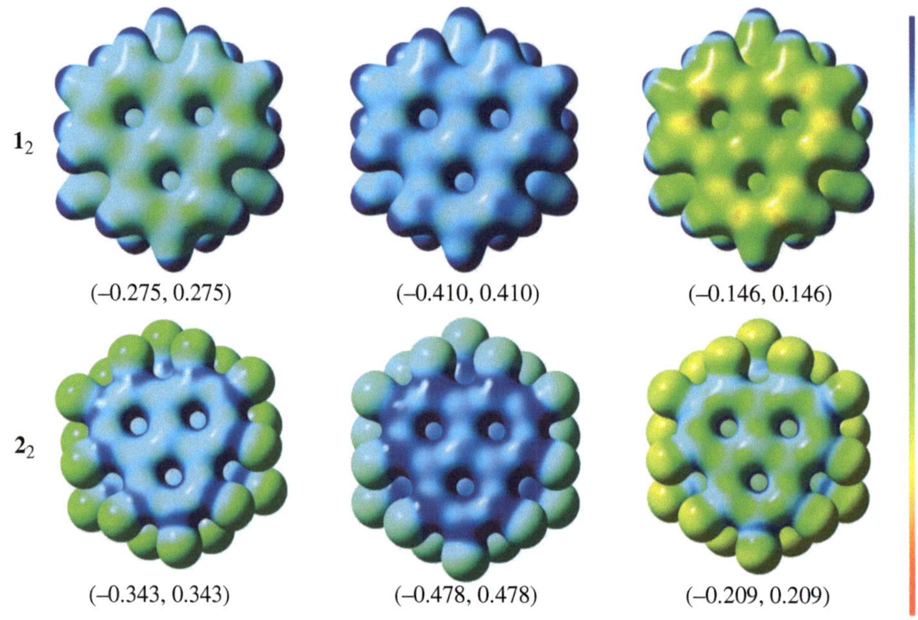

Figure 21.6 ESP isosurfaces at the M05-2X/6-311G(d) level for neutral (left), positive (middle), and negative (right) $\mathbf{1}_2$ and $\mathbf{2}_2$, mapping to the corresponding total electron density. Red means negative, and blue signifies positive in the color bar. The isovalue 0.04 a.u. was used.

To pinpoint the origin of molecular stability, we switch our gear to the total energy partition schemes. Collected in Table 21.1 are the interaction energies and their components obtained at the (U)M05-2X/Def2-TZVP level. We have observed a robust linear relationship between the quantum effect (ΔE_q) and steric hindrance (ΔE_s), which serves as quality control of our computations. It is clear from Table 21.1 that the formation of neutral and charged pancake dimers is energetically preferential. For the neutral dimers, the interaction energies (ΔE_{tot}) are very close to each other, ranging from −10.0 ($\mathbf{1}_2$) to −15.4 ($\mathbf{4}_2$) kcal/mol. If an electron is removed from the corresponding neutral dimer, the formation energies can be substantially reduced by ~12 kcal/mol, except for $\mathbf{2}_2$ (~2 kcal/mol). Moreover, adding an extra electron to the neutral dimer is energetically favorable. This process can gain an energy difference of ~12 kcal/mol for $\mathbf{2}_2$ and $\mathbf{4}_2$ and ~2 kcal/mol for $\mathbf{1}_2$ and $\mathbf{3}_2$. Then, what causes these considerable total energy differences? Is there a single energy component responsible for molecular stability? To answer these questions, we resort to the energy components, as shown in Table 21.1. One can see that while the steric effect makes positive contributions, they are compensated mainly by the quantum effect, as listed in the last two columns. For the exchange and correlation effect in the fourth and fifth columns, one can notice that though they make opposite contributions (negative for ΔE_x and positive for ΔE_c) to the molecular stability, their combination ΔE_{xc} governs the formation of a pancake dimer. A similar observation is discovered for triatomic clusters H_3^+ and Li_3^+ [72]. Moreover, the electrostatic potential is dominantly repulsive, especially for neutral dimers. Similar results have been observed for other systems in ref [12].

Finally, we explore the possible relationships between the ITA quantities, as shown in Table 21.2 and energetics in Table 21.1, to trace the origin of molecular stability of persistent radical-pair interactions in pancake dimers, as collected in Table 21.3. These density-based quantities, either from DFT or ITA, are intercorrelated, indicating that we can use one to approximate the other.

Table 21.1 Shown here are the interaction energy (in kcal/mol) and its components for all the systems in Figure 21.1.[a)]

Dimer	ΔE_{tot}	ΔT_s	ΔE_x	ΔE_c	ΔE_{xc}	ΔE_e	ΔE_s	ΔE_q
$\mathbf{1}_2$	−10.0	−18.1	21.1	−58.0	−36.9	45.0	−563.6	508.6
$\mathbf{1}_2^+$	−21.8	9.4	14.4	−40.7	−26.3	−4.8	−383.2	366.2
$\mathbf{1}_2^-$	−12.0	18.8	15.2	−41.4	−26.2	−4.7	−312.7	305.3
$\mathbf{2}_2$	−12.1	3.3	25.6	−63.0	−37.4	22.0	−601.6	567.5
$\mathbf{2}_2^+$	−14.5	14.9	23.8	−56.5	−32.7	3.3	−542.9	525.1
$\mathbf{2}_2^-$	−24.4	12.5	22.3	−59.8	−37.5	0.6	−608.6	583.5
$\mathbf{3}_2$	−11.1	−1.8	25.4	−63.3	−37.9	28.7	−564.4	524.7
$\mathbf{3}_2^+$	−23.3	20.5	20.7	−57.1	−36.5	−7.4	−533.0	517.0
$\mathbf{3}_2^-$	−12.6	20.5	21.9	−56.0	−34.1	1.0	−529.0	515.3
$\mathbf{4}_2$	−15.4	15.5	33.1	−80.1	−47.0	16.0	−771.6	740.2
$\mathbf{4}_2^+$	−17.1	23.4	32.5	−77.1	−44.5	4.0	−739.7	718.6
$\mathbf{4}_2^-$	−26.6	22.1	30.5	−78.8	−48.3	−0.3	−789.1	762.9

a) Single-point computations were performed at the (U)M05-2X/Def2-TZVP level.

Table 21.2 Shown here are the ITA quantities (in a.u.) for all the systems in Figure 21.1.[a)]

Dimer	ΔS_S	ΔI_F	ΔS_{GBP}	ΔE_2	ΔE_3	$\Delta^r R_2$	$\Delta^r R_3$	ΔG_1	ΔG_2	ΔG_3	ΔI_G
$\mathbf{1}_2$	−1.712	−7.185	−1.761	0.035	2.738	−0.084	−0.215	−0.183	0.318	6.235	−0.053
$\mathbf{1}_2^+$	−1.279	−4.886	−1.243	0.016	1.180	−0.077	−0.190	0.194	0.196	4.472	−0.049
$\mathbf{1}_2^-$	−1.279	−3.987	−1.190	0.016	0.989	−0.134	−0.342	0.270	0.008	4.468	−0.083
$\mathbf{2}_2$	−1.649	−7.670	−1.743	0.222	30.576	−0.092	−0.247	−0.459	−0.162	7.858	−0.053
$\mathbf{2}_2^+$	−1.442	−6.922	−1.553	0.300	43.235	−0.127	−0.338	−0.520	−0.071	7.020	−0.073
$\mathbf{2}_2^-$	−1.637	−7.758	−1.621	0.108	11.484	−0.097	−0.264	0.070	−0.349	7.295	−0.055
$\mathbf{3}_2$	−1.833	−7.196	−1.849	0.069	4.930	−0.103	−0.257	0.137	−0.147	6.876	−0.065
$\mathbf{3}_2^+$	−1.747	−6.795	−1.694	0.272	23.138	−0.092	−0.230	0.160	−1.482	6.191	−0.059
$\mathbf{3}_2^-$	−1.664	−6.744	−1.578	0.193	11.175	−0.148	−0.373	0.305	−0.522	6.173	−0.093
$\mathbf{4}_2$	−2.083	−9.838	−2.176	0.029	32.843	−0.118	−0.311	−0.330	0.315	9.806	−0.068
$\mathbf{4}_2^+$	−1.973	−9.430	−2.095	0.221	95.151	−0.153	−0.405	−0.489	−0.268	9.571	−0.089
$\mathbf{4}_2^-$	−2.120	−10.060	−2.112	−0.079	8.478	−0.115	−0.306	0.119	0.201	9.416	−0.066

a) Single-point computations were performed at the (U)M05-2X/Def2-TZVP level.

For example, the exchange–correlation effect (ΔE_{xc}) is in good line with ΔS_S, ΔI_F, and ΔS_{GBP}, with $R^2 = 0.952$, 0.960, and 0.962, respectively, indicating that one can use these quantities to develop better exchange–correlation functionals. A similar effort can be traced back to 1934 when Fermi and Amaldi [73] first proposed to use the electrostatic potential to approximate E_{xc}. Of note, Shannon entropy measures the spatial delocalization of the electron density, and Fisher

Table 21.3 Correlation coefficient (R^2) matrix between the total energy difference, its components, and the ITA quantities.

R^2	ΔE_{tot}	ΔT_s	ΔE_{xc}	ΔE_e	ΔE_s	ΔE_q	ΔS_S	ΔI_F	ΔS_{GBP}	ΔG_3	ΔI_G
ΔE_{tot}	1.000										
ΔT_s	0.257	1.000									
ΔE_{xc}	0.064	0.021	1.000								
ΔE_e	0.402	0.772	0.056	1.000							
ΔE_s	0.072	0.023	0.960	0.048	1.000						
ΔE_q	0.095	0.056	0.948	0.019	0.992	1.000					
ΔS_S	0.040	0.014	0.952	0.070	0.862	0.846	1.000				
ΔI_F	0.072	0.023	0.960	0.048	1.000	0.992	0.862	1.000			
ΔS_{GBP}	0.014	0.003	0.962	0.120	0.923	0.897	0.950	0.923	1.000		
ΔG_3	0.038	0.056	0.898	0.028	0.945	0.954	0.768	0.945	0.871	1.000	
ΔI_G	0.082	0.330	0.005	0.091	0.003	0.011	0.013	0.003	0.007	0.022	1.000

information measures the sharpness or localization of the same. These different but strongly correlated relationships provide effective measurements of the electron density distribution of the system; thus, they attribute valuable and novel insights into the nature and origin of various physicochemical phenomena, including the molecular stability of pancake dimers. Similar results have been found for fullerene buckyballs [74]. Other interesting results from Table 21.2 are the strong correlations between the steric effect (ΔE_s) and previous ΔS_{GBP} ($R^2 = 0.923$) and newly proposed ΔG_3 ($R^2 = 0.945$). This suggests that one can use these two simple density-based functions to quantify the steric effect.

21.5 Concluding Remarks

In summary, using a few theoretical tools from DFT, we have explored the origin and nature of the persistent radical–pair interactions in pancake dimers. Noncovalent interactions lucidly show that it is strong, attractive interactions that bind the two monomers together, which is in stark contrast to, for example, the dispersion-bound dimer of benzene. Local spin densities at the carbon atoms and global electrostatic potential maps fail to dictate the total molecular stability. From the total energy partition schemes, we have found that no single energy component can strongly correlate with the total energy difference. However, the exchange–correlation effect is highly positive in governing the formation of a dimer. We also found that quantities from information theory are in good correlation with the exchange–correlation effect, indicative of the significant role of the distribution of electron density that is responsible for molecular stability. We mention in passing that we have found in some intrinsically disordered proteins with low-complexity domains, there exist non-vanishing spin density maps. More work along this line is ongoing to verify whether it is a reality or just a fantasy.

Acknowledgment

Prof. Pratim Kumar Chattaraj is greatly acknowledged for this kind invitation. DBZ is supported by the startup funding of Yunnan University and the Yunnan Fundamental Research Projects (Grant No. 202101AU070012).

References

1 Howarth, O.W. and Fraenkel, G.K. (1966). Electron spin resonance study of mono- and dimeric cations of aromatic hydrocarbons. *JACS* 88 (19): 4514–4515.

2 Badger, B. and Brocklehurs, B. (1968). Formation of dimer cations of aromatic hydrocarbons. *Nature* 219 (5151): 263.

3 Cai, K., Zhang, L., Astumian, R.D., and Stoddart, J.F. (2021). Radical-pairing-induced molecular assembly and motion. *Nat. Rev. Chem.* 5 (7): 447–465.

4 Ito, S., Nagami, T., and Nakano, M. (2017). Singlet fission in pancake-bonded systems. *Phys. Chem. Chem. Phys.* 19 (8): 5737–5745.

5 Janesko, B.G. and Brothers, E.N. (2021). Half-pancake bonding in asphaltenes. *Energy Fuels* 35 (19): 15657–15662.

6 Zhang, Y., Siskin, M., Gray, M.R. et al. (2020). Mechanisms of asphaltene aggregation: puzzles and a new hypothesis. *Energy Fuels* 34 (8): 9094–9107.

7 Alili, A.S., Siddiquee, M.N., and de Klerk, A. (2020). Origin of free radical persistence in asphaltenes: cage effect and steric protection. *Energy Fuels* 34 (1): 348–359.

8 Liu, J., Zeng, J., Zhao, D. et al. (2022). Comprehending radicals, diradicals and their bondings in aggregates of imide-fused polycyclic aromatic hydrocarbons. *Chem. Sci.* 13 (34): 9985–9992.

9 Novoa, J.J., Lafuente, P., Del Sesto, R.E., and Miller, J.S. (2001). Exceptionally long (≥2.9 Å) C–C bonds between $[TCNE]^-$ ions: two-electron, four-center $\pi^*-\pi^*$ C–C bonding in π-$[TCNE]_2^{2-}$. *Angew. Chem. Int. Ed.* 40 (13): 2540–2545.

10 Jung, Y. and Head-Gordon, M. (2004). What is the nature of the long bond in the $TCNE_2^{2-}$ π-dimer? *Phys. Chem. Chem. Phys.* 6 (9): 2008–2011.

11 Jakowski, J. and Simons, J. (2003). Theoretical analysis of the electronic structure and bonding stability of the TCNE dimer di-anion $(TCNE)_2^{2-}$. *JACS* 125 (51): 16089–16096.

12 Mota, F., Miller, J.S., and Novoa, J.J. (2009). Comparative analysis of the multicenter, long bond in $[TCNE]^{\cdot-}$ and phenalenyl radical dimers: a unified description of multicenter, long bonds. *J. Am. Chem. Soc.* 131 (22): 7699–7707.

13 Goto, K., Kubo, T., Yamamoto, K. et al. (1999). A stable neutral hydrocarbon radical: synthesis, crystal structure, and physical properties of 2,5,8-tri-*tert*-butyl-phenalenyl. *JACS* 121 (7): 1619–1620.

14 Xiang, Q., Guo, J., Xu, J. et al. (2020). Stable olympicenyl radicals and their π-dimers. *JACS* 142 (25): 11022–11031.

15 Beneberu, H.Z., Tian, Y.-H., and Kertesz, M. (2012). Bonds or not bonds? Pancake bonding in 1,2,3,5-dithiadiazolyl and 1,2,3,5-diselenadiazolyl radical dimers and their derivatives. *Phys. Chem. Chem. Phys.* 14 (30): 10713–10725.

16 Cui, Z.-H., Lischka, H., Beneberu, H.Z., and Kertesz, M. (2014). Double pancake bonds: pushing the limits of strong $\pi-\pi$ stacking interactions. *JACS* 136 (37): 12958–12965.

17 Geraskina, M.R., Dutton, A.S., Juetten, M.J. et al. (2017). The viologen cation radical pimer: a case of dispersion-driven bonding. *Angew. Chem. Int. Ed.* 56 (32): 9435–9439.

18 Juneau, A. and Frenette, M. (2021). Exploring curious covalent bonding: Raman identification and thermodynamics of perpendicular and parallel pancake bonding (pimers) of ethyl viologen radical cation dimers. *J. Phys. Chem. B* 125 (38): 10805–10812.

19 Takamuku, S., Nakano, M., and Kertesz, M. (2017). Intramolecular pancake bonding in helical structures. *Chem. Eur. J.* 23 (31): 7474–7482.

20 Mou, Z. and Kertesz, M. (2017). Pancake bond orders of a series of π-stacked triangulene radicals. *Angew. Chem. Int. Ed.* 56 (34): 10188–10191.

21 Cui, Z.-H., Lischka, H., Beneberu, H.Z., and Kertesz, M. (2014). Rotational barrier in phenalenyl neutral radical dimer: separating pancake and van der Waals interactions. *JACS* 136 (15): 5539–5542.

22 Small, D., Zaitsev, V., Jung, Y. et al. (2004). Intermolecular π-to-π bonding between stacked aromatic dyads. experimental and theoretical binding energies and near-IR optical transitions for phenalenyl radical/radical versus radical/cation dimerizations. *JACS* 126 (42): 13850–13858.

23 Mou, Z., Tian, Y.-H., and Kertesz, M. (2017). Validation of density functionals for pancake-bonded π-dimers; dispersion is not enough. *Phys. Chem. Chem. Phys.* 19 (36): 24761–24768.

24 Lü, J.-M., Rosokha, S.V., and Kochi, J.K. (2003). Stable (long-bonded) dimers via the quantitative self-association of different cationic, anionic, and uncharged π-radicals: structures, energetics, and optical transitions. *JACS* 125 (40): 12161–12171.

25 Cui, Z.-H., Wang, M.-H., Lischka, H., and Kertesz, M. (2021). Unexpected charge effects strengthen π–stacking pancake bonding. *JACS Au* 1 (10): 1647–1655.

26 Suzuki, S., Morita, Y., Fukui, K. et al. (2006). Aromaticity on the pancake-bonded dimer of neutral phenalenyl radical as studied by MS and NMR spectroscopies and NICS analysis. *JACS* 128 (8): 2530–2531.

27 Devic, T., Yuan, M., Adams, J. et al. (2005). The maximin principle of π-radical packings. *JACS* 127 (42): 14616–14627.

28 Tian, Y.-H. and Kertesz, M. (2010). Is there a lower limit to the CC bonding distances in neutral radical π-dimers? The case of phenalenyl derivatives. *JACS* 132 (31): 10648–10649.

29 Uchida, K., Mou, Z., Kertesz, M., and Kubo, T. (2016). Fluxional σ-bonds of the 2,5,8-trimethylphenalenyl dimer: direct observation of the sixfold σ-bond shift via a π-dimer. *JACS* 138 (13): 4665–4672.

30 Mou, Z., Uchida, K., Kubo, T., and Kertesz, M. (2014). Evidence of σ- and π-dimerization in a series of phenalenyls. *JACS* 136 (52): 18009–18022.

31 Yang, W. and Parr, R.G. (1989). *Density Functional Theory of Atoms and Molecules*. New York: Oxford University.

32 Teale, A.M., Helgaker, T., Savin, A. et al. (2022). DFT exchange: sharing perspectives on the workhorse of quantum chemistry and materials science. *Phys. Chem. Chem. Phys.* 24 (47): 28700–28781.

33 Liu, S. (2007). Steric effect: a quantitative description from density functional theory. *J. Chem. Phys.* 126 (24): 244103.

34 Shannon, C.E. (1948). A mathematical theory of communication. *Bell Syst. Tech. J.* 27 (3): 379–423.

35 Fisher, R.A. (1925). Theory of statistical estimation. *Math. Proc. Cambridge Philos. Soc.* 22 (5): 700–725.

36 Ghosh, S.K., Berkowitz, M., and Parr, R.G. (1984). Transcription of ground-state density-functional theory into a local thermodynamics. *PNAS* 81 (24): 8028–8031.

37 Kullback, S. (1997). *Information Theory and Statistics*. Mineola, NY: Dover Publications.

38 Liu, S. (2019). Identity for Kullback-Leibler divergence in density functional reactivity theory. *J. Chem. Phys.* 151 (14): 141103.

39 Wang, B., Zhao, D., Lu, T. et al. (2021). Quantifications and applications of relative Fisher information in density functional theory. *J. Phys. Chem. A* 125 (17): 3802–3811.

40 Liu, S., Rong, C., Wu, Z., and Lu, T. (2015). Rényi entropy, Tsallis entropy and Onicescu information energy in density functional reactivity theory. *Acta Phys. Chim. Sin.* 31 (11): 2057–2063.

41 Johnson, E.R., Keinan, S., Mori-Sánchez, P. et al. (2010). Revealing noncovalent interactions. *JACS* 132 (18): 6498–6506.

42 Liu, S., Rong, C., Lu, T., and Hu, H. (2018). Identifying strong covalent interactions with Pauli energy. *J. Phys. Chem. A* 122 (11): 3087–3095.

43 Bader, R.F.W. and Preston, H.J.T. (1969). The kinetic energy of molecular charge distributions and molecular stability. *Int. J. Quantum Chem.* 3 (3): 327–347.

44 Tal, Y. and Bader, R.F.W. (1978). Studies of the energy density functional approach. 1. Kinetic energy. *Int. J. Quantum Chem.* 14 (S12): 153–168.

45 Cohen, L. (1979). Local kinetic energy in quantum mechanics. *J. Chem. Phys.* 70 (2): 788–789.

46 Cohen, L. (1984). Representable local kinetic energy. *J. Chem. Phys.* 80 (9): 4277–4279.

47 Yang, Z., Liu, S., and Wang, Y.A. (1996). Uniqueness and asymptotic behavior of the local kinetic energy. *Chem. Phys. Lett.* 258 (1–2): 30–36.

48 Ayers, P.W., Parr, R.G., and Nagy, Á. (2002). Local kinetic energy and local temperature in the density-functional theory of electronic structure. *Int. J. Quantum Chem.* 90 (1): 309–326.

49 Anderson, J.S.M., Ayers, P.W., and Hernandez, J.I.R. (2010). How ambiguous is the local kinetic energy? *J. Phys. Chem. A* 114 (33): 8884–8895.

50 Berkowitz, M. (1986). Exponential approximation for the density-matrix and the Wigner's distribution. *Chem. Phys. Lett.* 129 (5): 486–488.

51 Geerlings, P., De Proft, F., and Langenaeker, W. (2003). Conceptual density functional theory. *Chem. Rev.* 103 (5): 1793–1874.

52 Johnson, P.A., Bartolotti, L.J., Ayers, P.W. et al. (2012). Charge density and chemical reactivity: a unified view from conceptual DFT. In: *Modern Charge Density Analysis* (ed. C. Gatti and P. Macchi). New York: Springer.

53 Liu, S. (2009). Conceptual density functional theory and some recent developments. *Acta Phys. Chim. Sin.* 25 (03): 590–600.

54 Geerlings, P., Chamorro, E., Chattaraj, P.K. et al. (2020). Conceptual density functional theory: status, prospects, issues. *Theor. Chem. Acc.* 139 (2): 36.

55 Zhao, D., Liu, S., and Chen, D. (2022). A density functional theory and information-theoretic approach study of interaction energy and polarizability for base pairs and peptides. *Pharmaceuticals* 15 (8): 938.

56 Zhao, D., Zhao, Y., He, X. et al. (2023). Efficient and accurate density-based prediction of macromolecular polarizabilities. *Phys. Chem. Chem. Phys.* 25: 2131–2141. https://doi.org/10.1039/D2CP04690C.

57 Rong, C., Wang, B., Zhao, D., and Liu, S. (2020). Information-theoretic approach in density functional theory and its recent applications to chemical problems. *WIREs Comput. Mol. Sci.* 10 (4): e1461.

58 Rong, C., Zhao, D., He, X., and Liu, S. (2022). Development and applications of the density-based theory of chemical reactivity. *J. Phys. Chem. Lett.* 13 (48): 11191–11200.

59 Hirshfeld, F.L. (1977). Bonded-atom fragments for describing molecular charge densities. *Theor. Chim. Acta* 44 (2): 129–138.

60 Nalewajski, R.F. and Parr, R.G. (2000). Information theory, atoms in molecules, and molecular similarity. *PNAS* 97 (16): 8879–8882.

61 Ayers, P.W. (2006). Information theory, the shape function, and the Hirshfeld atom. *Theor. Chem. Acc.* 115 (5): 370–378.

62 Parr, R.G., Ayers, P.W., and Nalewajski, R.F. (2005). What is an atom in a molecule? *J. Phys. Chem. A* 109 (17): 3957–3959.

63 Heidar-Zadeh, F., Ayers, P.W., Verstraelen, T. et al. (2018). Information-theoretic approaches to atoms-in-molecules: Hirshfeld family of partitioning schemes. *J. Phys. Chem. A* 122 (17): 4219–4245.

64 Becke, A.D. (1988). A multicenter numerical integration scheme for polyatomic molecules. *J. Chem. Phys.* 88 (4): 2547–2553.

65 Bader, R.F.W. (1990). *Atoms in Molecules: A Quantum Theory*. Oxford, England: Oxford University Press.

66 Frisch, M.J., Trucks, G.W., Schlegel, H.B., et al. (2019). *Gaussian 16, Version C.01*. Gaussian, Inc.: Wallingford, CT.

67 Zhao, Y., Schultz, N.E., and Truhlar, D.G. (2006). Design of density functionals by combining the method of constraint satisfaction with parametrization for thermochemistry, thermochemical kinetics, and noncovalent interactions. *J. Chem. Theory Comput.* 2 (2): 364–382.

68 McLean, A.D. and Chandler, G.S. (1980). Contracted Gaussian-basis sets for molecular calculations. I. Second row atoms, $Z = 11–18$. *J. Chem. Phys.* 72 (10): 5639–5648.

69 Weigend, F. and Ahlrichs, R. (2005). Balanced basis sets of split valence, triple zeta valence and quadruple zeta valence quality for H to Rn: design and assessment of accuracy. *Phys. Chem. Chem. Phys.* 7 (18): 3297–3305.

70 Lu, T. and Chen, F. (2012). Multiwfn: a multifunctional wavefunction analyzer. *J. Comput. Chem.* 33 (5): 580–592.

71 Zhao, D., Rong, C., Jerkins, S. et al. (2013). Origin of the *cis*-effect: a density functional theory study of doubly substituted ethylenes. *Acta Phys. Chim. Sin.* 29 (01): 43–54.

72 He, X., Guo, C., Li, M. et al. (2022). Revisiting the trapping of noble gases (He–Kr) by the triatomic H_3^+ and Li_3^+ species: a density functional reactivity theory study. *J. Mol. Model.* 28 (5): 122.

73 Fermi, E. and Amaldi, E. (1934). Le orbite oos degli elementi. *Accad Ital. Rome* 6: 117–149.

74 Zhao, D., Liu, S., Rong, C. et al. (2018). Toward understanding the isomeric stability of fullerenes with density functional theory and the information-theoretic approach. *ACS Omega* 3 (12): 17986–17990.

22

Electron Spin Density and Magnetism in Organic Diradicals

Suranjan Shil[1], *Debojit Bhattacharya*[2] *and Anirban Misra*[2]

[1] *Manipal Centre for Natural Sciences, Manipal Academy of Higher Education, Manipal, Karnataka, India*
[2] *Department of Chemistry, University of North Bengal, Siliguri, West Bengal, India*

22.1 Introduction

Electron spin has fascinating properties with no classical counterpart. Electron spin plays an important role in designing modern electronics namely spintronics [1]. Electron spin is responsible for a system to become magnetic either Ferro-, Antiferro- or Ferri-magnet. The electron spin density (ESD) (s) is the difference between the electron densities of opposite spin of a particular atomic centre. $s(r) = \rho_\alpha(r) - \rho_\beta(r)$ where ρ_α and ρ_β are the electron densities of the α and β spin electrons [2]. Depending upon the value and sign of ESD, a system may behave differently. The importance of ESD and its distribution in molecular systems has incredible significance for their spin-related properties like spintronics and magnetism [1, 3–5]. The ESD plays an important role in magnetic interaction between the magnetic centres. If the sign of spin density of all the spin centres is similar (+/−), then the molecule may behave as a ferromagnet. On the other hand, the opposite sign between spin centers results from the antiferromagnetic coupling between the centres [6]. To identify a spin state of diradical systems, there is a robust thumb rule that states that generally, *m*-phenylene coupling tends to favour high spin states (this is called ferromagnetic coupling), whereas *p*-phenylene coupling tends to favour low spin states (called antiferromagnetic coupling), see Figure 22.1 [7–10].

The motive of writing this chapter is to sum up the spin density concepts applied to magnetic exchange interaction. The influence of ESD is the key factor of magnetic interaction in a molecule or complex because the magnetic exchange coupling constant is the result of the exchange interaction between unpaired electrons (spin centre).

There are several techniques by which the electron density can be measured, for example, Nuclear Magnetic Resonance (NMR) spectroscopy and X-ray [11–13]. However, it is not always very easy to measure the ESD and we need a theoretical method to calculate the ESD. Modern quantum chemical methods are quite capable of calculating the ESD [14].

Electron Density: Concepts, Computation and DFT Applications, First Edition.
Edited by Pratim Kumar Chattaraj and Debdutta Chakraborty.

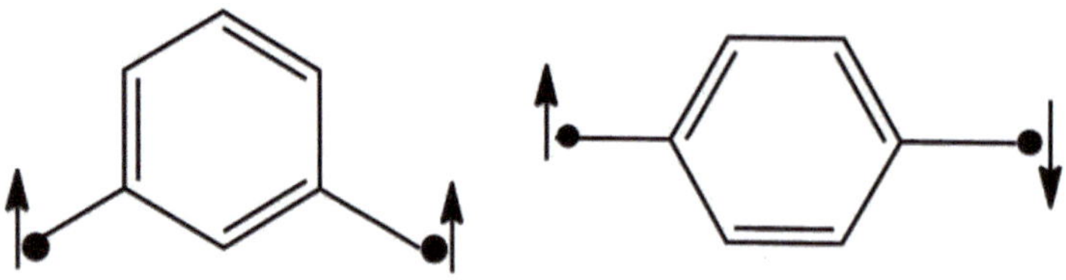

Figure 22.1 *m*-Phenylene and *p*-phenylene spin coupling.

22.2 Quantitative Relation Between Magnetic Exchange Coupling Constant and Spin Density

The relation between magnetic coupling and spin density was given by McConnell in early 1963 [15] based on the Heisenberg-Dirac-van Vleck (HDVV) Hamiltonian. The relation is as follows:

$$\hat{H} = -\hat{S}^{\mathrm{A}} \cdot \hat{S}^{\mathrm{B}} \sum_{ij} J_{ij}^{\mathrm{AB}} \rho_i^{\mathrm{A}} \rho_j^{\mathrm{B}} \tag{22.1}$$

The above Hamiltonian explains the exchange interaction between two aromatic radical fragments (A and B). For fragments A and B, $\hat{S}^{\mathrm{A}}$ and $\hat{S}^{\mathrm{B}}$ are the total spin operators, whereas ρ_i^{A} and ρ_j^{B} are the π-spin densities on atoms *i* and *j* of fragments A and B, respectively. In the above expression, the exchange coupling is generally considered as negative, and that is evaluated through Valence Bond Theory (VBT). As a consequence, ferromagnetic exchange interaction is found to be associated with the negative value of the spin-density product [5]. This model to predict the nature of magnetic interaction based on spin density has been popularly known as the McConnell-I model.

Following Eq. (22.1) Paul and Misra [16] developed a method to calculate the magnetic exchange coupling constant using the spin density of density functional theory (DFT)-based calculation, which is given by

$$J = -\frac{E_{\mathrm{BS}} - E_{\mathrm{HS}}}{\rho_{\mathrm{A}} \rho_{\mathrm{B}}} \tag{22.2}$$

where J is the magnetic exchange coupling constant between the magnetic centres, E_{BS} and E_{HS} are the energies of broken symmetry state and high spin state.

22.3 Spin Density Alternation

The ESD plays an important role in the magnetic exchange coupling constant, not only in the extent of the magnitude but also in the nature of interaction based on the atomic distance between magnetic centres. If the spin bearing centres are at an even atomic distance, then the molecule behaves as an antiferromagnet; on the other hand, if the distance is odd atomic, then the molecule behaves as ferromagnetic. Datta and Trindle formulate this thumb rule [6, 17].

To explain the spin alternation, we will discuss some examples of diradicals in the following.

22.3.1 Phenyl Nitroxide

A meta-phenylene coupler is used as a ferromagnetic coupler between two radical centres, which follow the spin alternation rule as demonstrated in Figure 22.2. The spin density of *meta*-phenylene coupled nitroxide diradicals is demonstrated in Figure 22.3. We can see from Figure 22.3 that in the

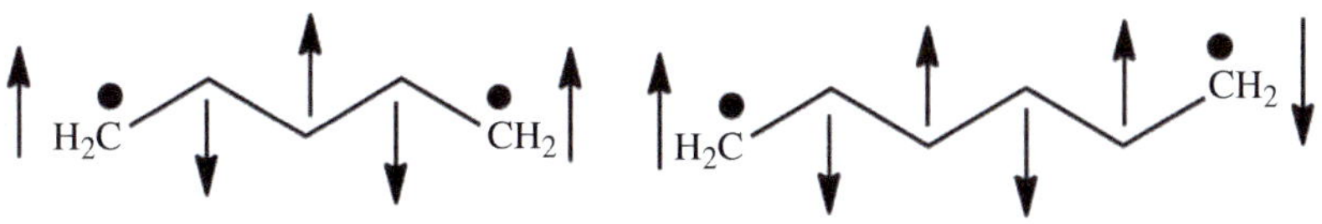

Figure 22.2 Spin alternation.

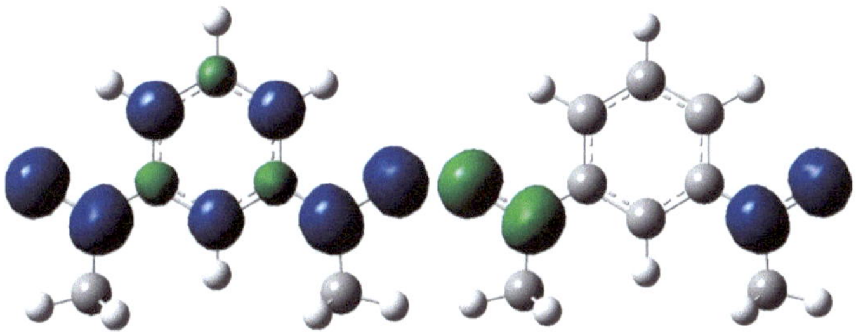

Figure 22.3 Spin density of phenyl nitroxide diradical (left: high spin state; right: broken symmetry state) (blue colour represents alpha spin and green colour represents beta spin).

high spin state, the spin density is delocalized over the entire molecule. In contrast, in the broken symmetry state, the spin density is localized only on the radical sites. The broken symmetry state is a weighted average of triplet and singlet state as the singlet state is not accountable in a single determinantal theory like DFT. The spin density calculation has been done using B3LYP hybrid DFT functional and 6-311++G(d,p) basis set by the Gaussian16 software package [18].

22.3.2 Methoxy Phenyl Nitroxide

In the example of the following diradicals phenylene group is substituted with two methoxy groups to increase their lifetime. We can see from Figure 22.4 that the spin density on the phenylene coupler is less compared to the radical centres. Also, in the broken symmetry state, there is negligible spin density on the coupler, which suggests that the high spin state is more stable compared to the broken symmetry state, resulting ferromagnetic ground state of the diradicals.

22.3.3 Phenyl Nitroxide Coupled Through Methylene

In this example the two *p*-phenylene diradicals coupled via a methylene group. The spin density plot shows that the spin density is distributed over the phenylene moiety. Still, there is no exchange between the two phenylenes as there is no π conjugation between them (Figure 22.5).

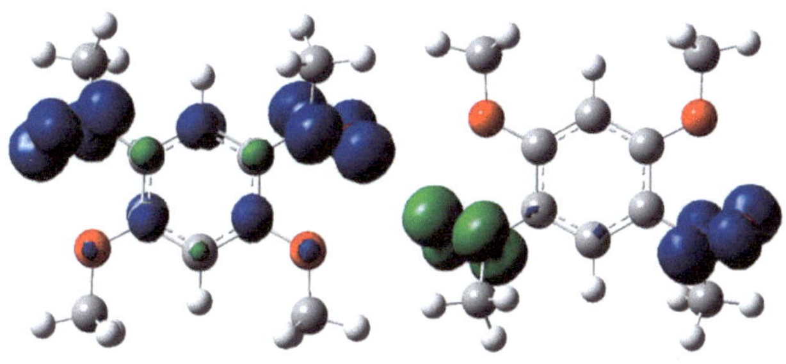

Figure 22.4 Spin density of methoxy phenyl nitroxide diradical (left: high spin state; right: broken symmetry state) (blue colour represents alpha spin and green colour represents beta spin).

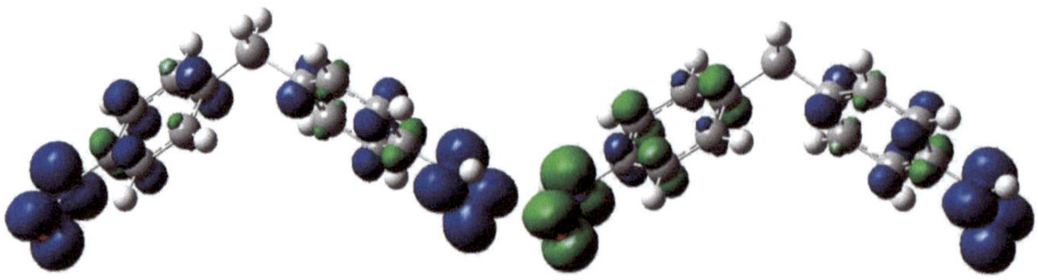

Figure 22.5 Spin density of nitroxide diradical (left: high spin state; right: broken symmetry state) (blue colour represents alpha spin and green colour represents beta spin).

22.3.4 Spin Density of Radical Systems

The magnetic properties of organic diradicals and polyradicals have been studied since long back experimentally as well as theoretically to get a desired single molecule organic magnet [19–23]. To look into the nature of magnetic properties and subsequently designing of magnetic radicals their spin density on radical centres as well as on the couplers are investigated [19, 21]. The extent of magnetic interaction depends on the spin density on the radical centres as well as on the bridging molecules or atoms. As the spin density is related to the charge density, it is influenced by the external substitution of the magnetic radicals. The spin density on the radical centres also depends on the geometry of the radicals. The planar molecule has more electron density, which is distributed over the whole molecule (radicals and coupler). If the radical and the coupler are in the same plane, the spin density distribution is better so that one will get higher values of the exchange coupling constant. On the contrary, the spin density distribution is disturbed if the coupler and radicals are not in the same plane [24, 25]. In the following few sections, we will discuss these effects in detail.

22.3.5 Distance Dependence of Spin Density

We know that the spin density is distributed through the pi conjugation path, but the question is how far the spin could travel, i.e. how long the spin density can delocalize. In one of our works [26], we discussed the effect of distance between radical centres and magnetic interaction between the radical centres. There are two hypothetical diradicals and their different variations based on the distance between the radical centres were investigated. The designed diradicals are shown in the following figures (Figures 22.6 and 22.7).

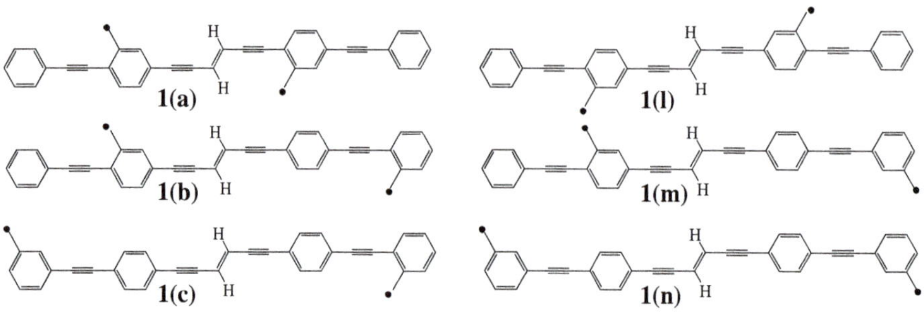

Figure 22.6 Designed diradical-based molecular wire 1 [26].

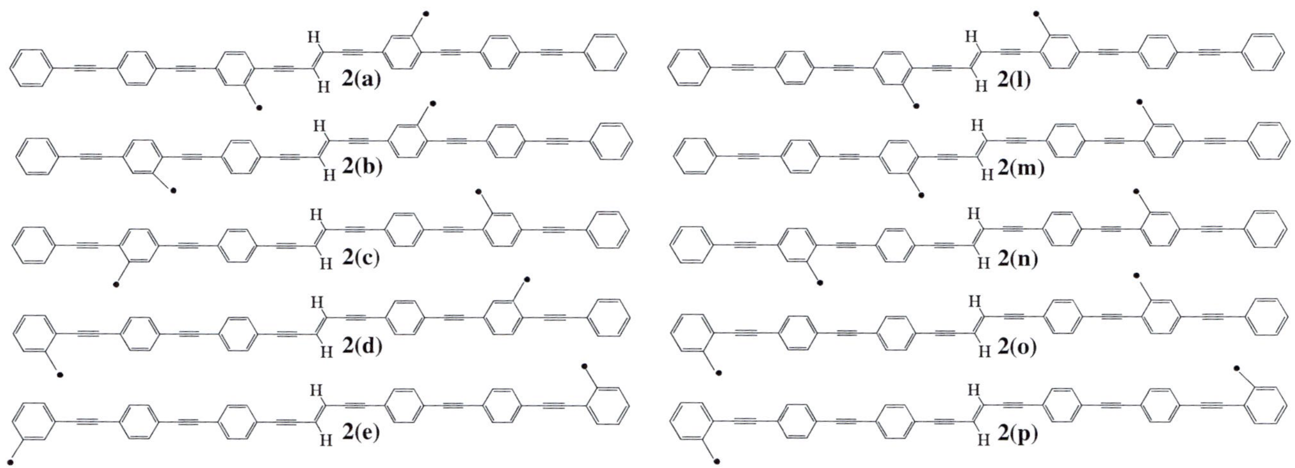

Figure 22.7 Designed diradical-based molecular wire 2 [26].

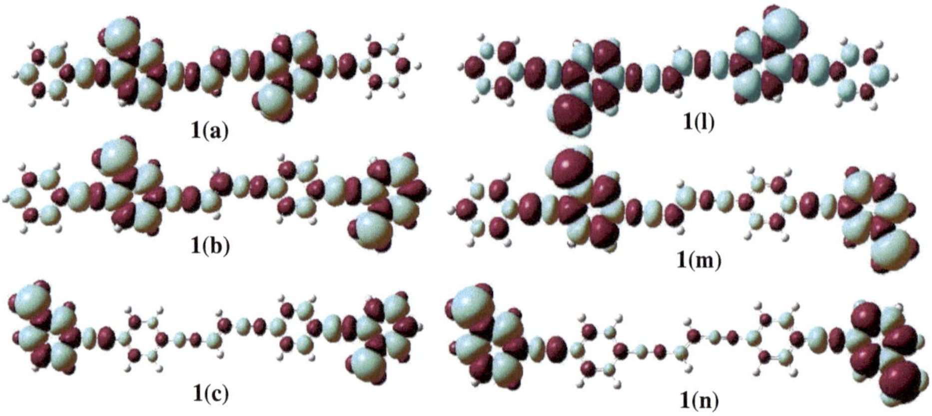

Figure 22.8 Spin density distribution of the ferromagnetically coupled diradicals (high spin state) and antiferromagnetically coupled diradicals (BS state) based on molecular wire 1 [26].

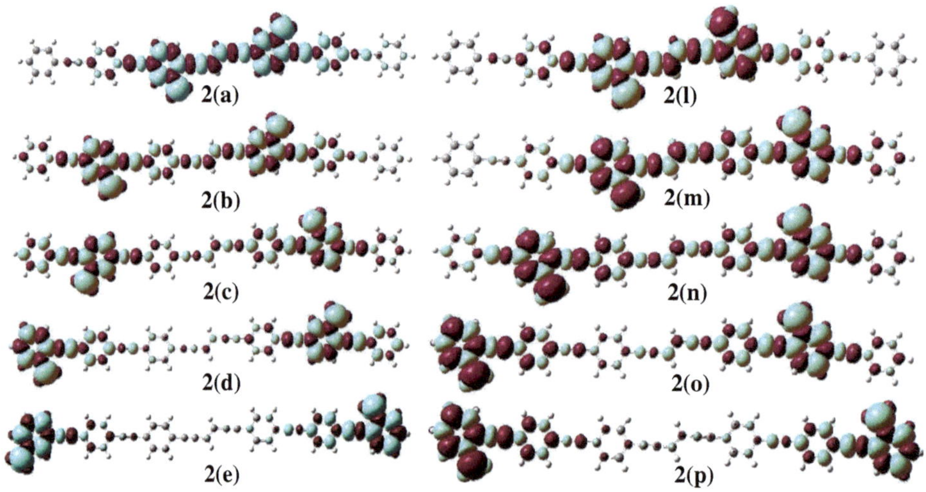

Figure 22.9 Spin density distribution of the ferromagnetically coupled diradicals (high spin state) and antiferromagnetically coupled diradicals (BS state) based on molecular wire 2 [26].

The spin density plots of the molecules are shown in Figures 22.8 and 22.9. It is evident from the figures that the spin density delocalization largely depends on the distance between the radicals. As the distance increases between the radical centres, the spin delocalization decreases. As a matter of fact, the magnetic exchange coupling constant between the radicals is affected, which is evident from Figure 22.10.

Now, we would like to discuss a few non-conventional π-conjugated systems, which are cumulene-based structures. The cumulene systems have fascinating properties like chirality, and they have two pi-conjugations between two carbon centres, which makes them unique [27]. As we have seen in the discussion, the spin density decreases when the distance between the radical centres increases. Subsequently, the magnetic exchange coupling constant decreases. However, in the case of allene and cumulene systems, the magnetic exchange coupling constant increases with

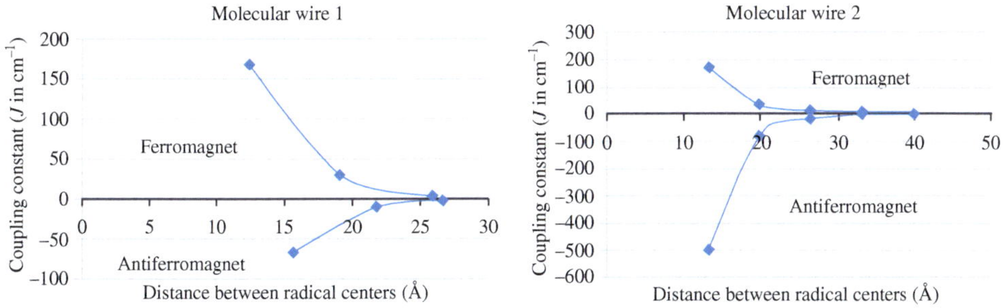

Figure 22.10 Plot of magnetic exchange coupling constants as a function of the distance between the radical centres [26].

Table 22.1 Spin density distribution on each atom of coupler and total spin density of coupler in designed diradicals in their triplet state (B3LYP/6-311++G(d,p)).

	Spin density on different atoms of coupler							
Systems	**C1**	**C2**	**C3**	**C4**	**C5**	**C6**	**C7**	**Total spin density**
1(a)	0.0768	−0.2655	0.0768	—	—	—	—	−0.1120
1(b)	0.1072	−0.2664	0.1572	−0.2665	0.1071	—	—	−0.1614
1(c)	0.1501	−0.3004	0.2041	−0.3196	0.2041	−0.3004	0.1501	−0.2120
2(a)	0.0616	−0.141	0.0616	—	—	—	—	−0.0177
2(b)	0.0831	−0.1341	0.0652	−0.1341	0.0831	—	—	−0.0366
2(c)	0.1223	−0.1659	0.0899	−0.1396	0.0899	−0.1659	0.1223	−0.0470
3(a)	0.0638	−0.1862	0.0683	—	—	—	—	−0.0541
3(b)	0.1025	−0.2095	0.114	−0.1969	0.0953	—	—	−0.0946
3(c)	0.1437	−0.2437	0.1711	−0.2488	0.1469	−0.2316	0.1362	−0.1262

Source: Reprinted with permission from Sarbadhikary et al. [27].

the increase in the distance between the radical centres. The diradicals are listed in Figure 22.11, along with the conjugated diradicals for comparison. The series 1 diradicals are constituted with an odd number of carbon atoms. They are chiral, and the series 2 diradicals are made up of an even number of carbons, which are achiral in nature. The spin density values of the carbon atoms in the coupler are listed in Table 22.1.

Figure 22.12 represents the spin density of the odd cumulene-based systems. It is evident from the figure that the π bonds in the two-end carbons are perpendicular to each other. The middle carbon has a spin density in each direction. Hence, the spin polarization from the radical to the coupler is possible from both sides, which makes the spin polarization into a greater extent as compared to the conjugated systems. We can see from Table 22.1 that the spin density on the coupler increases with the increase in the length for cumulene systems, as already said.

The increase in the spin density with the increasing length can be explained with the π electron sharing mechanism. In the cumulene systems, the end carbon has one pi-electron, and the middle carbon has two π-electrons, i.e. in total, for C-3 system, three carbons share four π electrons. Now, if we consider the C-5 system, here, five carbons share eight π-electrons. Therefore, with increasing

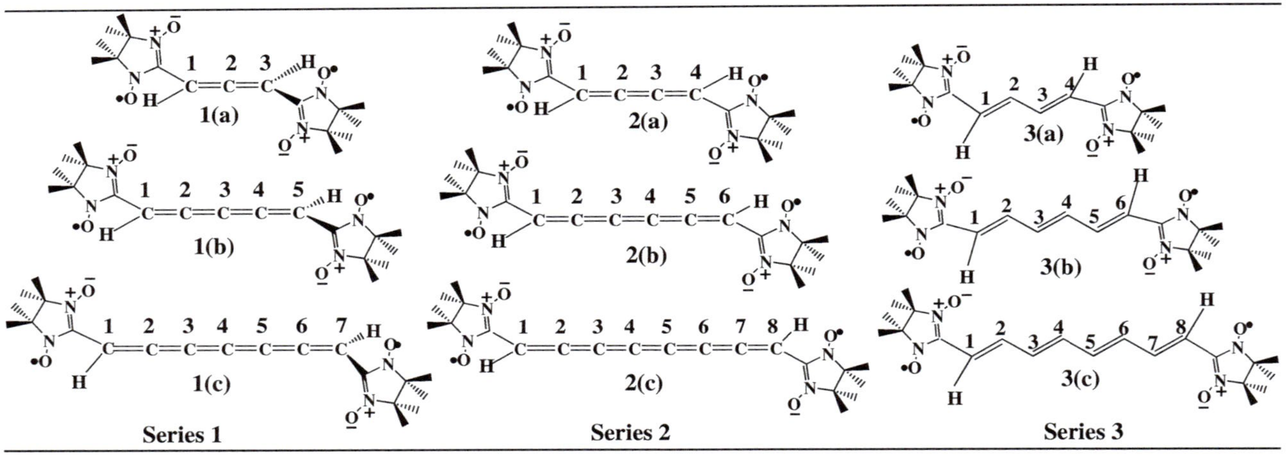

Figure 22.11 Designed diradicals based on allene, cumulene as well as conjugated couplers [28].

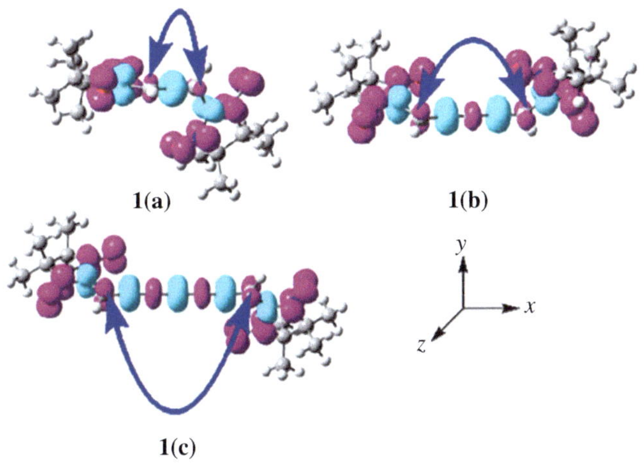

Figure 22.12 Spin density plot for the diradicals of series 1 in their triplet states (B3LYP/6-311++G(d,p)) (iso value 0.004), magenta and sky colours represent the α and β spins, respectively. The blue arrow represents two end carbons of a coupler. Source: Reprinted with permission from Sarbadhikary et al. [27].

length, the π-electron sharing increases, which facilitates more π conjugation than the conjugated systems. A pictorial view is given in Figure 22.13.

In contrast to the cumulene systems, the conjugated systems share one π-electron per carbon, and hence there is no such enhancement of spin polarization like cumulene systems. This spin polarization has a direct effect on the magnetic exchange coupling constant of the systems, which is shown in Figure 22.14.

22.3.6 Geometry Dependence of Spin Density

The spin density distribution along the π conjugation path is required for the magnetic exchange between two spin centres. However, not only the π conjugation but also the geometry of the radical site and the coupler is important. It has been shown that spin coupling happens when the radicals and the coupler are in the same plane. If the plane of the radical and the coupler is 90°, the spin propagation completely stopped [25]. The spin density of *meta*-phenylene coupled diradicals in different dihedral angles between the radicals and coupler is presented in Figure 22.15. We can see that when the radical and the coupler are in the same plane, the spin polarization is maximum, and the spin polarization reduces with the increase in the dihedral angle between the radical and the coupler. We have studied the effect of dihedral angle on the coupling constant in some polyacene coupled diradicals (Figure 22.16), which are shown in Figure 22.17.

22.3.7 Dependence on Connecting Atoms

The connecting atom between the radical and the coupler plays an important effect for determining the value of the magnetic exchange coupling constant. In Figure 22.18, we give an example of such effects. In the figure we have two similar diradicals connecting either in meta or in para positions. If we see the picture, we can find that one radical is connected by carbon and another one by phosphorus. It is evident that the carbon-connected diradicals have a higher value than the phosphorus-connected ones. It has been found that the spin density on the phosphorus

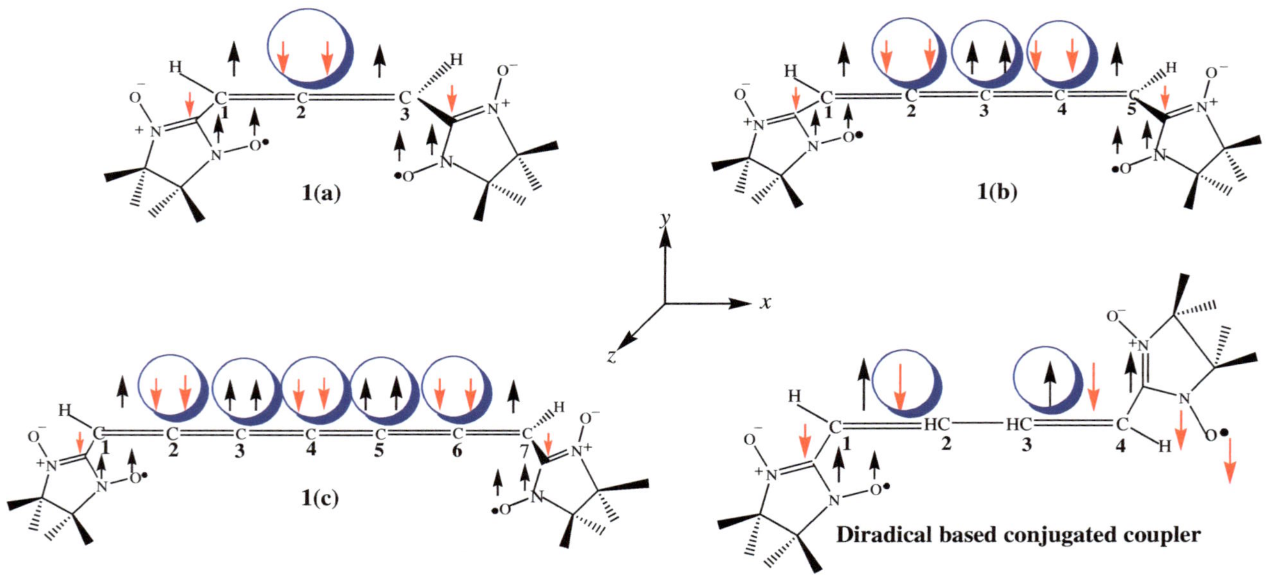

Figure 22.13 Spin polarization of π electron within the coupler for odd cumulene [27].

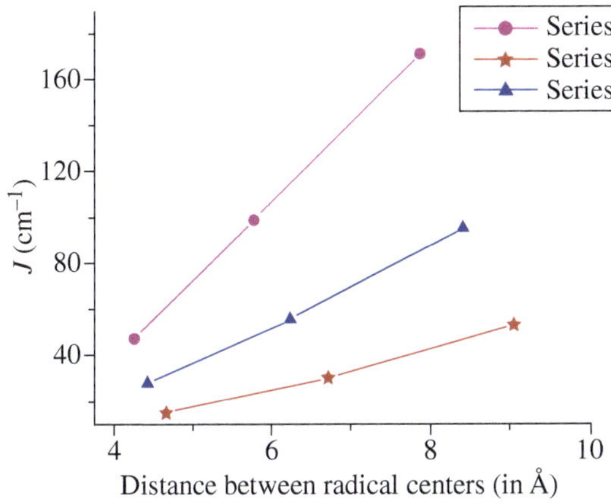

Figure 22.14 Plot of magnetic exchange coupling constant vs. distance (in Å) between radical centres (B3LYP/6-311++G(d,p)). Source: Reprinted with permission from Sarbadhikary et al. [27].

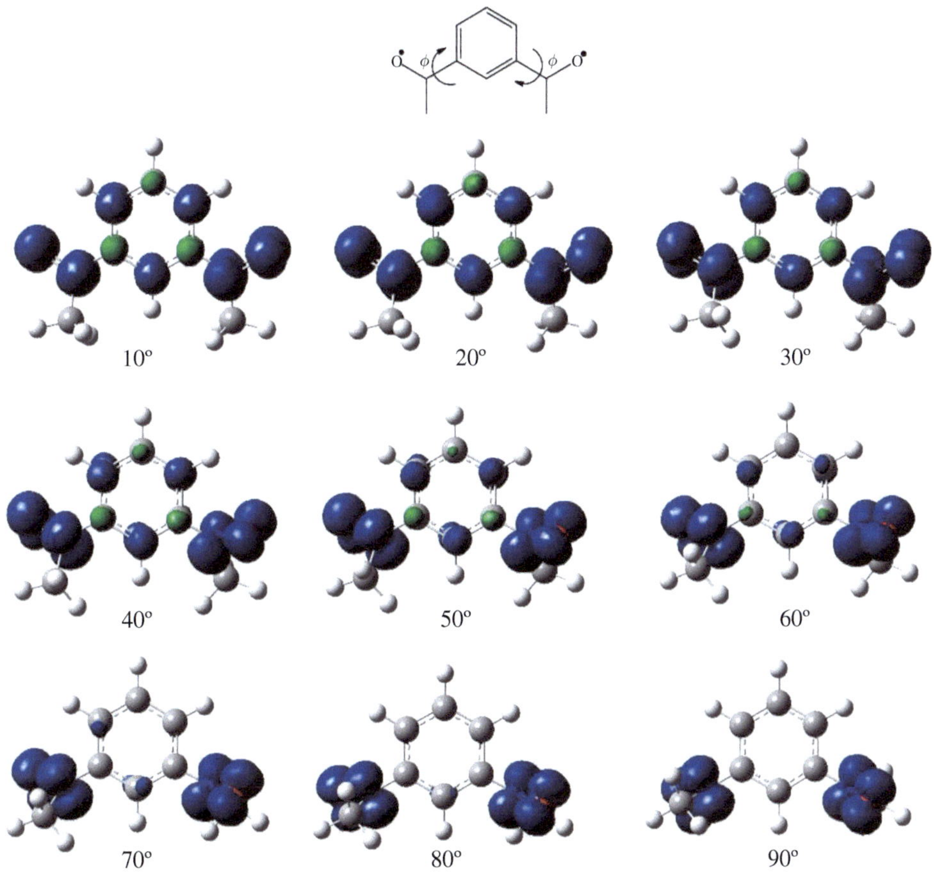

Figure 22.15 The spin density plot of the *meta*-phenylene coupled nitroxide diradical at different radical-coupler dihedral angles (ϕ).

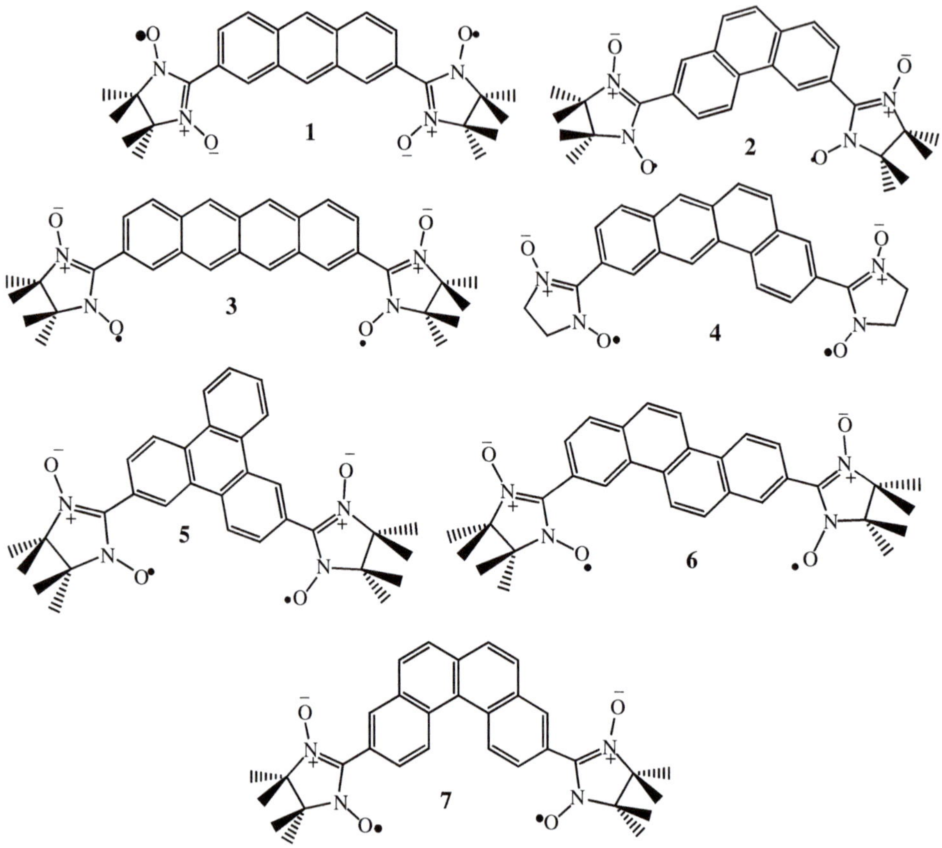

Figure 22.16 The polyacene coupled diradicals [25].

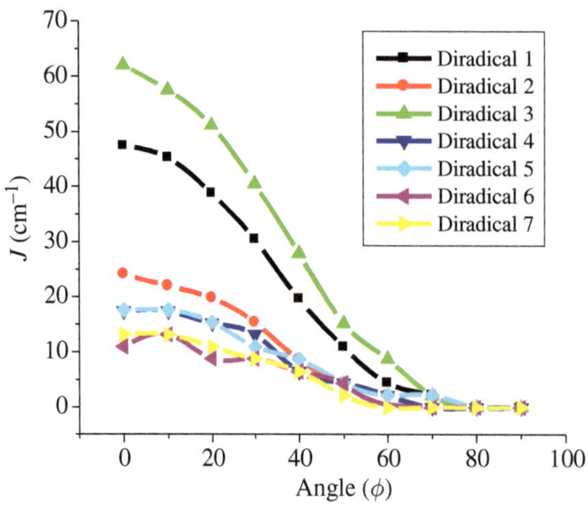

Figure 22.17 The magnetic exchange coupling constant vs. dihedral angle between the radical and the coupler [25].

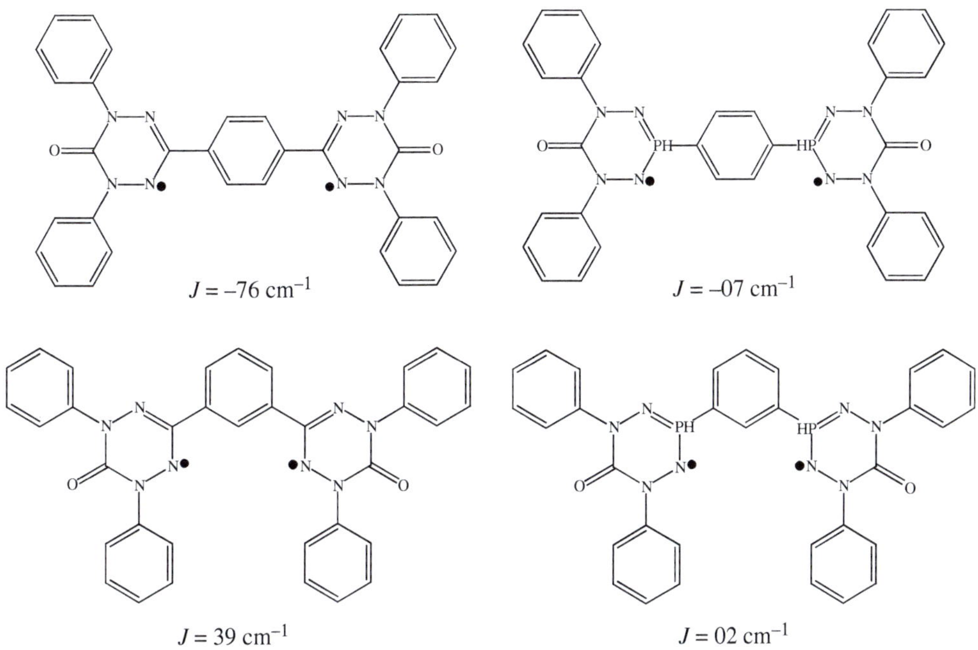

Figure 22.18 Diradicals showing spin leakage property.

atom (phosphoverdazyl system) is very low compared to the carbon-connected systems (oxoverdazyl) [29], which signifies that the phosphorus has eaten up the spin density, which is termed a spin leakage phenomenon. This resulted in the low value of the magnetic exchange coupling constant.

22.4 Concluding Remarks

The magnetic exchange coupling constant of diradicals is largely dependent on the spin polarization and spin density of the individual radicals and couplers. The increasing distance between radical centres resulted in a decreasing value of the magnetic exchange coupling constant in case of a conjugated coupler. However, the reverse is observed in the case of cumulene-coupled diradicals. Planarity largely controls the spin polarization, i.e. spin density and hence the magnetic exchange coupling constant. The connecting atoms between the radical and the coupler play a significant role in determining the magnetic exchange coupling constant.

Acknowledgements

S. Shil is thankful to SERB, India, for funding via Grant No. SRG/2022/000822. A. Misra is grateful to SERB, India (Grant No. CRG/2019/000262) for financial support.

References

1 Coronado, E. (2020). Molecular magnetism: from chemical design to spin control in molecules, materials and devices. *Nat. Rev. Mater.* 5 (2): 87–104. https://doi.org/10.1038/s41578-019-0146-8.

2 Bruno, G., Macetti, G., Presti, L.L., and Gatti, C. (2020). Spin density topology. *Molecules* 25 (15). https://doi.org/10.3390/molecules25153537.

3 Szulczewski, G., Sanvito, S., and Coey, M. (2009). A spin of their own. *Nat. Mater.* 8 (9): 693–695. https://doi.org/10.1038/nmat2518.

4 Anderson, P.W. (1963). Theory of magnetic exchange interactions:exchange in insulators and semiconductors. *J. Phys. C: Solid State Phys.* 99–214. https://doi.org/10.1016/S0081-1947(08)60260-X.

5 Yoshizawa, K., Yamabe, T., and Hoffmann, R. (1997). Orbital interaction analysis of McConnell's model for through-space magnetic coupling. *Mol. Cryst. Liq. Cryst. Sci. Technol., Sect. A* 305 (1): 157–166. https://doi.org/10.1080/10587259708045053.

6 Trindle, C. and Nath Datta, S. (1996). Molecular orbital studies on the spin states of nitroxide species: bis- and trisnitroxymetaphenylene, 1,1-bisnitroxyphenylethylene, and 4,6-dimethoxy-1,3-dialkylnitroxy-benzenes. *Int. J. Quantum Chem.* 57 (4): 781–799. https://doi.org/10.1002/(SICI)1097-461X(1996)57:4<781::AID-QUA26>3.0.CO;2-1.

7 Ovchinnikov, A.A. (1978). Multiplicity of the ground state of large alternant organic molecules with conjugated bonds (do organic ferromagnetics exist?). *Theor. Chim. Acta* 47: 297–304.

8 Longuet-Higgins, H.C. (1950). Some studies in molecular orbital theory. I. Resonance structures and molecular orbitals in unsaturated hydrocarbons. *J. Chem. Phys.* 18 (3): 265–274. https://doi.org/10.1063/1.1747618.

9 Borden, W.T. (1993). Qualitative methods for predicting the ground of heteroatom substitution on the ordering of the electronic states states of non-Kekule hydrocarbons and the effects. 232: 195–218.

10 Bhattacharya, D. and Misra, A. (2009). Density functional theory based study of magnetic interaction in bis-oxoverdazyl diradicals connected by different aromatic couplers. *J. Phys. Chem. A* 113 (18): 5470–5475. https://doi.org/10.1021/jp900341e.

11 Deutsch, M., Claiser, N., Pillet, S. et al. (2012). Experimental determination of spin-dependent electron density by joint refinement of X-ray and polarized neutron diffraction data. *Acta. Crystallogr. A* 68 (6): 675–686. https://doi.org/10.1107/S0108767312031996.

12 Campo, J., Luzon, J., Palacio, F., and Rawson, J. (2006). Spin density distribution and interaction mechanisms in thiazyl-based magnets. In: *Carbon Based Magnetism*, 159–188. Elsevier. https://doi.org/10.1016/B978-044451947-4/50008-6.

13 Gillon, B. Spin distributions in molecular systems with interacting transition metal ions. In: *Magnetism: Molecules to Materials I* (ed. J.S. Miller and M. Drillon), 357–378. Weinheim, FRG: Wiley-VCH Verlag GmbH & Co. KGaA. https://doi.org/10.1002/3527600841.ch11.

14 Boguslawski, K., Marti, K.H., Legeza, Ö., and Reiher, M. (2012). Accurate *ab initio* spin densities. *J. Chem. Theory Comput.* 8 (6): 1970–1982. https://doi.org/10.1021/ct300211j.

15 McConnell, H.M. (1963). Ferromagnetism in solid free radicals. *J. Chem. Phys.* 39 (7): 1910–1910. https://doi.org/10.1063/1.1734562.

16 Paul, S. and Misra, A. (2012). Interpretation and quantification of magnetic interaction through spin topology. *J. Chem. Theory Comput.* 8 (3): 843–853. https://doi.org/10.1021/ct2006506.

17 Trindle, C., Datta, S.N., and Mallik, B. (1997). Phenylene coupling of methylene sites. The spin states of bis(X−methylene)-*p*-phenylenes and bis(chloromethylene)-*m*-phenylene. *JACS* 119 (52): 12947–12951. https://doi.org/10.1021/ja9704139.

18 Frisch, M.J., Trucks, G.W., Schlegel, H.B., et al. (2016). *Gaussian 16, Revision C.01.*

19 Cho, D., Ko, K.C., and Lee, J.Y. (2016). Quantum chemical approaches for controlling and evaluating intramolecular magnetic interactions in organic diradicals. *Int. J. Quantum Chem.* 116 (8): 578–597. https://doi.org/10.1002/qua.25060.

20 Cui, H., Hu, Z.-B., Chen, C. et al. (2021). A high-spin diradical dianion and its bridged chemically switchable single-molecule magnet. *Chem. Sci.* 12 (29): 9998–10004. https://doi.org/10.1039/D1SC01932E.

21 Ma, J., Yuan, Y., Kang, B., and Lee, J.Y. (2021). Acetylene coupler builds strong and tunable diradical organic molecular magnets. *New J. Chem.* 45 (20): 9137–9142. https://doi.org/10.1039/D1NJ01140E.

22 Rajca, A. (1994). Organic diradicals and polyradicals: from spin coupling to magnetism? *Chem. Rev.* 94 (4): 871–893. https://doi.org/10.1021/cr00028a002.

23 Cho, D., Ko, K.C., and Lee, J.Y. (2014). Organic magnetic diradicals (radical–coupler–radical): standardization of couplers for strong ferromagnetism. *J. Phys. Chem. A* 118 (27): 5112–5121. https://doi.org/10.1021/jp5046639.

24 Ichimura, A.S., Koga, N., and Iwamura, H. (1993). Magnetic interaction between two triplet nitrene units through diphenylsilane and 1,2-diphenyldisilane couplers. *JACS* 115: 8928–8932.

25 Shil, S., Roy, M., and Misra, A. (2015). Role of the coupler to design organic magnetic molecules: LUMO plays an important role in magnetic exchange. *RSC Adv.* 5 (128): 105574–105582.

26 Shil, S., Sarbadhikary, P., and Misra, A. (2016). Effect of length on the transport and magnetic properties of diradical substituted molecular wires. *RSC Adv.* 6 (101): 99467–99474.

27 Sarbadhikary, P., Shil, S., Panda, A., and Misra, A. (2016). A perspective on designing chiral organic magnetic molecules with unusual behavior in magnetic exchange coupling. *J. Org. Chem.* 81 (13): 5623–5630.

28 Sarbadhikary, P., Shil, S., and Misra, A. (2018). Magnetic and transport properties of conjugated and cumulated molecules: the π-system enlightens part of the story. *Phys. Chem. Chem. Phys.* 20 (14): 9364–9375. https://doi.org/10.1039/C7CP06113G.

29 Bhattacharya, D., Shil, S., Panda, A., and Misra, A. (2010). A DFT study on the magnetostructural property of ferromagnetic heteroverdazyl diradicals with phenylene coupler. *J. Phys. Chem. A* 114 (43): 11833–11841.

23

Stabilization of Boron and Carbon Clusters with Transition Metal Coordination – An Electron Density and DFT Study

Amol B. Rahane[1], *Rudra Agarwal*[2], *Pinaki Saha*[2], *Nagamani Sukumar*[3] and *Vijay Kumar*[4,5]

[1] *Department of Physics, K. R. T. Arts, B. H. Commerce & A. M. Science (KTHM) College [Affiliated to Savitribai Phule Pune University (SPPU), Pune, India] Nashik, Maharashtra, India*
[2] *Department of Chemistry, School of Natural Sciences, Shiv Nadar University, Gautam Budhha Nagar, Uttar Pradesh, India*
[3] *Amrita School of AI, Amrita Vishwa Vidyapeetham, Coimbatore 641105, Tamil Nadu, India*
[4] *Center for Informatics, School of Natural Sciences, Shiv Nadar University, Gautam Budhha Nagar, Uttar Pradesh, India*
[5] *Dr. Vijay Kumar Foundation, Gurgaon, Haryana, India*

23.1 Introduction

Both carbon and boron form a rich variety of novel nanostructures including planar graphene [1], boron sheets (referred to as borophenes [2]), quasi-planar boron clusters stabilized with hexagonal holes in an otherwise triangular structure [3], fullerenes, such as C_{60} [4, 5] and fullerene-like boron cages [6], as well as many others. The proposition of an icosahedral structure for C_{60} was made in the 1970s [7, 8] but remained buried until the experimental findings of C_{60}, C_{70}, and other fullerenes by Kroto, Curl, Smalley and coworkers [4, 5]. Subsequently many other developments took place on fullerenes of carbon [9, 10]. The discovery of C_{60} and subsequent formation of a solid made of C_{60} fullerenes [11] fueled the growth of research on carbon clusters with less than a few hundred atoms. Among the many outcomes of this research was the discovery of a range of sizes for fullerene-like cages (in terms of the number of carbon atoms in the cage), fullerene onions, fullerene derivatives, fullerenes with endohedral encapsulation, and other small clusters with $n < 32$, where n is the number of carbon atoms involved in the structure. Here, $n = 32$ is the size of the smallest fullerene produced in wrap shrinking experiments, although C_{28} [12] and C_{20}, the smallest fullerene [13, 14], have also been produced in small quantities. Shrinking the size of the fullerene cage with the elimination of a pair of carbon atoms leads to the fragmentation of cages with less than 32 carbon atoms. However, more recently, theoretical studies have shown [15] that endohedral metal (M) complexation can stabilize the dodecahedral carbon 20 fullerene, C_{20}, while earlier experiments [12] on M complexation also produced TiC_{28} and UC_{28} in high abundance.

In more recent years, the findings of fullerene-like empty cage structure [6] of B_{40} and core–shell, bilayer, tubular, and quasi-planar structures for many boron clusters [16] have spurred a lot of interest in experimental and theoretical studies of boron clusters to understand their growth behavior and stability. Boron has one electron less than carbon, and this leads to its rich chemistry with multicenter bonding. Several studies on small boron clusters having up to around 36 atoms showed planar, quasi-planar, or tubular structures as their ground state [17–23] with the exception of B_{14} and B_{28}, for which a cage-like [24] and core-cage [25] structure, respectively, have been proposed. A double-ring tubular (DRT) structure has been suggested for neutral B_{20} [19] and B_{24} [26], while

Electron Density: Concepts, Computation and DFT Applications, First Edition.
Edited by Pratim Kumar Chattaraj and Debdutta Chakraborty.

a cage structure has been suggested [27] for B_{38}. A recent study by Chen et al. [28] has identified a cage structure with the axially chiral feature for B_{39}. Another recent theoretical study [25] on neutral B_{28} suggested a filled cage structure to be the lowest in energy, while a planar isomer was found to be nearly degenerate. Although for neutral B_{40}, a fullerene-like cage structure is the most stable one, a planar isomer is favored for B_{40} anion [6] and therefore, a quasi-planar isomer competes in energy for B_{40}. Following the findings of metal-encapsulated fullerenes and other polyhedral structures of silicon and other elements [15, 29–37], Rahane et al. [38] have investigated whether smaller cages of boron can be stabilized by encapsulation of a transition M atom and found that, depending upon the size of the M atom, cages of different sizes can be stabilized similarly to the earlier work on silicon and other systems [15, 37]. Also, for smaller clusters having ~10 boron atoms, even planar ring/wheel-shaped clusters can be stabilized [15, 39–44] with an M atom or a boron atom [45] at the center.

Interestingly, studies on a large number of bulk boron compounds show that an empty center icosahedral cage of B_{12} is the major building block of their structures [46]. However, an isolated B_{12} icosahedral cage is not stable due to the availability of 36 valence electrons, while 26 electrons are required by Wade's rules [47] to stabilize this cage. One way to stabilize the icosahedral cage is to endohedrally encapsulate an M atom that may interact with boron atoms strongly and stabilize cage structures. Such a strategy has been successfully used to stabilize non-carbon cage structures such as those of silicon and other elements [29–37]. In particular, exceptional stability has been obtained for $Zr@Si_{16}$ fullerene and $Ti@Si_{16}$ Frank–Kasper polyhedron structures [29] that have been subsequently realized in a laboratory [48] and even assemblies have been formed [49, 50]. The size of a B_{12} icosahedron is too small to encapsulate an M atom. As boron is smaller in size than silicon, the possibility of forming boron cages may lie around the size of 20 atoms. Elemental boron clusters in this size range are generally planar [51]. Still, the doping of an M atom transforms the most stable structures of elemental boron clusters into a bowl, drum, or cage structure [38]. Disk-like or wheel-shaped structures of boron have been predicted for clusters having up to about 10 boron atoms by encapsulation of a transition M atom [39–44, 51] as well as other elements [45]. The stability of some disk-shaped clusters has been correlated with electronic shell closing at 12 valence electrons [45]. Further, recently bowl-shaped [52–56] and drum-shaped [56, 57] structures of boron have been predicted to be stabilized by M atoms such as $M@B_{14}$ drum with M = Cr, Fe, Co, and Ni, and also with 16 boron atoms such as $Co@B_{16}^-$. For the larger sizes, boron cages B_n with $n = 28$, 36, 37, 38, 39, 40, and 44 have been explored theoretically by many groups [53–79] with encapsulation of various guest atoms, ions, or clusters. Extensive DFT calculations have also been carried out to explore the structure, stability, electronic properties, and potential applications of endohedral B_{40} cages doped with different atoms [67–79]. The stabilization of cage structures below the size of 24 boron atoms by M encapsulation has been performed in a recent systematic study, based on which fullerene-like cages $Cr@B_{20}$, $M@B_{22}$, and $M@B_{24}$ with M = Cr, Mo, and W atoms were predicted [38]. These M-doped boron species have enriched boron chemistry, and some of the clusters have been realized in the laboratory [15].

Carbon lies next to boron in the periodic table, and its nanostructures are generally different from those of boron. It is interesting to explore the kind of structures that small carbon clusters could form. Pure carbon clusters with less than 20 atoms form chain or ring structures compared with triangular networks of boron. The doping of an M atom in a chain has also been explored. However, recently, it has been found [80] that new ring structures of small carbon clusters could be stabilized by encapsulation of an M atom. These studies could open new possibilities of small carbon clusters and their applications. Here, we discuss these results of DFT calculations and further present analysis of the electron density to investigate the bonding behavior in boron clusters, transition

M atom encapsulated boron cages, monocyclic carbon rings, partially hydrogenated monocyclic carbon rings, and their complexation with an M atom.

As reported earlier [45], the chemical bond is not a uniquely defined concept. The electron pair bond, originally introduced by Lewis and still useful in valence bond theory [81, 82], was extended by Longuet–Higgins, Lipscomb, Massa et al. [83–85] to cover three-center, two-electron bonds in boranes, and subsequently, using adaptive natural density partitioning (AdNDP – itself an extension of natural bond orbitals analysis) [86], to include n-center, two-electron objects ($n \geq\geq 2$). In molecular orbital theory, a chemical bond is the force that holds pairs of atoms together, as manifested in the dissociation energy required to break the bond. Alternatively, bond paths emerge through topological analysis of the electron density scalar field and its critical points [87]. In spite of this ambiguity, the concept of the chemical bond has proven to be of enormous utility in chemistry. While bond paths are topological features of the electron density, and identification of bond paths with chemical bonds is often problematic (Shahbazian [87] has argued for the use of the term line path instead), there are strong correlations between electron density distributions, equilibrium molecular structures, and energetics, which make quantities derived from topological analysis of electron density distributions useful in discussions of chemical bonding [45, 88].

Detailed study of the chemical bonding in boron clusters as well as M atom encapsulated boron clusters with less than 25 boron atoms has been carried out [38, 45, 56] in order to find the smallest cage of free boron clusters besides the drum and wheel structures. In this size range, pure boron clusters generally favor quasi-planar or tubular structures. As the bonding in boron is stronger compared with silicon, Rahane et al. [38] considered strongly interacting transition M atoms with half-filled d states such as Cr, Mo, and W to explore the possibility of stabilizing boron cages in this size range. Based on a systematic study, they predicted fullerene-like cages $Cr@B_{20}$, $M@B_{22,}$ and $M@B_{24}$ with M = Cr, Mo, and W, in which one M atom is encapsulated in the cage. Also, they found dimer-capped drum structures for $Cr@B_{18}$ and $M@B_{20}$ (M = Mo and W) and a drum structure for $M@B_{18}$, M = Mo and W. The stability of these structures was studied by analyzing the MOs as well as the electronic charge density. These results suggest that Mo and W are well suited to produce endohedrally encapsulated novel cage structures of boron, as the encapsulation of the M atom has the effect of increasing the binding energy significantly. The encapsulation of the Cr atom also led to a cage formation, but the binding energy of the encapsulated clusters has similar values as for the pure boron clusters, suggesting that the encapsulation of the M atom does not lead to enhanced stability of the cage.

Further, in view of the stabilization of dodecahedral C_{20} fullerene using endohedral encapsulation [15, 89], Agarwal et al. have investigated the possibility of stabilizing novel structures of carbon with fewer than 20 atoms using endohedral encapsulation of small clusters. Based on an extensive body of research conducted at a lower level of theory through the 1990s and thereafter, it is known that for fewer than 20 carbon atoms, the system tends to form linear or 2-D (ring) structures [90]. This is rationalized in terms of strain from steric factors and excessive dangling bonds, raising the energy of any 3-D isomer to an unfavorable level. Among small clusters, studies have found that even-numbered clusters prefer a ring, whereas odd-numbered clusters prefer a linear chain (such as clusters with 7 and 9 carbon atoms). As mentioned earlier, a number of studies on small boron clusters have shown a preference for a ring when endohedrally encapsulated with an M atom. This also led Agarwal et al. [80] to explore if this is possible for carbon, and if so, how many atoms of carbon are required to form a ring with an M atom.

Agarwal et al. [80] studied the effects of transition M atom encapsulation in monocyclic carbon rings with 9, 10, and 12 carbon atoms, as well as on the corresponding rings with partial hydrogenation (2 or 3 hydrogen atoms) so as to alter the symmetry and the number of delocalized electrons on the C atoms – each H atom binds covalently with a C atom and effectively reduces one electron from the π bonded charge. Stabilization of the ring structure of C_9, as well as those of partially hydrogenated clusters C_9H_3, $C_{10}H_2$, and $C_{12}H_3$, was obtained with encapsulation of an M atom.

While the $C_{12}H_3$ ring has a puckered non-planar structure, complexation with certain 3d transition M atoms from the first half of the series was found to stabilize the planar D_{3h} structure to a minimum, leading to interesting structures such as $Cr@C_{12}H_3$ with a large HOMO–LUMO gap, planar geometry, and a large number of unpaired spins. This magic magnetic disk, with an open shell and a large HOMO–LUMO gap, is interesting as, generally, systems containing unpaired electrons have a slightly smaller HOMO–LUMO gap as compared to the spin-paired electronic state. One can also observe deviations from the D_{3h} geometry of the transition state structure on complexation with certain M atoms from the second half of the 3d series (distortion to C_{2v} in most cases, severe distortion in some cases, as elaborated further). Agarwal et al. [80] therefore carried out a detailed study on the effects of encapsulation of the $C_{12}H_3$ ring and the effects of changing the M atom, using charge and spin density analysis, HOMO–LUMO gaps and other metrics, in order to answer these questions.

23.2 Computational Details

We used the generalized gradient approximation (GGA) of Perdew, Burke, and Ernzerhof (PBE) [91] for the exchange–correlation functional and projector augmented wave (PAW) pseudopotential plane wave method [92, 93] in the Vienna *ab initio* simulation package (VASP) [94] to explore several isomers for the encapsulation of M atoms in boron and carbon clusters. The calculations were considered to be converged when the absolute value of the force on each ion was less than 0.005 eV/Å with convergence in the total energy of 10^{-5} eV. Further calculations were performed for the lowest energy isomers of different clusters using Gaussian 09 code [95] and PBE0 functional. Also, calculations of the vibrational modes of cation and neutral clusters were performed using the Gaussian 09 code, and in almost all cases, no imaginary frequency was obtained, suggesting the dynamical stability of the obtained atomic structures. For these calculations, B3PW91 hybrid exchange–correlation functional as well as PBE0 functional and 6-311+G basis set [96–100] for Cr encapsulation and LANL2DZ basis set for Mo and W while 6-311+G basis for B atoms have been used in all the cases. The atomic structures were optimized and used to calculate IR and Raman spectra for the neutral and cation clusters. Calculations using PBE0 exchange–correlation functional on the lowest energy isomers of the pure and M encapsulated boron clusters also suggest stabilization of boron cages with M encapsulation. The bonding characteristics have been studied by performing AdNDP [101] analysis at the PBE level of the theory and using the same basis set in the Gaussian 09 code and also from the Laplacian function

$$L = -\tfrac{1}{4}\nabla^2\rho(\mathbf{r}) \tag{23.1}$$

of the electronic charge density $\rho(\mathbf{r})$ using AIMALL [102, 103].

The electron localization function (ELF) [104] was calculated using the charge density distribution obtained from VASP for the lowest energy isomers of the neutrals and the electron localization–delocalization index using the AIMLDM script [105]. ELF is defined [104, 106] in terms of the local excess kinetic energy $\Delta(\mathbf{r})$ due to Pauli repulsion and the kinetic energy $\Delta_0(\mathbf{r})$ of a uniform electron gas with density $\rho(\mathbf{r})$:

$$\mathrm{ELF}(\mathbf{r}) = \left\{1 + [\Delta(\mathbf{r})/\Delta_0(\mathbf{r})]^2\right\}^{-1} \tag{23.2}$$

$\Delta(\mathbf{r})$ vanishes for one-electron systems and for regions within multi-electron systems dominated by a single, localized spin–orbital. Thus, ELF($\mathbf{r}$) varies between 0 and 1, with ELF = ½ corresponding to the uniform electron gas and ELF = 1 corresponding to perfect localization. The localization index λ(A) for atom A and the delocalization index δ(A,B) between two atoms A and B [103, 106] are defined in terms of the density of the Fermi hole (or exchange part of the exchange–correlation hole) ρ_X:

$$\lambda(\mathrm{A}) = \int_{\Omega_A}\int_{\Omega_A} \rho_X(\mathbf{r}_1,\ \mathbf{r}_2) d\mathbf{r}_1 d\mathbf{r}_2 \tag{23.3}$$

$$\delta(\mathrm{A},\mathrm{B}) = 2\int_{\Omega_A}\int_{\Omega_B} \rho_X(\mathbf{r}_1,\ \mathbf{r}_2) d\mathbf{r}_1 d\mathbf{r}_2 \tag{23.4}$$

where Ω_A and Ω_B are the basins of atoms A and B, respectively. The localization and delocalization indices for an atom add up to the atomic population:

$$N(\mathrm{A}) = \lambda(\mathrm{A}) + \frac{1}{2}\sum_{B \neq A} \delta(\mathrm{A},\mathrm{B}) \tag{23.5}$$

where the summation runs over all pairs of atoms.

Further calculations have been done on cation and anion clusters using PBE0 functional in Gaussian 09 code. Some results are also included for the isoelectronic anions of V, Nb, and Ta encapsulated clusters. The MOs have been analyzed using the Gaussian 09 code. We used VESTA 3 [107], Molekel 5.4 [108], XCrysden 1.5 [109], AIMALL [102], and Gaussview [110] for visualization. In all cases of charged and neutral clusters, the atomic structures were reoptimized when using the Gaussian 09 code. A similar analysis has been carried out for some carbon clusters.

23.3 Results and Discussion

23.3.1 Structures and Stability of Metal Atom Encapsulated Boron Clusters

Studies on doping of an M atom in small (10–20 atoms) boron clusters [45] showed stabilization of boron rings (planar wheel-shaped), as well as drum and bowl-shaped clusters [56]. Figure 23.1a shows the wheel structure of the Co@B_8 anion, and Figure 23.1b, the wheel structure of Ir@B_9. Isoelectronic neutral, anion, and cation species can also be obtained by doping of neighboring elements in the periodic table. The stability of these clusters can be understood with an electronic shell closure at 18 valence electrons, with each boron atom contributing one electron and the remaining two valence electrons being tightly bonded with neighboring atoms on the boron ring, as we shall discuss later. On the other hand, a drum-shaped DRT cluster (Figure 23.1c) is obtained for Ni@B_{14} and isoelectronic Co@B_{14}^- corresponding to an electronic shell closure at 24 valence electrons. Cr and Fe also form similar structures with spin-polarization, while larger M atoms, such as Mo, form bowl-shaped structures, as shown in Figure 23.1d for MoB_{14}. Drums of M@B_{16} (such as for M = Mn, Fe, Co, … shown in Figure 23.1e) and M@B_{18} (M = Mo, W) shown in Figure 23.1f encapsulating M atom, have also been obtained for many transition M atoms. Popov et al. [57] reported Co@B_{16}^- drum, but Saha et al. [56] showed this cluster to have 2 μ_B magnetic moments and that Co@B_{14} anion corresponded to a closed electronic shell cluster. Similarly, the anion of Mn@B_{16} and cation of Rh@B_{16} correspond to electronic shell closure with 24 valence electrons.

Further studies have been carried out [38] on the encapsulation of an M atom (M = Cr, Mo, and W) in several different structures of larger boron clusters in the size range of B_{18}–B_{24} in

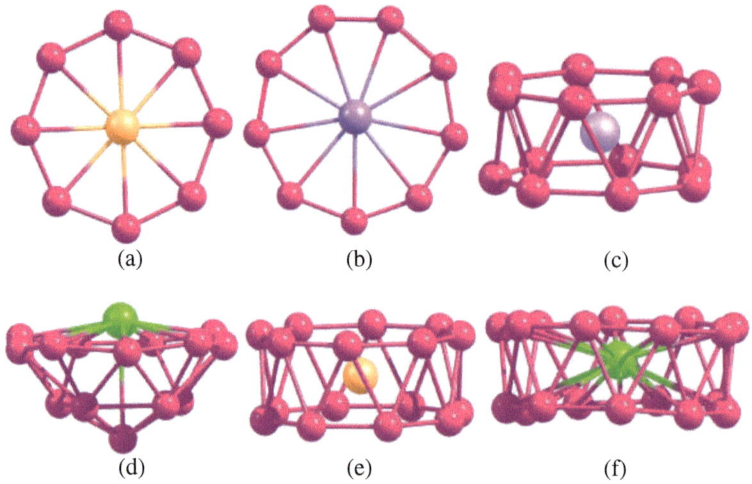

Figure 23.1 Atomic structures of (a) Co@B_8 anion, (b) Ir@B_9, (c) Ni@B_{14}, (d) Mo@B_{14}, (e) Co@B_{16}^- anion, and (f) Mo@B_{18}.

the hope of finding the smallest cage of boron. Figure 23.2a shows the lowest energy isomer for Cr@$B_{18,}$ which is a dimer-capped 16-atom drum-shaped structure. It has two pentagons arranged in a base-shared fashion. For slightly larger M atoms (Mo and W), a tubular drum-shaped structure is shown in Figure 23.2b,c has the lowest energy, and the dimer-capped drum-shaped isomer (Figure 23.2a) lies 0.36 and 0.42 eV higher in energy for Mo and W, respectively. On the other hand, the drum-shaped isomer of Cr@B_{18} lies 1.25 eV higher in energy compared with the lowest energy dimer-capped B_{16} drum isomer. The Cr@B_{18} structure remains open suggesting that 18 atoms of boron are not sufficient to form a cage with Cr encapsulation. A similar drum structure has been obtained for [Ru@$B_{18}]^{1-}$ by Jian et al. [111].

Continuing further with larger clusters, Cr is able to form a cage (Figure 23.2d) with 20 boron atoms, and it is the smallest cage of boron. This Cr@B_{20} cluster has a D_3 symmetric cage structure. There are three empty boron hexagons that are inter-linked through three 2-atom chains to two capped boron hexagons. However, for Mo@B_{20} and W@B_{20}, a dimer-capped drum structure with two boron atoms capping an 18-atom boron DRT with C_s symmetry and encapsulating M atom has the lowest energy (Figure 23.2e,f). This structure has one pentagon and one hexagon arranged in base shared fashion. For Cr@B_{20}, the dimer-capped drum isomer with C_s symmetry is 0.32 eV higher in energy than the cage structure. The D_3 fullerene structure for Mo@B_{20} and W@B_{20} is 1.79 and 2.06 eV, respectively, higher in energy than the lowest energy dimer-capped drum (C_s) structure. For TaB_{20} also, a dimer-capped structure similar to MoB_{20} and WB_{20} has been reported [112] for neutral and anionic species. These results suggest that for Mo, W, Ta, ... atoms, the cage needs to have more than 20 boron atoms.

The structure of the lowest energy isomer for Cr@B_{22}, obtained from a Gaussian 09 calculation, is shown in Figure 23.2g. The lowest energy configuration of neutral Mo@B_{22} has D_2 symmetry, as shown in Figure 23.2h. This structure has four heptagons that are interlinked through two 2-atom chains and three boron atoms. For W encapsulation, a D_2 structure was found to be of the lowest energy for W@B_{22}, as shown in Figure 23.2i. The lowest energy configuration for Cr@B_{24} is a somewhat distorted cage with C_1 symmetry, as shown in Figure 23.2j, in which the Cr atom does not interact with all the boron atoms optimally. This is because the size of a Cr atom is not

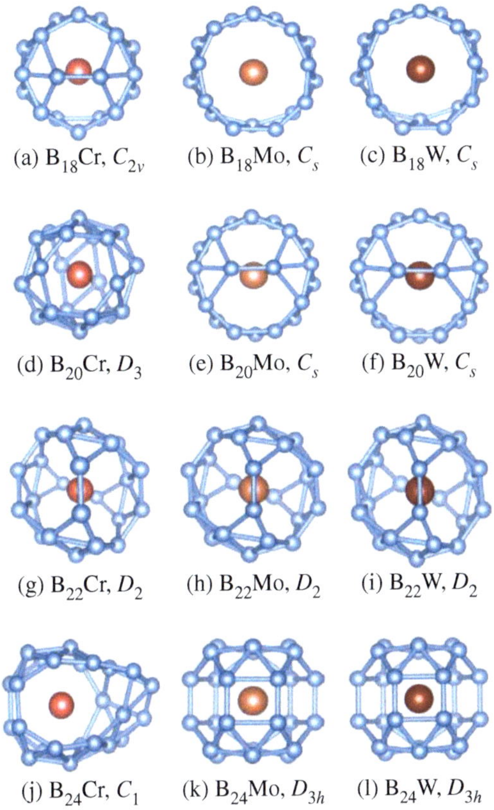

Figure 23.2 Atomic structures of the lowest energy configurations of clusters having 18, 20, 22, and 24 boron atoms encapsulation with a Cr, Mo, and W atom. In each case, the symmetry is also given. (a) $B_{18}Cr$, C_{2v} dimer-capped drum, (b) $B_{18}Mo$, C_s tubular drum, (c) $B_{18}W$, C_s tubular drum, (d) $B_{20}Cr$, D_3 cage, (e) $B_{20}Mo$, C_s dimer-capped drum, (f) $B_{20}W$, C_s dimer-capped drum, (g) $B_{22}Cr$, D_2 cage, (h) $B_{22}Mo$, D_2 cage, (i) $B_{22}W$, D_2 cage, (j) $B_{24}Cr$, C_1 distorted cage, (k) $B_{24}Mo$, D_{3h} cage, (l) $B_{24}W$, D_{3h} cage.

optimal for the B_{24} cage as it is slightly smaller than Mo and W atoms. The isomer with D_{3h} configuration for Cr lies 0.25 eV higher in energy using PBE in VASP. This result is in agreement with that reported by Lv et al. [113], except for the fact that Rahane et al. [38] found a triplet state to be lower in energy than a singlet state. For Mo@B_{24}, a very symmetric cage structure is obtained, as shown in Figure 23.2k. The lowest energy structure for Mo@B_{24} has D_{3h} symmetry and a large HOMO–LUMO gap (GGA) of 2.46 eV, indicating its high stability. The same isomer has also been independently obtained for Mo@B_{24} by Lv et al. [113] using a particle swarm optimization algorithm implemented in the CALYPSO package [114], which involved calculations of a few thousand structures. In these calculations, the hybrid PBE0 functional has been used for the structure optimization with a 6-311G* basis set in the Gaussian 09 package.

Further calculations [38] using cc-pVDZ basis set and PBE0 in Gaussian 09 code on Mo@B_{24} gave similar results. The lowest energy isomer for W@B_{24} is similar (Figure 23.2l) to that of Mo@B_{24}. Studies on encapsulation of a few other M atoms have led to the finding [38] of Mn@B_{20}^+ (isoelectronic with Cr@B_{20}), Ta@B_{22} anion cage [115] (isoelectronic with W@B_{22}), Ta@B_{24} anion cage which is isoelectronic with W@B_{24}, and also some larger cages [116] with 25–28 B atoms, as well as cages in size range of around 40 boron atoms [67–79].

It is to be noted that neutral pure boron clusters, B_n with $n = 18$–24, prefer quasi-planar or tubular structures in their ground states, and these become drum-shaped or form cages after encapsulation of a transition M atom. Calculations [38] of the energy of the drums/cages by removing the M atom and keeping the atomic positions of the boron atoms the same as in the optimized encapsulated cluster show that the empty cage structures of B_n ($n = 20$, 22, and 24) lie 1.5–3.5 eV higher in energy

Table 23.1 Calculated embedding energy (E_n(M)) of the M atom in n-atom boron cage, binding energy (BE), HOMO–LUMO gap (E_g), vertical electron affinity (VEA) and vertical ionization potential (VIP) for the lowest energy isomers of M@B_n (n = 18, 20, 22, and 24; M = Cr, Mo, W) clusters. BE w.r.t to B_n is the energy gained by doping an M atom with respect to the ground state of B_n using PBE0 in the Gaussian program, with 6-311+g basis set for B and Cr and LANL2DZ for Mo and W.

	PBE			PBE0				
Cluster	En(M) (eV)	BE per atom (eV)	E_g (eV)	BE w.r.t. B_n (eV)	BE per atom (eV)	E_g (eV)	VEA (eV)	VIP (eV)
B_{18}	—	5.48	0.47	—	4.93	1.39	—	—
B_{20}	—	5.56	1.36	—	5.05	2.72	—	—
B_{22}	—	5.57	0.29	—	5.06	1.35	—	—
B_{24}	—	5.63	1.30	—	5.10	2.18	—	—
Cr@B_{18}	6.67	5.55	1.24	5.65	4.97	2.73	2.72	7.63
Mo@B_{18}	8.28	5.63	0.40	7.53	5.07	1.71	3.33	7.37
W@B_{18}	9.61	5.70	0.47	9.51	5.17	1.71	3.37	7.44
Cr@B_{20}	5.62	5.57	2.14	4.28	5.02	3.85	1.88	7.94
Mo@B_{20}	7.76	5.67	1.76	6.54	5.12	3.21	2.06	6.99
W@B_{20}	9.22	5.74	1.79	8.72	5.23	3.31	2.00	7.06
Cr@B_{22}	5.29	5.57	2.26	3.99	5.03	3.78	2.05	7.05
Mo@B_{22}	8.25	5.70	2.51	7.59	5.17	3.87	1.97	8.01
W@B_{22}	9.67	5.76	2.50	9.76	5.26	3.85	1.99	8.03
Cr@B_{24}	4.59	5.59	0.89	3.32	5.03	2.35	3.44	7.11
Mo@B_{24}	7.33	5.71	2.46	7.42	5.20	4.40	1.59	7.86
W@B_{24}	8.69	5.76	2.45	9.55	5.28	4.37	1.56	7.89

VEA and VIP are calculated using Gaussian 09 program. For comparison, the binding energy and HOMO-LUMO gap calculated within PBE0 are also given. Results have also been given for the lowest energy isomers of pure B_n clusters (n = 18, 20, 22, and 24).

than their respective ground state isomers with the PBE functional. Therefore, the interaction of the M atom due to endohedral encapsulation lowers the energy of the cage significantly. Further, the gain in energy when an M atom is doped in a B_n cluster is obtained from:

$$E_d = E(B_n) + E(M) - E(M@B_n) \tag{23.6}$$

where $E(B_n)$ and $E(M@B_n)$ are the energies of B_n and M@B_n clusters in the lowest energy isomer, respectively. These values are given in Table 23.1. In general, large values (>8.0 eV) of E_d are obtained for M = Mo and W in B_{22} cluster. This also leads to significant enhancement in the binding energy of the encapsulated boron clusters compared with that of the undoped ones. The binding energy, defined as the energy per atom of the cluster with respect to the energy of the free atoms,

$$\Delta E = \{nE(B) + E(M) - E(M@B_n)\}/(n+1) \tag{23.7}$$

is shown in Figure 23.3 using PBE in VASP and PBE0 in the Gaussian 09 program for the lowest energy isomers of all sizes and with different transition M atoms. Here, E(B) and E(M) are the energies of the B atom and M atom, respectively. The overall trend of the binding energy in both the methods is similar, but the values are slightly lower when PBE0 is used, except for the encapsulation energy or the embedding energy, E_n(M) (defined as the energy gained by encapsulating an

M atom in the cage with the same structure as the encapsulated cluster) for W in B_{24} and Mo in B_{22} and B_{24}. The encapsulation energy increases in going from Cr to Mo and then to W, indicating that W encapsulation is the strongest among the three transition M atoms. The binding energy also increases upon going from Cr encapsulation to Mo and then to W. There is a slight increase in the binding energy on going from 18-atom to 20-atom boron clusters for all the TM atoms. For Cr encapsulation the binding energy remains nearly the same on going from Cr@B_{20} to Cr@B_{22}. However, it slightly increases to 5.59 eV for the Cr@B_{24} cluster in PBE. For Mo and W encapsulation, the trend shows that the increase in the binding energy from Mo@B_{20} to Mo@B_{22} is more than the increase from Mo@B_{22} to Mo@B_{24}. Similarly, for W encapsulation, the binding energy increases from W@B_{20} to W@B_{22}, but it remains nearly the same for W@B_{22} and W@B_{24} in PBE, suggesting the magic nature of these 22 boron atom clusters. The M@B_{24} (M = Mo and W) clusters also have good stability as the encapsulation energy continues to be large, and the HOMO–LUMO gap is also quite large (Table 23.1). Therefore, Mo and W are favorable to form endohedral cage structures of boron. However, for Cr encapsulation, the encapsulation energy is higher for Cr@B_{18} than the binding energy of a pure B_{18} cluster, while for Cr encapsulation in B_{20}, it is marginally higher in PBE and smaller in PBE0 compared with the binding energy of a pure B_{20} cluster (see Table 23.1). With the PBE0 functional, there is a large HOMO–LUMO gap for Cr@B_{20} and also in all three cases of M@B_{22}. In the following, we discuss Cr as well as Mo and W encapsulation to understand the cage stability and bonding characteristics in the encapsulated clusters.

From Table 23.1 the calculated HOMO–LUMO gaps (E_g) in PBE are 1.24, 0.40, and 0.47 eV for M@B_{18}, M = Cr, Mo, and W, respectively. GGA underestimates E_g and the values obtained using PBE0 are much larger (2.72, 1.71 and 1.71 eV, respectively). Note the large value of E_g for Cr encapsulation. For M@B_{20}, the HOMO–LUMO gaps increase to 2.14, 1.76, and 1.79 eV, respectively, with M = Cr, Mo, and W, and further to 2.26, 2.51, and 2.50 eV for M@B_{22} in PBE. These large values of E_g suggest the stability of the cage structures. However, for Cr@B_{24}, the HOMO–LUMO gap decreases to 0.89 eV as the cage becomes less symmetric and Cr is not bonded well with all the B atoms. Mo@B_{24} and W@B_{24} clusters also have large E_g (2.46 and 2.45 eV, respectively), but the values are slightly lower than for Mo@B_{22} and W@B_{22} cages. The calculated E_g for the M-encapsulated B_{22} and B_{24} cage structures using PBE0 are again much larger than the values obtained with PBE, and these are given in Table 23.1. The largest value (4.40 eV) of E_g in PBE0

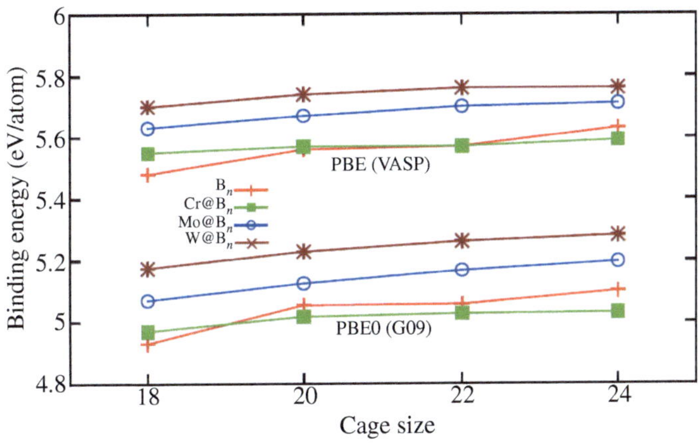

Figure 23.3 Calculated binding energies for the lowest energy pure boron clusters and isomers of clusters shown in Figure 23.2, as obtained by using PBE in VASP and PBE0 in Gaussian 09 program.

is obtained for Mo@B_{24} and a quite similar value for W@B_{24}. The largest value (5.28 eV/atom) of the binding energy in PBE0 is obtained for W@B_{24}, and a slightly smaller value (5.26 eV/atom) for W@B_{22}. This is in contrast to the values of 5.06 and 5.10 eV/atom for the binding energies of elemental boron clusters B_{22} and B_{24}, respectively, in PBE0. Therefore, M encapsulation enhances the stability of boron clusters and leads to the formation of their cage structures for smaller sizes compared with pure boron clusters. Table 23.1 also shows that the vertical ionization potential of Cr@B_{20}, Mo@B_{22}, W@B_{22}, Mo@B_{24}, and W@B_{24} is quite large, with values of about 8 eV, while the vertical electron affinity is low. These are also characteristics of magic/stable clusters.

23.3.2 Bonding Characteristics in M@B_{18}, M@B_{20}, M@B_{22}, and M@B_{24} Clusters

Studies on drum-shaped boron clusters [55] showed clear evidence of different kinds of bonding within a cluster: two-center covalent σ bonding along the two rims of a drum, three-center σ bonding on the drum faces between the two rims, and hybridization between the d orbitals of the transition M atom and the π-bonded MO of the drum.

In order to understand the bonding character in the M@B_{18}, M@B_{20}, M@B_{22}, and M@B_{24} structures, we have given the bond lengths in Table 23.2. It is noted that the shortest B—B bond lengths for Cr@B_{18}, Mo@B_{18}, and W@B_{18} are 1.61, 1.58, and 1.58 Å, respectively, whereas B—Cr, B—Mo, and B—W bond lengths are in the range of 2.12–2.36, 2.32–2.63, and 2.33–2.60 Å, respectively, using PBE. The shortest B—B bonds are the two-center (2c) sigma bonds. This is similar to the bonding in pure small clusters of boron with a ring [45]. As the cluster size increases, the B—B and B—Cr bond distances for Cr@B_{20} lie in the range of 1.60–1.83 and 2.08–2.30 Å, respectively, while for Mo@B_{20} and W@B_{20}, the shortest B—B bond length is 1.57 Å. The B—Mo and B—W bond lengths are in the range of 2.13–2.57 and 2.14–2.58 Å, respectively. These Mo—B and W—B bond lengths are shorter than those found in Mo@B_{18} and W@B_{18}, respectively. The shorter bonds are formed by the capping boron atoms with the M atom. For the M@B_{22} cage, some of the B—B bonds are shorter than those in the M@B_{20} cage. There are two shortest B—B bonds (1.56 Å) for the M@B_{22} cage for M = Cr and 1.57 Å for Mo and W encapsulation, but the M—B bond lengths are slightly elongated and lie in the range of 2.31–2.54 Å for M = Cr and 2.38–2.59 Å for M = Mo and W, as compared to those in the respective M@B_{20} cages. In the M@B_{24} cage, the shortest B—B bond length increases to 1.64 Å for M = Mo and W. Some of the M—B bond lengths are longer and lie in the range of 2.43–2.55 Å for Mo encapsulation and 2.44–2.55 Å for W encapsulation. For Cr@B_{24}, the B—B bond lengths are in the range of 1.54–1.84 Å, while the B—Cr bond lengths have wide

Table 23.2 Bond lengths for M doped boron clusters using PBE.

Size	M = Cr	Mo	W
MB_{18}	B–B = 1.61 Å B–Cr = 2.12–2.36 Å	B–B = 1.58 Å B–Mo = 2.32–2.63 Å	B–B = 1.58 Å B–W = 2.33–2.60 Å
M@B_{20}	B–B = 1.60–1.83 Å B–Cr = 2.08–2.30 Å	B–B = 1.57 Å B–Mo = 2.13–2.57 Å	B–B = 1.57 Å B–W = 2.14–2.58 Å
M@B_{22}	B–B = 1.56 Å B–Mo = 2.31–2.54 Å	B–B = 1.57 Å B–Mo = 2.38–2.59 Å	B–B = 1.57 Å B–Mo = 2.38–2.59 Å
M@B_{24}	B–B = 1.54–1.84 Å B–Cr = 2.18–2.66 Å.	B–B = 1.64 Å B–Mo = 2.43–2.55 Å	B–B = 1.64 Å B–Mo = 2.44–2.55 Å

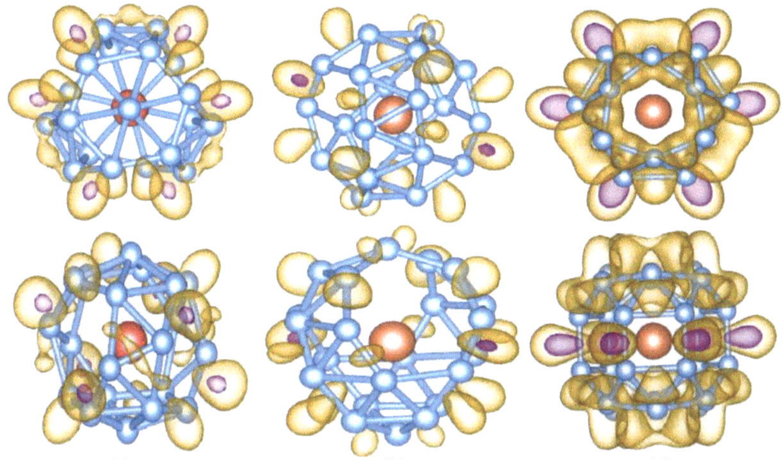

Figure 23.4 Isosurfaces of ELF for (a) $Cr@B_{20}$, (b) $Mo@B_{22}$, and (c) $Mo@B_{24}$. Two orientations are shown in each case. In (a) the lobes shown by purple color have strong localization of charge with a high value of ELF (0.94) representing 2c−2e bonds. As the value of the isosurface is decreased (0.84), these lobes become bigger and new lobes appear. These other lobes shown with yellowish color are 3c or higher center σ bonds. In (b), there are 2 purple color lobes with ELF = 0.96 (2c) and 16 yellowish color lobes with ELF = 0.88 (3c). In (c), there are six 2c (purple color) and eighteen 3c (yellowish) bonds.

variation and are in the range of 2.18–2.66 Å. This is because the Cr atom is not at the center of the cage.

Further, the results of the total charge density isosurfaces, ELF and AdNDP analysis for some of the encapsulated clusters are presented in Figures 23.4 and 23.5. Earlier studies on pure boron clusters have suggested the existence of multi-center bonds, namely 2c, 3c, 4c, 6c, and 7c bonds, using AdNDP analysis [101]. The total charge density and ELF, along with AdNDP analysis, help to understand 2c and multicenter bonds in the M-encapsulated boron clusters. Further, the calculation of Laplacian of the electron density and bond- as well as ring-critical points was done to analyze the bonding character.

Figure 23.4 shows ELF for the smallest cage, $Cr@B_{20}$ and also for $Mo@B_{22}$ and $Mo@B_{24}$ clusters. For $Cr@B_{20,}$ the bond length analysis shows that there are six short B—B bonds of 1.60 Å along the three 2-atom chains, joining the three empty heptagons. These bonds can be seen in the charge density and ELF plots at high isosurface values (Figure 23.4). A high value of ELF shows strong localization of charge as in covalent bonds. These six lobes are indicated by purple color in Figure 23.4a at the ELF value of 0.94. These bonds are identified as six 2c—2e bonds.

Further decreasing the ELF value to 0.84, these 6 lobes become bigger (shown by the yellowish color around the purple color lobes), and 12 more lobes can be seen. These 12 additional lobes (yellowish in Figure 23.4a) represent twelve 3c σ bonds. This identification was also supported by the AdNDP analysis, as discussed below.

AdNDP analysis of $Cr@B_{20}$, the smallest cage cluster, shows (Figure 23.5a) that there are six 2c—2e bonds, as also concluded from ELF. Further, there are twelve 3c bonds; six are along the three 2-atom chains, and the remaining six are placed alternately on the three edges of each capped hexagon, as shown in Figure 23.5b. These bonds were also detected in the total charge density and ELF analysis, as discussed above. In addition to these eighteen σ-bonds, there are four 5c–2e

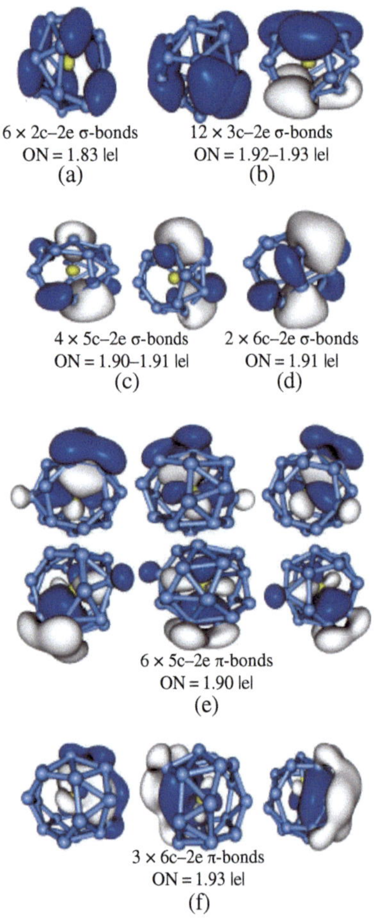

Figure 23.5 Different multicenter bonds in the lowest energy isomer for $Cr@B_{20}$ as obtained from the AdNDP analysis. (a) 6 2c-2e σ bonds, (b) 12 3c-2e σ bonds, (c) 4 5c-2e σ bonds, (d) 2 6c-2e σ bonds, (e) 6 5c-2e π bonds, (f) 3 6c-2e π bonds.

σ-bonds and two 6c–2e σ-bonds, as shown in Figure 23.5c,d, respectively. These bonds connect the atoms on the hexagons to the atoms on the three 2-atom chains. All these 24 σ-bonds involve bonding among boron atoms. There are six 5c–2e π-bonds, each connecting three B atoms on one capped hexagon to one B atom on the edge through the Cr atom. There are also three 6c–2e π-bonds along the three 2-atom chains, each involving six B atoms on the chain and the Cr atom. Thus, there are altogether nine π-bonds which involve the bonding of the boron atoms on the cage and the Cr atom. This indicates that the stability of the cage due to the Cr atom encapsulation is governed by the complete filling of an electronic shell with 18 valence electrons. It was also further shown from the analysis of the MOs of $Cr@B_{20}$ (see below).

An analysis of the electronic charge density $\rho(\mathbf{r})$ of $Cr@B_{22}$ was further performed [38] as a representative of the 22-atom boron cage by calculating contours of the Laplacian distribution of $\rho(\mathbf{r})$ as well as the Laplacian at bond critical points (BCPs) and ring critical points (RCPs). The method of calculation has been discussed earlier [45]. Figure 23.6a shows the atomic structure with BCPs (green dots) and RCPs (red dots), while Figure 23.6b shows a symmetrical view of $Cr@B_{22}$ along with the electrostatic potential mapped onto a $\rho(\mathbf{r}) = 0.01\ e/Bohr^3$ electron density isosurface. Blue regions indicate negative electrostatic potential associated with the boron atoms. Of the 22 boron atoms, there are 4 B atoms that have coordination 2 in the middle of the 2 boron chains, while 18 boron atoms are in two semicircular bands consisting of BBB triangles and/or BBBB

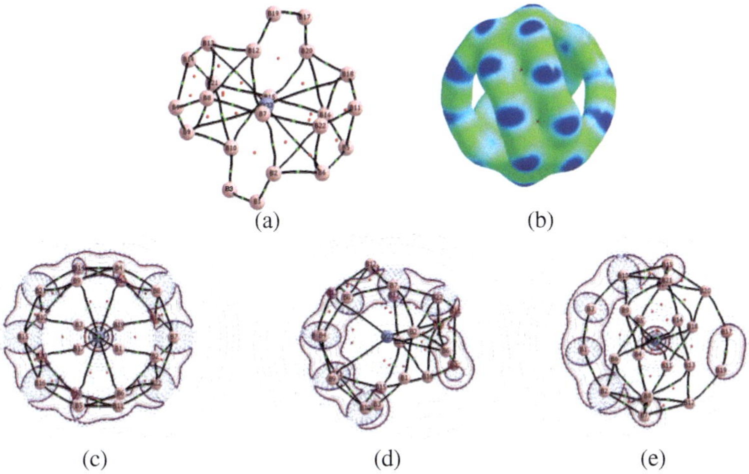

Figure 23.6 (a) BCPs (green dots) and RCPs of BBBB quadrilaterals (red dots) of $Cr@B_{22}$ cluster and (b) electrostatic potential mapped onto a $\rho(\mathbf{r}) = 0.001$ e/Bohr^3 electron density isosurface. Blue regions indicate negative electrostatic potentials associated with the boron atoms. (c–e) Contour plots of the Laplacian in different planes passing through atoms (c) B7, B8, B15, B16, B21, B22, and Cr, (d) B4, B7, B8, B9, B10, B22, and (e) B1, B2, B3, B10, and Cr. The contour values are 0.0, 0.001, 0.002, 0.004, 0.008, 0.02, 0.04, 0.08, 0.2, 0.4, 0.8, 2.0, 4.0, 8.0, 20.0, 40.0, 80.0, 200.0, 400.0, 800.0, −0.001, −0.002, −0.004, −0.008, −0.02, −0.04, −0.08, −0.2, −0.4, −0.8, −2.0, −4.0, −8.0, −20.0, −40.0, −80.0, −200.0, −400.0, −800.0. Pink (gray) balls show B (M) atoms.

quadrilaterals. The two boron chains appear at the left and right ends of the figure, while one of the semi-circular bands appear in the foreground, running diagonally from top left to bottom right (the other is partially occluded at the back, running from top right to bottom left). The two bands and the two chains loop around and encapsulate the M atom in a tetrahedral coordination. Such bands are a recurring motif in many kinds of boron clusters, including drums [45, 56] and quasi-planar structures [3]. The two bands join each other and the two chains at four tetra-coordinated boron atoms. The critical point analysis requires that the Poincare–Hopf relationship:

$$\text{NumNACP} + \text{NumNNACP} - \text{NumBCP} + \text{NumRCP} - \text{NumCCP} = 1 \tag{23.8}$$

be satisfied [104], where NumNACP is the number of nuclei (here 23), NumNNACP is the number of nonnuclear attractors (here 0) critical points, NumBCP is the number of BCPs (here 40), NumRCP is the number of RCPs (here 20), and NumCCP is the number of cage critical points (here 2). In Figure 23.6c–e red dots indicate the locations of the RCPs of the BBBB quadrilaterals. Of the 36 B–B BCPs (green dots), there are 6 that represent covalent bonds in the 2 boron chains.

Figure 23.4b,c show ELF plots for $Mo@B_{22}$ and $Mo@B_{24}$, respectively. For $Mo@B_{22}$, the two lobes shown by purple color are obtained at the ELF value of 0.96, and they represent two 2c bonds with a bond length of 1.57 Å. There are 16 more lobes shown by yellowish color at the ELF value of 0.88. Similar to the case of the B_{20} cage, they represent sixteen 3c bonds. For $Mo@B_{24}$, the analysis of the charge density and ELF shows that this cluster has six 2c σ-bonds (purple color lobes in ELF) and eighteen 3c σ-bonds. Further analysis using AdNDP shows six 5c σ-bonds (see Figure 23.5c). The remaining 18 valence electrons are involved in π-bonding as well as bonding with the M atom, and can be detected in the AdNDP analysis [38, 113].

Figure 23.6d shows a contour plot of the Laplacian distribution in a plane passing approximately through the rim of one of these bands of $Cr@B_{22}$. 12 BCPs correspond to bonds at the bases of

BBB triangles in the semicircular bands, and 10 correspond to bent bonds spanning the width of one or the other of the two bands. Figure 23.6e shows a contour plot of the Laplacian distribution in a plane passing approximately through one of the boron chains and the M atom of Cr@B_{22}. Eight BCPs represent covalent bonds along rims of the semicircular bands. The properties of these bonds are shown in Table 23.3 and Figure 23.7. Large values of the charge density and large positive Laplacian with low values of bond ellipticity correspond to covalent bonds, while high bond ellipticities, low positive values of the Laplacian $L(\mathbf{r})$ and delocalization index (off-diagonal localization–delocalization matrix [LDM] elements) are characteristic of multicenter bonding in the BBB triangles. On the other hand, negative values of Laplacian correspond to the M—Cr bonds. Similar results have been obtained for the Cr@B_{20} cage and are shown in Figure 23.7c.

In order to understand the electronic origin of the stability of these cages, the MOs were calculated for the M encapsulated clusters as well as for the corresponding boron cage by removing the M atom and keeping the other atomic positions fixed. For Cr@B_{20}, six electrons of the Cr atom and twelve π-bonded electrons from the B_{20} cage contribute to the stability of the electronic shell closing at eighteen π-bonded valence electrons. Figure 23.8 shows that the bare B_{20} cage has six occupied π-bonded MOs (doubly degenerate HOMO and HOMO-4, HOMO-5, and HOMO-9). The two HOMO levels have d angular momentum character. There are three unoccupied π-bonded MOs (LUMO and LUMO + 1) of d angular momentum character. For Cr@B_{20}, these five MOs of d angular momentum character hybridize with the d orbitals of Cr atom, forming bonding and antibonding MOs. The six electrons of the Cr atom are accommodated in the bonding MOs, and there is a large HOMO–LUMO gap, which results in the stability of the endohedral Cr@B_{20} cage cluster.

For the Mo@B_{22} cage, the π-bonded MOs for the bare B_{22} cage and for Mo@B_{22}, along with the corresponding eigenvalues, are shown in Figure 23.9, as obtained from the Gaussian calculation with the PBE0 functional. It is seen that the stability of the Mo-encapsulated B_{22} cage is governed by the 9 π-bonded MOs with an electronic shell closure at 18 valence electrons. For the bare B_{22}

Table 23.3 Different types of bonds, the number of bonds of each type, charge density (ρ) in units of e/Bohr^3 at the BCP, bond ellipticity, Laplacian (L) in units of e/Bohr^6 at the BCPs, and the delocalization index, for Cr@B_{22}.

Bond type	No. of bonds	ρ	Bond ellipticity	L	Delocalization index
B1–B3, B17–B19	2	0.174	0.02	0.104	1.183
B1–B2, B3–B10, B12–B19, B17–B20	4	0.162	0.15	0.085	1.023
B4–B8, B5–B16, B11–B22, B14–B21	4	0.160	0.12	0.083	1.009
B4–B9, B5–B6, B11–B18, B13–B14	4	0.151	0.22	0.069	0.959
B7–B8, B7–B22, B15–B16, B15–B21	4	0.145	0.47	0.061	0.886
B2–B6, B9–B10, B12–B13, B18–B20	4	0.128	1.80	0.033	0.686
B2–B22, B8–B12, B10–B21, B16–B20	4	0.125	2.87	0.028	0.555
B2–B7, B7–B12, B10–B15, B15–B20	4	0.124	1.66	0.030	0.747
B9–B21, B6–B22, B16–B18, B8–B13	4	0.118	5.10–5.20	0.018	0.555
B5–B11, B4–B14	2	0.115	1.295	0.019	0.572

Small value of ellipticity and large positive value of Laplacian indicates strong 2c covalent bonds while large values of ellipticity and small values of Laplacian indicates delocalization of charge (multi-center bonding). The numbering of boron atoms is given in Figure 23.6.

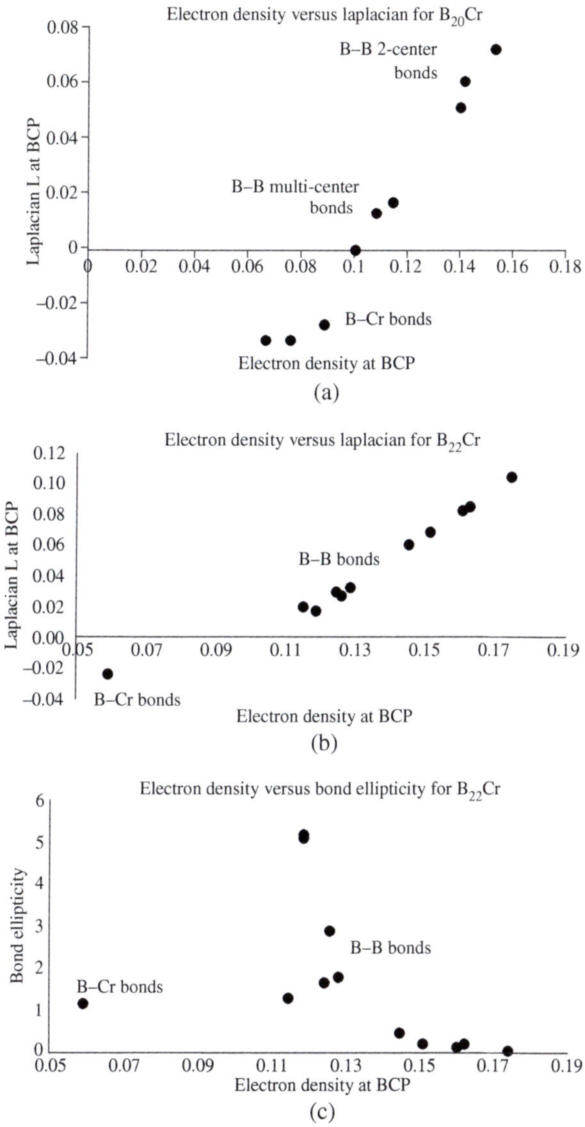

Figure 23.7 (a) Laplacian (in units of e/Bohr^6) versus electron density (e/Bohr^3) for $B_{20}Cr$ cage, (b) Laplacian versus electron density for $B_{22}Cr$ cluster and (c) bond ellipticity as a function of the electron density for $B_{22}Cr$ cluster.

cage, there are six occupied π-bonded MOs and three empty π-bonded MOs (not shown) of d angular momentum character just above the HOMO. The five π-bonded MOs with d angular momentum character hybridize strongly with the d orbitals of the Mo atom, forming five bonding MOs and five anti-bonding MOs. The bonding MOs are fully occupied and, similar to the case of 18 valence electron stability of the Cr@B_{20} cage, the stability of the Mo@B_{22} cluster also arises from 18 valence electrons. The ordering of the MOs can be seen in Figure 23.9. Note that the bare B_{22} cage has a very small HOMO-LUMO gap but it is very much increased by Mo encapsulation.

Further, the stability of the Mo@B_{24} cluster also arises from the occupation of 9 π-bonded MOs corresponding to 18 valence electrons. A similar result will hold for the case of W encapsulation. An

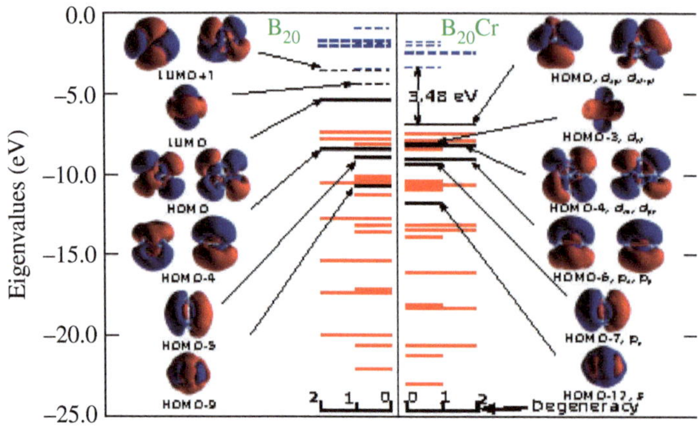

Figure 23.8 Molecular orbitals and MO energy level diagram of B_{20} cage without Cr atom and $B_{20}Cr$ cage.

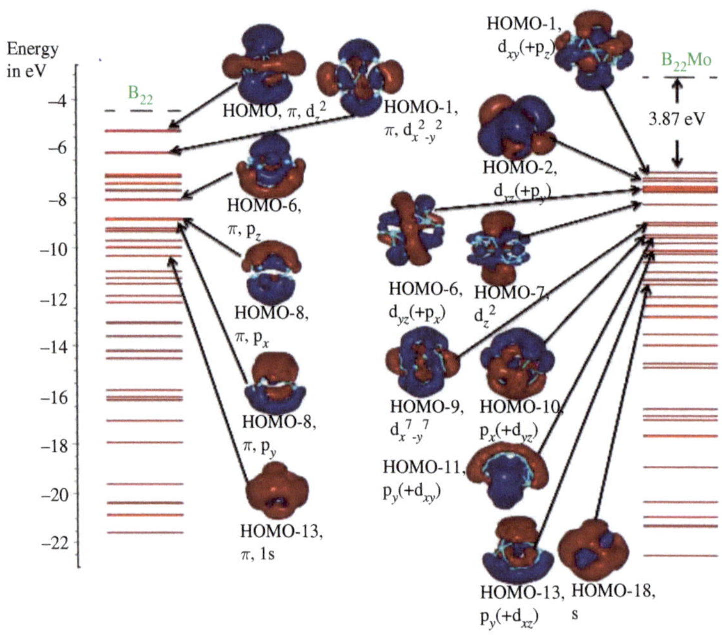

Figure 23.9 Molecular orbitals and MO energy level diagram of B_{22} cage without Mo atom and $B_{22}Mo$.

analysis of the MOs of the $Cr@B_{18}$ dimer-capped drum structure shows [38] that its stability is also associated with 18 π-bonded valence electrons. On the other hand, the stability of the disk-shaped $M@B_{20}$, for M = Mo and W, arises from 20 π-bonded valence electrons.

Calculations on cationic and anionic clusters were performed for the lowest energy and some low-lying isomers of neutral clusters using the Gaussian 09 program. Note that for the neutral and charged clusters, the calculations using the Gaussian 09 program gave almost the same energy order of the low-lying isomers as with the VASP calculations, but in some cases the lowest energy neutral isomer did not have the lowest energy when charged. As an example, the cation of $W@B_{22}$ has a double dimer-capped drum structure, while the anion of the lowest energy neutral isomer continues to have the lowest energy. Also, a cage structure of anions of $M@B_{20}$ (M = Mo and W)

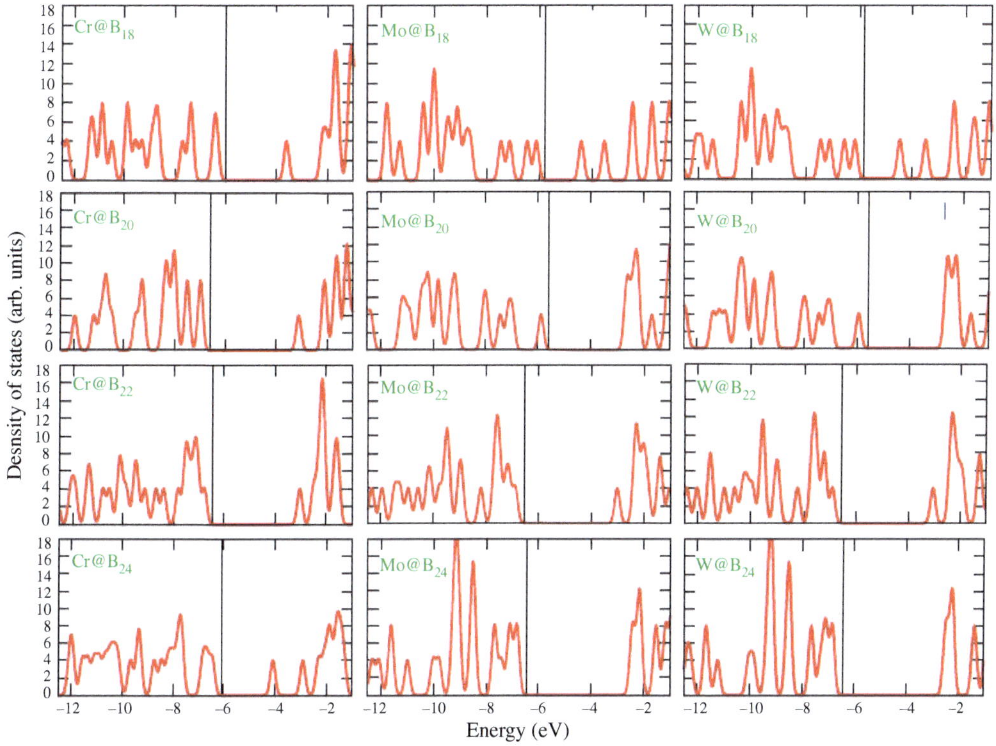

Figure 23.10 Calculated density of states (DOS) for neutral Cr, Mo, and W encapsulated B_{20}–B_{24} clusters. The vertical line shows the HOMO.

has the lowest energy compared with a dimer-capped drum for the neutral. More details can be found in Ref. [38]. The density of states (DOS) for the neutral and anionic clusters provides further information on the nature of the electronic states. The DOS for the neutral clusters with M = Cr, Mo, and W are shown in Figure 23.10. These results show that all the cages have a large HOMO–LUMO gap, as obtained from the PBE0 calculations, indicating their very good chemical stability. As seen from the MOs, all the energy levels near the HOMO (shown by the vertical line) arise from the hybridization of the boron cage orbitals with the d orbitals of the M atom. For Cr@B_{20}, very localized peaks can be seen below the HOMO, indicating the degeneracy arising from the symmetry of the structure. In going from Cr@B_{20} to Cr@B_{22} there is a spread in the distribution of the electronic energy levels. For M@B_{22} and M@B_{24}, some sharp peaks can be noticed. The overall nature of the DOS for Mo and W encapsulation remains similar.

23.3.3 Structures and Stability of Carbon Rings

The effects of transition metal encapsulation in monocyclic carbon rings with 9, 10, and 12 carbon atoms, as well as on the corresponding rings with partial hydrogenation (2 or 3 hydrogen atoms so as to alter the symmetry and the number of delocalized electrons on the C atoms), have been recently studied [80]. It is found that the stabilization of the ring structure of C_9, as well as those of partially hydrogenated clusters C_9H_3, $C_{10}H_2$, and $C_{12}H_3$, can be achieved with the complexation of an M atom. The lowest energy-optimized structures of 9-, 10- and 12-atom pure carbon clusters are shown in Figure 23.11a. While C_9 favors a chain structure, lying 0.634 eV (in PBE) lower in

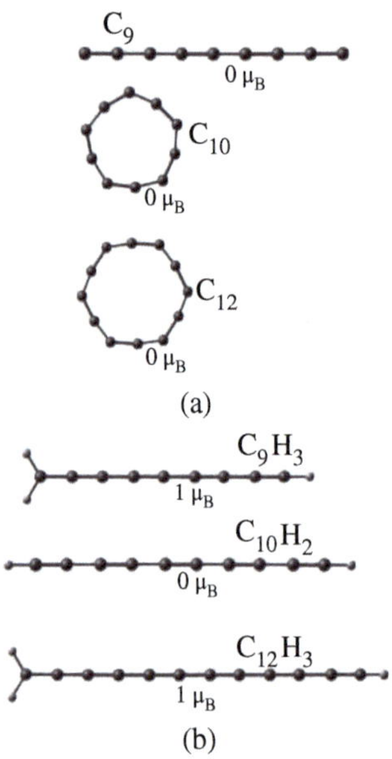

Figure 23.11 Lowest energy isomers of (a) pure and (b) partially hydrogenated carbon clusters. The magnetic moment is also given.

energy than a ring structure, C_{10} and C_{12} have fivefold and sixfold symmetric ring structures that are 2.607 and 1.304 eV (PBE) lower in energy, respectively, than their chain structures. The latter also has 2 μ_B magnetic moment. C_{10} is an especially stable ring with a large HOMO–LUMO gap of 3.750 eV in PBE (5.70 eV in PBE0). The stability of the D_{5h} symmetric structure of C_{10} can be understood by a straightforward application of Hückel MO theory. There are 15 occupied orbitals in the ring plane and 5 with a node in the ring plane (π-type involving carbon p_z orbitals), leading to π-aromaticity with $4n + 2 = 10$ electrons. C_{12} has sixfold symmetry, with alternating C—C bond lengths of 1.246 and 1.348 Å. The HOMO–LUMO gap (0.847 eV in PBE, 2.612 eV in PBE0) is much smaller compared to that of C_{10}.

The optimized structure of the transition M atom encapsulated C_9Ti is shown in Figure 23.12a, and a typical structure of $C_{10}M$ with M = Sc, Ti, and V in Figure 23.12b, showing distortions in the ring and the M atom above the ring. As C_{10} itself is a highly stable ring with a large HOMO–LUMO gap of 3.75 eV (in PBE), encapsulation of an M atom leads to only a small gain in energy with a significant reduction in the HOMO–LUMO gap (to ~0.5 eV in PBE) and distortions in the carbon

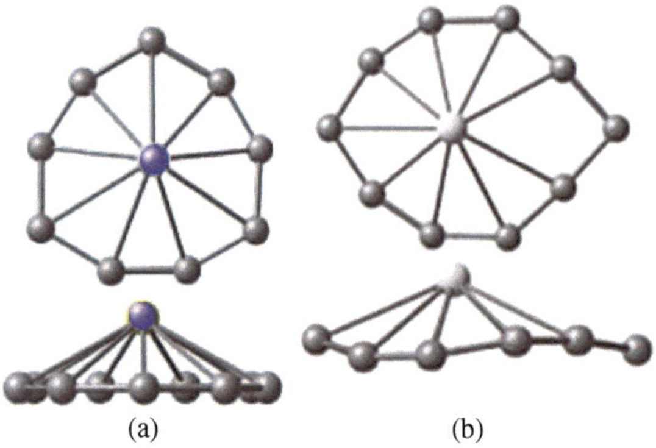

Figure 23.12 Lowest energy isomers of metal complexed carbon clusters. (a) The structure of C_9Ti, and (b) a typical structure of $C_{10}M$ with M = Sc, Ti, and V with distortions in the ring and the metal atom above the ring.

ring, although the ring structure itself is retained. For $C_{10}M$ with M = Cr, Mn, Fe, Co, and Ni, the M atom, being small, drifts toward the ring.

The hydrogenated clusters C_9H_3, $C_{10}H_2$, and $C_{12}H_3$ favor chain structures (Figure 23.11b) over ring structures by 0.498, 0.41, and 2.095 eV, respectively, but complexation of an M atom such as Cr or Sc makes the ring structures energetically favorable [80]. Encapsulation of a Cr atom within the C_9H_3 ring leads to a ring structure with the Cr atom capping and displaced from the center of the ring, having a 5 μ_B magnetic moment (Figure 23.13a). Complexation of a Sc atom with $C_{10}H_2$ leads to a bowl-shaped structure (Figure 23.13b) with 1 μ_B magnetic moment and a significant HOMO–LUMO gap of 2.449 eV in PBE0. A similar bowl-shaped structure is obtained for Ti with 0 μ_B magnetic moment in PBE. Another isomer of $C_{10}H_2Sc$ (1 μ_B) and $C_{10}H_2Ti$ (2 μ_B) with the M atom lying below the bowl is shown in Figure 23.13c. A similar structure is obtained for the case of V and Cr with 1 μ_B and 2 μ_B magnetic moments, respectively. For Mn atom complexation, an eye-shaped planar ring (Figure 23.13d) with Mn at the center and 1 μ_B magnetic moment has the lowest energy. Fe, Co, and Ni also lie in a slightly distorted planar $C_{10}H_2$ ring with 2, 1, and 0 μ_B magnetic moments.

The C_{12} ring has 24 π-bonded electrons and the resultant Hückel's aromaticity. Attaching three H atoms in a symmetric manner yields a threefold symmetric $C_{12}H_3$ ring, as shown in Figure 23.13e–g, reducing the number of available π electrons to 21, as three electrons on the carbon ring are used in σ bonding with three H atoms and sp^2 hybridization of the three corner carbon atoms. Encapsulation of a Sc or Ti atom within this structure leads to a slightly buckled

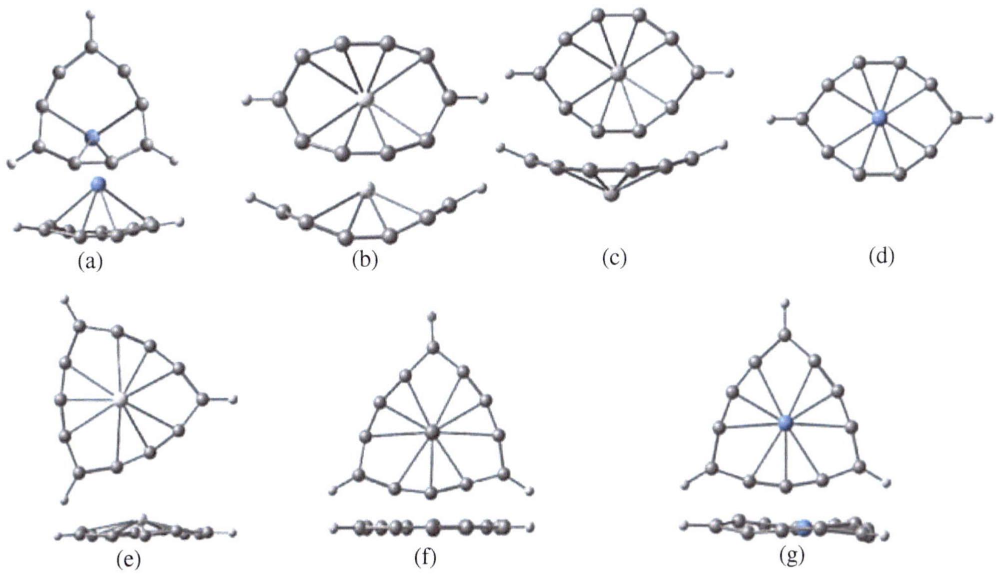

Figure 23.13 Lowest energy isomers of metal complexed partially hydrogenated carbon clusters. (a) The structure for C_9H_3Cr has 5 μ_B magnetic moment. (b) The bowl-shaped structure of $C_{10}H_2Sc$ with 1 μ_B magnetic moment. A similar structure has been obtained for the case of Ti doping with 0 μ_B magnetic moment. (c) Another isomer of $C_{10}H_2Sc$ (1 μ_B) and $C_{10}H_2Ti$ (2 μ_B) with the metal atom lying below the bowl. A similar structure is obtained for the case of V and Cr with 1 μ_B and 2 μ_B magnetic moments, respectively. (d) A nearly degenerate planar isomer of $C_{10}H_2Cr$. (e) The structure of $C_{12}H_3Sc$ with the Sc atom lying slightly above the ring. A similar structure is obtained for the doping of Ti with 1 μ_B magnetic moment. (f) The structure of $C_{12}H_3V$. The V atom lies in the plane of the ring, and it has 2 μ_B magnetic moments. (g) The structure of $C_{12}H_3Cr$ with 3 μ_B magnetic moment. The ring is slightly distorted.

ring with 1 μ_B magnetic moment and an M atom lying above the ring. $C_{12}H_3V$ has a planar structure (Figure 23.13f), with the V atom lying in the plane of the ring, and it has 2 μ_B magnetic moments. Figure 23.13g shows the structure for $C_{12}H_3Cr$; it is slightly distorted and has 3 μ_B magnetic moments. For smaller atoms such as Fe, the M atom drifts toward the ring and does not bind with all the C atoms. It has 5 μ_B magnetic moments. A similar structure has been obtained for $C_{12}H_3Co$ with 4 μ_B magnetic moments.

23.3.4 Bonding Characteristics in Carbon Rings

In this section, we examine the electron density distributions in the $C_{10}M$, $C_{10}H_2M$ and $C_{12}H_3M$ clusters, with M = Sc, Ti, V, and Cr, since these early transition M atoms are the ones that result in the best fit and the least distortion of the carbon rings upon complexation. Contours of the Laplacian of electron density (Eq. (23.1)) for $C_{10}Sc$, $C_{10}Ti$, $C_{10}V$, and $C_{10}Cr$ are plotted in Figure 23.14 in the plane of the carbon atoms. Positive contours of the Laplacian (Eq. (23.1)) (red continuous contour lines) identify regions of electron density accumulation, while negative contours (blue dotted contour lines) identify regions where the electron density is locally depleted. In all cases, we see strong electron density accumulation between adjacent carbon atoms, corresponding to covalent bonding, and depletion of electron density within the carbon ring, corresponding to noncovalent or coordinate bonding between the transition M atom and the ring. This is also borne out by the ELF plots, whose contours are shown in Figure 23.15. High contours of ELF (colored red in the figure) identify regions where electrons are strongly localized.

Contours of $L(\rho)$ for $C_{10}H_2Sc$, $C_{10}H_2Ti$, $C_{10}H_2V$, and $C_{10}H_2Cr$ are plotted in Figure 23.16, in a plane parallel to the carbon atoms, while ELF contours for the same systems are shown in Figure 23.17. Again, in all cases, we see strong electron density accumulation (as evidenced by positive Laplacian contours and high ELF contours), corresponding to covalent bonding between adjacent carbon atoms and depletion of electron density within the carbon ring (as evidenced

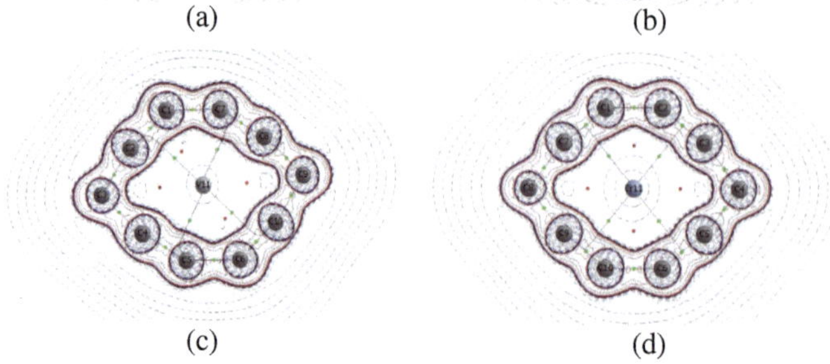

Figure 23.14 Contours of the Laplacian of electron density $L(\rho)$ for $C_{10}M$, where M is (a) Sc, (b) Ti, (c) V and (d) Cr in the plane of the carbon atoms. Blue dotted contours depict regions of depleting electron density, while red continuous contours represent regions of increasing electron density. Green dots identify the locations of bond critical points, while red dots identify the locations of ring critical points.

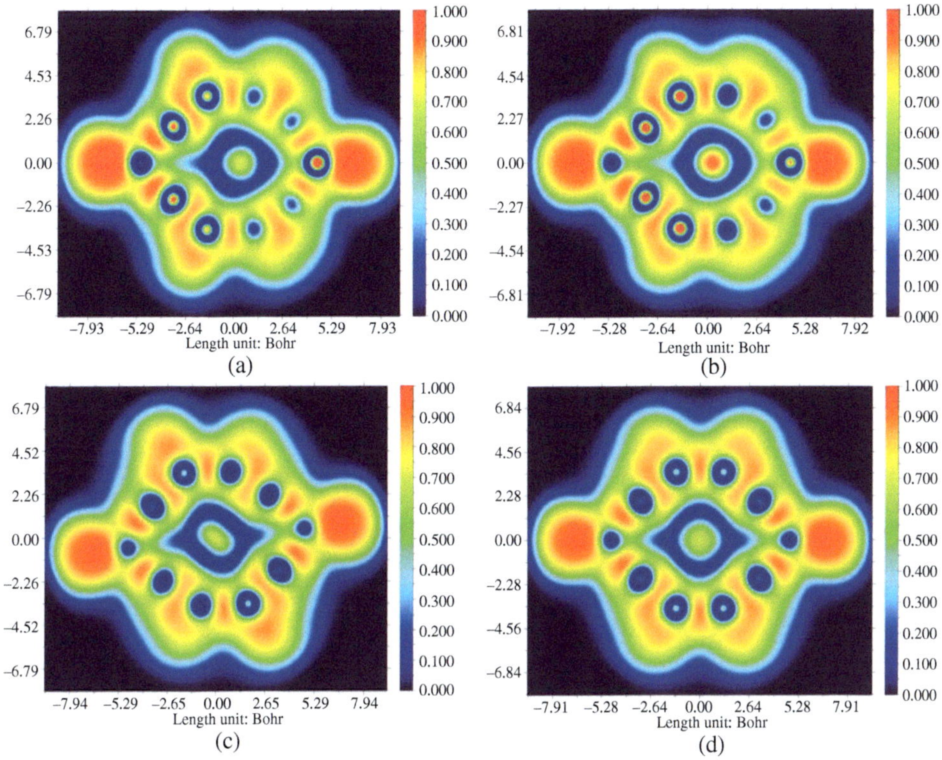

Figure 23.15 Contours of the electron localization function (ELF) for $C_{10}M$, where M is (a) Sc, (b) Ti, (c) V, and (d) Cr in the plane of the carbon atoms. Numbers along the left *y* axis are a measure of distance in Å, and the scale on the right describes the degree of electron density localization, with blue (≈ 0) being the least localization and red (≈ 1) being the maximum localization.

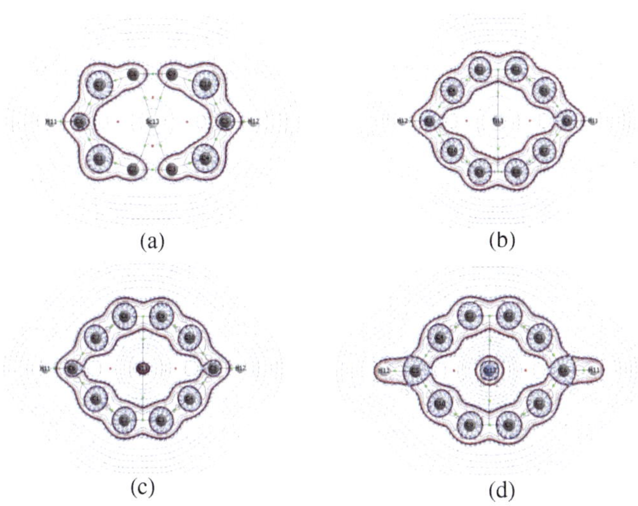

Figure 23.16 Contours of the Laplacian of electron density $L(\rho)$ for $C_{10}H_2M$, where M is (a) Sc, (b) Ti, (c) V and (d) Cr in a plane parallel to the carbon atoms. Blue dotted contours depict regions of depleting electron density, while red continuous contours represent regions of increasing electron density. Green dots identify the locations of bond critical points, while red dots identify the locations of ring critical points.

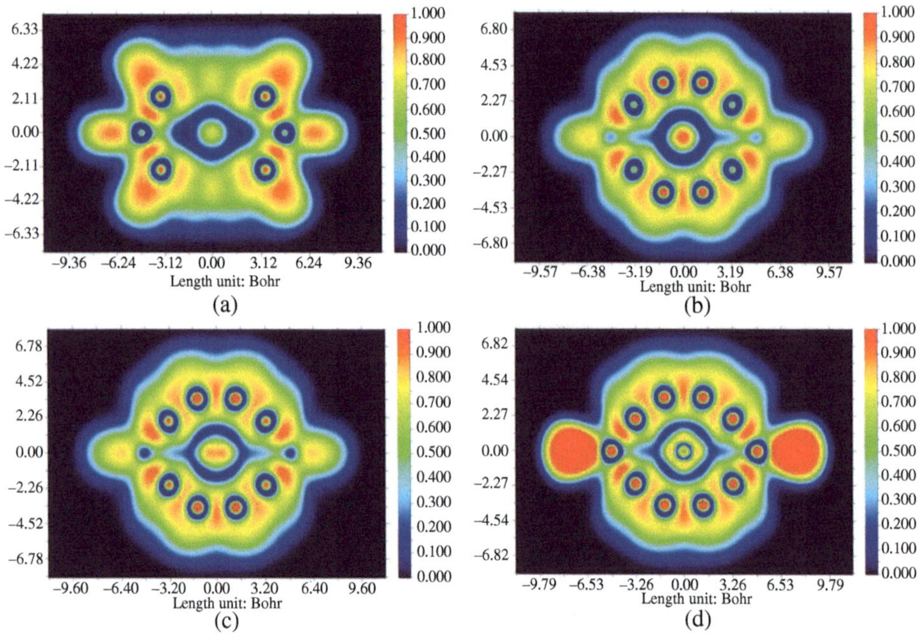

Figure 23.17 Contours of the electron localization function (ELF) for $C_{10}H_2M$, where M is (a) Sc, (b) Ti, (c) V and (d) Cr in a plane parallel to the carbon atoms. Numbers along the left *y* axis are a measure of distance in Å and the scale on the right describes the degree of electron density localization with blue (≈ 0) being the least localization and red (≈ 1), being the maximum localization.

by negative Laplacian contours and low ELF contours), corresponding to noncovalent bonding between the transition M atom and the ring.

The hydrogen atoms reduce the number of delocalizable electrons from the atomic basin of the carbons they bond to, thereby changing the electron density at the center of the ring and in the ring π cloud. As a result, the valence electron density of the $C_{10}H_2$ ring comprises an 18-electron system, as opposed to the 20 valence electrons in C_{10}. The electron density at the M-C BCP, ρ(BCP) increases steadily, and the Laplacian at the M-C BCP, L(BCP) becomes steadily more negative in the series Sc, Ti, V, Cr, Fe. The amount of electron density delocalized between the M atom and the carbon ring also steadily increases along the series, as seen in Table 23.4.

Figure 23.18 displays Laplacian contours for $C_{12}H_3Sc$, $C_{12}H_3Ti$, $C_{12}H_3V$, and $C_{12}H_3Cr$ that are plotted in the plane of the carbon atoms, while ELF contours for the same systems are shown in Figure 23.19. As before, in all cases, the Laplacian and ELF distributions indicate strong electron density accumulation or covalent bonding, between adjacent carbon atoms, and depletion of electron density within the carbon ring, corresponding to noncovalent or coordinate bonding between the transition M atom and the ring. $C_{12}H_3V$ has the most negative Laplacian L(BCP) at the M–C BCPs and the highest M—C bond ellipticity in the series $C_{12}H_3M$ with Sc, Ti, V, and Cr. $C_{12}H_3Cr$ has the highest electron density ρ(BCP) at the M–C BCPs and the lowest M—C bond ellipticity, as seen in Table 23.5.

The MO energy level diagram for $C_{12}H_3Sc$ is shown in Figure 23.20. In the odd-electron system $C_{12}H_3$, the uneven occupancy of α and β spin-orbital energy levels results in Jahn–Teller distortion of the ring, which is relieved in $C_{12}H_3Sc$ by the addition of an electron to the ring, resulting in a planar molecule, with uniform C—C bond lengths and $4n + 2$ π-electrons. In $C_{12}H_3Cr$ again, the ring is slightly distorted.

Table 23.4 Electron density ρ(BCP) and its Laplacian L(BCP) at the M−C bond critical points, bond ellipticity and delocalization in the M−C bonds for $C_{10}H_2M$ in the series M = Sc, Ti, V, Cr, Fe.

	$\rho(r)$						$L(r)$					
$C_{10}H_2M$	$C_{10}H_2$	Sc	Ti	V	Cr	Fe	$C_{10}H_2$	Sc	Ti	V	Cr	Fe
M—C BCP		0.07	0.089	0.104	0.107	0.121		−0.062	−0.064	−0.071	−0.083	−0.101
C—C BCP	0.344	0.312	0.312	0.309	0.287	0.314	0.121	0.199	0.195	0.189	0.16	0.196
C—H BCP	0.278	0.282	0.276	0.276	0.277	0.278	0.03	0.243	0.233	0.233	0.234	0.237
RCP	0.004	0.055	0.051	0.057	0.058	0.063		−0.051	−0.031	−0.036	−0.038	−0.036

	Ellipticity						Delocalization index						
$C_{10}H_2M$	$C_{10}H_2$	Sc	Ti	V	Cr	Fe		$C_{10}H_2$	Sc	Ti	V	Cr	Fe
M—C BCP	0.249	8.017	9.048	16.69	22.126	21.668	**Total**	21.526	22.902	23.392	23.697	23.631	24.312
C—C BCP	0.239	0.075	0.127	0.148	2.134	0.151	**C–DI**	1.588	1.560	1.580	1.570	1.570	1.56
C—H BCP	−0.004	0.045	0.055	0.061	0.061	0.053	**M–DI**		0.342	0.394	0.463	0.446	0.585

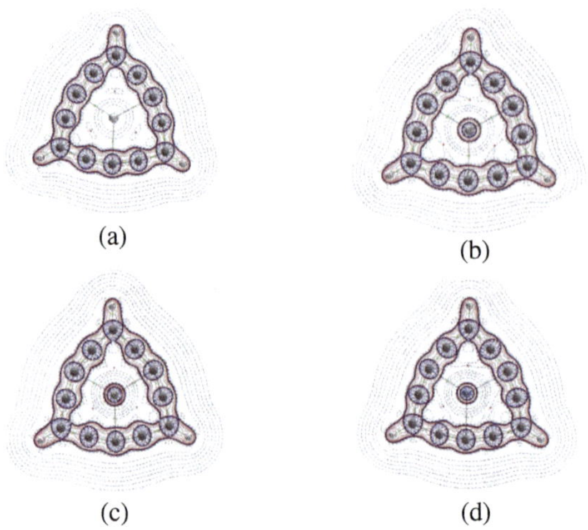

Figure 23.18 Contours of the Laplacian of electron density $L(\rho)$ for $C_{12}H_3M$, where M is (a) Sc, (b) Ti, (c) V and (d) Cr in the plane of the carbon atoms. Blue dotted contours depict regions of depleting electron density, while red continuous contours represent regions of increasing electron density. Green dots identify the locations of bond critical points, while red dots identify the locations of ring critical points.

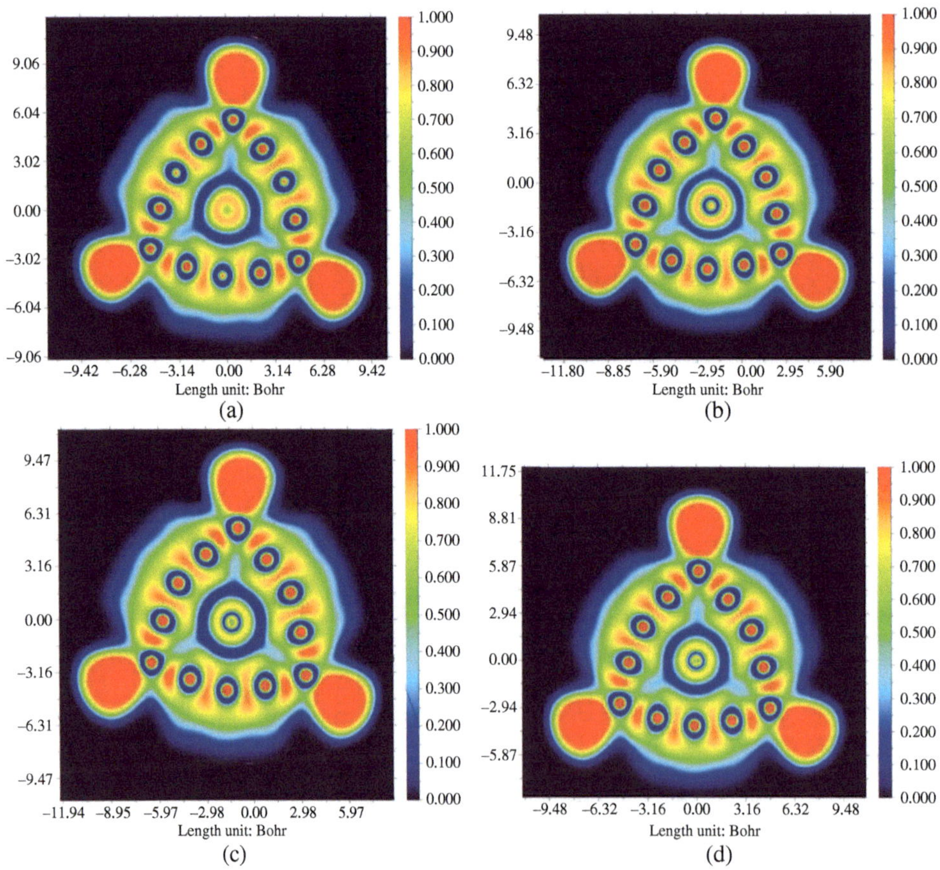

Figure 23.19 Contours of the electron localization function (ELF) for $C_{12}H_3M$, where M is (a) Sc, (b) Ti, (c) V and (d) Cr in the plane of the carbon atoms. Numbers along the left *y* axis is a measure of distance in Å and the scale on the right describes the degree of electron density localization with blue (≈ 0) being the least localization and red (≈ 1), being the maximum localization.

Table 23.5 Electron density ρ(BCP) and its Laplacian L(BCP) at the M−C bond critical points, bond ellipticity and delocalization in the M−C bonds for $C_{12}H_3M$ in the series M = Sc, Ti, V, Cr, Fe.

	$\rho(r)$				$L(r)$				Ellipticity			
$C_{12}H_3M$	Sc	Ti	V	Cr	Sc	Ti	V	Cr	Sc	Ti	V	Cr
M—C1 (BCP)	0.0614	0.0702	0.0641	0.0707	−0.0518	−0.0543	−0.064	−0.0599	5.2627	5.8557	6.0435	3.0926
C—H (BCP)	0.276	0.2764	0.2763	0.2762	0.2344	0.2357	0.2352	0.2357	0.0422	0.0415	0.0459	0.0409
M—C3 (RCP)	0.0385	0.0441	0.0428	0.0439	−0.0213	−0.0215	−0.0274	−0.021				

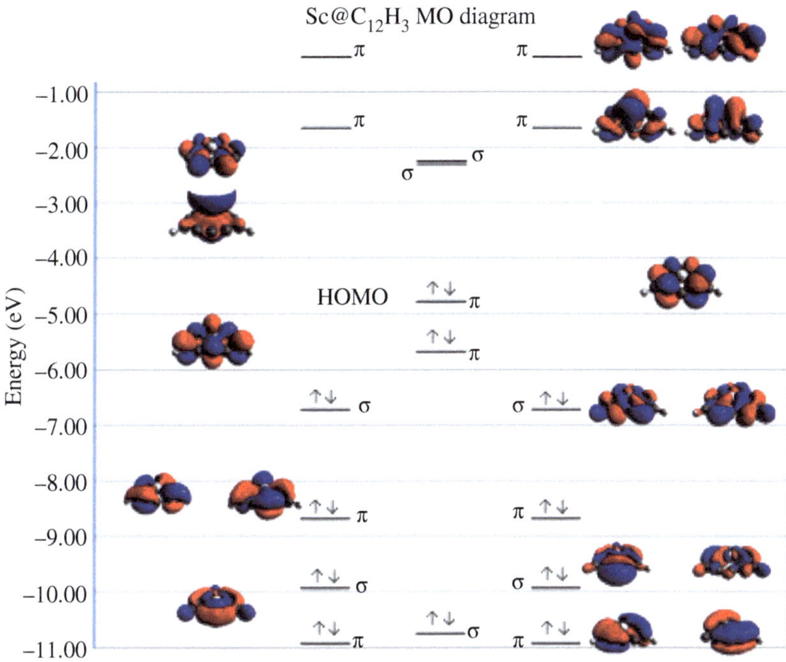

Figure 23.20 Frontier molecular orbitals and MO energy level diagram for $C_{12}H_3Sc$.

We have also computed the localization–delocalization indices, as defined in AIM theory, for all the systems. It is noteworthy that the delocalization indices for all the complexes studied fall on a single curve (Figure 23.21), which is differentiated into two regions – the C—C bonds lying at shorter internuclear distances and larger values of the delocalization index (corresponding to covalent bonding), and the carbon–metal bonds lying at longer internuclear distances and smaller values of the delocalization index. These data reinforce the trends seen in the Laplacian and ELF distributions.

In Table 23.6, we have collected the AIM atomic charges, average localization and delocalization indices on the carbons, and the delocalization indices on the M atom. Note that these delocalization indices count half the number of electron pairs shared between carbon atoms and between the atomic basins of carbon and the M atom, respectively.

Table 23.6 Average AIM atomic charges on carbon q(C), average localization indices LI(C) and delocalization DI(C) indices on the carbons, and the delocalization indices on the metal DI(M) for complexes of a transition metal with C_9H_3, C_{10}, $C_{10}H_2$, and $C_{12}H_3$ rings.

	q(C)	LI(C)	DI(C)	DI(M)
C_9H_3	−0.041	4.015	2.025	
C_9H_3Sc	−0.182	4.131	2.05	1.347
C_9H_3Ti	−0.184	4.134	2.05	1.588
C_9H_3V	−0.17	4.116	2.053	2.168
C_9H_3Cr	−0.158	4.106	2.052	1.882
C_9H_3Mn	−0.161	4.103	1.56	1.168
C_9H_3Fe	−0.133	4.073	2.06	2.016
C_{10}	0	3.986	2.014	
$C_{10}Sc$	−0.164	4.154	2.01	1.346
$C_{10}Ti$	−0.142	4.127	2.015	1.573
$C_{10}V$	−0.12	4.108	2.011	1.386
$C_{10}Cr$	−0.114	4.097	2.017	1.349
$C_{10}Mn$	−0.127	4.109	2.018	1.037
$C_{10}Fe$	−0.086	4.067	2.02	1.131
$C_{10}H_2$	−0.025	3.977	2.048	
$C_{10}H_2Sc$	−0.18	4.136	2.044	1.393
$C_{10}H_2Ti$	−0.152	4.082	2.069	1.638
$C_{10}H_2V$	−0.144	4.072	2.072	1.915
$C_{10}H_2Cr$	−0.132	4.062	2.069	1.874
$C_{10}H_2Fe$	−0.09	4.004	2.085	2.408
$C_{12}H_3Sc$	−0.168	4.089	2.079	1.284
$C_{12}H_3Ti$	−0.151	4.071	2.08	1.489
$C_{12}H_3V$	−0.114	4.042	2.072	1.462
$C_{12}H_3Cr$	−0.128	4.053	2.074	1.605
$C_{12}H_3Mn(III)$	−0.125	4.083	2.042	1.769
$C_{12}H_3Mn(V)$	−0.125	4.092	2.033	1.464
$C_{12}H_3Fe$	−0.112	4.069	2.044	1.651
$C_{12}H_3Co$	−0.094	4.069	2.025	1.696
$C_{12}H_3Ni$	−0.08	4.064	2.016	1.665
$C_{12}H_3Cu$	−0.085	4.084	2	1.19
$C_{12}H_3Zn$	−0.105	4.102	2.003	1.052

The numbers within $C_{12}H_3Mn$ parentheses refer to two different spin multiplicity states.

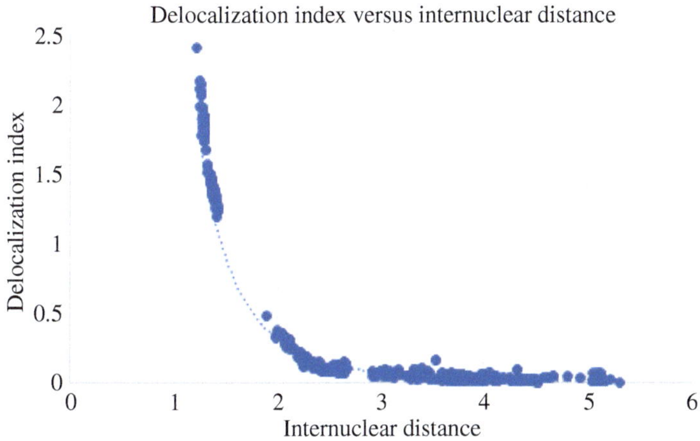

Figure 23.21 Delocalization Index plotted against internuclear distance (Å) for all bonds in the carbon clusters studied.

23.4 Conclusions

In summary, we have presented recent developments on metal-encapsulated boron and carbon clusters. We discussed in detail the results of a systematic study on M = Cr, Mo, and W encapsulation in boron clusters in the size range of 18–24 boron atoms and of transition metal encapsulation in monocyclic carbon rings of size 9–12 carbon atoms. Our results show that by M encapsulation, it is possible to have fullerene-like cage structures of boron with about 20 atoms, in contrast to dominantly planar, quasi-planar, or tubular structures of pure boron clusters for less than 40 atoms. These results suggest that encapsulation of Cr is suitable to produce symmetric small cage clusters of boron with B_{20}, whereas B_{22} is the smallest symmetric cage for Mo and W encapsulation, which is a magic cluster and is likely to be produced in high abundance in experiments. A symmetric B_{24} cage is also formed with Mo and W encapsulation, but the variation in the binding energy of the clusters suggests that the B_{22} cage has the optimal size for Mo and W encapsulation. There is a large gain in energy when a Mo or W atom is encapsulated in the cage, which also supports the strong stability of the encapsulated clusters. An analysis of the bonding nature in these clusters shows that the cage structures are stabilized by strong interaction between the M atom d orbitals and the π-bonded MOs of the bare boron cage. From this analysis, it is found that the stability of the Cr@B_{18}, Cr@B_{20}, M@B_{22}, M = Cr, Mo and W, and M@B_{24} clusters is associated with 18 π-bonded valence electrons, while for M@B_{18} (M = Mo and W) disk-shaped tubular clusters, the stability is associated with 20 π bonded valence electrons.

C_{10} is a stable ring with D_{5h} symmetry and a large HOMO–LUMO gap, due to the π-aromaticity of 10 π electrons. C_{12} has sixfold symmetry, with alternating C—C bond lengths and a much smaller HOMO–LUMO gap. The hydrogenated clusters C_9H_3, $C_{10}H_2$, and $C_{12}H_3$ prefer chain structures, but encapsulation of a metal atom such as Cr or Sc favors the ring structures. Attaching 3 H atoms to the 24 π-electron C_{12} ring gives a threefold symmetric $C_{12}H_3$ ring, reducing the number of available π electrons to 21; the stability can be regained by encapsulation of a suitable transition metal atom.

An analysis of the electron density distributions in the $C_{10}M$, $C_{10}H_2M$, and $C_{12}H_3M$ clusters, with M = Sc, Ti, V, and Cr, shows that in all cases the Laplacian and ELF distributions display strong electron density accumulation, indicating covalent bonding, between adjacent carbon atoms, and depletion of electron density within the carbon ring, corresponding to noncovalent or coordinate bonding between the transition metal atom and the ring. Delocalization indices for all the complexes studied fall on a single curve, with C—C bonds lying at shorter internuclear distances and larger values of the delocalization index (corresponding to covalent bonding) and carbon–metal bonds lying at longer internuclear distances and smaller values of the delocalization index, suggesting a similar nature of bonding in these systems.

These studies show how different electron density indices – Laplacian distributions, electron localization function, electron localization and delocalization indices, adaptive natural density partitioning and canonical molecular orbital theory – can complement each other and shed light on the nature of chemical bonding between transition metal atoms and carbon or boron clusters. We hope that our results will stimulate experimental work on these metal-encapsulated structures of boron and carbon.

Acknowledgments

We gratefully acknowledge the use of the high-performance computing facility *Magus* of Shiv Nadar University, where a part of the calculations has been performed. This work was done when ABR was at the Dr. Vijay Kumar Foundation. ABR and VK thankfully acknowledge financial support from the International Technology Center – Pacific. We thank Prof. Cherif Matta for providing access to AIMLDM software.

References

1 Novoselov, K.S., Geim, A.K., Morozov, S.V. et al. (2004). Electric field effect in atomically thin carbon films. *Science* 306: 666–669. https://doi.org/10.1126/science.1102896.
2 Mannix, A.J., Zhou, X.-F., Kiraly, B. et al. (2015). Synthesis of borophenes: anisotropic, two-dimensional boron polymorphs. *Science* 350 (6267): 1513–1516. https://doi.org/10.1126/science.aad1080.
3 Rahane, A.B. and Kumar, V. (2015). B_{84}: a quasi-planar boron cluster stabilized with hexagonal holes. *Nanoscale* 7: 4055. https://doi.org/10.1039/C4NR06026A.
4 Kroto, H.W., Heath, J.R., Obrien, S.C. et al. (1985). C_{60}: Buckminsterfullerene. *Nature* 318 (6042): 162–163. https://doi.org/10.1038/318162a0.
5 Curl, R.F. and Smalley, R.E. (1991). Fullerenes. *Sci. Am.* 265 (4): 54–63.
6 Zhai, H.-J., Zhao, Y.-F., Li, W.-L. et al. (2014). Observation of an all-boron fullerene. *Nat. Chem.* 6: 727–731. https://doi.org/10.1038/nchem.1999.
7 Osawa, E. (1970). Superaromaticity. *Kagaku* 25: 854–863.
8 Bochvar, D.A. and Galpern, E.G. (1973). On hypothetical systems: carbon dodecahedron, S-icosahedron and carbon-S-icosahedron. *Dokl. Akad. Nauk SSSR* 209: 610.
9 Thrower, P.A. (1999). Editorial. *Carbon* 37 (11): 1677–1678.
10 Kroto, H. (ed.) (1992). Special issue on fullerenes. *Carbon* 30 (8): 1139–1286.
11 Krätschmer, W., Lamb, L.D., Fostiropoulos, K., and Huffman, D.R. (1990). Solid C_{60}: a new form of carbon. *Nature* 347: 354–358. https://doi.org/10.1038/347354a0.
12 Dunk, P.W., Kaiser, N.K., Mulet-Gas, M. et al. (2012). The smallest stable fullerene, $M@C_{28}$ (M = Ti, Zr, U): stabilization and growth from carbon vapor. *J. Am. Chem. Soc.* 134: 9380–9389. https://doi.org/10.1021/ja302398h.

13 Prinzbach, H., Weiler, A., Landenberger, P. et al. (2000). Gas-phase production and photoelectron spectroscopy of the smallest fullerene, C_{20}. *Nature* 407: 60–63. https://doi.org/10.1038/35024037.

14 Iqbal, Z., Zhang, Y., Grebel, H. et al. (2003). Evidence for a solid phase of dodecahedral C_{20}. *Eur. Phys. J. B* 31: 509–515. https://doi.org/10.1140/epjb/e2003-00060-4.

15 Zhao, J., Du, Q., Zhou, S., and Kumar, V. (2020). Endohedrally doped cage clusters. *Chem. Rev.* 120 (17): 9021–9163. https://doi.org/10.1021/acs.chemrev.9b00651.

16 Wu, X., Sai, L., Zhou, S. et al. (2020). Competition between tubular, planar and cage geometries: a complete picture of structural evolution of B_n (n = 31–50) clusters. *Phys. Chem. Chem. Phys.* 22: 12959–12966. https://doi.org/10.1039/d0cp01256d.

17 Zhai, H.J., Alexandrova, A.N., Birch, K.A. et al. (2003). Hepta- and octacoordinate boron in molecular wheels of eight- and nine-atom boron clusters: observation and confirmation. *Angew. Chem. Int. Ed.* 42: 6004–6008. https://doi.org/10.1002/anie.200351874.

18 Zhai, H.J., Kiran, B., Li, J., and Wang, L.S. (2003). Hydrocarbon analogues of boron clusters—planarity, aromaticity and antiaromaticity. *Nat. Mater.* 2: 827–833. https://doi.org/10.1038/nmat1012.

19 Kiran, B., Bulusu, S., Zhai, H.-J. et al. (2005). Planar-to-tubular structural transition in boron clusters: B_{20} as the embryo of single-walled boron nanotubes. *PNAS* 102: 961–964. https://doi.org/10.1073/pnas.0408132102.

20 Lau, K.C., Deshpande, M., Pati, R., and Pandey, R. (2005). A theoretical study of electronic and vibrational properties of neutral, cationic, and anionic B_{24} cluster. *Int. J. Quantum Chem.* 102: 656–664. https://doi.org/10.1002/qua.20537.

21 Huang, W., Sergeeva, A.P., Zhai, H.-J. et al. (2010). A concentric planar doubly π-aromatic B_{19}^- cluster. *Nat. Chem.* 2: 202–206. https://doi.org/10.1038/nchem.534.

22 Popov, I.A., Piazza, Z.A., Li, W.L. et al. (2013). A combined photoelectron spectroscopy and ab initio study of the quasi-planar B_{24}^- cluster. *J. Chem. Phys.* 139: 144307. https://doi.org/10.1063/1.4824156.

23 Oger, E., Crawford, N.R.M., Kelting, R. et al. (2007). Boron cluster cations: transition from planar to cylindrical structures. *Angew. Chem. Int. Ed.* 46: 8503–8506. https://doi.org/10.1002/anie.200701915.

24 Cheng, L. (2012). B_{14}: an all-boron fullerene. *J. Chem. Phys.* 136: 104301–104304. https://doi.org/10.1063/1.3692183.

25 Zhao, J., Huang, X., Shi, R. et al. (2015). B_{28}: the smallest all-boron cage from an ab initio global search. *Nanoscale* 7: 15086–15090. https://doi.org/10.1039/C5NR04034E.

26 Chacko, S., Kanhere, D.G., and Boustani, I. (2003). Ab initio density functional investigation of B_{24} clusters: rings, tubes, planes, and cages. *Phys. Rev. B* 68: 035414. https://doi.org/10.1103/PhysRevB.68.035414.

27 Lv, J., Wang, Y., Zhua, L., and Ma, Y. (2014). B_{38}: an all-boron fullerene analogue. *Nanoscale* 6: 11692–11696. https://doi.org/10.1039/C4NR01846J.

28 Chen, Q., Li, W.-L., Zhao, Y.-F. et al. (2015). Experimental and theoretical evidence of an axially chiral borospherene. *ACS Nano* 9: 754–760. https://doi.org/10.1021/nn506262c.

29 Kumar, V. and Kawazoe, Y. (2001). Metal-encapsulated fullerenelike and cubic caged clusters of silicon. *Phys. Rev. Lett.* 87: 045503. https://doi.org/10.1103/PhysRevLett.87.045503.

30 Kumar, V. and Kawazoe, Y. (2002). Magic behavior of $Si_{15}M$ and $Si_{16}M$ (M = Cr, Mo, and W) clusters. *Phys. Rev. B* 65: 073404. https://doi.org/10.1103/PhysRevB.65.073404.

31 Kumar, V. and Kawazoe, Y. (2002). Metal-encapsulated icosahedral superatoms of germanium and tin with large gaps: Zn@Ge_{12} and Cd@Sn_{12}. *Appl. Phys. Lett.* 80: 859. https://doi.org/10.1063/1.1447315.

32 Kumar, V. and Kawazoe, Y. (2002). Metal-encapsulated caged clusters of germanium with large gaps and different growth behavior than silicon. *Phys. Rev. Lett.* 88: 235504. https://doi.org/10.1103/PhysRevLett.88.235504.

33 Kumar, V. (2003). Novel metal-encapsulated caged clusters of silicon and germanium. *Eur. Phys. J. D* 24: 227. https://doi.org/10.1140/epjd/e2003-00194-5.

34 Kumar, V. (2006). Alchemy at the nanoscale: magic heteroatom clusters and assemblies. *Comput. Mater. Sci.* 36 (1–2): 1–11. https://doi.org/10.1016/j.commatsci.2005.06.004.

35 Kumar, V. and Kawazoe, Y. (2003). Metal-doped magic clusters of Si, Ge, and Sn: the finding of a magnetic superatom. *Appl. Phys. Lett.* 83: 2677. https://doi.org/10.1063/1.1609661.

36 Kumar, V. and Kawazoe, Y. (2007). Hydrogenated caged clusters of Si, Ge, and Sn and their endohedral doping with atoms: ab initio calculations. *Phys. Rev. B* 75: 155425. https://doi.org/10.1103/PhysRevB.75.155425.

37 Kumar, V. (2008). Metal encapsulated clusters of silicon: silicon fullerenes and other polyhedral forms. In: *Nanosilicon* (ed. V. Kumar), 115–148. Elsevier.

38 Rahane, A.B., Saha, P., Sukumar, N., and Kumar, V. (2019). Smallest fullerene-like structures of boron with Cr, Mo, and W encapsulation. https://arxiv.org/abs/1907.12611 (accessed 29 Jul 2019).

39 Romanescu, C., Galeev, T.R., Li, W.L. et al. (2011). Aromatic metal-centered monocyclic boron rings: Co©B_8^- and Ru©B_9^-. *Angew. Chem. Int. Ed.* 50: 9334–9337. https://doi.org/10.1002/anie.201104166.

40 Li, W.L., Romanescu, C., Galeev, T.R. et al. (2012). Transition-metal-centered nine-membered boron rings: M©B_9 and M©B_9^- (M = Rh, Ir). *J. Am. Chem. Soc.* 134: 165–168. https://doi.org/10.1021/ja209808k.

41 Romanescu, C., Galeev, T.R., Sergeeva, A.P. et al. (2012). Experimental and computational evidence of octa- and nona-coordinated planar iron-doped boron clusters: Fe©B_8^- and Fe©B_9^-. *J. Organomet. Chem.* 721: 148–154. https://doi.org/10.1016/j.jorganchem.2012.07.050.

42 Galeev, T.R., Romanescu, C., Li, W.L. et al. (2012). Observation of the highest coordination number in planar species: decacoordinated Ta©B_{10}^- and Nb©B_{10}^- anions. *Angew. Chem. Int. Ed.* 51: 2101–2105. https://doi.org/10.1002/anie.201107880.

43 Romanescu, C., Galeev, T.R., Li, W.L. et al. (2013). Geometric and electronic factors in the rational design of transition-metal-centered boron molecular wheels. *J. Chem. Phys.* 138: 134315. https://doi.org/10.1063/1.4798935.

44 Romanescu, C., Galeev, T.R., Li, W.L. et al. (2013). Transition-metal-centered monocyclic boron wheel clusters (M©Bn): a new class of aromatic borometallic compounds. *Acc. Chem. Res.* 46: 350–358. https://doi.org/10.1021/ar300149a.

45 Saha, P., Rahane, A.B., Kumar, V., and Sukumar, N. (2016). Analysis of the electron density features of small boron ring clusters and the effects of doping. *Phys. Scr.* 91: 053005. https://doi.org/10.1088/0031-8949/91/5/053005.

46 Amin, D. (1987). Icosahedral boron-rich solids. *Phys. Today* 40: 55. https://doi.org/10.1063/1.881112.

47 Wade, K. (1971). The structural significance of the number of skeletal bonding electron-pairs in carboranes, the higher boranes and borane anions, and various transition-metal carbonyl cluster compounds. *J. Chem. Soc. D* 15: 792–793. https://doi.org/10.1039/C29710000792.

48 Koyasu, K., Akutsu, M., Mitsui, M., and Nakajima, A. (2005). Selective formation of MSi_{16} (M = Sc, Ti, and V). *J. Am. Chem. Soc.* 127: 4998–4999. https://doi.org/10.1021/ja045380t.

49 Du, Q., Zhen, W., Zhou, S. et al. (2021). Searching for cluster Lego blocks for three-dimensional and two-dimensional assemblies. *Phys. Rev. Mater.* 5: 066001. https://doi.org/10.1103/PhysRevMaterials.5.066001.

50 Kumar, V. (2021). Endohedrally doped superatoms and assemblies. In: *Superatoms: Principles, Synthesis and Applications* (ed. P. Jena and Q. Sun), 85–127. Wiley.

51 Jian, T., Chen, X., Li, S.-D. et al. (2019). Probing the structures and bonding of size-selected boron and doped-boron clusters. *Chem. Soc. Rev.* 48: 3550–3591. https://doi.org/10.1039/C9CS00233B.

52 Li, S.-D., Miao, C.-Q., Guo, J.-C., and Ren, G.-M. (2006). Transition metal-boron complexes BnM: from bowls (n = 8–14) to tires (n = 14). *J. Comput. Chem.* 27: 1858. https://doi.org/10.1002/jcc.20497.

53 Böyükata, M. and Güvenç, Z.B. (2011). Density functional study of AlB_n clusters for n = 1–14. *J. Alloys Compd.* 509: 4214. https://doi.org/10.1016/j.jallcom.2011.01.053.

54 Gu, J.-B., Yang, X.-D., Wang, H.-Q., and Li, H.-F. (2012). Structural, electronic, and magnetic properties of boron cluster anions doped with aluminum: B_nAl^- ($2 \leq n \leq 9$). *Chin. Phys. B* 21 (4): 043102. https://doi.org/10.1088/1674-1056/21/4/043102.

55 Pham, H.T. and Nguyen, M.T. (2015). Effects of bimetallic doping on small cyclic and tubular boron clusters: B_7M_2 and $B_{14}M_2$ structures with M = Fe, Co. *Phys. Chem. Chem. Phys.* 17: 17335–17345. https://doi.org/10.1039/C5CP01650A.

56 Saha, P., Rahane, A.B., Kumar, V., and Sukumar, N. (2017). Electronic origin of the stability of transition metal doped B_{14} drum shaped boron clusters and their assembly into a nanotube. *J. Phys. Chem. C* 121: 10728. https://doi.org/10.1021/acs.jpcc.6b10838.

57 Popov, I.A., Jian, T., Lopez, G.V. et al. (2015). Cobalt-centred boron molecular drums with the highest coordination number in the CoB_{16}^- cluster. *Nat. Commun.* 6: 8654. https://doi.org/10.1038/ncomms9654.

58 Xu, Q., Liu, C., Yang, L. et al. (2016). Computational investigation on MB_n (M = Li-Cs, Be-Ba, Sc-La and Ti; n = 28 and 38). *J. Mol. Model.* 22: 184. https://doi.org/10.1007/s00894-016-3055-4.

59 Wang, C.-Z., Bo, T., Lan, J.-H. et al. (2018). Ultrastable actinide endohedral borospherenes. *Chem. Commun.* 54: 2248–2251. https://doi.org/10.1039/C7CC09837E.

60 Chen, Q., Li, H.-R., Tian, W.-J. et al. (2016). Endohedral charge-transfer complex $Ca@B_{37}^-$: stabilization of a B_{37}^{3-} borospherene trianion by metal-encapsulation. *Phys. Chem. Chem. Phys.* 18: 14186–14190. https://doi.org/10.1039/C6CP02369J.

61 Lu, Q.L., Luo, Q.Q., De Li, Y., and Huang, S.G. (2015). DFT study on endohedral and exohedral B_{38} fullerenes: $M@B_{38}$ (M = Sc, Y, Ti) and $M\&B_{38}$ (M = Nb, Fe, Co, Ni). *Phys. Chem. Chem. Phys.* 17: 20897–20902. https://doi.org/10.1039/C5CP03378K.

62 Chen, Q., Li, H.-R., Miao, C.-Q. et al. (2016). Endohedral $Ca@B_{38}$: stabilization of a B_{38}^{2-} borospherene dianion by metal encapsulation. *Phys. Chem. Chem. Phys.* 18: 11610–11615. https://doi.org/10.1039/C5CP06169E.

63 Chen, Q., Gao, T., Tian, W.-J. et al. (2015). Endohedral C_3 $Ca@B_{39}^+$ and C_2 $Ca@B_{39}^+$: axially chiral metalloborospherenes based on B_{39}^-. *Phys. Chem. Chem. Phys.* 17: 19690–19694. https://doi.org/10.1039/C5CP03178H.

64 Stasyuk, A.J. and Sola, M. (2017). Does the endohedral borospherene supersalt $FLi_2@B_{39}$ maintain the super properties of its subunits? *Phys. Chem. Chem. Phys.* 19: 21276–21281. https://doi.org/10.1039/C7CP02550E.

65 Yang, L., Jin, P., Hou, Q., and Li, L. (2016). Endohedral metalloborofullerenes M@B_{44} (M = Ca, Sr, Ba): a computational investigation. *J. Mol. Model.* 22: 297. https://doi.org/10.1007/s00894-016-3170-2.

66 Zhao, R.-N., Yuan, Y.-H., Han, J.-G., and Duan, Y. (2016). Geometries, stabilities, and electronic properties of tungsten encapsulated nanosize irregular B_n (n = 20, 24, 28, and 32) fullerenes: a density functional investigation. *Chem. Phys. Lett.* 648: 41–46. https://doi.org/10.1016/j.cplett.2016.01.052.

67 Bai, H., Chen, Q., Zhai, H.J. et al. (2015). Exohedral metalloborospherenes: M@B_{40} (M = Ca, Sr) and M&B_{40} (M = Be, Mg). *Angew. Chem. Int. Ed.* 54: 941–945. https://doi.org/10.1002/anie.201408738.

68 Jin, P., Hou, Q., Tang, C., and Chen, Z. (2015). Computational investigation on the endohedral borofullerenes M@B_{40} (M = Sc, Y, La). *Theor. Chem. Acc.* 134: 13. https://doi.org/10.1007/s00214-014-1612-4.

69 Fa, W., Chen, S., Pande, S., and Zeng, X.C. (2015). Stability of metal-encapsulating boron fullerene B_{40}. *J. Phys. Chem. A* 119: 11208–11214. https://doi.org/10.1021/acs.jpca.5b07173.

70 Shah, E.V. and Roy, D.R. (2016). Sc_3N and Sc_2C_2 encapsulated B_{40}: smarter than its carbon analogue. *Phys. E* 84: 354–360. https://doi.org/10.1016/j.physe.2016.08.002.

71 Shakerzadeh, E., Biglari, Z., and Tahmasebi, E. (2016). M@B_{40} (M = Li, Na, K) serving as a potential promising novel NLO nanomaterial. *Chem. Phys. Lett.* 654: 76–80. https://doi.org/10.1016/j.cplett.2016.05.014.

72 Li, S.-X., Zhang, Z.-P., Long, Z.-W., and Qin, S.-J. (2017). Structures, stabilities and spectral properties of metalloborospherenes $MB^{0/-}{}_{40}$ (M = Cu, Ag, and Au). *RSC Adv.* 7: 38526–38537. https://doi.org/10.1039/C7RA05932A.

73 Jin, P., Yang, L., Liu, C. et al. (2017). Computational prediction of the endohedral metalloborofullerenes Ti_n@B_{40} (n = 1, 2). *Theor. Chem. Acc.* 136: 56. https://doi.org/10.1007/s00214-017-2087-x.

74 Xi, C., Yang, L., Liu, C. et al. (2018). Lanthanide metals in the boron cages: computational prediction of M@B_n (M = Eu, Gd; n = 38, 40). *Int. J. Quantum Chem.* 118: e25576. https://doi.org/10.1002/qua.25576.

75 Yong, Y., Su, X., Kuang, Y. et al. (2018). B_{40} and M@B_{40} (M = Li and Ba) fullerenes as potential molecular sensors for acetone detection: a first-principles study. *J. Mol. Liq.* 264: 1–8. https://doi.org/10.1016/j.molliq.2018.05.046.

76 Yu, T., Gao, Y., Xu, D., and Wang, Z. (2018). Actinide endohedral boron clusters: a closed-shell electronic structure of U@B_{40}. *Nano Res.* 11: 354–359. https://doi.org/10.1007/s12274-017-1637-9.

77 Chojecki, M., Yourdkhani, S., Rutkowska-Zbik, D., and Korona, T. (2018). Stability of endo- and exohedral complexes of all-boron fullerene B_{40}. *Comput. Theor. Chem.* 1133: 7–17. https://doi.org/10.1016/j.comptc.2018.04.007.

78 An, Y., Zhang, M., Wu, D. et al. (2016). Electronic transport properties of the first all-boron fullerene B_{40} and its metallofullerene Sr@B_{40}. *Phys. Chem. Chem. Phys.* 18: 12024–12028. https://doi.org/10.1039/C6CP01096B.

79 Wang, W., Guo, Y.-D., and Yan, X.-H. (2016). The spin-dependent transport of transition metal encapsulated B_{40} fullerene. *RSC Adv.* 6: 40155–40161. https://doi.org/10.1039/C6RA00179C.

80 Agarwal, R.; Sukumar, N.; Kumar, V.; *manuscript in preparation.*

81 Galbraith, J.M., Shaik, S., Danovich, D. et al. (2021). Valence bond and molecular orbital: two powerful theories that nicely complement one another. *J. Chem. Educ.* 98 (12): 3617–3620. https://doi.org/10.1021/acs.jchemed.1c00919.

82 Shaik, S., Danovich, D., and Hiberty, P.C. (2021). Valence bond theory—its birth, struggles with molecular orbital theory, its present state and future prospects. *Molecules* 26: 1624. https://doi.org/10.3390/molecules26061624.

83 Longuet-Higgins, H.C. and Roberts, M.D.V. (1954). The electronic structure of the borides MB_6. *Proc. R. Soc. A* 224: 336–347. https://doi.org/10.1098/rspa.1954.0162.

84 Lipscomb, W.N. (1973). Three-center bonds in electron-deficient compounds. Localized molecular orbital approach. *Acc. Chem. Res.* 6: 257–262. https://doi.org/10.1021/ar50068a001.

85 Wieczorek, C., Allwohn, J., Schmidt-Lukasch, G. et al. (1990). Structure of a non-classically bridged methyleneborane in the crystal. *Angew. Chem.* 29: 398–399. https://doi.org/10.1002/anie.199003981.

86 Reed, A.E. and Weinhold, F. (1983). Natural bond orbital analysis of near-Hartree–Fock water dimer. *J. Chem. Phys.* 78: 4066–4073. https://doi.org/10.1063/1.445134.

87 Shahbazian, S. (2018). Why bond critical points are not bond critical points. *Chem. Eur. J.* 24 (21): 5401–5405. https://doi.org/10.1002/chem.201705163.

88 Mierzwa, G., Gordon, A.J., and Berski, S. (2020). The nature of the triple BB, double B—B, single B—B, and one-electron, BB boron–boron bonds from the topological analysis of electron localisation function (ELF) perspective. *J. Mol. Struct.* 1221: 128530. https://doi.org/10.1016/j.molstruc.2020.128530.

89 Meng, F., Zhou, Z., Zhang, P. et al. (2015). Encapsulation of an f-block metal atom/ion to enhance the stability of C20 with the I_h symmetry. *Phys. Chem. Chem. Phys.* 17 (6): 4328–4336. https://doi.org/10.1039/C4CP03159H.

90 Handschuh, H., Ganteför, G., Kessler, B. et al. (1995). Stable configurations of carbon clusters: chains, rings, and fullerenes. *Phys. Rev. Lett.* 74: 1095. https://doi.org/10.1103/PhysRevLett.74.1095.

91 Perdew, J.P., Burke, K., and Ernzerhof, M. (1996). Generalized gradient approximation made simple. *Phys. Rev. Lett.* 77: 3865–3868. https://doi.org/10.1103/PhysRevLett.77.3865.

92 Blochl, P.E. (1994). Projector augmented-wave method. *Phys. Rev. B* 50: 17953. https://doi.org/10.1103/PhysRevB.50.17953.

93 Kresse, G. and Joubert, J. (1999). From ultrasoft pseudopotentials to the projector augmented-wave method. *Phys. Rev. B* 59: 1758. https://doi.org/10.1103/PhysRevB.59.1758.

94 Kresse, G. and Furthmuller, J. (1996). Efficient iterative schemes for ab initio total-energy calculations using a plane-wave basis set. *Phys. Rev. B* 54: 11169. https://doi.org/10.1103/PhysRevB.54.11169.

95 Frisch, M.J., Trucks, G.W., Schlegel, H.B. et al. (2013). *Gaussian 09, Revision D.01*. Gaussian, Inc.: Wallingford, CT.

96 Wachters, A.J.H. (1970). Gaussian basis set for molecular wavefunctions containing third-row atoms. *J. Chem. Phys.* 52: 1033. https://doi.org/10.1063/1.1673095.

97 Hay, P.J. (1977). Gaussian basis sets for molecular calculations. The representation of 3d orbitals in transition-metal atoms. *J. Chem. Phys.* 66: 4377–4384. https://doi.org/10.1063/1.433731.

98 Hay, P.J. and Wadt, W.R. (1985). Ab initio effective core potentials for molecular calculations. Potentials for the transition metal atoms Sc to Hg. *J. Chem. Phys.* 82: 270–283. https://doi.org/10.1063/1.448799.

99 Wadt, W.R. and Hay, P.J. (1985). Ab initio effective core potentials for molecular calculations. Potentials for main group elements Na to Bi. *J. Chem. Phys.* 82: 284–298. https://doi.org/10.1063/1.448800.

100 Dunning, T.H. Jr. (1989). Gaussian basis sets for use in correlated molecular calculations. I. The atoms boron through neon and hydrogen. *J. Chem. Phys.* 90: 1007–1023. https://doi.org/10.1063/1.456153.

101 Zubarev, D.Y. and Boldyrev, A.I. (2008). Developing paradigms of chemical bonding: adaptive natural density partitioning. *Phys. Chem. Chem. Phys.* 10: 5207–5217. https://doi.org/10.1039/B804083D.

102 Keith, T.A. (2012). AIMAll. TK Gristmill Software Overland Park KS, USA (Version 11.03.14). http://aim.tkgristmill.com.

103 Bader, R.F.W. (1990). *Atoms in Molecules*. New York: Wiley.

104 Becke, A.D. and Edgecombe, K.E. (1990). A simple measure of electron localization in atomic and molecular systems. *J. Chem. Phys.* 92: 5397–5403. https://doi.org/10.1063/1.458517.

105 Sumar, I., Cook, R., Ayers, P.W., and Matta, C.F. (2015). AIMLDM: a program to generate and manipulate electron localization-delocalization matrices (LDMs). *Comput. Theor. Chem.* 1070: 55–67.

106 Sukumar, N. (ed.) (2012). *A Matter of Density: Exploring the Electron Density Concept in Chemical, Biological, and Materials Sciences*. Hoboken, NJ: Wiley. ISBN: 978-0-470-76900-3.

107 Momma, K. and Izumi, F. (2011). VESTA 3 for three-dimensional visualization of crystal, volumetric and morphology data. *J. Appl. Crystallogr.* 44: 1272–1276. https://doi.org/10.1107/S0021889811038970.

108 Varetto, U. (2009). Molekel 5.4 (Swiss National Supercomputing Centre) (accessed 29 Jul 2019).

109 Kokalj, A. (2003). Computer graphics and graphical user interfaces as tools in simulations of matter at the atomic scale. *Comput. Mater. Sci.* 28: 155. https://doi.org/10.1016/S0927-0256(03)00104-6.

110 Dennington, R., Keith, T.A., and Millam, J.M. (2016). *GaussView*, Version 6. Shawnee Mission KS: Semichem Inc.

111 Jian, T., Li, W.L., Chen, X. et al. (2016). Competition between drum and quasi-planar structures in RhB_{18}^-: motifs for metalloboronanotubes and metallo-borophenes. *Chem. Sci.* 7: 7020–7027. https://doi.org/10.1039/C6SC02623K.

112 Li, W.-L., Jian, T., Chen, X. et al. (2017). Observation of a metal-centered B_2-Ta@B_{18} tubular molecular rotor and a perfect Ta@B_{20} boron drum with the record coordination number of twenty. *Chem. Commun.* 53: 1587–1590. https://doi.org/10.1039/c6cc09570d.

113 Lv, J., Wang, Y., Zhang, L. et al. (2015). Stabilization of fullerene-like boron cages by transition metal encapsulation. *Nanoscale* 7: 10482–10489. https://doi.org/10.1039/C5NR01659B.

114 Wang, Y., Lv, J., Zhu, L., and Ma, Y. (2012). CALYPSO: a method for crystal structure prediction. *Comput. Phys. Commun.* 183: 2063–2070. https://doi.org/10.1016/j.cpc.2012.05.008.

115 Li, H.-R., Liu, H., Tian, X.-X. et al. (2017). Structural transition in metal-centered boron clusters: from tubular molecular rotors Ta@B_{21} and Ta@B_{22}^+ to cage-like endohedral metalloborospherene Ta@B_{22}^-. *Phys. Chem. Chem. Phys.* 19: 27025–27030. https://doi.org/10.1039/c7cp05179d.

116 Li, H.-R., Liu, H., Lu, X.-Q. et al. (2018). Cage-like Ta@B^q_n complexes ($n = 23$–28, $q = -1$–$+3$) in 18-electron configurations with the highest coordination number of twenty-eight. *Nanoscale* 10: 7451–7456. https://doi.org/10.1039/c8nr01087k.

24

DFT-Based Computational Approach for Structure and Design of Materials: The Unfinished Story

Ravi Kumar[*1], *Mayank Khera*[*1,2], *Shivangi Garg*[*1], *and Neetu Goel*[1]

[1] *Computational and Theoretical Chemistry Group, Department of Chemistry & Centre for Advanced Studies in Chemistry, Panjab University, Chandigarh, India*
[2] *Department of Applied Science, Chandigarh College of Engineering, CGC-Jhanjheri, Mohali, India*

24.1 Introduction

> *The ground state of any interacting many-particle system with a given fixed inter-particle interaction is a unique functional of the electron density. The electron density that minimizes the energy of the overall functional is the true electron density corresponding to the full solution of the Schrödinger equation. Once the ground state electron density is known, all the properties can be calculated.*

The celebrated theorems by Hohnberg and Kohn in conjugation with Kohn Sham (KS) constructions laid the foundation for the build-up of a practical computational scheme, density functional theory (DFT).

Computational chemistry started its journey as a vital adjunct to experimental studies and has gained a firm footing as an independent research area owing to rigorous and advanced theoretical formulations that allow the near-accurate solutions of the Schrödinger equation for systems containing tens, or even hundreds of atoms. Translation of these formulations into efficient algorithms coupled with the incredible speed and capacity of computers has revolutionized the ability of theoretical chemistry to address critical problems in a wide range of disciplines. Assimilation of algorithms into a convenient computational package added computational chemistry as an indispensable tool in the arsenal of chemists. As a reflection of this observation, the efforts of Pople and Kohn to develop a DFT-based computational methodology were recognized by the award of the Nobel Prize in Chemistry in 1998.

Over the past decades, electronic structure calculations have become essential to obtain accurate and comprehensive information not only in the case of atoms and molecules but also in modeling complex systems such as those arising in biology and materials science. The underlying core technology of these calculations implies solving the electronic Schrödinger equation, given the positions of a collection of atomic nuclei, and the total number of electrons in the system, and thus calculating the electronic energy and other measurable properties by adopting well-defined theoretical formulations based on some approximations. The Born–Oppenheimer approximation assumes it reasonable to treat electron and nuclear motions separately and allows the solution

*These authors contributed equally to this work.

Electron Density: Concepts, Computation and DFT Applications, First Edition.
Edited by Pratim Kumar Chattaraj and Debdutta Chakraborty.

of the electronic Schrödinger equation at fixed nuclear coordinates. The electronic Schrödinger equation is a many-body problem; its computational complexity grows exponentially with the number of electrons, and hence, a brute-force solution is intractable. Hartree–Fock (HF) theory yields reasonable results by approximating the interaction between electrons by the Coulomb potential arising from a system of fixed electrons and also incorporates the exchange interactions by forcing the anti-symmetry of the electronic wave function. However, due to the neglect of correlations in the motion between two electrons with anti-parallel spins, it fails to provide a robust description of events in which electron correlation has a major role. Thus, the inclusion of electron correlation in the computational methodology that will remain tractable with the increase in the size of the system poses a key problem. Considerable progress made in this direction can be categorized as wave-function-based approach and DFT-based theoretical approach. The first one involves the expansion of electronic wavefunction as a sum of Slater determinants (the orbitals and coefficients are optimized by various numerical procedures). While HF theory is the simplest in this category, its usefulness is limited because of the complete neglect of electron correlation. Other popular wavefunction-based approaches include second-order Moller–Plesset perturbation (MP2), methods based on the coupled-cluster ansatz, and multi-reference perturbation methods, such as CASPT2 (complete active space with second-order perturbation theory). Each one scales differently with the number of electrons and is the method of choice keeping in view the complexity of the problem.

The basic framework of DFT approach considers all intricate motions and pair correlations of a many-body system in terms of total electron density. The theoretical foundations of DFT were laid in 1964–1965 by Hohenberg Kohn (HK) theorems [1, 2]. The period between 1980 and 2010 was the formative years of this engaging and stimulating methodology, DFT was the most active field during these 30 years as per web science citation data. Since its formal inception, KS-DFT has gained immense popularity owing to its computational elegance and simplicity as an electron structure method in physics and chemistry. In the mid-1980s, the rise of KS-DFT began with crucial developments in its exchange–correlation (XC) term, which gave it the predictive power competitive with the most developed wavefunction methods. Research in this area includes designing precise XC [3] functionals and developing new analytical methodologies as the estimated character of the XC functional is the basis of its success or failure in terms of implementation. DFT has the ability to correctly describe the quantum nature of matter if the exact XC functional is used [4]. However, framing the exact form of this functional is a continuously evolving process [5]. While excellent improvements have been made to suffice this limitation, unfortunately, no systematic has been formulated until [6]. Out of hundreds of different functional that have been proposed, the user is left bemused to ask "which functional should I use." Though there is no straightforward answer but there is an optimal answer for a particular problem. The story of DFT's success is the quest for the precise XC functional, that utilizes electron density to represent advanced many-body phenomena inside one element formalism.

24.2 Different Frameworks of DFT

24.2.1 Kohn Sham Density Functional Theory (KS-DFT)

The total energy of an electron system as a functional of charge density in KS-DFT is expressed as:

$$\mathrm{E}[\rho] = T_s[\rho] + \int drV_{\mathrm{ext}}(r)\rho(r) + \mathrm{E_H}[\rho] + \mathrm{E_{XC}}[\rho] \tag{24.1}$$

where T_s is the KS kinetic energy expressed in terms of KS orbitals as:

$$T_s[\rho] = \sum_{i=1}^{N} \phi_i^*(r) \left(-\frac{\hbar^2}{2m}\Delta^2\right) \phi_i(r) \tag{24.2}$$

and V_{ext} is the external potential acting on the interacting system and E_H is the Hartree (Coulomb) energy which can be expressed as:

$$E_H(\rho) = \frac{e^2}{2} \int dr \int dr' \frac{\rho(r)\rho(r')}{|r - r'|} \tag{24.3}$$

and E_{XC} is the XC energy.

To yield the KS potential, the KS equations are varied with respect to a set of orbitals, subject to constraints on those orbitals. The KS potential can be expressed as:

$$V_{eff}(r) = V_{ext}(r) + e^2 \int \frac{\rho(r')}{|r - r'|} dr' + \frac{\delta E_{XC}[\rho]}{\delta \rho(r)} \tag{24.4}$$

where the last term is V_{XC} is the XC potential:

$$V_{XC}(r) = \frac{\delta E_{XC}[\rho]}{\delta \rho(r)} \tag{24.5}$$

V_{XC} and the corresponding XC energy term are the only unknowns in the KS-DFT.

The KS orbital energies ϵ_i, generally have no physical significance. The sum of orbital energies is related to the total energy as:

$$E = \sum_{i=1}^{N} \epsilon_i - E_H[\rho] + E_{XC}[\rho] - \int \frac{\delta E_{XC}[\rho]}{\delta \rho(r)} \rho(r) dr \tag{24.6}$$

As the orbital energies are nonunique in restricted open-shell cases, this equation only holds true for specific choices of orbital energies.

24.2.2 Time-Dependent Density Functional Theory (TD-DFT)

TD-DFT is an approach to predict electronic excited states with high accuracy and efficiency. The formal foundation of the TD-DFT lies in the Runge–Gross (RG) theorem, which is a time-dependent analog of the HK theorem. It states that for any initial wave function there is a unique mapping between its time-dependent external potential and electron density. It means that the many-body wave function (dependent on $3N$ variables) is equivalent to density (dependent on three variables only) and thus all the properties of the system can be determined by the knowledge of density alone. Unlike DFT, there is no general minimization principle in time-dependent quantum mechanics (QM).

$$\hat{H}(t) = \hat{T} + \hat{V}_{ext}(t) + \hat{W} \tag{24.7}$$

where $\hat{T}$ is the kinetic operator, $\hat{W}$ is the e–e interaction, and $\hat{V}_{ext}(t)$ is the external potential, which along with the number of electrons defines the system. Generally, the external potential is the interaction of the electrons with nuclei of the system.

For nontrivial time dependence, an additional explicitly time-dependent potential is present that arises from a time-dependent electric or magnetic field. The many-body wave-function under a single initial condition can be written from time-dependent Schrödinger equation as:

$$\hat{H}(t) \langle \psi(t) = \imath\hbar \frac{\rho}{\rho(t)} |\psi(t)\rangle, |\psi(0)| = \psi(t)\rangle \tag{24.8}$$

Employing Schrödinger's equation as a starting point, the RG theorem shows that at any point in time, the density uniquely determines the external potential.

$$\hat{H}_s(t) = \hat{T} + \hat{V}_s(t) \tag{24.9}$$

which in turn determines a determinantal wavefunction.

$$\hat{H}_s(t)\,|\Phi(t)\rangle = i\frac{\delta}{\delta t}\,|\Phi(t)\rangle\,,\,|\Phi(0)\rangle = |\Phi\rangle \tag{24.10}$$

which is constructed in terms of a set of N orbitals that obey the equation:

$$\left(-\frac{1}{2}\nabla^2 + V_s(r,t)\right)\phi_j(r,t) = \imath\frac{\partial}{\partial t}\phi_j(r,t)\phi_j(r,0) = \psi_j(r) \tag{24.11}$$

and generate a time-dependent density

$$\rho_s(r,t) = \sum_{j=1}^{N_b} f_j(t)|\phi_j(r,t)|^2 \tag{24.12}$$

such that ρ_s is always equal to the density of the interacting system.

$$\rho_s(r,t) = |\rho(r,t)| \tag{24.13}$$

In the expression of the density above, the summation is over all bonding KS orbitals and $f_j(t)$ is the time-dependent occupation number for orbital j. If the potential $V_s(r,t)$ can be determined or well approximated, then Schrödinger's equation, a single partial differential equation in $3N$ variables is replaced by N differential equations of three variables, each differing only in one initial condition.

$$V_s(r,t) = V_{\text{ext}}(r,t) + V_{\text{J}}(r,t) + V_{\text{XC}}(r,t) \tag{24.14}$$

RG approached the definition of KS potential through an action-based argument starting from the Dirac equation.

$$A[\Psi] = \int dt\,\langle\Psi(t)|\,H - i\frac{\delta}{\delta t}\,|\Psi(t)\rangle \tag{24.15}$$

Treated as a functional of wavefunction, $A[\psi]$, variations of the wavefunction yield many-body Schrödinger's equation as the stationary point. Given the unique mapping between densities and wave functions, Runge and Gross then treated the Dirac action as a density functional

$$A[\rho] = A[\psi[\rho]] \tag{24.16}$$

and derived a formal expression that calculates the XC potential by functional differentiation. It was observed that the Dirac action-based approach results in paradoxical conclusions when considering the causality of the response function it generates. The density response function i.e. the functional derivative of the density with respect to the external potential should be causal. It means a change in potential at a given time should not affect the density at earlier times. As Dirac action-generated response functions are symmetric in time so they lack the required causal behavior. Keldysh's approach to complex path time integration later solved this problem. Recently, Vignale [7] proposed an alternative resolution of the causality paradox through the refinement of the action principle in real time.

However, its inability to well describe double, Rydberg charge transfer, and extended π-system excitations within its formal linear response formulation has seen its limitation in many practical applications.

24.2.3 Linear Response Time-Dependent Density-Functional Theory (LR-TDDFT)

LR-TDDFT is used where the external perturbation is so small that it cannot completely destroy the structure of the ground state of the system. In this case, the linear response of the system can be analyzed. It is of great advantage as the first-order variation to the system will depend only on the ground state wave function such that one can simply utilize all the properties of DFT.

Consider a small time-dependent external perturbation, $\delta V^{\text{ext}}(t)$. It gives:

$$H'(t) = H + \delta V^{\text{ext}}(t) \tag{24.17}$$

$$H'_{\text{KS}}[\rho](t) = H_{\text{KS}}[\rho] + \delta V_{\text{H}}[\rho](t) + \delta V_{\text{XC}}[\rho](t) + \delta V^{\text{ext}}(t) \tag{24.18}$$

and looking at the linear response of the density

$$\delta\rho(rt) = \chi(rt, r't')\delta V^{\text{ext}}(r't') \tag{24.19}$$

$$\delta\rho(rt) = \chi_{\text{KS}}(rt, r't')\delta V^{\text{eff}}[\rho](r't') \tag{24.20}$$

where,

$$\delta V^{\text{eff}}[\rho](r't') = \delta V^{\text{ext}}(t) + \delta V_{\text{H}}[\rho](t) + \delta V_{\text{XC}}[\rho](t) \tag{24.21}$$

(Here all the prime variables are integrated).

In the linear response domain, variation of Hartree and XC potential to linear order may be expanded with respect to density variation.

$$\delta V_{\text{H}}[\rho](r) = \frac{\delta V_{\text{H}}[\rho]}{\delta\rho}\delta\rho = \frac{1}{|r - r'|}\delta\rho(r') \tag{24.22}$$

and

$$\delta V_{\text{XC}}[\rho](r) = \frac{\delta V_{\text{XC}}[\rho]}{\delta\rho}\delta\rho = f_{\text{XC}}(rt, r't')\delta\rho(r') \tag{24.23}$$

After, adding this relation to the response equation of the KS system and comparing both the resultant equation and the response equation for the physical system, the Dyson equation of TD-DFT is obtained:

$$\begin{aligned}\chi(r_1t_1, r_2t_2) &= \chi_{\text{KS}}(r_1t_1, r_2t_2) + \chi_{KS}(r_1t_1, r'_2t'_2)\left(\frac{1}{|r'_2 - r'_1|}\right. \\ &\left. + f_X C(r'_2t'_2, r'_1t'_1)\right)\chi(r'_1t'_1, r_2t_2)\end{aligned} \tag{24.24}$$

From this equation, it is possible to derive the excitation energies of the system, as these are simply the poles of the response function. Other linear-response approaches include the Casida formalism (an expansion in electron–hole pairs) and the Sternheimer equation (density-functional perturbation theory).

24.2.4 Discontinuous Galerkin Density Functional Theory (DGDFT)

The DGDFT method is based on the solution of KS equations by making use of adaptive local basis (ALB) functions generated on the fly during the self-consistent field (SCF) iterations with high precision in comparison to that of plane wave (PW) basis sets. In general, a two-level parallelization strategy is adopted in the DGDFT method to deal with different types of data distribution, task scheduling, and data communication schemes. The accuracy of the approximation gets improved systematically by using the ALB set. We can make the computational complexity of the DGDFT scale quadratically with respect to the number of electrons for both insulating and metallic systems, by making use of pole expansion and selected inversion technique to compute electron density, energy, and atomic forces. In DGDFT each KS orbital can be expanded in the form of ALB sets as:

$$\phi_i(x) = \sum_{K=1}^{M}\sum_{j=1}^{J_K} c_i; \quad K, j\phi_{Kj}(x) \tag{24.25}$$

where J_K is the total number of ALBs, and $\phi_{K,j}(x)$ is the jth ALB in the element E_K, which has nonzero support only in E_K. The basis function $\phi_{K,j}$ is not necessarily continuous across the boundary of E_K, as it is assumed to be square integrable within E_K and zero outside its boundary.

The minimized total energy functional with respect to N KS orbitals $\psi(i)$ subjected to the ortho-normal condition can be written as:

$$\begin{aligned} E_{\text{DG}}(\{\psi(i)\}) = {} & \frac{1}{2}\sum_{l=1}^{N}\langle\Delta\psi_i, \Delta\psi_i\rangle_T + \langle V_{\text{eff}}, \rho\rangle_T + \sum_{I=1}^{N_A}\sum_{l=1}^{L_I}\gamma_{I,l}\sum_{i=1}^{N}|\langle b_{I,l}(\cdot - R_1), \psi_i\rangle_T|^2 \\ & - \sum_{i=1}^{N}\langle\Delta\{\{\psi_i\}\}, [[\psi_i]]\rangle_S + \alpha\sum_{i=1}^{N}\langle[[\psi_i]], [[\psi_i]]\rangle_S \end{aligned} \tag{24.26}$$

where V_{eff} is the effective one body potential, which includes local pseudo-potential V_{loc}, Hartree potential V_{H}, and the XC potential $V_{\text{XC}}[\rho]$. The term containing $\{b_{I,l}\}$ and $\{\gamma_{I,l}\}$ corresponds to the nonlocal pseudo-potential in the Kleinman–Bylander form [8] and α is the adjustable penalty parameter to ensure that the equation gives well-defined ground-state energy. The parameters $\{\gamma_{\text{I},l}\}$ are the real-valued scalars. For each atom I, there are L_{I} functions $\{b_{\text{I},l}\}$, called the projection vectors of the nonlocal pseudo-potentials.

The rigorous yet scalable methodology has established a benchmark in design and engineering of novel materials to be used as in electronics [9, 10], photonics [11, 12], adsorption [13, 14], catalysis [15–17], and energy storage devices [18, 19] via size scaling and structural modification. A number of up-to-date surveys and a wealth of monographs are available on the success of DFT. This chapter attempts to make an addition to this treasure by providing instances of its recent practical applications in the design of two-dimensional (2D) nano-materials, understanding of noncovalent interactions, and tailor-made photovoltaic devices in a form that is rigorous yet concise.

24.3 DFT Implemented Computational Packages

Significant advances in computational packages largely contributed to the infiltration of computational chemistry in almost every branch of science. The development of DFT-based methods that lowered the computational cost without compromising on accuracy facilitated its wide applicability. The software packages can be categorized as commercial and freeware. The commercial packages that are widely popular among industrial as well as academic circles include Amsterdam density functional (ADF) [20], Atomistix ToolKit (ATK) [21], DMol^3 [22], Cambridge Serial Total Energy Package (CASTEP) [23], General Atomic and Molecular Electronic Structure System (GAMESS-UK) [24], Gaussian [25], Orca [26], Q-Chem [27], Turbomole [28], Vienna *ab initio* Simulation Package (VASP) [29], and WIEN2k [30]. Their popularity arise from the fact that they are not only highly efficient in performing accurate DFT calculations but the current versions of most of these commercial packages can perform molecular mechanics (MM) [31], semi-empirical [32], SCF [33], QM/MM [34], Moller–Plosset (MPn) [35], and coupled-cluster (CC) [36] calculations in a very sophisticated manner. These packages have also assimilated electrical, magnetic, and optical property calculations with post-KS calculations within the random-phase approximation (RPA) [37]. These software packages can be accessed through graphical user interface (GUI) that has either been built separately or integrated within the package. GUI built separately for the packages mentioned above include GaussView [38] for Gaussian, VESTA [39] for VASP, TmoleX [40] for Turbolmole, and Avagadro [41] for ORCA. Software packages that come with integrated GUI include ADF and Q-chem. Apart from performing regular calculations, these packages have achieved excellence in solving specific problems, for instance, ADF can calculate spectroscopic, transition metals, and heavy elements problems very efficiently [42]. Similarly, Q-Chem has integrated configuration interaction (CI), equation-of-motion coupled-cluster (EOM-CC), and algebraic diagrammatic construction (ADC) methods [43]. GAMESS-UK has the capability to calculate valence bond wave functions using the TURTLE code [44] provided in the package. Dmol3 has pioneered the use of the COSMO solvation model for in-silico simulations of solvated molecules and surfaces whereas ADF has championed conductor-like screening model for real solvents (COSMO-RS) [45] method that predicts chemical potential (μ) for liquids by screening charge density on the surface of molecules of each species in the solution. Similarly, Turbomole was initially spearheaded to perform large-scale QM simulations, it has evolved itself to perform treatment of relativistic effects like spin–orbit coupling and exact two-component Hamiltonian calculations [46].

Over the period of time, some open-source programs have been developed to decentralize access to quantum calculations. Some of the popular freeware for performing QM calculations include ABINIT [47], AIMAll [48] BigDFT [49], CONQUEST [50], CrystalExplorer (CE) [51], Dalton [52], PARSEC [53], Quantum ESPRESSO [54], SIESTA [55], and Orca [26]. Each of these packages has a specific task-solving capability as its unique feature. ABINIT computes electronic density and derived properties of materials using a PW basis set and pseudo-potentials [56]. AIMAll is a quantum chemistry software package with the primary purpose of performing quantitative atoms in molecules (AIM) analysis and then providing the output in visual mode. It requires molecular wavefunction files from *ab initio* or DFT calculations as input [57]. BigDFT, another modern-day package, has expertise in atomistic simulation of challenging materials and biological systems. It also offers calculations for periodic systems, surfaces, wires, or isolated molecules. The outstanding property of BigDFT is to use Daubechies wavelets basis set that forms a flexible, systematic, and accurate basis set providing an adaptive environment for simulation [58]. CONQUEST is another open-source code available for performing DFT calculations on very large systems containing thousands of atoms [59]. Its levels of precision can be tuned ranging from *ab initio* tight binding to full DFT with PW accuracy. It has been successfully applied to the study of 3D reconstructions formed by Ge on Si (001) that contain 20,000 atoms. Another freeware CrystalExplorer (CE) is designed to specifically analyze the crystal structure using crystallographic information file (cif) by obtaining Hirshfeld surface and calculating inter-molecular interactions. It can also be used to study the effect of pressure and temperature on crystal structure, single-crystal to single-crystal reactions, the voids present in the crystal, and structure-property relationships [60]. Quantum ESPRESSO performs first-principle electronic-structure calculations and has achieved excellence in materials modeling based on DFT, PW basis sets, and pseudo-potentials (both norm-conserving and ultra-soft). The core PW DFT functions used for simulations are provided by the PW SCF [61]. SIESTA is a very popular open-source package that is capable in terms of tuning its accuracy and cost along a wide range. It can perform quick exploratory calculations as well as highly accurate simulations to match the quality of PW and all-electron methods [62]. However, to augment the program, post-processing tools for SIESTA have been developed that are helpful in processing the output.

The research in computational chemistry lead to the development of software packages that were able to perform quantum-chemical calculations on solids such as semiconductors and metals. In this arena, both commercial and freeware packages are available to execute calculations. CASTEP is a commercially available program that employs DFT-based PW pseudo-potential method to explore the properties of crystals and surfaces in materials such as semiconductors, ceramics, metals, minerals, and zeolites [63]. Its most prevalent use is to conduct studies on surface chemistry, structural properties, band structure, density of states (DOS), and vibrational and optical properties [64]. In addition, it is used to calculate second-order elastic constants and to study properties of point defects and extended defects in semiconductors and other materials [65]. CASTEP is also used widely to calculate the properties required to analyze the chemical shifts and electronic field gradient obtained from the results of solid-state nuclear magnetic resonance (NMR) experiments [66]. ATK, popularly known as QuantumATK, is another commercial package for atomic-scale modeling and simulation of materials including nano-systems. It has achieved excellence in large-scale real-time material computations, thus accelerating research into semiconductor and material chemistry [67]. WIEN2k also performs DFT to calculate the electronic structure of a solid and its properties such as bond energy, DOS, electron density, spin density, X-ray structure factor, and optical properties [68]. Some open source codes are also available for performing simulations in material sciences such as FHI-aims (Fritz Haber Institute *ab initio* molecular simulations) [69] and PARSEC [53].

24.4 DFT as Backbone of Electronic Structure Calculations

24.4.1 Design of 2D Nano-Materials

The crystalline solids comprising a single layer of atoms are termed 2D materials and are considered promising candidates for various applications in material science [70]. The experimental realization of graphene [71] sparked the interest of the scientific community in the field of 2D materials owing to its remarkable properties [72]. DFT emerged as a strong workforce in the field of novel material design and has become a popular method used by the material science community [73]. Subsequently, numerous density functional studies were reported to account for the estimation of various properties of graphene in its pristine form [74], with different kinds of defects [75, 76], doped [77, 78], and in functionalized form [79]. DFT studies have played a pivotal role in understanding the adsorption [80] and catalysis [81] over the 2D material surface.

Here, we present the efficacy of graphene-based 2D materials investigated for some critically important reactions using frameworks of DFT. Oxidation of water is the key step for the large-scale production of molecular hydrogen, a clean and efficient fuel. It involves two crucial reactions viz Hydrogen evolution reaction (HER) and Oxygen evolution reaction (OER). The HER is a facile two-electron process that occurs at the cathode of the electrolyzer [82] while the OER is a four-electron process that occurs at the anode. OER is kinetically sluggish in comparison to HER and governs the overall efficiency of oxidation of water for H_2 production [83]. A number of electro-catalysts have been designed in order to overcome the over-potential associated with the uncatalyzed OER [84–88]. Our group proposed a graphene-based electro-catalyst (Sulphur doped graphene in the environment of Boron (B_3S-G) for OER) and calculated the over-potential using spin polarised DFT with PBE XC functional as instigated in Quantum Espresso. The study provided evidence of a four-step reaction mechanism with a very low over-potential of 0.44 eV and proved that there exists a synergy between the dopants [89]. The OER mechanism (Figure 24.1) initiates

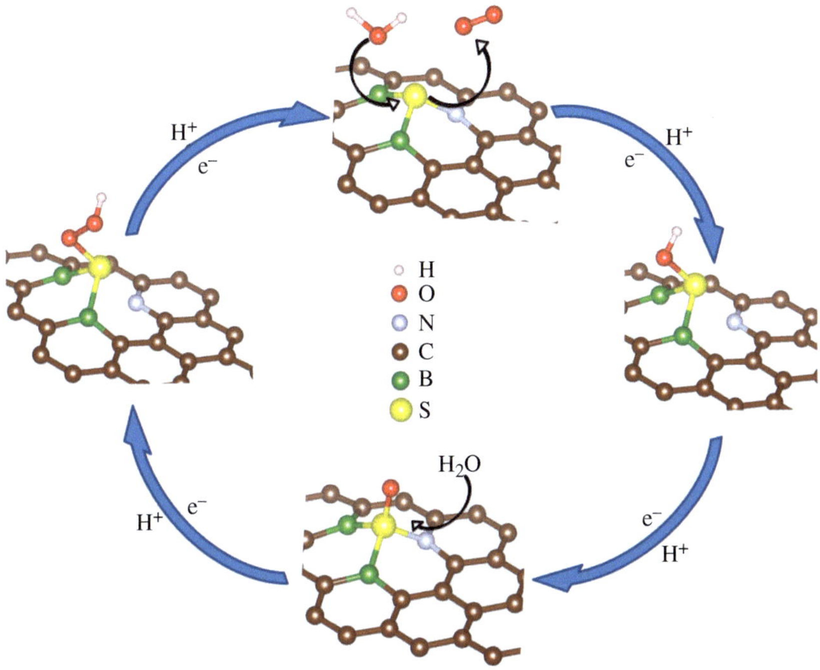

Figure 24.1 Schematic representation of the reaction mechanism for OER on B_3S-G catalyst.

with the oxidation of water into OH* intermediate on an active site. This OH* intermediate is then oxidized to O* in the second step followed by a reaction of another water molecule with O* in the third step to form a super-oxide intermediate (HOO*), which then oxidizes to release O_2 molecule in the final step. Every step releases a proton and an electron as side products as shown in the equations below:

$$H_2O + * \rightarrow OH^* + (H^+ + e^-) \tag{24.27}$$

$$OH^* \rightarrow O^* + (H^+ + e^-) \tag{24.28}$$

$$O^* + H_2O \rightarrow OOH^* + (H^+ + e^-) \tag{24.29}$$

$$OOH^* \rightarrow * + O_2 + (H^+ + e^-) \tag{24.30}$$

where * represents the adsorption site.

The change in Gibbs free energy (ΔG) on going from one step to another for the above mechanism was calculated as:

$$\Delta G = \Delta E + \Delta ZPE - T\Delta S \tag{24.31}$$

where ΔE is the adsorption energy, ΔZPE is the zero point correction to the total energy, and ΔS represents entropy change.

The over-potential (η) for the reaction is determined as:

$$\eta = (G^{max}/e) - 1.23 \tag{24.32}$$

where G^{max} must be the highest ΔG value obtained from the four steps.

Table 24.1 Table showing the maximum number of Mg atom over each model, the adsorption energy per atom, and corresponding magnetization potential.

Model	Doping percentage (%)	Number of Mg atom adsorbed	E_{ad}/n (eV)	Magnetization potential
BG	1.9	2	−1.39	0.70
B_2G	3.7	4	−0.87	0.44
B_3G	5.6	6	−0.42	0.21
B_4G	7.4	8	0.09	−0.05
B_5G	9.3	5	−0.95	0.48
B_6G	11.1	6	0.04	0.02

Rechargeable metal-ion batteries are an attractive alternative as a clean energy source to address the environmental issues posed by the excessive use of fossil fuels. The bivalent nature of Mg covers the possibility of storing more charge per ion that could allow the building of batteries with higher energy density. Currently, the promised potential of Mg ion battery (MIB) is far to achieve owing to the lack of an appropriate electrolyte, since the Mg metal used as anode in MIB is not compatible with common battery electrolytes. Carbon-based electrodes function well with conventional battery electrolytes, therefore, it was proposed that B and N-doped graphene intercalated with Mg could be an efficient anode material for MIB [90]. Density functional calculations were carried out to study interaction of Mg with the doped graphene surface by calculating adsorption energy (E_{ad}), magnesiation potential (V_{Mg}) [91], and specific capacity (C_c) (Table 24.1) of the doped surface, respectively, using the equations:

$$E_{ad} = E_{(gra+Mg)} - (E_{(Gra)} + nE_{Mg}) \tag{24.33}$$

where $E_{(Gra)}$ and $E_{(gra+Mg)}$ are the total energy of doped graphene before and after adsorption of Mg and E_{Mg} is the energy of the Mg atom in the bulk cluster.

$$V_{Mg} = -\Delta G_f/zF \tag{24.34}$$

where ΔG_f is the change in free energy, z is the Mg valency, and F is the Faraday's constant (26.8 Ah/mol)

$$C_c = n_{Mg}zF/M_G \tag{24.35}$$

where n_{Mg} is the number of Mg adsorbed and M_G mass of doped graphene substrate. The calculations predicted a B-doped surface as an efficient anode material for MIB. Moreover, it shows good electrical and ionic conductivity and high cycling stability.

Though graphene has proved itself to be a wonderful material having enormous applications in the chemical world, its direct application in the field of electronics is restricted due to its zero band gap [92]. Functionalization of graphene into graphane (hydrogenated graphene) [93] has added new avenues to the use of graphene-based devices. Graphane has further been functionalized and doped to enhance its properties such as hydrogen storage capacity [94], fuel cell applications [95], bio-sensing capabilities [96], electronic, and mechanical properties [97]. Our group has investigated the effect of B and N doping on the electronic and mechanical properties of graphane using PBE/6-31G. The cohesive energy (E), strain (ϵ), and in-plane stiffness (S) of the pure and

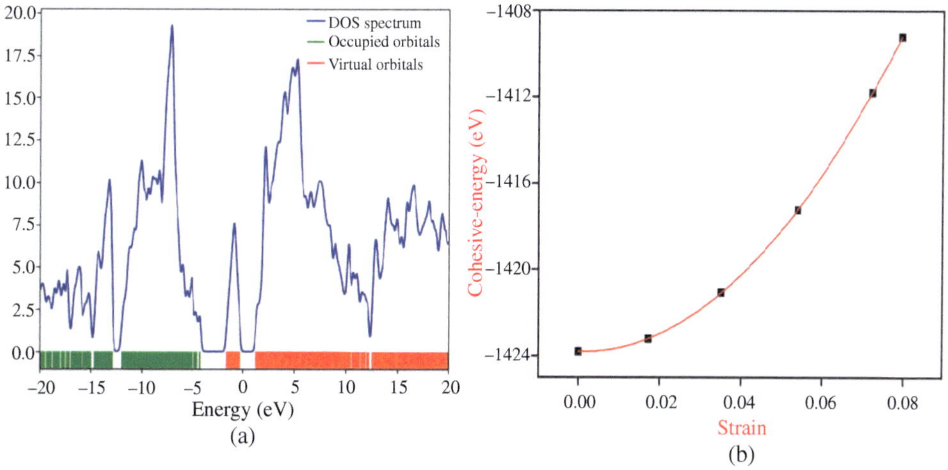

Figure 24.2 (a) DOS plot and (b) Strain versus Cohesive energy curve of B doped (8%) graphane.

doped sheets were calculated, respectively, according to the equations:

$$E = [E_S - (n_C E_C + n_H E_H + n_D E_D)] \tag{24.36}$$

where E_S is the total electronic energy of the sheet, n_C, n_H, and n_D are the numbers, and E_C, E_H, and E_D are the electronic energies of carbon, hydrogen, and doped atoms, respectively.

$$\epsilon = \frac{L_s - L_u}{L_u} \tag{24.37}$$

where L_u and L_s are the lengths of the unstretched and stretched sheets, respectively.

$$S = \frac{1}{A}\left(\frac{\partial^2 E}{\partial \epsilon^2}\right) \tag{24.38}$$

where A is the area of the unstretched sheet.

The results were realized in the form of strain versus cohesive energy curve and DOS plots of graphane with 8% B doping (Figure 24.2). It was concluded that B doping has a greater influence on the electronic and mechanical properties in comparison to that of N doping (Table 24.2). The B and N centered bond angles are at 120° and 112.2°, respectively, while the carbon atom is centered 111.44° very close to N centered bond angles [98].

Table 24.2 Band gap (eV) and In-plane stiffness (Nm^{-1}) values of graphane at various doping concentrations of B and N.

	Boron		Nitrogen	
Doping (%)	**Stiffness (Nm^{-1})**	**Band-Gap (eV)**	**Stiffness (Nm^{-1})**	**Band-Gap (eV)**
2	188.5	3.65	210.8	4.90
4	185.6	3.45	217.2	5.05
6	185.8	2.84	236.8	5.17
8	183.6	2.71	248.1	5.33

24.4.2 Non-covalent Interactions and Crystal Packing

DFT has played a remarkable role in providing assistance for the discovery and rediscovery of covalent [99] and noncovalent forces [100] in nature. Electronic structure calculations are particularly successful in gaining insight on the interactions involved in the molecular system as well as crystal structure [101]. Experimental techniques are at a disadvantage to derive reliable averaged binding energies while computational approaches especially Hirshfeld surface [102] and DFT in collaboration with Bader's AIM theory [103] provide a suitable methodology to quantify noncovalent interactions [104]. Quantification of σ-hole [105] on the covalently bonded halogen atoms has led to the discovery of halogen bonding [106] and redefining of hydrogen bonding [107] as σ-hole interaction [108] supported by the QM Hellmann–Feynman theorem [109, 110]. A series of noncovalent interactions such as Alkali bonds [111, 112], Alkaline Earth Bonds [113–115], Regium Bonds [116], Sodium Bonds [117], Triel bond [118], Tetrel bond [119], Pnictogen Bond [120], Chalcogen Bond [121], and Aerogen Bond [122] have been discovered with the help of QM calculations. Recently, Daolio et al. in two separate studies provided evidence of σ-hole on tetroxide of osmium that lead to the coining of the term "osme bond" (OmB, Om = Fe, Ru, Os) for noncovalent interactions of group eight elements [123] and gold atom can form Au···Cl and Au···O short contacts in the solid state that enables gold to engage in Anion···Anion Coinage bonds [124]. The interactions were rationalized by using the anisotropic distribution of the electron density in the anions giving a valuable extension to the already known aurophilic bonds [125].

Our group highlighted the role of noncovalent interactions in shaping a crystal structure using DFT linked with CE. For this purpose, nine halogen-substituted benzene derivatives were considered and experimental geometries were extracted from neutron diffraction studies submitted to Cambridge Crystallographic Data Centre (CCDC). The total interaction energy was calculated as the sum of electrostatic (E_{ele}), polarization (E_{pol}), dispersion (E_{dis}), and exchange repulsion (E_{rep}) by the energy model (CE-B3LYP) employed in CE17 using QM charge distribution with Gaussian09 according to the equation:

$$E_{total} = E_{ele} + E_{pol} + E_{dis} + E_{rep} \quad (24.39)$$

The noncovalent interactions were observed and analyzed based on overlapped Hirshfeld surfaces and it was proved that out of these nine molecules, 4-bromo benzoic acid (4BBA) attributes its crystal packing according to halogen bonding present between in its crystal structure and a similar compound, 4-chloro benzoic acid (4CBA), does not exhibit halogen bonding (Figure 24.3). Other conventional and nonconventional interactions such as hydrogen bonding, halogen bonding, chlorine–hydrogen, bromine–hydrogen, and bromine–oxygen interactions were also deciphered in the nine halogen crystal structures. [126].

Not only these interactions define a crystal structure, they do cooperate with themselves [127]. A recent investigation by our group has proved strong reinforcement in tetrel bonding between halogenated silanes and water, orchestrated by H-bonding owing to their cooperativity. The percentage strengthening of tetrel bond was studied by optimizing 16 binary complexes of halogenated silane and water of general formula $SiX_nH_{4-n} - H_2O$ and 16 ternary complexes, of general formula $NaX - SiX_nH_{4-n} - H_2O$, where X = F, Cl, Br, and I and $n = 1, 2, 3$ and 4 using DFT. An average of 30% reinforcement with some exceptions of tetrel bond length shortening up to 49% was observed indicating the effectiveness of cooperativity (Figure 24.4). The results were confirmed by electron density values calculated using QTAIM analysis. An unexpected reinforcement due to cooperativity despite the low σ-hole potential has proved that σ-hole is just a bystander when other strong noncovalent interaction, such as H-bonding, takes rein of the atomic arrangement [128].

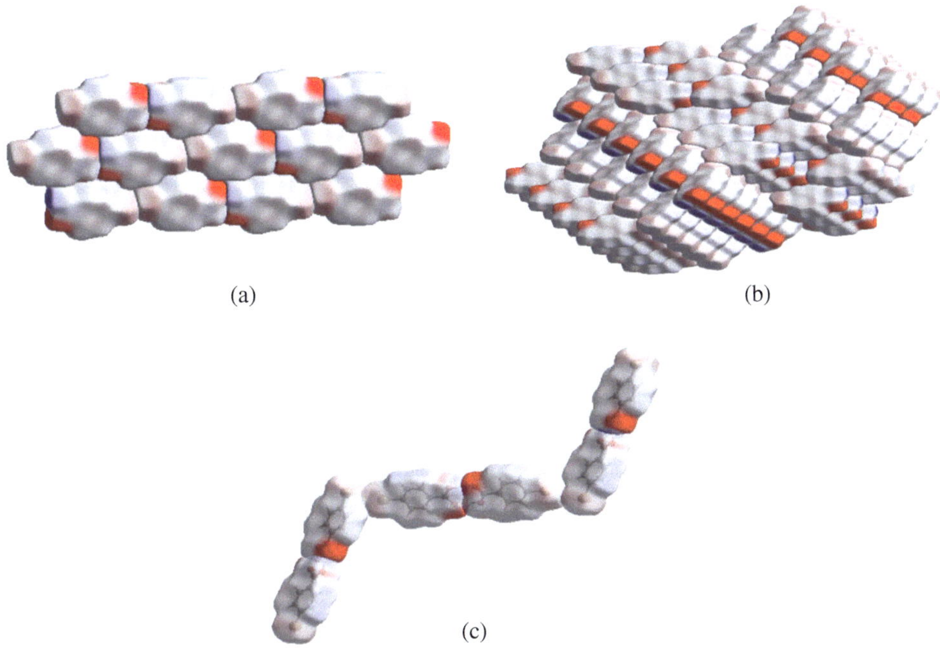

Figure 24.3 Crystal structure of (a) 4-chloro benzoic acid, (b) 4-bromo benzoic acid, and (c) evidence of halogen bonding in 4-bromo benzoic acid.

24.4.3 Designing of Organic Solar Cell

Organic solar cells (OSCs) are one of the emerging photovoltaic technology that is an attractive alternative to resolve future energy problems by maximizing solar energy production [129, 130]. It employs light-absorbing functional layers based on organic conductive polymers or small organic molecules to convert sunlight into electricity [131, 132]. In the arena of organic photovoltaic technology, Bulk hetero-junction solar cell is making tremendous advances, it is a blend of donor material (conjugated polymer) and a fullerene-based acceptor material ([6, 6]-phenylC_{61}-butyric acid methyl ester; PCBM) [133–135]. Figure 24.5 shows the mechanism of charge transport in an OSC. The organic photovoltaic technology mostly employs polymer solar cells (PSCs) [136] because, in comparison to inorganic counterparts (silicon-based solar cells), PSCs are lightweight, flexible, potentially disposable, inexpensive, customizable on the molecular level, and potentially have less carbon footprint [137, 138]. Recent developments in PSCs have shown a 16.5 % power conversion efficiency (PCE) in single-junction devices [139] and a 17.3 % PCE in tandem devices [140], amounting to a threefold improvement in 15 years. However, significant work is required to translate the progress made at the lab scale toward large-scale industrial production.

Theoretical approaches are generally employed to overcome practical synthetic issues and investigate cost-saving manufacturing processes. DFT and TD-DFT have evolved as rigorous and efficient QM methods to be employed for qualitative and quantitative analyses of OSCs [141–144]. DFT has the ability to resolve many problems as it examines the structural, electronic, and optical properties of materials such as PSCs at a low computational cost [145–147] and TD-DFT is an essential tool for computing excited-state characteristics and charge–transfer excitations [148] in large systems such as organic light emitting devices [149], organic field effect transistors [150], and

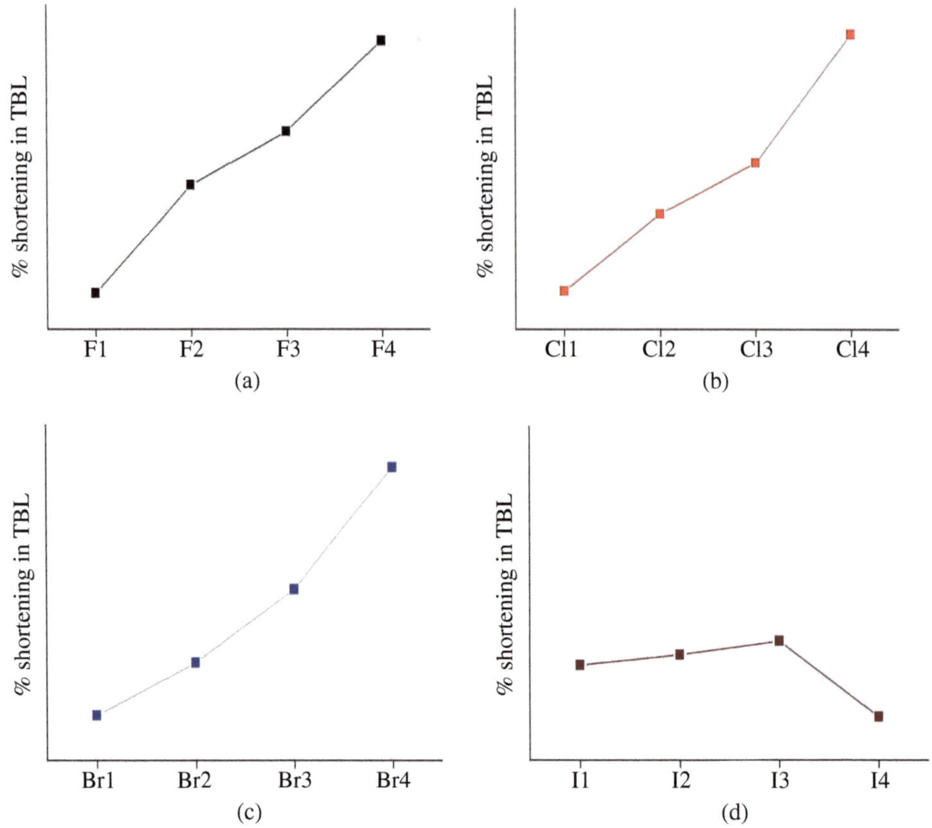

Figure 24.4 Percentage shortening in TBL (a) NaF – SiF_nH_{4-n} – H_2O, (b) NaCl – $SiCl_nH_{4-n}$ – H_2O, (c) NaBr – $SiBr_nH_{4-n}$ – H_2O, and (d) NaI – SiI_nH_{4-n} – H_2O.

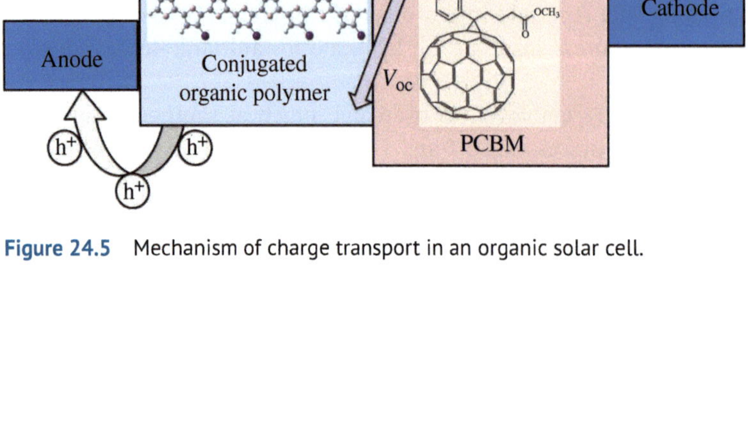

Figure 24.5 Mechanism of charge transport in an organic solar cell.

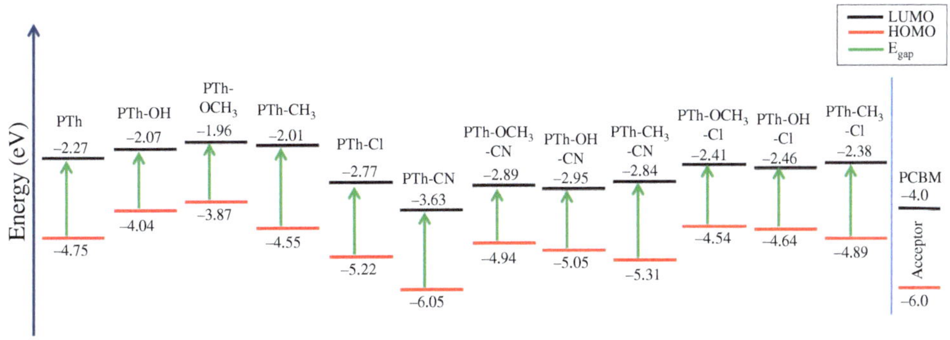

Figure 24.6 Energy ordering of frontier orbitals of polythiophene and its derivatives with respect to $PC_{70}BM$ acceptor.

OSCs [144]. DFT uses frontier molecular orbitals (highest occupied molecular orbital [HOMO] and lowest unoccupied molecular orbital [LUMO]), band gap, charge mobility, open-circuit voltage (V_{OC}) and reorganization energy (λ) as key electronic parameters to determine the optoelectronic properties of PSCs [151, 152].

The best performing polymeric solar cells are made with regioregular poly(3-hexylthiophene) as donor material and achieve an efficiency surpassing 6% [153]. However, P3HT has a relatively wide band-gap (~ 2 eV) that prevents the absorption of photons, therefore, the methods such as doping or chemical modification [154–156] were subsequently developed to reduce the band gap of conjugated polymers for maximum photon harvesting that leads to higher PCE [157, 158].

Scientific research efforts are now being directed toward exploring polymeric materials to enhance the performance of PSCs by various techniques such as functionalization of polymer [159], donor–acceptor–donor (D–A–D) polymer backbone approach [160], all-polymer solar cell approach [161], and asymmetric molecular strategy [162]. Various recent studies have explored the effect of functionalization on optoelectronic properties of polymers such as polypyrrole [163] and oligothiophenes [164]. The results have indicated that the Hammett parameters of the side group are sufficient to enable the design of improved donor materials. The research on PSCs has mostly been focused on various polymers and their derivatives, whereas their close analogs were largely ignored. A recent study by our group explored the influence of different electron donor/acceptor side groups on structural and optoelectronic properties of three polymers (polythiophene (PTh), polyfuran (PFu), and polyselenophene (PSe)) via DFT and TD-DFT calculations. The embodiment of different electron releasing/withdrawing groups significantly shifted the energies of frontier orbitals (Figure 24.6 shows the frontier orbitals' energy spacing for PTh and its derivatives). The study showed that the substitution of the polymer chains by electron donating groups (like –OH and $–OCH_3$) resulted in a lower band gap. While electron-withdrawing groups (such as –Cl and –CN) stabilized the frontier orbitals in all the studied polymers and these functionalized polymers were then suitably aligned with the HOMO (−4.0 eV) and LUMO (−6.0 eV) levels of commonly used fullerene-based ($PC_{70}BM$) acceptor. All the processes starting from the excitation of electrons from HOMO to LUMO of the donor, breaking of excitons into free charge carriers and diffusion of excitons have been investigated thoroughly. The CN substituted derivatives exhibited the maximum efficiency with V_{OC} of about 1.63–1.75 eV [165].

24.5 Concluding Remarks

Triumphs of DFT are now well documented, but one must be cautious about using it as a "black box" and apply it mindlessly to the problem at hand. The unavailability of universal XC functionals renders it approximate in application though it is exact in construction. Since an XC is used twice in a calculation, first in the approximation of electron density and then in finding the energy, thus its choice becomes very crucial to the electronic structure calculations. DFT considers energy as functional of electron density [1], and the error in self-consistent KS calculation can be estimated as:

$$\Delta \mathrm{E} = \mathrm{E}'[\rho'] - \mathrm{E}[\rho] \tag{24.40}$$

where, $\rho(\mathrm{r})$ is the exact density and $\mathrm{E}[\rho]$ is the exact functional, while $\mathrm{E}'[\rho']$ is the approximated energy.

Most practical calculations employ XC approximations that yield excellent approximate densities [166], which leads to a very slight error in calculating energy. Since XC is the only approximated term, so the functional error is because of XC approximation:

$$\Delta \mathrm{E}_\mathrm{F} = \mathrm{E}'[\rho'] - \mathrm{E}[\rho] = \mathrm{E}'_\mathrm{XC}[\rho] - \mathrm{E}_\mathrm{XC}[\rho] \tag{24.41}$$

and the remaining energy error is the density-driven error, and it is given by the difference in the approximate functional on exact and the self-consistent densities:

$$\Delta \mathrm{E}_\mathrm{D} = \mathrm{E}'[\rho'] - \mathrm{E}'[\rho] \tag{24.42}$$

For any calculation, the density-driven error is always negative. So, in KS formalism the error is the sum of functional and the density-driven error ($\Delta \mathrm{E}_\mathrm{F} + \Delta \mathrm{E}_\mathrm{D}$), will be there, irrespective of employed functional or the system at the hand [167].

As a result of these approximations, DFT sometimes underestimates the chemical reaction energy barrier, the band gap of the materials, the dissociation energy of molecular ions, and the charge–transfer excitation energies. Amazingly, these diversified errors originate from the delocalization error of the approximate functionals, because of the dominating Coulomb term that pushes electrons apart. This error can be understood from a perspective that invokes fractional charges [168, 169]. DFT calculation suffers from the limitation of the description of degenerate states that arise in transition metal systems, the breaking of chemical bonds, and the strongly correlated materials. This problem occurs because of difficulty in using the electron density to describe the interaction of degenerate states and can be understood from the perspective of fractional spins [169]. Paradoxically, it is the approximate nature of the XC functional that is the reason both for the success and the failure of DFT applications. Obtaining the exact form of this functional is one of the most active areas of research. While excellent improvements have been made to design functionals that are optimal for a given physical or chemical system, there is no well-defined approach for choosing a functional from the existing range of functionals that continues to grow fast.

Starting from the Thomas Fermi Dirac (TFD) theory to wavefunctional approaches like HF and post-HF methodology to electron density-based DFT, electronic calculations have come a long way. The fascinating formalism of DFT has gained huge popularity, reflected in its applications in various fields, the development of different DFT-based computational tools, and the citations of DFT-based articles in recent years. DFT not only provides useful insight about the material properties but is also an excellent tool for providing motivation as well as direction to experimentalists for the synthesis of new materials. A large number of publications include DFT-based computations either to pre or post-validate their experimental findings. Despite the limitation resulting from the unavailability of accurate XC functional, DFT has set a high benchmark among all the prevailing computational approaches. Its reliability and predictive power have resulted in close nexus between the theoretical and experimental communities. With the advancement in theoretical formalism, this nexus will be further strengthened to work for the ultimate goal of science; the service of humankind.

Acknowledgment

NG is grateful to Prof. Pratim Kumar Chattaraj and Prof. Debdutta Chakraborty for their kind invitation to contribute to this chapter. MK is grateful to the Council of Scientific and Industrial Research, India for the Senior Research Fellowship (SRF). RK and SG acknowledge the University Grants Commission, India for the SRF.

References

1 Hohenberg, P. and Kohn, W. (1964). Density functional theory (DFT). *Phys. Rev.* 136: B864.
2 Sham, L.J. and Kohn, W. (1966). One-particle properties of an inhomogeneous interacting electron gas. *Phys. Rev.* 145: 561.
3 Perdew, J.P. (1986). Density-functional approximation for the correlation energy of the inhomogeneous electron gas. *Phys. Rev. B*33: 8822.
4 Eschrig, H. (2003). *The Fundamentals of Density Functional Theory*. Springer.
5 Perdew, J.P. and Schmidt, K. (2001). Jacob's ladder of density functional approximations for the exchange-correlation energy. *AIP Conf. Proc.* 577: 1–20.
6 Perdew, J.P., Ruzsinszky, A., Tao, J. et al. (2005). Prescription for the design and selection of density functional approximations: more constraint satisfaction with fewer fits. *J. Chem. Phys.* 123: 062201.
7 Vignale, G. (2008). Real-time resolution of the causality paradox of time-dependent density-functional theory. *Phys. Rev. A* 77: 062511.
8 Kleinman, L. and Bylander, D. (1982). Efficacious form for model pseudopotentials. *Phys. Rev. Lett.* 48: 1425.
9 Gusain, A., Faria, R.M., and Miranda, P.B. (2019). Polymer solar cells-interfacial processes related to performance issues. *Front. Chem.* 7: 61.
10 Manoharan, K. and Subramanian, V. (2018). Exploring multifunctional applications of hexagonal boron arsenide sheet: a DFT study. *ACS Omega* 3: 9533–9543.
11 Zhai, X.-P., Ma, B., Wang, Q., and Zhang, H.-L. (2020). 2D materials towards ultrafast photonic applications. *Phys. Chem. Chem. Phys.* 22: 22140–22156.
12 Ahmed, H. and Hashim, A. (2022). Exploring the characteristics of new structure based on silicon doped organic blend for photonics and electronics applications. *Silicon* 14: 4907–4914.
13 Giovannetti, G., Khomyakov, P.A., Brocks, G. et al. (2008). Doping graphene with metal contacts. *Phys. Rev. Lett.* 101: 026803.
14 Lousada, C.M., Johansson, A.J., Brinck, T., and Jonsson, M. (2012). Mechanism of H_2O_2 decomposition on transition metal oxide surfaces. *J. Phys. Chem. C* 116: 9533–9543.
15 Nößler, M., Mitrić, R., Bonačić-Koutecký, V. et al. (2010). Generation of oxygen radical centers in binary neutral metal oxide clusters for catalytic oxidation reactions. *Angew. Chem. Int. Ed.* 49: 407–410.

16 Azofra, L.M., MacFarlane, D.R., and Sun, C. (2016). A DFT study of planar *vs.* corrugated graphene-like carbon nitride (g-C_3N_4) and its role in the catalytic performance of CO_2 conversion. *Phys. Chem. Chem. Phys.* 18: 18507–18514.
17 Garg, S., Kaur, N., and Goel, N. (2020). Conceptual DFT and TDDFT study on electronic structure and reactivity of pure and sulfur doped (CrO_3) n (n = 1–10) clusters. *J. Mol. Graph. Model.* 99: 107617.
18 Kamila, S., Chakraborty, B., Basu, S., and Jena, B.K. (2019). Combined experimental and theoretical insights into energy storage applications of a VO_2(D)–graphene hybrid. *J. Phys. Chem. C* 123: 24280–24288.
19 Adekoya, D., Qian, S., Gu, X. et al. (2021). DFT-guided design and fabrication of carbon-nitride-based materials for energy storage devices: a review. *Nano-Micro Lett.* 13: 1–44.
20 van Gisbergen, S., Snijders, J., and Ziegler, T. (2001). Chemistry with ADF. *J. Comput. Chem.* 22: 931–967.
21 Taylor, J., Guo, H., and Wang, J. (2001). *Ab initio* modeling of quantum transport properties of molecular electronic devices. *Phys. Rev. B* 63: 245407.
22 Delley, B. (2000). From molecules to solids with the $DMol^3$ approach. *J. Chem. Phys.* 113: 7756–7764.
23 Clark, S.J., Segall, M.D., Pickard, C.J. et al. (2005). First principles methods using CASTEP. *Z. Kristallogr. –Cryst. Mater.* 220: 567–570.
24 Guest, M.F., Bush, I.J., Van Dam, H.J. et al. (2005). The GAMESS-UK electronic structure package: algorithms, developments and applications. *Mol. Phys.* 103: 719–747.
25 Frisch, M.J., Trucks, G.W., and Schlegel, H.B. (2009). *Gaussian09 Revision E.01*. Wallingford, CT: Gaussian Inc.
26 Neese, F. (2012). The ORCA program system. *Wiley Interdiscip. Rev. Comput. Mol. Sci.* 2: 73–78.
27 Kong, J., White, C.A., Krylov, A.I. et al. (2000). Q-Chem 2.0: a high-performance *ab initio* electronic structure program package. *J. Comput. Chem.* 21: 1532–1548.
28 Furche, F. (2001). Molecular tests of the random phase approximation to the exchange-correlation energy functional. *Phys. Rev. B* 64: 195120.
29 Hafner, J. (2008). *Ab-initio* simulations of materials using VASP: density-functional theory and beyond. *J. Comput. Chem.* 29: 2044–2078.
30 Blaha, P., Schwarz, K., Tran, F. et al. (2020). WIEN2k: an APW+lo program for calculating the properties of solids. *J. Chem. Phys.* 152: 074101.
31 Allinger, N.L. (1982). *Molecular Mechanics*. American Chemical Society.
32 Stewart, J.J. (1989). Optimization of parameters for semiempirical methods I. Method. *J. Comput. Chem.* 10: 209–220.
33 Echenique, P. and Alonso, J.L. (2007). A mathematical and computational review of Hartree–Fock SCF methods in quantum chemistry. *Mol. Phys.* 105: 3057–3098.
34 Senn, H.M. and Thiel, W. (2009). QM/MM methods for biomolecular systems. *Angew. Chem. Int. Ed.* 48: 1198–1229.
35 Binkley, J. and Pople, J. (1975). Møller–Plesset theory for atomic ground state energies. *Int. J. Quantum Chem.* 9: 229–236.
36 Bartlett, R.J. and Musiał, M. (2007). Coupled-cluster theory in quantum chemistry. *Rev. Mod. Phys.* 79: 291.
37 Ren, X., Rinke, P., Joas, C., and Scheffler, M. (2012). Random-phase approximation and its applications in computational chemistry and materials science. *J. Mater. Sci.* 47: 7447–7471.
38 Dennington, R., Keith, T., and Millam, J. (2009). *Gauss View, Version 5*. Semichem Inc.

39 Momma, K. and Izumi, F. (2008). VESTA: a three-dimensional visualization system for electronic and structural analysis. *J. Appl. Crystallogr.* 41: 653–658.

40 Steffen, C., Thomas, K., Huniar, U. et al. (2010). TmoleX—a graphical user interface for TURBOMOLE. *J. Comput. Chem.* 31: 2967–2970.

41 Hanwell, M.D., Curtis, D.E., Lonie, D.C. et al. (2012). Avogadro: an advanced semantic chemical editor, visualization, and analysis platform. *J. Cheminf.* 4: 1–17.

42 Novotný, J., Vícha, J., Bora, P.L. et al. (2017). Linking the character of the metal–ligand bond to the ligand NMR shielding in transition-metal complexes: NMR contributions from spin–orbit coupling. *J. Chem. Theory Comput.* 13: 3586–3601.

43 Epifanovsky, E., Gilbert, A.T., Feng, X. et al. (2021). Software for the frontiers of quantum chemistry: an overview of developments in the Q-Chem 5 package. *J. Chem. Phys.* 155: 084801.

44 van Lenthe, J.H., Dijkstra, F., and Havenith, R.W. (2002). TURTLE–A gradient VBSCF program theory and studies of aromaticity. *Comput. Theor. Chem.* 10: 79.

45 Klamt, A. (1995). Conductor-like screening model for real solvents: a new approach to the quantitative calculation of solvation phenomena. *J. Phys. Chem.* 99: 2224–2235.

46 Liu, J. and Cheng, L. (2018). An atomic mean-field spin-orbit approach within exact two-component theory for a non-perturbative treatment of spin-orbit coupling. *J. Chem. Phys.* 148: 144108.

47 Gonze, X., Amadon, B., Anglade, P.-M. et al. (2009). ABINIT: First-principles approach to material and nanosystem properties. *Comput. Phys. Commun.* 180: 2582–2615.

48 Keith, T.A. (2019). *AIMAll (Version 19.10. 12).* Overland Park, KS: TK Gristmill Software.

49 Genovese, L., Neelov, A., Goedecker, S. et al. (2008). Daubechies wavelets as a basis set for density functional pseudopotential calculations. *J. Chem. Phys.* 129: 014109.

50 Bruno, I.J., Cole, J.C., Edgington, P.R. et al. (2002). New software for searching the Cambridge structural database and visualizing crystal structures. *Acta Crystallogr., Sect. B. Struct. Sci. Cryst. Eng. Mater.* 58: 389–397.

51 Turner, M., McKinnon, J., Wolff, S. et al. (2017). CrystalExplorer17.

52 Olsen, J.M.H., Reine, S., Vahtras, O. et al. (2020). Dalton Project: a Python platform for molecular-and electronic-structure simulations of complex systems. *J. Chem. Phys.* 152: 214115.

53 Bienia, C., Kumar, S., Singh, J.P., and Li, K. (2008). The PARSEC benchmark suite: Characterization and architectural implications. In: *Proceedings of the 17th International Conference on Parallel Architectures and Compilation Techniques*, 72–81. ACM.

54 Giannozzi, P., Baroni, S., Bonini, N. et al. (2009). QUANTUM ESPRESSO: a modular and open-source software project for quantum simulations of materials. *J. Phys. Condens. Matter* 21: 395502.

55 Soler, J.M., Artacho, E., Gale, J.D. et al. (2002). The SIESTA method for *ab initio* order-*N* materials simulation. *J. Phys. Condens. Matter* 14: 2745.

56 Wang, S. and Ye, H. (2002). Plane-wave pseudopotential study on mechanical and electronic properties for IV and III-V crystalline phases with zinc-blende structure. *Phys. Rev. B* 66: 235111.

57 Klein, J., Khartabil, H., Boisson, J.-C. et al. (2020). New way for probing bond strength. *J. Phys. Chem. A* 124: 1850–1860.

58 Ratcliff, L.E., Dawson, W., Fisicaro, G. et al. (2020). Flexibilities of wavelets as a computational basis set for large-scale electronic structure calculations. *J. Chem. Phys.* 152: 194110.

59 Nakata, A., Baker, J.S., Mujahed, S.Y. et al. (2020). Large scale and linear scaling DFT with the CONQUEST code. *J. Chem. Phys.* 152: 164112.

60 Spackman, P.R., Turner, M.J., McKinnon, J.J. et al. (2021). CrystalExplorer: A program for Hirshfeld surface analysis, visualization and quantitative analysis of molecular crystals. *J. Appl. Crystallogr.* 54: 1006–1011.

61 Scandolo, S., Giannozzi, P., Cavazzoni, C. et al. (2005). First-principles codes for computational crystallography in the Quantum-ESPRESSO package. *Z. Kristallogr. –Cryst. Mater.* 220: 574–579.

62 Garcia, A., Papior, N., Akhtar, A. et al. (2020). S<sc>IESTA</sc>: Recent developments and applications. *J. Chem. Phys.* 152: 204108.

63 Coquet, R., Howard, K.L., and Willock, D.J. (2008). *Theory: Periodic Electronic Structure Calculations*, 323–389. Wiley Online Library.

64 Naeem, M., Rizwan, M., Bakhtawar, R. et al. (2022). Oxynitride perovskite: computational approach to correlate structural, electronic, and optical properties of c-$BiAlO_3/N_3$. *ACS Appl. Electron. Mater.* 4: 375–385.

65 Rahaman, M.Z. and Hossain, A.M.A. (2018). Effect of metal doping on the visible light absorption, electronic structure and mechanical properties of non-toxic metal halide $CsGeCl_3$. *RSC Adv.* 8: 33010–33018.

66 Bryce, D.L. and Bultz, E.B. (2007). Alkaline earth chloride hydrates: chlorine quadrupolar and chemical shift tensors by solid-state NMR spectroscopy and plane wave pseudopotential calculations. *Chem. Eur. J.* 13: 4786–4796.

67 Gaur, T., Sharma, R., and Chaujar, R. (2022). Quantum ATK analysis of silicon nanowire FET with a cylindrical metallic wrap-around gate varied with dielectrics. *Mater. Today: Proc.* 62 (Part 6): 3823–3826.

68 Schwarz, K. and Blaha, P. (2003). Solid state calculations using WIEN2k. *Comput. Mater. Sci.* 28: 259–273.

69 Blum, V., Gehrke, R., Hanke, F. et al. (2009). *Ab initio* molecular simulations with numeric atom-centered orbitals. *Comput. Phys. Commun.* 180: 2175–2196.

70 Singh, A.K., Mathew, K., Zhuang, H.L., and Hennig, R.G. (2015). Computational screening of 2D materials for photocatalysis. *J. Phys. Chem. Lett.* 6: 1087–1098.

71 Novoselov, K.S., Geim, A.K., Morozov, S.V. et al. (2004). Electric field effect in atomically thin carbon films. *Science* 306: 666–669.

72 Novoselov, K., Morozov, S., Mohinddin, T. et al. (2007). Electronic properties of graphene. *Phys. Status Solidi B* 244: 4106–4111.

73 Kent, P.R. and Kotliar, G. (2018). Toward a predictive theory of correlated materials. *Science* 361: 348–354.

74 Tawfik, S.A., Isayev, O., Stampfl, C. et al. (2019). Efficient prediction of structural and electronic properties of hybrid 2D materials using complementary DFT and machine learning approaches. *Adv. Theory Simul.* 2: 1800128.

75 Sun, T., Zhang, G., Xu, D. et al. (2019). Defect chemistry in 2D materials for electrocatalysis. *Mater. Today Energy* 12: 215–238.

76 Wang, D. and Sundararaman, R. (2019). Substrate effects on charged defects in two-dimensional materials. *Phys. Rev. Mater.* 3: 083803.

77 Zhang, K. and Robinson, J. (2019). Doping of two-dimensional semiconductors: a rapid review and outlook. *MRS Adv.* 4: 2743–2757.

78 Shinde, P.V., Tripathi, A., Thapa, R., and Rout, C.S. (2022). Nanoribbons of 2D materials: a review on emerging trends, recent developments and future perspectives. *Coord. Chem. Rev.* 453: 214335.

79 Lavini, F., Rejhon, M., and Riedo, E. (2022). Two-dimensional diamonds from sp^2-to-sp^3 phase transitions. *Nat. Rev. Mater.* 7: 814–832.

80 Zhou, Q., Ju, W., Su, X. et al. (2017). Adsorption behavior of SO_2 on vacancy-defected graphene: a DFT study. *J. Phys. Chem. Solids* 109: 40–45.

81 Sinthika, S., Waghmare, U.V., and Thapa, R. (2018). Structural and electronic descriptors of catalytic activity of graphene-based materials: first-principles theoretical analysis. *Small* 14: 1703609.

82 Yang, T.T. and Saidi, W.A. (2020). Graphene activation explains the enhanced hydrogen evolution on graphene-coated molybdenum carbide electrocatalysts. *J. Phys. Chem. Lett.* 11: 2759–2764.

83 Zhang, Y.-C., Afzal, N., Pan, L. et al. (2019). Structure-activity relationship of defective metal-based photocatalysts for water splitting: experimental and theoretical perspectives. *Adv. Sci.* 6: 1900053.

84 Li, Y., Li, Y.-L., Sa, B., and Ahuja, R. (2017). Review of two-dimensional materials for photocatalytic water splitting from a theoretical perspective. *Catal. Sci. Technol.* 7: 545–559.

85 Zhao, X., Liu, X., Huang, B. et al. (2019). Hydroxyl group modification improves the electrocatalytic ORR and OER activity of graphene supported single and bi-metal atomic catalysts (Ni, Co, and Fe). *J. Mater. Chem. A* 7: 24583–24593.

86 Xue, Z., Zhang, X., Qin, J., and Liu, R. (2021). TMN_4 complex embedded graphene as bifunctional electrocatalysts for high efficiency OER/ORR. *J. Energy Chem.* 55: 437–443.

87 Zhang, X., Yang, Z., Lu, Z., and Wang, W. (2018). Bifunctional CoN_x embedded graphene electrocatalysts for OER and ORR: a theoretical evaluation. *Carbon* 130: 112–119.

88 Liang, Z., Luo, M., Chen, M. et al. (2020). Exploring the oxygen electrode bi-functional activity of Ni–N–C-doped graphene systems with N, C co-ordination and OH ligand effects. *J. Mater. Chem. A* 8: 20453–20462.

89 Riyaz, M., Gupta, S., and Goel, N. (2021). First principle studies to tailor graphene through synergistic effect as a highly efficient electrocatalyst for oxygen evolution reaction. *ChemPhysChem* 22: 1141–1147.

90 Riyaz, M., Garg, S., Kaur, N., and Goel, N. (2022). Boron doped graphene as anode material for Mg ion battery: a DFT study. *Comput. Theor. Chem.* 1214: 113757.

91 Datta, D., Li, J., and Shenoy, V.B. (2014). Defective graphene as a high-capacity anode material for Na-and Ca–ion batteries. *ACS Appl. Mater. Interfaces* 6: 1788–1795.

92 Gui, G., Li, J., and Zhong, J. (2008). Band structure engineering of graphene by strain: first-principles calculations. *Phys. Rev. B* 78: 075435.

93 Sofo, J.O., Chaudhari, A.S., and Barber, G.D. (2007). Graphane: a two-dimensional hydrocarbon. *Phys. Rev. B* 75: 153401.

94 Hussain, T., De Sarkar, A., and Ahuja, R. (2012). Strain induced lithium functionalized graphane as a high capacity hydrogen storage material. *Appl. Phys. Lett.* 101: 103907.

95 Hussain, T., Pathak, B., Ramzan, M. et al. (2012). Calcium doped graphane as a hydrogen storage material. *Appl. Phys. Lett.* 100: 183902.

96 Poh, H.L., Sofer, Z., and Pumera, M. (2012). Graphane electrochemistry: electron transfer at hydrogenated graphenes. *Electrochem. Commun.* 25: 58–61.

97 Topsakal, M., Cahangirov, S., and Ciraci, S. (2010). The response of mechanical and electronic properties of graphane to the elastic strain. *Appl. Phys. Lett.* 96: 091912.

98 Kumar, R., Goel, N., Riyaz, M., and Gupta, S. (2021). Effect of boron and nitrogen doping on mechanical and electronic properties of graphane under uni-axial strain conditions: a DFT study. *Comput. Theor. Chem.* 1200: 113195.

99 Pauling, L. (1931). The nature of the chemical bond. II. The one-electron bond and the three-electron bond. *J. Am. Chem. Soc.* 53: 3225–3237.

100 Kollman, P.A. (1977). Noncovalent interactions. *Acc. Chem. Res.* 10: 365–371.

101 Zhang, M. and Li, T. (2014). Intermolecular interactions in organic crystals: gaining insight from electronic structure analysis by density functional theory. *CrystEngComm* 16: 7162–7171.

102 Spackman, M.A. and Jayatilaka, D. (2009). Hirshfeld surface analysis. *CrystEngComm* 11: 19–32.

103 Tognetti, V. and Joubert, L. (2014). Density functional theory and Bader's atoms-in-molecules theory: towards a vivid dialogue. *Phys. Chem. Chem. Phys.* 16: 14539–14550.

104 Schneider, H.-J. (2019). Quantification of noncovalent interactions–promises and problems. *New J. Chem.* 43: 15498–15512.

105 Clark, T., Hennemann, M., Murray, J.S., and Politzer, P. (2007). Halogen bonding: the σ-hole. *J. Mol. Model.* 13: 291–296.

106 Desiraju, G.R., Ho, P.S., Kloo, L. et al. (2013). Definition of the halogen bond (IUPAC Recommendations 2013). *Pure Appl. Chem.* 85: 1711–1713.

107 Kollman, P.A. and Allen, L.C. (1972). Theory of the hydrogen bond. *Chem. Rev.* 72: 283–303.

108 Murray, J.S., Riley, K.E., Politzer, P., and Clark, T. (2010). Directional weak intermolecular interactions: σ-hole bonding. *Aust. J. Chem.* 63: 1598–1607.

109 Hellman, H. (1937). *Einführung in die Quantenchemie*, 285. Leipzig: Franz Deuticke.

110 Feynman, R.P. (1939). Forces in molecules. *Phys. Rev.* 56: 340.

111 Del Bene, J.E., Alkorta, I., and Elguero, J. (2009). Characterizing complexes with F–Li…N, H–Li…N, and CH_3Li…N Lithium bonds: structures, binding energies, and spin–spin coupling constants. *J. Phys. Chem. A* 113: 10327–10334.

112 Esrafili, M.D. and Mohammadirad, N. (2014). Halogen bond interactions enhanced by sodium bonds—theoretical evidence for cooperative and substitution effects in NCX…NCNa…NCY complexes (X = F, Cl, Br, I; Y = H, F, OH). *Can. J. Chem.* 92: 653–658.

113 Montero-Campillo, M.M., Mó, O., Yáņez, M. et al. (2019). Chapter three - The beryllium bond. *Adv. Inorg. Chem.* 73: 73–121.

114 Yang, X., Li, Q., Cheng, J., and Li, W. (2013). A new interaction mechanism of $LiNH_2$ with MgH_2: magnesium bond. *J. Mol. Model.* 19: 247–253.

115 Montero-Campillo, M.M., Sanz, P., Mó, O. et al. (2018). Alkaline-earth (Be, Mg and Ca) bonds at the origin of huge acidity enhancements. *Phys. Chem. Chem. Phys.* 20: 2413–2420.

116 Frontera, A. and Bauzá, A. (2018). Regium–π bonds: an unexplored link between noble metal nanoparticles and aromatic surfaces. *Chem. Eur. J.* 24: 7228–7234.

117 Wang, S.R., Arrowsmith, M., Braunschweig, H. et al. (2017). Monomeric 16-electron π-diborene complexes of Zn(II) and Cd(II). *J. Am. Chem. Soc.* 139: 10661–10664.

118 Grabowski, S.J. (2014). Boron and other triel Lewis acid centers: from hypovalency to hypervalency. *ChemPhysChem* 15: 2985–2993.

119 Bauzá, A., Mooibroek, T.J., and Frontera, A. (2013). Tetrel-bonding interaction: rediscovered supramolecular force? *Angew. Chem. Int. Ed.* 52: 12317–12321.

120 Zahn, S., Frank, R., Hey-Hawkins, E., and Kirchner, B. (2011). Pnicogen bonds: a new molecular linker? *Chem. Eur. J.* 17: 6034–6038.

121 Wang, W., Ji, B., and Zhang, Y. (2009). Chalcogen bond: a sister noncovalent bond to halogen bond. *J. Phys. Chem. A* 113: 8132–8135.

122 Bauzá, A. and Frontera, A. (2015). Aerogen bonding interaction: a new supramolecular force? *Angew. Chem. Int. Ed.* 54: 7340–7343.

123 Daolio, A., Pizzi, A., Calabrese, M. et al. (2021). Molecular electrostatic potential and noncovalent interactions in derivatives of group 8 elements. *Angew. Chem.* 133: 20891–20895.

124 Daolio, A., Pizzi, A., Terraneo, G. et al. (2021). Anion· · ·anion coinage bonds: the case of tetrachloridoaurate. *Angew. Chem.* 133: 14506–14510.

125 Bachman, R.E., Fioritto, M.S., Fetics, S.K., and Cocker, T.M. (2001). The structural and functional equivalence of aurophilic and hydrogen bonding: evidence for the first examples of rotator phases induced by aurophilic bonding. *J. Am. Chem. Soc.* 123: 5376–5377.

126 Khera, M. and Goel, N. (2020). Qualitative as well as quantitative analysis of interactions present in chlorine and bromine substituted aromatic organic crystals: a DFT linked crystal explorer study. *J. Mol. Graph. Model.* 95: 107503.

127 Alkorta, I., Montero-Campillo, M.M., Mó, O. et al. (2019). Weak interactions get strong: synergy between tetrel and alkaline-earth bonds. *J. Phys. Chem. A* 123: 7124–7132.

128 Khera, M. and Goel, N. (2022). Cooperative effect of noncovalent interactions on tetrel bonding in halogenated silanes. *ChemPhysChem* 23: e202100776.

129 Hoppe, H. and Sariciftci, N.S. (2004). Organic solar cells: an overview. *J. Mater. Res.* 19: 1924–1945.

130 Ameri, T., Dennler, G., Lungenschmied, C., and Brabec, C.J. (2009). Organic tandem solar cells: a review. *Energy Environ. Sci.* 2: 347–363.

131 Wöhrle, D. and Meissner, D. (1991). Organic solar cells. *Adv. Mater.* 3: 129–138.

132 Clarke, T.M. and Durrant, J.R. (2010). Charge photogeneration in organic solar cells. *Chem. Rev.* 110: 6736–6767.

133 Arbogast, J.W., Darmanyan, A.P., Foote, C.S. et al. (1991). Photophysical properties of sixty atom carbon molecule C_{60}. *J. Phys. Chem.* 95: 11–12.

134 Allemand, P., Koch, A., Wudl, F. et al. (1991). Two different fullerenes have the same cyclic voltammetry. *J. Am. Chem. Soc.* 113: 1050–1051.

135 Sariciftci, N.S., Smilowitz, L., Heeger, A.J., and Wudl, F. (1992). Photoinduced electron transfer from a conducting polymer to buckminsterfullerene. *Science* 258: 1474–1476.

136 Günes, S., Neugebauer, H., and Sariciftci, N.S. (2007). Conjugated polymer-based organic solar cells. *Chem. Rev.* 107: 1324–1338.

137 Deibel, C. and Dyakonov, V. (2010). Polymer–fullerene bulk heterojunction solar cells. *Rep. Prog. Phys.* 73: 096401.

138 Zhou, E., Hashimoto, K., and Tajima, K. (2013). Low band gap polymers for photovoltaic device with photocurrent response wavelengths over 1000 nm. *Polymer* 54: 6501–6509.

139 You, J., Dou, L., Yoshimura, K. et al. (2013). A polymer tandem solar cell with 10.6% power conversion efficiency. *Nat. Commun.* 4: 1–10.

140 Meng, L., Zhang, Y., Wan, X. et al. (2018). Organic and solution-processed tandem solar cells with 17.3% efficiency. *Science* 361: 1094–1098.

141 Sholl, D.S. and Steckel, J.A. (2011). *Density Functional Theory: A Practical Introduction*. Wiley.

142 Jain, A., Shin, Y., and Persson, K.A. (2016). Computational predictions of energy materials using density functional theory. *Nat. Rev. Mater.* 1: 1–13.

143 Ku, J., Lansac, Y., and Jang, Y.H. (2011). Time-dependent density functional theory study on benzothiadiazole-based low-band-gap fused-ring copolymers for organic solar cell applications. *J. Phys. Chem. C* 115: 21508–21516.

144 Babu, N.S. (2021). *Density Functional Theory-Recent Advances, New Perspectives and Applications*. IntechOpen.

145 Sham, L. and Schlüter, M. (1985). Density-functional theory of the band gap. *Phys. Rev. B* 32: 3883.

146 Vaschetto, M.E., Monkman, A.P., and Springborg, M. (1999). First-principles studies of some conducting polymers: PPP, PPy, PPV, PPyV, and PANI. *J. Mol. Struct. THEOCHEM* 468: 181–191.

147 Scharber, M.C., Koppe, M., Gao, J. et al. (2010). Influence of the bridging atom on the performance of a low-bandgap bulk heterojunction solar cell. *Adv. Mater.* 22: 367–370.

148 Autschbach, J. (2009). Charge-transfer excitations and time-dependent density functional theory: problems and some proposed solutions. *ChemPhysChem* 10: 1757–1760.

149 Raftani, M., Abram, T., Kacimi, R. et al. (2020). Organic compounds based on pyrrole and terphenyl for organic light-emitting diodes (OLED) applications: design and electro-optical properties. *J. Mater. Environ. Sci.* 11: 933–946.

150 Chen, H., Cui, Q., Yu, G. et al. (2011). Synthesis and characterization of novel semiconductors based on thieno[3,2-*b*][1]benzothiophene cores and their applications in the organic thin-film transistors. *J. Phys. Chem. C* 115: 23984–23991.

151 Koster, L.J.A., Mihailetchi, V., and Blom, P.W. (2006). Ultimate efficiency of polymer/fullerene bulk heterojunction solar cells. *Appl. Phys. Lett.* 88: 093511.

152 Scharber, M.C. (2016). On the efficiency limit of conjugated polymer: fullerene-based bulk heterojunction solar cells. *Adv. Mater.* 28: 1994–2001.

153 Reyes-Reyes, M., Kim, K., and Carroll, D.L. (2005). High-efficiency photovoltaic devices based on annealed poly(3-hexylthiophene) and 1-(3-methoxycarbonyl)-propyl-1-phenyl-(6, 6)C_{61} blends. *Appl. Phys. Lett.* 87: 083506.

154 Xu, T. and Yu, L. (2014). How to design low bandgap polymers for highly efficient organic solar cells. *Mater. Today Commun.* 17: 11–15.

155 Khokhar, D., Jadoun, S., Arif, R., and Jabin, S. (2021). Functionalization of conducting polymers and their applications in optoelectronics. *Polym. Plast. Technol. Eng.* 60: 463–485.

156 Ullah, H., Shah, A.-u.-H.A., Bilal, S., and Ayub, K. (2014). Doping and dedoping processes of polypyrrole: DFT study with hybrid functionals. *J. Phys. Chem. C* 118: 17819–17830.

157 Yeh, N. and Yeh, P. (2013). Organic solar cells: their developments and potentials. *Renewable Sustainable Energy Rev.* 21: 421–431.

158 Brabec, C.J., Winder, C., Sariciftci, N.S. et al. (2002). A low-bandgap semiconducting polymer for photovoltaic devices and infrared emitting diodes. *Adv. Funct. Mater.* 12: 709–712.

159 Namsheer, K. and Rout, C.S. (2021). Conducting polymers: a comprehensive review on recent advances in synthesis, properties and applications. *RSC Adv.* 11: 5659–5697.

160 Babu, N.S. (2022). DFT and TD-DFT studies of new triphenylamine-based (D–A–D) donor materials for high-efficiency organic solar cells. *Mater. Adv.* 3: 3526–3535.

161 Yang, Q., Yu, W., Lv, J. et al. (2022). Effects of fluorination position on all-polymer organic solar cells. *Dyes Pigm.* 200: 110180.

162 Liang, Z., Yan, L., Si, J. et al. (2021). Rational design and characterization of symmetry-breaking organic semiconductors in polymer solar cells: a theory insight of the asymmetric advantage. *Materials* 14: 6723.

163 Coleone, A.P., Lascane, L.G., and Batagin-Neto, A. (2019). Polypyrrole derivatives for optoelectronic applications: a DFT study on the influence of side groups. *Phys. Chem. Chem. Phys.* 21: 17729–17739.

164 Franco, F.C. Jr. (2020). Tuning the optoelectronic properties of oligothiophenes for solar cell applications by varying the number of cyano and fluoro substituents for solar cell applications: a theoretical study. *J. Chem. Res.* 44: 235–242.

165 Garg, S. and Goel, N. (2022). Optoelectronic applications of conjugated organic polymers: influence of donor/acceptor groups through density functional studies. *J. Phys. Chem. C* 126 (22): 9313–9323.

166 Kim, M.-C., Sim, E., and Burke, K. (2013). Understanding and reducing errors in density functional calculations. *Phys. Rev. Lett.* 111: 073003.

167 Sim, E., Song, S., Vuckovic, S., and Burke, K. (2022). Improving results by improving densities: density-corrected density functional theory. *J. Am. Chem. Soc.* 144: 6625–6639.

168 Mori-Sánchez, P., Cohen, A.J., and Yang, W. (2008). Localization and delocalization errors in density functional theory and implications for band-gap prediction. *Phys. Rev. Lett.* 100: 146401.

169 Cohen, A.J., Mori-Sánchez, P., and Yang, W. (2008). Fractional charge perspective on the band gap in density-functional theory. *Phys. Rev. B* 77: 115123.

25

Structure, Stability and Bonding in Ligand Stabilized C_3 Species

Sudip Pan and Zhong-hua Cui

Institute of Atomic and Molecular Physics, Jilin University, Changchun, China

25.1 Introduction

Ligand-supported transient species, small clusters or even single atoms become a fascinating arena of research, particularly because of their catalytic property or acting as precursors for different transformations [1]. Persistent carbenes like cyclic (alkyl)(amino) carbene (cAAC) [1] and N-heterocyclic carbene (NHC) [2], in addition to the phosphine, turn out to be excellent ligands for this purpose [3]. Emphasizing the carbon atom(s) to be stabilized by ligands, in 1961, Ramirez et al. [4] reported the experimental synthesis of the first divalent carbon(0) complex, carbodiphosphorane $C(PPh_3)_2$, the structure of which was later properly elucidated by X-ray analysis in 1978 by Hardy et al. [5]. Later, Frenking and coworkers analyzed the bonding situation and found that the bonding in $C(PPh_3)_2$ is best described as the donor–acceptor interactions, $Ph_3P{\rightarrow}C(0){\leftarrow}PPh_3$ between the phosphine ligands and a bare carbon atom in the excited ^1D state [6]. Subsequent search for other carbon(0) complexes led to the in silico report of carbodicarbene $C(NHC)_2$ in 2007 [7], a benzoannulated derivative $C(NHC^{Bz})_2$ of which was experimentally isolated by the groups of Bertrand and Fürstner [8] in 2009. With time, the number of complexes $L{\rightarrow}C(0){\leftarrow}L$ with carbone [9] kind of bonding situation was extended for other donor ligands L [10]. However, the dative bonding notion in carbones was not accepted by all, initiating debate [11], but it has now been well-accepted to be a dominant Lewis form [12].

Regarding its next higher analogue, C_2, although long back in 1989, Bestmann et al. [13] prepared a push-pull molecule, $R_3P{-}C_2{-}BR_3$ with a C≡C core, the possibility of stabilization by employing a carbene as $NHC{\rightarrow}C_2{\leftarrow}NHC$ (**1**) was theoretically explored only in 2012 [14]. More recently, attempts to synthesize cAAC stabilized C_2 species as $cAAC{\rightarrow}C_2{\cdot}^{+}{\leftarrow}cAAC$ and $cAAC{\rightarrow}C_2{\leftarrow}cAAC$ (**2**) by the Roesky and Bertrand groups separately (even $\mathbf{2}^{2+}$ was reported by the former group) have become successful [15]. A similar effort to synthesize NHC stabilized C_2 via the deprotonation of a doubly protonated precursor, $[\mathbf{1}\text{-}\mathbf{H_2}]^{2+}$, failed presumably because of its small HOMO–LUMO energy gap and high proton affinity (PA) of **1** [16]. The C_4 core in **2** slightly deviates from the linearity (by 1.1–2.3°), which can be explained by a donor–acceptor description instead of an electron-sharing cumulene unit [16]. Later on, Wu et al. [17] isolated the 4-pyridylidene supported C_2 unit (**3**), which has an even more significant deviation from linearity (by 24.2°) than that in **2**. Recently, a single phosphine ligand stabilized C_2, $(L)_2P{\rightarrow}C_2$ was isolated by the groups of Ong and Frenking [18]. We also recently carried out bonding analysis

Electron Density: Concepts, Computation and DFT Applications, First Edition.
Edited by Pratim Kumar Chattaraj and Debdutta Chakraborty.

of $(L)_2C_2$ (L = NHC, cAAC, SNHC and DAC) compounds which shows that $(cAAC)_2C_2$ and $(DAC)_2C_2$ possess genuine cumulene C_4 moieties with electron-sharing $C{=}C_L$ σ and π bonds, whereas the bonding in core C_4 in $(NHC)_2C_2$ and $(SNHC)_2C_2$ is a combination of dative and electron-sharing interactions [19]. More recently, using $Me_3Si–C_2–I(Ph)FBF_3$ precursor reacting with CsF, Miyamoto et al. [20] used $Me_3Si–C_2–I(Ph)FBF_3$ as precursor to generate elusive C_2, which can be further employed to form graphene, carbon nanotubes and fullerenes. A follow-up study was also devoted to the bonding analysis of $Me_3E–C_2–I(Ph)FBF_3$ (E = C, Si, Ge) [21].

There is a scarcity of investigations towards the ligand-stabilized compounds beyond the C_2 unit. Only a single report was made by Bestmann et al. [22], where they synthesized $Ph_3P–C{=}C{=}C–PPh_3$ and $Ph_2(Me)P–C{\equiv}C–C{\equiv}C–B(CH_2Ph)_3$ based on the assumption that a linear C_n unit can be stabilized by attaching two donor ligands, and one donor and one acceptor ligands at either end of the C_n unit for n = odd and even number, respectively.

Here, we review the results of our recent density functional theory (DFT) study on $(L)_2C_3$ complexes where L = PPh_3, NHC^{Me} and $cAAC^{Me}$ [23]. The bonding situation is studied by employing natural bond orbital (NBO) and charge and energy decomposition analyses (EDA-NOCV). Here, we showed that the phosphine complex $Ph_3P–C_3–PPh_3$ unexpectedly has a synclinal (*gauche*) conformation as the most stable isomer and not a trans conformation as originally proposed [22]. The present results also showed that the $L–C_3–L$ with L = NHC^{Me} and $cAAC^{Me}$ are also viable for experimental synthesis at ambient conditions. While $(NHC^{Me})_2C_3$ complex has a trans-bent equilibrium geometry, $(cAAC^{Me})_2C_3$ has a nearly linear arrangement of the central C_4 core. The bonding in the $(cAAC^{Me})_2C_3$ complex is particularly different from the other two complexes.

25.2 Computational Details

The geometry optimizations and harmonic vibrational frequencies calculations of $(L)_2C_3$ complexes where L = NHC^{Me}, $cAAC^{Me}$ and PPh_3 in their respective singlet and triplet electronic states were performed at the BP86-D3(BJ)/def2-TZVPP level [24]. All the complexes are minima on the potential energy surfaces, as indicted by the absence of any imaginary frequency. For some complexes, additional computations were studied with the M06-2X functional [25]. These calculations were carried out with the Gaussian 16 program package [26]. The NBO analysis [27] was performed to evaluate the partial charge, Wiberg bond indices (WBI) and natural bond orbitals using the NBO 6.0 program [28]. Unless otherwise stated, we discussed geometries and energies at the BP86-D3(BJ)/def2-TZVPP level. Some *ab initio* computations were also performed with the DLPNO-CCSD(T) method [29] in conjunction with the cc-pVTZ basis set [30] using the program package ORCA [31].

The bonding situations were analyzed by means of an energy decomposition analysis (EDA) [32] together with the natural orbitals for chemical valence (NOCV) [33] method by using the ADF 2018.105 program package [34]. The EDA-NOCV calculations were carried out at the BP86-D3(BJ)/TZ2P level [35] using the BP86-D3(BJ)/def2-TZVPP optimized geometries. In this analysis, the intrinsic interaction energy (ΔE_{int}) between two fragments can be divided into four energy components as follows:

$$\Delta E_{int} = \Delta E_{elstat} + \Delta E_{Pauli} + \Delta E_{orb} + \Delta E_{disp} \quad (25.1)$$

While the electrostatic ΔE_{elstat} term represents the quasiclassical electrostatic interaction between the unperturbed charge distributions of the prepared fragments, the Pauli repulsion ΔE_{Pauli} corresponds to the energy change associated with the transformation from the superposition of the unperturbed electron densities of the isolated fragments to the wavefunction, which properly obeys the Pauli principle through explicit antisymmetrization and renormalization of the production wavefunction. Since we included D3(BJ), it provides us with the dispersion interaction energy between two interacting fragments ΔE_{disp}. The orbital term ΔE_{orb} is originated from the mixing of orbitals, charge transfer and polarization between the isolated fragments, which can be further decomposed into contributions from each irreducible representation of the point group of the interacting system as follows:

$$\Delta E_{\text{orb}} = \sum_r \Delta E_r \tag{25.2}$$

The combination of the EDA with NOCV enables the partition of the total orbital interactions into pairwise contributions of the orbital interactions, which is very vital to get a complete picture of the bonding. The charge deformation $\Delta\rho_k(r)$, resulting from the mixing of the orbital pairs $\psi_k(r)$ and $\psi_{-k}(r)$ of the interacting fragments, presents the amount and the shape of the charge flow due to the orbital interactions (Eq. (25.3)), and the associated energy term ΔE_{orb} provides with the size of stabilizing orbital energy originated from such interaction (Eq. (25.4)).

$$\Delta\rho_{\text{orb}}(r) = \sum_k \Delta\rho_k(r) = \sum_{k=1}^{N/2} v_k \left[-\psi_{-k}^2(r) + \psi_k^2(r)\right] \tag{25.3}$$

$$\Delta E_{\text{orb}} = \sum_k \Delta E_{\text{orb}}^k = \sum_k v_k \left[-F_{-k,-k}^{\text{TS}} + F_{k,k}^{\text{TS}}\right] \tag{25.4}$$

More details about the EDA-NOCV method and its application are given in recent review articles [36].

25.3 Structures and Energetics

Figure 25.1 displays the minimum energy geometries of $(L)_2C_3$ complexes where L = NHC^{Me}, $cAAC^{Me}$ and PPh_3 in their respective singlet and triplet electronic states. While the ground state geometry of $(NHC^{Me})_2C_3$ (**2**) corresponds to the C_2 point group and ^1A electronic state, the same for $(cAAC^{Me})_2C_3$ (**3**) and $(PPh_3)_2C_3$ (**1**) belongs to C_1 point group and ^1A electronic state. The triplet state for $(NHC^{Me})_2C_3$ also has a C_2 point group and ^3A electronic state and is 41.7 kcal/mol higher in energy than the corresponding singlet state. On the other hand, in cases of $(PPh_3)_2C_3$ and $(cAAC^{Me})_2C_3$, the triplet state is 38.5 and 30.8 kcal/mol higher in energy, respectively, than the singlet state and has C_1 symmetry and ^3A electronic state. Three complexes show distinct differences in their equilibrium geometry. The orientation of two ligands in **2** is trans-bent with respect to almost linear C_3 moiety in both their singlet and triplet electronic states. On the other hand, $cAAC^{Me}$ ligands in **3** bind with C_3, orienting themselves almost linearly along the C_3 axis as reflected from the L—C1—C2 and L—C3—C2 bond angles of 171.5° and 171.8°, respectively. Another interesting structural difference is that while in **2** two NHC^{Me} ligands lie in the same plane as that of C_3, in **3**, two $cAAC^{Me}$ ligands are perpendicular to each other, which can be understood from the torsional angle, *t*(N′–C′–C″–N′) of 178.7° in the former complex and of 99.9° in the latter one.

In the case of **1**, in the lowest energy form, the phosphine ligands are orientated around the central C_3 unit in a synclinal (*gauche*) conformation and not a trans conformation as originally proposed [22]. At the BP86-D3(BJ)/def2-TZVPP level, the trans conformation **1a** is 8.9 kcal/mol higher in energy than the gauche form **1**. Interestingly, exclusion of the dispersion correction leads to **1** only 0.4 kcal/mol lower in energy than **1a**, which indicates that dispersion interaction between –Ph ligands in the former conformer is the reason to make it lower in energy than

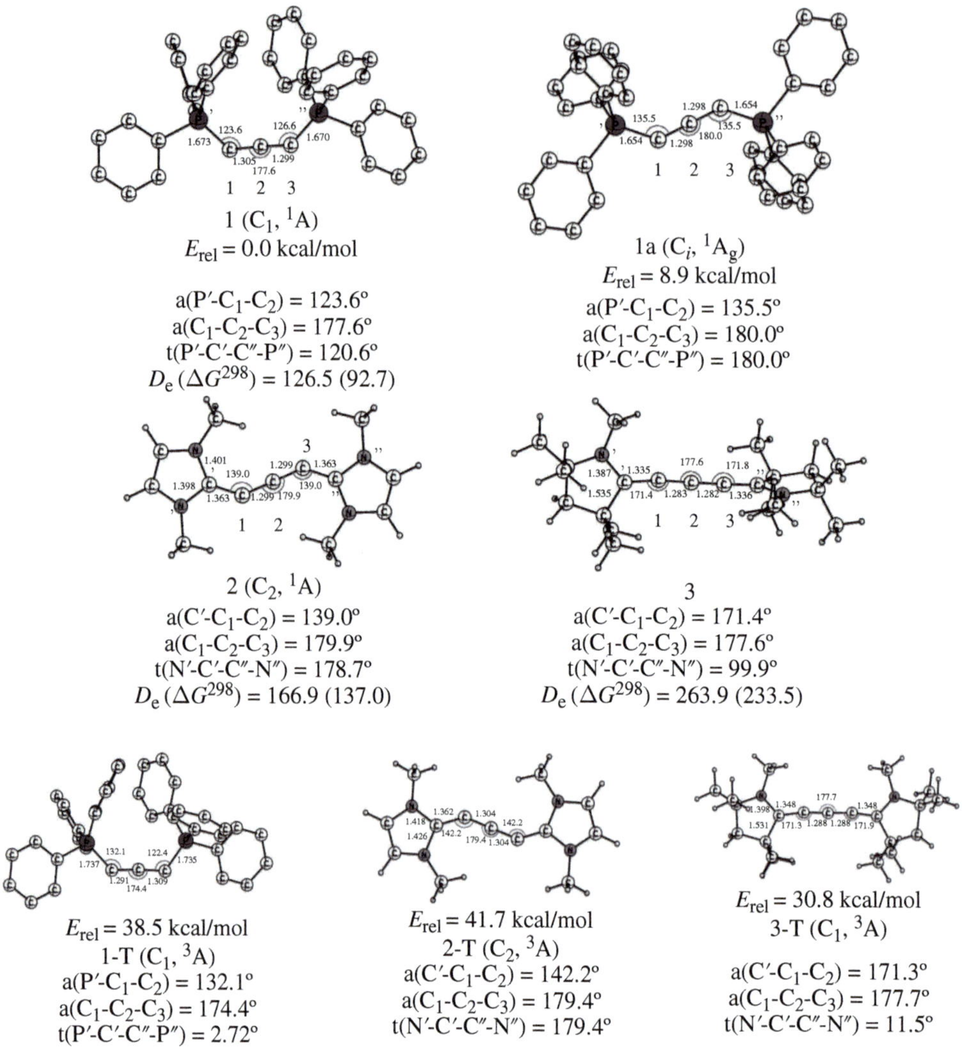

Figure 25.1 The optimized geometries of $(L)_2C_3$ (L = NHC^{Me}, $cAAC^{Me}$ and PPh_3) complexes at the BP86-D3(BJ)/def2-TZVPP level. Source: This figure is reproduced from Gorantla et al. [23].

the latter conformer. To rule out any artifact of the level of theory, we also minimized both the isomers at the M06-2X-D3/def2-TZVPP level and found that **1** is 7.0 kcal/mol lower in energy than **1a**. A similar observation is noted at the DLPNO-CCSD(T)/cc-pVTZ//BP86-D3(BJ)/def2-TZVPP level, which gives the *gauche* isomer **1** 7.7 kcal/mol lower in energy than the trans isomer **1a**. Therefore, there is no doubt that our gas-phase calculations unequivocally confirm that *gauche* form for **1** is the preferred one to adopt. The earlier experimental work also reported ^{31}P NMR spectrum of $(PPh_3)_2C_3$, which gives a signal at $\delta = -10.4$ ppm [22]. We also computed NMR signals of **1** and **1a** using the GIAO approach at the BP86/def2-TZVPP level to check which matches better with the experimental one. The computed chemical shifts for the ^{31}P NMR spectrum are $\delta = +11.3$ and $+17.5$ ppm for two nondegenerate P atoms of **1**, whereas it is $\delta = -5.6$ ppm for the degenerate P of **1a**. Therefore, **1a** gives better agreement with the experimental result, which was recorded in C_6D_6. It seems that in solution **1a** predominates over **1**, or there is a fast equilibrium between two gauche conformations, which mimics a trans form. A further rechecking of the relative energies after inclusion of the solvent effect of C_6D_6 using the CPCM [37] approach at the BP86-D3(BJ)/def2-TZVPP level still gives energetic preference of **1** over **1a** by 9.2 kcal/mol.

The C—C bond distances in the central C_3 core range from 1.282 to 1.301 Å, which fall in between the typical C≡C (1.20 Å) and C=C (1.34 Å) bond lengths, being closer to the latter bond length. The C–L bond distances are 1.67 (**1**), 1.36 (**2**) and 1.34 (**3**) Å, which agree with the C=C and P=C (1.69 Å) bond description [38]. The computed C—C and C—L bond lengths in $(cAAC^{Me})_2C_3$ are comparable with the reported C—C (1.249 Å) and C—L (1.324 Å) bond lengths in $(cAAC^{Dipp})_2C_2$ [13].

Figure 25.1 shows the computed C—L bond dissociation energies (BDEs) at 0 K and at 298 K with respect to dissociation, $(L)_2C_3 \rightarrow C_3 + 2L$. The BDE value, D_e is quite large and is the maximum for L = $cAAC^{Me}$ (263.9 kcal/mol) followed by NHC^{Me} (166.9 kcal/mol) and PPh_3 (126.5 kcal/mol). Moreover, even after the consideration of thermal correction and entropy correction, the BDE values at 298 K, ΔG^{298} still remain highly positive (92.7–233.5 kcal/mol), which implies that the bond dissociation is highly endergonic in nature at room temperature. Apart from BDE, the HOMO–LUMO energy gap (Δ_{H-L}) is also an important descriptor to understand the electronic stability of the complex. A complex with a small Δ_{H-L} value becomes too reactive to isolate experimentally. Previously, the small Δ_{H-L} value of $(NHC^{Bz})_2C_2$ (1.70 eV) was considered as the main culprit for the failure to get $(NHC^{Bz})_2C_2$ via the deprotonation of a doubly protonated precursor, $[(NHC^{Bz})_2C_2H_2]^{2+}$ [16]. On the other hand, $(cAAC^{Dipp})C_2$, having a reasonably large Δ_{H-L} value of 2.60 eV, was synthesized without any hurdle [15, 16]. In the present cases, Δ_{H-L} values are computed as 1.90 (**1**), 2.08 (**2**) and 2.90 (**3**) eV. Therefore, both $(NHC^{Me})_2C_3$ and $(CAAC^{Me})_2C_3$ have a larger Δ_{H-L} value than experimentally synthesized $(Ph_3P)_2C_3$ complex [22].

25.4 Bonding

We have first performed an NBO analysis to shed light into the bonding situation on the complexes. Table 25.1 shows the bonding composition of the natural bond orbital found in between L and C1,

Table 25.1 NBO results of the complexes $(L)_2C_3$ at the BP86/def2-TZVPP level of theory.

Complex	Bond	ON	$L-C_1$ polarization and hybridization (%)		*q* C1	*q* C2	*q* C3
1	P—C1 σ	1.96	P: 40.6 s(30.0), p(69.4)	C1: 59.4 s(49.0), p(51.0)	−0.80	0.09	−0.79
2	C_L—C1 σ	1.97	C_L: 52.6 s(42.1), p(57.9)	C1: 47.4 s(47.6), p(52.4)	−0.42	0.05	−0.42
	C_L—C1 π	1.65	C_L: 46.7 s(0.1), p(99.9)	C1: 53.3 s(0.1), p(99.9)			
3	C_L—C1 σ	1.97	C_L: 51.2 s(38.6), p(61.4)	C1: 48.8 s(50.1), p(49.9)	−0.19	−0.01	−0.19
	C_L—C1 π	1.80	C_L: 46.1 s(0.1), p(99.9)	C1: 53.9 s(0.1), p(99.9)			

Occupation number ON, polarization and hybridization of the L—C1 bonds and partial charges *q*.
Source: This table is reproduced from Gorantle et al. [23].

WBI values and partial charges on C in C_3 moiety. For **2** and **3**, the NBO method identifies σ- and π-natural bond orbitals either between C_L and C1 or between any two carbon atoms in the C_3 unit, giving a C_L=C=C=C=C_L Lewis description, which corroborates with their bond distances. In the case of **1**, only P—C1 σ bond is identified, indicating that the π-bond is weaker. A look at the % contribution reveals that while in **2** and **3**, both the σ- and π-bonds formed between C_L and C1 are almost nonpolarized, in the case of **3**, the σ-bond is more polarized towards C1. The WBI values ranging from 1.25 (**3**) to 1.60 (**2**) also support the double bond description. A lower WBI value for P—C1 bond of **3** corroborates with the less effective π-bond than the others. The net amount of charge on the C_3 unit is −1.5 (**1**), −0.79 (**2**) and −0.39 (**3**) e^-, showing a significant amount of electron transfer from ligand to C3 unit, where C1 and C3 centers carry the negative charge and the middle C2 centers are either positively charged or only minutely negatively charged. Such electron distribution hints about the possible Lewis basic property of the title complexes, which is evaluated by calculating their proton affinity (PA). As obvious from the charge distribution, C1 and C3 centers act as the Lewis basic centers. The computed first PA values are very large, being 295.0 (**2**) and 289.3 (**3**) kcal/mol, whereas the second PA is comparatively smaller (183.9 and 170.7 kcal/mol, respectively).

Now, the question is: what would be the best Lewis form between L:⇄:C=C=C:⇆:L and L=C=C=C=L to describe the actual bonding situation in the title complexes? This can be addressed by the EDA-NOCV method, which is able to dictate the most suitable fragments for the formation of a chemical bond. In general, when there is more than one possible Lewis structure available, the best one can be decided from the smallest orbital interaction value ΔE_{orb}. This is because those fragments are better prepared to generate the electronic structure of the complex. The utility of this criterion has been proved for a variety of chemical bonds [39]. We have performed EDA-NOCV on **1**, **2** and **3** considering $(L)_2$ and C_3 fragments in singlet electronic states (dative) and $(L)_2$ and C_3 fragments in quintet states (electron sharing). In the dative scheme, the C_3 fragment is in an excited singlet state where two σ-orbitals are vacant to act as acceptors for (+,+) and (+,−) σ-donation of $(L)_2$ and two π* orbitals are filled up that participate in L←C_3→L π/σ backdonation. Table 25.2 presents the numerical results of these two schemes.

The results show that for **2** and **3**, the electron-shared L=C=C=C=L Lewis form is better suited to describe the bonding between L and C_3, whereas **3** is best described as donor–acceptor Ph_3P:⇄:C=C=C:⇆:PPh_3. Interestingly, in addition to these neutral schemes, other possibilities are ionic schemes, as given in Table 25.2, which are not generally considered. In the ionic scheme, both dative and electron-sharing bonds are involved. A comparison of ΔE_{orb} value in all these four schemes now gives that **1** and **2** complexes can be best described as the interaction between mono-charged $(L)_2{}^+$ and $[C_3]^-$ in doublet states engaging both electron-sharing and dative bonding. On the other hand, in the case of **3**, still $(cAAC^{Me})_2$ and C_3 in quintet states forming electron-sharing σ and π bonds are the most reasonable scheme.

More details about EDA-NOCV results taking the best energy partitioning scheme are given in Table 25.3. Irrelevant of the electron-sharing (neutral) or mixed bonding (charged), in all cases, the orbital interaction (covalent) is dominant over the Coulombic interaction. The dispersion contribution is negligible. The most important information comes from the breakdown of the total ΔE_{orb} term into pair-wise orbital interactions, which are five orbital interactions, $\Delta E_{orb(1)} - \Delta E_{orb(5)}$ for **1** and **2**, and four, $\Delta E_{orb(1)} - \Delta E_{orb(4)}$ for **3**. The qualitative and quantitative information about the bonding becomes clear if we scrutinize the orbital terms, the involved fragments orbitals, the associated deformation densities $\Delta\rho_n$ and connected MOs, which are shown in Figures 25.2–25.4.

The strongest orbital interaction $\Delta E_{orb(1)}$ in **1** and **2** comes from the σ donation of the HOMO of the ligands $L_2{}^+$ into the LUMO of $C_3{}^-$. This is counter-intuitive that the cation can act as a donor

Table 25.2 EDA-NOCV results of $(L)_2C_3$ complexes using four different sets of fragments with different charges and electronic states (S = singlet, D = doublet, T = triplet, Q = quintet) and associated bond types at the BP86-D3(BJ)/TZ2P level.

Molecule	Bond type[a)]	Fragments	ΔE_{int}	ΔE_{Pauli}	ΔE_{elstat}	ΔE_{disp}	ΔE_{orb}
$(PPh_3)_2C_3$	D	$(PPh_3)_2$ (S) + C_3 (S)	−488.5	674.6	−428.1	−16.3	−718.6
	E	$(PPh_3)_2$ (Q) + C_3 (Q)	−461.1	630.4	−371.4	−16.3	−703.7
	D + E	$[(PPh_3)_2]^+$ (D) + $[C_3]^-$ (D)	−493.8	766.5	−565.5	−16.3	**−678.5**
	D + E	$[(PPh_3)_2]^{2+}$ (T) + $[C_3]^{2-}$ (T)	−824.4	967.8	−918.8	−16.3	−857.4
$(NHC^{Me})_2C_3$	D	$(NHC^{Me})_2$ (S) + C_3 (S)	−541.0	634.7	−466.2	−9.4	−700.1
	E	$(NHC^{Me})_2$ (Q) + C_3 (Q)	−544.8	506.4	−373.8	−9.4	−667.9
	D + E	$[(NHC^{Me})_2]^+$ (D) + $[C_3]^-$ (D)	−551.9	621.5	−517.2	−9.4	**−646.7**
	D + E	$[(NHC^{Me})_2]^{2+}$ (T) + $[C_3]^{2-}$ (T)	−875.9	659.2	−798.2	−9.4	−727.5
$(cAAC^{Me})_2C_3$	D	$(cAAC^{Me})_2$ (S) + C_3 (S)	−594.2	419.5	−359.6	−10.6	−643.5
	E	$(cAAC^{Me})_2$ (Q) + C_3 (Q)	−407.7	519.1	−348.5	−10.6	**−567.6**
	D + E	$[(cAAC^{Me})_2]^+$ (D) + $[C_3]^-$ (D)	−586.8	555.8	−445.4	−10.6	−686.5
	D + E	$[(cAAC^{Me})_2]^{2+}$ (T) + $[C_3]^{2-}$ (T)	−930.1	873.0	−819.7	−10.6	−972.7

Energies are in kcal/mol. The most favourable fragmentation scheme and bond type is given by the smallest ΔE_{orb} value written in bold.
a) D = Dative bond; E = Electron-sharing bond.
Source: This table is reproduced from Gorantle et al. [23].

Table 25.3 The EDA-NOCV results at the BP86-D3(BJ)/TZ2P level of L_2C_3 complexes **1–3** using $(L_2)^+$ and $(C_3)^-$ in the electronic doublet (D) states as interacting fragments for **1** and **2** and neutral L_2 and C_3 in the quintet (Q) states as interacting fragments for **3**.

Energy	Interaction[c]	$[(PPh_3)_2]^+$ (D) + $[C_3]^-$ (D)	$[(NHC^{Me})_2]^+$ (D) + $[C_3]^-$ (D)	$(cAAC^{Me})_2$ (Q) + C_3 (Q)
ΔE_{int}		−493.8	−551.9	−407.7
ΔE_{Pauli}		766.5	621.5	519.1
ΔE_{disp}[a]		−16.3 (1.2%)	−9.4 (0.8%)	−10.6 (1.1%)
ΔE_{elstat}[a]		−565.5 (44.8%)	−517.2 (44.0%)	−348.5 (37.6%)
ΔE_{orb}[a]		−678.5 (53.8%)	−646.7 (55.1%)	−567.6 (61.2%)
$\Delta E_{orb(1)}$[b]	L–C_3–L σ e^- sharing (+, +)			−187.6 (33.1%)
	L→C_3←L σ donation (+, +)	−286.3 (42.1%)	−235.1 (36.3%)	—
$\Delta E_{orb(2)}$[b]	L–C_3–L σ e^- sharing (+, −)	−231.3 (34.1%)	−208.7 (32.2%)	−180.3 (31.8%)
	L→C_3←L σ donation (+, −)			—
$\Delta E_{orb(3)}$[b]	L–C_3–L π/σ e^- sharing			−77.7 (13.7%)
	L←C_3→L σ back donation	−51.8 (7.6%)	−88.3 (13.6%)	—
$\Delta E_{orb(4)}$[b]	L–C_3–L π e^- sharing			−76.7 (13.5%)
	L←C_3→L π back donation	−40.4 (5.9%)	−52.9 (8.1%)	—
$\Delta E_{orb(5)}$[b]	L←C_3→L π back donation	−25.0 (3.6%)	−18.4 (2.8%)	—
$\Delta E_{orb(rest)}$[b]		−43.6 (6.4%)	−43.3 (6.6%)	−45.3 (8.0%)

Energies are in kcal/mol.
a) The values in the parentheses show the contribution to the total attractive interaction $\Delta E_{elstat} + \Delta E_{orb} + \Delta E_{disp}$.
b) The values in parentheses show the contribution to the total orbital interaction ΔE_{orb}.
Source: This table is reproduced from Gorantla et al. [23].

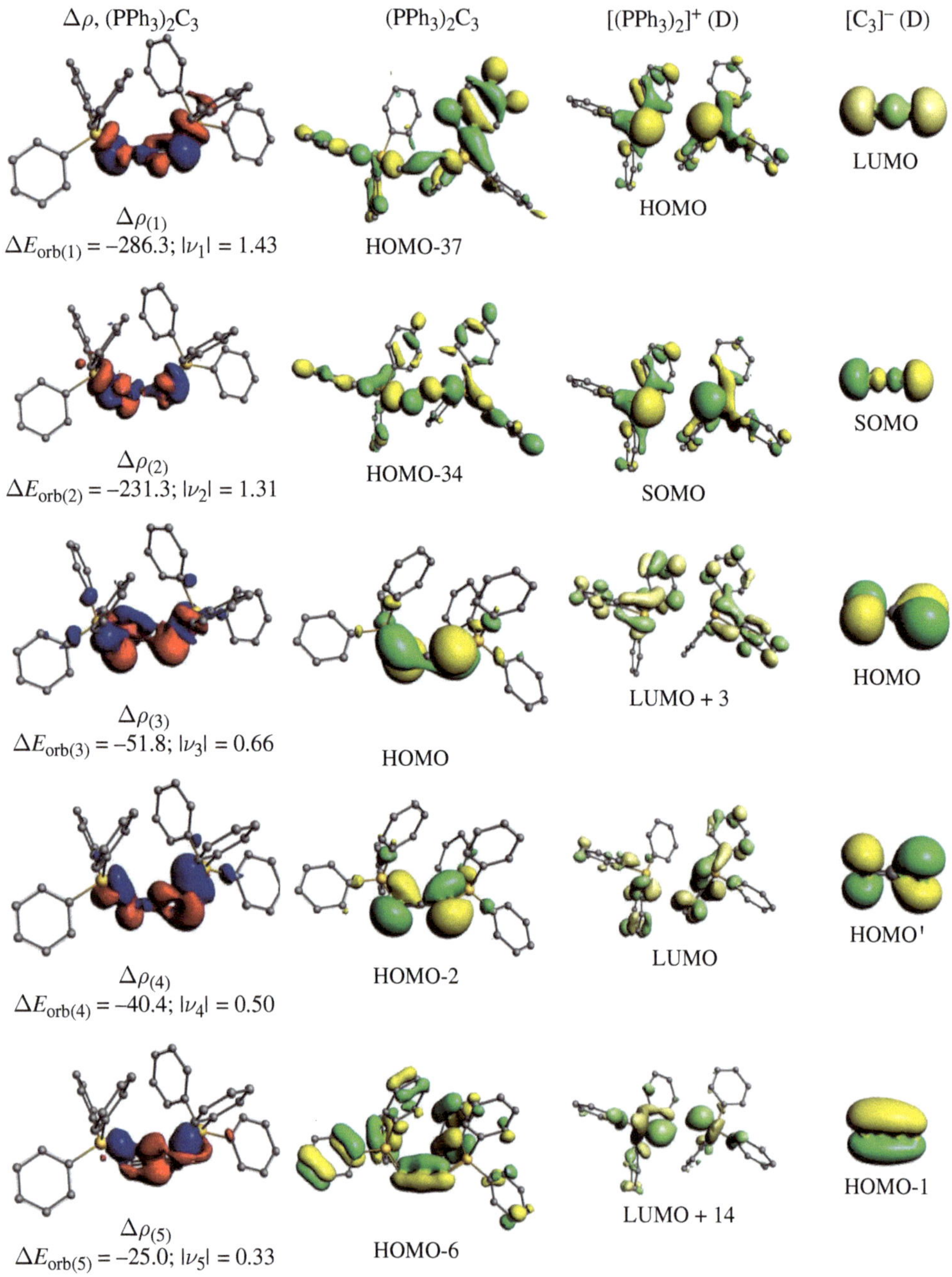

Figure 25.2 The shape of the deformation densities $\Delta\rho_{(1)-(5)}$ that correspond to $\Delta E_{\mathrm{orb}(1)-(5)}$, and the associated MOs of $(PPh_3)_2C_3$ and the fragments orbitals of $[(PPh_3)_2]^+$ and $[C_3]^-$ in the doublet state at the BP86-D3(BJ)/TZ2P level. Isosurface values are 0.003 a.u. The eigenvalues $|\nu_n|$ give the size of the charge migration in *e*. The direction of the charge flow of the deformation densities is red → blue. Source: This figure is reproduced from Gorantla et al. [23].

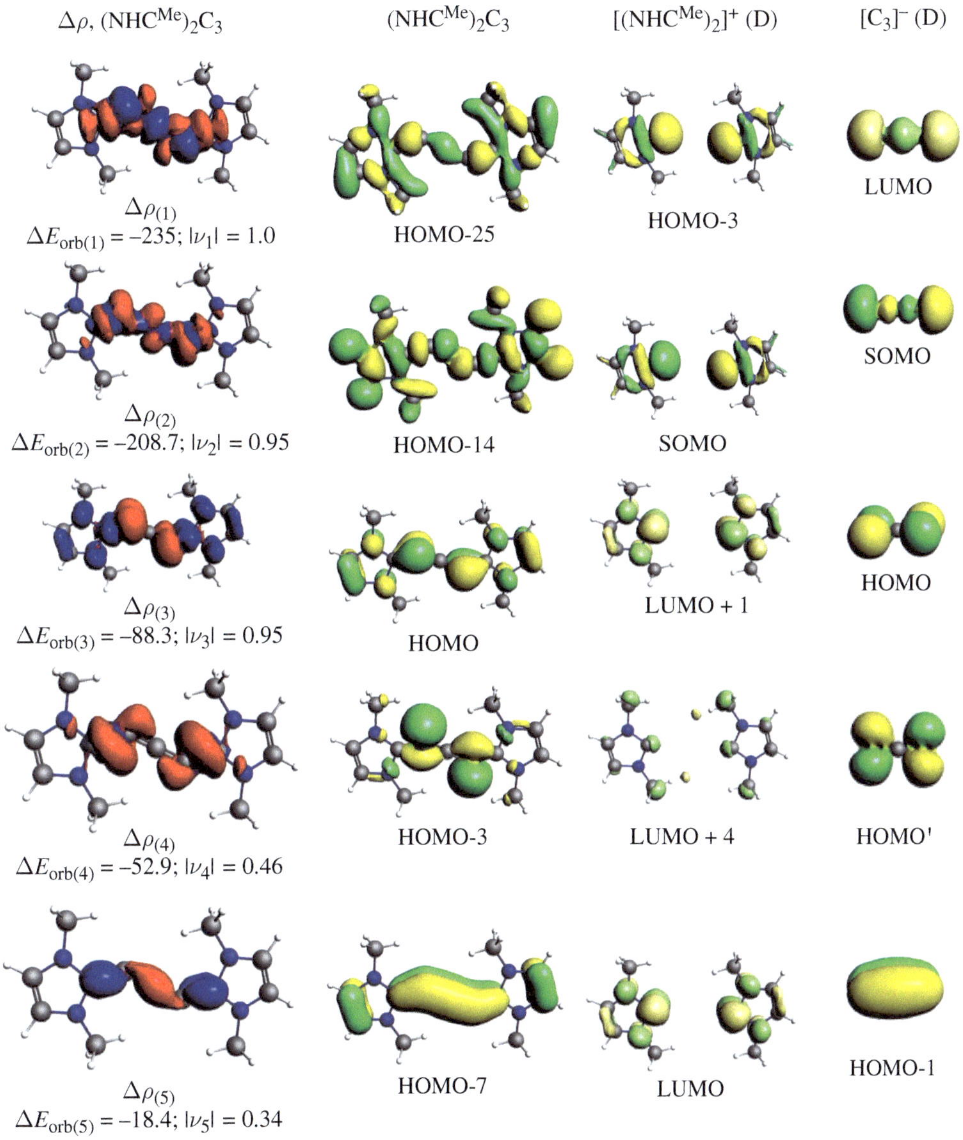

Figure 25.3 The shape of the deformation densities $\Delta\rho_{(1)-(5)}$ that correspond to $\Delta E_{orb(1)-(5)}$, and the associated MOs of $(NHC^{Me})_2C_3$ and the fragments orbitals of $[(NHC^{Me})_2]^+$ and $[C_3]^-$ in the doublet state at the BP86-D3(BJ)/TZ2P level. Isosurface values are 0.003 a.u. The eigenvalues $|v_n|$ give the size of the charge migration in *e*. The direction of the charge flow of the deformation densities is red → blue. Source: This figure is reproduced from Gorantla e al. [23].

and the anion as an acceptor. Somewhat weaker orbital term $\Delta E_{orb(2)}$ in **1** and **2** is originated from the electron-sharing interactions of the unpaired electrons in the SOMO of the fragments. The remaining three orbital terms $\Delta E_{orb(3)} - \Delta E_{orb(5)}$ in these complexes come from the π backdonation of the degenerate HOMO and the HOMO-1 of C_3^- to vacant orbitals of L_2^+. It may also be noted that although the strength of the π backdonation is significantly lower than the σ donation $\Delta E_{orb(1)}$ but the amount of charge migration that is provided by the eigenvalues υ_n, indicates that the degree of π backdonation has a similar magnitude as the σ donation. On the other hand, the four orbital terms

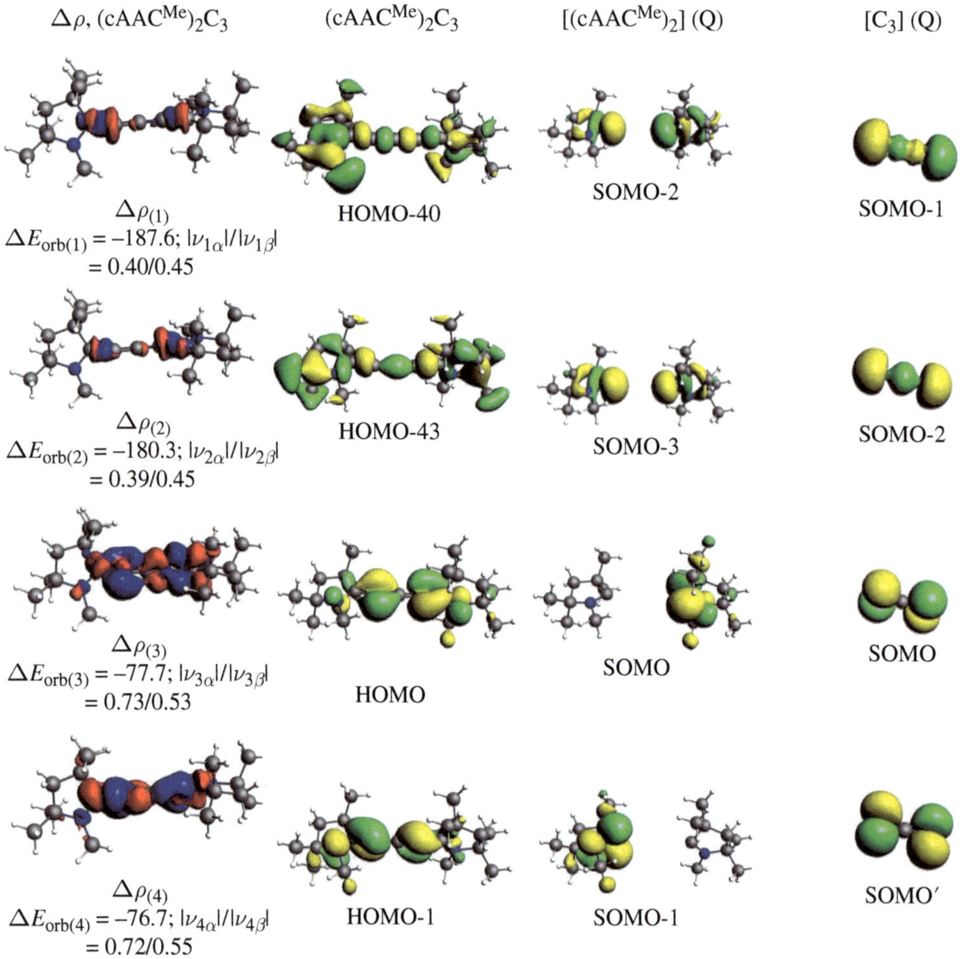

Figure 25.4 The shape of the deformation densities $\Delta\rho_{(1)-(5)}$ that correspond to $\Delta E_{orb(1)-(5)}$, and the associated MOs of $(cAAC^{Me})_2C_3$ and the fragments orbitals of $(cAAC^{Me})$ and C_3 in the quintet state at the BP86-D3(BJ)/TZ2P level. Isosurface values are 0.003 a.u. The eigenvalues $|\nu_n|$ give the size of the charge migration in *e*. The direction of the charge flow of the deformation densities is red → blue. Source: This figure is reproduced from Gorantla et al. [23].

$\Delta E_{orb(1)} - \Delta E_{orb(4)}$ between the neutral fragments in the quintet states in complex **3** are originated from the strong in-phase (+,+) and out-of-phase (+,−) electron-sharing σ-interactions $\Delta E_{orb(1)}$ and $\Delta E_{orb(2)}$ and relatively weaker but still significant π interactions $\Delta E_{orb(3)}$ and $\Delta E_{orb(4)}$.

25.5 Conclusions

The quantum chemical calculations show that in the equilibrium geometries of L–C_3–L with L = PPh_3 (**1**) and NHC^{Me} (**2**), the ligands are connected with rather acute bonding angles at the linear C_3 moiety. While **1** exhibits a synclinal (*gauche*) conformation, **2** has a trans conformation of the ligands around central C_3. On the other hand, the complex with L = $cAAC^{Me}$ (**3**) has a nearly linear arrangement of the $cAAC^{Me}$ ligands at the C_3 fragment. The BDEs of the ligands have the order **1** < **2** < **3**. The EDA-NOCV analysis suggests that **3** is best described as a cumulene with electron-sharing double bonds between neutral fragments $(cAAC^{Me})=C_3=(cAAC^{Me})$, whereas **1** and **2** possess electron-sharing and dative bonds between positively charged ligands $[(PPh_3)_2]^+$ and $[(NHC^{Me})_2]^+$ and negatively charged $[C_3]^-$.

Acknowledgements

This work was funded by the National Natural Science Foundation of China (No. 11874178, 11922405, 91961204). The partial calculations in this work were supported by the High-Performance Computing Center of Jilin University, China.

References

1 (a) Bourissou, D., Guerret, O., Gabbai, F.P., and Bertrand, G. (2000). *Chem. Rev.* 100 (1): 39–92. (b) Lavallo, V., Canac, Y., Prasang, C. et al. (2005). *Angew. Chem.* 117: 5851; (2005). *Angew. Chem. Int. Ed.* 44: 5705. (c) Soleilhavoup, M. and Bertrand, G. (2015). *Acc. Chem. Res.* 48: 256.

2 Arduengo, A.J. III, Harlow, R.L., and Kline, M. (1991). *JACS* 113: 2801.

3 (a) Mondal, K.C., Roesky, H.W., Schwarzer, M.C. et al. (2013). *Angew. Chem. Int. Ed.* 52: 2963. (b) Li, Y., Mondal, K.C., Roesky, H.W. et al. (2013). *JACS* 135: 12422. (c) Chu, T., Belding, L., van der Est, A. et al. (2014). *Angew. Chem. Int. Ed.* 53: 2711. (d) Mondal, K.C., Samuel, P.P., Roesky, H.W. et al. (2014). *JACS* 136: 8919. (e) Holzmann, N., Dange, D., Jones, C., and Frenking, G. (2013). *Angew. Chem. Int. Ed.* 52: 3004. (f) Holzmann, N., Stasch, A., Jones, C., and Frenking, G. (2011). *Chem. Eur. J.* 17: 13517. (g) Scheschkewitz, D. (2005). *Angew. Chem. Int. Ed.* 44: 2954; (2005). *Angew. Chem.* 117: 3014;. (h) Scheschkewitz, D. (2009). *Chem. Eur. J.* 15: 2476. (i) Bejan, I. and Scheschkewitz, D. (2007). *Angew. Chem. Int. Ed.* 46: 5783. (j) Abersfelder, K., White, A.J.P., Rzepa, H.S., and Scheschkewitz, D. (2010). *Science* 327: 564. (k) Abersfelder, K., White, A.J.P., Berger, R.J.F. et al. (2011). *Angew. Chem. Int. Ed.* 50: 7936. (l) Mondal, K.C., Roy, S., Dittrich, B. et al. (2016). *Angew. Chem. Int. Ed.* 55: 3158. (m) Pan, S., Saha, R., Osorio, E. et al. (2017). *Chem. Eur. J.* 23: 7463. (n) Saha, R., Pan, S., and Chattaraj, P.K. (2018). *ACS Omega* 3: 13720. (o) Ghara, M., Pan, S., and Chattaraj, P.K. (2019). *J. Phys. Chem. A* 123 (50): 10764. (p) Arrowsmith, M., Braunschweig, H., Celik, M. et al. (2016). *Nat. Chem.* 8: 890. (q) Wang, G., Walley, J.E., Dickie, D.A. et al. (2020). *JACS* 142 (10): 4560.

4 Ramirez, F., Desai, N.B., Hansen, B., and McKelvie, N. (1961). *JACS* 83: 3539.

5 Hardy, G.E., Zink, J.I., Kaska, W.C., and Baldwin, J.C. (1978). *JACS* 100: 8001.

6 Tonner, R., Öxler, F., Neumüller, B. et al. (2006). *Angew. Chem. Int. Ed. Engl.* 45: 8038.

7 Tonner, R. and Frenking, G. (2007). *Angew. Chem. Int. Ed.* 46: 8695.

8 (a) Dyker, C.A., Lavallo, V., Donnadieu, B., and Bertrand, G. (2008). *Angew. Chem. Int. Ed.* 47: 3206. (b) Fürstner, A., Alcarazo, M., Goddard, R., and Lehmann, C.W. (2008). *Angew. Chem. Int. Ed.* 47: 3210.

9 Frenking, G. and Tonner, R. (2009). *Pure Appl. Chem.* 81: 597.

10 (a) Tonner, R. and Frenking, G. (2008). *Chem. Eur. J.* 14: 326; (b) Tonner, R. and Frenking, G. (2008). *Chem. Eur. J.* 14: 3273; (c) Tonner, R., Heydenrych, G., and Frenking, G. (2008). *ChemPhysChem* 9: 1474; (d) Klein, S., Tonner, R., and Frenking, G. (2010). *Chem. Eur. J.* 16: 10160; (e) Esterhuysen, C. and Frenking, G. (2011). *Chem. Eur. J.* 17: 9944. (f) Klein, S. and Frenking, G. (2010). *Angew. Chem. Int. Ed.* 49: 7106. (g) Esterhuysen, C. and Frenking, G.

(2013). *Dalton Trans.* 42: 13349. (h) Kaufhold, O. and Hahn, F.E. (2008). *Angew. Chem. Int. Ed.* 47: 4057; (i) Phukan, A.K. and Guha, A.K. (2012). *Dalton Trans.* 41: 8973.

11 (a) Himmel, D., Krossing, I., and Schnepf, A. (2014). *Angew. Chem. Int. Ed.* 53: 370–374; (b) Frenking, G. (2014). *Angew. Chem. Int. Ed.* 53: 6040; (c) Himmel, D., Krossing, I., and Schnepf, A. (2014). *Angew. Chem. Int. Ed.* 53: 6047.

12 (a) Frenking, G., Tonner, R., Klein, S. et al. (2014). *Chem. Soc. Rev.* 43: 5106. (b) Frenking, G., Hermann, M., Andrada, D.M., and Holzmann, N. (2016). *Chem. Soc. Rev.* 45: 1129. (c) Zhao, L., Hermann, M., Holzmann, N., and Frenking, G. (2017). *Coord. Chem. Rev.* 344: 163.

13 Bestmann, H.J., Behl, H., and Bremer, M. (1989). *Angew. Chem. Int. Ed.* 28: 1219.

14 Dutton, J.L. and Wilson, D.J.D. (2012). *Angew. Chem. Int. Ed.* 51: 1477.

15 (a) Li, Y., Mondal, K.C., Samuel, P.P. et al. (2014). *Angew. Chem. Int. Ed.* 53: 4168. (b) Jin, L., Melaimi, M., Liu, L., and Bertrand, G. (2014). *Org. Chem. Front.* 1: 351.

16 Georgiou, D.C., Stringer, B.D., Hogan, C.F. et al. (2015). *Chem. Eur. J.* 21: 3377.

17 Wu, D., Li, Y., Ganguly, R., and Kinjo, R. (2014). *Chem. Commun.* 50: 12378.

18 Leung, T.-F., Jiang, D., Wu, M.-C. et al. (2021). *Nat. Chem.* 13: 89.

19 Gorantla, S.M.N.V.T., Pan, S., Mondal, K.C., and Frenking, G. (2021). *J. Phys. Chem. A* 125: 291.

20 Miyamoto, K., Narita, S., Masumoto, Y. et al. (2020). *Nat. Commun.* 11: 2134.

21 Gorantla, S.M.N.V.T., Pan, S., Mondal, K.C., and Frenking, G. (2022). *Pure Appl. Chem.* 94: 767–781.

22 Bestmann, H.J., Hadawi, D., Behl, H. et al. (1993). *Angew. Chem. Int. Ed.* 32: 1205.

23 Gorantla, S.M.N.V.T., Pan, S., Mondal, K.C., and Frenking, G. (2020). *Chem. Eur. J.* 26: 14211.

24 (a) Becke, A.D. (1988). *Phys. Rev. A* 38: 3098. (b) Perdew, J.P. (1986). *Phys. Rev. B* 33: 8822. (c) Grimme, S., Ehrlich, S., and Goerigk, L. (2011). *J. Comput. Chem.* 32: 1456. (d) Grimme, S., Antony, J., Ehrlich, S., and Krieg, H. (2010). *J. Chem. Phys.* 132: 154104. (e) Weigend, F. and Ahlrichs, R. (2005). *Phys. Chem. Chem. Phys.* 7: 3297. (f) Weigend, F. (2006). *Phys. Chem. Chem. Phys.* 8: 1057.

25 Zhao, Y. and Truhlar, D.G. (2006). *Theor. Chem. Acc.* 120: 215–241.

26 Frisch, M.J., Trucks, G.W., Schlegel, H.B. et al. (2016). *Gaussian 16, Revision A.03*. Gaussian, Inc.: Wallingford, CT.

27 (a) Weinhold, F. and Landis, C. (2005). *Valency and Bonding, a Natural Bond Orbital Donor–Acceptor Perspective*. Cambridge: Cambridge University Press. (b) Landis, C.R. and Weinhold, F. (2014). The NBO view of chemical bonding. In: *The Chemical Bond: Fundamental Aspects of Chemical Bonding* (ed. G. Frenking and S. Shaik), 91–120. Wiley.

28 Glendening, E.D., Landis, C.R., and Weinhold, F. (2013). *J. Comput. Chem.* 34: 1429.

29 (a) Riplinger, C. and Neese, F. (2013). *J. Chem. Phys.* 138: 034106; (b) Riplinger, C., Sandhoefer, B., Hansen, A., and Neese, F. (2013). *J. Chem. Phys.* 139: 134101.

30 Dunning, T.H. Jr. (1989). *J. Chem. Phys.* 90: 1007.

31 Neese, F. (2018). *WIREs Comput. Mol. Sci.* 8: e1327.

32 Ziegler, T. and Rauk, A. (1977). *Theor. Chim. Acta* 46: 1.

33 (a) Mitoraj, M. and Michalak, A. (2007). *Organometallics* 26: 6576. (b) Mitoraj, M. and Michalak, A. (2008). *J. Mol. Model.* 14: 681.

34 (a) Baerends, E. J., Ziegler, T., Autschbach, J. (2014). ADF2017, SCM, Theoretical Chemistry, Vrije Universiteit, Amsterdam, The Netherlands; (b) te Velde, G., Bickelhaupt, F.M., Baerends, E.J. et al. (2001). *J. Comput. Chem.* 22: 931.

35 (a) van Lenthe, E. and Baerends, E.J. (2003). *J. Comput. Chem.* 24: 1142. (b) van Lenthe, E., Baerends, E.J., and Snijders, J.G. (1993). *J. Chem. Phys.* 99: 4597. (c) van Lenthe, E., Baerends, E.J., and Snijders, J.G. (1994). *J. Chem. Phys.* 101: 9783.

36 (a) Frenking, G. and Bickelhaupt, F.M. (2014). The EDA perspective of chemical bonding. In: *The Chemical Bond 1. Fundamental Aspects of Chemical Bonding* (ed. G. Frenking and S. Shaik), 121. Weinheim: Wiley-VCH. (b) Zhao, L.M., von Hopffgarten, D.M., and Andrada, G.F. (2018). *WIREs Comput. Mol. Sci.* 8: 1345. (c) Zhao, L., Hermann, M., Schwarz, W.H.E., and Frenking, G. (2019). *Nat. Rev. Chem.* 3: 48. (d) Zhao, L., Pan, S., Holzmann, N. et al. (2019). *Chem. Rev.* 119: 8781.

37 (a) Klamt, A. and Schüürmann, G. (1993). *J. Chem. Soc., Perkin Trans.* 2: 799. (b) Andzelm, J., Kölmel, C., and Klamt, A. (1995). *J. Chem. Phys.* 103: 9312. (c) Barone, V. and Cossi, M. (1998). *J. Phys. Chem. A* 102, 1995. (d) Cossi, M., Rega, N., Scalmani, G., and Barone, V. (2003). *J. Comput. Chem.* 24: 669.

38 Pyykkö, P. and Atsumi, M. (2009). *Chem. Eur. J.* 15: 12770–12779.

39 (a) Saha, R., Pan, S., Merino, G., and Chattaraj, P.K. (2019). *Angew. Chem. Int. Ed.* 58: 8372–8377. (b) Chi, C., Pan, S., Meng, L. et al. (2019). *Angew. Chem. Int. Ed.* 58: 1732–1738. (c) Pan, S., Zhao, L., Dias, H.V.R., and Frenking, G. (2018). *Inorg. Chem.* 57: 7780–7791. (d) Pecher, L., Pan, S., and Frenking, G. (2019). *Theor. Chem. Acc.* 138: 47. (e) Pan, S. and Frenking, G. (2020). *Angew. Chem. Int. Ed.* 59: 8836–8839. (f) Zhao, L., Pan, S., Zhou, M., and Frenking, G. (2019). *Science* 365: eaay5021. (g) Chi, C., Pan, S., Jin, J. et al. (2019). *Chem. Eur. J.* 25: 11772–11784. (h) Deng, G., Pan, S., Wang, G. et al. (2020). *Angew. Chem. Int. Ed.* 59: 10603–10609. (i) Bickelhaupt, F.M., Fonseca Guerra, C., Mitoraj, M. et al. (2022). *Phys. Chem. Chem. Phys.* 24: 15726. (j) Zhou, Y., Pan, S., Dong, X. et al. (2022). *JACS* 144: 8355–8361.

26

The Role of Electronic Activity Toward the Analysis of Chemical Reactions

Swapan Sinha[1,2] *and Santanab Giri*[1]

[1] *School of Applied Science and Humanities, Haldia Institute of Technology, Haldia, West Bengal, India*
[2] *Department of Applied Chemistry, Maulana Abul Kalam Azad University of Technology, Haringhata, West Bengal, India*

26.1 Introduction

In chemical reactions, bond formations and deformations occur due to the accumulation and depletion of electron density between the systems. Understanding how the electron density in molecules is redistributed during a chemical reaction is thus one of the main research interests of chemistry [1]. Density functional theory (DFT) is one of the most powerful methods to explain and predict the reaction mechanism for a variety of reactions. In the last two decades, several DFT-based global and local reactivity descriptors have been popularly used for analyzing the chemical reactivity of molecules involved in the chemical reaction [1–10]. Not only the descriptors but different electronic structure principles like maximum hardness and minimum polarizability principles [11, 12] provide relationships between processes of bond making/breaking and electron transfer [13, 14]. Here, we have discussed the mechanism of three popular chemical reactions with the help of an electronic activity. In organic chemistry, there exist several types of reactions that are very important. These are substitution, elimination, addition, and rearrangement reactions [15]. Out of these, bimolecular substitution reactions (S_N2) are very common in most chemical transformations. In this chapter, we have discussed the two different types of bimolecular substitution reactions (S_N2) where nucleophiles i.e. OH^- and Zintl clusters, react with normal alkyl halide [16, 17]. As per the literature, with the help of high-level *ab initio* and DFT calculations, several S_N2-type reactions are fully characterized in terms of energetic and structural parameters [18–20]. There are different types of approaches like Berlin function, electronic stress, and second derivative tensor of electron density that have been used to analyze these reactions [21–23]. Expect S_N2 reaction, and we have also discussed another two types of reactions, i.e. double proton transfer (DPT) reaction of formic acid dimer and water activations by intermolecular frustrated Lewis pair (FLP) [24, 25]. DPT plays an important role in biology and chemistry [26]. DPT in DNA base pair adenine–thymine is one of the fundamental examples of such a biological process [27]. To understand these complicated biological proton reactions, DPT in the formic acid dimer (FAD) has been studied extensively as a typical example of a multiple proton transfer reaction.

On the other hand, water activation by FLP is getting attention because it is a metal-free activation. Experimentally [28–30], there are some examples regarding the inter or intramolecular water activation by FLPs, but none of these looks into the detailed mechanism in terms of electronic activity. Here, we mainly focus on electronic activity along with the intrinsic reaction coordinate

Electron Density: Concepts, Computation and DFT Applications, First Edition.
Edited by Pratim Kumar Chattaraj and Debdutta Chakraborty.

(IRC), with a special emphasis on the reaction electronic flux (REF) and reaction force (RF) [31–37]. The detailed equations about the REF, and force are given in the theoretical background and computational details section.

26.2 Theoretical Backgrounds and Computational Details

To understand any chemical reaction, the generation of a minimum energy profile ($E(\xi)$), which links the reactant and product via a transition state, is very important. It helps to understand the physical nature of the reaction. The IRC, which is the condensed form of the three-dimensional atomic motion, monitors the minimum energy profile. The reaction force $F(\xi)$ is an important parameter to know the amount of work required to complete a reaction. It is defined as the negative derivative of $E(\xi)$ with respect to reaction coordinate ξ by the following expression [34, 38–46]

$$F(\xi) = -\frac{dE}{d\xi} \tag{26.1}$$

For any elementary step of a chemical reaction, the reaction force is characterized by a minimum and a maximum located at ξ_1 and ξ_2 along IRC (see Figure 26.1).

It is already established that, for any single-step reaction, there are three distinct reaction regions along ξ: (i) the reactant region ($\xi_R \leq \xi \leq \xi_{min}$) where the reactants are prepared for the initiation of the reaction through structural rearrangements, (ii) the transition state region ($\xi_{min} \leq \xi \leq \xi_{max}$) characterized mainly by electronic reordering, and (iii) the product region ($\xi_{max} \leq \xi \leq \xi_p$) where mainly structural relaxation takes place to form the desired product. This feature is also very important to characterize different properties along the IRC within the reaction regions. In particular, the activation energy ($\Delta E^{\ddagger}$) and the reaction energy (ΔE_R) can be obtained through the analysis of reaction force by the following expression.

$$\Delta E^{\ddagger} = [E(\xi_{TS}) - E(\xi_R)] = W_1 + W_2 \tag{26.2}$$

$$\Delta E_R = [E(\xi_p) - E(\xi_R)] = W_1 + W_2 + W_3 + W_4 \tag{26.3}$$

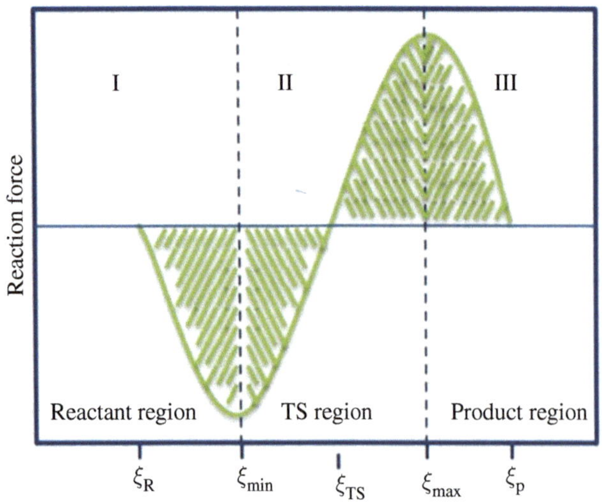

Figure 26.1 General reaction force profile for a single-step reaction. Source: Adapted with permission from Giri et al. [25]. Copyright 2017 American Chemical Society.

where W_1, W_2, W_3, and W_4 are the reaction works involved in the reaction and are defined as

$$W_1 = -\int_{\xi_R}^{\xi_{min}} F(\xi)d\xi > 0;\ W_2 = -\int_{\xi_{min}}^{\xi_{TS}} F(\xi)d\xi > 0$$

$$W_3 = -\int_{\xi_{TS}}^{\xi_{max}} F(\xi)d\xi < 0;\ W_4 = -\int_{\xi_{max}}^{\xi_p} F(\xi)d\xi < 0 \tag{26.4}$$

Chemical potential (μ) is a global descriptor, which also helps to understand the electronic activity of the chemical reaction [47–50]. It was calculated within the DFT framework for an N electron system having total energy E as

$$\mu = \left(\frac{\partial E}{\partial N}\right)_{v(r)} = -\chi \tag{26.5}$$

Here, χ and $v(r)$ are the electronegativity and external potential, respectively. It can also be calculated by using the finite difference approximation from ionization potential (IP) and electron affinity (EA) through the following expression [2],

$$\mu \cong -\left(\frac{\mathrm{IP}+\mathrm{EA}}{2}\right) \tag{26.6}$$

IP and EA can be further approximated in terms of the highest occupied molecular orbital (HOMO) and the lowest unoccupied molecular orbital (LUMO). Thus, expression (26.6) can be written as

$$\mu \cong \left(\frac{E_{\mathrm{LUMO}}+E_{\mathrm{HOMO}}}{2}\right) \tag{26.7}$$

where $E_{\mathrm{LUMO}} \approx -\mathrm{EA}$ and $E_{\mathrm{HOMO}} \approx -\mathrm{IP}$, as stated by Koopmans' theorem [51]. After getting the μ in each step of the reaction, the REF [32–37, 52] can be represented as

$$J(\xi) = -\left(\frac{d\mu}{d\xi}\right) \tag{26.8}$$

It was stated that positive $J(\xi)$ represents bond-strengthening or bond-forming processes, whereas bond-weakening or bond-breaking processes are associated with a negative $J(\xi)$ profile. The $J(\xi)$ can also be separated in two terms, e.g. electronic polarization $J_p(\xi)$ and electronic transfer $J_t(\xi)$ flux as

$$J(\xi) = J_p(\xi) + J_t(\xi) \tag{26.9}$$

The polarization contribution $J_p(\xi)$ can be calculated numerically from a partition of the reactive complex into molecular fragments; for an n-fragment system, $J_p(\xi)$ can be expressed as

$$J_p(\xi) = \sum_{i=1}^{n} J_p^i(\xi) \tag{26.10}$$

By use of the counterpoise method, the ith fragment of $J_p(\xi)$, $J_p^i(\xi)$, can be calculated separately along the IRC by using the expression

$$J_p^i(\xi) = -\left(\frac{N_i}{N}\right)\frac{d\mu_i}{d\xi} \tag{26.11}$$

where μ_i and N_i are the chemical potential and the number of electrons of fragment i, respectively, and N is the total number of electrons for the supramolecular system. The flux associated with the electronic transfer, $J_t(\xi)$, can be obtained from the difference of the total REF, calculated from the supramolecular system, and $J_p(\xi)$ as

$$J_t(\xi) = J(\xi) - J_p(\xi) = -\frac{d\mu}{d\xi} + \sum_i^n \left(\frac{N_i}{N}\right)\frac{d\mu_i}{d\xi} \tag{26.12}$$

By use of the equalization principle, the chemical potential of the supramolecule (μ) can be written in terms of the chemical potential of hypothetical fragments (μ_i^0) such that $\mu_i^0 \equiv \mu$ for all i^{th} fragments,

$$\mu(\xi) = \sum_i^n \left(\frac{N_i}{N}\right) \mu_i^0(\xi) \tag{26.13}$$

and thus, Eq. (26.8) can be written as

$$J(\xi) = -\left(\frac{d\mu}{d\xi}\right) = -\sum_i^n \left(\frac{N_i}{N}\right) \frac{d\mu_i^0}{d\xi} \equiv \sum_i^n \left(\frac{N_i}{N}\right) \frac{d\mu}{d\xi} = \sum_i^n J^i(\xi) \tag{26.14}$$

Finally, the REF associated with electron transfer for each fragment can be expressed as

$$\begin{aligned} J_{\text{t}}(\xi) &= J(\xi) - J_{\text{p}}(\xi) \\ &= \sum_i^n \left(\frac{N_i}{N}\right) \frac{d\mu_i}{d\xi} - \sum_i^n \left(\frac{N_i}{N}\right) \frac{d\mu_i^0}{d\xi} \\ &= \sum_i^n \left(\frac{N_i}{N}\right) \frac{d}{d\xi}(\mu_i - \mu_i^0) \equiv \sum_i^n \left(\frac{N_i}{N}\right) \frac{d}{d\xi}(\mu_i - \mu) \\ &= \sum_i^n J_{\text{t}}^i(\xi) \end{aligned} \tag{26.15}$$

To explore the electronic activity in detail, i.e. whether the reaction is nucleophilic or electrophilic in nature, the REF can be separated in terms of nucleophilic reactivity flux $J_{\text{N}}(\xi)$ and electrophilic reactivity flux $J_{\text{E}}(\xi)$ along the IRC. Therefore, the REF can be expressed in terms of $J_{\text{N}}(\xi)$ and $J_{\text{E}}(\xi)$ as

$$J(\xi) = J_{\text{N}}(\xi) + J_{\text{E}}(\xi) \tag{26.16}$$

where $J_{\text{N}}(\xi)$ and $J_{\text{E}}(\xi)$ are defined as the negative derivative of HOMO and LUMO, respectively, along the reaction coordinate, ξ.

Fukui function ($f(r)$), which can describe the selectivity in terms of electrophilic and nucleophilic sites inside the molecule [53], is one of the most popular local reactivity descriptors. It is defined as the change in electron density $\rho(r)$ concerning the total number of electrons N at constant external potential, $v(r)$:

$$f(r) = \left(\frac{\partial \rho(r)}{\partial N}\right)_{v(r)} \tag{26.17}$$

Two different types of Fukui functions [54–56], i.e. for nucleophilic ($f^+(r)$) or electrophilic ($f^-(r)$) attacks, have been introduced to know the selectivity of a particular center of a system. Morell et al. introduced a new reactivity descriptor, known as a dual descriptor, which can easily identify the nucleophilic and electrophilic regions of a molecule simultaneously [57]. It can be expressed as

$$\Delta f(r) = \rho_{\text{LUMO}(r)} - \rho_{\text{HOMO}(r)} \tag{26.18}$$

where $\Delta f(r)$ is positive in electrophilic regions and negative in the nucleophilic region of a molecule. This dual descriptor is quite useful and reliable for analyzing the reactivity and selectivity inside a molecule [58].

All the geometries presented in this chapter are optimized using hybrid functionals (B3LYP, M06-2X) and 6-311+G(d), 6-31+G (d, p), 6-311+G (d, p), and 6-311++G (d, p) basis sets [59–61]. For the solvent phase study, the polarizable continuum model (PCM) has been used. Frequency calculation was done to know the true nature of the stationary states. IRCs were performed to get the minimum energy pathway from the reactant complex to the product complex via transition state (TS). To know about the synchronic and asynchronic nature of the reaction, the Wiberg bond index, along with reaction coordinates, has been determined. All the calculations have been performed in *Gaussian 09 and Multiwfn* software [62, 63].

26.3 Results and Discussions

As discussed earlier, in this chapter, we are going to three types of reactions, namely (i) Bimolecular nucleophilic substitution (S_N2) reaction, (ii) DPT reaction in the formic acid dimer, and (iii) Water activation by FLPs.

26.3.1 Bimolecular Nucleophilic Substitution (S_N2) Reaction

Substitution reaction is one of the most popular reactions in organic chemistry. In substitution reactions, the bimolecular nucleophilic substitution (S_N2) reactions are common. Here, we are going to analyze the electronic activity throughout the reaction in terms of reaction electronic force and flux [17] of simple S_N2 reaction, i.e. $OH^- + CH_3F \rightarrow CH_3OH + F^-$. It is well known that in this reaction, the nucleophile OH^- attacks the carbon center of CH_3F and F^- leaves as a leaving group, which is depicted in Scheme 26.1.

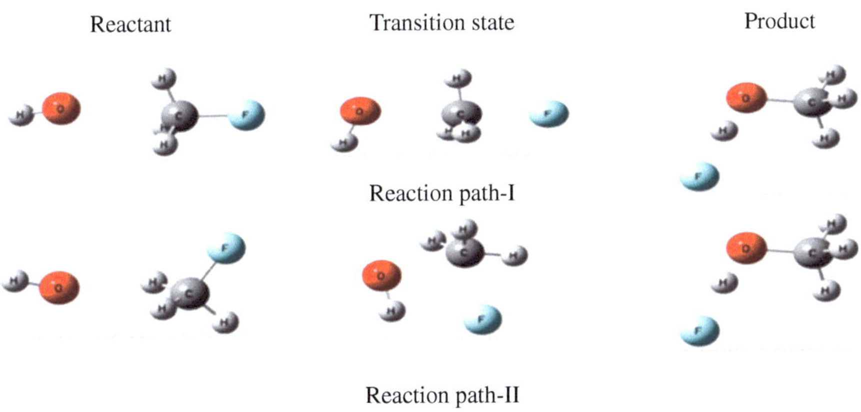

Scheme 26.1 Possible reaction mechanism pathway of the reaction $OH^- + CH_3F \rightarrow CH_3OH + F^-$. Source: Adapted with permission from Giri et al. [17]. Copyright 2012 American Chemical Society.

Scheme 26.1 represents the two possible gas phase S_N2 reaction pathways, wherein both the reaction paths and the product are similar. Still, the reactants and transition states are different depending on the orientation of the groups. The minimum energy profile, along with reaction coordinates for both reaction paths, is given in Figure 26.2.

From Figure 26.2, it has been observed that both the reaction pathways are exothermic in nature. The reaction path-I is more favorable than path II because of the lower activation energy barrier. The linear form TS of reaction path-I is also more familiar with conventional S_N2 reaction whereas TS of path-II is inconsistent with S_N2 reaction, which proceeds via three-membered cyclic TS.

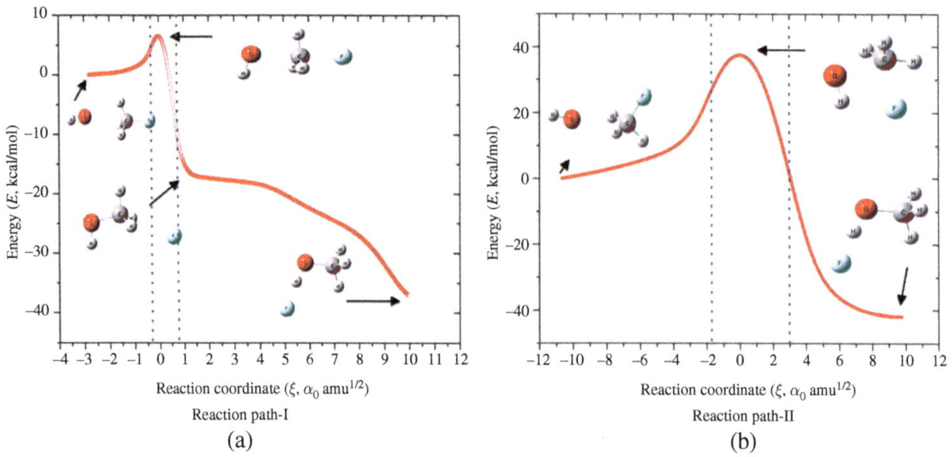

Figure 26.2 Reaction energy profile for a simple S_N2 reaction, viz., $OH^- + CH_3F \rightarrow CH_3OH + F^-$, in two possible pathways. Vertical lines define the reaction regions obtained from the reaction force analysis. Source: Adapted with permission from Giri et al. [17]. Copyright 2012 American Chemical Society.

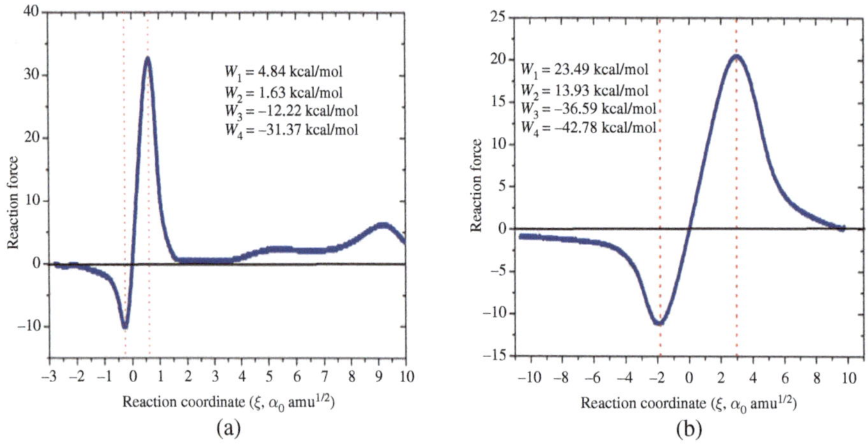

Figure 26.3 Reaction force profile of S_N2 reaction, viz., $OH^- + CH_3F \rightarrow CH_3OH + F^-$, for path-I (a) and path-II (b). Source: Adapted with permission from Giri et al. [17]. Copyright 2012 American Chemical Society.

The plots of reaction force are depicted in Figure 26.3. The red dashed line indicates the three reaction regions. From Figure 26.3 it is observed that less amount of work is required for the rearrangement of the reactants in path-I than path-II. This supports the feasibility of path-I over path-II.

As in this S_N2 reaction, path-I is favorable, our further discussion about the electronic activity of this reaction in terms of REF is only for reaction path-I. The REF and its separation are shown in Figure 26.4.

The zero-flux profile in the initial stage of the reaction implies no electronic activity but rather structural orientation among the reactants to initiate the electronic activity. After this, the positive increase of REF indicates that the electronic activity starts with the formation of an O—C bond. After reaching the maximum, the REF shows a decreasing trend in the TS region, which indicates the weakening of the C—F bond. Both positive and negative flux profile in the TS region implies that the spontaneous electronic activity associated with the bond-making and braking process drives the electronic activity in this part of the reaction. From the separation of REF (J)

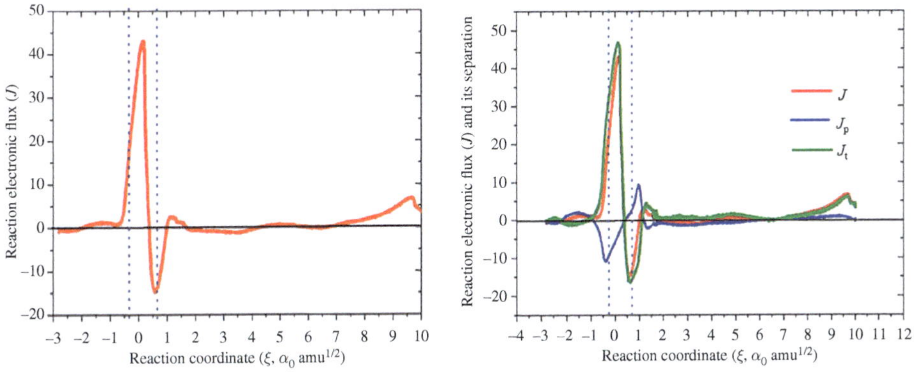

Figure 26.4 Reaction flux (REF) and its separation profile of $OH^- + CH_3F \rightarrow CH_3OH + F^-$, S_N2 reaction for reaction path-I. Source: Adapted with permission from Giri et al. [17]. Copyright 2012 American Chemical Society.

(see Figure 26.4), it is observed that the electronic transfer (J_t) gives the major contribution toward the total J. At the same time, J_p has some contribution in reactant and product regions.

Further, we looked into the contribution of the induvial fragments (OH^-, CH_3^+, and F^-) towards the total electronic transfer and polarization flux. The corresponding profiles are plotted in Figure 26.5. From Figure 26.5, it can be noticed that all three fragments contribute towards $J_{t,}$ but it is CH_3^+ whose contribution is less than the other two moieties, F^- and OH^-. A close look at the transition state region reveals that the contribution of OH^- is more, which indicates the initiation of the reaction through nucleophilic attack. After that, it is the F^- moiety whose contribution is more important than those of the OH^- and CH_3^+ moieties. This situation is different when we look at the electronic polarization contribution, as depicted in Figure 26.5 (right side view). It is observed that, in the transition state region, the electronic polarization (J_p) contribution for CH_3^+ is more significant than F^- and OH^-. Therefore, we can say that, while OH^- and F^- are the two most crucial moieties throughout the reaction paths, CH_3^+ plays an important role in the transition state region. This can be explained nicely if we have a look at the transition state structure of path I. Here, F^-, CH_3^+, and OH^- moieties exist in a linear geometry, and the electronegative F^- and OH^- make the CH_3^+ highly polarizable.

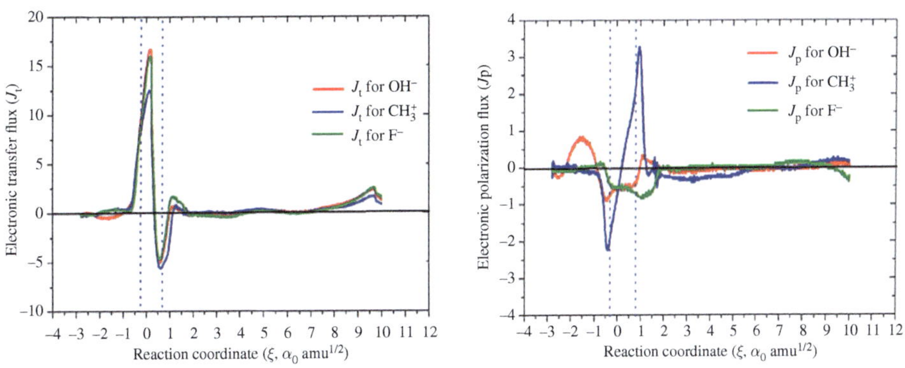

Figure 26.5 Individual components of electronic transfer flux (J_t) and polarization flux (J_p) profile in S_N2 reaction, viz., $OH^- + CH_3F = CH_3OH + F^-$ for path I. Source: Adapted with permission from Giri et al. [17] Copyright 2012 American Chemical Society.

26.3.2 Alkylation of Zintl Cluster

Previously, we explained that the DFT-based electronic activity parameter, i.e. reaction force and flux can nicely explain the mechanism of simple S_N2 reaction, viz., $OH^- + CH_3F = CH_3OH + F^-$. Now, we have looked into another S_N2 reaction, where the functionalized Zintl cluster acts as a nucleophile in the alkylation of the Zintl cluster. The alkylation process is important to get a new functionalized deltahedral organo-Zintl cluster. It has been found that the alkylation reaction between the Zintl ion and the alkyl halide follows an S_N2-like mechanism. A recent study [16] showed that the Zintl cluster having an electron-donating like methyl group as a ligand, $[Ge_9(CH_3)_3^-]$, acts better nucleophile than the electron-withdrawing cyanide (–CN) ligand substituted cluster $[Ge_9(CN)_3^-]$. Here, we will thoroughly discuss the alkylation process of the Zintl cluster ((i) $Ge_9(CH_3)_3^- + CH_3Cl \rightarrow Ge_9(CH_3)_4 + Cl^-$ and (ii) $Ge_9(CN)_3^- + CH_3Cl \rightarrow Ge_9(CN)_3CH_3 + Cl^-$) in terms of different thermodynamic parameters like reaction energy, free energy, and enthalpy of activation and electronic activity in terms of REF. The reaction force, Wiberg bond indices, and dual descriptor lend additional support to the reaction mechanism.

For both reactions, the ground state structures of the reactant complex (RC), transition structure (TS), and product complex (PC) are given in Figure 26.6. Here, both the Zintl nucleophiles,

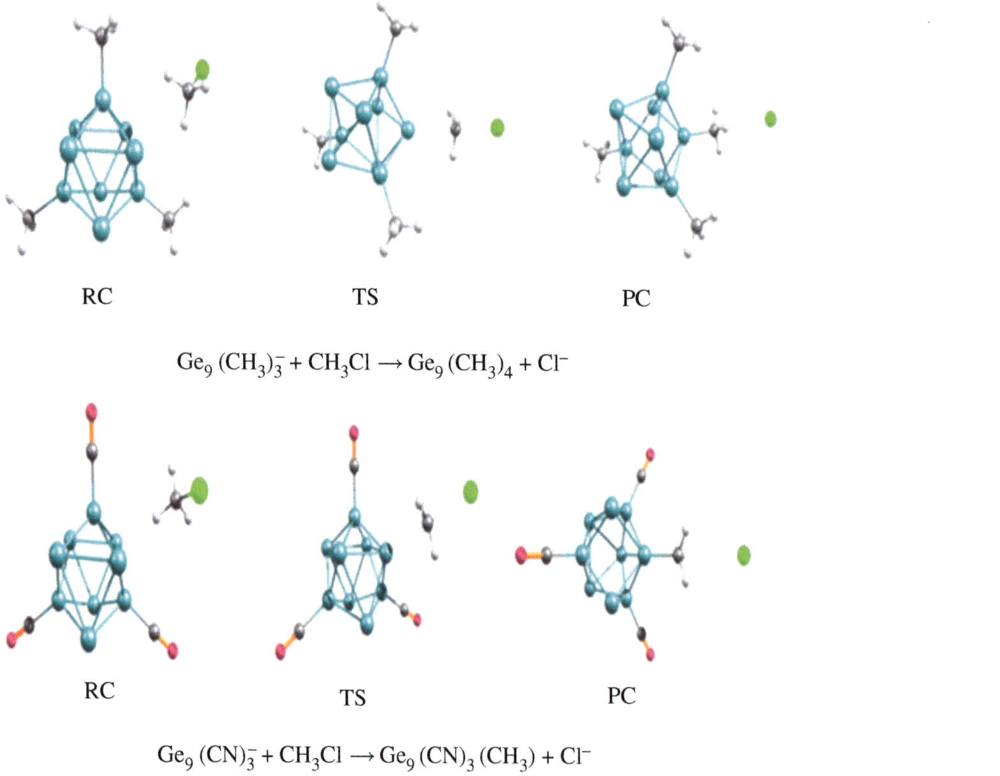

Figure 26.6 Optimized structures of reactant complex (RC), transition structures (TS), and product complex (PC) in first reaction i.e. $Ge_9(CH_3)_3^- + CH_3Cl \rightarrow Ge_9(CH_3)_4 + Cl^-$ and second reaction i.e. $Ge_9(CN)_3^- + CH_3Cl \rightarrow Ge_9(CN)_3CH_3 + Cl^-$ reactions in B3LYP/6-31+G(d, p) level of theory calculation. Source: Adapted with permission from Parida et al. [16]. Copyright 2020 American Chemical Society.

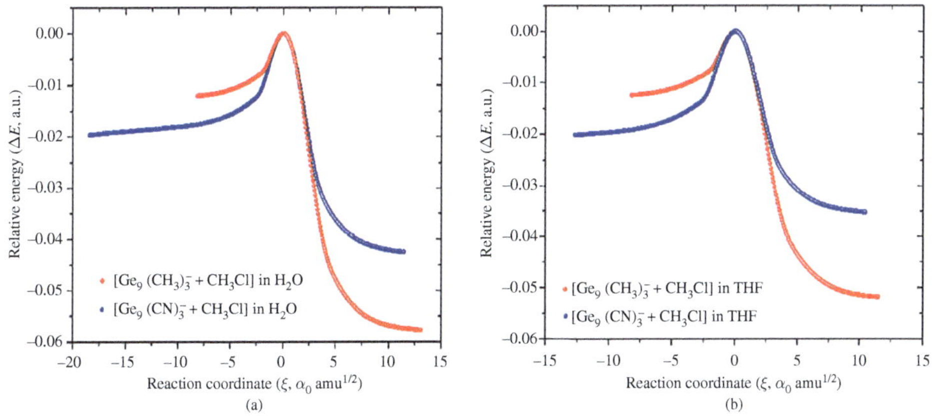

Figure 26.7 Minimum energy pathway along the IRC for the reactions involving the nucleophiles $Ge_9(CH_3)_3^-$ and $Ge_9(CN)_3^-$ in (a) H_2O and (b) THF solvent medium obtained from the B3LYP/6-31+G(d, p) level of theory calculation. Source: Adapted with permission from Parida et al. [16]. Copyright 2020 American Chemical Society.

i.e. $Ge_9(CH_3)_3^-$ and $Ge_9(CN)_3^-$ contain 40 electrons and are stable according to the jellium electronic shell closer. It is expected that the $Ge_9(CH_3)_3^-$ acts as a better nucleophile than the $Ge_9(CN)_3^-$ because of the electron donating group ($-CH_3$) present in $Ge_9(CH_3)_3^-$ whereas an electron-withdrawing group (–CN) in $Ge_9(CN)_3^-$. The lower electron affinity of $Ge_9(CH_3)_3^-$ (2.8 eV) than the $Ge_9(CN)_3^-$ (4.2 eV) also tells us the same finding. To know about the true TS of these two reactions, the IRC calculation has been performed in both the gas and solvent phases (Water and THF). The thermodynamics parameters like Gibbs Free Energy ($\Delta G^{\ddagger}$), Enthalpy of Activation ($\Delta H^{\ddagger}$), and Reaction Energies (ΔG_R, ΔH_R) are obtained from the IRC calculation (Figure 26.7), which is tabulated in Table 26.1. From Table 26.1 data, we can see that the first reaction shows lower activation energy than the second reaction in both the gas phase and different solvent phase calculations.

To know the amount of work required to complete the reaction, the reaction force ($F(\xi)$) has been calculated from the minimum energy pathway in both water and THF solvent phases. The corresponding $F(\xi)$ profiles for the two reactions are depicted in Figure 26.8. The force profiles in both H_2O and THF solvents are found to be similar. The force profile can be separated into three regions, namely, reactant, TS, and product, depending on the appearance of the minimum and maximum. The calculated reaction works information starting from the reactant to the product

Table 26.1 Calculated Gibbs free energy ($\Delta G^{\ddagger}$), enthalpy of activation ($\Delta H^{\ddagger}$), and reaction energies (ΔG_R, ΔH_R) obtained from the B3LYP/6-31+G(d, p) level of theory calculation in kcal/mol unit.

Reaction in a different medium	$\Delta G^{\ddagger}$	ΔG_R	$\Delta H^{\ddagger}$	ΔH_R
$Ge_9(CH_3)_3^- + CH_3Cl \rightarrow Ge_9(CH_3)_4 + Cl^-$ (Gas)	14.09	−0.72	14.33	0.43
$Ge_9(CN)_3^- + CH_3Cl \rightarrow Ge_9(CN)_3CH_3 + Cl^-$ (Gas)	21.73	17.01	20.85	16.44
$Ge_9(CH_3)_3^- + CH_3Cl \rightarrow Ge_9(CH_3)_4 + Cl^-$ (Water)	11.71	−25.32	12.36	−24.06
$Ge_9(CN)_3^- + CH_3Cl \rightarrow Ge_9(CN)_3CH_3 + Cl^-$ (Water)	15.43	−11.90	15.86	−10.63
$Ge_9(CH_3)_3^- + CH_3Cl \rightarrow Ge_9(CH_3)_4 + Cl^-$ (THF)	11.98	−19.32	12.46	−18.17
$Ge_9(CN)_3^- + CH_3Cl \rightarrow Ge_9(CN)_3CH_3 + Cl^-$ (THF)	16.53	−5.11	16.51	−5.20

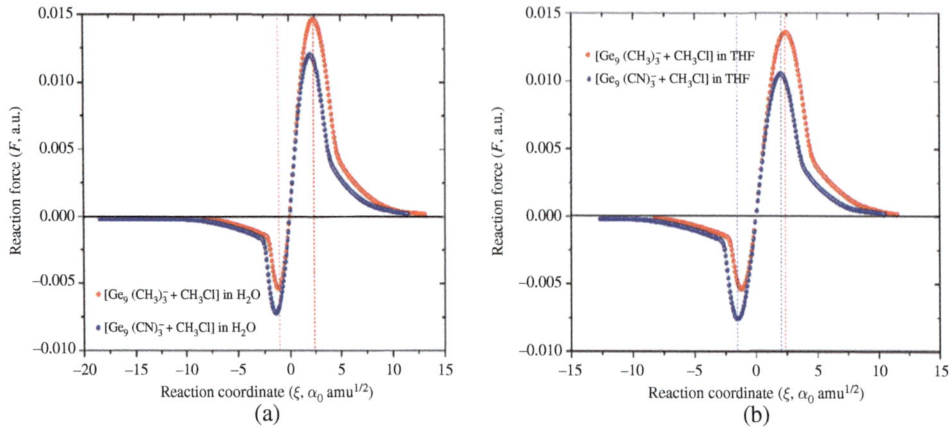

Figure 26.8 Reaction force ($F(\xi)$) profile of these two reactions in (a) H_2O and (b) THF solvent medium obtained from the B3LYP/6-31+G(d, p) level of theory calculation. Source: Adapted with permission from Parida et al. [16]. Copyright 2020 American Chemical Society.

Table 26.2 Calculated work done (W1–W4) in a different region of our two studied reactions in different solvent phases (in kcal/mol unit).

Reactions	**W1**	**W2**	**W3**	**W4**
First reaction (H_2O)	10.59	3.17	−13.38	−23.42
Second reaction (H_2O)	12.73	3.87	−10.48	−16.40
First reaction (THF)	11.23	2.63	−13.68	−17.34
Second reaction (THF)	13.87	4.20	−7.97	−14.29

region, W1, W2, W3, and W4, are tabulated in Table 26.2. From W1 values in Table 26.2 suggest that almost 75% of the activation energy is required for the structural orientation of the reactants to achieve the reactant complex (force minimum). The Lower W2 values indicate that only 25% is necessary for electronic activity to reach the TS.

After having the physical interpretation from the reaction force profile of the alkylation reactions, for a better understanding of an electronic activity throughout the reaction pathway, the REF, $J(\xi)$ has been generated (Figure 26.9). A positive $J(\xi)$ indicates the electronic activity is associated with the bond formation or strengthening process, whereas a negative $J(\xi)$ profile indicates the bond breaking or bond weakening process. The REF $J(\xi)$ and its separation into nucleophilic, $J_N(\xi)$, and electrophilic, $J_E(\xi)$ flux profile in the water solvent of these two-alkylation reactions are given in Figure 26.9. A zero-flux profile for both reactions at the initial stage is associated with structural orientation. The electronic activity starts with a positive value before entering the TS region. This electronic activity is associated with the formation of the Ge—C bond. After reaching the maximum $J(\xi)$ in the TS region, it gradually decreases and becomes negative, which indicates the bond weakening of the C—Cl bond. In the product region, we could see again a zero $J(\xi)$, which suggests the formation as well as relaxation of the product complex. From Figure 26.9, it is evident that in the reactant region both $J_N(\xi)$ and $J_E(\xi)$ have the same behavior because of no electronic activity. When electronic activity starts, the contributions of $J_N(\xi)$ and $J_E(\xi)$ are found to be different. For both reactions, it is the $J_E(\xi)$ which have a similar profile as the total $J(\xi)$.

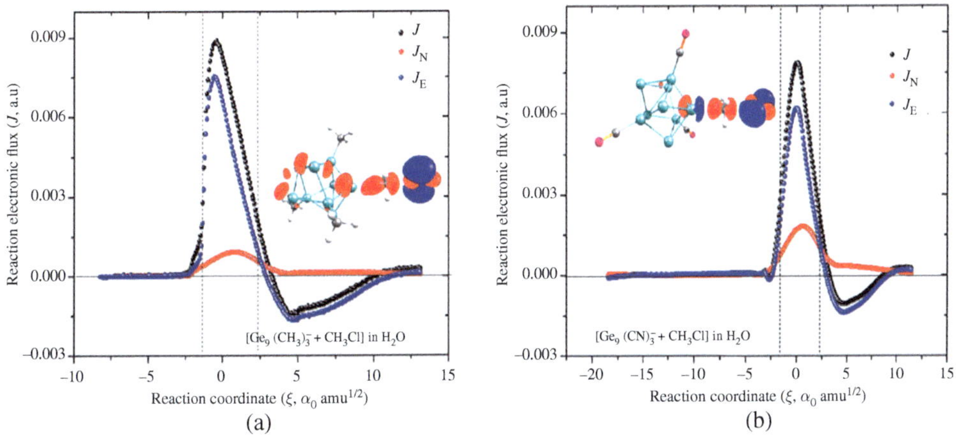

Figure 26.9 (a) Reaction electronic flux ($J(\xi)$) and its separation $J_N(\xi)$ and $J_E(\xi)$ profile for first reaction (left side view) and (b) second reaction (right side view) in H_2O medium obtained from the B3LYP/6-31+G(d, p) level of theory calculation. (Inset) Dual descriptor ($\Delta f(r)$) has been depicted for the transition state. Source: Adapted with permission from Parida et al. [16]. Copyright 2020 American Chemical Society.

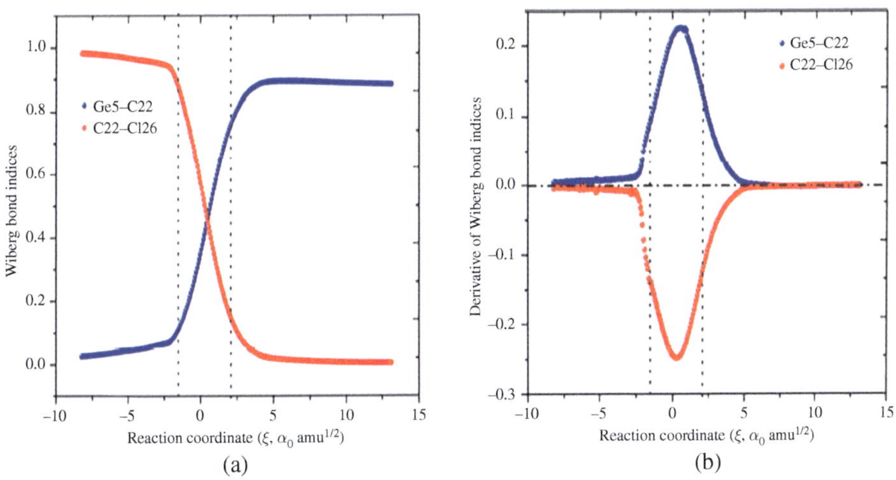

Figure 26.10 (a) Wiberg bond indices profile (b) their derivatives for first reactions in H_2O solvent medium obtained from the B3LYP/6-31+G(d, p) level of theory calculation. Source: Adapted with permission from Parida et al. [16]. Copyright 2020 American Chemical Society.

This suggests that these reactions are nucleophilic substitution reactions. To understand the bond formation and deformation processes, more specifically, whether they occur synchronously, the Wiberg bond indices (WBI) and their derivative have been calculated along the IRC. Here, we have shown (Figure 26.10) the Wiberg bond indices profile and their derivatives) for the first reaction. From Figure 26.10, it is observed that bond formation (Ge—C) and deformation (C—Cl) are taking place simultaneously.

26.3.3 Proton Transfer Reaction

Here, we will discuss the mechanism of ground and excited state DPT reaction in formic acid dimer with the help of reaction force and the REF [25]. The structural information for both the ground (GS) and excited state (ES) DPT reaction of the formic acid dimer is given in Figure 26.11.

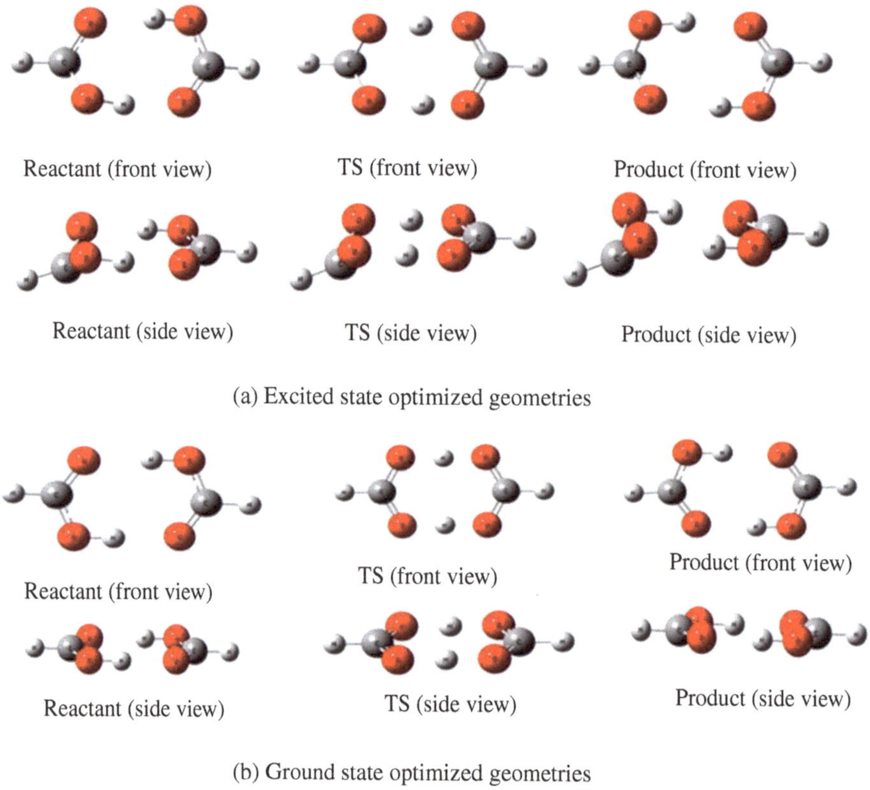

Figure 26.11 (a) Excited and (b) ground state optimized geometries of reactant complex, transition state (TS), and product complex for the double proton transfer in the formic acid dimer. Source: Adapted with permission from Giri et al. [25]. Copyright 2017 American Chemical Society.

From Figure 26.11, it is observed that the ground state structures of the formic acid dimer are planer, whereas in excited states the carbon atom in one of the formic acids of the dimer is out of a plane. But the O and H atoms in both acids in the dimer, which are responsible for the proton transfer, lie within the plane. Next, from the IRC calculation, it was found that both minimum energy profiles are thermoneutral and concerted in nature. The activation energy barrier for the ground state and excited state are 7.35 and 7.43 kcal/mol, respectively. Reaction force calculation has been done to know about the work done in each elementary step of the reactions. The ground state and excited state reaction force profile for the DPT in the formic acid dimer are given in Figure 26.12 and work done data is given in Table 26.3.

Both the force profile for GS and ES is very similar. From Table 26.3 data, it is observed that almost 70% of activation energy is related to structural orientation to start the electronic activity. The W3 and W4 values are the same as the W1 and W2 values since these reactions are thermoneutral in nature. Although the physical interpretation for both ground and excited state reactions is the same, the excited state reaction needs for work to complete.

For a better understanding of the electronic activity throughout the reaction, an electronic flux study has been carried out. The flux profiles of the GS and ES DPT reactions are depicted in Figure 26.13. Although for both reactions, we have zero flux profile at the beginning and end of the reaction, the initiation of the electronic activity differs. The starting point of the electronic activity is also found to be different for ground and excited state reactions. From Figure 26.13, it

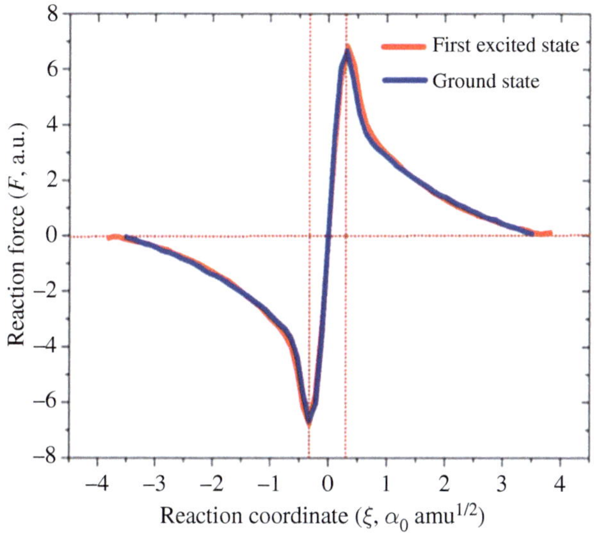

Figure 26.12 Ground state and excited state reaction force profile (F, ξ) for the double proton transfer in the formic acid dimer. Source: Adapted with permission from Giri et al. [25]. Copyright 2017 American Chemical Society.

Table 26.3 The work done calculated from the force profile of each elementary step in GS and ES double proton transfer reaction in the formic acid dimer (in kcal/mol).

Reaction state	W1	W2	W3	W4
Ground state (GS)	5.93	1.39	−1.39	−5.93
Excited state (ES)	6.07	1.32	−1.32	−6.07

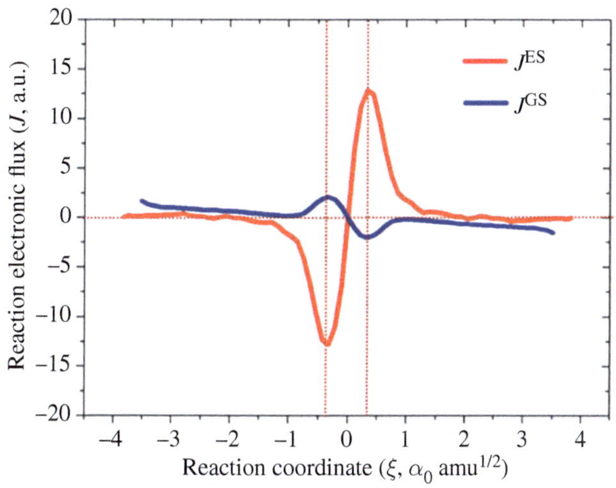

Figure 26.13 Profiles of reaction electronic flux, $J(\xi)$, in the ground and excited states for the double proton transfer in formic acid dimer along the reaction path. Source: Adapted with permission from Giri et al. [25]. Copyright 2017 American Chemical Society.

can be noted that in the ground state, such activity starts at $\xi = -0.5$ whereas it starts earlier, i.e. at $\xi = -1.0$, for the excited state. It is further interesting to note that at $\xi = -0.4$ the GS, REF profile exhibits maxima, whereas the ES, REF profile shows minima. This indicates that the ground state electronic activity is associated with the formation or strengthening of the bond. At the same time, the opposite is true for ES reaction.

To get more microscopic knowledge about the electronic activity of DPT reactions, the REF is further separated in terms of electronic transfer (J_t) and polarization (J_p) contribution using the counterpoise method. The electronic flux and its separation profile of GS and ES DPT reactions are given in Figure 26.14. From this figure, it is shown that the contribution of J_t in total J of GS DPT reaction is found to be more than J_p. However, in the ES DPT reaction, the contribution of polarization flux (J_p) is more than the J_t.

Next, to know which formic acid monomer has more contribution toward the electronic transfer (J_t) and electronic polarization (J_p) during the DPT reaction in both GS and ES, the total system is further separated into two fragments. The monomer contributions for both the ground and excited state reaction toward the total electronic transfer and polarization flux are presented in Figure 26.15. It has been observed that the monomer contribution is different for ground and excited states. In the ground state, both monomers fragment (Frags 1 and 2) has equal contributions to the total electronic transfer flux (J_t) and polarization flux (J_p). However, for excited state reaction, it is found to be different. For the ES reaction initially in the reactant region, the maximum contribution appears from the monomer, which has a distorted planner structure (J_t Frag 2). After that, the planar monomer (J_t Frag 1) is found to have more contribution toward the total J_t. In the TS region also the planar monomer (Frag 1) is the only contributor to the total electronic transfer flux J_t. For separation in polarization flux (J_p), it is found that initially, both fragments have the same contribution in the reactant region (up to $\xi = -0.8$). After that monomer Frag 1 has more contribution in reactant as well as TS region toward the total J_p. Then, in the product region, both fragments have a similar contribution.

Therefore, this study concludes that the DPT reaction in the ground state starts with spontaneous electronic activity with a strengthening of the bond. Whereas in an excited state it begins with the weakening of the bond. Electronic flux and its separation and the contribution of different fragments into separation flux infer that the electronic activity in the ground state is exactly opposite to that in the excited state. Also, the monomer contributions toward the electronic polarization

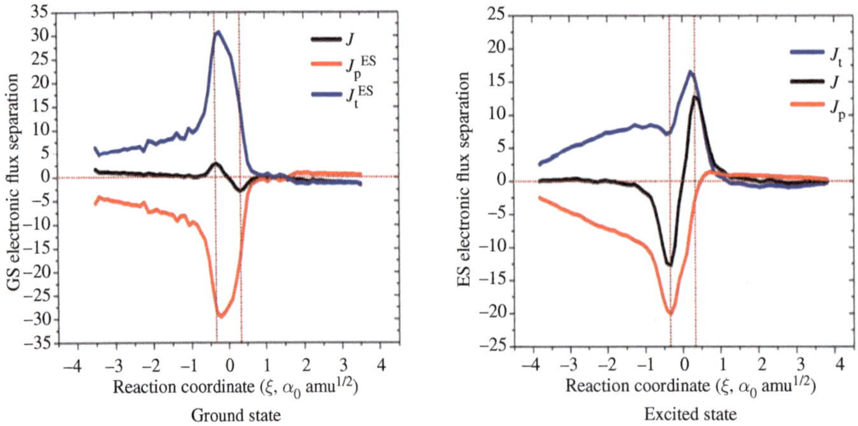

Figure 26.14 Profiles of the separation of reaction electronic flux in the ground and excited states for the double proton transfer in formic acid dimer along the reaction path. Source: Adapted with permission from Gii et al. [25]. Copyright 2017 American Chemical Society.

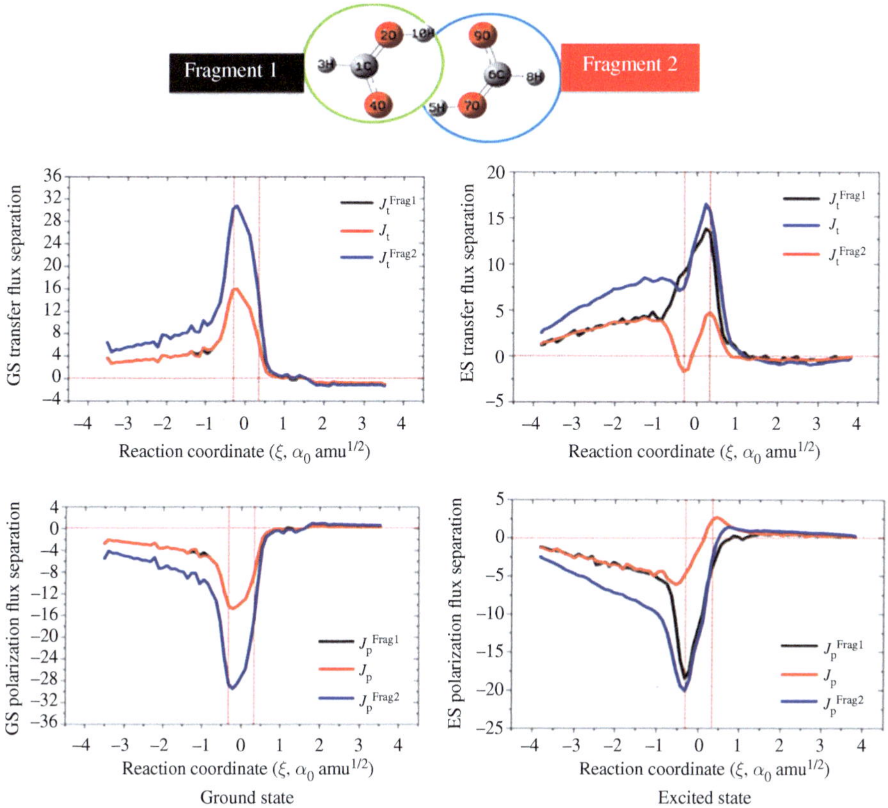

Figure 26.15 Profiles of the electronic transfer (J_t) and polarization (J_p) contribution for individual monomer toward total reaction electronic flux in the ground and excited states for the double proton transfer in formic acid dimer along the reaction path. Source: Adapted with permission from Giri et al. [25]. Copyright 2017 American Chemical Society.

and transfer flux are different. In the ground state, the contribution of each monomer fragment is almost similar and the electronic activity is spontaneous due to having planar geometry. This is quite different in the excited state. The nonplanarity of one of the formic acid monomers changes the electronic activity.

26.3.4 Water Activation by Frustrated Lewis Pairs (FLPs)

FLPs belong to a special class of complex molecules where both sterically encumbered Lewis acid and base do not form a classical adduct [64]. These complex molecules are well-known for metal-free small molecule activation like H_2, CO, CO_2, H_2O, NO, and N_2O [30, 64–71]. Here, the detailed mechanism of water activation will be discussed in terms of electron density and electronic activity. From the ground state geometries of the reactant, TS, intermediates, and product (Figure 26.16), it has been observed that water activation by the FLP is a two-step process having two transition structures (TS) and two intermediates (INTs). The first step is associated with H_2O addition to Lewis acid (LA), and then the hydrogen transfer to Lewis base occurs in the second step.

First, the mechanism is explained in terms of the electron transfer (ET) model with the help of natural bonding orbital (NBO) analyses. The orbital's involvement in the ET process of reactant complex, TS, INT, and product are given in Figure 26.17. Starting from the reactant complex, it

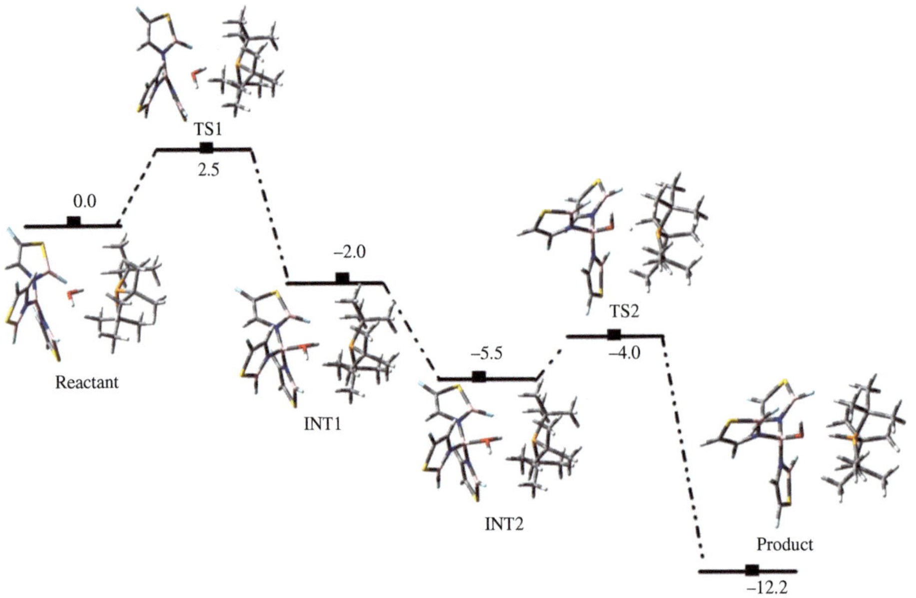

Figure 26.16 Potential energy diagram for H_2O dissociation by FLP, $(C_2HBNSF_2)_3B{-}P(^tBu)_3$ along with their ground state geometries of reactant complex, intermediate, and product complex. Source: Adapted with permission from Sinha and Giri [24] Copyright 2022 Elsevier.

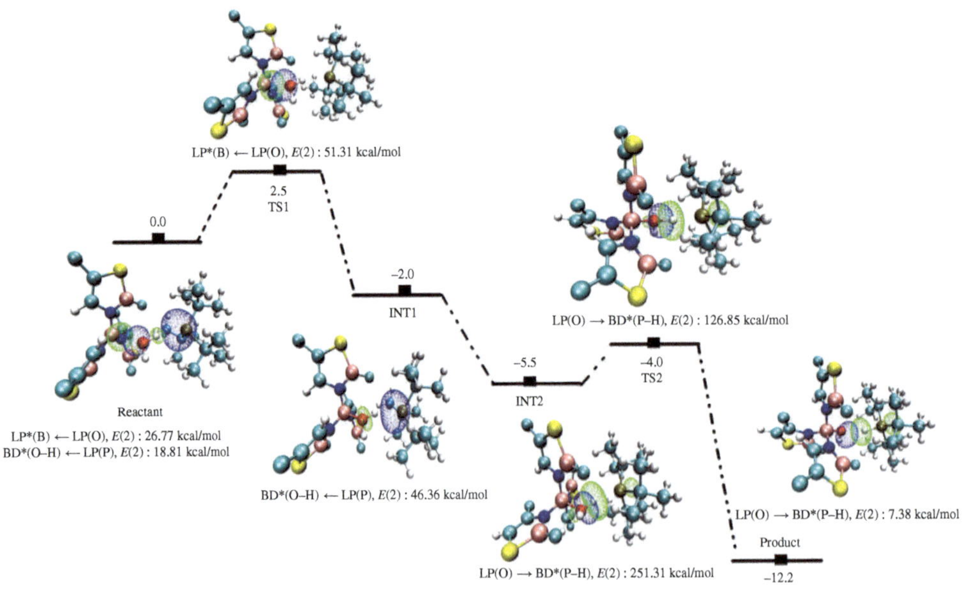

Figure 26.17 The interacting donor (blue color) and acceptor (green color) orbitals of the reactant, transitions states, intermediates, and products from the second-order perturbation theory-based NBO analysis through the reaction energy profile. Source: Adapted with permission from Sinha and Giri [24] Copyright 2022 Elsevier.

shows two major interactions. First, ET occurs from the lone pair (LP) of oxygen in the water to antibonding LP of boron (B) in LA, and second is LP of phosphorous (P) in LB to σ* or BD* of O–H in water. The corresponding estimated second-order perturbation theory (PT) level $E(2)$ for both interactions are 26.77 and 18.81 kcal/mol, respectively. So, in the reactant complex, both LA and LB interact with H_2O. However, the interaction with LA is more predominant, which is also reflected in TS1 having a higher $E(2)$ value of 51.31 kcal/mol for the same LP(O) → LP*(B) ET. Coming to the intermediate state, in INT1, H_2O is coordinated with LA, and the ET occurs from LP(P) → σ*(O–H) in the water-LA complex. In INT2, we have observed different ET. Here, since the O—H bond is weakening and P–H is going to be formed, the charge transfer occurs from LP(O) → σ*(P–H) with an $E(2)$ value of 251.31 kcal/mol. The same ET [LP(O) → σ*(P–H)] has been observed in TS2 with a lower $E(2)$ value of 126.85 kcal/mol. In product (LA–OH/LB–H), where water is completely dissociated and H makes a bond with P of LB shows less ET [LP(O) → σ*(P–H)] with a very small $E(2)$ value in comparison to INT2 and TS2.

Next, the mechanism of water activation is explained by electronic activity such as REF (J). The REF (J) and its separation, i.e. electronic transfer (J_t) and polarization (J_p) for the first step of the reaction, is given in Figure 26.18.

The dashed line in Figure 26.18 indicates three different reaction regions, namely, reactant, TS, and product. It has been taken from the appearance of the maximum and minimum on the reaction force profile (which is not shown here). The total REF begins with a zero-flux profile at the start of the reactant region, which indicates that the structural orientation is predominant over the electronic process. After that, the REF shows a negative profile in the TS region. This is due to the weakening of the O—H bond in the H_2O molecule. After the TS region, the REF has an increasing nature, which is due to the start of the formation of B—O bond. The appearance of the intermediate is supported by the zero-flux profile in the product region. If we look at the separation of REF, we see that J is predominantly by the electronic transfer process, J_t. Polarization flux (J_p) has a reverse trend. To see which fragment is contributing more towards the total J_t and J_p, we have performed a counterpoise calculation [72, 73]. The total system has been separated into three parts, namely, LA, LB, and H_2O. From Figure 26.18b, it is evident that all three components contribute toward the total J_t. However, it is the LA that contributes more than LB and H_2O. In the polarization flux also (Figure 26.18c), LA is the main contributor, while LB and H_2O do not have any such contribution.

In the second step of the activation process, the REF starts with a zero-flux profile due to the formation of the intermediate (INT2) (see Figure 26.19). Here, the electronic activity starts in the TS2 region, and it begins with the weakening of the O—H bond as H is detached from H_2O. In the product region, the increasing nature of the REF shows the gradual formation of the P—H bond.

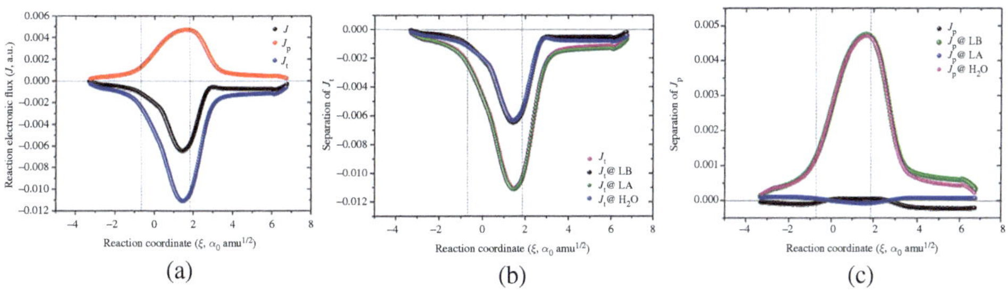

Figure 26.18 (a) Variation of total reaction electronic flux and its separation into electronic transfer and polarization components, (b) individual fragment contributions toward the total electronic transfer flux, and (c) electronic polarization flux for the first step of the reaction. Source: Adapted with permission from Sinha and Giri [24]. Copyright 2022 Elsevier.

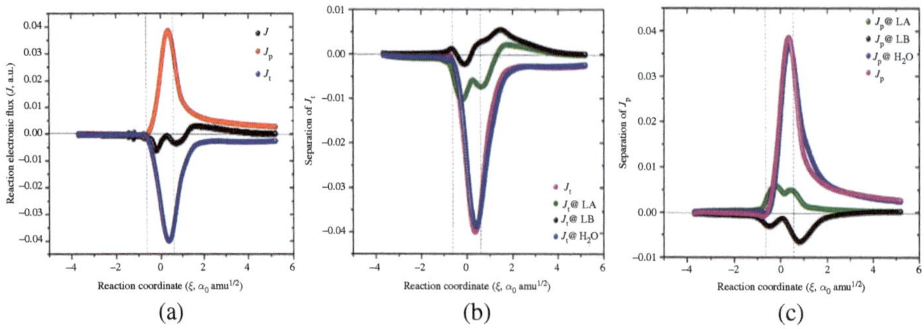

Figure 26.19 (a) Variation of total reaction electronic flux and its separation into electronic transfer and polarization components, (b) Individual fragment contributions toward the total electronic transfer flux, and (c) electronic polarization flux for the second step of the reaction. Source: Adapted with permission from Sinha and Giri [24]. Copyright 2022 Elsevier.

Both electronic transfer processes (J_t) and polarization flux (J_p) have the same kind of contribution toward total J. The second step is to associate the hydrogen transfer from the water to LB; therefore, for both cases, further separation of J_t and $J_{p,}$ the contribution of water (H_2O) is more predominant than the LA and LB.

26.4 Concluding Remarks

Within the conceptual DFT, in this chapter, we have demonstrated the detailed mechanistic view of three different types of reactions in terms of electronic activity. The reaction force (RF), REF, and its separations give the detailed electronic activity of each step in the chemical reaction. Natural bonding orbitals (NBO) and WBI lend additional support regarding charge transfer and the spontaneity of the chemical reactions. This kind of study helps to understand the reaction mechanism of several reactions better.

Acknowledgments

We are thankful to the Editors, Prof. Pratim Kumar Chattaraj and Dr. Debdutta Chakraborty, for inviting us to contribute a chapter to the book entitled "The Fundamentals of Electron Density." SG and SS gratefully acknowledge the DST-SERB grant CRG/2019/001125, and the Haldia Institute of Technology for providing funding and computational facilities.

References

1 Ayers, P.W., Yang, W.T., Bartolotti, L.J., and Chattaraj, P.K. (2009). *Chemical Reactivity Theory: A Density Functional View*, 255. CRC Press.

2 Chattaraj, P.K., Giri, S., and Duley, S. (2011). Update 2 of: electrophilicity index. *Chem. Rev.* 111 (2): PR43–PR75.

3 Stuyver, T., De Proft, F., Geerlings, P., and Shaik, S. (2020). How do local reactivity descriptors shape the potential energy surface associated with chemical reactions? The valence bond delocalization perspective. *JACS* 142 (22): 10102–10113.

4 Cárdenas, C., Rabi, N., Ayers, P.W. et al. (2009). Chemical reactivity descriptors for ambiphilic reagents: dual descriptor, local hypersoftness, and electrostatic potential. *J. Phys. Chem. A* 113 (30): 8660–8667.

5 Gupta, K., Giri, S., and Chattaraj, P.K. (2013). Charge-based DFT descriptors for Diels–Alder reactions. *J. Phys. Org. Chem.* 26 (2): 187–193.

6 Chakraborty, A., Das, R., Giri, S., and Chattaraj, P.K. (2011). Net reactivity index ($\Delta\omega$). *J. Phys. Org. Chem.* 24 (9): 854–864.

7 Chattaraj, P.K., Chakraborty, A., and Giri, S. (2009). Net electrophilicity. *J. Phys. Chem. A* 113 (37): 10068–10074.

8 Padmanabhan, J., Parthasarathi, R., Elango, M. et al. (2007). Multiphilic descriptor for chemical reactivity and selectivity. *J. Phys. Chem. A* 111: 9130.

9 Padmanabhan, J., Parthasarathi, R., Subramanian, V., and Chattaraj, P.K. (2007). Electrophilicity-based charge transfer descriptor. *J. Phys. Chem. A* 111 (7): 1358–1361.

10 Chattaraj, P.K., Maiti, B., and Sarkar, U. (2003). Philicity: a unified treatment of chemical reactivity and selectivity. *J. Phys. Chem. A* 107 (25): 4973–4975.

11 Chattaraj, P.K. and Sengupta, S. (1996). Popular electronic structure principles in a dynamical context. *J. Phys. Chem.* 100 (40): 16126–16130.

12 Pearson, R.G. (1987). Recent advances in the concept of hard and soft acids and bases. *J. Chem. Educ.* 64 (7): 561.

13 Chattaraj, P.K., Cedillo, A., Parr, R.G., and Arnett, E.M. (1995). Appraisal of chemical bond making, bond breaking, and electron transfer in solution in the light of the principle of maximum hardness. *J. Org. Chem.* 60 (15): 4707–4714.

14 Chattaraj, P.K. and Maiti, B. (2003). HSAB principle applied to the time evolution of chemical reactions. *JACS* 125 (9): 2705–2710.

15 Clayden, J., Greeves, N., and Warren, S. (2012). *Organic Chemistry*. Oxford University Press.

16 Parida, R., Ganguly, S., Das, G., and Giri, S. (2020). Density functional treatment on alkylation of a functionalized deltahedral Zintl cluster. *J. Phys. Chem. A* 124 (36): 7248–7258.

17 Giri, S., Echegaray, E., Ayers, P.W. et al. (2012). Insights into the mechanism of an S_N2 reaction from the reaction force and the reaction electronic flux. *J. Phys. Chem. A* 116 (40): 10015–10026.

18 Parthiban, S., Oliveria, G., and Martin, J.M.L. (2001). Benchmark ab initio energy profiles for the gas-phase S_N2 reactions $Y^- + CH_3X \rightarrow CH_3Y + X^-$ (X, Y = F, Cl, Br). Validation of hybrid DFT methods. *J. Phys. Chem.* 105: 895.

19 Gonzales, J., Allen, W., and Schaefer, H.F. (2005). Model identity S_N2 reactions $CH_3X + X^-$ (X = F, Cl, CN, OH, SH, NH_2, PH_2): Marcus theory analyzed. *J. Phys. Chem. A* 109: 10613.

20 Gonzales, J., Pack, C., Cox, R.S. et al. (2003). Definitive ab initio studies of model S_N2 reactions $CH_3X + F^-$ (X = F, Cl, CN, OH, SH, NH_2, PH_2). *Chem. - Eur. J.* 9: 2173.

21 Chakraborty, D., Cardenas, C., Echegaray, E. et al. (2012). Understanding chemical binding using the Berlin function and the reaction force. *Chem. Phys. Lett.* 539: 168.

22 Guevara-Garcia, A., Ayers, P.W., Jenkins, S. et al. (2011). Electronic stress as a guiding force for chemical bonding. *Top. Curr. Chem.* 128: 193.

23 Guevara-Garcia, A., Echegaray, E., Toro-Labbe, A. et al. (2011). Pointing the way to the products? Comparison of the stress tensor and the second-derivative tensor of the electron density. *J. Chem. Phys.* 134: 234106.

24 Sinha, S. and Giri, S. (2022). Unraveling H_2O activation by intermolecular frustrated Lewis pair. *Chem. Phys. Lett.* 799: 139634.

25 Giri, S., Parida, R., Jana, M. et al. (2017). Insights into the mechanism of ground and excited state double proton transfer reaction in formic acid dimer. *J. Phys. Chem. A* 121 (49): 9531–9543.

26 Belevich, I., Verkhovsky, M.I., and Wikström, M. (2006). Proton-coupled electron transfer drives the proton pump of cytochrome c oxidase. *Nature* 440: 829.

27 Florian, J., Hrouda, V., and Hobza, P. (1994). Proton transfer in the adenine-thymine base pair. *JACS* 116: 1457.

28 Klahn, M., Spannenberg, A., and Rosenthal, U. (2012). Tri-*tert*-butylphosphonium hydroxytris (pentafluorophenyl)borate. *Acta Crystallogr., Sect. E* 68 (5): o1549.

29 Roesler, R., Piers, W.E., and Parvez, M. (2003). Synthesis, structural characterization and reactivity of the amino borane 1-(NPh_2)-2-$[B(C_6F_5)_2]C_6H_4$. *J. Organomet. Chem.* 680 (1): 218–222.

30 Mo, Z., Szilvási, T., Zhou, Y.-P. et al. (2017). An intramolecular silylene borane capable of facile activation of small molecules, including metal-free dehydrogenation of water. *Angew. Chem. Int. Ed.* 56 (13): 3699–3702.

31 Amaral, L. and Viveiros, M. (2012). Why thioridazine in combination with antibiotics cures extensively drug-resistant *Mycobacterium tuberculosis* infections. *Int. J. Antimicrob. Agents* 39 (5): 376–380.

32 Cerón, M.L., Echegaray, E., Gutiérrez-Oliva, S. et al. (2011). The reaction electronic flux in chemical reactions. *Sci. China Chem.* 54 (12): 1982–1988.

33 Ceron, M.L., Echegaray, E., Gutierrez-Oliva, S. et al. (2011). The reaction electronic flux in chemical reactions. *Sci. China Chem.* 54, 1982.

34 Cerón, M.L., Herrera, B., Araya, P. et al. (2011). The mechanism of methanol decomposition by CuO. A theoretical study based on the reaction force and reaction electronic flux analysis. *J. Mol. Model.* 17: 1625.

35 Flores-Morales, P., Gutierrez-Oliva, S., Silva, E., and Toro-Labbé, A. (2010). The reaction electronic flux: a new descriptor of the electronic activity taking place during a chemical reaction. Application to the characterization of the mechanism of the Schiff's base formation in the Maillard reaction. *J. Mol. Struct. THEOCHEM* 943: 121.

36 Vogt-Geisse, S. and Toro-Labbé, A. (2009). The mechanism of the interstellar isomerization reaction $HOC^+ \rightarrow HCO^+$ catalyzed by H_2: new insights from the reaction electronic flux. *J. Chem. Phys.* 130: 244308.

37 Echegaray, E. and Toro-Labbé, A. (2008). Reaction electronic flux: a new concept to get insights into reaction mechanisms. Study of model symmetric nucleophilic substitutions. *J. Phys. Chem. A* 112 (46): 11801–11807.

38 Herrera, B. and Toro-Labbe, A. (2004). The role of the reaction force to characterize local specific interactions that activate the intramolecular proton transfers in DNA basis. *J. Chem. Phys.* 121: 7096.

39 Politzer, P., Toro-Labbé, A., Gutierrez-Oliva, S. et al. (2005). The reaction force: three key points along an intrinsic reaction coordinate. *J. Chem. Sci.* 117: 467.

40 Rincon, E., Jaque, P., and Toro-Labbé, A. (2006). Reaction force analysis of the effect of Mg(II) on the 1,3 intramolecular hydrogen transfer in thymine. *J. Phys. Chem. A* 110: 9478.

41 Politzer, P., Burda, J.V., Concha, M.C. et al. (2006). Analysis of the reaction force for a gas phase S_N2 process: $CH_3Cl + H_2O \rightarrow CH_3OH + HCl$. *J. Phys. Chem. A* 110: 756.

42 Herrera, B. and Toro-Labbé, A. (2007). The role of reaction force and chemical potential in characterizing the mechanism of double proton transfer in the adenine–uracil complex. *J. Phys. Chem. A* 111 (26): 5921–5926.

43 Herrera, B. and Toro-Labbe, A. (2007). The role of reaction force and chemical potential in characterizing the mechanism of double proton transfer in the adenine-uracil complex. *J. Phys. Chem. A* 111: 5921.

44 Burda, J.V., Toro-Labbé, A., Gutierrez-Oliva, S. et al. (2007). Reaction force decomposition of activation barriers to elucidate solvent effects. *J. Phys. Chem. A* 111: 2455.

45 Rincon, E. and Toro-Labbe, A. (2007). Reaction force and electron localization function analysis of the metal chelation process in Mg(II)–thymine complex. *Chem. Phys. Lett.* 438: 93.

46 Toro-Labbé, A., Gutierrez-Oliva, S., Murray, J.S., and Politzer, P. (2007). A new perspective on chemical and physical processes: the reaction force. *Mol. Phys.* 105: 2619.

47 Parr, R.G., Donnelly, R.A., Levy, M., and Palke, W.E. (1978). Electronegativity: the density functional viewpoint. *J. Chem. Phys.* 68: 3801.

48 Parr, R.G. (1980). Density functional theory of atoms and molecules. In: *Horizons of Quantum Chemistry* (ed. K. Fukui and B. Pullman), 5–15. Springer.

49 Parr, R.G. and Pearson, R.G. (1983). Absolute hardness: companion parameter to absolute electronegativity. *JACS* 105 (26): 7512–7516.

50 Parr, R.G. and Yang, W. (1997). *Density Functional Theory of Atoms and Molecules.* Oxford University Press.

51 Koopmans, T. (1993). Theoretical aspects of chemical reactivity. *Physica* 91: 651.

52 Echegaray, E. and Toro-Labbé, A. (2008). Reaction electronic flux: a new concept to get insights into reaction mechanisms. Study of model symmetric nucleophilic substitutions. *J. Phys. Chem. A* 112: 11801.

53 Parr, R.G. and Yang, W. (1984). Density functional approach to the frontier-electron theory of chemical reactivity. *JACS* 106: 4049.

54 Fukui, K. (1970). Formulation of the reaction coordinate. *J. Phys. Chem.* 74: 4161.

55 Fukui, K. (1981). The path of chemical reactions – the IRC approach. *Acc. Chem. Res.* 14: 363.

56 Fukui, K. (1982). Role of frontier orbitals in chemical reactions. *Science* 218 (4574): 747–754.

57 Morell, C., Grand, A., and Toro-Labbe, A. (2005). New dual descriptor for chemical reactivity. *J. Phys. Chem. A* 109: 205.

58 Martínez, J. (2009). Local reactivity descriptors from degenerate frontier molecular orbitals. *Chem. Phys. Lett.* 478 (4): 310–322.

59 Pople, J.A., Gill, P.M.W., and Johnson, B.G. (1992). Kohn–Sham density-functional theory within a finite basis set. *Chem. Phys. Lett.* 199 (6): 557–560.

60 Krishnan, R., Binkley, J.S., Seeger, R., and Pople, J.A. (1980). Self-consistent molecular orbital methods. XX. A basis set for correlated wave functions. *J. Chem. Phys.* 72 (1): 650–654.

61 Lee, C., Yang, W., and Parr, R.G. (1988). Development of the Colle–Salvetti correlation-energy formula into a functional of the electron density. *Phys. Rev. B: Condens. Matter Mater. Phys.* 37: 785.

62 Frisch, M.J., Trucks, G.W., Schlegel, H.B. et al. (2009). *Gaussian 09*; Revision A.1. Wallingford, CT: Gaussian, Inc.

63 Lu, T. and Chen, F. (2012). Multiwfn: a multifunctional wavefunction analyzer. *J. Comput. Chem.* 33: 580.

64 Stephan, D.W. and Erker, G. (2015). Frustrated Lewis pair chemistry: development and perspectives. *Angew. Chem. Int. Ed.* 54 (22): 6400–6441.

65 Grimme, S., Kruse, H., Goerigk, L., and Erker, G. (2010). The mechanism of dihydrogen activation by frustrated Lewis pairs revisited. *Angew. Chem. Int. Ed.* 49 (8): 1402–1405.

66 Cardenas, A.J.P., Culotta, B.J., Warren, T.H. et al. (2011). Capture of NO by a frustrated Lewis pair: a new type of persistent N-oxyl radical. *Angew. Chem. Int. Ed.* 50 (33): 7567–7571.

67 Neu, R.C., Otten, E., Lough, A., and Stephan, D.W. (2011). The synthesis and exchange chemistry of frustrated Lewis pair–nitrous oxide complexes. *Chem. Sci.* 2 (1): 170–176.

68 Sajid, M., Elmer, L.-M., Rosorius, C. et al. (2013). Facile carbon monoxide reduction at intramolecular frustrated phosphane/borane Lewis pair templates. *Angew. Chem. Int. Ed.* 52 (8): 2243–2246.

69 Sajid, M., Klose, A., Birkmann, B. et al. (2013). Reactions of phosphorus/boron frustrated Lewis pairs with SO_2. *Chem. Sci.* 4 (1): 213–219.

70 Holthausen, M.H., Mahdi, T., Schlepphorst, C. et al. (2014). Frustrated Lewis pair-mediated C–O or C–H bond activation of ethers. *Chem. Commun.* 50 (70): 10038–10040.

71 Parida, R., Inostroza Rivera, R., Sinha, S., and Giri, S. (2021). Can superhalogen ligand make more reactive frustrated Lewis pairs? *ChemistrySelect* 6 (31): 8068–8073.

72 Boys, S. and Bernardi, F. (1970). The calculation of small molecular interactions by the differences of separate total energies. Some procedures with reduced errors. *Mol. Phys.* 19: 553.

73 Simon, S., Duran, M., and Dannenberg, J. (1996). How does basis set superposition error change the potential surfaces for hydrogen-bonded dimers? *J. Chem. Phys.* 105: 11024.

27

Prediction of Radiative Efficiencies and Global Warming Potential of Hydrofluoroethers and Fluorinated Esters Using Various DFT Functionals

Kanika Guleria, Suresh Tiwari, Dali Barman, Snehasis Daschakraborty, and Ranga Subramanian

Department of Chemistry, Indian Institute of Technology Patna, Patna, India

27.1 Introduction

Chlorofluorocarbons (CFCs), mainly used as refrigerants, cleaning solvents, foam-blowing agents, aerosol propellants, and fire-extinguishing agents, were being phased out before the mid-1990s. CFCs alter the reflection of infrared radiation from the earth's surface and induce global climate change [1]. Due to the harmful impact of CFCs on stratospheric ozone and global warming potential, several international efforts have been made to find environmentally acceptable alternatives such as hydrofluorocarbons and hydrofluorochlorocarbons. Hydrofluoroethers (HFEs) have been invented recently as third-generation CFC substitutes and are used in various industrial applications. Because of the inclusion of H atoms in the structure, HFEs were intended to react quickly with OH radicals in the troposphere, reducing their contribution to global warming [2]. Still, these fluorinated chemicals may absorb strongly in the earth's infrared "atmospheric transparency window" between 8 and 12 μm. Thermal radiation is absorbed by water in the atmosphere below $800\,cm^{-1}$, but water and carbon dioxide absorb beyond $1400\,cm^{-1}$. As a result, there is an infrared zone between 800 and $1400\,cm^{-1}$ where heat radiation from the earth can escape [3]. Therefore, HFEs may be considered potent greenhouse gases [4].

It is well known that the tropospheric oxidation of HFEs results in the formation of fluorinated esters (FESs). FESs are also potential greenhouse gases [2] as they absorb infrared radiation between 1000 and $1400\,cm^{-1}$. With individuals becoming more concerned about environmental preservation, the topic of global warming potential (GWP) has received greater attention. The GWP of a molecule is defined as the time-integrated radiative forcing resulting from the instantaneous emission of a kilogram of trace gas expressed relative to that of a kilogram of a reference gas, usually CO_2. It is essential to understand the effects of HFEs and FES on the atmosphere because they are both powerful greenhouse gases.

We have estimated the radiative efficiency and GWP values of some significant HFEs and FES using *ab initio* calculation. We have considered the following HFEs: CF_3OCH_3 (HFE-143a), $CF_3CF_2OCH_3$ (HFE-245cb2), CH_3OCHF_2 (HFE-152a), CH_2FOCH_2F, $CF_3CH_2OCH_2CF_3$ (HFE-365pcf2), and CF_3OCHF_2 (HFE-125). The studied FES are $CHF_2C(O)OCH_3$, $CF_3C(O)OCH_2CH_3$, $CF_3C(O)OCH_2CF_3$, $FC(O)OCF_2CH_3$, and $CF_3C(O)OCH_3$. The infrared absorption spectra and the reaction rate are needed to compute the molecule's GWP and atmospheric lifetime. HFEs and FESs' interaction with hydroxyl radicals mostly regulates the troposphere's lifetime of these hydrogen-containing trace gases [2].

Electron Density: Concepts, Computation and DFT Applications, First Edition.
Edited by Pratim Kumar Chattaraj and Debdutta Chakraborty.

27.2 Computational Methodology

We have calculated the radiative efficiencies (REs), a measure of the radiative forcing for a unit change in the atmospheric concentration of gases, and GWPs values of HFEs and FESs at various theoretical methods and basis sets. Geometrical optimization and frequency calculations were performed using different density functional theory (DFT) methods. The DFT methods are Becke-3-Lee–Yang–Parr (B3LYP), Becke three-parameter Perdew/Wang 91 (B3PW91), wB97X-D, Austin–Petersson–Frisch-Dispersion (APF-D), M06-2X, Tao–Perdew–Staroverov–Scuseria hybrid (TPSSh), Slater–Vosvo–Wilk–Nusair (SVWN), and Perdew–Burke–Ernzerhof (PBE). Pople basis sets, namely 6-31g(d,p), 6-31+g(d,p), and 6-311g(d,p), were used for geometrical and frequency calculations. We visualized molecular geometries and vibrational motions using Gaussview software [5]. The Gaussian 16 software package was used to perform the quantum mechanical calculations [6]. Anharmonic vibrations were considered for calculating radiative forcing values because potential energy surfaces are more accurate [7].

27.3 RE and GWP Calculation Methodology

Pinnock et al. developed a simple radiative forcing model that allows us to immediately translate absorption cross-sections to radiative forcing without the need of complex atmospheric modeling [8]. We used Pinnock's model for calculating REs and GWPs of two fluorinated esters ($CF_3CF_2C(O)OCH_3$ and $CHF_2CF_2C(O)OCH_3$) in our previous study [9]. In this approach, the REs is given by:

$$F_{\text{tot}} = \sum_{i=1}^{250} 10\,\sigma_i F(\boldsymbol{\nu}_i) \tag{27.1}$$

where F_{tot} is the RE (a_i) in W m^{-2} ppbv^{-1}. $F(\boldsymbol{\nu}_i)$ is the RE per unit cross-section in Wm^{-2} (cm^{-1})$^{-1}$ (cm^2/mol)$^{-1}$, and σ_i is the cross-sectional area of species I in cm^2/mol.

In subsequent studies, Papasavva et al. [10] developed an extension that employed *ab initio* data to substitute experimentally measured absorption cross-sections [8], and is given below.

$$F_{\text{tot}} = \sum_{k} A_k F(\boldsymbol{\nu}_k) \tag{27.2}$$

where A_k is the intensity of infrared radiation evaluated from peak height, in km/mol and $F(\boldsymbol{\nu}_k)$ is the radiative efficiency per unit cross-section, in W m^{-2} (cm^{-1})$^{-1}$ (cm^2/mol)$^{-1}$, F_{tot} is the RE (a_i) in W m^{-2} ppbv^{-1}.

The expression to experimentally compute GWPs utilizing information about species degradation rates in the troposphere and infrared vibrational spectroscopy is given as:

$$\text{GWP} = \frac{\int_0^{\text{TH}} a_i[x_i(t)]dt}{\int_0^{\text{TH}} a_{\text{ref}}[x_{\text{ref}}(t)]dt} \tag{27.3}$$

where TH is the time horizon for GWP, a_i is the RE due to unit increase in the atmospheric concentration of species, i, $[x_i(t)]$ is the time-dependent concentration of a pulse of species i. The term in the denominator is reference gas species' (CO_2) absolute global warming potential (AGWP). Instead of integrating $x_i(t)$, a modification of Eq. (27.3) employs the atmospheric lifetime of a species,

$$\text{GWP} = \frac{a_i \int_0^{TH} e - \frac{t}{\tau} dt}{\text{AGWP CO}_2(\text{TH})} \tag{27.4}$$

where AGWP is the absolute global warming potential of reference gas CO_2 at three different time horizons (20, 100, and 500 years). The World Meteorological Organization released a current set of AGWP of CO_2 in 2007, with values of 0.192, 0.676, and 2.223 W yr m^{-2} $ppmv^{-1}$ over the same periods [11]. τ is the lifetime, and t is the time horizon. These values of AGWP of CO_2 are the most validated values available in the literature. To calculate radiative efficiency and GWP values of HFEs and FES, the lowest energy rotamers have been selected.

27.4 Results and Discussions

The optimized structures, calculated RE, and GWP of various HFEs and FES are shown below.

27.4.1 (Difluoromethoxy)trifluoromethane (CF_3OCHF_2)

The CF_3OCHF_2 molecule is found in two different rotamers: a syn-periplanar (sp) and an antiperiplanar (ap) rotamer. Good and Francisco discovered that the syn-periplanar (sp) form is the most stable rotamer [3]. The energy difference between the two conformers of the CF_3OCHF_2 molecule is 1.080 kcal/mol. We used the lowest energy structure for the computation of REs. The optimized structure of the syn-periplanar CF_3OCHF_2 molecule is shown in Figure 27.1. Our calculated GWPs values using M06-2X, TPSSh, B3LYP, B3PW91, APF-D, wB97X-D, SVWN, and PBE DFT functionals are tabulated in Table 27.1. There were experimental reported GWP values for the CF_3OCHF_2 molecule. They are 13,400, 15,600, and 10,200 over 20, 100, and 500 years [12]. Our calculated GWP values at the PBE/6-31g(d,p), TPSSh/6-31g(d,p), and APFD/6-311g(d,p) level of theory are close to experimental GWP literature values.

27.4.2 Difluoro(methoxy)methane (CH_3OCHF_2)

There are two possible rotamers of the CH_3OCHF_2 molecule. Syn-clinal (sc) rotamer is one kind, whereas antiperiplanar rotamer is another (ap). The antiperiplanar (ap) rotamer is the most stable conformer of CH_3OCHF_2 [13]. The CH_3OCHF_2 molecule has a 1.98 kcal/mol energy difference between its two conformers. Therefore, the antiperiplanar form was then used for further calculations and its optimized structure is shown in Figure 27.2. The GWPs of the CH_3OCHF_2 molecule have previously been published. The experimentally reported GWP values are 1025, 291, and 89 for

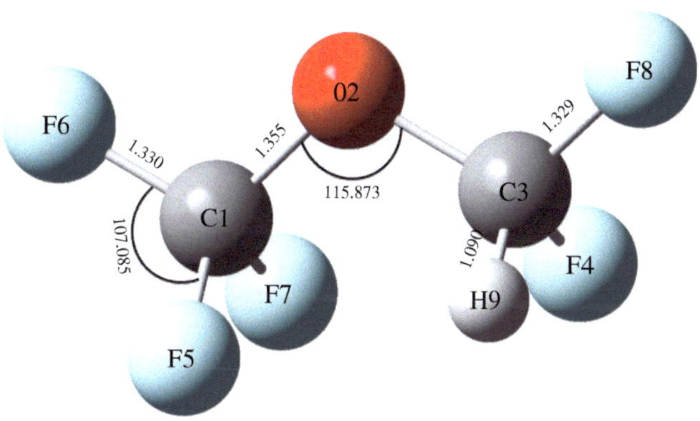

Figure 27.1 Optimized structure of CF_3OCHF_2 molecule at M06-2X/6-311G(d,p) level of theory.

Table 27.1 Radiative efficiency (RE) (a_i) and global warming potential (GWP) value of CF_3OCHF_2 at different DFT methods and Pople basis sets.

				This work (GWP)			Experimental (GWP)		
DFT approximation	**Functional**	**Basis set**	**REs**	**20 years**	**100 years**	**500 years**	**20 years**	**100 years**	**500 years**
Meta-Hybrid-GGA	M06-2X	6-31g(d,p)	0.2321	7,370	8,334	5,310	**13,400**	**15,600**	**10,200**
		6-31+g(d,p)	0.2818	8,946	10,116	6,446			
		6-311g(d,p)	0.2534	8,044	9,097	5,797			
	TPSSh	6-31g(d,p)	0.4136	13,128	14,846	9,456			
		6-31+g(d,p)	0.5445	17,283	19,545	12,449			
		6-311g(d,p)	0.5250	16,664	18,844	12,003			
Hybrid-GGA	B3LYP	6-31g(d,p)	0.3733	11,850	13,400	8,536			
		6-31+g(d,p)	0.5206	16,524	18,686	11,902			
		6-311g(d,p)	0.5142	16,323	18,458	11,758			
	B3PW91	6-31g(d,p)	0.3362	10,672	12,068	7,687			
		6-31+g(d,p)	0.4741	15,051	17,020	10,841			
		6-311g(d,p)	0.4730	15,013	16,977	10,815			
	APF-D	6-31g(d,p)	0.2961	9,398	10,628	6,770			
		6-31+g(d,p)	0.4616	14,653	16,750	10,555			
		6-311g(d,p)	0.4404	13,980	15,809	10,071			
	wB97X-D	6-31g(d,p)	0.2461	7,813	8,835	5,629			
		6-31+g(d,p)	0.3830	12,157	13,748	8,759			
		6-311g(d,p)	0.3803	12,074	13,653	8,698			
Pure	SVWN	6-31g(d,p)	0.3224	10,233	11,572	7,371			
		6-31+g(d,p)	0.4737	15,037	17,004	10,831			
		6-311g(d,p)	0.4571	14,508	16,406	10,450			
	PBE	6-31g(d,p)	0.4359	13,838	15,649	9,967			
		6-31+g(d,p)	0.6087	19,320	21,848	13,915			
		6-311g(d,p)	0.5607	17,797	20,125	12,818			

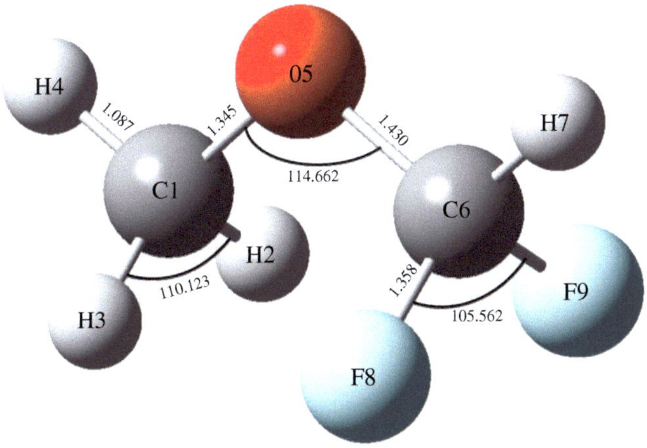

Figure 27.2 Optimized structure of CH_3OCHF_2 molecule at M06-2X/6-311G(d,p) level of theory.

time horizons of 20, 100, and 500 years [4]. From Table 27.2, We can infer that our computed GWPs at the M06-2X/6-31g+(d,p), B3PW91/6-31g+(d,p), and SVWN/6-311g(d,p) levels are comparable to the reported values.

27.4.3 Trifluoro(methoxy)methane (CF_3OCH_3)

The optimized geometry of the CF_3OCH_3 molecule has C_s symmetry [13]. No other conformers of this molecule exist. The optimized structure of CF_3OCH_3 is shown below in Figure 27.3:

The experimental GWP values for the CF_3OCH_3 molecule have already been reported. Previously reported experimental GWP values are 2550, 739, and 225 for periods of 20, 100, and 500 years [4]. Table 27.3 shows that the obtained results at the APF-D/6-31g+(d,p) level, SVWN/6-311g(d,p) level, and PBE/6-31g(d,p) of the theory are comparable to the previously published GWPs values.

27.4.4 Bis(2,2,2-trifluoroethyl)ether ($CF_3CH_2OCH_2CF_3$)

The hydrogen atoms in each of the two $-CH_2$ groups in the $CF_3CH_2OCH_2CF_3$ molecule are not identical because of different stereographical orientations of hydrogen atoms bonded to the same

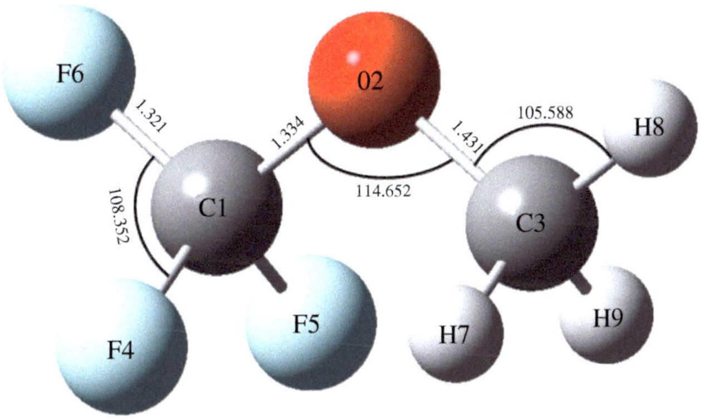

Figure 27.3 Optimized structure of CF_3OCH_3 molecule at M06-2X/6-311G(d,p) level of theory.

Table 27.2 Radiative efficiency (a_i) and global warming potential value of CH_3OCHF_2 at different DFT methods and different Pople basis sets.

DFT approximation	Functional	Basis set	REs	This work (GWP)			Experimental (GWP)		
				20 years	100 years	500 years	20 years	100 years	500 years
Meta-Hybrid-GGA	M06-2X	6-31g(d,p)	0.1566	783	222	67	**1025**	**291**	**89**
		6-31+g(d,p)	0.1995	997	283	86			
		6-311g(d,p)	0.1969	985	279	84			
	TPSSh	6-31g(d,p)	0.2355	1178	334	101			
		6-31+g(d,p)	0.2467	1233	350	106			
		6-311g(d,p)	0.2139	1069	303	92			
Hybrid-GGA	B3LYP	6-31g(d,p)	0.2268	1134	322	97			
		6-31+g(d,p)	0.2273	1136	322	98			
		6-311g(d,p)	0.2254	1127	320	97			
	B3PW91	6-31g(d,p)	0.1981	990	281	85			
		6-31+g(d,p)	0.2119	1059	300	91			
		6-311g(d,p)	0.1892	946	268	81			
	APF-D	6-31g(d,p)	0.1795	897	254	77			
		6-31+g(d,p)	0.1712	856	243	73			
		6-311g(d,p)	0.2143	1071	304	92			
	wB97X-D	6-31g(d,p)	0.1765	883	250	76			
		6-31+g(d,p)	0.1744	872	247	75			
		6-311g(d,p)	0.2149	1075	305	92			
Pure	SVWN	6-31g(d,p)	0.1901	950	270	81			
		6-31+g(d,p)	0.1811	905	257	78			
		6-311g(d,p)	0.1993	996	283	85			
	PBE	6-31g(d,p)	0.1902	951	270	82			
		6-31+g(d,p)	0.2814	1407	400	121			
		6-311g(d,p)	0.2551	1275	362	110			

Table 27.3 Radiative efficiency (a_i) and global warming potential value of CF_3OCH_3 at different DFT methods and Pople basis sets.

DFT approximation	Functional	Basis set	REs	This work (GWP) 20 years	This work (GWP) 100 years	This work (GWP) 500 years	Experimental (GWP) 20 years	Experimental (GWP) 100 years	Experimental (GWP) 500 years
Meta-Hybrid-GGA	M06-2X	6-31g(d,p)	0.0861	995	288	88			
		6-31+g(d,p)	0.1583	1,829	530	161			
		6-311g(d,p)	0.1466	1,695	492	149			
	TPSSh	6-31g(d,p)	0.1796	2,075	602	183			
		6-31+g(d,p)	0.2717	3,140	910	277			
		6-311g(d,p)	0.2543	2,939	852	259			
Hybrid-GGA	B3LYP	6-31g(d,p)	0.1829	2,114	613	186			
		6-31+g(d,p)	0.2587	2,990	867	263			
		6-311g(d,p)	0.2528	2,921	847	257			
	B3PW91	6-31g(d,p)	0.1498	1,731	502	153			
		6-31+g(d,p)	0.2368	2,737	793	241			
		6-311g(d,p)	0.2119	2,449	710	216	**2550**	**739**	**225**
	APF-D	6-31g(d,p)	0.1431	1,654	480	146			
		6-31+g(d,p)	0.2279	2,634	764	232			
		6-311g(d,p)	0.1968	2,275	660	200			
	wB97X-D	6-31g(d,p)	0.1276	1,474	427	129			
		6-31+g(d,p)	1.5141	17,499	5075	1543			
		6-311g(d,p)	0.1931	2,232	647	197			
Pure	SVWN	6-31g(d,p)	0.1591	1,839	533	162			
		6-31+g(d,p)	0.2413	2,789	808	246			
		6-311g(d,p)	0.2277	2,631	763	232			
	PBE	6-31g(d,p)	0.2289	2,645	767	233			
		6-31+g(d,p)	0.3485	4,028	1168	355			

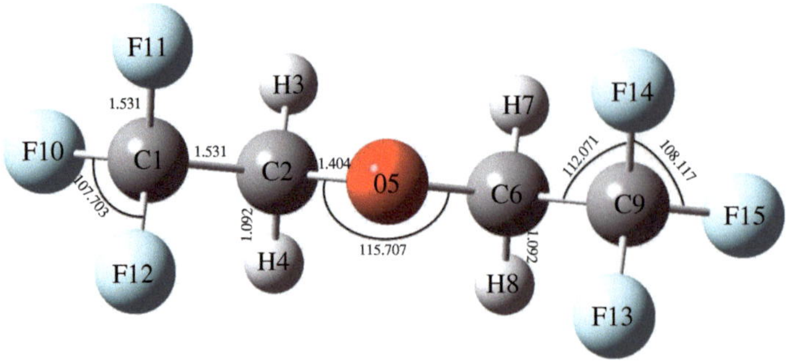

Figure 27.4 Optimized structure of $CF_3CH_2OCH_2CF_3$ molecule at M06-2X/6-311G(d,p) level of theory.

carbon atom. Two stable conformers were reported for the $CF_3CH_2OCH_2CF_3$ molecule. We chose the lower energy (more stable) conformer out of two feasible conformers to further calculate RE values [14]. The optimized structure of the stable conformer of the $CF_3CH_2OCH_2CF_3$ molecule is shown in Figure 27.4.

The $CF_3CH_2OCH_2CF_3$ molecule has previously published GWP values. Previous experimental reported GWP values are 134, 39, and 12 for 20, 100, and 500 years time horizons [12]. From Table 27.4, we can deduce that our predicted GWPs at the wB97X-D/6-31g+(d,p) level, wB97X-D/6-311g(d,p) level, and SVWN/6-311g(d,p) level are comparable to these experimental values.

27.4.5 1,1,1,2,2-Pentafluoro-2-Methoxyethane ($CF_3CF_2OCH_3$)

There are two distinct conformations of the $CF_3CF_2OCH_3$ molecule. The $CF_3CF_2OCH_3$ molecule has a 2.44 kcal/mol energy difference between its two conformers. We used the lowest energy (more stable) conformer to calculate the RE, and its optimized structure is shown in Figure 27.5 [14]. The $CF_3CF_2OCH_3$ compound has previously published GWP values. Hodenbrog et al. employed experimental approaches to derive GWP values and reported a GWP value of 784 across the 100-year time frame [15]. Table 27.5 depicts that at the B3PW91/6-311g(d,p) level, our anticipated GWPs are close to the literature values.

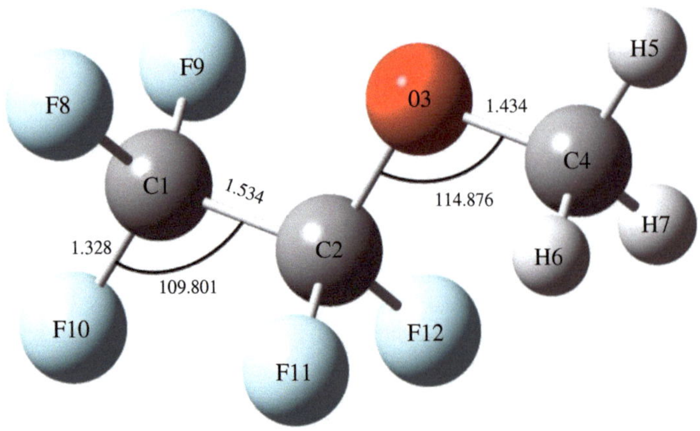

Figure 27.5 Optimized $CF_3CF_2OCH_3$ structure at M06-2X/6-311G(d,p) level of theory.

Table 27.4 Radiative efficiency (a_i) and global warming potential value of $CF_3CH_2OCH_2CF_3$ at different DFT and Pople basis sets.

				This work (GWP)			Experimental (GWP)		
DFT approximation	**Functional**	**Basis set**	**REs**	**20 years**	**100 years**	**500 years**	**20 years**	**100 years**	**500 years**
Meta-Hybrid-GGA	M06-2X	6-31g(d,p)	0.1731	63	17	5	**134**	**39**	**12**
		6-31+g(d,p)	0.2647	97	27	8			
		6-311g(d,p)	0.2641	97	27	8			
Hybrid-GGA	TPSSh	6-31g(d,p)	0.3384	125	35	10			
		6-31+g(d,p)	0.4689	173	49	14			
		6-311g(d,p)	0.4602	170	49	15			
	B3LYP	6-31g(d,p)	0.3157	116	33	10			
		6-31+g(d,p)	0.4419	163	46	14			
		6-311g(d,p)	0.4100	151	43	13			
	B3PW91	6-31g(d,p)	0.2759	102	29	9			
		6-31+g(d,p)	0.3829	141	40	12			
		6-311g(d,p)	0.3837	142	40	12			
	APF-D	6-31g(d,p)	0.2752	101	29	9			
		6-31+g(d,p)	0.3463	128	36	11			
		6-311g(d,p)	0.3550	131	37	11			
	wB97X-D	6-31g(d,p)	0.2490	92	26	7			
		6-31+g(d,p)	0.3616	133	38	11			
		6-311g(d,p)	0.3601	133	37	11			
Pure	SVWN	6-31g(d,p)	0.2325	86	24	7			
		6-31+g(d,p)	0.3651	135	38	11			
		6-311g(d,p)	0.3730	138	39	12			
	PBE	6-31g(d,p)	0.3942	145	41	12			
		6-31+g(d,p)	0.5329	197	56	17			
		6-311g(d,p)	0.5089	188	53	16			

Table 27.5 Radiative efficiency (a_i) and global warming potential value of $CF_3CF_2OCH_3$ at different DFT methods and different Pople basis sets.

DFT approximation	Functional	Basis set	REs	This work (GWP)			Experimental (GWP)		
				20 years	100 years	500 years	20 years	100 years	500 years
Meta-Hybrid-GGA	M06-2X	6-31g(d,p)	0.1924	1443	417	126	—	**784**	—
		6-31+g(d,p)	0.2197	1647	476	144			
		6-311g(d,p)	0.2177	1632	472	143			
	TPSSh	6-31g(d,p)	0.3295	2470	714	217			
		6-31+g(d,p)	0.4511	3382	978	297			
		6-311g(d,p)	0.4858	3643	1054	320			
Hybrid-GGA	B3LYP	6-31g(d,p)	0.3069	2301	665	202			
		6-31+g(d,p)	0.4169	3126	904	275			
		6-311g(d,p)	0.4078	3057	884	269			
	B3PW91	6-31g(d,p)	0.2502	1876	542	165			
		6-31+g(d,p)	0.5970	4476	1295	393			
		6-311g(d,p)	0.3691	2768	800	243			
	APF-D	6-31g(d,p)	0.2321	1740	503	153			
		6-31+g(d,p)	0.3134	2350	670	206			
		6-311g(d,p)	0.3128	2346	678	206			
	wB97X-D	6-31g(d,p)	0.2243	1682	486	147			
		6-31+g(d,p)	0.2976	2231	645	196			
		6-311g(d,p)	0.3186	2389	691	210			
Pure	SVWN	6-31g(d,p)	0.2204	1653	478	145			
		6-31+g(d,p)	0.3336	2501	724	220			
		6-311g(d,p)	0.3416	2561	741	225			
	PBE	6-31g(d,p)	0.4118	3087	893	271			
		6-31+g(d,p)	0.3986	2989	864	262			
		6-311g(d,p)	0.4177	3132	906	275			

The optimized lowest energy structure of the $CF_3CF_2OCH_3$ molecule is given below:

27.4.6 Fluoro(fluoromethoxy)methane (CH_2FOCH_2F)

According to the literature studies, two distinct conformations of the CH_2FOCH_2F molecule exist. One conformer corresponds to C_2 symmetry, and the other corresponds to C_s symmetry [16]. The conformation with C_2 symmetry was reported to be more stable as having lower energy than the other conformer. So, we have calculated GWP for the C_2 symmetry conformer, and its optimized structure is shown in Figure 27.6. The CH_2FOCH_2F compound has previously reported GWP values. Hodenbrog and coworkers employed experimental approaches to derive GWP values and reported 470 and 127 over 20 and 100-year time frames [17]. From Table 27.6, we can conclude that at the M06-2X/6-311g(d,p) level and wB97X-D/6-311g(d,p), our anticipated GWPs are close to the literature values.

27.4.7 Methyl 2,2,2-Difluoroacetate ($CHF_2C(O)OCH_3$)

The $CHF_2C(O)OCH_3$ molecule has several structures, out of which two stable conformers, one with C_s symmetry and the other with C_1 symmetry, are identified [18]. The C_s conformer is found to be more stable by 0.39 kcal/mol. We have optimized the conformer having the C_s symmetry, and its GWP has been calculated. The optimized structure of stable $CHF_2C(O)OCH_3$ molecule is shown in Figure 27.7. The GWPs of the $CHF_2C(O)OCH_3$ molecule have previously been published. The experimental GWP reported values are 35 and 9 for time horizons of 20 and 100 years [17]. In Table 27.7, we can see that our computed GWPs at the B3LYP/6-31g(d,p) level and B3PW91/6-31g(d,p) are comparable to these reported experimental values.

27.4.8 Ethyl 2,2,2-Trifluoroacetate ($CF_3C(O)OCH_2CH_3$)

The $CF_3C(O)OCH_2CH_3$ molecule exists in different conformations, among which two conformers are stable [18]. The conformer with C_s symmetry was found to be more stable by 0.30 kcal/mol than the conformer with C_1 symmetry, hence used for optimization and GWP calculation. Figure 27.8 shows the optimized structure of stable C_s conformer of $CF_3C(O)OCH_2CH_3$ molecule.

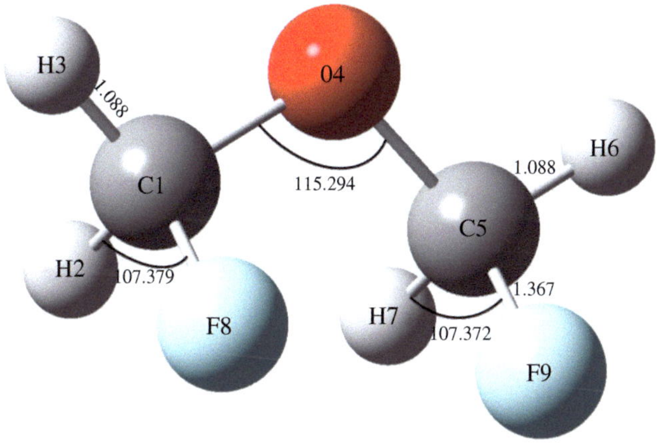

Figure 27.6 Optimized structure of CH_2FOCH_2F molecule at M06-2X/6-311G(d,p) level of theory.

Table 27.6 Radiative efficiency (a_i) and global warming potential value of CH_2FOCH_2F at different DFT methods and Pople basis sets.

				This work (GWP)			Experimental (GWP)		
DFT approximation	**Functional**	**Basis set**	**REs**	**20 years**	**100 years**	**500 years**	**20 years**	**100 years**	**500 years**
Meta-Hybrid-GGA	M06-2X	6-31g(d,p)	0.1526	383	108	33	**470**	**127**	—
		6-31+g(d,p)	0.2015	506	143	43			
		6-311g(d,p)	0.1838	462	131	39			
	TPSSh	6-31g(d,p)	0.2227	560	159	48			
		6-31+g(d,p)	0.2501	629	178	54			
		6-311g(d,p)	0.2438	613	174	52			
Hybrid-GGA	B3LYP	6-31g(d,p)	0.2139	537	152	46			
		6-31+g(d,p)	0.2224	559	158	48			
		6-311g(d,p)	0.2278	572	162	49			
	B3PW91	6-31g(d,p)	0.2045	514	146	44			
		6-31+g(d,p)	0.2333	586	166	50			
		6-311g(d,p)	0.2413	606	172	52			
	APF-D	6-31g(d,p)	0.1836	461	131	39			
		6-31+g(d,p)	0.2224	559	158	48			
		6-311g(d,p)	0.2108	530	150	45			
	wB97X-D	6-31g(d,p)	0.1526	383	108	33			
		6-31+g(d,p)	0.2015	506	143	43			
		6-311g(d,p)	0.1838	462	131	39			
Pure	SVWN	6-31g(d,p)	0.2227	560	159	48			
		6-31+g(d,p)	0.2501	629	178	54			
		6-311g(d,p)	0.2438	613	174	52			
	PBE	6-31g(d,p)	0.2139	537	152	46			
		6-31+g(d,p)	0.2224	559	158	48			
		6-311g(d,p)	0.2278	572	162	49			

Table 27.7 Radiative efficiency (a_i) and global warming potential value of $CHF_2C(O)OCH_3$ at different DFT methods and different Pople basis sets.

DFT approximation	Functional	Basis set	REs	This work (GWP)			Experimental (GWP)		
				20 years	100 years	500 years	20 years	100 years	500 years
		6-31g(d,p)	0.099	22	6	1	**35**	**9**	—
	M06-2X	6-31+g(d,p)	0.137	31	8	2			
Meta-Hybrid-GGA		6-311g(d,p)	0.147	33	9	2			
		6-31g(d,p)	0.179	41	11	3			
	TPSSh	6-31+g(d,p)	0.243	55	15	4			
		6-311g(d,p)	0.236	54	15	4			
		6-31g(d,p)	0.154	35	10	3			
	B3LYP	6-31+g(d,p)	0.223	51	14	4			
		6-311g(d,p)	0.216	49	14	4			
		6-31g(d,p)	0.154	35	10	3			
Hybrid-GGA	B3PW91	6-31+g(d,p)	0.192	44	12	3			
		6-311g(d,p)	0.191	52	14	4			
		6-31g(d,p)	0.141	31	8	2			
	APF-D	6-31+g(d,p)	0.150	33	9	2			
		6-311g(d,p)	0.158	36	11	3			
		6-31g(d,p)	0.129	29	8	2			
	wB97X-D	6-31+g(d,p)	0.175	40	11	3			
		6-311g(d,p)	0.171	39	11	3			
		6-31g(d,p)	0.115	26	7	2			
	SVWN	6-31+g(d,p)	0.185	42	12	3			
Pure		6-311g(d,p)	0.173	39	11	3			
		6-31g(d,p)	0.211	48	13	4			
	PBE	6-31+g(d,p)	0.215	49	14	4			
		6-311g(d,p)	0.226	50	14	4			

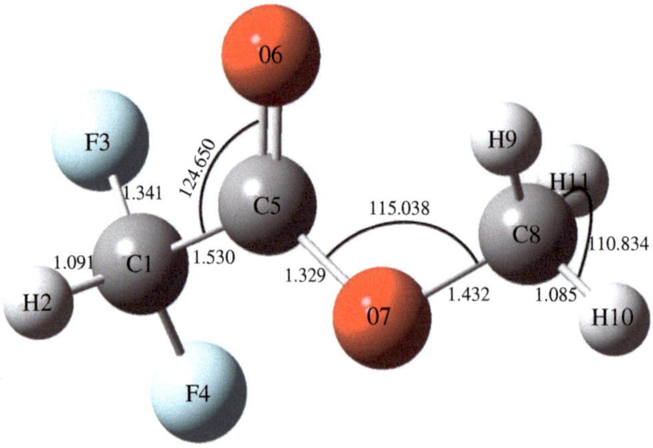

Figure 27.7 Optimized structure of $CHF_2C(O)OCH_3$ molecule at M06-2X/6-311G(d,p), level of theory.

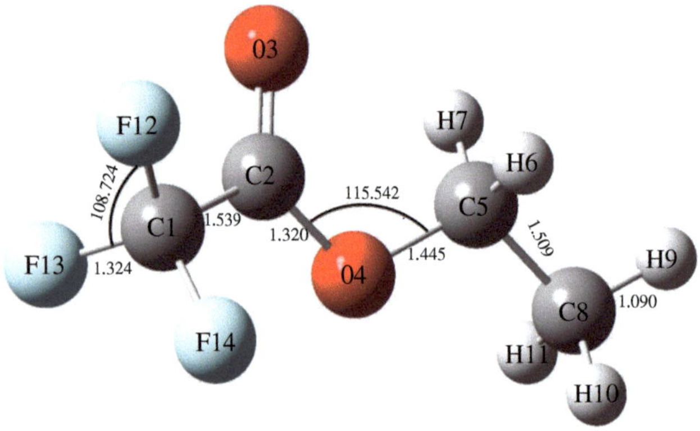

Figure 27.8 Optimized structure of $CF_3C(O)OCH_2CH_3$ molecule at M06-2X/6-311G(d,p) level of theory.

The $CF_3C(O)OCH_2CH_3$ molecule has previously published GWP values. Previous experimental GWP literature reported values are 5 and 1 over time horizons of 20 and 100 years [17]. Our predicted GWPs at the M06-2X/6-31g(d,p) level, M06-2X/6-31g+(d,p), and wB97XD/6-31g(d,p) level are closest to the experimental values reported in Table 27.8.

27.4.9 2,2,2-Trifluoroethyl 2,2,2-trifluoroacetate ($CF_3C(O)OCH_2CF_3$)

Geometrical optimization of $CF_3C(O)OCH_2CF_3$ molecule predicts two possible conformers. The lowest energy conformer is stable by 0.65 kcal/mol from its higher energy conformer [19]. To calculate the RE and GWP, the conformer of the lowest energy is considered for optimization and frequency calculation, and its optimized structure is shown in Figure 27.9. There were previously reported GWP values for the $CF_3C(O)OCH_2CF_3$ molecule. The experimental GWP values are 85 and 23 over 20 and 100 years [17]. From Table 27.9, we can glean that our calculated GWP values at the M06-2X/6-311g(d,p) level and B3LYP/6-311g(d,p) level of theory are close to experimental GWP literature values.

Table 27.8 Radiative efficiency (a_i) and global warming potential value of $CF_3C(O)OCH_2CH_3$ at different DFT methods and different Pople basis sets.

DFT approximation	Functional	Basis set	REs	This work (GWP)			Experimental (GWP)		
				20 years	100 years	500 years	20 years	100 years	500 years
Meta-Hybrid-GGA	M06-2X	6-31g(d,p)	0.174	17	4	1	**5**	**1**	—
		6-31+g(d,p)	0.193	18	5	1			
		6-311g(d,p)	0.147	33	9	2			
	TPSSh	6-31g(d,p)	0.270	26	7	2			
		6-31+g(d,p)	0.336	32	9	2			
		6-311g(d,p)	0.348	23	9	2			
Hybrid-GGA	B3LYP	6-31g(d,p)	0.236	22	6	1			
		6-31+g(d,p)	0.317	30	8	2			
		6-311g(d,p)	0.339	32	9	2			
	B3PW91	6-31g(d,p)	0.237	22	6	1			
		6-31+g(d,p)	0.274	26	7	2			
		6-311g(d,p)	0.298	28	8	2			
	APF-D	6-31g(d,p)	0.219	21	6	1			
		6-31+g(d,p)	0.299	29	8	2			
		6-311g(d,p)	0.308	29	8	2			
	wB97X-D	6-31g(d,p)	0.190	18	5	2			
		6-31+g(d,p)	0.361	34	9	3			
		6-311g(d,p)	0.291	28	8	2			
Pure	SVWN	6-31g(d,p)	0.215	20	5	1			
		6-31+g(d,p)	0.275	26	7	2			
		6-311g(d,p)	0.266	25	7	2			
	PBE	6-31g(d,p)	0.365	35	9	3			
		6-31+g(d,p)	0.378	36	10	3			
		6-311g(d,p)	0.373	36	10	3			

Table 27.9 Radiative efficiency (a_i) and global warming potential value of $CF_3C(O)OCH_2CF_3$ at different DFT methods and Pople basis sets.

				This work (GWP)			Experimental (GWP)		
DFT approximation	**Functional**	**Basis set**	**REs**	**20 years**	**100 years**	**500 years**	**20 years**	**100 years**	**500 years**
Meta-Hybrid-GGA	M06-2X	6-31g(d,p)	0.260	45	12	3	**85**	**23**	—
		6-31+g(d,p)	0.303	53	15	4			
		6-311g(d,p)	0.474	83	23	7			
	TPSSh	6-31g(d,p)	0.374	65	18	5			
		6-31+g(d,p)	0.505	88	25	7			
		6-311g(d,p)	0.519	91	25	7			
Hybrid-GGA	B3LYP	6-31g(d,p)	0.352	61	17	5			
		6-31+g(d,p)	0.501	88	25	7			
		6-311g(d,p)	0.495	86	24	7			
	B3PW91	6-31g(d,p)	0.326	57	16	4			
		6-31+g(d,p)	0.436	76	21	6			
		6-311g(d,p)	0.450	78	22	6			
	APF-D	6-31g(d,p)	0.305	53	15	4			
		6-31+g(d,p)	0.409	72	20	5			
		6-311g(d,p)	0.434	76	21	6			
	wB97X-D	6-31g(d,p)	0.386	68	18	5			
		6-31+g(d,p)	0.390	68	18	5			
		6-311g(d,p)	0.411	73	20	6			
Pure	SVWN	6-31g(d,p)	0.281	49	14	1			
		6-31+g(d,p)	0.396	69	19	6			
		6-311g(d,p)	0.399	70	19	6			
	PBE	6-31g(d,p)	0.408	74	20	5			
		6-31+g(d,p)	0.575	100	28	6			
		6-311g(d,p)	0.386	68	18	5			

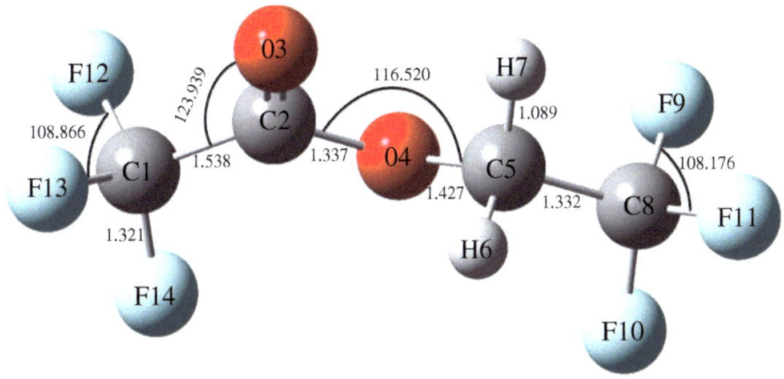

Figure 27.9 Optimized structure of $CF_3C(O)OCH_2CF_3$ molecule at M06-2X/6-311G(d,p) level of theory.

27.4.10 1,1-Difluoroethyl Carbonofluoridate ($FC(O)OCF_2CH_3$)

The molecule $FC(O)OCF_2CH_3$ exists in only one conformation. We have geometrically optimized this structure, as shown in Figure 27.10, and further determined RE. Using the RE value, we have calculated the GWP value. The $FC(O)OCF_2CH_3$ compound has previously published GWP values. Hodenbrog et al. employed experimental approaches to derive GWP values and reported 90 and 24 over 20 and 100 years time frames [17]. Table 27.10 shows us that at the M06-2X/6-31g(d,p) level, our anticipated GWPs are close to the literature values.

27.4.11 Methyl 2,2,2-trifluoroacetate ($CF_3C(O)OCH_3$)

This molecule exists in two conformations, anti and syn-conformer. As both are found to be stable [20], both the isomers are considered for optimization and further REs and GWP studies. The optimized structures of anti and syn $CF_3C(O)OCH_3$ molecules are shown in Figures 27.11 and 27.12, respectively. The $CF_3C(O)OCH_3$ compound has previously reported GWP values. Hodenbrog et al. employed experimental approaches to derive GWP values and reported 315, 86 over 20 and 100 years of time horizons [17]. In Tables 27.11 and 27.12, we can see that for anti-conformer and syn-conformer, our anticipated GWPs at the TPSSh/6-31+g(d,p) level are close to the literature values.

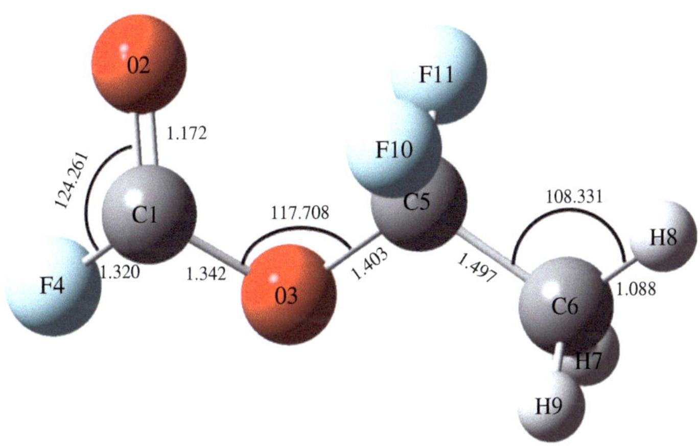

Figure 27.10 Optimized structure of $FC(O)OCF_2CH_3$ molecule at M06-2X/6-311G(d,p) level of theory.

Table 27.10 Radiative efficiency (a_i) and global warming potential value of $FC(O)OCF_2CH_3$ at different DFT methods and different Pople basis sets.

DFT approximation	Functional	Basis set	REs	This work (GWP)			Experimental (GWP)		
				20 years	100 years	500 years	20 years	100 years	500 years
Meta-Hybrid-GGA	M06-2X	6-31g(d,p)	0.223	167	47	14	**90**	**24**	—
		6-31+g(d,p)	0.249	186	53	16			
		6-311g(d,p)	0.273	204	58	17			
	TPSSh	6-31g(d,p)	0.316	236	67	20			
		6-31+g(d,p)	0.456	340	96	29			
		6-311g(d,p)	0.446	333	94	28			
Hybrid-GGA	B3LYP	6-31g(d,p)	0.306	228	65	19			
		6-31+g(d,p)	0.414	309	87	26			
		6-311g(d,p)	0.418	312	87	26			
	B3PW91	6-31g(d,p)	0.298	218	60	19			
		6-31+g(d,p)	0.277	206	58	17			
		6-311g(d,p)	0.280	210	59	17			
	APF-D	6-31g(d,p)	0.414	312	88	27			
		6-31+g(d,p)	0.391	290	81	22			
		6-311g(d,p)	0.406	299	84	24			
	wB97X-D	6-31g(d,p)	0.238	177	50	15			
		6-31+g(d,p)	0.341	254	72	22			
		6-311g(d,p)	0.349	261	74	22			
Pure	SVWN	6-31g(d,p)	0.251	187	53	16			
		6-31+g(d,p)	0.313	233	66	20			
		6-311g(d,p)	0.393	293	83	25			
	PBE	6-31g(d,p)	0.408	304	86	26			
		6-31+g(d,p)	0.514	384	109	33			
		6-311g(d,p)	0.490	366	104	31			

Table 27.11 Radiative efficiency (a_i) and global warming potential value of *anti*-$CF_3C(O)OCH_3$ at different DFT methods and different Pople basis sets.

				This work (GWP)			Experimental (GWP)		
DFT approximation	**Functional**	**Basis set**	**REs**	**20 years**	**100 years**	**500 years**	**20 years**	**100 years**	**500 years**
		6-31g(d,p)	0.147	87	24	7	**315**	**86**	—
	M06-2X	6-31+g(d,p)	0.295	174	49	15			
		6-311g(d,p)	0.288	172	48	14			
		6-31g(d,p)	0.259	153	43	13			
Meta-Hybrid-GGA	TPSSh	6-31+g(d,p)	0.312	188	57	19			
		6-311g(d,p)	0.323	186	56	18			
		6-31g(d,p)	0.205	121	34	10			
	B3LYP	6-31+g(d,p)	0.295	174	49	15			
		6-311g(d,p)	0.301	178	50	15			
		6-31g(d,p)	0.293	173	48	15			
Hybrid-GGA	B3PW91	6-31+g(d,p)	0.289	172	47	13			
		6-311g(d,p)	0.285	171	47	13			
		6-31g(d,p)	0.204	120	34	10			
	APF-D	6-31+g(d,p)	0.255	150	42	13			
		6-311g(d,p)	0.265	157	44	13			
		6-31g(d,p)	0.168	99	28	8			
	wB97X-D	6-31+g(d,p)	0.198	117	31	7			
		6-311g(d,p)	0.176	104	25	5			
		6-31g(d,p)	0.168	99	28	8			
	SVWN	6-31+g(d,p)	0.236	139	40	11			
		6-311g(d,p)	0.224	133	36	9			
		6-31g(d,p)	0.284	170	45	12			
Pure	PBE	6-31+g(d,p)	0.276	167	44	12			
		6-311g(d,p)	0.168	99	28	8			

Table 27.12 Radiative efficiency (a_i) and global warming potential value of *syn*-$CF_3C(O)OCH_3$ at different DFT methods and different Pople basis sets.

				This work (GWP)			Experimental (GWP)		
DFT approximation	**Functional**	**Basis set**	**REs**	**20 years**	**100 years**	**500 years**	**20 years**	**100 years**	**500 years**
Meta-Hybrid-GGA	M06-2X	6-31g(d,p)	0.138	81	23	7	**315**	**86**	—
		6-31+g(d,p)	0.181	106	30	9			
		6-311g(d,p)	0.176	103	27	6			
	TPSSh	6-31g(d,p)	0.259	153	43	13			
		6-31+g(d,p)	0.304	170	41	13			
		6-311g(d,p)	0.298	162	39	12			
Hybrid-GGA	B3LYP	6-31g(d,p)	0.189	111	31	9			
		6-31+g(d,p)	0.258	152	43	13			
		6-311g(d,p)	0.250	148	42	12			
	B3PW91	6-31g(d,p)	0.226	132	36	7			
		6-31+g(d,p)	0.244	142	35	7			
		6-311g(d,p)	0.240	140	34	6			
	APF-D	6-31g(d,p)	0.203	120	33	10			
		6-31+g(d,p)	0.199	114	34	10			
		6-311g(d,p)	0.218	129	37	10			
	wB97X-D	6-31g(d,p)	0.228	132	38	2			
		6-31+g(d,p)	0.246	146	39	12			
		6-311g(d,p)	0.223	131	38	10			
Pure	SVWN	6-31g(d,p)	0.141	83	23	7			
		6-31+g(d,p)	0.205	121	34	10			
		6-311g(d,p)	0.207	122	34	10			
	PBE	6-31g(d,p)	0.258	152	43	13			
		6-31+g(d,p)	0.250	147	38	8			
		6-311g(d,p)	0.228	132	38	2			

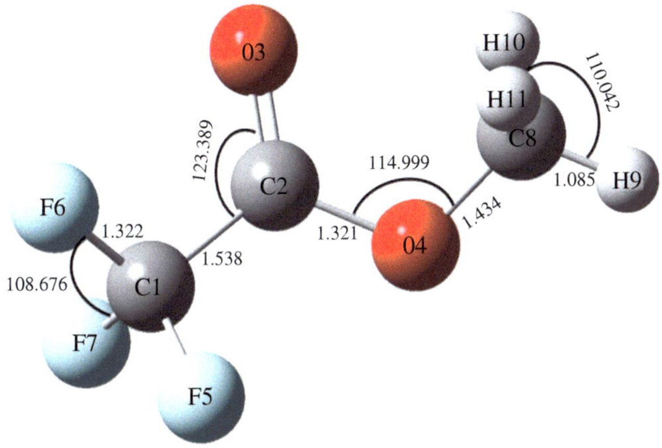

Figure 27.11 Optimized structure of *anti*-$CF_3C(O)OCH_3$ molecule at M06-2X/6-311G(d,p) level of theory.

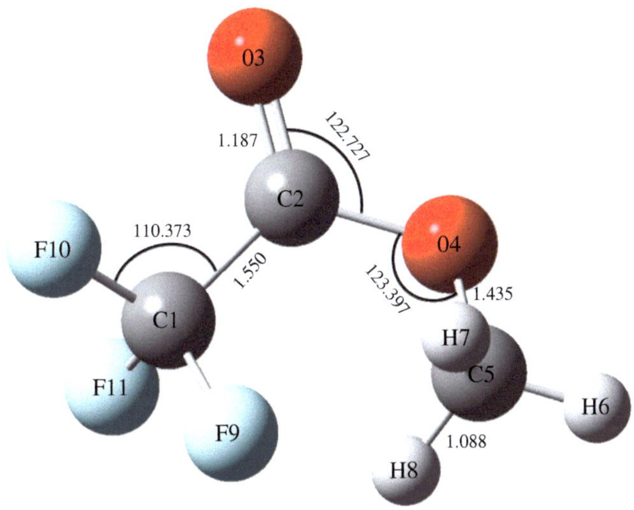

Figure 27.12 Optimized structure of *syn*-$CF_3C(O)OCH_3$ molecule at M06-2X/6-311G(d,p) level of theory.

27.5 Concluding Remarks

We have calculated radiative forcing and projected GWP values for six molecules of HFEs and five FES. It may be gleaned from the smaller GWP values; the molecules get oxidized by OH radicals rather than influencing the atmosphere for 20, 100, and 500 years. Overall, no method or basis set combination consistently predicts the experimental values of REs and GWPs. It is also clear that conformers do not affect REs and GWPs evaluation.

Acknowledgment

The authors are thankful to the Indian Institute of Technology Patna for providing research facilities and financial support to carry out this work.

References

1 Tsai, W.-T. (2005). Environmental risk assessment of hydrofluoroethers (HFEs). *J. Hazard. Mater.* 119: 69–78.

2 Bravo, I., Díaz-de-Mera, Y., Aranda, A. et al. (2011). Radiative efficiencies for fluorinated esters: indirect global warming potentials of hydrofluoroethers. *Phys. Chem. Chem. Phys.* 13: 17185–17193.

3 Good, D., Francisco, J., Jain, A., and Wuebbles, D.J. (1998). Lifetimes and global warming potentials for dimethyl ether and for fluorinated ethers: CH_3OCF_3 (E143a), CHF_2OCHF_2 (E134), CHF_2OCF_3 (E125). *J. Geophys. Res. Atmos.* 103: 28181–28186.

4 Orkin, V.L., Khamaganov, V.G., and Guschin, A.G. (2014). Photochemical properties of hydrofluoroethers CH_3OCHF_2, CH_3OCF_3, and $CH_2FCCF_2OCHF_2$: reactivity toward OH, IR absorption cross-sections, atmospheric lifetimes, and global warming potentials. *J. Phys. Chem. A* 118: 10770–10777.

5 Dennington, R., Keith, T., and Millam, J.G. (2009). *GaussView*, Version 5. Shawnee Mission KS: Semichem Inc.

6 Frisch, M., Trucks, G., Schlegel, H. et al. (2016). *Gaussian 16, Revision A.03*. Gaussian, Inc.: Wallingford, CT.

7 Jacobsen, R.L., Johnson, R.D. III, Irikura, K.K., and Kacker, R.N. (2013). Anharmonic vibrational frequency calculations are not worthwhile for small basis sets. *J. Chem. Theory Comput.* 9: 951–954.

8 Pinnock, S., Hurley, M.D., Shine, K.P. et al. (1995). Radiative forcing of climate by hydrochlorofluorocarbons and hydrofluorocarbons. *J. Geophys. Res. Atmos.* 100: 23227–23238.

9 Guleria, K. and Subramanian, R. (2022). Quantum chemical and chemical kinetic investigation on hydrogen abstraction reactions of $CF_3CF_2C(O)OCH_3$ and $CHF_2CF_2C(O)OCH_3$ with OH radicals and fate of haloalkoxy radicals. *ACS Earth Space Chem.* 6: 1596–1611.

10 Papasavva, S., Tai, S., Illinger, K.H., and Kenny, J.E. (1997). Infrared radiative forcing of CFC substitutes and their atmospheric reaction products. *J. Geophys. Res.* 102 (D12): 13643–13650.

11 WMO (World Meteorological Organisation) Scientific Assessment of Ozone Depletion (2006). Global ozone research and monitoring project – Report No. 50, 572pp. Geneva, 2007.

12 Orkin, V.L., Villenave, E., Huie, R.E., and Kurylo, M.J. (1999). Atmospheric lifetimes and global warming potentials of hydrofluoroethers: reactivity toward OH, UV spectra, and IR absorption cross sections. *J. Phys. Chem. A* 103 (48): 9770–9779.

13 Blowers, P., Moline, D.M., Tetrault, K.F. et al. (2007). Prediction of radiative forcing values for hydrofluoroethers using density functional theory methods. *J. Geophys. Res. Atmos.* 112: D15108.

14 Lily, M., Mishra, B.K., and Chandra, A.K. (2014). Kinetics, mechanism and thermochemistry of the gas phase reactions of $CF_3CH_2OCH_2CF_3$ with OH radicals: a theoretical study. *J. Fluorine Chem.* 161: 51–59.

15 Hodnebrog, Ø., Aamaas, B., Fuglestvedt, J.S. et al. (2020). Updated global warming potentials and radiative efficiencies of halocarbons and other weak atmospheric absorbers. *Rev. Geophys.* 58 (3).

16 Song, G., Jia, X., Gao, Y. et al. (2010). Theoretical studies on the mechanisms and dynamics of OH radicals with $CH_2FCF_2OCHF_2$ and CH_2FOCH_2F. *J. Phys. Chem. A* 114: 9057–9068.

17 Burkholder, J., Hodnebrog, Ø., and Orkin, V. (2018). Appendix A – Summary of abundances, lifetimes, ozone depletion potentials (ODPs), radiative efficiencies (REs), global warming potentials (GWPs), and global temperature change potentials (GTPs), global ozone research and monitoring project – Report No. 58. *World Meteorological Organization, Geneva, Switzerland.*

18 Zhu, P., Ai, L.-l., Wang, H., and Liu, J.-y. (2014). Theoretical studies on mechanism and kinetics of the hydrogen-abstraction reaction of $CF_2HCOOCH_3$ with OH radicals. *Comput. Theor. Chem.* 1029: 91–98.

19 Gour, N.K., Deka, R.C., Singh, H.J., and Mishra, B.K. (2014). A computational perspective on mechanism and kinetics of the reactions of $CF_3C(O)OCH_2CF_3$ with OH radicals and Cl atoms at 298 K. *J. Fluorine Chem.* 160: 64–71.

20 Kuze, N., Ishikawa, A., Kono, M. et al. (2015). Molecular structure and internal rotation of CF_3 group of methyl trifluoroacetate: gas electron diffraction, microwave spectroscopy, and quantum chemical calculation studies. *J. Phys. Chem. A* 119: 1774–1786.

28

Density Functional Theory-Based Study on Some Natural Products

Abhishek Kumar[1,3], *Ambrish K. Srivastava*[2], *Ratnesh Kumar*[1], *and Neeraj Misra*[1]

[1] *Department of Physics, University of Lucknow, Lucknow, Uttar Pradesh, India*
[2] *Department of Physics, Deen Dayal Upadhyaya Gorakhpur University, Gorakhpur, Uttar Pradesh, India*
[3] *Department of Geology, Babasaheb Bhimrao Ambedkar University (A Central University), Lucknow, Uttar Pradesh, India*

28.1 Introduction

A natural compound, phomarin, is isolated from plant sources. This kind of compound usually exhibits antimalarial activity, i.e. increasing occurrence of *Plasmodium falciparum* resistance against current anti-malarial therapy [1–3]. Synthetic *antiplasmodial* drugs have no longer effects after the first use as a natural compound. But It is interesting to note that for many synthetic antiplasmodial drugs, *P. falciparum* resistance isolated was observed 1–12 years after the first use whereas, it was longer for natural compounds. Drawing from the rich plant biodiversity phomarin may be helpful in the fight against Malaria.

A biologically relevant phytosterol, Poriferasterol, is isolated and identified from Cassiasophera Linn. The steroids occupy a significant position among bioactive natural products, and have contributed a lot in supplying useful and promising leads in drug discovery programs. This type of natural compound exhibits a variety of biological activities [4–8]. Poriferasterol was first reported from the marine sponge *Spheciospongia vesparia* by Valentine et al. [9]. Later on, the phytosterol was isolated from a few algae species, such as *Chlorella ellipsoidea* [10, 11], *C. saccharophila* [12], *Chloromorum toxicum* [13], *Dunaliella tertiolecta* [14], and also from some plants [15–18].

The natural product Gerberinol was isolated from *Gerbera lanuginose* Benth. (Compositae) and its structure is 5,5′-dimethyl dicoumarol based on physiochemical evidence [19]. Natural and synthetic dicoumarols are established as anticoagulant drugs in use, and they act as competitive inhibitors of Vitamin K epoxide reductase. This enzyme recycles Vitamin K, thus causing the depletion of active Vitamin K in blood [20, 21]. This type of compound prevents the formation of the active form of prothrombin and several other coagulant enzymes. Dicoumarol and coumarins are a new structural class for synthetic elaboration that could lead to improved antineoplastic drugs [22].

A member of the steroidal class of natural products, stigmasterol has significant biological and pharmaceutical promise [6, 23–27]. This phytosterol is reported to act as a precursor in the synthesis of progesterone [28–33], and as an intermediate in the biosynthesis of hormones such as corticoids, androgens, and estrogens [34], as well as in the synthesis of vitamin D3 [35]. Stigmasterol, with a wide range of beneficial metabolic and therapeutic actions on living systems, has been evaluated to exhibit many pharmacological properties [36–43]. Recently, It has been found that *Peltophorum pterocarpum* (DC.) Backer ex K. Heyne (Family: Fabaceae) is a new natural source

Electron Density: Concepts, Computation and DFT Applications, First Edition.
Edited by Pratim Kumar Chattaraj and Debdutta Chakraborty.

of this biologically potent phytosterol. *P. pterocarpum* is an important plant used in traditional systems of medicine to treat many diseases, such as stomatitis, insomnia, skin disorders, and ringworm infections [44–46].

Density functional theory (DFT)-based methods use the description of the electronic density of an atom or molecule to calculate important properties, which are not easily obtained via experimental methods. These properties include molecular geometry, vibrational frequencies, dipole moments, higher-order moments, and thermo-chemical properties. There are several variants of DFT whose applications depend on the nature of molecular systems under study and the properties to be explored. DFT offers a better compromise between computational cost and accuracy for medium-size molecules, and hence, it has been successfully employed in many previous studies [47–49]. A comprehensive coverage of the topic can be found in recent literature [50].

This chapter deals with a detailed study of various aspects of phomarin, poriferasterol, gerberinol, and stigmasterol revealed by DFT. The geometrical and vibrational features of these compounds have been discussed. The molecular properties of the natural products are explored with the help of various chemical reactivity surfaces such as HOMO, LUMO, and molecular electrostatic potential (MESP), as well as a number of reactivity parameters.

28.2 Computational Details

All DFT-based calculations in this chapter have been employed using Beckes three-parameter hybrid exchange functional [51] with Lee–Yang–Parr correlation functional [52] (B3LYP) functional. The B3LYP functional is a gradient-corrected hybrid functional and is most popular for studying a variety of systems of biological interest [53–56]. The standard split-valence basis set added with diffuse as well as polarization functions 6-311++G(d,p) is used throughout these calculations. The Gaussian 09 program [57] is used to perform all these computations with the OPT+FREQ keyword. The equilibrium structures of molecules are obtained by geometry optimization followed by frequency calculations.

The aim of geometry optimization is to find an atomic arrangement that makes the molecule most stable, i.e. its energy becomes the lowest. The lowest energy structures are ensured by all positive real values of frequencies. In addition, the frequency calculation provides information about the position and relative intensity of vibrational bands and this information is independent of an experiment. Therefore, it can be used as a tool to predict peak positions and intensities when experimental data is not available. These calculated frequencies are based on the harmonic model, while real vibrational frequencies are anharmonic. This partially explains discrepancies between calculated and experimental frequencies, so the calculated frequencies are generally scaled by some factor [58] (depending on the method of computation) prior to comparison with experimental data obtained from FT-IR, etc.

28.3 Results and Discussion

28.3.1 Geometrical Properties

The optimized geometries of compounds are shown in Figure 28.1. These optimized geometries are further ensured by the absence of any imaginary normal mode frequency at the same level of theory.

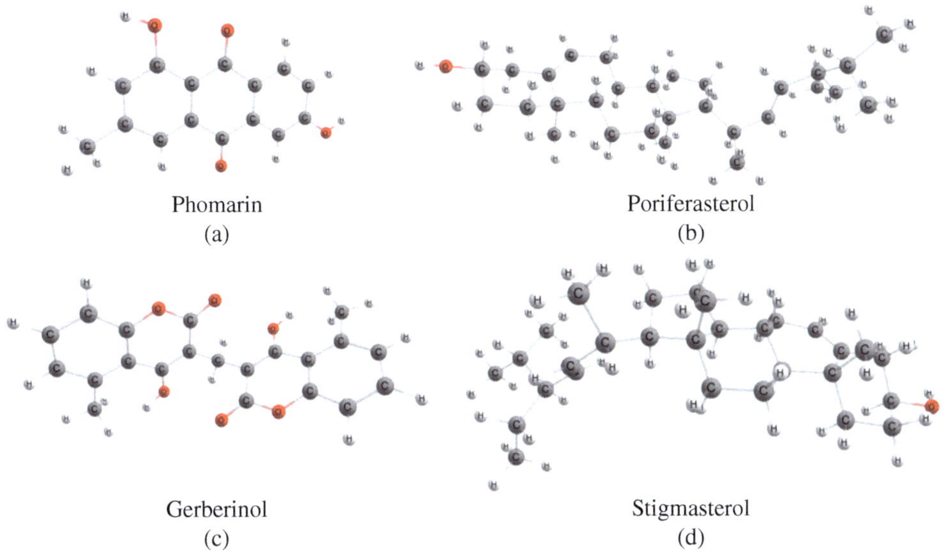

Figure 28.1 Optimized geometry of natural compounds.

The most common feature of these molecules includes the presence of the -OH group. Phomarin and gerberimol contain two -OH and =O groups; however, there is a substitution of O atom in two rings of gerberimol. Poriferasterol and stigmasterol contain only one -OH group. The calculated geometrical parameters show a good proximity when compared with the experimental results. The average value for C—C bonds in rings is found to be in the range of 1.515–1.567 Å, and that for C—H bonds in the range of 1.090–1.100 Å, which are in good agreement with those reported for similar structures [59, 60]. The C—O and O—H bond lengths of 1.43 and 0.96 Å, respectively, are also in good agreement with the literature data [61, 62].

28.3.2 Vibrational Properties

The vibrational analysis is performed along with their assignments and potential energy distribution (PED). The PED analysis is carried out with the VEDA 4 program [63]. Generally, the PED contribution is based on parameters such as the quadratic force constants matrix, diagonal mass matrix, its eigenvalues, and eigenvectors expressed in Cartesian coordinates. The contribution of the *i*th valence-force coordinate in the *k*th normal mode is expressed as the percent of potential energy distribution, %PED(*i*,*k*) [64]. The VEDA program attempts to enhance the PED contributions by the composition of local coordinates. These local coordinates are constructed by combining only the same type of movements, i.e. stretching, bending, or torsion. The movements of heavy atoms are combined separately from the movements engaging the light atoms. The DFT hybrid B3LYP functional tends to overestimate the fundamental normal modes of vibration. A scaling factor of 0.9614 is applied [58]. The descriptions of selected stretching modes of vibration of natural compounds carried out in terms of their PED contribution are given in Table 28.1. Note that PEDs of less than 15% are not included in the assignment. The calculated IR spectra of these compounds have been shown in Figure 28.2.

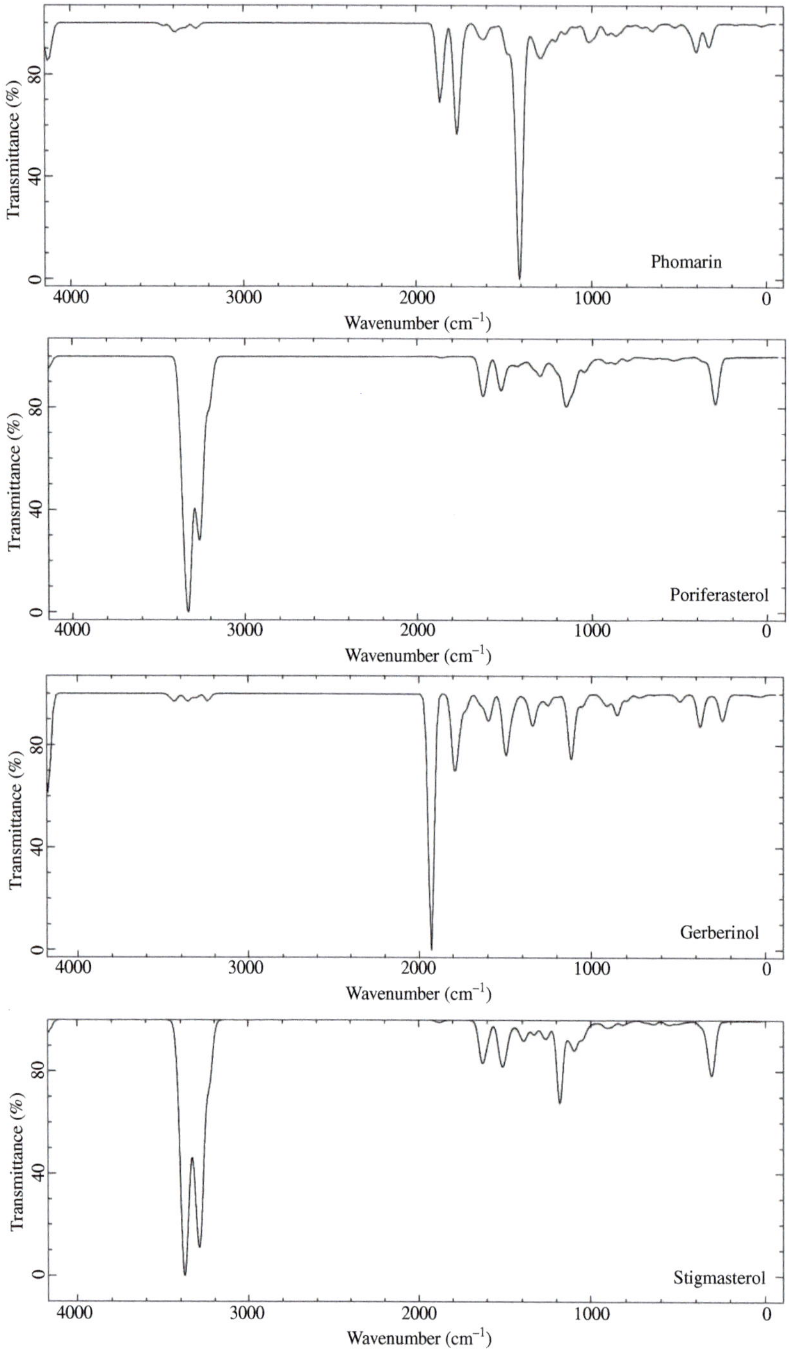

Figure 28.2 The calculated vibrational IR spectra of natural compounds.

Table 28.1 Vibrational analysis of some selected stretching modes of natural compounds.

Vibrational frequency in cm^{-1} (rel. intensity)				
Phoamrin	**Poriferasterol**	**Gerberinol**	**Stigmasterol**	**Assignment (%PED)**
3699 (s)	3710 (w)	3741 (s)	3715 (w)	ν_s(H) (100–O)
3103 (w)	3025 (m)	2986 (w)	3051 (w)	ν(C–H) (98)
1678 (s)	—	1713 (w)	—	ν(O–C) (85)
1594 (m)	1668 (w)	1616 (w)	1678 (w)	ν(C–C) (58)
1557 (w)	1664 (w)	1552 (w)	1670 (w)	R[ν(C–C) (41)]

28.3.2.1 O–H Vibration

A free hydroxyl group (O–H) absorbs in the range 3700–3500 cm^{-1} [65]. The O–H stretching vibrations of natural compounds are observed in the range 3699–3741 cm^{-1} with 100% PED. These results are found to be in accordance with the value reported in the literature. In-plane, out-of-plane bending and torsion-based O–H modes are observed at lower frequency ranges.

28.3.2.2 C–H Vibration

The C–H stretching modes of vibrations are observed as mixed modes in the frequency range 3051–2986 cm^{-1}. In general, the aromatic C–H stretching modes are found in the region 3100–3000 cm^{-1} [66]. The other modes of C–H vibration of natural compounds, such as scissoring, rocking, and torsion, are obtained below the frequency 1500 cm^{-1}.

28.3.2.3 C–C Vibration

The bands observed between the spectral ranges from 1650 to 1200 cm^{-1} in the aromatic compounds are assigned to C–C stretching modes. The C–C stretching vibrations of the compounds are observed in the range 1152–1678 cm^{-1} and these modes are confirmed by their PED values. The C–C in-plane and out-of-plane bending vibrations are also observed below 1000 cm^{-1}.

28.3.2.4 C=O Vibration

The vibrational frequencies of the carbonyl group are the significant characteristic bands for extensive studies [67] and this frequency of the stretch mainly depends on the band strength. The modes calculated at 1678 cm^{-1} for phomarin and 1713 cm^{-1} for geberinol have been identified as the stretching modes of C=O groups. The scissoring and torsion modes of vibration are obtained at lower frequencies. These are also in good agreement with the values reported in the literature [68].

28.3.3 HOMO–LUMO and MESP Plots

The frontier orbital highest occupied molecular orbital (HOMO) and lowest unoccupied molecular orbital (LUMO) determine the way the molecule interacts with other species and the gap between these helps characterizes the kinetic stability and chemical reactivity of the molecule. The frontier molecular orbital gap has been used as a measure for bioactivity, since the intra-molecular charge transfer determining the bioactivity depends on this energy gap. The HOMO and LUMO plots of the natural compounds are shown in Figure 28.3. A molecule having a small frontier orbital gap is more polarizable. It is generally associated with a high chemical reactivity and low kinetic stability and is

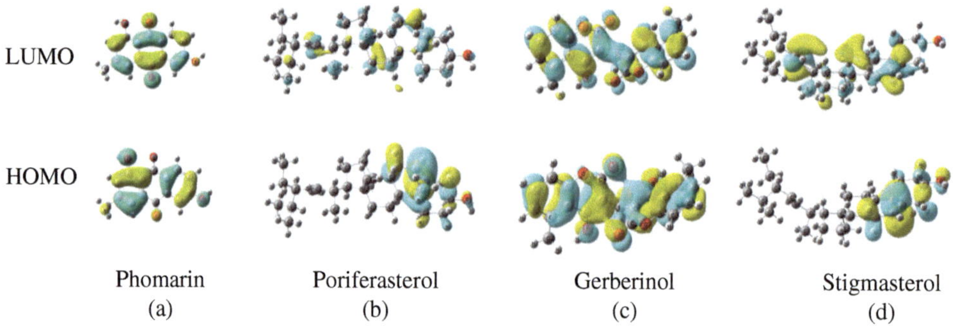

Figure 28.3 HOMO and LUMO surfaces of natural compounds.

also termed a soft molecule [69, 70]. The HOMO primarily acts as an electron donor and the LUMO acts as an electron acceptor. The frontier orbital gap of phomarin, poriferasterol, gerberinol and stigmasterol are 3.82, 6.27, 4.39 and 6.32 eV, respectively. This result suggested that in the studies of all natural products, phomarin is the most polarizable and soft compound. Further, poriferasterol and stigmasterol possess nearly equal frontier orbital gaps, as expected from their structures, as shown in Figure 28.1.

The importance of MESP lies in the fact that it simultaneously displays molecular size and shape as well as positive, negative, and neutral electrostatic potential regions in terms of color grading. This surface is very useful in the investigation of molecular structure with its physiochemical property relationship [71]. The different values of the electrostatic potential at the surface are represented by different colors and potential increases in the order red < orange < yellow < green < blue, where blue indicates the most electropositive, i.e. electron-poor region and red indicates the most electronegative region, i.e. electron-rich region.

The MESP plots of natural compounds are shown in Figure 28.4. In the MESP plots of compounds, the molecular surface clearly suggests that the most electropositive region (blue region,

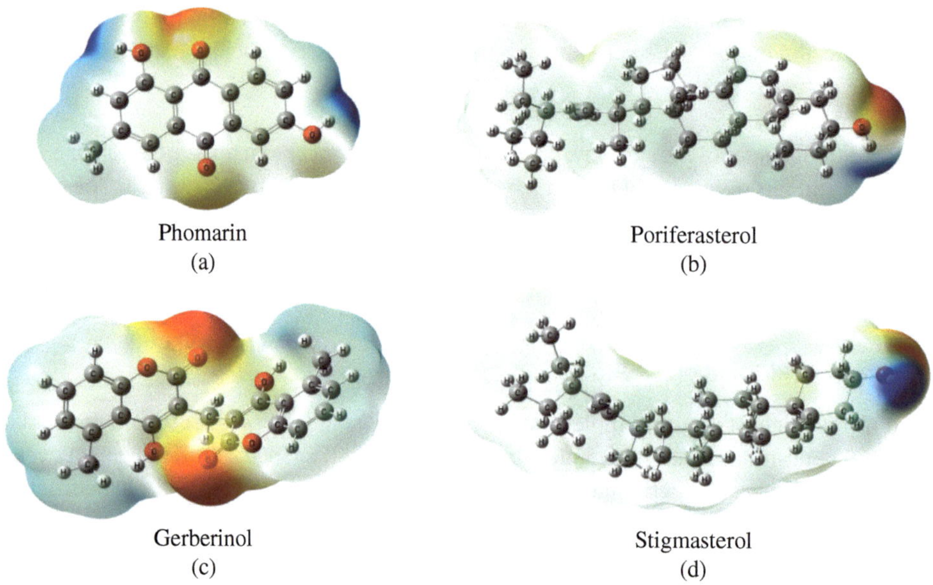

Figure 28.4 MESP plots of natural compounds.

electron-poor) of phomarin, poriferasterol, and stigmasterol are located over the side where the H-atom attached by O-atom and the electronegative region (red region, electron-rich) is located over the carbonyl oxygen atom which essentially, acts as an electron acceptor region and electron donor region, respectively. However, the MESP surface of gerberinol indicates that the most electronegative region (red region, electron-rich) is located over the carbonyl oxygen atom, which makes it the most reactive part of the molecule.

28.3.4 Chemical Reactivity

DFT enables to obtain some electronic parameters, which are used to describe the chemical reactivity of compounds. These electronic parameters include ionization potential (I), electron affinity (A), absolute electronegativity (χ), and chemical hardness (η). These electronic parameters are, sometimes, referred to as chemical reactivity descriptors, molecular descriptors, or global reactivity descriptors.

The calculated electronic parameters of the natural products are shown in Table 4. The energies of HOMO and LUMO have been used to obtain the ionization potential (I) and electron affinity (A) using the Koopmans' theorem [72]. According to this theorem, the negative of the highest occupied molecular orbital energy ($-E_{\text{HOMO}}$) and the lowest unoccupied molecular orbital energy ($-E_{\text{LUMO}}$) corresponds to ionization potential and electron affinity, respectively. Note, however, that I and A values, thus obtained, are only "approximate" values as the Koopmans' theorem is valid only for the HF method. For DFT, the "derivative discontinuity" should be taken into account, as suggested by Perdew et al. [73].

The electronegativity is the power of an atom in a molecule to attract an electron toward it [74], and the principle of electronegativity equalization was introduced by Sanderson [75]. The chemical hardness (η) is an important quantity in chemical reactivity theory, which was first put forward by Pearson [76]. Chemical hardness is a measure of the resistance of a chemical species to change its electronic configuration. The practical method used to calculate chemical hardness and electronegativity was given by Parr and Pearson [77, 78] as,

$$\chi = \tfrac{1}{2}(I + A)$$

$$\eta = \tfrac{1}{2}(I - A)$$

where the ionization potential (I) and the electron affinity (A) can be calculated through various techniques through experimental and theoretical means. Initially, experimental ionization potential and electron affinity values have been used to calculate the absolute chemical hardness and absolute electronegativity [79]. Electrophilicity index (ω) [80] is calculated as,

$$\omega = \frac{\chi^2}{2\eta}$$

From Table 28.2, it is clear that phomarin is chemically softer than other studied natural compounds as mentioned earlier. It is because of the fact the chemical hardness is directly proportional to the frontier orbital gap. Further, phomarin possesses higher electron affinity and absolute electronegativity. These parameters also indicate its more chemically reactive nature. The electronic parameters of poriferasterol and stigmasterol are almost equal due to their structural similarities (see Figure 28.1). This further supports the fact that the chemical reactivity of molecules largely depends on their molecular structures. According to the principles of minimum electrophilicity [81, 82] and maximum hardness [83, 84], the minimum electrophilicity and maximum hardness

Table 28.2 The electronic parameters of natural compounds.

Parameters	Phomarin	Poriferasterol	Gerberinol	Stigmasterol
I (eV)	6.73	6.45	6.30	6.43
A (eV)	2.91	0.18	1.92	0.11
χ (eV)	4.82	3.31	4.11	3.32
η (eV)	1.91	3.13	2.19	3.16
ω (eV)	6.08	1.75	3.85	1.74

should correspond to the equilibrium geometry along a reaction path or in a system. In terms of these two parameters, phomarin comes with higher electrophilicity and lower hardness. This further suggests the enhanced chemical reactivity feature of phomarin among all these four natural products.

28.4 Conclusion

In this chapter, we have discussed the DFT results of four natural products, namely, phomarin, poriferasterol, gerberinol, and stigmasterol, using the B3LYP/6-311++G(d,p) level. Their structural features have been highlighted. Normal mode analyses have been performed, and assignments are given to the selected stretching modes of vibrations. The chemical reactivity of these compounds has been discussed by frontier orbitals and MESP surfaces along with various electronic parameters, viz ionization potential, electron affinity, absolute electronegativity, chemical harness, and electrophilicity. It has been found that the transition from HOMO to LUMO leads to the transfer of electron density and the smaller energy gap of phomarin along with electronic parameters suggests its chemically more reactive behavior. The study should provide a firsthand example of how DFT can be applied to assess the chemical reactivity pattern of compounds.

Acknowledgments

AKS is thankful to CSIR, SERB, and UGC for financial assistance in the form of a senior research fellowship (SRF), national postdoctoral fellowship (NPDF) and start-up grant, respectively.

References

1 Brew, E.J.C. and Thomson, R.H. (1971). *J. Chem. Soc. C* 2007.
2 Arrebola, M.L., Ringbom, T., and Verpoorte, R. (1999). *Phytochemistry* 52: 1283.
3 Bick, I.R.C. and Rhee, C. (1966). *Biochem. J.* 98: 112.
4 Newman, D.J. and Cragg, G.M. (2012). *J. Nat. Prod.* 75: 311.
5 Kaur, K., Jain, M., Kaur, T., and Jain, R. (2009). *Bioorg. Med. Chem.* 17: 3229.
6 Shukla, V., Joshi, G.P., and Rawat, M.S.M. (2010). *Phytochem. Rev.* 9: 303.
7 Wang, Z., Tang, H., Wang, P. et al. (2013). *Mar. Drugs* 11: 775.

8 Singariya, P., Kumar, P., and Mourya, K.K. (2012). *Int. J. Res. Pharm. Sci.* 3: 678.
9 Valentine, F.R., Bergmann, J., and Bergmann, W. (1941). *J. Org. Chem.* 6: 452.
10 Patterson, G.W. and Krauss, R.W. (1965). *Plant Cell Physiol.* 6: 211.
11 Thompson, M.J., Dutky, S.R., Patterson, G.W., and Gooden, E.L. (1972). *Phytochemistry* 11: 1781.
12 Giner, J.L., Zhao, H., and Tomas, C. (2008). *Phytochemistry* 69: 2167.
13 Wright, J.L.C. (1979). *Can. J. Chem.* 57: 2569.
14 Esa, H.S.A., Kornprobst, J.M., and Rizk, A.M. (1995). *Phytochemistry* 39: 373.
15 Misra, L. and Siddiqi, S.A. (2005). *Indian J. Chem.* 44B: 1915.
16 Rontani, J.F., Beker, B., and Volkman, J.K. (2004). *Phytochemistry* 65: 117.
17 Amatyaa, S. and Tuladhar, S.M. (2005). *Z. Naturforsch.* 60b: 1006.
18 Akihisa, T., Matsubara, Y., Ghosh, P. et al. (1988). *Phytochemistry* 27: 1169.
19 Sengupta, P., Sen, M., Karuri, P. et al. (1985). *J. Indian Chem. Soc.* 52: 916.
20 Wardrop, D. and Keeling, D. (2008). *Br. J. Haematol.* 141: 757.
21 Paknikar, S.K., Pai Fondekar, K.P., Kirtany, J.K., and Natori, S. (1996). *Phytochemistry* 41: 931.
22 Madari, H., Panda, D., Wilson, L., and Jacobs, R.S. (2003). *Cancer Res.* 63: 1214.
23 Shabana, M.M., Salama, M.M., Ezzat, S.M., and Ismail, L.R. (2013). *J. Carcinog. Mutagen.* 4: 1.
24 Zidorn, C. (2016). *Phytochemistry* 124: 5.
25 Sun, Y., Lin, Y., Cao, X. et al. (2014). *Int. J. Mol. Sci.* 15: 21660.
26 Berger, A., Jones, P., and Abumweis, S.S. (2004). *Lipids Health Dis.* 3: 1.
27 Goldberg, A.C., Ostlund, R.E., Bateman, J.H. et al. (2006). *Am. J. Cardiol.* 97: 376.
28 Sali, V.K., Mansingh, D.P., and Vasanthi, H.R. (2016). *Med. Chem. Commun.* 7: 1429.
29 Tanemossu, S.A.F., Franke, K., Arnold, N. et al. (2014). *Phytochemistry* 105: 171.
30 Nkengfack, A.E., Mkounga, P., Fomum, Z.T. et al. (2002). *J. Nat. Prod.* 65: 734.
31 Tapia, E.B., Bucio, M.A., Rivera, A., and Nathan, P.J. (2001). *J. Nat. Prod.* 64: 518.
32 Tasdemir, D. and Sticher, O. (1997). *J. Nat. Prod.* 60: 874.
33 Jamaluddin, F., Mohamed, S., and Lajis, M. (1994). *Food Chem.* 49: 339.
34 Sundararaman, P. and Djerassi, C. (1977). *J. Org. Chem.* 42: 3633.
35 Kametani, T. and Furuyama, H. (1987). *Med. Res. Rev.* 7: 147.
36 Gabay, O., Sanchez, C., Salvat, C. et al. (2010). *Osteoarthritis Cartilage* 18: 106.
37 Shanthakumar, B., Sathish, M., and Suresh, A.J. (2013). *Res. J. Pharm. Biol. Chem. Sci.* 4: 1411.
38 Jeon, G., Park, M., Yoon, D. et al. (2005). *Exp. Mol. Med.* 37: 111.
39 Chowdhury, R., Rashid, R.B., Sohrab, M.H., and Hasan, C.M. (2003). *Pharmazie* 58: 272.
40 Yin, G., Zeng, H., He, M., and Wang, M. (2009). *Int. J. Mol. Sci.* 10: 4330.
41 Sheu, J.H., Wang, G.H., Sung, J., and Duh, C.Y. (1999). *J. Nat. Prod.* 62: 224.
42 Sadish, S.K., Kumar, Y., Khan, M.S.Y. et al. (2009). *Pharmacologyonline* 1: 1104.
43 He, F., Wang, M., Gao, M. et al. (2014). *Molecules* 19: 4046.
44 Rastogi, R.P. and Mehrotra, B.N. (1999). *Compendium of Indian Medicinal Plants*, vol. 1–4. New Delhi: CDRI, Lucknow and NISC.
45 Ambasta, S.P. (1956). *The Useful Plants of India*. New Delhi: PID, CSIR.
46 Raj, M.K., Balachandran, C., Duraipandiyan, V. et al. (2012). *Med. Chem. Res.* 21: 11.
47 Srivastava, A.K., Pandey, S.K., Kumar, A., and Misra, N. (2016). *Med. Chem. Res.* 25: 2832.
48 Srivastava, A.K., Pandey, A.K., Jain, S., and Misra, N. (2015). *Spectr. Chim. Acta Part A* 136: 682.
49 Kumar, A., Srivastava, A.K., Gangwar, S. et al. (2015). *J. Mol. Struct.* 1096: 94.
50 Srivastava, A.K. and Misra, N. (2021). *DFT Based Studies on Bioactive Molecules*. Singapore: Bentham Science Publishers.
51 Becke, A.D. (1988). *Phys. Rev. A: Gen. Phys.* 38: 3098.
52 Lee, C., Yang, W., and Parr, R.G. (1988). *Phys. Rev. B* 1337: 785.

53 Kumar, A., Dwivedi, A., Srivastava, A.K. et al. (2017). *Polycyclic Aromat. Compd.* 37: 1.

54 Brahmachari, G., Mondal, A., Nayek, N. et al. (2017). *J. Mol. Struct.* 1143: 184.

55 Dwivedi, A. and Kumar, A. (2019). *Polycyclic Aromat. Compd.* 41: 387.

56 Brahmachari, G., Majhi, S., Mandal, B. et al. (2018). *J. Indian Chem. Soc.* 95: 1231.

57 Frisch, M.J., Trucks, G.W., Schlegel, H.B., et al. (2009). *Gaussian 09, Revision A.1*. Gaussian Inc.: Wallingford, CT.

58 Merrick, J.P., Moran, D., and Radom, L. (2007). *J. Phys. Chem. A* 111: 11683.

59 Bernhard, S. (1995). *Infrared & Raman Spectroscopy*. New York: Wiley VCH Inc.

60 Arnett, E.M. and Larsen, J.W. (1969). *JACS* 91: 1438.

61 Kumar, A., Srivastava, A.K., Gangwar, S.K. et al. (2020). *Polycyclic Aromatic Compound* 40: 540.

62 Tiwari, G., Srivastava, A.K., Kumar, R., and Kumar, A. (2018). *Main Grp. Chem.* 18: 107.

63 Jamróz, M.H. (2004). Vibrational Energy Distribution Analysis: VEDA 4, Program, Warsaw.

64 Jamroz, M.H. (2013). *Spectrochim. Acta Part A* 114: 220.

65 Srivastava, A.K., Kumar, A., Misra, N. et al. (2016). *J. Mol. Struct.* 1107: 137.

66 Silverstein, R.M. and Webster, F.X. (1998). *Spectrometric Identification of Organic Compounds*, 6e. New York: Wiley.

67 Smith, B. (1999). *Infrared Spectral Interpretation – A Systematic Approach*. Washington, DC: CRC Press.

68 Arivazhagan, M. and Jeyavijayan, S. (2011). *Spectrochim. Acta A* 79: 161.

69 Fleming, I. (1976). *Frontier Orbitals and Organic Chemical Reactions*. New York: Wiley.

70 Belletete, M., Morin, J.F., Leclere, M., and Durocher, G. (2005). *J. Phys. Chem.* 109A: 6953.

71 Murray, J.S. and Sen, K. (1996). *Molecular Electrostatic Potential; Concepts and Applications*. Amsterdam, The Netherlands: Elsevier.

72 Koopmans, T. (1934). *Physica* 1: 104.

73 Perdew, J.P., Parr, R.G., Levy, M., and Balduz, J.L. (1982). *Phys. Rev. Lett.* 49: 1691.

74 Pauling, L. (1960). *The Nature of the Chemical Bond*, 3e. Ithaca, NY: Cornell University Press.

75 Sanderson, R.T. (1983). *Polar Covalence*. New York: Academic Press.

76 Pearson, R.G. (1973). *Hard and Soft Acids and Bases*. Stroudsberg: Dowen, Hutchinson and Ross.

77 Parr, R.G. and Pearson, R.G. (1983). *JACS* 105: 7512.

78 Pearson, R.G. (1988). *JACS* 110: 7684.

79 Zhan, C.G., Nichols, J.A., and Dixon, D.A. (2003). *J. Phys. Chem. A* 107: 4184.

80 Parr, R.G., Szentpaly, L.V., and Liu, S. (1999). *JACS* 121: 1922.

81 Chamorro, E., Chattaraj, P.K., and Fuentealba, P.J. (2003). *Phys. Chem.* 107: 7068.

82 Pan, S., Sola, M., and Chattaraj, P.K. (2013). *J. Phys. Chem.* 117: 1843.

83 Pearson, R.G. (1993). *Acc. Chem. Res.* 26: 250.

84 Pearson, R.G. (1999). *J. Chem. Educ.* 76: 267.

Index

Note: *Italic* page numbers refer to figures and tables.

a

Electron Density: Concepts, Computation and DFT Applications, First Edition.
Edited by Pratim Kumar Chattaraj and Debdutta Chakraborty.

c

e

f

g

n

p